S0-BPI-620

Springer Series in Statistics

Springer Series in Statistics

L. A. Goodman and W. H. Kruskal, Measures of Association for Cross Classifications. x, 146 pages, 1979.

J. O. Berger, Statistical Decision Theory: Foundations, Concepts, and Methods. xiv, 425 pages, 1980.

R. G. Miller, Jr., Simultaneous Statistical Inference, 2nd edition. xvi, 299 pages, 1981.

P. Brémaud, Point Processes and Queues: Martingale Dynamics. xviii, 354 pages, 1981.

E. Seneta, Non-Negative Matrices and Markov Chains. xv, 279 pages, 1981.

F. J. Anscombe, Computing in Statistical Science through APL. xvi, 426 pages, 1981.

J. W. Pratt and J. D. Gibbons, Concepts of Nonparametric Theory. xvi, 462 pages, 1981.

V. Vapnik, Estimation of Dependences based on Empirical Data. xvi, 399 pages, 1982.

H. Heyer, Theory of Statistical Experiments. x, 289 pages, 1982.

L. Sachs, Applied Statistics: A Handbook of Techniques. xxviii, 706 pages, 1982.

M. R. Leadbetter, G. Lindgren and H. Rootzen, Extremes and Related Properties of Random Sequences and Processes. xii, 336 pages, 1983.

H. Kres, Statistical Tables for Multivariate Analysis. xxii, 504 pages, 1983.

J. A. Hartigan, Bayes Theory. xii, 145 pages, 1983.

Heinz Kres

Statistical Tables for Multivariate Analysis

A Handbook with References to Applications

Translated by Peter Wadsack

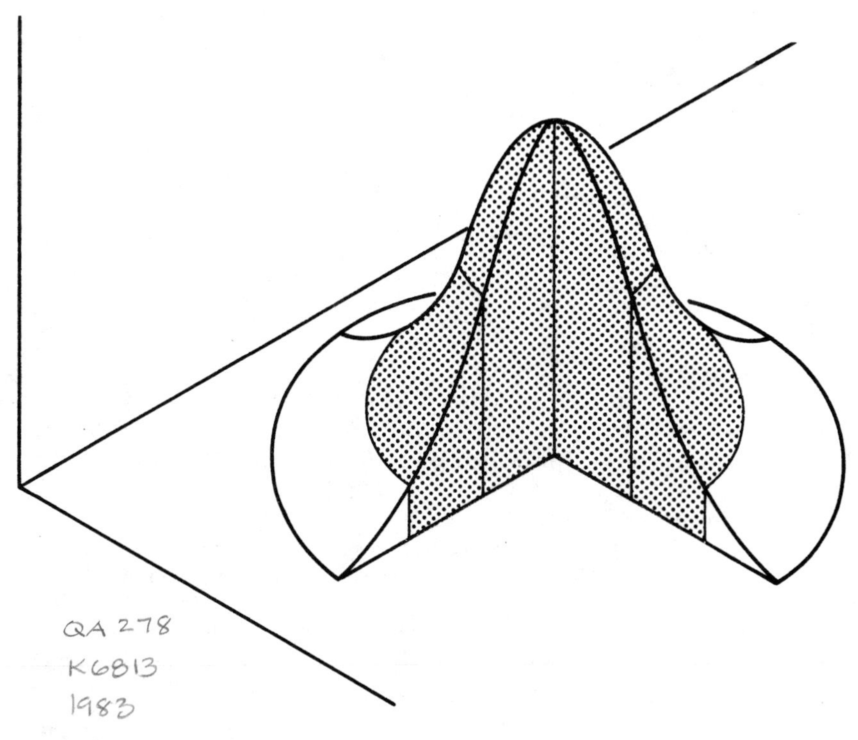

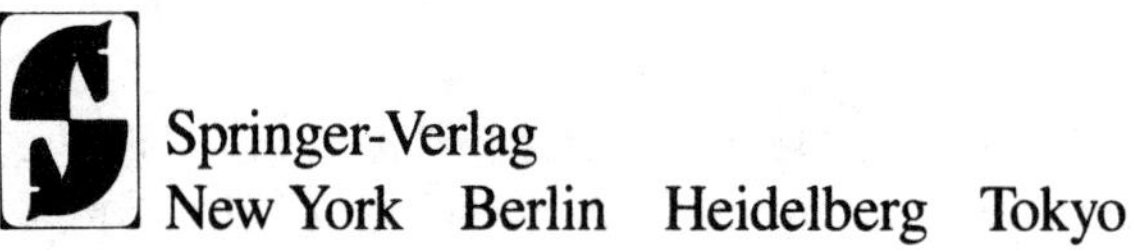

Springer-Verlag
New York Berlin Heidelberg Tokyo

AMS Classification: 62HXX

Library of Congress Cataloging in Publication Data
Kres, Heinz, 1929–
Statistical tables for multivariate analysis.
(Springer series in statistics)
Translation of: Statistische Tafeln zur multivariaten Analysis.
1. Multivariate analysis—Tables. 2. Mathematical statistics—Tables. I. Title. II. Series.
QA278.K6813 1983 519.5'35'0212 83-14688

With one illustration.

Title of the original German edition: *Statistische Tafeln zur multivariaten Analysis*.

Printed and bound by R.R. Donnelley & Sons, Harrisonburg, VA.
Printed in the United States of America.

9 8 7 6 5 4 3 2 1

ISBN 0-387-90909-5 Springer-Verlag New York Berlin Heidelberg Tokyo
ISBN 3-540-90909-5 Springer-Verlag Berlin Heidelberg New York Tokyo

To GISELA

Preface to the English Edition

An English translation of my "Statistische Tafeln zur multivariaten Analysis - Ein Handbuch mit Hinweisen zur Anwendung" was planned already in 1975 when I prepared the German volume. The tables were immediately supplied with German and English headings for the intended photo-offset printing.

In the meantime, new and important tables for multivariate statistical hypotheses and procedures have been compiled and published. Only four of them have been incorporated in the present volume. The selection of these tables must be on an individual basis for reasons of space. Let me mention only the new tables for sample size determination in MANOVA.

The instructions for using the tables are all organized in the same way. They are kept short since it is not the task of such a work to provide an introduction to the theory and practice of multivariate analysis. I have renounced giving examples for the same reason. I wish instead to refer the reader to the many good textbooks that are available, as well as to my own works on methods that are in preparation. Furthermore, I am of the opinion that statistical tables should accompany the textbook rather than be included in it.

Finally, my thanks go to the translator, Mr. Peter R. Wadsack, as well as to the ladies and gentlemen of Springer-Verlag for their pleasant collaboration and their indulgence of my numerous requests.

Baden-Baden / Black Forest,
Brenner's Park-Hotel, Easter 1983

H.K.

Preface to the German Edition

This collection of "Statistical Tables for Multivariate Analysis" is part of a multivolume survey of this branch of mathematical statistics which is growing in significance. The tables, as well as the complementary volumes on methods, arose out of my consulting work for various clients while I was at the University of Freiburg in Üchtland and the Swiss Federal Institute of Technology Zürich. For technical reasons, the volume of tables appears in print before the volumes on methods.

The desirability or even the necessity for such a collection of tables can be seen from the fact that authors of English language textbooks invariably include several of the most-used tables as appendices. The use of computers does not make such a collection obsolete, for the approximations used on machines generally do not reach sufficient accuracy until cases of high dimension or large sample size are encountered.

The present tables, with few exceptions, consider only parameter values of dimensions less than 10. In making the selections I was limited by the size of existing tables. At this point it is worth noting that well-designed studies, perhaps following preliminary investigations, with a number of variates p less than 10 hold often greater promise of success than investigations with 50 to 200 variates whose evaluation and commentary is the product of a routine computer program. Both for reasons of cost and error avoidance, the tables selected to follow the commentary are reproduced through offset printing. The quality of the reproductions may be described as satisfactory : in any case, technical printing efforts can only lead to inconsequential improvements over the original copy.

At this point I wish to thank all the publishers and editors, as well as colleagues, who were kind enough to permit the reproduction of their tables and assist me with their help and advice.

The motivation for the choice of tables is noted in the various preliminary remarks to the sections. In general, I also chose those tables which are in inaccessible technical literature, graduate texts, and company reports.

In writing the explanations for the tables I was guided by the thought that they should in no way attempt to replace a textbook; this is especially true for the tables in Part I. The tables in Part II occasionally have a more extensive explanation, especially where the application of the table is restricted to a single problem. By using the same format for all of the explanations to the tables, I hope to expedite the user's progress.

I have in general decided against inclusion of accuracy estimates for the tabulated values, since these are of secondary importance to the practitioner. The circle of those users who feel that an effect is "assured" when $\alpha = 4.99$ % *, while nothing "noteworthy" has been found when* $\alpha = 5.01$ % *should be shrinking in spite of routine computer evaluations. For the theoretically inclined, there are various citations from the literature.*

Baden-Baden, October 1975

H.K.

Contents

Preface to the English Edition VII

From the Preface to the German Edition IX

Permissions and Acknowledgments XV

Remarks on Notation XXI

Part I : The Primary Tables for Testing Multivariate Statistical Hypotheses 1

Preliminary Remarks 1

Chapter 0 : A Brief Summary of the Test Criteria for the Multivariate General Linear Hypothesis 2

Table 1 : The Likelihood Ratio Criterion Λ of S. S. WILKS : Tables of F. J. WALL 14

Table 2 : The Θ_{max} - Criterion of K. C. S. PILLAI : A Version of the Maximum Root Criterion of S. N. ROY 52

Table 3 : The Generalized F-Criterion of R. D. BOCK : A Version of the Maximal Root Criterion of S. N. ROY 62

Table 4 : The Nomograms of D. L. HECK for the Distribution of the Θ_{max} - Criterion of S. N. ROY 87

Table 5 : The Θ_{max} - Criterion of F. G. FOSTER and D. H. REES : A Version of the Maximum Root Criterion of S. N. ROY 105

Table 6 : The Trace Criterion T_o^2 of H. HOTELLING and D. N. LAWLEY in the Version of K. C. S. PILLAI 118

Table 7 : The Trace Criterion of H. HOTELLING and D. N. LAWLEY in the Version $V^{(s)}$ of K. C. S. PILLAI 136

Table 8 : The T^2-Criterion of H. HOTELLING : Tables of D. R. JENSEN and R. B. HOWE 154

Part II : Tables Related to the Multivariate Normal Distribution 183

Preliminary Remarks 183

Table 9 : The Multivariate Normal Distribution with Equal Correlations : Tables of S. S. GUPTA 184

Table 10 : The Distribution of the Maximum of N Equally Correlated Normal Standardised Random Variables : Tables of S. S. GUPTA, K. NAGEL, and S. PANCHAPAKESAN 204

Table 11 : The Sphericity Test of J. W. MAUCHLY : Tables of B. N. NAGARSENKER and K. C. S. PILLAI 212

Table 12 : The Test Criteria L_{mvc} , L_{vc} , and L_m of S. S. WILKS : Tables of S. S. WILKS and also of J. ROY and V. K. MURTHY 225

Table 13 : The Multivariate Outlier Criteria of S. S. WILKS 235

Table 14 : Multivariate Tolerance Regions with β-Expectation (Type 2) : Tables of D. A. S. FRASER and I. GUTTMAN 248

Table 15 : Multivariate Tolerance Regions with β-Content (Type 1): Tables of V. CHEW 258

Table 16 : Testing a Single Covariance Matrix : Tables of B. P. KORIN 263

Table 17 : Testing the Equality of k Covariance Matrices : Tables of B. P. KORIN 268

Table 18 : Distribution of the Extreme Roots of a WISHART Matrix : Tables of R. Ch. HANUMARA and W. A. THOMPSON 274

Table 19 : The Multivariate t-Distribution : Tables of P. R. KRISHNAIAH and J. V. ARMITAGE 281

Part III : Further Tables for Multivariate Problems 305

Preliminary Remarks 305

Table 20 : The Gamma Distribution : Tables of M. B. WILK, R. GNANADESIKAN, and M. J. HUYETTE 306

Table 21 : The BARGMANN Test for Simple Structure of a Factor Pattern : Tables of R. BARGMANN 314

Table 22 : Upper Percentage Points of the BONFERRONI Chi-Square Statistic : Tables of G. B. BEUS and D. R. JENSEN 329

Table 23 : Lower Percentage Points of the BONFERRONI Chi-Square Statistic : Tables of G. B. BEUS and D. R. JENSEN 358

Table 24 : The Sequential Chi-Square Criterion for Multivariate Comparisons of Means : Tables of R. J. FREUND and J. E. JACKSON 381

Table 25 : The Sequential T^2-Criterion for Multivariate Testing for Means : Tables of R. J. FREUND and J. E. JACKSON 394

Part IV (Appendix) : Supplement 419

Preliminary Remarks 419

Table 26 : The MARDIA-Test for Multivariate Normality, Skewness, and Kurtosis : Tables by K. V. MARDIA 420

Table 27 : Sample Size Requirements for the T^2-Test of MANOVA in One-Way Classifications : Tables of J. LÄUTER 432

Table 28 : Critical Values for Simultaneous and Sequential BONFERRONI z-Tests : Tables of G. A. LIENERT, O. LUDWIG, and K. ROCKENFELLER 452

Table 29 : Upper Percentage Points of the BONFERRONI t-Statistic : Tables of B. J. R. BAILEY 467

Table 30 : Upper Percentage Points of Statistics for Testing Covariance Matrices : Tables of J. C. LEE, T. C. CHANG, and P. R. KRISHNAIAH 478

Final Remarks : Univariate Test Distributions as a Special Case of Their Multivariate Analogs 503

Permissions and Acknowledgments

Acknowledgment is made to the authors, editors, and publishers of the following publications, whose tables have been used in this " STATISTICAL TABLES for MULTIVARIATE ANALYSIS - A HANDBOOK with REFERENCES to APPLICATIONS " , for which permissions have been received. (Numbering corresponds to that of the tables) :

(1) WALL, F. J. : The generalized variance ratio or U-statistic.
Albuquerque / New Mexico : The Dikewood Corporation, 1967 .

(2) PILLAI, K. C. S. : Statistical Tables for Tests of Multivariate Hypotheses. (Table 1).
Manila / Philippines : The Statistical Center, The University of the Philippines, 1960 .
PEARSON, E. S. + HARTLEY, H. O. (eds.) : Biometrika Tables for Statisticians. Vol. II. (Table 48).
Cambridge : The Cambridge University Press, 1972 .
(Published for the Biometrika Trustees).

(3) BOCK, R. D. : Multivariate Statistical Methods in Behavioral Research. (Table : The generalized F Statistic).
New York : McGraw-Hill, 1975 .
(And : Preprints 1969/1971) .

(4) HECK, D. L. : Charts of some upper percentage points of the distribution of the largest characteristic root.
The Annals of Mathematical Statistics 31, 625 - 642 (1960).

(5) FOSTER, F. G. + REES, D. H. : Upper percentage points of the generalized beta distribution. I.
Biometrika 44, 237 - 247 (1957) .

FOSTER, F. G. : Upper percentage points of the generalized beta distribution. II.
Biometrika 44, 441 -453 (1957).

FOSTER, F. G. : Upper percentage points of the generalized beta distribution. III.
Biometrika 45, 492 - 503 (1958) .

PEARSON, E. S. + HARTLEY, H. O. (eds.) : Biometrika Tables for Statisticians. Vol. II. (Table 49).
Cambridge : The Cambridge University Press, 1972 .
(Published for the Biometrika Trustees).

(6) PILLAI, K. C. S. : Statistical Tables for Tests of Multivariate Hypotheses. (Table 3).
Manila / Philippines : The Statistical Center, The University of the Philippines, 1960 .

(7) PILLAI, K. C. S. : Statistical Tables for Tests of Multivariate Hypotheses. (Table 2).
Manila / Philippines : The Statistical Center, The University of the Philippines, 1960 .

(8) JENSEN, D. R. + HOWE, R. B. : Tables of HOTELLING's T^2-Distribution.
Blacksburg / Virginia : Virginia Polytechnic Institute (Technical Report No. 9)., March 1968.
(Revised Edition : August 1972).

(9) GUPTA, S. S. : Probability integrals of multivariate normal and multivariate t . (Table II).
The Annals of Mathematical Statistics 34, 792 - 828 (1963).

(10) GUPTA, S. S. + NAGEL, K. + PANCHAPAKESAN, S. : On the order statistics from equally correlated normal random variables.
Biometrika 60, 403 - 413 (1973).

(11) NAGARSENKER, B. N. + PILLAI, K. C. S. : The distribution of the sphericity test criterion. Mimeograph Series No. 284 , Lafayette / Indiana : Department of Statistics , Purdue University, 1972 .

(12) WILKS, S. S. : Sample criteria for testing equality of means , equality of variances , and equality of covariances in a normal multivariate distribution. (Table I, Table II, Table III, Table IV).
The Annals of Mathematical Statistics 17, 257 - 281 (1946).

ROY, J. + MURTHY, V. K. : Percentage points of WILKS' L_{mvc} and L_{vc} criteria. (Table 3 , Table 4).
Psychometrika 25. 243 - 250 (1960).

(13) WILKS, S. S. : Multivariate Statistical Outliers.
Sankhya, Series A, Vol. 25, 407 - 426 (1963).

(14) FRASER, D. A. S. + GUTTMAN, I. : Tolerance regions.
The Annals of Mathematical Statistics 27, 162 - 179(1956).

(15) CHEW, V. : Confidence, prediction, and tolerance regions for the multivariate normal distribution. (Table 2).
Journal of the American Statistical Association 61, 605 - 617 (1966) .

(16) KORIN, B. P. : On the distribution of a statistic used for testing a covariance matrix. (Table 3).
Biometrika 55, 171 - 178 (1968) .

PEARSON, E. S. + HARTLEY, H. O. (eds.) : Biometrika Tables for Statisticians. Vol. ii. (Table 53).
Cambridge : The Cambridge University Press, 1972 .
(Published for the Biometrika Trustees) .

(17) KORIN, B. P. : On testing the equality of k covariance matrices. (Table 2).
Biometrika 56, 216 218 (1969) .

PEARSON, E. S. + HARTLEY, H. O. (eds.) : Biometrika Tables for Statisticians. Vol. II. (Table 50).
Cambridge : The Cambridge University Press, 1972 .
(Published for the Biometrika Trustees).

(18) HANUMARA, R. Ch. + THOMPSON, Jr. W. A. : Percentage points of the extreme roots of a WISHART matrix.
Biometrika 55, 503 - 512 (1968).

(19) KRISHNAIAH, P. R. + Armitage, J. V. : Tables for multivariate t-distribution.
Sankhya, Series B, Vol. 28, 31 - 56 (1966).

(20) WILK, M. B. + GNANADESIKAN, R. + HUYETTE, M. J. : Probability plots for the gamma distribution.
Technometrics 4, 1 - 20 (1962) .

(21) BARGMANN, R. : Signifikanzuntersuchungen der einfachen Struktur in der Faktoren-Analyse.
Mitteilungsblatt für mathematische Statistik 7, 1 - 24 (1955) .

(22) BEUS, G. B. + JENSEN, D. R. : Percentage points of the Bonferroni chi-square statistic. (Technical Report No. 3). (Table 1) .
Blacksburg / Virginia : Department of Statistics , Virginia Polytechnic Institute and State University , September 1967 .

(23) BEUS, G. B. + JENSEN, D. R. : Percentage points of the Bonferroni chi-square statistic. (Technical Report No. 3). (Table 2).
Blacksburg / Virginia : Department of Statistics, Virginia Polytechnic Institute and State University , September 1967 .

(24) FREUND, R. J. + JACKSON, J. E. : Tables to facilitate multivariate sequential testing for means. (Technical Report No. 12). (Table 7).
Blacksburg / Virginia : Department of Statistics and Statistical Laboratory, Virginia Agricultural Experiment Station, Virginia Polytechnic Institute , September 1960 .

(25) FREUND, R. J. + JACKSON, J. E. : Tables to facilitate multivariate sequential testing for means. (Technical Report No. 12). (Table 8).
Blacksburg / Virginia : Department of Statistics and Statistical Laboratory, Virginia Agricutltural Experiment Station, Virginia Polytechnic Institute , September 1960 .

(26) MARDIA, K. V. : 'Personal Communication'. August 1975. Department of Statistics, School of Mathematics. The University of Leeds, Leeds / England .

(27) LÄUTER, J. : Sample Size Requirements for the T^2 Rest of MANOVA (Tables for One-way Classification).
Biometrica Journal 20, 389 - 406 (1978) .

(28) LIENERT, G. A. + LUDWIG, O. + ROCKENFELLER, K. : Tables of the critical values for simultaneous and sequential Bonferroni-z-tests.
Biometrical Journal 24, 239 255 (1982) .

(29) BAILEY, B. J. R. : Tables of the Bonferroni t Statistic. Journal of the American Statistical Association 72, 469 - 478 (1977) .

(30) LEE, J. C. + CHANG, T. C. + KRISHNAIAH, P. R. : Approximations to the distributions of the likelihood ratio statistics for testing ceetain structures on the covariance matrices of real multivariate normal populations.
ARL-Report 75 / 0167 (1975) .
(Table III ; Table VII ; Table IX) .

LEE, J. C. + CHANG, T. C. + KRISHNAIAH, P. R. : Approximations to the distributions of the likelihood ratio statistics for testing certain structures on the covariance matrices of real multivariate normal populations.
= pp. 105 - 118 in
KRISHNAIAH, P. R. (ed.) : Multivariate Analysis-IV .
Amsterdam : North Holland Publ. Comp., 1977 .

Remarks on Notation

(1) Vectors and Matrices :

In lieu of the usual printing convention of boldface type , we use underlined lower case letters for vectors and underlined upper case letters for matrices .

For example :

$$\underline{x}' = (x_1, x_2, x_3, \dots , x_p)$$

for the vector $\underline{x}'$,

$$\underline{A} = (a_{ik})$$

for the matrix $\underline{A}$.

(2) Exponential Notation for the Floting Point Representation of Numbers :

The number given in decimal notation is multiplied by a power of ten , with the exponent being the number following the letter E .

For example :

.135 E+3 for $0.135 \times 10^3 = 135$.

.248 E-2 for $0.248 \times 10^{-2} = 0.00248$.

(3) Condensed Notation for a Sequence of Zeros :

In order to abbreviate unneccessarily long expressions , a sequence of zeros following the decimal points is indicated by an exponential notation.

For example : $.0^{4}321$ for 0.0000321 .

PART I

The Primary Tables for Testing Multivariate Statistical Hypotheses

Preliminary Remarks :

The tables contained in this part of the collection constitute the main content of this volume.
The test statistics included in this are usually used together in the English speaking world, so that it seems appropriate to print them together here.

On the one hand there is insufficient research into the test theoretical properties of the various test criteria at the present time to permit a preferential or exclusive choice - in analogy to the F-distribution in univariate analysis - of any one of these.

On the other hand the development of special test methods - as e.g. simultaneous comparisons - based on the less common test statistics, provides a reason for considering these.

The various user customs even make it worthwhile to reproduce together the various tabulated versions and analogs for the maximal root criterion of S. N. ROY.
The test criteria for the multivariate general linear hypothesis are briefly reviewed in chapter 0 .

CHAPTER 0

A Brief Summary of the Test Criteria for the Multivariate General Linear Hypothesis

In the following we consider the multivariate general linear hypothesis in the form

(1) $$E(\underline{X}) = \underline{A} \cdot \underline{\Phi}$$

with a view to testing a corresponding multivariate general linear hypothesis in the form

(2) $$H_0 : \underline{C} \cdot \underline{\Phi} \cdot \underline{M} = \underline{0}$$

against the alternative

(3) $$H_1 : \underline{C} \cdot \underline{\Phi} \cdot \underline{M} \neq \underline{0} \quad .$$

where :

E = "expected value" operator

$\underline{X}$ = matrix of observed values

$\underline{A}$ = sampling method matrix, design matrix

$\underline{\Phi}$ = parameter matrix

$\underline{C}$ = hypothesis matrix for comparisons of elements in the column of the parameter matrix

$\underline{M}$ = hypothesis matrix for comparisons between components of the p-dimensional variate. ($\underline{M}$ plays a role in profile analysis, for example; in ordinary hypothesis tests it is taken to be the identity matrix $\underline{I}$.)

$\underline{0}$ = zero matrix.

If a test criterion for a corresponding multivariate general linear hypothesis is to satisfy an invariance condition with respect to a certain class of transformations (namely, changes in the origin or the unit of measure should not change the value of the test criterion!), then it can be shown that no "uniformly most selective" test of this sort exists.
This otherwise desirable condition is satisfied, however, by the F-test in the univariate case.

As a result, various authors have proposed a number of different criteria for the multivariate case :

Let $\underline{S}_h$ denote the (pxp)-dimensional SP-matrix for the hypothesis ($\underline{S}_h$ is WISHART-distributed with n_h degrees of freedom) and let $\underline{S}_e$ denote the SP-matrix for the error or the remainder, also (pxp)-dimensional ($\underline{S}_e$ is also WISHART-distributed, with n_e degrees of freedom).

Then all of the test statistics may be represented as functions of the $s = \min(p,n_h)$ non-zero roots of the characteristic (determinant) equation

(4) $$\left| \underline{S}_h - \lambda \underline{S}_e \right| = 0$$

that is, as functions of the eigenvalues λ_i of the matrix

(4') $$\underline{S}_e^{-1} \cdot \underline{S}_h \quad \text{oder} \quad \underline{S}_h \cdot \underline{S}_e^{-1} \quad .$$

Instead of (4) and (4') one can also use

(5) $$\left| \underline{S}_h - \theta(\underline{S}_e + \underline{S}_h) \right| = 0$$

and

(5') $$(\underline{S}_e + \underline{S}_h)^{-1} \cdot \underline{S}_h \quad \text{or} \quad \underline{S}_h \cdot (\underline{S}_e + \underline{S}_h)^{-1} \quad .$$

Equally useful is

$$(6) \qquad |\underline{S}_e - \mu(\underline{S}_e + \underline{S}_h)| = 0$$

and

$$(6') \qquad (\underline{S}_e + \underline{S}_h)^{-1} \cdot \underline{S}_e \quad \text{or} \quad \underline{S}_e \cdot (\underline{S}_e + \underline{S}_h)^{-1} .$$

The connection between the various eigenvalues is given by the following relations :

$$(7) \qquad \lambda_i = \frac{\theta_i}{1 - \theta_i} = \frac{1 - \mu_k}{\mu_k} , \quad (i = 1, 2, \ldots, s ; \; k = s, s-1, \ldots, 2, 1)$$

$$(8) \qquad \theta_i = \frac{\lambda_i}{1 + \lambda_i} = 1 - \mu_k \quad \text{and}$$

$$(9) \qquad \mu_k = \frac{1}{1 + \lambda_i} = 1 - \theta_i .$$

Occasionelly one also uses the characteristic equation in the form

$$(10) \qquad | \underline{M}_h - \xi \cdot \underline{M}_e | = 0$$

This contains the "mean" product matrices defined by

$$(11) \qquad \underline{M}_h = (1/n_h) \cdot \underline{S}_h \quad \text{and} \quad \underline{M}_e = (1/n_e) \cdot \underline{S}_e ,$$

where n_h = degrees of freedom of the hypothesis and

n_e = degrees of freedom of the error.

The connection with the eigenvalues defined farther above is given by the relations

$$(12) \qquad \lambda_i = (n_h/n_e) \cdot \xi_i$$

$$(13) \qquad \theta_i = \frac{(n_h/n_e) \cdot \xi_i}{1 + (n_h/n_e) \cdot \xi_i} \quad \text{and}$$

$$(14) \qquad \mu_k = \frac{1}{1 + (n_h / n_e) \cdot \xi_i} .$$

We now present a listing of variously proposed test criteria based on the characteristic determinant equations just listed, with no claims of completeness
(the most important of them are considered in this collection of tables) :

1). The likelihood ratio criterion Λ of S.S.WILKS :

(S. S. WILKS (1932) and P. L. HSU (1940)).

$$K_1 \equiv \Lambda = \frac{|\underline{S}_e|}{|\underline{S}_e + \underline{S}_h|} = \frac{1}{|\underline{I} + \underline{S}_e^{-1} \cdot \underline{S}_h|}$$

$$= \prod_{i=1}^{s} \frac{1}{1+\lambda_i} = \prod_{i=1}^{s} (1-\theta_i) = \prod_{i=1}^{s} \mu_i \quad .$$

The test statistic is tabulated in Table 1.

Remark : In computing Λ as the quotient of the determinants $|\underline{S}_e|$ and $|\underline{S}_h + \underline{S}_e|$ one ordinarily uses the GAUSS-DOOLITTLE- reduction. The pivot elements which arise thereby permit a simple computation of the initial values for the step-down test of J. ROY (1958) and of S. N. ROY & R. E. BARGMANN (1958). This test is a multivariate process wherein the results of the test depend on the order in which the individual components of the multivariate vectors of the GAUSS DOOLITTLE reduction are considered.

2). WILKS' U-criterion :

(S. S. WILKS (1932) and P. L. HSU (1940)).

$$K_2 \equiv U = \frac{S_h}{|S_e + S_h|} = \prod_{i=1}^{s} (1 + \lambda_i) \text{ , or}$$

when defined by the other eigenvalues :

$$= \prod_{i=1}^{s} \frac{1}{1 - \theta_i} = \prod_{i=1}^{s} \frac{1}{\mu_i} .$$

3). <u>The trace criterion of H. HOTELLING and D.N. LAWLEY</u> :

(H. HOTELLING (1951) and D.N. LAWLEY (1938)).

$$K_3 \equiv V = \text{Trace}\ (S_h \cdot S_e^{-1}) = \sum_{i=1}^{s} \lambda_i$$

$$= \sum_{i=1}^{s} \frac{\theta_i}{1 - \theta_i} = \sum_{i=1}^{s} \frac{1 - \mu_i}{\mu_i} = T_o^2 / n_e .$$

(Notation : V in LAWLEY ; T_o^2 in HOTELLING) .

The test statistic is tabulated in <u>Table 6</u>.

In addition, Table 8 contains for the special case of T_o^2 for $n_h = 1$ a tabulation of the test statistic $T^2 = n_e \cdot V$ of H. HOTELLING. Here we have direct multivariate extension of the univariate t-test.

4). <u>The trace criterion of H. HOTELLING, D.N. LAWLEY, K. C. S. PILLAI and M. S. BARTLETT</u> :

(H. HOTELLING (1951), D. N. LAWLEY (1938), K. C. S. PILLAI (1955), M. S. BARTLETT (1939)).

$$K_4 \equiv V^{(s)} = \text{Trace}\ \{S_h \cdot (S_e + S_h)^{-1}\} = \sum_{i=1}^{s} \theta_i$$

$$= \sum_{i=1}^{s} \frac{\lambda_i}{1 + \lambda_i} = \sum_{i=1}^{s} (1 - \mu_i) .$$

This criterion is tabulated in <u>Table 7</u>.

5). <u>The maximal root criterion of S. N. ROY</u> :

(S. N. ROY (1957)).

$$K_5 \equiv \lambda_{max} = \frac{\theta_{max}}{1 - \theta_{max}} = \frac{1 - \mu_{min}}{\mu_{min}}$$

<u>Remark</u> : S. N. ROY (1957) suggested the use of λ_{max} as a test criterion and derived the zero distribution of $\theta_{max} = (\lambda_{max}) / (1 + \lambda_{max})$.

This criterion is tabulated in <u>Table 3</u> in the form of the generalized F-criterion of R. D. BOCK (1969/71 , 1975) .

6). <u>A further maximal root criterion of K. C. S. PILLAI and S. N. ROY</u> :

(K. C. S. PILLAI (1960), S. N. ROY (1957)).

$$K_6 \equiv \theta_{max} = \frac{\lambda_{max}}{1 + \lambda_{max}} = 1 - \mu_{min} \quad .$$

This criterion is tabulated in <u>Table 2</u> in the version due to K. C. S. PILLAI, in <u>Table 4</u> in the form of nomograms due to D. L. HECK and in <u>Table 5</u> in a version due to F. G. FORSTER & D. H. REES.

7). <u>The minimal root criterion of T. W. ANDERSON</u> :

(T. W. ANDERSON (1958)).

$$K_7 \equiv \lambda_{min} = \frac{\theta_{min}}{1 - \theta_{min}} = \frac{1 - \mu_{max}}{\mu_{max}}$$

8). <u>The H-criterion of K. C. S. PILLAI</u> :

(K. C. S. PILLAI (1955)).

$$K_8 \equiv H^{(s)} = s \cdot \left\{ \sum_{i=1}^{s} (1-\theta_i)^{-1} \right\}^{-1} \quad .$$

9). <u>The R-criterion of K. C. S. PILLAI</u> :

(K. C. S. PILLAI (1955)).

$$K_9 \equiv R^{(s)} = s \cdot \left\{ \sum_{i=1}^{s} \theta_i^{-1} \right\}^{-1} .$$

10). <u>The T-criterion of K. C. S. PILLAI</u> :

(K. C. S. PILLAI (1955)).

$$K_{10} \equiv T^{(s)} = s \cdot \left\{ \sum_{i=1}^{s} \lambda_i^{-1} \right\}^{-1} .$$

11). <u>The U-criterion of R. GNANADESIKAN</u> :

(R. GNANADESIKAN et al. (1965), S. N. ROY et al (1971)).

$$K_{11} \equiv U = \prod_{i=1}^{s} \theta_i = \prod_{i=1}^{s} \frac{\lambda_i}{1 + \lambda_i} .$$

12). <u>The S-criterion of CH. L. OLSON</u> :

(CH. L. OLSON (1974)).

$$K_{12} \equiv S = \prod_{i=1}^{s} \lambda_i = \prod_{i=1}^{s} \frac{\theta_i}{1 - \theta_i} .$$

13). <u>The T^2-criterion of H. HOTELLING</u> :

(H. HOTELLING (1931)).

$$K_{13} \equiv T^2(p, \nu) = k \cdot D .$$

This criterion is a special case of the generalized T_o^2-criterion of H. HOTELLING (see Nr. 3). - for $n_h = 1$; it is a direct multivariate extension of the univariate t-test.

<u>Table 8</u> contains a tabulation of the T^2-criterion.

As already noted (and in contrast to univariate analysis, where the F-test is almost exclusively used), there is a variety of test statistics available in multivariate analysis. Nevertheless, in practice the number of significant tests reduces itself to the following three :

I. WILKS' Λ-criterion ,

II. HOTELLING's trace criterion ,

III. S. N. ROY's maximal root criterion ,

the latter in several tabulations with varying initial parameters.

This multiplicity of current test statistics is primarily due to the fact that all presently available investigations into the quality (power) and robustness of the test methods (see, e.g.: K. ITO (1962, 1969), K.V. MARDIA (1974, 1975), CH.L. OLSON (1974), K.C.S. PILLAI & K. JAYACHANDRA (1967), H.O. POSTEN & R.E. BARGMANN (1964)) have not led to a preference for any particular test statistic.

In addition, for several test statistics special test methods have been developed - for example for simultaneous comparisons - which alone would lead one to expect a wider application of these test statistics. In conclusion, let us note that in addition to the statistics described above (which are based on the distribution of eigenvalues), other criteria have been proposed, based on the multivariate t-distribution, the multivariate chi-square distribution, and the multivariate E-distribution, and the multivariate F-distribution.

See e.g. P.R. KRISHNAIAH (1969) and the survey of K.R. GABRIEL (1969).

References :

ANDERSON, T. W. : An introduction to multivariate statistical analysis. New York : Wiley, 1958.

BARTLETT, M. S. : A note on tests of significance in multivariate analysis. Proceedings of the Cambridge Philosophical Society 35, 180 - 185 (1939).

BOCK, R. D. : Multivariate statistical methods in behavioral research. New York : McGraw-Hill, 1975. (+ Preprints 1969/ 1971).

DAS GUPTA, S. + PERLMAN, M. D. : On the power of WILKS' U-test for MANOVA. Journal of Multivariate Analysis 3, 220 - 225 (1973).

FOSTER, F. G. : Upper percentage points of the generalized beta distribution II. Biometrika 44, 441 - 453 (1957).

FOSTER, F. G. : Upper percentage points of the generalized beta distribution III. Biometrika 45, 492 - 503 (1958).

FOSTER, F. G. + REES, D. H. : Upper percentage points of the generalized beta distribution I. Biometrika 44, 237 - 247 (1957).

GABRIEL, K. R. : A comparison of some methods of simultaneous inference in MANOVA. = pp. 67 - 86 in : KRISHNAIAH, P. R. (ed.) : Multivariate Analysis II. New York & London : Academic Press, 1969.

GNANADESIKAN, R. + LAUH, E. + SNYDER, M. + YAO, Y : Efficiency comparisons of certain multivariate analysis of variance test procedures. Paper presented at the Central Regional Meeting of the Institute of Mathematical Statistics, Chicago, December 1964. Abstract in : The Annals of Mathematical Statistics 36, 356 - 357 (February 1965).

HECK, D. L. : Charts of some upper percentage points of the largest characteristic root. Ann. Math. Statist. 31, 625 - 642 (1960).

HOTELLING, H. : The generalization of Student's ratio. The Annals of Mathematical Statistics 2, 360 - 378 (1931).

HOTELLING, H. : A generalized T test and measure of multivariate dispersion. Proc. Second Berk. Symp. on Math. Stat. and Probability 1, 23 - 41 (1951).

HSU, P. L. : On generalized analysis of variance.
Biometrika 31, 221 - 237 (1940).

ITO, K. : On multivariate analysis of variance tests.
Bulletin de l'Institut International de Statistique,
Vol. 38, 87 - 98 (1961).

ITO, K. : A comparison of powers of two multivariate analysis
of variance tests.
Biometrika 49, 445 - 462 (1962).

ITO, K. : On the effect of heteroscedasticity and non-normality
upon some multivariate test procedures. = pp. 87 - 120 in:
KRISHNAIAH, P. R. (ed.) : Multivariate Analysis II.
New York & London : Academic Press, 1969.

KRES, H. : Multivariate Statistische Methoden. Band I.
(Unveröffentlichtes Manuskript. Zürich: ETH, 1969 /
Basel, 1977). (To be published in 1983/84).

KRES, H. : Multivariate Statistische Methoden. Band II.
(Unveröffentlichtes Manuskript. Zürich: ETH, 1969 /
Basel, 1979). (To be published in 1983/84).

KRES, H. : Statistische Tafeln zur multivariaten Analysis.
Ein Handbuch mit Hinweisen zur Anwendung.
Berlin - Heidelberg - New York : Springer , 1975.

KRES, H. : Stochastische Mathematik.
- Eine Siegener Einführungsvorlesung -
(Unveröffentlichtes Vorlesungsmanuskript. Siegen, 1979).

KRES, H. : Multivariate Analysis im Komplexen.
- Eine Siegener Einführungsvorlesung -
(Unveröffentlichtes Vorlesungsmanuskript. Siegen, 1980).
(To be published 1983).

KRES, H. : Stochastische Funktionen (= Prozesse) im Komplexen.
- Eine Siegener Einführungsvorlesung -
(Unveröffentlichtes Vorlesungsmanuskript. Siegen, 1980).
(In preparation for publication).

KRISHNAIAH, P. R. (ed.) : Multivariate Analysis.
(Proceedings of an International Symposium held in Dayton, Ohio, June 14 - 19, 1965).
New York & London : Academic Press, 1966.

KRISHNAIAH, P. R. : Simultaneous test procedures under general MANOVA models. = pp. 121 - 143 in : KRISHNAIAH, P.R.(ed.): Multivariate Analysis II.
New York & London : Academic Press, 1969.

KRISHNAIAH, P. R. : Multivariate Analysis II.
(Proceedings of the Second International Symposium on Multivariate Analysis held at Wright State University, Dayton/Ohio, June 17 - 22, 1968).
New York & London : Academic Press, 1969.

KRISHNAIAH, P. R. (ed.) : Multivariate Analysis III.
(Proceedings of the Third International Symposium on Multivariate Analysis held at Wright State University, Dayton/Ohio, June 19 - 24, 1972).
New York & London : Academic Press, 1973.

LAWLEY, D. N. : A generalization of Fisher's z test.
Biometrika 30, 180 - 187 (1938).

MARDIA, K. V. : Applications of some measures of multivariate skewness and kurtosis in testing normality and robustness studies.
Sankhya, Series B, Vol. 36, 115 - 128 (1974).

MARDIA, K. V. : Assessment of multinormality and robustness of HOTELLING's T^2-test.
Applied Statistics 24, 163 - 171 (1975).

OLSON, CH. L. : Comparative robustness of six tests in multivariate analysis of variance.
Journal of the American Statistical Association 69, 894 - 908 (1974).

PILLAI, K. C. S. : Some new test criteria in multivariate analysis.The Annals of Mathematical Statistics 26, 117 - 121 (1955).

PILLAI, K. C. S. : Statistical Tables for Tests of Multivariate Hypotheses.
Manila/Philippines : The Statistical Center. The University of the of the Philippines, 1960.

PILLAI, K. C. S. + JAYACHANDRA, K. : Power comparisons of tests of two multivariate hypotheses based on four criteria.
Biometrika 54, 195 - 210 (1967).

POSTEN, H. O. + BARGMANN, R. E. : Power of the likelihood ratio test of the general linear hypothesis in multivariate analysis.
Biometrika 51, 467 - 480 (1964).

ROY, J. : Step-down procedure in multivariate analysis.
The Annals of Mathematical Statistics 29,1177-1187(1958).

ROY, S. N. : Some aspects of multivariate analysis.
New York : Wiley, 1957. + Calcutta : Indian Statistical Institute, 1957.

Roy, S. N. + BARGMANN, R. E. : Tests of multiple independence and the associated confidence bounds.
The Annals of Mathematical Statistics 29,491 - 503(1958).

ROY, S. N. + GNANADESIKAN, R. + SRIVASTAVA, J. N. : Analysis and design of certain quantitative multiresponse experiments.
Oxford : Pergamon Press, 1971.

SEBER, G. A. F. : The Linear Hypothesis : A General Theory.
(= No. 19 of Griffin's Statistical Monographs & Courses).
London : Griffin, 1966.

WILKS, S. S. : Certain generalizations in the analysis of variance.
Biometrika 24, 471 - 494 (1932).

TABLE 1

The Likelihood Ratio Criterion Λ of S. S. Wilks: Tables of F. J. Wall

(a) Content of the tables and definition of the test statistic :

The tables contain the lower percentage points of the test statistic (the likelihood ratio criterion)

$$\Lambda \equiv U_{p,q,n} = \frac{|\underline{S}_e|}{|\underline{S}_e + \underline{S}_h|} = \frac{1}{|\underline{I} + \underline{S}_e^{-1} \cdot \underline{S}_h|} = \prod_{i=1}^{s} \frac{1}{1+\lambda_i} .$$

Here

$$\underline{S}_h = \text{SP matrix of the hypothesis}$$

$$\underline{S}_e = \text{SP matrix of the error.}$$

The λ_i may be determined as the roots of the determinant, $|\underline{S}_h - \lambda \underline{S}_e| = 0$, or as the eigenvalues of the matrix $\underline{S}_e^{-1} \cdot \underline{S}_h$.

The relationship with the maximum root criterion of S. N. ROY is given by the expressions

$$\lambda = \frac{\theta}{1-\theta} \quad \text{or} \quad \theta = \frac{\lambda}{1+\lambda} \quad \text{or} \quad \frac{1}{1+\lambda} = 1 - \theta .$$

Similarly, Λ can be computed from the μ_k in the relation

$$\frac{1}{1+\lambda_i} = \mu_k .$$

(b) Extent of the tables and definition of the parameters :

(1) The parameter α :

α = the error probability

$\alpha \equiv C = 0.05$ and 0.01

$= 5\%$ and 1%.

(2) The parameter p :

p = dimension of the variates

for p = 1(1)8. Condition : $p \leq n_e$.

(3) The parameter q :

$q = n_h$ = degrees of freedom of the hypothesis

for q = 1(1)15,18(3)30,40(20)120 .

(4) The parameter n :

$n = n_e$ = degrees of freedom of the error

for n = 1(1)30,40(2)140,170,200,240,320,440,

600,800,1000,∞

Condition : $n = n_e \geq p$.

(c) Suggestions for applications :

(1) Multivariate general linear hypotheses, especially MANOVA problems, multivariate regression and correlation analyses, multivariate confidence regions, linear contrasts, canonical discriminant analyses.

(2) For testing the independence of p (dependent) from q (independent) variates (multivariate regression analysis)

where

$$\underline{S}_h = \underline{R}_{yx} \cdot \underline{R}_x^{-1} \cdot \underline{R}_{xy} \quad \text{and}$$

$$\underline{S}_e = \underline{R}_y - \underline{R}_{yx} \cdot \underline{R}_x^{-1} \cdot \underline{R}_{xy} \quad \text{(Standardized}$$

samples ; $\underline{R}_x$, $\underline{R}_y$, $\underline{R}_{xy}$, $\underline{R}_{yx}$ = correlation matrices)

using the initial parameters

p = dimension of the (dependent) vector $\underline{y}$

$q = n_h$ = dimension of the (independent) vector $\underline{x}$

$n = N - q - 1$, where N = sample size.

(d) References :

(1) For the test criterion :

WILKS, S. S. : Certain generalizations in the analysis of variance. Biometrika 24,471 - 494(1932).

(2) For a copy of the tables :

WALL, F. J. : The generalized variance ratio or U-statistic. Albuquerque / New Mexico : The Dikewood Corporation, 1967 .

(e) Further remarks :

(1) The table for $p = 1$ is identical with the table for the beta distribution, which may be applied as an (although rarely used !!) alternative to the F-distribution to test hypotheses of univariate statistics.

(2) For special cases of the parameter p and $n_h = q$, the distribution of Λ is identical with certain F-distributions. The following table provides the transformation of Λ to exact right-sided tests (i.e., upper percentage points, corresponding to lower ones for Λ)used in

applications of the F-distribution.

The parameters p and n_h	Statistics with F-distribution	Degrees of freedom of the F-distribution f_1 ; f_2
$n_h=1$;p arbitrary	$\frac{1-\Lambda}{\Lambda} \cdot \frac{n_e + n_h - p}{p}$	p ; $n_e + n_h - p$
$n_h=2$;p arbitrary	$\frac{1-\sqrt{\Lambda}}{\sqrt{\Lambda}} \cdot \frac{n_e + n_h - p - 1}{p}$	$2p$; $2.(n_e + n_h - p - 1)$
$p=1$;n_h arbitrary	$\frac{1-\Lambda}{\Lambda} \cdot \frac{n_e}{n_h}$	n_h ; n_e
$p=2$;n_h arbitrary	$\frac{1-\sqrt{\Lambda}}{\sqrt{\Lambda}} \cdot \frac{n_e - 1}{n_h}$	$2.n_h$; $2.(n_e - 1)$

(3) Approximations of Λ :

Outside the tubulated range, the Λ-distribution can be approximated by the χ^2- and the F-distribution.

(3.1) The χ^2 -approximation of M. S. BARTLETT :

In this case,

$$\chi_B^2 \sim V = -(n_e + n_h - \frac{p + n_h + 1}{2}) \cdot \ln \Lambda$$

$$= (n_e + n_h - \frac{p + n_h + 1}{2}) \cdot \prod_{i=1} \ln(1+\lambda_i)$$

with $f = p \cdot n_h$ degrees of freedom.

When n_e is small in comparison to n_h and p, the following F-approximation of C.R. RAO provides more exact values.

(3.2) The F-approximation of C. R. RAO :

$$F_R \sim R = \frac{1 - \Lambda^{1/s}}{\Lambda^{1/s}} \cdot \frac{m.s - (p.n_h)/2 + 1}{p \cdot n_h}$$

with parameters

$$m = n_e + n_h - \frac{(p + n_h + 1)}{2} \qquad \text{and}$$

$$s = \sqrt{\frac{(p \cdot n_h)^2 - 4}{p^2 + n_h^2 - 5}}$$

and degrees of freedom

$$f_1 = p \cdot n_h \qquad \text{and}$$

$$f_2 = m \cdot s - (p \cdot n_h)/2 + 1 \quad .$$

(3.3) To test for canonical correlation use in particular

(aa) for M. S. BARTLETT :

$$\chi_B^2 = V = - \left[(N - 1) - \frac{1}{2} \cdot (p + q + 1) \right] \cdot \ln \Lambda$$

with $f = p.q$ degrees of freedom.
(N = sample size).

(bb) for C. R. RAO :

in this case simply substitute q for n_e in the expressions given in (3.2) .

(3.4) References for the approximations :

BARTLETT, M. S. : Multivariate Analysis.
J. Roy. Statist. Soc. , Series B, Vol. 9 ,
176 - 197 (1947).

RAO, C. R. : Advanced Statistical Methods in
Biometric Research.
New York : Wiley, 1952.

(4) Remarks on different forms of notations :

In referring to the technical literature, one, one should be awere of the multiplicity of notation for the SP-matrices and the parameters, which is a steady source of confusion.

Herewith a very small selection :

Here :	F.J. WALL :	T.W. ANDERSON* :	G.A.F. SEBER* :
N	(N)	N	n
p	p	p	p
S_h	B - A	H	A
S_e	A	G	A
n_h	q	q_1	q
n_e	n	n	n - p

(*) See references in chapter 0 .

Tafel 1 Table 1

$U_{p,q,n}$

p=1 C=0.05

q

n	1	2	3	4	5	6	7	8	9	10	11	12	n
1	0.006157	0.002501	0.001543	0.001112	0.000868	0.000712	0.000603	0.000523	0.000462	0.000413	0.000374	0.000341	1
2	0.097504	0.050003	0.033615	0.025322	0.020309	0.016953	0.014549	0.012741	0.011333	0.010208	0.009281	0.008512	2
3	0.228516	0.135712	0.097321	0.076019	0.062408	0.052963	0.046005	0.040672	0.036446	0.033020	0.030182	0.027794	3
4	0.341614	0.223602	0.168243	0.135345	0.113373	0.097610	0.085724	0.076447	0.068985	0.062851	0.057724	0.053375	4
5	0.430725	0.301697	0.235535	0.194031	0.165283	0.144073	0.127777	0.114822	0.104279	0.095505	0.088120	0.081787	5
6	0.500549	0.368408	0.295990	0.248596	0.214783	0.189255	0.169266	0.153168	0.139893	0.128754	0.119278	0.111115	6
7	0.555908	0.424896	0.349304	0.298096	0.260620	0.231812	0.208893	0.190186	0.174606	0.161423	0.150116	0.140289	7
8	0.600708	0.472870	0.396057	0.342590	0.302612	0.271332	0.246124	0.225311	0.207825	0.192902	0.180008	0.168747	8
9	0.637512	0.513916	0.437164	0.382446	0.340790	0.307770	0.280823	0.258362	0.239288	0.222931	0.208679	0.196182	9
10	0.668243	0.549286	0.473389	0.418213	0.375519	0.341248	0.313019	0.289246	0.268936	0.251373	0.235992	0.222443	10
11	0.694275	0.580017	0.505463	0.450317	0.407104	0.372040	0.342834	0.318054	0.296768	0.278229	0.261932	0.247467	11
12	0.716553	0.606964	0.534027	0.479309	0.435913	0.400299	0.370453	0.344940	0.322876	0.303528	0.286469	0.271240	12
13	0.735840	0.630737	0.559570	0.505524	0.462189	0.426361	0.396057	0.369995	0.347321	0.327362	0.309662	0.293823	13
14	0.752686	0.651825	0.582581	0.529327	0.486267	0.450348	0.419800	0.393372	0.370239	0.349823	0.331589	0.315247	14
15	0.767548	0.670715	0.603333	0.551025	0.508362	0.472534	0.441864	0.415222	0.391754	0.370941	0.352325	0.335541	15
16	0.780701	0.687653	0.622162	0.570862	0.528717	0.493103	0.462433	0.435638	0.411957	0.390869	0.371918	0.354797	16
17	0.792480	0.702972	0.639343	0.589081	0.547516	0.512177	0.481598	0.454742	0.430939	0.409637	0.390472	0.373077	17
18	0.803070	0.716858	0.655029	0.605835	0.564911	0.529907	0.499481	0.472687	0.448807	0.427368	0.408020	0.390411	18
19	0.812622	0.729553	0.669434	0.621307	0.581024	0.546448	0.516235	0.489502	0.465637	0.444138	0.424652	0.406891	19
20	0.821320	0.741135	0.682709	0.635651	0.596039	0.561890	0.531952	0.505341	0.481506	0.459991	0.440430	0.422546	20
21	0.829224	0.751770	0.694977	0.648941	0.610046	0.576355	0.546692	0.520264	0.496521	0.475006	0.455414	0.437469	21
22	0.836472	0.761597	0.706329	0.661316	0.623108	0.589905	0.560562	0.534332	0.510712	0.489258	0.469635	0.451660	22
23	0.843140	0.770660	0.716858	0.672867	0.635361	0.602631	0.573639	0.547638	0.524139	0.502762	0.483185	0.465179	23
24	0.849274	0.779083	0.726685	0.683655	0.646851	0.614609	0.585968	0.560211	0.536896	0.515594	0.496078	0.478088	24
25	0.854950	0.786896	0.735870	0.693771	0.657639	0.625900	0.597626	0.572128	0.548981	0.527817	0.508362	0.490402	25
26	0.860199	0.794189	0.744446	0.703278	0.667786	0.636566	0.608643	0.583435	0.560486	0.539459	0.520081	0.502167	26
27	0.865112	0.800995	0.752487	0.712189	0.677383	0.646637	0.619080	0.594147	0.571411	0.550537	0.531281	0.513428	27
28	0.869675	0.807373	0.760040	0.720612	0.686432	0.656174	0.628998	0.604370	0.581833	0.561127	0.541962	0.524200	28
29	0.873947	0.813339	0.767151	0.728546	0.694992	0.665222	0.638428	0.614075	0.591766	0.571228	0.552200	0.534515	29
30	0.877945	0.818970	0.773865	0.736053	0.703110	0.673798	0.647385	0.623322	0.601242	0.580872	0.561996	0.544418	30
40	0.907349	0.860886	0.824463	0.793274	0.765594	0.740540	0.717575	0.696365	0.676636	0.658188	0.640884	0.624603	40
60	0.937485	0.904968	0.878807	0.855911	0.835175	0.816055	0.798233	0.781494	0.765686	0.750702	0.736420	0.722809	60
80	0.952827	0.927841	0.907471	0.889450	0.872940	0.857590	0.843124	0.829437	0.816391	0.803925	0.791962	0.780464	80
100	0.962128	0.941845	0.925179	0.910324	0.896637	0.883835	0.871696	0.860153	0.849083	0.838455	0.828201	0.818314	100
120	0.968363	0.951297	0.937200	0.924578	0.912894	0.901916	0.891475	0.881501	0.871901	0.862660	0.853706	0.845045	120
140	0.972836	0.958107	0.945890	0.934921	0.924731	0.915131	0.905971	0.897200	0.888734	0.880563	0.872625	0.864929	140
170	0.977588	0.965370	0.955195	0.946025	0.937476	0.929401	0.921669	0.914245	0.907057	0.900101	0.893324	0.886738	170
200	0.980926	0.970487	0.961768	0.953893	0.946532	0.939564	0.932877	0.926443	0.920200	0.914149	0.908239	0.902486	200
240	0.984086	0.975345	0.968024	0.961396	0.955187	0.949296	0.943631	0.938171	0.932861	0.927705	0.922660	0.917740	240
320	0.988046	0.981451	0.975907	0.970876	0.966145	0.961649	0.957311	0.953121	0.949035	0.945058	0.941155	0.937344	320
440	0.991295	0.986475	0.982411	0.978715	0.975232	0.971914	0.968704	0.965599	0.962561	0.959605	0.956692	0.953846	440
600	0.993610	0.990064	0.987067	0.984337	0.981759	0.979301	0.976917	0.974611	0.972349	0.970144	0.967969	0.965842	600
800	0.995204	0.992539	0.990282	0.988225	0.986279	0.984422	0.982619	0.980873	0.979158	0.977487	0.975834	0.974218	800
1000	0.996161	0.994026	0.992216	0.990566	0.989003	0.987512	0.986062	0.984658	0.983276	0.981931	0.980598	0.979296	1000
INF	1.000000	1.000000	1.000000	1.000000	1.000000	1.000000	1.000000	1.000000	1.000000	1.000000	1.000000	1.000000	INF

$U_{p,q,n}$

p=1 C=0.01

n	q=1	2	3	4	5	6	7	8	9	10	11	12	n
1	0.000247	0.000100	0.000062	0.000044	0.000035	0.000028	0.000024	0.000021	0.000018	0.000017	0.000015	0.000014	1
2	0.019900	0.010000	0.006678	0.005013	0.004012	0.003344	0.002867	0.002509	0.002231	0.002008	0.001826	0.001674	2
3	0.080827	0.046416	0.032834	0.025458	0.020806	0.017599	0.015251	0.013458	0.012043	0.010898	0.009951	0.009157	3
4	0.158742	0.100000	0.073959	0.058903	0.049014	0.041999	0.036755	0.032682	0.029427	0.026763	0.024544	0.022665	4
5	0.235203	0.158489	0.121418	0.098877	0.083563	0.072430	0.063947	0.057265	0.051854	0.047390	0.043634	0.040434	5
6	0.303867	0.215443	0.169784	0.140867	0.120651	0.105640	0.094010	0.084728	0.077134	0.070801	0.065439	0.060839	6
7	0.363705	0.268270	0.216358	0.182355	0.158006	0.139585	0.125112	0.113417	0.103749	0.095628	0.088696	0.082716	7
8	0.415397	0.316227	0.259967	0.222073	0.194363	0.173070	0.156116	0.142270	0.130724	0.120944	0.112552	0.105261	8
9	0.460089	0.359381	0.300242	0.259453	0.229097	0.205430	0.186374	0.170658	0.157452	0.146189	0.136452	0.127957	9
10	0.498896	0.398108	0.337189	0.294313	0.261901	0.236323	0.215512	0.198202	0.183548	0.170965	0.160030	0.150442	10
11	0.532793	0.432877	0.370993	0.326670	0.292708	0.265602	0.243349	0.224692	0.208793	0.195061	0.183068	0.172501	11
12	0.562582	0.464159	0.401904	0.356635	0.321526	0.293230	0.269804	0.250027	0.233063	0.218338	0.205413	0.193976	12
13	0.588936	0.492388	0.430204	0.384373	0.348450	0.319237	0.294872	0.274166	0.256310	0.240729	0.226996	0.214791	13
14	0.612381	0.517948	0.456147	0.410058	0.373579	0.343685	0.318575	0.297116	0.278506	0.262202	0.247764	0.234893	14
15	0.633365	0.541170	0.479986	0.433867	0.397056	0.366662	0.340981	0.318908	0.299682	0.282756	0.267719	0.254259	15
16	0.652233	0.562342	0.501931	0.455967	0.418982	0.388257	0.362133	0.339583	0.319844	0.302404	0.286850	0.272887	16
17	0.669300	0.581709	0.522195	0.476513	0.439507	0.408566	0.382133	0.359198	0.339049	0.321175	0.305186	0.290785	17
18	0.684789	0.599484	0.540936	0.495647	0.458725	0.427677	0.401023	0.377807	0.357322	0.339101	0.322737	0.307970	18
19	0.698917	0.615849	0.558319	0.513499	0.476742	0.445681	0.418900	0.395470	0.374735	0.356217	0.339555	0.324463	19
20	0.711843	0.630958	0.574471	0.530184	0.493661	0.462657	0.435811	0.412241	0.391308	0.372565	0.355644	0.340290	20
21	0.723730	0.644947	0.589523	0.545805	0.509577	0.478683	0.451836	0.428178	0.407108	0.388183	0.371060	0.355477	21
22	0.734669	0.657933	0.603568	0.560456	0.524563	0.493830	0.467022	0.443332	0.422166	0.403108	0.385819	0.370054	22
23	0.744795	0.670019	0.616713	0.574221	0.538693	0.508161	0.481441	0.457752	0.436534	0.417377	0.399965	0.384048	23
24	0.754176	0.681293	0.629026	0.587173	0.552034	0.521736	0.495132	0.471485	0.450247	0.431029	0.413515	0.397487	24
25	0.762902	0.691831	0.640594	0.599381	0.564657	0.534611	0.508160	0.484576	0.463349	0.444097	0.426523	0.410397	25
26	0.771028	0.701704	0.651468	0.610905	0.576603	0.546834	0.520546	0.497064	0.475866	0.456613	0.438989	0.422804	26
27	0.778625	0.710971	0.661723	0.621798	0.587931	0.558452	0.532362	0.508986	0.487854	0.468607	0.450974	0.434734	27
28	0.785730	0.719686	0.671391	0.632109	0.598682	0.569507	0.543615	0.520379	0.499314	0.480110	0.462471	0.446211	28
29	0.792406	0.727896	0.680539	0.641884	0.608900	0.580037	0.554370	0.531274	0.510313	0.491149	0.473532	0.457257	29
30	0.798670	0.735642	0.689191	0.651161	0.618619	0.590076	0.564636	0.541702	0.520842	0.501748	0.484160	0.467894	30
40	0.845412	0.794328	0.755603	0.723155	0.694813	0.669500	0.646550	0.625549	0.606163	0.588188	0.571417	0.555726	40
60	0.894480	0.857696	0.828970	0.804330	0.782305	0.762272	0.743738	0.726513	0.710318	0.695108	0.680672	0.667012	60
80	0.919918	0.891251	0.868522	0.848784	0.830928	0.814526	0.799185	0.784809	0.771162	0.758235	0.745861	0.734069	80
100	0.935478	0.912011	0.893219	0.876803	0.861820	0.847989	0.834952	0.822679	0.810943	0.799788	0.789035	0.778749	100
120	0.945976	0.926119	0.910119	0.896070	0.883183	0.871238	0.859925	0.849237	0.838971	0.829183	0.819700	0.810616	120
140	0.953532	0.936329	0.922402	0.910129	0.898828	0.888325	0.878338	0.868886	0.859771	0.851065	0.842600	0.834476	140
170	0.961595	0.947263	0.935601	0.925292	0.915755	0.906869	0.898385	0.890335	0.882539	0.875082	0.867790	0.860800	170
200	0.967270	0.954993	0.944964	0.936079	0.927834	0.920134	0.912760	0.905757	0.898951	0.892434	0.886056	0.879907	200
240	0.972661	0.962351	0.953904	0.946399	0.939414	0.932883	0.926606	0.920640	0.914819	0.909247	0.903766	0.898492	240
320	0.979433	0.971628	0.965202	0.959483	0.954136	0.949127	0.944294	0.939692	0.935191	0.930865	0.926597	0.922489	320
440	0.985001	0.979285	0.974556	0.970342	0.966383	0.962678	0.959082	0.955661	0.952291	0.949067	0.945865	0.942783	440
600	0.988980	0.984767	0.981267	0.978151	0.975211	0.972459	0.969781	0.967231	0.964708	0.962302	0.959899	0.957590	600
800	0.991723	0.988553	0.985913	0.983561	0.981337	0.979256	0.977220	0.975291	0.973377	0.971545	0.969713	0.967956	800
1000	0.993372	0.990832	0.988711	0.986824	0.985033	0.983362	0.981722	0.980169	0.978621	0.977148	0.975664	0.974250	1000
INF	1.000000	1.000000	1.000000	1.000000	1.000000	1.000000	1.000000	1.000000	1.000000	1.000000	1.000000	1.000000	INF

Tafel 1 (Forts.) Table 1 (cont.)

$U_{p,q,n}$

p=1 C=0.05

q

n	13	14	15	18	21	24	27	30	40	60	80	100	120	n
1	0.000314	0.000291	0.000271	0.000225	0.000192	0.000167	0.000148	0.000133	0.000100	0.000066	0.000049	0.000040	0.000033	1
2	0.007860	0.007301	0.006817	0.005684	0.004873	0.004265	0.003792	0.003414	0.002562	0.001708	0.001282	0.001025	0.000854	2
3	0.025757	0.023998	0.022465	0.018852	0.016239	0.014263	0.012718	0.011473	0.008652	0.005799	0.004362	0.003495	0.002916	3
4	0.049637	0.046387	0.043541	0.036774	0.031822	0.028053	0.025082	0.022678	0.017191	0.011585	0.008736	0.007012	0.005857	4
5	0.076309	0.071533	0.067307	0.057198	0.049736	0.043991	0.039444	0.035748	0.027241	0.018459	0.013960	0.011225	0.009386	5
6	0.104004	0.097748	0.092209	0.078819	0.068832	0.061104	0.054932	0.049896	0.038223	0.026043	0.019751	0.015907	0.013317	6
7	0.131683	0.124100	0.117325	0.100861	0.088455	0.078781	0.071014	0.064651	0.049782	0.034103	0.025940	0.020931	0.017542	7
8	0.158829	0.150024	0.142151	0.122849	0.108185	0.096657	0.087357	0.079697	0.061676	0.042480	0.032402	0.026188	0.021976	8
9	0.185120	0.175247	0.166382	0.144501	0.127747	0.114487	0.103729	0.094826	0.073746	0.051071	0.039066	0.031631	0.026577	9
10	0.210373	0.199585	0.189850	0.165649	0.146988	0.132111	0.119980	0.109909	0.085884	0.059784	0.045860	0.037197	0.031288	10
11	0.234558	0.222931	0.212433	0.186203	0.165787	0.149429	0.136017	0.124832	0.098007	0.068581	0.052750	0.042862	0.036098	11
12	0.257599	0.245300	0.234131	0.206070	0.184082	0.166367	0.151779	0.139557	0.110054	0.077393	0.059692	0.048588	0.040966	12
13	0.279572	0.266663	0.254913	0.225250	0.201843	0.182877	0.167206	0.154007	0.121994	0.086212	0.066681	0.054367	0.045895	13
14	0.300476	0.287048	0.274811	0.243729	0.219055	0.198944	0.182266	0.168175	0.133774	0.094994	0.073662	0.060165	0.050846	14
15	0.320343	0.306488	0.293823	0.261505	0.235687	0.214569	0.196945	0.182037	0.145386	0.103737	0.080658	0.065987	0.055832	15
16	0.339233	0.325027	0.311981	0.278595	0.251770	0.229721	0.211258	0.195557	0.156815	0.112396	0.087616	0.071800	0.060822	16
17	0.357208	0.342712	0.329361	0.295044	0.267303	0.244415	0.225174	0.208771	0.168053	0.121002	0.094566	0.077621	0.065826	17
18	0.374329	0.359558	0.345947	0.310822	0.282303	0.258652	0.238693	0.221634	0.179077	0.129494	0.101456	0.083412	0.070816	18
19	0.390625	0.375656	0.361832	0.326004	0.296783	0.272446	0.251846	0.234177	0.189896	0.137909	0.108322	0.089195	0.075817	19
20	0.406143	0.391022	0.377014	0.340607	0.310745	0.285797	0.264618	0.246384	0.200500	0.146217	0.115112	0.094940	0.080788	20
21	0.420944	0.405685	0.391541	0.354630	0.324234	0.298737	0.277008	0.258286	0.210892	0.154427	0.121872	0.100670	0.085762	21
22	0.435059	0.419708	0.405441	0.368118	0.337250	0.311264	0.289063	0.269852	0.221054	0.162521	0.128555	0.106354	0.090698	22
23	0.448547	0.433121	0.418762	0.381104	0.349823	0.323395	0.300751	0.281113	0.231018	0.170517	0.135193	0.112015	0.095631	23
24	0.461426	0.445953	0.431534	0.393585	0.361954	0.335144	0.312103	0.292084	0.240768	0.178383	0.141747	0.117619	0.100517	24
25	0.473755	0.458252	0.443787	0.405609	0.373688	0.346527	0.323135	0.302750	0.250320	0.186150	0.148254	0.123199	0.105400	25
26	0.485535	0.470032	0.455536	0.417206	0.385010	0.357559	0.333847	0.313141	0.259659	0.193787	0.154671	0.128716	0.110237	26
27	0.496826	0.481339	0.466827	0.428375	0.395966	0.368256	0.344254	0.323242	0.268799	0.201340	0.161041	0.134209	0.115059	27
28	0.507645	0.492188	0.477692	0.439133	0.406555	0.378616	0.354370	0.333084	0.277740	0.208755	0.167313	0.139641	0.119835	28
29	0.518036	0.502594	0.488113	0.449524	0.416809	0.388672	0.364197	0.342682	0.286499	0.216064	0.173538	0.145042	0.124596	29
30	0.528000	0.512604	0.498154	0.459549	0.426727	0.398422	0.373749	0.352005	0.295059	0.223251	0.179672	0.150372	0.129303	30
40	0.609207	0.594650	0.580826	0.543274	0.510559	0.481750	0.456146	0.433212	0.371368	0.289360	0.237228	0.201080	0.174522	40
60	0.709793	0.697327	0.685349	0.652084	0.622238	0.595215	0.570602	0.548080	0.484772	0.394585	0.333054	0.288269	0.254158	60
80	0.769379	0.758698	0.748367	0.719315	0.692764	0.668335	0.645737	0.624741	0.564194	0.473579	0.408539	0.359406	0.320923	80
100	0.808723	0.799446	0.790421	0.764862	0.741211	0.719208	0.698639	0.679352	0.622650	0.534698	0.469109	0.418106	0.377228	100
120	0.836619	0.828441	0.820463	0.797725	0.776501	0.756609	0.737860	0.720157	0.667378	0.583258	0.518637	0.467198	0.425191	120
140	0.857422	0.850122	0.842982	0.822544	0.803342	0.785239	0.768078	0.751784	0.702681	0.622729	0.559824	0.508785	0.466444	140
170	0.880295	0.874014	0.867852	0.850140	0.833378	0.817471	0.802288	0.787793	0.743560	0.669737	0.610002	0.560404	0.518457	170
200	0.896845	0.891341	0.885930	0.870316	0.855463	0.841302	0.827722	0.814694	0.774583	0.706377	0.649978	0.602294	0.561335	200
240	0.912906	0.908181	0.903526	0.890057	0.877178	0.864839	0.852950	0.841498	0.805911	0.744248	0.692117	0.647197	0.607973	240
320	0.933589	0.929911	0.926278	0.915722	0.905561	0.895769	0.886275	0.877077	0.848147	0.796735	0.751929	0.712279	0.676831	320
440	0.951033	0.948276	0.945543	0.937582	0.929874	0.922410	0.915132	0.908049	0.885531	0.844610	0.807956	0.774709	0.744313	440
600	0.963734	0.961668	0.959616	0.953623	0.947797	0.942134	0.936589	0.931175	0.913827	0.881761	0.852426	0.825297	0.800044	600
800	0.972614	0.971041	0.969476	0.964900	0.960437	0.956088	0.951817	0.947638	0.934170	0.908972	0.885562	0.863600	0.842879	800
1000	0.978000	0.976731	0.975466	0.971765	0.968148	0.964620	0.961149	0.957746	0.946744	0.926003	0.906550	0.888133	0.870607	1000
INF	1.000000	1.000000	1.000000	1.000000	1.000000	1.000000	1.000000	1.000000	1.000000	1.000000	1.000000	1.000000	1.000000	INF

Tafel 1 (Forts.) Table 1 (cont.)

$U_{p,q,n}$

p=1 C=0.01

n	q = 13	14	15	18	21	24	27	30	40	60	80	100	120	n
1	0.000013	0.000012	0.000011	0.000009	0.000008	0.000007	0.000006	0.000005	0.000004	0.000003	0.000002	0.000002	0.000001	1
2	0.001545	0.001435	0.001339	0.001116	0.000957	0.000837	0.000744	0.000670	0.000502	0.000335	0.000251	0.000201	0.000167	2
3	0.008480	0.007896	0.007387	0.006192	0.005329	0.004678	0.004168	0.003759	0.002832	0.001897	0.001426	0.001143	0.000953	3
4	0.021055	0.019658	0.018436	0.015538	0.013429	0.011824	0.010562	0.009544	0.007223	0.004860	0.003662	0.002938	0.002453	4
5	0.037671	0.035267	0.033148	0.028095	0.024378	0.021534	0.019283	0.017460	0.013276	0.008976	0.006780	0.005448	0.004554	5
6	0.056844	0.053346	0.050255	0.042822	0.037309	0.033056	0.029675	0.026921	0.020565	0.013972	0.010580	0.008513	0.007123	6
7	0.077492	0.072899	0.068817	0.058944	0.051552	0.045818	0.041228	0.037482	0.028768	0.019642	0.014914	0.012022	0.010070	7
8	0.098865	0.093210	0.088172	0.075891	0.066628	0.059388	0.053573	0.048795	0.037623	0.025814	0.019649	0.015861	0.013299	8
9	0.120464	0.113822	0.107870	0.093287	0.082192	0.073474	0.066426	0.060626	0.046959	0.032379	0.024714	0.019986	0.016778	9
10	0.141951	0.134382	0.127589	0.110830	0.097993	0.087835	0.079593	0.072771	0.056619	0.039226	0.030015	0.024306	0.020423	10
11	0.163101	0.154701	0.147127	0.128348	0.113852	0.102325	0.092923	0.085118	0.066516	0.046310	0.035530	0.028823	0.024251	11
12	0.183767	0.174601	0.166326	0.145676	0.129640	0.116804	0.106299	0.097539	0.076542	0.053537	0.041175	0.033454	0.028171	12
13	0.203865	0.194024	0.185100	0.162745	0.145258	0.131206	0.119643	0.109973	0.086663	0.060908	0.046967	0.038225	0.032230	13
14	0.223324	0.212877	0.203377	0.179459	0.160647	0.145440	0.132887	0.122343	0.096797	0.068328	0.052818	0.043051	0.036332	14
15	0.242132	0.231142	0.221131	0.195800	0.175751	0.159480	0.145984	0.134623	0.106935	0.075835	0.058771	0.047983	0.040546	15
16	0.260261	0.248799	0.238326	0.211711	0.190540	0.173273	0.158902	0.146751	0.117013	0.083335	0.064736	0.052931	0.044769	16
17	0.277736	0.265847	0.254967	0.227203	0.204996	0.186806	0.171608	0.158737	0.127041	0.090879	0.070772	0.057962	0.049083	17
18	0.294542	0.282290	0.271041	0.242248	0.219095	0.200049	0.184095	0.170521	0.136964	0.098377	0.076786	0.062980	0.053383	18
19	0.310718	0.298139	0.286575	0.256850	0.232839	0.213015	0.196336	0.182125	0.146800	0.105889	0.082859	0.068064	0.057764	19
20	0.326267	0.313409	0.301561	0.271013	0.246215	0.225664	0.208334	0.193503	0.156498	0.113333	0.088875	0.073112	0.062106	20
21	0.341222	0.328120	0.316031	0.284742	0.259244	0.238029	0.220071	0.204684	0.166095	0.120776	0.094936	0.078222	0.066523	21
22	0.355595	0.342291	0.329985	0.298047	0.271902	0.250074	0.231554	0.215628	0.175537	0.128117	0.100925	0.083275	0.070880	22
23	0.369429	0.355942	0.343454	0.310936	0.284220	0.261841	0.242783	0.226368	0.184861	0.135451	0.106953	0.088380	0.075313	23
24	0.382723	0.369095	0.356446	0.323424	0.296179	0.273293	0.253751	0.236865	0.194019	0.142673	0.112888	0.093414	0.079675	24
25	0.395530	0.381770	0.368989	0.335521	0.307817	0.284459	0.264467	0.247161	0.203056	0.149881	0.118858	0.098496	0.084106	25
26	0.407843	0.393989	0.381079	0.347240	0.319114	0.295338	0.274928	0.257218	0.211920	0.156963	0.124722	0.103500	0.088456	26
27	0.419715	0.405771	0.392781	0.358595	0.330101	0.305936	0.285153	0.267075	0.220667	0.164024	0.130620	0.108547	0.092873	27
28	0.431140	0.417136	0.404060	0.369598	0.340768	0.316261	0.295123	0.276695	0.229234	0.170959	0.136407	0.113508	0.097202	28
29	0.442160	0.428103	0.414975	0.380263	0.351147	0.326320	0.304867	0.286127	0.237682	0.177867	0.142220	0.118511	0.101595	29
30	0.452778	0.438691	0.425510	0.390602	0.361225	0.336119	0.314365	0.295337	0.245954	0.184643	0.147915	0.123413	0.105890	30
40	0.540979	0.527095	0.513973	0.478594	0.448079	0.421443	0.397921	0.376998	0.321106	0.248187	0.202472	0.171057	0.148107	40
60	0.653960	0.641568	0.629677	0.596941	0.567805	0.541669	0.517993	0.496465	0.436552	0.352577	0.296104	0.255388	0.224572	60
80	0.722713	0.711848	0.701349	0.672069	0.645513	0.621290	0.598999	0.578429	0.519625	0.433012	0.371737	0.325879	0.290207	80
100	0.768775	0.759214	0.749910	0.723767	0.699743	0.677593	0.656969	0.637765	0.581777	0.496261	0.433390	0.384961	0.346405	100
120	0.801768	0.793256	0.784937	0.761455	0.739667	0.719423	0.700414	0.682593	0.629875	0.547092	0.484366	0.434905	0.394790	120
140	0.826535	0.818887	0.811381	0.790125	0.770260	0.751693	0.734146	0.717613	0.668144	0.588748	0.527114	0.477565	0.436741	140
170	0.853927	0.847304	0.840772	0.822208	0.804722	0.788268	0.772613	0.757765	0.712782	0.638740	0.579594	0.530926	0.490040	170
200	0.873859	0.868023	0.862255	0.845798	0.830206	0.815474	0.801379	0.787952	0.746884	0.677977	0.621704	0.574538	0.534291	200
240	0.893281	0.888248	0.883255	0.868987	0.855388	0.842479	0.830061	0.818187	0.781511	0.718771	0.666360	0.621578	0.582721	240
320	0.918408	0.914471	0.910551	0.899290	0.888491	0.878163	0.868168	0.858556	0.828488	0.775695	0.730198	0.690253	0.654754	320
440	0.939709	0.936743	0.933777	0.925238	0.916990	0.909074	0.901356	0.893910	0.870341	0.827990	0.790448	0.756647	0.725916	440
600	0.955275	0.953047	0.950805	0.944357	0.938091	0.932063	0.926154	0.920436	0.902183	0.868805	0.838567	0.810796	0.785080	600
800	0.966190	0.964489	0.962772	0.957837	0.953019	0.948379	0.943812	0.939385	0.925163	0.898826	0.874586	0.851993	0.830782	800
1000	0.972817	0.971448	0.970056	0.966059	0.962145	0.958376	0.954654	0.951045	0.939401	0.917670	0.897470	0.878467	0.860467	1000
INF	1.000000	1.000000	1.000000	1.000000	1.000000	1.000000	1.000000	1.000000	1.000000	1.000000	1.000000	1.000000	1.000000	INF

Tafel 1 (Forts.) Table 1 (cont.)

$U_{p,q,n}$

p=2 C=0.05

n	q=1	2	3	4	5	6	7	8	9	10	11	12	n
1	0.000000	0.000000	0.000000	0.000000	0.000000	0.000000	0.000000	0.000000	0.000000	0.000000	0.000000	0.000000	1
2	0.002500	0.000641	0.000287	0.000162	0.000104	0.000072	0.000053	0.000041	0.000032	0.000026	0.000022	0.000018	2
3	0.049998	0.018318	0.009528	0.005844	0.003950	0.002849	0.002152	0.001683	0.001352	0.001110	0.000928	0.000787	3
4	0.135725	0.061800	0.035817	0.023460	0.016578	0.012346	0.009555	0.007615	0.006212	0.005165	0.004362	0.003734	4
5	0.223606	0.117368	0.073621	0.050765	0.037211	0.028476	0.022507	0.018244	0.015092	0.012695	0.010826	0.009343	5
6	0.301715	0.174902	0.116450	0.083663	0.063188	0.049481	0.039834	0.032772	0.027440	0.023320	0.020068	0.017453	6
7	0.368405	0.229737	0.160239	0.118984	0.092129	0.073571	0.060172	0.050155	0.042465	0.036426	0.031600	0.027678	7
8	0.424876	0.280187	0.202813	0.154741	0.122376	0.099380	0.082397	0.069475	0.059404	0.051386	0.044908	0.039579	8
9	0.472866	0.325883	0.243151	0.189781	0.152779	0.125881	0.105643	0.089993	0.077615	0.067661	0.059515	0.052772	9
10	0.513885	0.367036	0.280802	0.223433	0.182644	0.152421	0.129282	0.111138	0.096610	0.084797	0.075044	0.066901	10
11	0.549281	0.404052	0.315720	0.255369	0.211592	0.178545	0.152898	0.132506	0.116013	0.102453	0.091178	0.081680	11
12	0.580029	0.437339	0.347988	0.285511	0.239373	0.203997	0.176155	0.153782	0.135511	0.120356	0.107656	0.096885	12
13	0.606971	0.467384	0.377744	0.313837	0.265838	0.228568	0.198874	0.174774	0.154909	0.138311	0.124284	0.112321	13
14	0.630737	0.494599	0.405216	0.340396	0.291016	0.252171	0.220930	0.195325	0.174061	0.156149	0.140923	0.127849	14
15	0.651851	0.519281	0.430564	0.365263	0.314863	0.274786	0.242249	0.215357	0.192837	0.173755	0.157442	0.143350	15
16	0.670711	0.541775	0.454003	0.388530	0.337412	0.296391	0.262763	0.234782	0.211185	0.191059	0.173755	0.158740	16
17	0.687662	0.562317	0.475724	0.410322	0.358763	0.316990	0.282502	0.253583	0.229036	0.208000	0.189807	0.173946	17
18	0.702982	0.581146	0.495888	0.430784	0.378964	0.336632	0.301430	0.271723	0.246366	0.224530	0.205530	0.188918	18
19	0.716866	0.598489	0.514629	0.449961	0.398041	0.355335	0.319573	0.289225	0.263169	0.240614	0.220915	0.203611	19
20	0.729531	0.614483	0.532092	0.467968	0.416109	0.373163	0.336951	0.306072	0.279429	0.256249	0.235937	0.218013	20
21	0.741124	0.629283	0.548399	0.484925	0.433211	0.390129	0.353609	0.322287	0.295147	0.271437	0.250565	0.232083	21
22	0.751776	0.643011	0.563622	0.500886	0.449429	0.406286	0.369555	0.337873	0.310325	0.286147	0.264800	0.245821	22
23	0.761598	0.655775	0.577893	0.515922	0.464800	0.421699	0.384810	0.352883	0.324978	0.300409	0.278639	0.259224	23
24	0.770680	0.667666	0.591286	0.530135	0.479373	0.436391	0.399429	0.367295	0.339116	0.314213	0.292087	0.272280	24
25	0.779088	0.678783	0.603884	0.543551	0.493227	0.450412	0.413436	0.381165	0.352775	0.327593	0.305127	0.285006	25
26	0.786893	0.689182	0.615752	0.556269	0.506409	0.463802	0.426867	0.394506	0.365946	0.340539	0.317798	0.297372	26
27	0.794192	0.698945	0.626937	0.568306	0.518951	0.476588	0.439744	0.407337	0.378645	0.353047	0.330095	0.309407	27
28	0.800992	0.708108	0.637517	0.579727	0.530891	0.488822	0.452093	0.419700	0.390911	0.365171	0.342019	0.321110	28
29	0.807354	0.716737	0.647497	0.590582	0.542291	0.500519	0.463948	0.431586	0.402753	0.376900	0.353591	0.332484	29
30	0.813343	0.724899	0.656962	0.600899	0.553155	0.511722	0.475325	0.443028	0.414182	0.388244	0.364807	0.343537	30
40	0.857594	0.786433	0.729818	0.681627	0.639419	0.601870	0.568076	0.537426	0.509476	0.483873	0.460296	0.438550	40
60	0.903437	0.852599	0.810662	0.773804	0.740586	0.710190	0.682157	0.656096	0.631804	0.609029	0.587643	0.567501	60
80	0.926967	0.887496	0.854347	0.824736	0.797636	0.772490	0.748974	0.726849	0.705927	0.686107	0.667279	0.649328	80
100	0.941272	0.909051	0.881684	0.856993	0.834186	0.812834	0.792697	0.773596	0.755405	0.738034	0.721395	0.705440	100
120	0.950898	0.923673	0.900382	0.879233	0.859569	0.841056	0.823491	0.806739	0.790700	0.775302	0.760485	0.746201	120
140	0.957812	0.934247	0.913983	0.895493	0.878224	0.861896	0.846339	0.831442	0.817125	0.803326	0.789999	0.777105	140
170	0.965169	0.945562	0.928606	0.913057	0.898465	0.884603	0.871338	0.858581	0.846267	0.834352	0.822797	0.811574	170
200	0.970341	0.953554	0.938982	0.925569	0.912940	0.900904	0.889349	0.878202	0.867412	0.856939	0.846755	0.836834	200
240	0.975243	0.961158	0.948887	0.937554	0.926848	0.916613	0.906758	0.897224	0.887968	0.878959	0.870174	0.861593	240
320	0.981393	0.970741	0.961415	0.952766	0.944563	0.936691	0.929082	0.921692	0.914493	0.907461	0.900579	0.893835	320
440	0.986445	0.978644	0.971788	0.965408	0.959337	0.953491	0.947824	0.942303	0.936908	0.931623	0.926435	0.921337	440
600	0.990047	0.984298	0.979233	0.974507	0.969998	0.965648	0.961420	0.957293	0.953251	0.949283	0.945380	0.941537	600
800	0.992529	0.988203	0.984384	0.980814	0.977404	0.974108	0.970900	0.967763	0.964687	0.961662	0.958683	0.955744	800
1000	0.994021	0.990552	0.987487	0.984620	0.981877	0.979224	0.976640	0.974110	0.971627	0.969184	0.966775	0.964397	1000
INF	1.000000	1.000000	1.000000	1.000000	1.000000	1.000000	1.000000	1.000000	1.000000	1.000000	1.000000	1.000000	INF

Tafel 1 (Forts.) Table 1 (cont.)

$U_{p,q,n}$

p=2 C=0.01

n	q=1	2	3	4	5	6	7	8	9	10	11	12	n
1	0.000000	0.000000	0.000000	0.000000	0.000000	0.000000	0.000000	0.000000	0.000000	0.000000	0.000000	0.000000	1
2	0.000100	0.000025	0.000011	0.000006	0.000004	0.000003	0.000002	0.000002	0.000001	0.000001	0.000000	0.000000	2
3	0.010000	0.003470	0.001764	0.001068	0.000716	0.000514	0.000386	0.000301	0.000241	0.000198	0.000165	0.000140	3
4	0.046416	0.019844	0.011160	0.007179	0.005013	0.003701	0.002846	0.002257	0.001834	0.001520	0.001280	0.001093	4
5	0.099999	0.049316	0.029953	0.020241	0.014627	0.011080	0.008688	0.006999	0.005760	0.004824	0.004099	0.003527	5
6	0.158490	0.086620	0.055849	0.039284	0.029229	0.022633	0.018059	0.014752	0.012283	0.010389	0.008903	0.007715	6
7	0.215444	0.127189	0.085984	0.062513	0.047671	0.037627	0.030485	0.025222	0.021222	0.018110	0.015639	0.013643	7
8	0.268270	0.168148	0.118119	0.088278	0.068750	0.055175	0.045317	0.037913	0.032206	0.027708	0.024097	0.021153	8
9	0.316228	0.207906	0.150743	0.115317	0.091448	0.074467	0.061901	0.052316	0.044821	0.038849	0.034006	0.030024	9
10	0.359382	0.245666	0.182908	0.142738	0.114989	0.094846	0.079687	0.067961	0.058684	0.051208	0.045092	0.040019	10
11	0.398107	0.281095	0.214051	0.169943	0.138805	0.115797	0.098225	0.084458	0.073448	0.064492	0.057100	0.050924	11
12	0.432876	0.314111	0.243868	0.196543	0.162496	0.136940	0.117163	0.101490	0.088832	0.078445	0.069807	0.062537	12
13	0.464159	0.344773	0.272209	0.222298	0.185786	0.157996	0.136231	0.118805	0.104603	0.092857	0.083018	0.074689	13
14	0.492388	0.373205	0.299027	0.247072	0.208495	0.178764	0.155227	0.136206	0.120576	0.107553	0.096574	0.087224	14
15	0.517947	0.399561	0.324338	0.270794	0.230506	0.199104	0.174003	0.153543	0.136603	0.122393	0.110341	0.100021	15
16	0.541170	0.424011	0.348190	0.293441	0.251751	0.218924	0.192450	0.170701	0.152570	0.137265	0.124211	0.112976	16
17	0.562341	0.446714	0.370654	0.315019	0.272197	0.238163	0.210492	0.187598	0.168387	0.152079	0.138096	0.126003	17
18	0.581709	0.467823	0.391807	0.335555	0.291830	0.256785	0.228078	0.204169	0.183989	0.166764	0.151924	0.139032	18
19	0.599484	0.487482	0.411733	0.355085	0.310657	0.274771	0.245174	0.220372	0.199322	0.181266	0.165639	0.152007	19
20	0.615848	0.505819	0.430515	0.373654	0.328695	0.292119	0.261761	0.236178	0.214352	0.195544	0.179196	0.164879	20
21	0.630957	0.522953	0.448231	0.391312	0.345965	0.308831	0.277829	0.251565	0.229052	0.209566	0.192561	0.177614	21
22	0.644947	0.538990	0.464956	0.408104	0.362497	0.324920	0.293378	0.266524	0.243403	0.223308	0.205707	0.190182	22
23	0.657933	0.554026	0.480762	0.424082	0.378320	0.340403	0.308412	0.281051	0.257394	0.236756	0.218614	0.202560	23
24	0.670019	0.568146	0.495715	0.439293	0.393467	0.355296	0.322939	0.295146	0.271020	0.249897	0.231267	0.214731	24
25	0.681292	0.581428	0.509875	0.453782	0.407970	0.369622	0.336971	0.308812	0.284279	0.262726	0.243657	0.226681	25
26	0.691831	0.593939	0.523299	0.467592	0.421860	0.383403	0.350522	0.322057	0.297172	0.275239	0.255776	0.238402	26
27	0.701704	0.605746	0.536040	0.480765	0.435169	0.396661	0.363607	0.334890	0.309703	0.287436	0.267621	0.249886	27
28	0.710971	0.616902	0.548144	0.493339	0.447927	0.409418	0.376242	0.347322	0.321877	0.299318	0.279190	0.261129	28
29	0.719686	0.627457	0.559655	0.505352	0.460163	0.421696	0.388443	0.359363	0.333703	0.310890	0.290483	0.272130	29
30	0.727896	0.637459	0.570615	0.516835	0.471903	0.433519	0.400227	0.371026	0.345187	0.322156	0.301504	0.282889	30
40	0.789652	0.714476	0.656673	0.608581	0.567185	0.530850	0.498541	0.469542	0.443323	0.419481	0.397694	0.377702	40
60	0.855467	0.799984	0.755573	0.717315	0.683328	0.652617	0.624558	0.598723	0.574795	0.552535	0.531746	0.512271	60
80	0.889953	0.846188	0.810436	0.779081	0.750765	0.724783	0.700697	0.678213	0.657114	0.637235	0.618444	0.600635	80
100	0.911163	0.875081	0.845239	0.818780	0.794644	0.772286	0.751373	0.731681	0.713048	0.695352	0.678495	0.662399	100
120	0.925522	0.894844	0.869263	0.846415	0.825431	0.805865	0.787452	0.770011	0.753414	0.737564	0.722385	0.707815	120
140	0.935886	0.909213	0.886840	0.866750	0.848208	0.830840	0.814421	0.798803	0.783877	0.769567	0.755808	0.742551	140
170	0.946959	0.924659	0.905836	0.888839	0.873070	0.858224	0.844122	0.830646	0.817710	0.805252	0.793224	0.781586	170
200	0.954772	0.935614	0.919375	0.904652	0.890941	0.877988	0.865642	0.853805	0.842407	0.831395	0.820731	0.810382	200
240	0.962196	0.946071	0.932347	0.919856	0.908184	0.897119	0.886539	0.876363	0.866534	0.857011	0.847760	0.838758	240
320	0.971540	0.959297	0.948819	0.939239	0.930247	0.921688	0.913469	0.905533	0.897838	0.890354	0.883057	0.875930	320
440	0.979238	0.970243	0.962512	0.955416	0.948732	0.942346	0.936194	0.930234	0.924437	0.918780	0.913248	0.907828	440
600	0.984741	0.978097	0.972369	0.967098	0.962118	0.957350	0.952745	0.948273	0.943913	0.939649	0.935470	0.931366	600
800	0.988539	0.983531	0.979204	0.975215	0.971440	0.967819	0.964316	0.960909	0.957581	0.954322	0.951123	0.947977	800
1000	0.990823	0.986804	0.983329	0.980120	0.977081	0.974162	0.971336	0.968584	0.965894	0.963257	0.960665	0.958115	1000
INF	1.000000	1.000000	1.000000	1.000000	1.000000	1.000000	1.000000	1.000000	1.000000	1.000000	1.000000	1.000000	INF

Tafel 1 (Forts.) Table 1 (cont.)

$U_{p,q,n}$

p=2 C=0.05

n	q=13	14	15	18	21	24	27	30	40	60	80	100	120	n
1	0.000000	0.000000	0.000000	0.000000	0.000000	0.000000	0.000000	0.000000	0.000000	0.000000	0.000000	0.000000	0.000000	1
2	0.000016	0.000013	0.000012	0.000008	0.000006	0.000005	0.000004	0.000003	0.000002	0.000001	0.000000	0.000000	0.000000	2
3	0.000676	0.000587	0.000514	0.000362	0.000269	0.000207	0.000165	0.000134	0.000076	0.000034	0.000019	0.000012	0.000009	3
4	0.003232	0.002825	0.002490	0.001778	0.001333	0.001037	0.000829	0.000678	0.000390	0.000177	0.000101	0.000065	0.000045	4
5	0.008146	0.007164	0.006352	0.004598	0.003482	0.002728	0.002195	0.001805	0.001050	0.000483	0.000276	0.000179	0.000125	5
6	0.015318	0.013555	0.012080	0.008856	0.006772	0.005345	0.004327	0.003574	0.002103	0.000979	0.000564	0.000366	0.000257	6
7	0.024443	0.021747	0.019476	0.014450	0.011149	0.008865	0.007217	0.005990	0.003563	0.001678	0.000973	0.000634	0.000445	7
8	0.035156	0.031438	0.028283	0.021226	0.016519	0.013223	0.010823	0.009024	0.005426	0.002585	0.001507	0.000986	0.000695	8
9	0.047120	0.042333	0.038242	0.029003	0.022755	0.018333	0.015088	0.012633	0.007677	0.003699	0.002170	0.001425	0.001007	9
10	0.060023	0.054162	0.049122	0.037618	0.029743	0.024110	0.019943	0.016769	0.010293	0.005015	0.002959	0.001950	0.001381	10
11	0.073613	0.066688	0.060705	0.046922	0.037367	0.030469	0.025321	0.021380	0.013251	0.006527	0.003874	0.002562	0.001820	11
12	0.087673	0.079721	0.072820	0.056771	0.045524	0.037323	0.031162	0.026413	0.016526	0.008226	0.004911	0.003260	0.002321	12
13	0.102024	0.093104	0.085313	0.067051	0.054112	0.044601	0.037403	0.031820	0.020092	0.010104	0.006066	0.004042	0.002884	13
14	0.116544	0.106697	0.098057	0.077658	0.063051	0.052233	0.043988	0.037553	0.023923	0.012152	0.007336	0.004905	0.003510	14
15	0.131110	0.120388	0.110945	0.088498	0.072277	0.060157	0.050861	0.043579	0.027994	0.014360	0.008717	0.005849	0.004195	15
16	0.145624	0.134099	0.123908	0.099505	0.081706	0.068321	0.057988	0.049841	0.032282	0.016719	0.010204	0.006870	0.004939	16
17	0.160032	0.147763	0.136862	0.110610	0.091297	0.076674	0.065316	0.056321	0.036768	0.019219	0.011792	0.007967	0.005741	17
18	0.174277	0.161304	0.149752	0.121758	0.101003	0.085170	0.072812	0.062974	0.041430	0.021851	0.013477	0.009136	0.006599	18
19	0.188308	0.174698	0.162544	0.132906	0.110772	0.093785	0.080446	0.069781	0.046246	0.024608	0.015253	0.010375	0.007512	19
20	0.202099	0.187911	0.175196	0.144033	0.120579	0.102473	0.088180	0.076708	0.051199	0.027480	0.017118	0.011681	0.008477	20
21	0.215626	0.200907	0.187673	0.155078	0.130387	0.111209	0.095999	0.083729	0.056277	0.030458	0.019067	0.013053	0.009494	21
22	0.228875	0.213675	0.199965	0.166045	0.140168	0.119965	0.103875	0.090832	0.061459	0.033536	0.021095	0.014487	0.010560	22
23	0.241843	0.226196	0.212056	0.176886	0.149917	0.128734	0.111780	0.097990	0.066735	0.036706	0.023198	0.015981	0.011676	23
24	0.254506	0.238448	0.223909	0.187607	0.159593	0.137478	0.119701	0.105197	0.072089	0.039962	0.025372	0.017532	0.012837	24
25	0.266861	0.250458	0.235551	0.198195	0.169193	0.146195	0.127630	0.112424	0.077513	0.043293	0.027614	0.019139	0.014043	25
26	0.278913	0.262184	0.246958	0.208613	0.178700	0.154861	0.135550	0.119669	0.082994	0.046697	0.029917	0.020798	0.015294	26
27	0.290671	0.273651	0.258122	0.218883	0.188109	0.163481	0.143443	0.126906	0.088520	0.050169	0.032282	0.022507	0.016585	27
28	0.302134	0.284843	0.269041	0.228977	0.197407	0.172023	0.151303	0.134144	0.094085	0.053698	0.034703	0.024264	0.017918	28
29	0.313290	0.295777	0.279735	0.238911	0.206583	0.180498	0.159111	0.141358	0.099679	0.057281	0.037176	0.026067	0.019289	29
30	0.324161	0.306427	0.290177	0.248659	0.215640	0.188878	0.166879	0.148544	0.105296	0.060916	0.039697	0.027914	0.020698	30
40	0.418386	0.399660	0.382240	0.336544	0.298822	0.267264	0.240547	0.217714	0.161390	0.099068	0.067027	0.048374	0.036555	40
60	0.548467	0.530490	0.513448	0.467280	0.427406	0.392630	0.362089	0.335076	0.264298	0.177001	0.126977	0.095579	0.074560	60
80	0.632204	0.615847	0.600190	0.556963	0.518600	0.484309	0.453479	0.425632	0.349700	0.248903	0.186478	0.145019	0.116044	80
100	0.690119	0.675368	0.661161	0.621452	0.585574	0.552962	0.523181	0.495877	0.419344	0.312163	0.241828	0.193007	0.157681	100
120	0.732410	0.719079	0.706177	0.669807	0.636526	0.605913	0.577638	0.551435	0.476447	0.367086	0.292018	0.238044	0.197858	120
140	0.764612	0.752493	0.740725	0.707333	0.676486	0.647856	0.621186	0.596267	0.523805	0.414717	0.337092	0.279642	0.235843	140
170	0.800658	0.790028	0.779668	0.750059	0.722421	0.696512	0.672146	0.649171	0.581140	0.474899	0.396041	0.335619	0.288193	170
200	0.827159	0.817711	0.808479	0.781957	0.757012	0.733457	0.711150	0.689974	0.626420	0.524383	0.446147	0.384542	0.335049	200
240	0.853202	0.844988	0.836939	0.813703	0.791686	0.770750	0.750785	0.731707	0.673679	0.577877	0.501945	0.440430	0.389771	240
320	0.887217	0.880717	0.874327	0.855759	0.837999	0.820954	0.804555	0.788748	0.739807	0.655921	0.586347	0.527690	0.477647	320
440	0.916320	0.911379	0.906507	0.892275	0.878553	0.865280	0.852412	0.839916	0.800628	0.731019	0.670879	0.618275	0.571867	440
600	0.937746	0.934006	0.930311	0.919473	0.908962	0.898737	0.888768	0.879034	0.848077	0.791839	0.741703	0.696563	0.655657	600
800	0.952843	0.949975	0.947138	0.938794	0.930669	0.922734	0.914967	0.907353	0.882940	0.837787	0.796604	0.758723	0.723698	800
1000	0.962047	0.959722	0.957420	0.950640	0.944020	0.937539	0.931180	0.924932	0.904798	0.867144	0.832309	0.799833	0.769417	1000
INF	1.000000	1.000000	1.000000	1.000000	1.000000	1.000000	1.000000	1.000000	1.000000	1.000000	1.000000	1.000000	1.000000	INF

Tafel 1 (Forts.) Table 1 (cont.)

$U_{p,q,n}$

p=2 C=0.01

n	q=13	14	15	18	21	24	27	30	40	60	80	100	120	n
1	0.000000	0.000000	0.000000	0.000000	0.000000	0.000000	0.000000	0.000000	0.000000	0.000000	0.000000	0.000000	0.000000	1
2	0.000000	0.000000	0.000000	0.000000	0.000000	0.000000	0.000000	0.000000	0.000000	0.000000	0.000000	0.000000	0.000000	2
3	0.000120	0.000104	0.000091	0.000064	0.000047	0.000037	0.000029	0.000024	0.000013	0.000006	0.000003	0.000002	0.000002	3
4	0.000944	0.000823	0.000725	0.000516	0.000386	0.000299	0.000239	0.000195	0.000112	0.000051	0.000029	0.000019	0.000013	4
5	0.003067	0.002691	0.002381	0.001715	0.001294	0.001011	0.000812	0.000666	0.000386	0.000177	0.000101	0.000065	0.000046	5
6	0.006750	0.005957	0.005296	0.003861	0.002939	0.002312	0.001867	0.001539	0.000901	0.000417	0.000240	0.000155	0.000109	6
7	0.012008	0.010652	0.009514	0.007014	0.005386	0.004266	0.003462	0.002866	0.001695	0.000794	0.000458	0.000298	0.000209	7
8	0.018719	0.016686	0.014968	0.011154	0.008634	0.006882	0.005615	0.004669	0.002790	0.001320	0.000767	0.000500	0.000352	8
9	0.026704	0.023911	0.021536	0.016213	0.012648	0.010145	0.008318	0.006945	0.004191	0.002004	0.001171	0.000767	0.000541	9
10	0.035765	0.032160	0.029077	0.022098	0.017368	0.014012	0.011545	0.009678	0.005896	0.002850	0.001674	0.001100	0.000778	10
11	0.045710	0.041265	0.037444	0.028712	0.022725	0.018438	0.015263	0.012844	0.007899	0.003857	0.002279	0.001503	0.001065	11
12	0.056361	0.051069	0.046495	0.035954	0.028650	0.023371	0.019434	0.016414	0.010186	0.005024	0.002985	0.001975	0.001404	12
13	0.067566	0.061436	0.056105	0.043735	0.035069	0.028756	0.024012	0.020356	0.012744	0.006348	0.003792	0.002518	0.001793	13
14	0.079191	0.072228	0.066161	0.051963	0.041918	0.034541	0.028963	0.024637	0.015556	0.007824	0.004699	0.003131	0.002235	14
15	0.091109	0.083356	0.076560	0.060561	0.049132	0.040678	0.034243	0.029227	0.018607	0.009448	0.005703	0.003814	0.002729	15
16	0.103228	0.094712	0.087224	0.069456	0.056656	0.047119	0.039813	0.034093	0.021879	0.011213	0.006804	0.004564	0.003274	16
17	0.115466	0.106223	0.098066	0.078589	0.064440	0.053821	0.045643	0.039204	0.025355	0.013113	0.007998	0.005383	0.003869	17
18	0.127753	0.117822	0.109026	0.087902	0.072431	0.060746	0.051696	0.044536	0.029021	0.015141	0.009281	0.006267	0.004515	18
19	0.140033	0.129453	0.120053	0.097351	0.080594	0.067855	0.057943	0.050061	0.032859	0.017292	0.010652	0.007216	0.005210	19
20	0.152260	0.141072	0.131100	0.106887	0.088888	0.075122	0.064353	0.055755	0.036855	0.019559	0.012107	0.008228	0.005953	20
21	0.164395	0.152637	0.142127	0.116479	0.097277	0.082515	0.070902	0.061597	0.040995	0.021936	0.013643	0.009300	0.006745	21
22	0.176407	0.164119	0.153104	0.126093	0.105746	0.090005	0.077569	0.067563	0.045266	0.024414	0.015257	0.010432	0.007583	22
23	0.188272	0.175491	0.164003	0.135704	0.114254	0.097568	0.084332	0.073637	0.049652	0.026992	0.016945	0.011622	0.008465	23
24	0.199971	0.186732	0.174804	0.145289	0.122783	0.105190	0.091170	0.079802	0.054141	0.029660	0.018704	0.012868	0.009392	24
25	0.211489	0.197826	0.185486	0.154827	0.131315	0.112853	0.098066	0.086036	0.058730	0.032412	0.020531	0.014167	0.010362	25
26	0.222812	0.208758	0.196037	0.164303	0.139834	0.120530	0.105008	0.092335	0.063400	0.035248	0.022425	0.015517	0.011374	26
27	0.233933	0.219519	0.206444	0.173701	0.148323	0.128212	0.111977	0.098679	0.068142	0.038154	0.024378	0.016919	0.012427	27
28	0.244846	0.230101	0.216698	0.183011	0.156770	0.135887	0.118971	0.105056	0.072952	0.041131	0.026393	0.018368	0.013519	28
29	0.255546	0.240498	0.226792	0.192222	0.165165	0.143542	0.125965	0.111463	0.077810	0.044174	0.028463	0.019865	0.014650	29
30	0.266032	0.250705	0.236719	0.201326	0.173498	0.151169	0.132955	0.117882	0.082723	0.047275	0.030585	0.021406	0.015818	30
40	0.359292	0.342285	0.326530	0.285564	0.252121	0.224393	0.201107	0.181337	0.133119	0.080710	0.054195	0.038914	0.029296	40
60	0.493980	0.476759	0.460514	0.416847	0.379494	0.347198	0.319026	0.294269	0.230087	0.152294	0.108433	0.081184	0.063085	60
80	0.583720	0.567620	0.552272	0.510201	0.473197	0.440372	0.411055	0.384722	0.313647	0.220850	0.164264	0.127080	0.101284	80
100	0.646999	0.632239	0.618071	0.578731	0.543482	0.511671	0.482799	0.456470	0.383356	0.282574	0.217423	0.172663	0.140506	100
120	0.693805	0.680311	0.667295	0.630823	0.597710	0.567455	0.539672	0.514057	0.441398	0.337024	0.266399	0.216121	0.178966	120
140	0.729755	0.717386	0.705413	0.671632	0.640658	0.612092	0.585628	0.561021	0.490073	0.384795	0.310906	0.256745	0.215750	140
170	0.770307	0.759360	0.748723	0.718487	0.690459	0.664342	0.639906	0.616969	0.549593	0.445801	0.369757	0.312023	0.267021	170
200	0.800323	0.790533	0.780992	0.753726	0.728251	0.704332	0.681790	0.660484	0.597023	0.496447	0.420280	0.360830	0.313385	200
240	0.829984	0.821419	0.813051	0.789009	0.766374	0.744963	0.724642	0.705301	0.646895	0.551644	0.477024	0.417073	0.368018	240
320	0.868958	0.862130	0.855434	0.836067	0.817651	0.800064	0.783215	0.767035	0.717271	0.632927	0.563710	0.505798	0.456679	320
440	0.902510	0.897287	0.892149	0.877205	0.862875	0.849077	0.835754	0.822860	0.782567	0.711906	0.651439	0.598912	0.552818	440
600	0.927331	0.923359	0.919444	0.908009	0.896976	0.886290	0.875911	0.865808	0.833860	0.776371	0.725566	0.680108	0.639113	600
800	0.944878	0.941823	0.938808	0.929975	0.921417	0.913093	0.904974	0.897039	0.871735	0.825349	0.783384	0.745005	0.709677	800
1000	0.955601	0.953120	0.950669	0.943476	0.936488	0.929674	0.923012	0.916485	0.895561	0.856762	0.821144	0.788118	0.757316	1000
INF	1.000000	1.000000	1.000000	1.000000	1.000000	1.000000	1.000000	1.000000	1.000000	1.000000	1.000000	1.000000	1.000000	INF

p=3 $U_{p,q,n}$ C=0.05

n	q=1	2	3	4	5	6	7	8	9	10	11	12	n
1	0.000000	0.000000	0.000000	0.000000	0.000000	0.000000	0.000000	0.000000	0.000000	0.000000	0.000000	0.000000	1
2	0.000000	0.000000	0.000000	0.000000	0.000000	0.000001	0.000002	0.000004	0.000005	0.000008	0.000010	0.000013	2
3	0.001698	0.000354	0.000179	0.000127	0.000105	0.000095	0.000091	0.000090	0.000091	0.000092	0.000095	0.000098	3
4	0.033740	0.009612	0.004205	0.002314	0.001479	0.001052	0.000809	0.000659	0.000562	0.000496	0.000449	0.000416	4
5	0.097355	0.035855	0.017521	0.010010	0.006357	0.004369	0.003195	0.002458	0.001971	0.001636	0.001397	0.001222	5
6	0.168271	0.073634	0.039672	0.024047	0.015792	0.011018	0.008067	0.006148	0.004849	0.003939	0.003281	0.002793	6
7	0.235525	0.116476	0.067711	0.043226	0.029433	0.021043	0.015642	0.012012	0.009485	0.007674	0.006345	0.005347	7
8	0.295976	0.160244	0.098932	0.065947	0.046378	0.033966	0.025706	0.019990	0.015911	0.012927	0.010697	0.008997	8
9	0.349277	0.202814	0.131378	0.090794	0.065660	0.049161	0.037855	0.029838	0.023995	0.019637	0.016323	0.013763	9
10	0.396084	0.243139	0.163846	0.116701	0.086448	0.066012	0.051643	0.041238	0.033514	0.027654	0.023135	0.019593	10
11	0.437147	0.280808	0.195556	0.142927	0.108110	0.083979	0.066659	0.053876	0.044225	0.036801	0.030993	0.026391	11
12	0.473377	0.315719	0.226090	0.168939	0.130131	0.102644	0.082534	0.067443	0.055894	0.046882	0.039757	0.034049	12
13	0.505452	0.347981	0.255220	0.194414	0.152160	0.121656	0.098973	0.081704	0.068298	0.057724	0.049278	0.042437	13
14	0.534018	0.377735	0.282849	0.219113	0.173959	0.140775	0.115736	0.096413	0.081246	0.069166	0.059407	0.051442	14
15	0.559570	0.405221	0.308951	0.242944	0.195322	0.159796	0.132619	0.111416	0.094593	0.081052	0.070029	0.060954	15
16	0.582577	0.430566	0.333588	0.265812	0.216138	0.178574	0.149493	0.126564	0.108178	0.093264	0.081026	0.070875	16
17	0.603338	0.454006	0.356777	0.287689	0.236338	0.197017	0.166236	0.141728	0.121917	0.105704	0.092299	0.081109	17
18	0.622168	0.475728	0.378631	0.308599	0.255858	0.215044	0.182762	0.156827	0.135694	0.118273	0.103768	0.091588	18
19	0.639337	0.495908	0.399223	0.328552	0.274710	0.232604	0.199009	0.171789	0.149446	0.130904	0.115361	0.102241	19
20	0.655028	0.514622	0.418629	0.347546	0.292843	0.249666	0.214918	0.186544	0.163097	0.143521	0.127018	0.113012	20
21	0.669437	0.532101	0.436898	0.365676	0.310304	0.266216	0.230467	0.201077	0.176620	0.156088	0.138689	0.123835	21
22	0.682712	0.548393	0.454182	0.382934	0.327083	0.282253	0.245626	0.215325	0.189969	0.168561	0.150321	0.134680	22
23	0.694960	0.563637	0.470473	0.399402	0.343191	0.297740	0.260397	0.229291	0.203123	0.180907	0.161896	0.145521	23
24	0.706310	0.577895	0.485889	0.415077	0.358665	0.312738	0.274743	0.242939	0.216044	0.193091	0.173370	0.156313	24
25	0.716875	0.591311	0.500491	0.430041	0.373523	0.327222	0.288709	0.256276	0.228718	0.205103	0.184720	0.167023	25
26	0.726681	0.603899	0.514336	0.444332	0.387790	0.341199	0.302238	0.269280	0.241137	0.216929	0.195944	0.177651	26
27	0.735837	0.615757	0.527453	0.457946	0.401488	0.354711	0.315386	0.281968	0.253300	0.228535	0.206998	0.188160	27
28	0.744404	0.626944	0.539914	0.470981	0.414658	0.367742	0.328131	0.294313	0.265188	0.239935	0.217899	0.198546	28
29	0.752437	0.637514	0.551741	0.483431	0.427307	0.380334	0.340477	0.306326	0.276805	0.251110	0.228615	0.208809	29
30	0.759984	0.647501	0.563023	0.495347	0.439475	0.392490	0.352461	0.318033	0.288158	0.262062	0.239155	0.218912	30
40	0.816139	0.723938	0.651356	0.590773	0.538846	0.493686	0.453976	0.418785	0.387401	0.359271	0.333940	0.311045	40
60	0.874843	0.807778	0.752424	0.704238	0.661334	0.622640	0.587440	0.555224	0.525598	0.498272	0.472957	0.449477	60
80	0.905160	0.852653	0.808266	0.768805	0.732964	0.700027	0.669520	0.641124	0.614572	0.589678	0.566281	0.544236	80
100	0.923660	0.880557	0.843610	0.810333	0.779746	0.751296	0.724666	0.699598	0.675935	0.653520	0.632235	0.611999	100
120	0.936178	0.899588	0.867973	0.839253	0.812632	0.787686	0.764150	0.741841	0.720623	0.700389	0.681054	0.662546	120
140	0.945137	0.913391	0.885776	0.860534	0.836998	0.814820	0.793780	0.773732	0.754565	0.736197	0.718557	0.701592	140
170	0.954680	0.928199	0.904999	0.883652	0.863624	0.844636	0.826518	0.809156	0.792465	0.776383	0.760857	0.745847	170
200	0.961395	0.938685	0.918687	0.900202	0.882782	0.866197	0.850307	0.835018	0.820262	0.805990	0.792160	0.778739	200
240	0.967765	0.948679	0.931793	0.916116	0.901281	0.887100	0.873459	0.860284	0.847521	0.835131	0.823081	0.811346	240
320	0.975762	0.961296	0.948422	0.936405	0.924972	0.913987	0.903369	0.893064	0.883033	0.873250	0.863692	0.854341	320
440	0.982336	0.971725	0.962235	0.953337	0.944835	0.936632	0.928671	0.920913	0.913333	0.905910	0.898630	0.891482	440
600	0.987028	0.979198	0.972173	0.965563	0.959229	0.953099	0.947133	0.941302	0.935589	0.929978	0.924461	0.919029	600
800	0.990261	0.984364	0.979060	0.974060	0.969257	0.964600	0.960057	0.955610	0.951243	0.946947	0.942713	0.938538	800
1000	0.992204	0.987475	0.983215	0.979193	0.975326	0.971571	0.967905	0.964310	0.960776	0.957296	0.953863	0.950473	1000
INF	1.000000	1.000000	1.000000	1.000000	1.000000	1.000000	1.000000	1.000000	1.000000	1.000000	1.000000	1.000000	INF

$U_{p,q,n}$

p=3   C=0.01

n	q=1	2	3	4	5	6	7	8	9	10	11	12	n
1	0.000000	0.000000	0.000000	0.000000	0.000000	0.000000	0.000000	0.000000	0.000000	0.000000	0.000000	0.000000	1
2	0.000000	0.000000	0.000000	0.000000	0.000000	0.000000	0.000000	0.000000	0.000001	0.000002	0.000003	0.000004	2
3	0.000080	0.000021	0.000016	0.000015	0.000015	0.000017	0.000019	0.000021	0.000023	0.000026	0.000029	0.000031	3
4	0.006763	0.001829	0.000824	0.000484	0.000335	0.000258	0.000215	0.000188	0.000172	0.000161	0.000154	0.000149	4
5	0.032882	0.011211	0.005326	0.003037	0.001959	0.001383	0.001047	0.000837	0.000698	0.000602	0.000533	0.000483	5
6	0.073980	0.029981	0.015536	0.009229	0.006018	0.004211	0.003116	0.002414	0.001943	0.001614	0.001376	0.001200	6
7	0.121426	0.055863	0.031196	0.019423	0.013027	0.009244	0.006864	0.005293	0.004214	0.003450	0.002892	0.002474	7
8	0.169788	0.085991	0.051041	0.033146	0.022897	0.016575	0.012459	0.009664	0.007703	0.006286	0.005237	0.004444	8
9	0.216359	0.118124	0.073700	0.049627	0.035223	0.026018	0.019845	0.015549	0.012468	0.010203	0.008501	0.007198	9
10	0.259966	0.150746	0.098030	0.068087	0.049501	0.037260	0.028840	0.022851	0.018472	0.015200	0.012705	0.010772	10
11	0.300240	0.182909	0.123161	0.087852	0.065237	0.049951	0.039203	0.031408	0.025614	0.021217	0.017821	0.015158	11
12	0.337186	0.214052	0.148469	0.108380	0.081998	0.063758	0.050682	0.041037	0.033759	0.028161	0.023785	0.020317	12
13	0.370989	0.243868	0.173524	0.129256	0.099424	0.078384	0.063039	0.051550	0.042762	0.035922	0.030516	0.026188	13
14	0.401904	0.272209	0.198043	0.150167	0.117224	0.093576	0.076062	0.062771	0.052482	0.044385	0.037922	0.032700	14
15	0.430202	0.299027	0.221841	0.170888	0.135171	0.109125	0.089567	0.074542	0.062783	0.053438	0.045911	0.039779	15
16	0.456147	0.324338	0.244809	0.191257	0.153091	0.124859	0.103396	0.086724	0.073546	0.062978	0.054394	0.047349	16
17	0.479984	0.348191	0.266888	0.211160	0.170848	0.140644	0.117419	0.099197	0.084662	0.072908	0.063289	0.055338	17
18	0.501932	0.370654	0.288051	0.230524	0.188345	0.156371	0.131529	0.111859	0.096037	0.083144	0.072519	0.063680	18
19	0.522191	0.391807	0.308300	0.249300	0.205509	0.171955	0.145640	0.124624	0.107589	0.093611	0.082016	0.072312	19
20	0.540934	0.411734	0.327644	0.267462	0.222288	0.187334	0.159680	0.137421	0.119251	0.104243	0.091719	0.081178	20
21	0.558316	0.430515	0.346122	0.284999	0.238647	0.202457	0.173594	0.150193	0.130962	0.114983	0.101574	0.090228	21
22	0.574470	0.448231	0.363762	0.301910	0.254564	0.217287	0.187337	0.162890	0.142676	0.125784	0.111534	0.099418	22
23	0.589519	0.464956	0.380593	0.318203	0.270024	0.231801	0.200875	0.175473	0.154348	0.136602	0.121558	0.108708	23
24	0.603567	0.480761	0.396664	0.333888	0.285024	0.245977	0.214182	0.187912	0.165948	0.147403	0.131611	0.118064	24
25	0.616709	0.495715	0.412006	0.348987	0.299564	0.259807	0.227238	0.200181	0.177443	0.158158	0.141663	0.127455	25
26	0.629025	0.509875	0.426661	0.363513	0.313646	0.273282	0.240029	0.212260	0.188816	0.168840	0.151686	0.136855	26
27	0.640592	0.523299	0.440664	0.377492	0.327281	0.286402	0.252545	0.224135	0.200042	0.179430	0.161661	0.146241	27
28	0.651469	0.536040	0.454050	0.390942	0.340476	0.299164	0.264779	0.235795	0.211110	0.189910	0.171566	0.155593	28
29	0.661719	0.548144	0.466858	0.403887	0.353244	0.311575	0.276730	0.247231	0.222009	0.200265	0.181387	0.164895	29
30	0.671391	0.559656	0.479116	0.416348	0.365597	0.323637	0.288394	0.258438	0.232727	0.210485	0.191110	0.174132	30
40	0.744674	0.649620	0.577483	0.518712	0.469272	0.426891	0.390088	0.357822	0.329317	0.303979	0.281338	0.261014	40
60	0.823683	0.751990	0.654679	0.645816	0.602970	0.564801	0.530443	0.499282	0.470857	0.444810	0.420849	0.398737	60
80	0.865422	0.808282	0.761397	0.720482	0.683828	0.650513	0.619945	0.591715	0.565509	0.541095	0.518272	0.496881	80
100	0.891201	0.843804	0.804298	0.769332	0.737595	0.708389	0.681275	0.655947	0.632179	0.609801	0.588667	0.568664	100
120	0.908698	0.868241	0.834163	0.803715	0.775833	0.749959	0.725744	0.702949	0.681398	0.660958	0.641519	0.622994	120
140	0.921350	0.886074	0.856136	0.829206	0.804388	0.781216	0.759404	0.738756	0.719127	0.700413	0.682523	0.665388	140
170	0.934886	0.905306	0.880001	0.857075	0.835802	0.815812	0.796876	0.778843	0.761600	0.745064	0.729169	0.713861	170
200	0.944448	0.918986	0.897083	0.877137	0.858541	0.840986	0.824284	0.808309	0.792971	0.778202	0.763939	0.750169	200
240	0.953545	0.932073	0.913505	0.896514	0.880601	0.865514	0.851099	0.837256	0.823911	0.811012	0.798488	0.786374	240
320	0.965006	0.948662	0.934435	0.921337	0.909000	0.897239	0.885943	0.875039	0.864473	0.854210	0.844184	0.834449	320
440	0.974459	0.962428	0.951897	0.942156	0.932937	0.924109	0.915592	0.907335	0.899302	0.891467	0.883774	0.876280	440
600	0.981223	0.972324	0.964504	0.957246	0.950354	0.943732	0.937324	0.931093	0.925012	0.919063	0.913203	0.907480	600
800	0.985892	0.979179	0.973264	0.967760	0.962522	0.957478	0.952586	0.947819	0.943158	0.938588	0.934077	0.929662	800
1000	0.988702	0.983312	0.978556	0.974124	0.969900	0.965827	0.961872	0.958013	0.954234	0.950525	0.946859	0.943268	1000
INF	1.000000	1.000000	1.000000	1.000000	1.000000	1.000000	1.000000	1.000000	1.000000	1.000000	1.000000	1.000000	INF

Tafel 1 (Forts.) Table 1 (cont.)

$U_{p,q,n}$

p=3 C=0.05

n	q = 13	14	15	18	21	24	27	30	40	60	80	100	120	n
1	0.000000	0.000000	0.000000	0.000000	0.000000	0.000000	0.000000	0.000000	0.000000	0.000000	0.000000	0.000000	0.000000	1
2	0.000017	0.000020	0.000024	0.000036	0.000049	0.000063	0.000078	0.000093	0.000142	0.000236	0.000318	0.000389	0.000453	2
3	0.000102	0.000106	0.000110	0.000124	0.000138	0.000153	0.000168	0.000183	0.000231	0.000318	0.000394	0.000460	0.000519	3
4	0.000391	0.000372	0.000358	0.000333	0.000324	0.000322	0.000324	0.000330	0.000357	0.000422	0.000484	0.000541	0.000592	4
5	0.001089	0.000987	0.000907	0.000750	0.000662	0.000609	0.000576	0.000555	0.000530	0.000549	0.000589	0.000632	0.000673	5
6	0.002423	0.002138	0.001913	0.001467	0.001214	0.001056	0.000953	0.000882	0.000761	0.000703	0.000710	0.000733	0.000761	6
7	0.004583	0.003987	0.003516	0.002578	0.002041	0.001707	0.001486	0.001332	0.001059	0.000887	0.000848	0.000846	0.000858	7
8	0.007680	0.006644	0.005818	0.004156	0.003198	0.002600	0.002204	0.001928	0.001434	0.001105	0.001005	0.000971	0.000964	8
9	0.011755	0.010157	0.008872	0.006256	0.004727	0.003769	0.003133	0.002691	0.001896	0.001358	0.001182	0.001109	0.001078	9
10	0.016783	0.014525	0.012690	0.008905	0.006661	0.005242	0.004296	0.003637	0.002454	0.001651	0.001381	0.001261	0.001203	10
11	0.022699	0.019703	0.017248	0.012112	0.009014	0.007036	0.005709	0.004781	0.003116	0.001986	0.001602	0.001428	0.001337	11
12	0.029419	0.025631	0.022501	0.015861	0.011789	0.009160	0.007383	0.006136	0.003888	0.002366	0.001847	0.001609	0.001481	12
13	0.036845	0.032229	0.028385	0.020129	0.014983	0.011618	0.009325	0.007707	0.004778	0.002793	0.002118	0.001806	0.001636	13
14	0.044875	0.039412	0.034837	0.024881	0.018579	0.014406	0.011537	0.009500	0.005790	0.003270	0.002415	0.002019	0.001802	14
15	0.053418	0.047107	0.041785	0.030079	0.022552	0.017512	0.014016	0.011515	0.006928	0.003799	0.002739	0.002249	0.001980	15
16	0.062389	0.055234	0.049161	0.035675	0.026884	0.020927	0.016754	0.013750	0.008195	0.004381	0.003092	0.002497	0.002170	16
17	0.071695	0.063716	0.056904	0.041633	0.031543	0.024629	0.019744	0.016201	0.009591	0.005019	0.003474	0.002762	0.002371	17
18	0.081280	0.072492	0.064957	0.047906	0.036501	0.028602	0.022974	0.018862	0.011118	0.005714	0.003886	0.003046	0.002585	18
19	0.091073	0.081504	0.073265	0.054455	0.041730	0.032828	0.026430	0.021725	0.012775	0.006466	0.004330	0.003349	0.002812	19
20	0.101022	0.090702	0.081774	0.061242	0.047197	0.037285	0.030101	0.024780	0.014560	0.007278	0.004805	0.003672	0.003052	20
21	0.111071	0.100036	0.090442	0.068235	0.052884	0.041951	0.033970	0.028016	0.016471	0.008149	0.005313	0.004015	0.003306	21
22	0.121181	0.109464	0.099236	0.075397	0.058764	0.046812	0.038023	0.031426	0.018506	0.009081	0.005853	0.004377	0.003573	22
23	0.131325	0.118959	0.108124	0.082699	0.064801	0.051844	0.042246	0.034995	0.020662	0.010073	0.006427	0.004760	0.003854	23
24	0.141462	0.128475	0.117066	0.090118	0.070988	0.057032	0.046624	0.038715	0.022934	0.011125	0.007035	0.005165	0.004149	24
25	0.151565	0.137996	0.126039	0.097633	0.077298	0.062355	0.051142	0.042577	0.025319	0.012239	0.007678	0.005591	0.004459	25
26	0.161625	0.147506	0.135022	0.105215	0.083707	0.067800	0.055789	0.046561	0.027812	0.013412	0.008354	0.006039	0.004783	26
27	0.171597	0.156968	0.143992	0.112840	0.090204	0.073350	0.060553	0.050665	0.030406	0.014646	0.009065	0.006508	0.005122	27
28	0.181487	0.166372	0.152935	0.120500	0.096770	0.078995	0.065419	0.054879	0.033100	0.015938	0.009812	0.007000	0.005476	28
29	0.191279	0.175712	0.161832	0.128178	0.103391	0.084715	0.070373	0.059190	0.035888	0.017289	0.010593	0.007513	0.005844	29
30	0.200957	0.184972	0.170668	0.135863	0.110051	0.090499	0.075410	0.063587	0.038763	0.018697	0.011409	0.008049	0.006229	30
40	0.290289	0.271393	0.254158	0.210687	0.176777	0.149889	0.128283	0.110711	0.071321	0.035678	0.021451	0.014654	0.010933	40
60	0.427617	0.407240	0.388210	0.338067	0.296483	0.261629	0.232145	0.207024	0.145767	0.080871	0.050340	0.034274	0.025030	60
80	0.523426	0.503760	0.485147	0.434837	0.391568	0.354082	0.321390	0.292697	0.218934	0.132425	0.086752	0.060564	0.044578	80
100	0.592729	0.574335	0.556781	0.508552	0.466092	0.428463	0.394949	0.364969	0.284954	0.184110	0.126282	0.090850	0.068058	100
120	0.644805	0.627778	0.611418	0.565955	0.525245	0.488594	0.455452	0.425373	0.342876	0.233125	0.166148	0.122945	0.093923	120
140	0.685253	0.669499	0.654295	0.611677	0.573032	0.537823	0.505620	0.476074	0.393308	0.278453	0.204836	0.155372	0.120955	140
170	0.731318	0.717238	0.703583	0.664944	0.629419	0.596622	0.566244	0.538032	0.457087	0.339172	0.259166	0.202772	0.161864	170
200	0.765701	0.753021	0.740680	0.705519	0.672869	0.642434	0.613982	0.587320	0.509452	0.391785	0.308397	0.247403	0.201696	200
240	0.799905	0.788741	0.777838	0.746568	0.717247	0.689659	0.663633	0.639031	0.565893	0.451223	0.366280	0.301723	0.251685	240
320	0.845185	0.836210	0.827406	0.801944	0.777768	0.754743	0.732764	0.711747	0.647789	0.542419	0.459503	0.393040	0.339013	320
440	0.884456	0.877544	0.870739	0.850916	0.831895	0.813592	0.795944	0.778901	0.725974	0.634882	0.559202	0.495492	0.441337	440
600	0.913675	0.908393	0.903180	0.887916	0.873155	0.858845	0.844944	0.831422	0.788790	0.712951	0.647266	0.589805	0.539189	600
800	0.934414	0.930339	0.926310	0.914467	0.902955	0.891734	0.880777	0.870065	0.835924	0.773733	0.718217	0.668254	0.623049	800
1000	0.947121	0.943805	0.940522	0.930853	0.921423	0.912202	0.903169	0.894310	0.865891	0.813357	0.765571	0.721789	0.681497	1000
INF	1.000000	1.000000	1.000000	1.000000	1.000000	1.000000	1.000000	1.000000	1.000000	1.000000	1.000000	1.000000	1.000000	INF

$U_{p,q,n}$

p=3 C=0.01

q

n	13	14	15	18	21	24	27	30	40	60	80	100	120	n
1	0.000000	0.000000	0.000000	0.000003	0.000006	0.000010	0.000015	0.000022	0.000048	0.000113	0.000181	0.000245	0.000304	1
2	0.000005	0.000006	0.000008	0.000014	0.000021	0.000030	0.000039	0.000048	0.000084	0.000158	0.000228	0.000293	0.000352	2
3	0.000035	0.000038	0.000041	0.000052	0.000063	0.000074	0.000086	0.000098	0.000138	0.000215	0.000285	0.000348	0.000404	3
4	0.000147	0.000145	0.000145	0.000148	0.000154	0.000163	0.000172	0.000182	0.000218	0.000288	0.000352	0.000410	0.000463	4
5	0.000445	0.000416	0.000393	0.000350	0.000329	0.000318	0.000314	0.000314	0.000328	0.000378	0.000431	0.000481	0.000527	5
6	0.001067	0.000963	0.000881	0.000718	0.000626	0.000569	0.000533	0.000510	0.000478	0.000487	0.000521	0.000560	0.000598	6
7	0.002155	0.001907	0.001710	0.001315	0.001088	0.000945	0.000851	0.000786	0.000674	0.000620	0.000626	0.000649	0.000676	7
8	0.003833	0.003354	0.002972	0.002204	0.001757	0.001477	0.001289	0.001158	0.000924	0.000777	0.000745	0.000747	0.000761	8
9	0.006184	0.005382	0.004740	0.003436	0.002674	0.002192	0.001870	0.001644	0.001236	0.000962	0.000881	0.000856	0.000854	9
10	0.009252	0.008039	0.007061	0.005054	0.003869	0.003117	0.002613	0.002259	0.001617	0.001176	0.001033	0.000976	0.000955	10
11	0.013042	0.011340	0.009955	0.007084	0.005367	0.004272	0.003535	0.003017	0.002074	0.001424	0.001204	0.001108	0.001064	11
12	0.017533	0.015275	0.013423	0.009540	0.007187	0.005673	0.004649	0.003928	0.002615	0.001706	0.001394	0.001253	0.001181	12
13	0.022683	0.019815	0.017448	0.012423	0.009335	0.007331	0.005967	0.005003	0.003244	0.002025	0.001604	0.001410	0.001308	13
14	0.028436	0.024921	0.021998	0.015725	0.011815	0.009251	0.007496	0.006249	0.003967	0.002383	0.001835	0.001581	0.001444	14
15	0.034732	0.030543	0.027037	0.019430	0.014621	0.011435	0.009239	0.007672	0.004787	0.002783	0.002089	0.001766	0.001589	15
16	0.041510	0.036630	0.032521	0.023514	0.017743	0.013881	0.011198	0.009273	0.005710	0.003226	0.002367	0.001966	0.001745	16
17	0.048705	0.043129	0.038406	0.027955	0.021170	0.016582	0.013371	0.011053	0.006737	0.003713	0.002668	0.002180	0.001911	17
18	0.056261	0.049988	0.044648	0.032722	0.024884	0.019531	0.015754	0.013013	0.007870	0.004247	0.002994	0.002411	0.002088	18
19	0.064121	0.057160	0.051204	0.037789	0.028868	0.022716	0.018343	0.015149	0.009111	0.004828	0.003347	0.002657	0.002276	19
20	0.072235	0.064597	0.058031	0.043125	0.033102	0.026126	0.021129	0.017457	0.010460	0.005458	0.003725	0.002919	0.002475	20
21	0.080556	0.072257	0.065093	0.048704	0.037568	0.029748	0.024104	0.019933	0.011918	0.006137	0.004131	0.003199	0.002685	21
22	0.089042	0.080101	0.072351	0.054498	0.042245	0.033568	0.027260	0.022569	0.013482	0.006867	0.004565	0.003496	0.002907	22
23	0.097657	0.088094	0.079775	0.060481	0.047114	0.037571	0.030586	0.025361	0.015153	0.007648	0.005027	0.003811	0.003141	23
24	0.106365	0.096205	0.087334	0.066628	0.052157	0.041745	0.034072	0.028300	0.016928	0.008480	0.005518	0.004143	0.003387	24
25	0.115139	0.104405	0.095000	0.072919	0.057355	0.046075	0.037708	0.031379	0.018805	0.009364	0.006038	0.004494	0.003647	25
26	0.123952	0.112668	0.102750	0.079330	0.062691	0.050547	0.041483	0.034590	0.020781	0.010300	0.006587	0.004863	0.003918	26
27	0.132781	0.120971	0.110560	0.085844	0.068150	0.055148	0.045388	0.037925	0.022855	0.011288	0.007166	0.005252	0.004202	27
28	0.141606	0.129296	0.118412	0.092442	0.073715	0.059867	0.049411	0.041377	0.025022	0.012328	0.007776	0.005659	0.004499	28
29	0.150409	0.137623	0.126288	0.099108	0.079373	0.064691	0.053544	0.044937	0.027280	0.013419	0.008416	0.006086	0.004810	29
30	0.159175	0.145938	0.134172	0.105827	0.085111	0.069607	0.057777	0.048598	0.029624	0.014562	0.009086	0.006533	0.005134	30
40	0.242696	0.226130	0.211096	0.173514	0.144569	0.121868	0.103795	0.089223	0.057055	0.028647	0.017457	0.012094	0.009137	40
60	0.378272	0.359289	0.341637	0.295496	0.257636	0.226191	0.199806	0.177475	0.123700	0.068014	0.042401	0.029052	0.021379	60
80	0.476788	0.457876	0.440044	0.392205	0.351460	0.316437	0.286099	0.259651	0.192380	0.115013	0.075031	0.052435	0.038740	80
100	0.549692	0.531670	0.514521	0.467724	0.426875	0.390945	0.359148	0.330860	0.256121	0.163561	0.111482	0.080033	0.060002	100
120	0.605307	0.588394	0.572199	0.527469	0.487735	0.452209	0.420274	0.391446	0.313105	0.210557	0.149024	0.109865	0.083839	120
140	0.648946	0.633150	0.617953	0.575599	0.537482	0.502978	0.471597	0.442948	0.363394	0.254657	0.186033	0.140487	0.109120	140
170	0.699095	0.684833	0.671043	0.632228	0.596790	0.564269	0.534303	0.506601	0.427763	0.314520	0.238754	0.185923	0.147949	170
200	0.736825	0.723889	0.711333	0.675743	0.642910	0.612477	0.584164	0.557749	0.481186	0.367006	0.287138	0.229286	0.186289	200
240	0.774600	0.763133	0.751962	0.720079	0.690368	0.662559	0.636443	0.611856	0.539283	0.426889	0.344628	0.282663	0.234984	240
320	0.824943	0.815652	0.806561	0.780364	0.755626	0.732179	0.709889	0.688652	0.624439	0.519812	0.438357	0.373572	0.321238	320
440	0.868933	0.861725	0.854644	0.834102	0.814487	0.795690	0.777632	0.760252	0.706588	0.615136	0.539871	0.476944	0.423743	440
600	0.901853	0.896316	0.890862	0.874954	0.859647	0.844866	0.830551	0.816669	0.773133	0.696377	0.630457	0.573141	0.522892	600
800	0.925314	0.921026	0.916795	0.904405	0.892417	0.880776	0.869448	0.858400	0.823361	0.760060	0.703985	0.653797	0.608582	800
1000	0.939725	0.936229	0.932774	0.922635	0.912790	0.903199	0.893834	0.884674	0.855420	0.801763	0.753305	0.709136	0.668647	1000
INF	1.000000	1.000000	1.000000	1.000000	1.000000	1.000000	1.000000	1.000000	1.000000	1.000000	1.000000	1.000000	1.000000	INF

Tafel 1 (Forts.) Table 1 (cont.)

$U_{p,q,n}$

p=4 C=0.05

n	q=1	2	3	4	5	6	7	8	9	10	11	12	n
1	0.000000	0.000000	0.000000	0.000000	0.000000	0.000000	0.000000	0.000000	0.000000	0.000000	0.000000	0.000000	1
2	0.000000	0.000000	0.000000	0.000000	0.000000	0.000000	0.000000	0.000000	0.000000	0.000000	0.000000	0.000000	2
3	0.000000	0.000000	0.000000	0.000000	0.000000	0.000001	0.000001	0.000001	0.000002	0.000002	0.000002	0.000003	3
4	0.001378	0.000292	0.000127	0.000075	0.000052	0.000040	0.000033	0.000029	0.000026	0.000025	0.000023	0.000022	4
5	0.025529	0.006091	0.002314	0.001128	0.000647	0.000416	0.000292	0.000218	0.000172	0.000141	0.000120	0.000105	5
6	0.076071	0.023604	0.010010	0.005073	0.002903	0.001818	0.001223	0.000872	0.000652	0.000508	0.000409	0.000338	6
7	0.135374	0.050839	0.024047	0.013014	0.007737	0.004938	0.003338	0.002365	0.001745	0.001333	0.001050	0.000848	7
8	0.194043	0.083695	0.043226	0.024857	0.015415	0.010129	0.006975	0.004994	0.003698	0.002819	0.002206	0.001766	8
9	0.248619	0.118995	0.065947	0.039919	0.025729	0.017408	0.012249	0.008907	0.006664	0.005112	0.004009	0.003208	9
10	0.298130	0.154758	0.090794	0.057378	0.038260	0.026586	0.019107	0.014130	0.010706	0.008288	0.006542	0.005254	10
11	0.342593	0.189778	0.116701	0.076502	0.052524	0.037385	0.027402	0.020589	0.015806	0.012365	0.009839	0.007948	11
12	0.382448	0.223411	0.142927	0.096664	0.068077	0.049495	0.036933	0.028170	0.021899	0.017314	0.013895	0.011302	12
13	0.418181	0.255376	0.168939	0.117377	0.084546	0.062632	0.047493	0.036731	0.028895	0.023075	0.018675	0.015303	13
14	0.450335	0.285511	0.194414	0.138286	0.101586	0.076537	0.058886	0.046115	0.036676	0.029572	0.024133	0.019917	14
15	0.479286	0.313829	0.219113	0.159131	0.118954	0.090983	0.070925	0.056188	0.045140	0.036722	0.030208	0.025101	15
16	0.505512	0.340400	0.242944	0.179688	0.136434	0.105779	0.083443	0.066806	0.054181	0.044440	0.036830	0.030804	16
17	0.529312	0.365253	0.265812	0.199832	0.153891	0.120780	0.096316	0.077856	0.063688	0.052645	0.043936	0.036980	17
18	0.551035	0.388530	0.287689	0.219490	0.171171	0.135856	0.109411	0.089236	0.073577	0.061263	0.051456	0.043568	18
19	0.570858	0.410325	0.308599	0.238570	0.188209	0.150905	0.122643	0.100843	0.083764	0.070213	0.059338	0.050514	19
20	0.589077	0.430766	0.328552	0.257052	0.204926	0.165853	0.135926	0.112607	0.094180	0.079441	0.067513	0.057782	20
21	0.605832	0.449947	0.347546	0.274909	0.221288	0.180626	0.149180	0.124462	0.104757	0.088877	0.075938	0.065315	21
22	0.621318	0.467988	0.365676	0.292142	0.237242	0.195197	0.162364	0.136342	0.115440	0.098474	0.084565	0.073068	22
23	0.635634	0.484922	0.382934	0.308765	0.252783	0.209511	0.175434	0.148204	0.126185	0.108191	0.093352	0.081008	23
24	0.648934	0.500883	0.399402	0.324767	0.267896	0.223535	0.188341	0.160009	0.136950	0.117977	0.102254	0.089100	24
25	0.661320	0.515918	0.415077	0.340175	0.282568	0.237277	0.201067	0.171726	0.147695	0.127818	0.111240	0.097305	25
26	0.672864	0.530124	0.430041	0.355004	0.296810	0.250710	0.213597	0.183333	0.158399	0.137656	0.120274	0.105608	26
27	0.683663	0.543561	0.444332	0.369254	0.310608	0.263809	0.225900	0.194794	0.169017	0.147483	0.129346	0.113968	27
28	0.693769	0.556262	0.457946	0.382979	0.323980	0.276602	0.237971	0.206105	0.179569	0.157274	0.138418	0.122368	28
29	0.703259	0.568303	0.470981	0.396197	0.336947	0.289051	0.249798	0.217241	0.189991	0.167006	0.147478	0.130785	29
30	0.712188	0.579734	0.483431	0.408914	0.349488	0.301188	0.261373	0.228198	0.200311	0.176673	0.156516	0.139205	30
40	0.778877	0.668158	0.582817	0.513297	0.455181	0.405867	0.363565	0.326959	0.295085	0.267163	0.242600	0.220888	40
60	0.849044	0.767047	0.700066	0.642556	0.592126	0.547349	0.507256	0.471148	0.438462	0.408771	0.381699	0.356960	60
80	0.885442	0.820705	0.766251	0.718260	0.675124	0.635912	0.600023	0.566986	0.536460	0.508176	0.481887	0.457414	80
100	0.907714	0.854312	0.808614	0.767700	0.730354	0.695928	0.663968	0.634166	0.606280	0.580112	0.555487	0.532298	100
120	0.922736	0.877325	0.838018	0.802443	0.769650	0.739118	0.710513	0.683595	0.658183	0.634132	0.611324	0.589657	120
140	0.933554	0.894066	0.859605	0.828176	0.798994	0.771635	0.745829	0.721386	0.698162	0.676045	0.654943	0.634778	140
170	0.945088	0.912072	0.883006	0.856283	0.831279	0.807662	0.785224	0.763821	0.743347	0.723717	0.704865	0.686733	170
200	0.953211	0.924848	0.899727	0.876499	0.854647	0.833900	0.814087	0.795095	0.776838	0.759251	0.742281	0.725885	200
240	0.960919	0.937047	0.915781	0.896012	0.877319	0.859482	0.842366	0.825881	0.809961	0.794554	0.779622	0.765130	240
320	0.970605	0.952477	0.936212	0.920990	0.906503	0.892593	0.879164	0.866153	0.853513	0.841211	0.829220	0.817517	320
440	0.978571	0.965253	0.953233	0.941922	0.931100	0.920655	0.910522	0.900654	0.891022	0.881602	0.872376	0.863331	440
600	0.984259	0.974422	0.965507	0.957084	0.948995	0.941160	0.933530	0.926075	0.918772	0.911606	0.904563	0.897634	600
800	0.988181	0.980767	0.974028	0.967644	0.961498	0.955529	0.949702	0.943994	0.938390	0.932877	0.927446	0.922092	800
1000	0.990538	0.984589	0.979173	0.974034	0.969078	0.964257	0.959545	0.954922	0.950376	0.945898	0.941481	0.937120	1000
INF	1.000000	1.000000	1.000000	1.000000	1.000000	1.000000	1.000000	1.000000	1.000000	1.000000	1.000000	1.000000	INF

Tafel 1 (Forts.) Table 1 (cont.)

$U_{p,q,n}$

p=4 C=0.01

n	q = 1	2	3	4	5	6	7	8	9	10	11	12	n
1	0.000000	0.000000	0.000000	0.000000	0.000000	0.000000	0.000000	0.000000	0.000000	0.000000	0.000000	0.000000	1
2	0.000000	0.000000	0.000000	0.000000	0.000000	0.000000	0.000000	0.000000	0.000000	0.000000	0.000000	0.000000	2
3	0.000000	0.000000	0.000000	0.000000	0.000000	0.000000	0.000000	0.000000	0.000000	0.000000	0.000000	0.000000	3
4	0.000090	0.000026	0.000015	0.000011	0.000009	0.000008	0.000007	0.000007	0.000007	0.000007	0.000007	0.000007	4
5	0.005218	0.001224	0.000484	0.000250	0.000153	0.000106	0.000079	0.000063	0.000052	0.000045	0.000040	0.000037	5
6	0.025586	0.007345	0.003037	0.001538	0.000893	0.000574	0.000398	0.000293	0.000227	0.000183	0.000152	0.000130	6
7	0.058962	0.020352	0.009229	0.004891	0.002885	0.001846	0.001259	0.000906	0.000681	0.000531	0.000427	0.000353	7
8	0.098904	0.039349	0.019423	0.010860	0.006623	0.004315	0.002966	0.002131	0.001590	0.001225	0.000971	0.000789	8
9	0.140881	0.062551	0.033146	0.019474	0.012300	0.008211	0.005732	0.004155	0.003111	0.002396	0.001891	0.001527	9
10	0.182362	0.088300	0.049627	0.030445	0.019865	0.013591	0.009662	0.007095	0.005357	0.004145	0.003278	0.002644	10
11	0.222076	0.115330	0.068087	0.043357	0.029128	0.020392	0.014763	0.010993	0.008388	0.006539	0.005197	0.004202	11
12	0.259456	0.142746	0.087852	0.057777	0.039832	0.028478	0.020973	0.015836	0.012218	0.009607	0.007685	0.006242	12
13	0.294315	0.169948	0.108380	0.073308	0.051709	0.037679	0.028192	0.021570	0.016825	0.013349	0.010755	0.008785	13
14	0.326670	0.196547	0.129256	0.089607	0.064505	0.047814	0.036299	0.028118	0.022164	0.017742	0.014400	0.011834	14
15	0.356636	0.222301	0.150167	0.106392	0.077992	0.058712	0.045168	0.035392	0.028175	0.022747	0.018597	0.015378	15
16	0.384374	0.247074	0.170888	0.123435	0.091973	0.070212	0.054675	0.043297	0.034790	0.028315	0.023313	0.019396	16
17	0.410058	0.270796	0.191257	0.140556	0.106281	0.082172	0.064703	0.051742	0.041937	0.034394	0.028510	0.023860	17
18	0.433867	0.293442	0.211160	0.157615	0.120777	0.094469	0.075148	0.060641	0.049546	0.040928	0.034143	0.028739	18
19	0.455967	0.315021	0.230524	0.174505	0.135348	0.106995	0.085915	0.069912	0.057551	0.047861	0.040170	0.033996	19
20	0.476513	0.335555	0.249300	0.191144	0.149903	0.119660	0.096921	0.079482	0.065888	0.055142	0.046546	0.039596	20
21	0.495648	0.355086	0.267462	0.207474	0.164368	0.132389	0.108094	0.089287	0.074500	0.062720	0.053228	0.045503	21
22	0.513499	0.373655	0.284999	0.223450	0.178686	0.145118	0.119371	0.099266	0.083333	0.070548	0.060177	0.051683	22
23	0.530184	0.391312	0.301910	0.239044	0.192810	0.157797	0.130700	0.109371	0.092342	0.078585	0.067354	0.058103	23
24	0.545805	0.408104	0.318203	0.254237	0.206707	0.170381	0.142036	0.119555	0.101484	0.086790	0.074725	0.064730	24
25	0.560457	0.424083	0.333888	0.269015	0.220349	0.182837	0.153340	0.129781	0.110720	0.095129	0.082255	0.071537	25
26	0.574221	0.439293	0.348987	0.283374	0.233718	0.195137	0.164581	0.140015	0.120018	0.103570	0.089917	0.078495	26
27	0.587173	0.453781	0.363513	0.297314	0.246799	0.207259	0.175732	0.150228	0.129349	0.112085	0.097684	0.085579	27
28	0.599381	0.467592	0.377492	0.310838	0.259584	0.219187	0.186772	0.160396	0.138688	0.120648	0.105530	0.092768	28
29	0.610904	0.480765	0.390942	0.323953	0.272068	0.230906	0.197681	0.170499	0.148013	0.129238	0.113435	0.100038	29
30	0.621798	0.493340	0.403887	0.336665	0.284247	0.242408	0.208447	0.180518	0.157304	0.137834	0.121378	0.107372	30
40	0.704846	0.593044	0.510028	0.444079	0.390022	0.344862	0.306628	0.273929	0.245739	0.221270	0.199908	0.181164	40
60	0.795314	0.709205	0.641042	0.583746	0.534292	0.490946	0.452558	0.418305	0.387562	0.359837	0.334734	0.311927	60
80	0.843446	0.774138	0.717496	0.668503	0.625080	0.586056	0.550669	0.518371	0.488748	0.461472	0.436276	0.412939	80
100	0.873280	0.815461	0.767296	0.724909	0.686729	0.651890	0.619831	0.590158	0.562571	0.536836	0.512760	0.490184	100
120	0.893573	0.844036	0.802240	0.765030	0.731147	0.699908	0.670876	0.643745	0.618288	0.594325	0.571711	0.550325	120
140	0.908268	0.864962	0.828089	0.794989	0.764613	0.736396	0.709985	0.685132	0.661655	0.639410	0.618283	0.598178	140
170	0.924010	0.887597	0.856293	0.827944	0.801710	0.777147	0.753978	0.731988	0.711113	0.691172	0.672101	0.653829	170
200	0.935142	0.903739	0.876559	0.851792	0.828739	0.807033	0.786448	0.766791	0.748045	0.730070	0.712780	0.696132	200
240	0.945741	0.919212	0.896104	0.874924	0.855100	0.836335	0.818447	0.801269	0.784813	0.768957	0.753648	0.738836	240
320	0.959108	0.938870	0.921100	0.904693	0.889230	0.874495	0.860356	0.846684	0.833511	0.820740	0.808335	0.796267	320
440	0.970142	0.955217	0.942028	0.929777	0.918164	0.907036	0.896302	0.885863	0.875757	0.865908	0.856294	0.846895	440
600	0.978043	0.966989	0.957176	0.948021	0.939309	0.930927	0.922811	0.914887	0.907187	0.899657	0.892279	0.885041	600
800	0.983500	0.975154	0.967720	0.960765	0.954127	0.947724	0.941507	0.935422	0.929493	0.923681	0.917972	0.912357	800
1000	0.986785	0.980081	0.974098	0.968491	0.963130	0.957951	0.952914	0.947976	0.943158	0.938427	0.933773	0.929189	1000
INF	1.000000	1.000000	1.000000	1.000000	1.000000	1.000000	1.000000	1.000000	1.000000	1.000000	1.000000	1.000000	INF

Tafel 1 (Forts.) Table 1 (cont.)

$U_{p,q,n}$

p=4 C=0.05

n	q=13	14	15	18	21	24	27	30	40	60	80	100	120	n
1	0.000000	0.000000	0.000000	0.000000	0.000000	0.000000	0.000000	0.000000	0.000000	0.000000	0.000000	0.000000	0.000000	1
2	0.000000	0.000000	0.000000	0.000000	0.000000	0.000000	0.000000	0.000000	0.000000	0.000000	0.000000	0.000000	0.000000	2
3	0.000003	0.000004	0.000004	0.000005	0.000007	0.000008	0.000010	0.000011	0.000017	0.000026	0.000035	0.000043	0.000050	3
4	0.000022	0.000022	0.000021	0.000021	0.000022	0.000023	0.000024	0.000025	0.000029	0.000038	0.000046	0.000053	0.000060	4
5	0.000093	0.000084	0.000078	0.000065	0.000058	0.000054	0.000052	0.000050	0.000050	0.000054	0.000059	0.000065	0.000071	5
6	0.000287	0.000248	0.000219	0.000162	0.000131	0.000113	0.000101	0.000093	0.000080	0.000074	0.000076	0.000079	0.000083	6
7	0.000701	0.000591	0.000508	0.000349	0.000265	0.000215	0.000182	0.000161	0.000123	0.000101	0.000096	0.000095	0.000097	7
8	0.001444	0.001202	0.001017	0.000669	0.000485	0.000377	0.000308	0.000262	0.000183	0.000134	0.000119	0.000114	0.000113	8
9	0.002614	0.002165	0.001820	0.001167	0.000821	0.000619	0.000492	0.000408	0.000265	0.000175	0.000147	0.000136	0.000131	9
10	0.004287	0.003549	0.002978	0.001886	0.001303	0.000962	0.000749	0.000608	0.000372	0.000226	0.000181	0.000161	0.000151	10
11	0.006512	0.005404	0.004539	0.002864	0.001957	0.001426	0.001093	0.000873	0.000509	0.000288	0.000219	0.000189	0.000173	11
12	0.009309	0.007756	0.006532	0.004129	0.002809	0.002029	0.001539	0.001215	0.000681	0.000362	0.000264	0.000221	0.000198	12
13	0.012679	0.010614	0.008971	0.005704	0.003878	0.002788	0.002100	0.001644	0.000893	0.000450	0.000315	0.000256	0.000225	13
14	0.016601	0.013970	0.011857	0.007599	0.005178	0.003716	0.002787	0.002169	0.001149	0.000553	0.000374	0.000296	0.000255	14
15	0.021050	0.017805	0.015181	0.009820	0.006720	0.004825	0.003610	0.002799	0.001454	0.000672	0.000441	0.000341	0.000289	15
16	0.025985	0.022093	0.018921	0.012363	0.008508	0.006122	0.004578	0.003540	0.001813	0.000810	0.000516	0.000391	0.000325	16
17	0.031371	0.026805	0.023059	0.015222	0.010543	0.007611	0.005697	0.004401	0.002230	0.000967	0.000601	0.000446	0.000365	17
18	0.037155	0.031902	0.027563	0.018386	0.012823	0.009295	0.006969	0.005384	0.002708	0.001146	0.000696	0.000507	0.000409	18
19	0.043304	0.037352	0.032405	0.021840	0.015342	0.011173	0.008398	0.006493	0.003251	0.001347	0.000801	0.000573	0.000456	19
20	0.049764	0.043114	0.037557	0.025567	0.018092	0.013242	0.009985	0.007732	0.003861	0.001572	0.000918	0.000646	0.000507	20
21	0.056506	0.049160	0.042988	0.029547	0.021065	0.015499	0.011728	0.009101	0.004542	0.001822	0.001046	0.000726	0.000563	21
22	0.063489	0.055455	0.048670	0.033767	0.024248	0.017938	0.013626	0.010598	0.005295	0.002099	0.001187	0.000813	0.000623	22
23	0.070678	0.061962	0.054569	0.038208	0.027631	0.020552	0.015673	0.012224	0.006122	0.002403	0.001341	0.000907	0.000688	23
24	0.078032	0.068654	0.060666	0.042846	0.031202	0.023336	0.017870	0.013978	0.007024	0.002737	0.001508	0.001008	0.000757	24
25	0.085529	0.075510	0.066938	0.047665	0.034947	0.026277	0.020207	0.015856	0.008002	0.003100	0.001690	0.001118	0.000831	25
26	0.093145	0.082494	0.073350	0.052652	0.038856	0.029373	0.022681	0.017854	0.009056	0.003494	0.001887	0.001236	0.000911	26
27	0.100843	0.089592	0.079888	0.057787	0.042918	0.032610	0.025286	0.019971	0.010187	0.003920	0.002099	0.001363	0.000996	27
28	0.108619	0.096779	0.086533	0.063050	0.047114	0.035983	0.028017	0.022202	0.011394	0.004379	0.002328	0.001498	0.001086	28
29	0.116436	0.104032	0.093266	0.068437	0.051441	0.039481	0.030870	0.024542	0.012676	0.004870	0.002572	0.001643	0.001183	29
30	0.124279	0.111336	0.100064	0.073921	0.055884	0.043097	0.033831	0.026986	0.014032	0.005396	0.002833	0.001797	0.001285	30
40	0.201607	0.184450	0.169118	0.131916	0.104542	0.084000	0.068327	0.056193	0.031375	0.012585	0.006462	0.003925	0.002676	40
60	0.334285	0.313468	0.294307	0.245203	0.206160	0.174731	0.149156	0.128162	0.080505	0.036886	0.019733	0.011905	0.007882	60
80	0.434576	0.413240	0.393266	0.340456	0.296513	0.259620	0.228418	0.201851	0.137432	0.070586	0.040321	0.025089	0.016764	80
100	0.510415	0.489723	0.470157	0.417334	0.372048	0.332968	0.299025	0.269408	0.194209	0.108880	0.066064	0.042728	0.029180	100
120	0.569046	0.549415	0.530697	0.479453	0.434588	0.395082	0.360128	0.329073	0.247486	0.148493	0.094784	0.063600	0.044539	120
140	0.615483	0.596999	0.579277	0.530236	0.486620	0.447637	0.412653	0.381146	0.296172	0.187523	0.124835	0.086543	0.062113	140
170	0.669273	0.652445	0.636210	0.590753	0.549624	0.512254	0.478182	0.447029	0.360473	0.242932	0.170096	0.122861	0.091136	170
200	0.710024	0.694669	0.679790	0.637777	0.599288	0.563895	0.531252	0.501070	0.415341	0.293507	0.213784	0.159629	0.121753	200
240	0.751052	0.737363	0.724044	0.686122	0.650961	0.618249	0.587735	0.559211	0.476396	0.353195	0.267985	0.207264	0.162968	240
320	0.806085	0.794908	0.783973	0.752514	0.722888	0.694906	0.668419	0.643304	0.568236	0.449483	0.360908	0.293445	0.241214	320
440	0.854453	0.845735	0.837167	0.812296	0.788567	0.765866	0.744108	0.723223	0.659206	0.552289	0.466957	0.397864	0.341299	440
600	0.890812	0.884090	0.877462	0.858099	0.839447	0.821437	0.804016	0.787142	0.734440	0.642740	0.565658	0.500182	0.444139	600
800	0.916807	0.911588	0.906430	0.891294	0.876617	0.862351	0.848463	0.834925	0.792076	0.715289	0.648271	0.589294	0.537102	800
1000	0.932809	0.928545	0.924326	0.911912	0.899825	0.888030	0.876502	0.865221	0.829223	0.763530	0.704835	0.652022	0.604282	1000
INF	1.000000	1.000000	1.000000	1.000000	1.000000	1.000000	1.000000	1.000000	1.000000	1.000000	1.000000	1.000000	1.000000	INF

$U_{p,q,n}$

p=4 C=0.01

n	q=13	14	15	18	21	24	27	30	40	60	80	100	120	n
1	0.000000	0.000000	0.000000	0.000000	0.000000	0.000000	0.000000	0.000000	0.000002	0.000007	0.000014	0.000020	0.000026	1
2	0.000000	0.000000	0.000000	0.000000	0.000000	0.000001	0.000002	0.000002	0.000005	0.000011	0.000018	0.000025	0.000032	2
3	0.000000	0.000001	0.000001	0.000002	0.000003	0.000004	0.000005	0.000006	0.000009	0.000017	0.000025	0.000032	0.000038	3
4	0.000007	0.000008	0.000008	0.000009	0.000010	0.000011	0.000012	0.000013	0.000017	0.000025	0.000032	0.000039	0.000045	4
5	0.000034	0.000032	0.000031	0.000028	0.000027	0.000026	0.000026	0.000027	0.000029	0.000035	0.000042	0.000048	0.000054	5
6	0.000114	0.000101	0.000092	0.000073	0.000063	0.000057	0.000053	0.000051	0.000048	0.000049	0.000054	0.000059	0.000064	6
7	0.000299	0.000257	0.000226	0.000165	0.000131	0.000111	0.000098	0.000089	0.000075	0.000068	0.000068	0.000071	0.000074	7
8	0.000655	0.000554	0.000477	0.000329	0.000249	0.000201	0.000170	0.000149	0.000113	0.000091	0.000086	0.000085	0.000087	8
9	0.001256	0.001052	0.000895	0.000596	0.000434	0.000339	0.000278	0.000236	0.000165	0.000120	0.000106	0.000102	0.000101	9
10	0.002170	0.001809	0.001531	0.000997	0.000709	0.000539	0.000431	0.000358	0.000234	0.000155	0.000131	0.000121	0.000117	10
11	0.003452	0.002876	0.002428	0.001564	0.001094	0.000817	0.000641	0.000524	0.000324	0.000199	0.000160	0.000143	0.000134	11
12	0.005143	0.004292	0.003625	0.002324	0.001611	0.001187	0.000919	0.000740	0.000439	0.000252	0.000193	0.000167	0.000154	12
13	0.007268	0.006083	0.005147	0.003301	0.002276	0.001664	0.001275	0.001016	0.000581	0.000315	0.000232	0.000195	0.000175	13
14	0.009838	0.008265	0.007013	0.004514	0.003107	0.002260	0.001721	0.001360	0.000755	0.000389	0.000276	0.000226	0.000200	14
15	0.012850	0.010842	0.009230	0.005975	0.004117	0.002988	0.002264	0.001778	0.000965	0.000476	0.000327	0.000261	0.000226	15
16	0.016293	0.013807	0.011798	0.007694	0.005316	0.003856	0.002914	0.002279	0.001214	0.000577	0.000384	0.000300	0.000255	16
17	0.020147	0.017150	0.014711	0.009672	0.006710	0.004871	0.003677	0.002868	0.001506	0.000693	0.000449	0.000343	0.000287	17
18	0.024388	0.020854	0.017957	0.011910	0.008303	0.006041	0.004559	0.003551	0.001844	0.000825	0.000521	0.000391	0.000322	18
19	0.028990	0.024897	0.021521	0.014402	0.010097	0.007367	0.005565	0.004333	0.002231	0.000974	0.000602	0.000443	0.000360	19
20	0.033924	0.029256	0.025386	0.017141	0.012090	0.008853	0.006699	0.005216	0.002671	0.001142	0.000692	0.000501	0.000402	20
21	0.039160	0.033909	0.029531	0.020119	0.014279	0.010498	0.007962	0.006204	0.003166	0.001330	0.000791	0.000564	0.000447	21
22	0.044668	0.038829	0.033936	0.023325	0.016660	0.012302	0.009354	0.007299	0.003718	0.001539	0.000901	0.000633	0.000495	22
23	0.050420	0.043992	0.038580	0.026746	0.019226	0.014262	0.010877	0.008502	0.004329	0.001770	0.001021	0.000708	0.000548	23
24	0.056388	0.049375	0.043443	0.030370	0.021970	0.016374	0.012528	0.009813	0.005001	0.002025	0.001152	0.000790	0.000604	24
25	0.062546	0.054954	0.048504	0.034183	0.024885	0.018635	0.014307	0.011232	0.005736	0.002303	0.001295	0.000877	0.000665	25
26	0.068869	0.060707	0.053743	0.038172	0.027962	0.021039	0.016210	0.012759	0.006533	0.002606	0.001450	0.000972	0.000730	26
27	0.075335	0.066612	0.059142	0.042325	0.031192	0.023581	0.018235	0.014390	0.007396	0.002935	0.001617	0.001074	0.000800	27
28	0.081921	0.072650	0.064682	0.046628	0.034567	0.026256	0.020378	0.016126	0.008322	0.003291	0.001798	0.001184	0.000874	28
29	0.088609	0.078804	0.070347	0.051068	0.038077	0.029057	0.022635	0.017963	0.009314	0.003674	0.001992	0.001301	0.000953	29
30	0.095380	0.085055	0.076120	0.055633	0.041713	0.031977	0.025002	0.019899	0.010371	0.004086	0.002201	0.001426	0.001037	30
40	0.164643	0.150023	0.137036	0.105837	0.083189	0.066392	0.053706	0.043978	0.024372	0.009844	0.005138	0.003171	0.002192	40
60	0.291144	0.272154	0.254761	0.210577	0.175838	0.148142	0.125801	0.107595	0.066806	0.030332	0.016279	0.009900	0.006614	60
80	0.391270	0.371112	0.352322	0.303018	0.262398	0.228588	0.200205	0.176198	0.118648	0.060147	0.034227	0.021346	0.014332	80
100	0.468972	0.449007	0.430186	0.379741	0.336885	0.300175	0.268515	0.241053	0.172040	0.095172	0.057356	0.037036	0.025336	100
120	0.530063	0.510838	0.492570	0.442871	0.399718	0.361987	0.328812	0.299500	0.223230	0.132247	0.083760	0.055994	0.039189	120
140	0.579014	0.560721	0.543239	0.495142	0.452696	0.415010	0.381384	0.351257	0.270734	0.169395	0.111877	0.077187	0.055284	140
170	0.636297	0.619453	0.603253	0.558135	0.517603	0.480998	0.447801	0.417590	0.334350	0.222942	0.154920	0.111307	0.082290	170
200	0.680082	0.664590	0.649622	0.607569	0.569298	0.534304	0.502188	0.472625	0.389301	0.272481	0.197072	0.146376	0.111223	200
240	0.724484	0.710570	0.697065	0.658801	0.623539	0.590905	0.560604	0.532394	0.451077	0.331611	0.250009	0.192401	0.150697	240
320	0.784512	0.773050	0.761856	0.729786	0.699754	0.671521	0.644906	0.619759	0.545079	0.428243	0.342046	0.276925	0.226831	320
440	0.837696	0.828682	0.819844	0.794287	0.770016	0.746894	0.724814	0.703687	0.639298	0.532806	0.448617	0.380924	0.325812	440
600	0.877932	0.870942	0.864065	0.844046	0.824853	0.806384	0.788579	0.771383	0.717950	0.625788	0.548962	0.484100	0.428850	600
800	0.906829	0.901380	0.896006	0.880291	0.865119	0.850426	0.836162	0.822294	0.778609	0.700946	0.633667	0.574782	0.522893	800
1000	0.924669	0.920208	0.915802	0.902879	0.890350	0.878166	0.866293	0.854703	0.817878	0.751178	0.691995	0.639008	0.591297	1000
INF	1.000000	1.000000	1.000000	1.000000	1.000000	1.000000	1.000000	1.000000	1.000000	1.000000	1.000000	1.000000	1.000000	INF

$U_{p,q,n}$

p=5 C=0.05

q

n	1	2	3	4	5	6	7	8	9	10	11	12	n
1	0.000000	0.000000	0.000000	0.000000	0.000000	0.000000	0.000000	0.000000	0.000000	0.000000	0.000000	0.000000	1
2	0.000000	0.000000	0.000000	0.000000	0.000000	0.000000	0.000000	0.000000	0.000000	0.000000	0.000000	0.000000	2
3	0.000000	0.000000	0.000000	0.000000	0.000000	0.000000	0.000000	0.000000	0.000000	0.000000	0.000000	0.000000	3
4	0.000000	0.000000	0.000000	0.000000	0.000001	0.000001	0.000001	0.000001	0.000001	0.000001	0.000001	0.000001	4
5	0.001598	0.000291	0.000105	0.000052	0.000031	0.000021	0.000015	0.000012	0.000010	0.000008	0.000007	0.000007	5
6	0.021145	0.004391	0.001479	0.000647	0.000335	0.000197	0.000126	0.000087	0.000064	0.000049	0.000039	0.000032	6
7	0.062771	0.016898	0.006357	0.002903	0.001514	0.000872	0.000544	0.000361	0.000253	0.000185	0.000141	0.000110	7
8	0.113526	0.037390	0.015792	0.007737	0.004208	0.002479	0.001557	0.001032	0.000716	0.000516	0.000385	0.000296	8
9	0.165351	0.063279	0.029433	0.015415	0.008787	0.005348	0.003433	0.002304	0.001607	0.001159	0.000861	0.000657	9
10	0.214794	0.092191	0.046378	0.025729	0.015321	0.009639	0.006343	0.004335	0.003062	0.002225	0.001660	0.001267	10
11	0.260635	0.122403	0.065660	0.038260	0.023674	0.015360	0.010358	0.007216	0.005173	0.003802	0.002858	0.002192	11
12	0.302608	0.152793	0.086448	0.052524	0.033618	0.022418	0.015467	0.010980	0.007991	0.005946	0.004512	0.003486	12
13	0.340813	0.182662	0.108110	0.068077	0.044878	0.030680	0.021607	0.015611	0.011530	0.008685	0.006659	0.005187	13
14	0.375528	0.211602	0.130131	0.084546	0.057198	0.039965	0.028683	0.021061	0.015774	0.012024	0.009313	0.007317	14
15	0.407128	0.239373	0.152160	0.101586	0.070324	0.050117	0.036584	0.027266	0.020687	0.015949	0.012475	0.009885	15
16	0.435899	0.265851	0.173959	0.118954	0.084048	0.060965	0.045199	0.034145	0.026219	0.020428	0.016129	0.012885	16
17	0.462173	0.291015	0.195322	0.136434	0.098187	0.072367	0.054409	0.041618	0.032312	0.025427	0.020252	0.016307	17
18	0.486266	0.314859	0.216138	0.153891	0.112582	0.084178	0.064111	0.049602	0.038909	0.030904	0.024819	0.020133	18
19	0.508362	0.337418	0.236338	0.171171	0.127108	0.096308	0.074209	0.058024	0.045951	0.036810	0.029790	0.024339	19
20	0.528714	0.358776	0.255858	0.188209	0.141662	0.108634	0.084619	0.066805	0.053373	0.043100	0.035137	0.028896	20
21	0.547516	0.378956	0.274710	0.204926	0.156176	0.121083	0.095254	0.075885	0.061122	0.049724	0.040817	0.033782	21
22	0.564905	0.398038	0.292843	0.221288	0.170563	0.133590	0.106063	0.085203	0.069149	0.056652	0.046803	0.038962	22
23	0.581036	0.416105	0.310304	0.237242	0.184782	0.146095	0.116974	0.094699	0.077408	0.063832	0.053052	0.044411	23
24	0.596032	0.433216	0.327083	0.252783	0.198795	0.158544	0.127948	0.104337	0.085849	0.071231	0.059537	0.050103	24
25	0.610030	0.449429	0.343191	0.267896	0.212568	0.170898	0.138945	0.114058	0.094444	0.078809	0.066222	0.056005	25
26	0.623126	0.464800	0.358665	0.282568	0.226071	0.183129	0.149909	0.123843	0.103144	0.086535	0.073084	0.062103	26
27	0.635368	0.479382	0.373523	0.296810	0.239294	0.195207	0.160826	0.133657	0.111931	0.094385	0.080093	0.068358	27
28	0.646832	0.493247	0.387790	0.310608	0.252224	0.207116	0.171667	0.143454	0.120766	0.102323	0.087220	0.074761	28
29	0.657645	0.506421	0.401488	0.323980	0.264873	0.218828	0.182408	0.153240	0.129630	0.110335	0.094455	0.081283	29
30	0.667803	0.518945	0.414658	0.336947	0.277200	0.230347	0.193043	0.162971	0.138499	0.118393	0.101767	0.087901	30
40	0.744010	0.617178	0.521747	0.446045	0.384424	0.333492	0.290896	0.254963	0.224433	0.198322	0.175874	0.156480	40
60	0.824764	0.729155	0.652037	0.586878	0.530670	0.481578	0.438367	0.400085	0.365997	0.335520	0.308193	0.283593	60
80	0.866847	0.790730	0.727186	0.671775	0.622536	0.578316	0.538319	0.501966	0.468774	0.438392	0.410497	0.384827	80
100	0.892643	0.829563	0.775817	0.728040	0.684827	0.645343	0.609037	0.575509	0.544420	0.515540	0.488629	0.463515	100
120	0.910071	0.856268	0.809790	0.767957	0.729656	0.694256	0.661341	0.630608	0.601822	0.574793	0.549362	0.525395	120
140	0.922634	0.875748	0.834850	0.797705	0.763400	0.731431	0.701466	0.673268	0.646653	0.621477	0.597616	0.574968	140
170	0.936039	0.896748	0.862122	0.830370	0.800777	0.772953	0.746649	0.721687	0.697934	0.675284	0.653648	0.632953	170
200	0.945486	0.911680	0.881674	0.853973	0.827989	0.803406	0.780024	0.757705	0.736343	0.715856	0.696177	0.677251	200
240	0.954455	0.925960	0.900496	0.876838	0.854512	0.833264	0.812938	0.793426	0.774647	0.756540	0.739054	0.722148	240
320	0.965732	0.944055	0.924519	0.906224	0.888827	0.872146	0.856074	0.840535	0.825476	0.810855	0.796641	0.782805	320
440	0.975013	0.959064	0.944590	0.930949	0.917894	0.905302	0.893096	0.881226	0.869655	0.858357	0.847311	0.836500	440
600	0.981642	0.969850	0.959096	0.948913	0.939124	0.929642	0.920411	0.911396	0.902572	0.893921	0.885429	0.877084	600
800	0.986214	0.977320	0.969181	0.961450	0.953996	0.946753	0.939682	0.932756	0.925957	0.919273	0.912693	0.906209	800
1000	0.988963	0.981823	0.975277	0.969047	0.963029	0.957171	0.951441	0.945820	0.940292	0.934848	0.929480	0.924182	1000
INF	1.000000	1.000000	1.000000	1.000000	1.000000	1.000000	1.000000	1.000000	1.000000	1.000000	1.000000	1.000000	INF

$U_{p,q,n}$

p=5 C=0.01

n	q=1	2	3	4	5	6	7	8	9	10	11	12	n
1	0.000000	0.000000	0.000000	0.000000	0.000000	0.000000	0.000000	0.000000	0.000000	0.000000	0.000000	0.000000	1
2	0.000000	0.000000	0.000000	0.000000	0.000000	0.000000	0.000000	0.000000	0.000000	0.000000	0.000000	0.000000	2
3	0.000000	0.000000	0.000000	0.000000	0.000000	0.000000	0.000000	0.000000	0.000000	0.000000	0.000000	0.000000	3
4	0.000000	0.000000	0.000000	0.000000	0.000000	0.000000	0.000000	0.000000	0.000000	0.000000	0.000000	0.000000	4
5	0.000164	0.000036	0.000015	0.000009	0.000006	0.000004	0.000003	0.000003	0.000003	0.000002	0.000002	0.000002	5
6	0.004668	0.000962	0.000335	0.000153	0.000084	0.000052	0.000035	0.000025	0.000020	0.000016	0.000013	0.000011	6
7	0.021333	0.005332	0.001959	0.000893	0.000472	0.000277	0.000177	0.000121	0.000088	0.000066	0.000052	0.000042	7
8	0.049302	0.014879	0.006018	0.002885	0.001557	0.000918	0.000582	0.000390	0.000275	0.000202	0.000154	0.000121	8
9	0.083710	0.029395	0.013027	0.006623	0.003709	0.002237	0.001432	0.000964	0.000677	0.000493	0.000371	0.000287	9
10	0.120729	0.047777	0.022897	0.012300	0.007165	0.004443	0.002899	0.001974	0.001394	0.001017	0.000763	0.000587	10
11	0.158044	0.068815	0.035223	0.019865	0.012007	0.007658	0.005103	0.003528	0.002519	0.001850	0.001392	0.001072	11
12	0.194389	0.091490	0.049501	0.029128	0.018203	0.011922	0.008113	0.005702	0.004122	0.003055	0.002315	0.001790	12
13	0.229107	0.115016	0.065237	0.039832	0.025645	0.017209	0.011946	0.008535	0.006252	0.004681	0.003576	0.002781	13
14	0.261911	0.138822	0.081998	0.051709	0.034186	0.023452	0.016584	0.012034	0.008929	0.006758	0.005207	0.004077	14
15	0.292711	0.162507	0.099424	0.064505	0.043665	0.030560	0.021981	0.016183	0.012158	0.009299	0.007228	0.005702	15
16	0.321529	0.185794	0.117224	0.077992	0.053925	0.038428	0.028076	0.020951	0.015926	0.012305	0.009648	0.007667	16
17	0.348449	0.208501	0.135171	0.091973	0.064813	0.046952	0.034797	0.026294	0.020207	0.015764	0.012465	0.009978	17
18	0.373583	0.230510	0.153091	0.106281	0.076195	0.056028	0.042071	0.032161	0.024971	0.019658	0.015670	0.012633	18
19	0.397053	0.251754	0.170848	0.120777	0.087948	0.065559	0.049825	0.038499	0.030179	0.023963	0.019247	0.015623	19
20	0.418988	0.272198	0.188345	0.135348	0.099970	0.075458	0.057989	0.045254	0.035793	0.028649	0.023178	0.018937	20
21	0.439505	0.291832	0.205509	0.149903	0.112170	0.085645	0.066497	0.052374	0.041771	0.033687	0.027441	0.022558	21
22	0.458721	0.310659	0.222288	0.164368	0.124471	0.096050	0.075288	0.059809	0.048073	0.039045	0.032012	0.026471	22
23	0.476739	0.328696	0.238647	0.178686	0.136809	0.106612	0.084307	0.067511	0.054660	0.044692	0.036865	0.030655	23
24	0.493662	0.345966	0.254564	0.192810	0.149131	0.117275	0.093504	0.075438	0.061497	0.050597	0.041977	0.035091	24
25	0.509575	0.362498	0.270024	0.206707	0.161392	0.127995	0.102837	0.083549	0.068546	0.056730	0.047322	0.039758	25
26	0.524560	0.378321	0.285024	0.220349	0.173556	0.138732	0.112264	0.091808	0.075777	0.063064	0.052876	0.044637	26
27	0.538691	0.393468	0.299564	0.233718	0.185593	0.149451	0.121752	0.100183	0.083160	0.069572	0.058617	0.049707	27
28	0.552034	0.407970	0.313646	0.246799	0.197480	0.160123	0.131271	0.108643	0.090666	0.076229	0.064523	0.054951	28
29	0.564655	0.421860	0.327281	0.259584	0.209196	0.170725	0.140795	0.117163	0.098272	0.083013	0.070573	0.060349	29
30	0.576601	0.435170	0.340476	0.272068	0.220728	0.181235	0.150300	0.125719	0.105955	0.089902	0.076748	0.065885	30
40	0.668249	0.542257	0.451107	0.380670	0.324502	0.278831	0.241173	0.209788	0.183401	0.161054	0.142005	0.125677	40
60	0.769057	0.669979	0.592611	0.528633	0.474345	0.427572	0.386853	0.351131	0.319602	0.291636	0.266723	0.244448	60
80	0.823038	0.742525	0.677232	0.621368	0.572440	0.529007	0.490107	0.455043	0.423278	0.394390	0.368029	0.343906	80
100	0.856606	0.789069	0.733048	0.684125	0.640446	0.600959	0.564970	0.531981	0.501610	0.473552	0.447557	0.423417	100
120	0.879484	0.821412	0.772515	0.729225	0.690073	0.654241	0.621196	0.590558	0.562038	0.535406	0.510473	0.487083	120
140	0.896071	0.845177	0.801862	0.763137	0.727789	0.695150	0.664792	0.636413	0.609782	0.584719	0.561076	0.538731	140
170	0.913861	0.870955	0.834024	0.800662	0.769907	0.741242	0.714313	0.688960	0.664941	0.642146	0.620465	0.599807	170
200	0.926453	0.889383	0.857222	0.827957	0.800792	0.775304	0.751191	0.728357	0.706614	0.685839	0.665962	0.646914	200
240	0.938451	0.907082	0.879663	0.854540	0.831068	0.808906	0.787800	0.767701	0.748450	0.729967	0.712175	0.695028	240
320	0.953594	0.929613	0.908456	0.888903	0.870485	0.852956	0.836125	0.819982	0.804406	0.789342	0.774746	0.760584	320
440	0.966104	0.948390	0.932642	0.917985	0.904086	0.890771	0.877902	0.865484	0.853429	0.841699	0.830267	0.819109	440
600	0.975067	0.961932	0.950192	0.939210	0.928747	0.918676	0.908897	0.899419	0.890178	0.881148	0.872309	0.863646	600
800	0.981261	0.971335	0.962430	0.954071	0.946080	0.938365	0.930849	0.923543	0.916398	0.909395	0.902519	0.895761	800
1000	0.984990	0.977013	0.969840	0.963094	0.956632	0.950381	0.944279	0.938337	0.932515	0.926799	0.921177	0.915641	1000
INF	1.000000	1.000000	1.000000	1.000000	1.000000	1.000000	1.000000	1.000000	1.000000	1.000000	1.000000	1.000000	INF

Tafel 1 (Forts.)

Table 1 (cont.)

$U_{p,q,n}$

p=5 C=0.05

n	q = 13	14	15	18	21	24	27	30	40	60	80	100	120	n
1	0.000000	0.000000	0.000000	0.000000	0.000000	0.000000	0.000000	0.000000	0.000000	0.000000	0.000000	0.000000	0.000000	1
2	0.000000	0.000000	0.000000	0.000000	0.000000	0.000000	0.000000	0.000000	0.000000	0.000000	0.000000	0.000000	0.000000	2
3	0.000000	0.000000	0.000000	0.000000	0.000000	0.000000	0.000000	0.000000	0.000000	0.000000	0.000000	0.000000	0.000000	3
4	0.000001	0.000001	0.000001	0.000001	0.000001	0.000001	0.000002	0.000002	0.000002	0.000003	0.000004	0.000005	0.000006	4
5	0.000006	0.000006	0.000005	0.000005	0.000004	0.000004	0.000004	0.000004	0.000004	0.000005	0.000006	0.000007	0.000007	5
6	0.000027	0.000023	0.000020	0.000015	0.000012	0.000011	0.000010	0.000009	0.000008	0.000008	0.000008	0.000008	0.000009	6
7	0.000089	0.000074	0.000062	0.000041	0.000030	0.000024	0.000020	0.000018	0.000014	0.000011	0.000011	0.000011	0.000011	7
8	0.000233	0.000188	0.000155	0.000096	0.000066	0.000050	0.000040	0.000033	0.000022	0.000016	0.000014	0.000013	0.000013	8
9	0.000513	0.000409	0.000333	0.000197	0.000130	0.000094	0.000072	0.000058	0.000035	0.000022	0.000018	0.000016	0.000016	9
10	0.000987	0.000784	0.000634	0.000366	0.000235	0.000164	0.000122	0.000096	0.000054	0.000030	0.000023	0.000020	0.000019	10
11	0.001713	0.001361	0.001099	0.000627	0.000396	0.000271	0.000198	0.000151	0.000080	0.000041	0.000030	0.000025	0.000022	11
12	0.002738	0.002183	0.001765	0.001005	0.000629	0.000424	0.000304	0.000229	0.000115	0.000054	0.000037	0.000030	0.000026	12
13	0.004100	0.003285	0.002666	0.001524	0.000950	0.000636	0.000451	0.000336	0.000161	0.000071	0.000046	0.000036	0.000031	13
14	0.005825	0.004694	0.003827	0.002203	0.001374	0.000915	0.000645	0.000476	0.000221	0.000092	0.000057	0.000043	0.000036	14
15	0.007926	0.006426	0.005266	0.003062	0.001915	0.001275	0.000894	0.000656	0.000297	0.000117	0.000070	0.000051	0.000042	15
16	0.010408	0.008491	0.006994	0.004113	0.002587	0.001724	0.001207	0.000881	0.000391	0.000147	0.000085	0.000061	0.000049	16
17	0.013263	0.010889	0.009018	0.005368	0.003400	0.002272	0.001590	0.001159	0.000506	0.000183	0.000103	0.000072	0.000056	17
18	0.016484	0.013615	0.011335	0.006832	0.004361	0.002926	0.002051	0.001493	0.000644	0.000226	0.000123	0.000084	0.000064	18
19	0.020056	0.016661	0.013943	0.008509	0.005478	0.003694	0.002595	0.001890	0.000809	0.000277	0.000147	0.000098	0.000074	19
20	0.023958	0.020011	0.016832	0.010400	0.006754	0.004581	0.003228	0.002354	0.001003	0.000335	0.000174	0.000113	0.000084	20
21	0.028168	0.023653	0.019990	0.012501	0.008191	0.005589	0.003954	0.002889	0.001229	0.000402	0.000204	0.000131	0.000096	21
22	0.032665	0.027566	0.023407	0.014810	0.009790	0.006723	0.004777	0.003499	0.001488	0.000479	0.000239	0.000150	0.000108	22
23	0.037428	0.031737	0.027066	0.017320	0.011550	0.007984	0.005699	0.004187	0.001784	0.000567	0.000278	0.000172	0.000122	23
24	0.042429	0.036140	0.030953	0.020025	0.013469	0.009372	0.006722	0.004955	0.002117	0.000666	0.000321	0.000196	0.000138	24
25	0.047648	0.040764	0.035051	0.022915	0.015542	0.010887	0.007847	0.005805	0.002491	0.000777	0.000370	0.000223	0.000155	25
26	0.053064	0.045582	0.039344	0.025981	0.017767	0.012527	0.009075	0.006739	0.002908	0.000902	0.000424	0.000252	0.000173	26
27	0.058656	0.050579	0.043820	0.029216	0.020139	0.014291	0.010406	0.007756	0.003368	0.001040	0.000483	0.000284	0.000193	27
28	0.064402	0.055743	0.048458	0.032611	0.022652	0.016174	0.011837	0.008859	0.003873	0.001193	0.000549	0.000320	0.000215	28
29	0.070283	0.061047	0.053249	0.036153	0.025300	0.018177	0.013369	0.010046	0.004424	0.001361	0.000621	0.000358	0.000239	29
30	0.076280	0.066480	0.058171	0.039834	0.028076	0.020294	0.015001	0.011317	0.005023	0.001545	0.000700	0.000400	0.000265	30
40	0.139655	0.124987	0.112171	0.082271	0.061543	0.046832	0.036187	0.028354	0.013714	0.004417	0.001940	0.001050	0.000655	40
60	0.261398	0.241318	0.223118	0.177808	0.143281	0.116603	0.095736	0.079241	0.044377	0.016790	0.007725	0.004136	0.002485	60
80	0.361157	0.339305	0.319076	0.266766	0.224675	0.190458	0.162409	0.139243	0.086263	0.037606	0.018737	0.010401	0.006314	80
100	0.440041	0.418059	0.397456	0.342846	0.297298	0.259010	0.226615	0.199050	0.132448	0.064417	0.034575	0.020110	0.012528	100
120	0.502775	0.481400	0.461177	0.406654	0.360003	0.319840	0.285080	0.254860	0.178816	0.094664	0.054113	0.032928	0.021152	120
140	0.553442	0.532961	0.513455	0.460182	0.413736	0.373026	0.337187	0.305516	0.223286	0.126421	0.076154	0.048253	0.031936	140
170	0.613135	0.594138	0.575912	0.525431	0.480502	0.440337	0.404302	0.371871	0.284633	0.174214	0.111771	0.074528	0.051378	170
200	0.659028	0.641467	0.624530	0.577145	0.534336	0.495515	0.460198	0.427988	0.339029	0.220159	0.148384	0.103127	0.073593	200
240	0.705785	0.689936	0.674573	0.631168	0.591386	0.554802	0.521069	0.489895	0.401520	0.276806	0.196318	0.142561	0.105662	240
320	0.769325	0.756183	0.743362	0.706681	0.672432	0.640364	0.610271	0.581981	0.498956	0.372884	0.283803	0.219357	0.171845	320
440	0.825909	0.815527	0.805344	0.775901	0.747978	0.721430	0.696144	0.672021	0.599068	0.480889	0.390326	0.319815	0.264235	440
600	0.868876	0.860799	0.852845	0.829672	0.807444	0.786074	0.765494	0.745650	0.684278	0.579879	0.494751	0.424557	0.366192	600
800	0.899814	0.893504	0.887273	0.869023	0.851377	0.834280	0.817688	0.801567	0.750910	0.661658	0.585530	0.520043	0.463373	800
1000	0.918948	0.913774	0.908657	0.893621	0.879012	0.864791	0.850924	0.837389	0.794442	0.717113	0.649282	0.589358	0.536168	1000
INF	1.000000	1.000000	1.000000	1.000000	1.000000	1.000000	1.000000	1.000000	1.000000	1.000000	1.000000	1.000000	1.000000	INF

$U_{p,q,n}$

p=5 C=0.01

q

n	13	14	15	18	21	24	27	30	40	60	80	100	120	n
1	0.000000	0.000000	0.000000	0.000000	0.000000	0.000000	0.000000	0.000000	0.000000	0.000000	0.000000	0.000002	0.000002	1
2	0.000000	0.000000	0.000000	0.000000	0.000000	0.000000	0.000000	0.000000	0.000000	0.000000	0.000001	0.000002	0.000003	2
3	0.000000	0.000000	0.000000	0.000000	0.000000	0.000000	0.000000	0.000000	0.000000	0.000001	0.000002	0.000003	0.000004	3
4	0.000000	0.000000	0.000000	0.000000	0.000000	0.000000	0.000000	0.000000	0.000001	0.000002	0.000003	0.000004	0.000004	4
5	0.000002	0.000002	0.000002	0.000002	0.000002	0.000002	0.000002	0.000002	0.000002	0.000003	0.000004	0.000005	0.000005	5
6	0.000010	0.000009	0.000008	0.000006	0.000006	0.000005	0.000005	0.000005	0.000004	0.000005	0.000005	0.000006	0.000007	6
7	0.000035	0.000029	0.000026	0.000018	0.000014	0.000012	0.000010	0.000009	0.000008	0.000007	0.000007	0.000008	0.000008	7
8	0.000097	0.000080	0.000067	0.000044	0.000032	0.000025	0.000021	0.000018	0.000013	0.000010	0.000010	0.000010	0.000010	8
9	0.000227	0.000184	0.000152	0.000094	0.000065	0.000049	0.000039	0.000032	0.000021	0.000015	0.000013	0.000012	0.000012	9
10	0.000462	0.000371	0.000303	0.000181	0.000121	0.000087	0.000067	0.000054	0.000033	0.000020	0.000016	0.000015	0.000014	10
11	0.000843	0.000674	0.000549	0.000322	0.000210	0.000148	0.000110	0.000087	0.000049	0.000027	0.000021	0.000018	0.000017	11
12	0.001409	0.001128	0.000917	0.000533	0.000341	0.000236	0.000173	0.000134	0.000071	0.000037	0.000026	0.000022	0.000020	12
13	0.002198	0.001764	0.001436	0.000832	0.000529	0.000361	0.000262	0.000199	0.000101	0.000048	0.000033	0.000027	0.000024	13
14	0.003240	0.002611	0.002130	0.001237	0.000783	0.000531	0.000381	0.000286	0.000140	0.000063	0.000041	0.000032	0.000028	14
15	0.004558	0.003690	0.003021	0.001765	0.001116	0.000753	0.000536	0.000400	0.000190	0.000080	0.000051	0.000038	0.000032	15
16	0.006168	0.005019	0.004127	0.002429	0.001539	0.001037	0.000735	0.000545	0.000252	0.000102	0.000062	0.000046	0.000037	16
17	0.008078	0.006608	0.005459	0.003242	0.002063	0.001390	0.000984	0.000725	0.000330	0.000128	0.000075	0.000054	0.000043	17
18	0.010291	0.008464	0.007024	0.004215	0.002696	0.001820	0.001287	0.000947	0.000424	0.000158	0.000090	0.000063	0.000050	18
19	0.012804	0.010588	0.008828	0.005354	0.003446	0.002333	0.001651	0.001213	0.000537	0.000195	0.000108	0.000074	0.000057	19
20	0.015611	0.012976	0.010870	0.006665	0.004319	0.002936	0.002081	0.001529	0.000672	0.000237	0.000128	0.000086	0.000065	20
21	0.018701	0.015624	0.013148	0.008150	0.005321	0.003634	0.002581	0.001897	0.000829	0.000286	0.000151	0.000099	0.000074	21
22	0.022062	0.018522	0.015656	0.009810	0.006454	0.004430	0.003155	0.002322	0.001012	0.000342	0.000177	0.000115	0.000084	22
23	0.025680	0.021661	0.018388	0.011645	0.007721	0.005328	0.003808	0.002808	0.001222	0.000407	0.000207	0.000131	0.000096	23
24	0.029540	0.025028	0.021334	0.013652	0.009123	0.006330	0.004542	0.003357	0.001461	0.000480	0.000240	0.000150	0.000108	24
25	0.033624	0.028611	0.024485	0.015827	0.010658	0.007439	0.005358	0.003971	0.001731	0.000563	0.000277	0.000171	0.000121	25
26	0.037917	0.032397	0.027830	0.018166	0.012328	0.008654	0.006260	0.004652	0.002034	0.000656	0.000318	0.000194	0.000136	26
27	0.042403	0.036371	0.031358	0.020664	0.014128	0.009975	0.007247	0.005403	0.002372	0.000759	0.000364	0.000220	0.000152	27
28	0.047064	0.040521	0.035058	0.023314	0.016057	0.011403	0.008322	0.006224	0.002745	0.000874	0.000415	0.000248	0.000170	28
29	0.051885	0.044831	0.038918	0.026109	0.018112	0.012937	0.009483	0.007117	0.003156	0.001002	0.000471	0.000278	0.000189	29
30	0.056851	0.049291	0.042926	0.029044	0.020289	0.014573	0.010731	0.008082	0.003605	0.001142	0.000532	0.000311	0.000210	30
40	0.111611	0.099438	0.088858	0.064448	0.047766	0.036077	0.027710	0.021611	0.010380	0.003378	0.001512	0.000833	0.000528	40
60	0.224469	0.206496	0.190284	0.150280	0.120164	0.097134	0.079283	0.065284	0.036099	0.013539	0.006261	0.003384	0.002054	60
80	0.321777	0.301433	0.282690	0.234610	0.196315	0.165467	0.140379	0.119801	0.073307	0.031510	0.015649	0.008715	0.005321	80
100	0.400952	0.380010	0.360454	0.308993	0.266464	0.231002	0.201211	0.176021	0.115821	0.055501	0.029582	0.017188	0.010736	100
120	0.465099	0.444405	0.424898	0.372638	0.328306	0.290419	0.257842	0.229682	0.159530	0.083262	0.047197	0.028619	0.018380	120
140	0.517574	0.497518	0.478479	0.426792	0.382085	0.343167	0.309110	0.279174	0.202176	0.112955	0.067452	0.042527	0.028094	140
170	0.580095	0.561262	0.543248	0.493627	0.449781	0.410828	0.376070	0.344941	0.261919	0.158407	0.100775	0.066812	0.045899	170
200	0.628636	0.611075	0.594187	0.547176	0.504991	0.466954	0.432526	0.401267	0.315620	0.202761	0.135584	0.093682	0.066583	200
240	0.678484	0.662504	0.647056	0.603615	0.564046	0.527850	0.494628	0.464055	0.378012	0.258141	0.181777	0.131272	0.096882	240
320	0.746821	0.733434	0.720404	0.683283	0.648815	0.616691	0.586669	0.558550	0.476551	0.353436	0.267416	0.205705	0.160515	320
440	0.808208	0.797546	0.787111	0.757043	0.728666	0.701798	0.676298	0.652050	0.579131	0.462158	0.373379	0.304752	0.250964	440
600	0.855146	0.846800	0.838597	0.814783	0.792037	0.770248	0.749334	0.729225	0.667345	0.562988	0.478602	0.409451	0.352233	600
800	0.889112	0.882563	0.876109	0.857267	0.839125	0.821606	0.804653	0.788225	0.736838	0.647002	0.570941	0.505867	0.449794	800
1000	0.910185	0.904801	0.899487	0.883920	0.868856	0.854240	0.840028	0.826188	0.782461	0.704298	0.636199	0.576335	0.523404	1000
INF	1.000000	1.000000	1.000000	1.000000	1.000000	1.000000	1.000000	1.000000	1.000000	1.000000	1.000000	1.000000	1.000000	INF

Tafel 1 (Forts.) Table 1 (cont.)

$U_{p,q,n}$

p=6 C=0.05

q

n	1	2	3	4	5	6	7	8	9	10	11	12	n
1	0.000000	0.000000	0.000000	0.000000	0.000000	0.000000	0.000000	0.000000	0.000000	0.000000	0.000000	0.000000	1
2	0.000000	0.000000	0.000000	0.000000	0.000000	0.000000	0.000000	0.000000	0.000000	0.000000	0.000000	0.000000	2
3	0.000000	0.000000	0.000000	0.000000	0.000000	0.000000	0.000000	0.000000	0.000000	0.000000	0.000000	0.000000	3
4	0.000000	0.000000	0.000000	0.000000	0.000000	0.000000	0.000000	0.000000	0.000000	0.000000	0.000000	0.000000	4
5	0.000007	0.000002	0.000001	0.000001	0.000001	0.000000	0.000000	0.000000	0.000000	0.000000	0.000000	0.000000	5
6	0.002045	0.000315	0.000095	0.000040	0.000021	0.000012	0.000008	0.000006	0.000004	0.000003	0.000003	0.000002	6
7	0.018804	0.003479	0.001052	0.000416	0.000197	0.000106	0.000063	0.000040	0.000027	0.000020	0.000015	0.000011	7
8	0.053911	0.012883	0.004369	0.001818	0.000872	0.000465	0.000270	0.000168	0.000111	0.000076	0.000055	0.000041	8
9	0.098038	0.028824	0.011018	0.004938	0.002479	0.001358	0.000798	0.000497	0.000325	0.000222	0.000157	0.000115	9
10	0.144274	0.049685	0.021043	0.010129	0.005348	0.003035	0.001826	0.001155	0.000762	0.000521	0.000369	0.000269	10
11	0.189355	0.073697	0.033966	0.017408	0.009639	0.005672	0.003507	0.002263	0.001514	0.001046	0.000744	0.000543	11
12	0.231866	0.099450	0.049161	0.026586	0.015360	0.009348	0.005940	0.003915	0.002664	0.001865	0.001338	0.000983	12
13	0.271356	0.125933	0.066012	0.037385	0.022418	0.014071	0.009172	0.006173	0.004273	0.003033	0.002200	0.001630	13
14	0.307797	0.152453	0.083979	0.049495	0.030680	0.019795	0.013205	0.009066	0.006381	0.004592	0.003370	0.002520	14
15	0.341285	0.178581	0.102644	0.062632	0.039965	0.026433	0.018012	0.012593	0.009005	0.006568	0.004877	0.003682	15
16	0.372033	0.204010	0.121656	0.076537	0.050117	0.033893	0.023544	0.016741	0.012147	0.008974	0.006740	0.005137	16
17	0.400304	0.228568	0.140775	0.090983	0.060965	0.042061	0.029737	0.021472	0.015794	0.011811	0.008966	0.006898	17
18	0.426364	0.252176	0.159796	0.105779	0.072367	0.050834	0.036522	0.026746	0.019924	0.015070	0.011554	0.008971	18
19	0.450349	0.274785	0.178574	0.120780	0.084178	0.060119	0.043825	0.032520	0.024510	0.018734	0.014503	0.011356	19
20	0.472562	0.296393	0.197017	0.135856	0.096308	0.069818	0.051576	0.038739	0.029518	0.022785	0.017796	0.014049	20
21	0.493091	0.316990	0.215044	0.150905	0.108634	0.079840	0.059715	0.045350	0.034906	0.027193	0.021418	0.017040	21
22	0.512182	0.336628	0.232604	0.165853	0.121083	0.090122	0.068178	0.052311	0.040646	0.031936	0.025354	0.020317	22
23	0.529913	0.355328	0.249666	0.180626	0.133590	0.100596	0.076899	0.059574	0.046695	0.036988	0.029582	0.023864	23
24	0.546452	0.373143	0.266216	0.195197	0.146095	0.111189	0.085836	0.067090	0.053016	0.042316	0.034078	0.027670	24
25	0.561889	0.390109	0.282253	0.209511	0.158544	0.121873	0.094944	0.074824	0.059586	0.047895	0.038825	0.031716	25
26	0.576348	0.406285	0.297740	0.223535	0.170898	0.132587	0.104168	0.082735	0.066362	0.053696	0.043795	0.035986	26
27	0.589899	0.421688	0.312738	0.237277	0.183129	0.143309	0.113485	0.090793	0.073318	0.059697	0.048977	0.040460	27
28	0.602633	0.436379	0.327222	0.250710	0.195207	0.153998	0.122849	0.098970	0.080420	0.065867	0.054339	0.045123	28
29	0.614602	0.450416	0.341199	0.263809	0.207116	0.164629	0.132250	0.107224	0.087654	0.072196	0.059866	0.049957	29
30	0.625896	0.463794	0.354711	0.276602	0.218828	0.175171	0.141648	0.115539	0.094994	0.078649	0.065542	0.054951	30
40	0.710937	0.569976	0.466792	0.387183	0.324162	0.273470	0.232192	0.198251	0.170132	0.146678	0.126985	0.110367	40
60	0.801604	0.693451	0.607528	0.536153	0.475641	0.423707	0.378774	0.339636	0.305361	0.275238	0.248638	0.225098	60
80	0.849063	0.762264	0.690479	0.628610	0.574313	0.526153	0.483144	0.444543	0.409736	0.378269	0.349725	0.323787	80
100	0.878218	0.805945	0.744748	0.690824	0.642495	0.598763	0.558956	0.522538	0.489125	0.458377	0.430004	0.403784	100
120	0.897944	0.836112	0.782919	0.735354	0.692128	0.652489	0.615927	0.582063	0.550602	0.521300	0.493955	0.468392	120
140	0.912172	0.858176	0.811198	0.768751	0.729786	0.693709	0.660119	0.628724	0.599296	0.571649	0.545628	0.521100	140
170	0.927365	0.882016	0.842092	0.805615	0.771776	0.740119	0.710350	0.682254	0.655667	0.630455	0.606507	0.583730	170
200	0.938078	0.899001	0.864314	0.832375	0.802523	0.774395	0.747758	0.722444	0.698328	0.675308	0.653300	0.632233	200
240	0.948255	0.915270	0.885761	0.858391	0.832628	0.808187	0.784886	0.762599	0.741229	0.720701	0.700953	0.681935	240
320	0.961056	0.935919	0.913212	0.891956	0.871772	0.852459	0.833892	0.815985	0.798676	0.781916	0.765666	0.749894	320
440	0.971597	0.953076	0.936212	0.920308	0.905097	0.890438	0.876249	0.862471	0.849063	0.835995	0.823242	0.810784	440
600	0.979129	0.965422	0.952870	0.940969	0.929529	0.918448	0.907669	0.897152	0.886868	0.876798	0.866924	0.857233	600
800	0.984325	0.973979	0.964469	0.955420	0.946689	0.938203	0.929921	0.921812	0.913858	0.906042	0.898354	0.890785	800
1000	0.987450	0.979142	0.971487	0.964187	0.957129	0.950256	0.943532	0.936937	0.930455	0.924073	0.917783	0.911578	1000
INF	1.000000	1.000000	1.000000	1.000000	1.000000	1.000000	1.000000	1.000000	1.000000	1.000000	1.000000	1.000000	INF

Tafel 1 (Forts.) Table 1 (cont.)

$U_{p,q,n}$

p=6 C=0.01

n	q=1	2	3	4	5	6	7	8	9	10	11	12	n
1	0.000000	0.000000	0.000000	0.000000	0.000000	0.000000	0.000000	0.000000	0.000000	0.000000	0.000000	0.000000	1
2	0.000000	0.000000	0.000000	0.000000	0.000000	0.000000	0.000000	0.000000	0.000000	0.000000	0.000000	0.000000	2
3	0.000000	0.000000	0.000000	0.000000	0.000000	0.000000	0.000000	0.000000	0.000000	0.000000	0.000000	0.000000	3
4	0.000000	0.000000	0.000000	0.000000	0.000000	0.000000	0.000000	0.000000	0.000000	0.000000	0.000000	0.000000	4
5	0.000000	0.000000	0.000000	0.000000	0.000000	0.000000	0.000000	0.000000	0.000000	0.000000	0.000000	0.000000	5
6	0.000295	0.000050	0.000017	0.000008	0.000004	0.000003	0.000002	0.000001	0.000001	0.000000	0.000000	0.000000	6
7	0.004608	0.000839	0.000258	0.000106	0.000052	0.000029	0.000018	0.000012	0.000008	0.000006	0.000005	0.000004	7
8	0.018808	0.004182	0.001383	0.000574	0.000277	0.000150	0.000089	0.000057	0.000038	0.000027	0.000020	0.000015	8
9	0.042762	0.011508	0.004211	0.001846	0.000918	0.000503	0.000297	0.000187	0.000124	0.000086	0.000062	0.000046	9
10	0.072861	0.022948	0.009244	0.004315	0.002237	0.001257	0.000754	0.000477	0.000317	0.000219	0.000156	0.000115	10
11	0.105882	0.037842	0.016575	0.008211	0.004443	0.002575	0.001578	0.001014	0.000678	0.000470	0.000336	0.000247	11
12	0.139723	0.055318	0.026018	0.013591	0.007658	0.004578	0.002873	0.001878	0.001272	0.000889	0.000639	0.000471	12
13	0.173151	0.074563	0.037260	0.020392	0.011922	0.007339	0.004715	0.003140	0.002158	0.001525	0.001105	0.000818	13
14	0.205478	0.094910	0.049951	0.028478	0.017209	0.010886	0.007151	0.004851	0.003384	0.002420	0.001769	0.001320	14
15	0.236354	0.115841	0.063758	0.037679	0.023452	0.015207	0.010199	0.007041	0.004984	0.003608	0.002664	0.002004	15
16	0.265622	0.136971	0.078384	0.047814	0.030560	0.020265	0.013856	0.009723	0.006980	0.005114	0.003815	0.002894	16
17	0.293243	0.158016	0.093576	0.058712	0.038428	0.026009	0.018099	0.012897	0.009383	0.006954	0.005240	0.004009	17
18	0.319245	0.178778	0.109125	0.070212	0.046952	0.032373	0.022896	0.016549	0.012192	0.009135	0.006950	0.005362	18
19	0.343692	0.199115	0.124859	0.082172	0.056028	0.039291	0.028206	0.020658	0.015398	0.011658	0.008952	0.006963	19
20	0.366666	0.218933	0.140644	0.094469	0.065559	0.046693	0.033983	0.025196	0.018987	0.014517	0.011245	0.008816	20
21	0.388260	0.238169	0.156371	0.106995	0.075458	0.054514	0.040181	0.030131	0.022939	0.017701	0.013827	0.010921	21
22	0.408567	0.256789	0.171955	0.119660	0.085645	0.062689	0.046751	0.035431	0.027233	0.021197	0.016688	0.013275	22
23	0.427679	0.274775	0.187334	0.132389	0.096050	0.071160	0.053649	0.041060	0.031843	0.024988	0.019819	0.015873	23
24	0.445681	0.292121	0.202457	0.145118	0.106612	0.079874	0.060830	0.046985	0.036744	0.029056	0.023208	0.018708	24
25	0.462657	0.308833	0.217287	0.157797	0.117275	0.088781	0.068253	0.053173	0.041911	0.033382	0.026841	0.021769	25
26	0.478684	0.324922	0.231801	0.170381	0.127995	0.097838	0.075880	0.059592	0.047319	0.037946	0.030703	0.025046	26
27	0.493829	0.340404	0.245977	0.182837	0.138732	0.107006	0.083676	0.066212	0.052941	0.042728	0.034778	0.028527	27
28	0.508160	0.355297	0.259807	0.195137	0.149451	0.116251	0.091609	0.073004	0.058756	0.047709	0.039052	0.032201	28
29	0.521737	0.369623	0.273282	0.207259	0.160123	0.125542	0.099650	0.079944	0.064740	0.052871	0.043508	0.036054	29
30	0.534611	0.383404	0.286402	0.219187	0.170725	0.134852	0.107773	0.087007	0.070872	0.058193	0.048132	0.040075	30
40	0.633971	0.495984	0.398981	0.326182	0.269778	0.225181	0.189401	0.160362	0.136572	0.116924	0.100582	0.086906	40
60	0.744292	0.633481	0.548313	0.479114	0.421438	0.372626	0.330877	0.294885	0.263658	0.236424	0.212564	0.191578	60
80	0.803733	0.712825	0.639827	0.578119	0.524751	0.477984	0.436634	0.399837	0.366925	0.337368	0.310733	0.286658	80
100	0.840806	0.764135	0.700945	0.646245	0.597871	0.554570	0.515501	0.480046	0.447731	0.418175	0.391064	0.366135	100
120	0.866117	0.799962	0.744489	0.695702	0.651910	0.612148	0.575775	0.542326	0.511446	0.482846	0.456292	0.431584	120
140	0.884492	0.826371	0.777033	0.733148	0.693333	0.656808	0.623066	0.591738	0.562544	0.535260	0.509701	0.485712	140
170	0.904218	0.855095	0.812851	0.774824	0.739929	0.707546	0.677354	0.649014	0.622338	0.597164	0.573355	0.550798	170
200	0.918192	0.875677	0.838782	0.805291	0.774313	0.745329	0.718117	0.692405	0.668017	0.644841	0.622774	0.601727	200
240	0.931516	0.895480	0.863940	0.835082	0.808188	0.782831	0.758862	0.736059	0.714299	0.693473	0.673511	0.654352	240
320	0.948345	0.920741	0.896322	0.873758	0.852531	0.832324	0.813059	0.794571	0.776777	0.759613	0.743028	0.726975	320
440	0.962259	0.941834	0.923609	0.906632	0.890537	0.875096	0.860269	0.845937	0.832046	0.818554	0.805427	0.792640	440
600	0.972232	0.957070	0.943458	0.930704	0.918546	0.906818	0.895499	0.884501	0.873788	0.863330	0.853106	0.843097	600
800	0.979127	0.967660	0.957322	0.947597	0.938291	0.929281	0.920553	0.912044	0.903725	0.895577	0.887583	0.879732	800
1000	0.983279	0.974060	0.965726	0.957869	0.950332	0.943020	0.935921	0.928985	0.922190	0.915520	0.908964	0.902511	1000
INF	1.000000	1.000000	1.000000	1.000000	1.000000	1.000000	1.000000	1.000000	1.000000	1.000000	1.000000	1.000000	INF

Tafel 1 (Forts.) Table 1 (cont.)

$U_{p,q,n}$

p=6 C=0.05

q

n	13	14	15	18	21	24	27	30	40	60	80	100	120	n
1	0.000000	0.000000	0.000000	0.000000	0.000000	0.000000	0.000000	0.000000	0.000000	0.000000	0.000000	0.000000	0.000000	1
2	0.000000	0.000000	0.000000	0.000000	0.000000	0.000000	0.000000	0.000000	0.000000	0.000000	0.000000	0.000000	0.000000	2
3	0.000000	0.000000	0.000000	0.000000	0.000000	0.000000	0.000000	0.000000	0.000000	0.000000	0.000000	0.000000	0.000000	3
4	0.000000	0.000000	0.000000	0.000000	0.000000	0.000000	0.000000	0.000000	0.000000	0.000000	0.000000	0.000000	0.000000	4
5	0.000000	0.000000	0.000000	0.000000	0.000000	0.000000	0.000000	0.000000	0.000000	0.000000	0.000001	0.000001	0.000001	5
6	0.000002	0.000002	0.000002	0.000001	0.000001	0.000001	0.000001	0.000001	0.000001	0.000001	0.000001	0.000001	0.000001	6
7	0.000009	0.000007	0.000006	0.000004	0.000003	0.000002	0.000002	0.000002	0.000001	0.000001	0.000001	0.000001	0.000001	7
8	0.000032	0.000025	0.000020	0.000012	0.000008	0.000006	0.000005	0.000004	0.000003	0.000002	0.000002	0.000002	0.000002	8
9	0.000087	0.000067	0.000053	0.000029	0.000019	0.000013	0.000010	0.000008	0.000004	0.000003	0.000002	0.000002	0.000002	9
10	0.000201	0.000154	0.000120	0.000064	0.000039	0.000026	0.000019	0.000014	0.000007	0.000004	0.000003	0.000003	0.000002	10
11	0.000406	0.000310	0.000241	0.000126	0.000074	0.000048	0.000033	0.000025	0.000012	0.000006	0.000004	0.000003	0.000003	11
12	0.000737	0.000563	0.000438	0.000226	0.000131	0.000083	0.000057	0.000041	0.000019	0.000008	0.000005	0.000004	0.000003	12
13	0.001229	0.000944	0.000736	0.000380	0.000218	0.000136	0.000091	0.000065	0.000028	0.000011	0.000007	0.000005	0.000004	13
14	0.001915	0.001479	0.001158	0.000600	0.000344	0.000213	0.000142	0.000099	0.000041	0.000015	0.000009	0.000006	0.000005	14
15	0.002822	0.002193	0.001726	0.000903	0.000518	0.000320	0.000211	0.000147	0.000058	0.000020	0.000011	0.000008	0.000006	15
16	0.003970	0.003107	0.002460	0.001303	0.000750	0.000464	0.000305	0.000211	0.000081	0.000026	0.000014	0.000009	0.000007	16
17	0.005375	0.004236	0.003376	0.001811	0.001050	0.000651	0.000427	0.000294	0.000111	0.000034	0.000017	0.000011	0.000009	17
18	0.007046	0.005593	0.004485	0.002440	0.001427	0.000887	0.000582	0.000400	0.000149	0.000044	0.000022	0.000014	0.000010	18
19	0.008989	0.007185	0.005796	0.003200	0.001888	0.001180	0.000775	0.000533	0.000196	0.000056	0.000027	0.000016	0.000012	19
20	0.011201	0.009013	0.007315	0.004097	0.002441	0.001534	0.001012	0.000695	0.000254	0.000070	0.000032	0.000020	0.000014	20
21	0.013680	0.011079	0.009044	0.005138	0.003092	0.001957	0.001295	0.000892	0.000324	0.000087	0.000039	0.000023	0.000016	21
22	0.016421	0.013378	0.010982	0.006327	0.003847	0.002452	0.001630	0.001125	0.000409	0.000108	0.000047	0.000027	0.000019	22
23	0.019411	0.015906	0.013128	0.007666	0.004710	0.003025	0.002021	0.001399	0.000508	0.000132	0.000057	0.000032	0.000022	23
24	0.022641	0.018656	0.015476	0.009156	0.005682	0.003678	0.002471	0.001717	0.000625	0.000160	0.000068	0.000038	0.000025	24
25	0.026098	0.021620	0.018021	0.010797	0.006769	0.004415	0.002982	0.002080	0.000761	0.000192	0.000080	0.000044	0.000029	25
26	0.029769	0.024784	0.020755	0.012589	0.007969	0.005237	0.003559	0.002493	0.000916	0.000229	0.000094	0.000051	0.000033	26
27	0.033641	0.028140	0.023671	0.014527	0.009283	0.006147	0.004202	0.002956	0.001094	0.000272	0.000110	0.000059	0.000037	27
28	0.037700	0.031679	0.026761	0.016608	0.010712	0.007146	0.004915	0.003473	0.001294	0.000321	0.000128	0.000068	0.000042	28
29	0.041932	0.035388	0.030015	0.018830	0.012254	0.008236	0.005697	0.004044	0.001520	0.000376	0.000149	0.000078	0.000048	29
30	0.046324	0.039254	0.033424	0.021186	0.013908	0.009415	0.006550	0.004672	0.001771	0.000437	0.000172	0.000089	0.000054	30
40	0.096273	0.084268	0.073997	0.050990	0.035974	0.025909	0.019010	0.014183	0.005938	0.001538	0.000579	0.000280	0.000160	40
60	0.204183	0.185557	0.168929	0.128714	0.099372	0.077624	0.061279	0.048846	0.024372	0.007611	0.003014	0.001433	0.000782	60
80	0.300152	0.278572	0.258843	0.208964	0.170157	0.139625	0.115377	0.095954	0.054063	0.019993	0.008687	0.004304	0.002376	80
100	0.379515	0.357007	0.336105	0.281723	0.237593	0.201482	0.171721	0.147038	0.090287	0.038074	0.018074	0.009457	0.005376	100
120	0.444459	0.422023	0.400966	0.345065	0.298338	0.259015	0.225735	0.197426	0.129202	0.060328	0.030878	0.017042	0.010042	120
140	0.497948	0.476069	0.455375	0.399607	0.351955	0.311006	0.275650	0.244997	0.168388	0.085232	0.046452	0.026902	0.016420	140
170	0.562041	0.541367	0.521643	0.467616	0.420327	0.378743	0.342036	0.309524	0.224863	0.124979	0.073463	0.045220	0.028972	170
200	0.612046	0.592685	0.574100	0.522592	0.476716	0.435695	0.398895	0.365791	0.276904	0.165227	0.103036	0.066648	0.044501	200
240	0.663600	0.645910	0.628829	0.580949	0.537582	0.498168	0.462249	0.429439	0.338625	0.217072	0.143901	0.098110	0.068544	240
320	0.734572	0.719676	0.705186	0.663971	0.625825	0.590422	0.557490	0.526805	0.438387	0.309538	0.223317	0.164081	0.122503	320
440	0.798604	0.786688	0.775023	0.741435	0.709781	0.679875	0.651571	0.624744	0.544702	0.418963	0.326469	0.257243	0.204704	440
600	0.847715	0.838361	0.829163	0.802441	0.776922	0.752499	0.729088	0.706619	0.637816	0.523422	0.432962	0.360569	0.302103	600
800	0.883327	0.875974	0.868721	0.847519	0.827084	0.807349	0.788261	0.769778	0.712129	0.612293	0.529096	0.459149	0.399965	800
1000	0.905452	0.899401	0.893421	0.875875	0.858869	0.842354	0.826293	0.810659	0.761338	0.673746	0.598332	0.532935	0.475938	1000
INF	1.000000	1.000000	1.000000	1.000000	1.000000	1.000000	1.000000	1.000000	1.000000	1.000000	1.000000	1.000000	1.000000	INF

$U_{p,q,n}$

p=6 C=0.01

n	q=13	14	15	18	21	24	27	30	40	60	80	100	120	n
1	0.000000	0.000000	0.000000	0.000000	0.000000	0.000000	0.000000	0.000000	0.000000	0.000000	0.000000	0.000000	0.000000	1
2	0.000000	0.000000	0.000000	0.000000	0.000000	0.000000	0.000000	0.000000	0.000000	0.000000	0.000000	0.000000	0.000000	2
3	0.000000	0.000000	0.000000	0.000000	0.000000	0.000000	0.000000	0.000000	0.000000	0.000000	0.000000	0.000000	0.000000	3
4	0.000000	0.000000	0.000000	0.000000	0.000000	0.000000	0.000000	0.000000	0.000000	0.000000	0.000000	0.000000	0.000000	4
5	0.000000	0.000000	0.000000	0.000000	0.000000	0.000000	0.000000	0.000000	0.000000	0.000000	0.000000	0.000000	0.000000	5
6	0.000000	0.000000	0.000000	0.000000	0.000000	0.000000	0.000000	0.000000	0.000000	0.000000	0.000000	0.000000	0.000000	6
7	0.000003	0.000003	0.000002	0.000002	0.000001	0.000001	0.000000	0.000000	0.000000	0.000000	0.000000	0.000000	0.000000	7
8	0.000012	0.000010	0.000008	0.000005	0.000004	0.000003	0.000002	0.000002	0.000001	0.000001	0.000001	0.000001	0.000001	8
9	0.000036	0.000028	0.000023	0.000013	0.000009	0.000006	0.000005	0.000004	0.000003	0.000002	0.000001	0.000001	0.000001	9
10	0.000087	0.000068	0.000054	0.000030	0.000019	0.000013	0.000010	0.000008	0.000004	0.000003	0.000002	0.000002	0.000002	10
11	0.000186	0.000144	0.000113	0.000061	0.000037	0.000025	0.000018	0.000014	0.000007	0.000004	0.000003	0.000002	0.000002	11
12	0.000355	0.000273	0.000214	0.000113	0.000068	0.000044	0.000031	0.000023	0.000011	0.000005	0.000004	0.000003	0.000003	12
13	0.000619	0.000477	0.000374	0.000197	0.000116	0.000074	0.000051	0.000037	0.000017	0.000007	0.000005	0.000004	0.000003	13
14	0.001003	0.000776	0.000609	0.000320	0.000187	0.000118	0.000080	0.000057	0.000025	0.000010	0.000006	0.000005	0.000004	14
15	0.001533	0.001191	0.000938	0.000495	0.000288	0.000181	0.000122	0.000086	0.000036	0.000013	0.000008	0.000006	0.000005	15
16	0.002229	0.001741	0.001378	0.000733	0.000427	0.000268	0.000179	0.000125	0.000051	0.000018	0.000010	0.000007	0.000005	16
17	0.003110	0.002444	0.001944	0.001044	0.000610	0.000382	0.000254	0.000177	0.000070	0.000023	0.000012	0.000008	0.000006	17
18	0.004190	0.003314	0.002650	0.001439	0.000845	0.000530	0.000352	0.000245	0.000095	0.000030	0.000015	0.000010	0.000008	18
19	0.005481	0.004362	0.003507	0.001927	0.001139	0.000716	0.000475	0.000330	0.000126	0.000038	0.000019	0.000012	0.000009	19
20	0.006988	0.005595	0.004523	0.002517	0.001499	0.000947	0.000629	0.000436	0.000165	0.000048	0.000023	0.000015	0.000011	20
21	0.008715	0.007020	0.005706	0.003215	0.001931	0.001225	0.000816	0.000566	0.000212	0.000060	0.000028	0.000017	0.000012	21
22	0.010662	0.008640	0.007059	0.004029	0.002441	0.001557	0.001040	0.000722	0.000270	0.000075	0.000034	0.000021	0.000014	22
23	0.012828	0.010454	0.008585	0.004962	0.003033	0.001947	0.001305	0.000908	0.000338	0.000092	0.000041	0.000024	0.000017	23
24	0.015208	0.012461	0.010284	0.006017	0.003712	0.002398	0.001614	0.001126	0.000419	0.000112	0.000049	0.000028	0.000019	24
25	0.017797	0.014657	0.012155	0.007199	0.004482	0.002914	0.001970	0.001378	0.000514	0.000136	0.000059	0.000033	0.000022	25
26	0.020586	0.017039	0.014196	0.008506	0.005344	0.003498	0.002376	0.001668	0.000624	0.000163	0.000069	0.000039	0.000025	26
27	0.023568	0.019600	0.016401	0.009940	0.006300	0.004153	0.002835	0.001997	0.000750	0.000194	0.000081	0.000045	0.000029	27
28	0.026733	0.022333	0.018768	0.011500	0.007353	0.004880	0.003348	0.002367	0.000893	0.000230	0.000095	0.000051	0.000033	28
29	0.030071	0.025231	0.021290	0.013185	0.008503	0.005682	0.003919	0.002781	0.001055	0.000270	0.000110	0.000059	0.000037	29
30	0.033573	0.028287	0.023961	0.014992	0.009750	0.006559	0.004548	0.003240	0.001238	0.000316	0.000128	0.000068	0.000042	30
40	0.075396	0.065659	0.057384	0.039057	0.027278	0.019488	0.014207	0.010550	0.004390	0.001152	0.000443	0.000218	0.000127	40
60	0.173054	0.156653	0.142088	0.107188	0.082048	0.063617	0.049897	0.039549	0.019465	0.006030	0.002404	0.001156	0.000638	60
80	0.264842	0.245026	0.226988	0.181756	0.146946	0.119819	0.098453	0.081464	0.045282	0.016499	0.007153	0.003559	0.001977	80
100	0.343162	0.321953	0.302340	0.251678	0.210964	0.177928	0.150903	0.128634	0.078018	0.032377	0.015263	0.007983	0.004553	100
120	0.408550	0.387043	0.366934	0.313896	0.269953	0.233257	0.202408	0.176324	0.114115	0.052461	0.026615	0.014640	0.008630	120
140	0.463160	0.441926	0.421910	0.368298	0.322860	0.284090	0.250823	0.222141	0.151148	0.075397	0.040708	0.023456	0.014296	140
170	0.529397	0.509066	0.489729	0.437055	0.391289	0.351301	0.316201	0.285269	0.205428	0.112687	0.065631	0.040153	0.025636	170
200	0.581627	0.562408	0.544014	0.493293	0.448426	0.408543	0.372948	0.341076	0.256192	0.151069	0.093396	0.060035	0.039915	200
240	0.635938	0.618223	0.601164	0.553568	0.510727	0.472001	0.436875	0.404924	0.317146	0.201195	0.132334	0.089680	0.062368	240
320	0.711420	0.696337	0.681699	0.640237	0.602075	0.566823	0.534167	0.503849	0.417058	0.292021	0.209316	0.152989	0.113733	320
440	0.780170	0.767998	0.756103	0.721982	0.689980	0.659871	0.631476	0.604648	0.525056	0.401252	0.311074	0.244075	0.193528	440
600	0.833289	0.823670	0.814231	0.786898	0.760904	0.736119	0.712436	0.689770	0.620709	0.506873	0.417609	0.346622	0.289578	600
800	0.872013	0.864417	0.856938	0.835145	0.814226	0.794087	0.774666	0.755908	0.697664	0.597570	0.514770	0.445540	0.387215	800
1000	0.896154	0.889886	0.883703	0.865613	0.848149	0.831244	0.814845	0.798921	0.748892	0.660673	0.585230	0.520128	0.463612	1000
INF	1.000000	1.000000	1.000000	1.000000	1.000000	1.000000	1.000000	1.000000	1.000000	1.000000	1.000000	1.000000	1.000000	INF

Tafel 1 (Forts.) Table 1 (cont.)

$U_{p,q,n}$

p=7 C=0.05

n	q=1	2	3	4	5	6	7	8	9	10	11	12	n
1	0.000000	0.000000	0.000000	0.000000	0.000000	0.000000	0.000000	0.000000	0.000000	0.000000	0.000000	0.000000	1
2	0.000000	0.000000	0.000000	0.000000	0.000000	0.000000	0.000000	0.000000	0.000000	0.000000	0.000000	0.000000	2
3	0.000000	0.000000	0.000000	0.000000	0.000000	0.000000	0.000000	0.000000	0.000000	0.000000	0.000000	0.000000	3
4	0.000000	0.000000	0.000000	0.000000	0.000000	0.000000	0.000000	0.000000	0.000000	0.000000	0.000000	0.000000	4
5	0.000000	0.000000	0.000000	0.000000	0.000000	0.000000	0.000000	0.000000	0.000000	0.000000	0.000000	0.000000	5
6	0.000043	0.000006	0.000002	0.000001	0.000001	0.000000	0.000000	0.000000	0.000000	0.000000	0.000000	0.000000	6
7	0.002625	0.000350	0.000091	0.000033	0.000015	0.000008	0.000005	0.000003	0.000002	0.000002	0.000001	0.000001	7
8	0.017612	0.002953	0.000809	0.000292	0.000126	0.000063	0.000034	0.000020	0.000013	0.000009	0.000006	0.000005	8
9	0.047835	0.010329	0.003195	0.001223	0.000543	0.000270	0.000147	0.000086	0.000053	0.000035	0.000024	0.000017	9
10	0.086645	0.023060	0.008067	0.003338	0.001558	0.000798	0.000440	0.000259	0.000160	0.000104	0.000070	0.000049	10
11	0.128234	0.040186	0.015642	0.006974	0.003433	0.001826	0.001035	0.000619	0.000387	0.000252	0.000170	0.000119	11
12	0.169506	0.060396	0.025707	0.012249	0.006343	0.003508	0.002048	0.001252	0.000796	0.000525	0.000357	0.000249	12
13	0.209026	0.082538	0.037857	0.019109	0.010357	0.005940	0.003571	0.002234	0.001448	0.000967	0.000665	0.000468	13
14	0.246203	0.105734	0.051646	0.027402	0.015466	0.009172	0.005668	0.003628	0.002395	0.001625	0.001131	0.000804	14
15	0.280861	0.129346	0.066659	0.036933	0.021607	0.013206	0.008371	0.005476	0.003682	0.002537	0.001787	0.001285	15
16	0.313032	0.152929	0.082533	0.047494	0.028684	0.018013	0.011688	0.007801	0.005337	0.003733	0.002664	0.001936	16
17	0.342842	0.176179	0.098971	0.058884	0.036586	0.023544	0.015606	0.010611	0.007379	0.005235	0.003782	0.002778	17
18	0.370455	0.198894	0.115731	0.070921	0.045199	0.029736	0.020096	0.013900	0.009814	0.007057	0.005159	0.003829	18
19	0.396050	0.220944	0.132623	0.083445	0.054409	0.036520	0.025122	0.017653	0.012640	0.009204	0.006805	0.005102	19
20	0.419802	0.242252	0.149498	0.096315	0.064111	0.043824	0.030640	0.021845	0.015847	0.011676	0.008725	0.006605	20
21	0.441876	0.262777	0.166240	0.109415	0.074209	0.051579	0.036603	0.026450	0.019422	0.014469	0.010921	0.008342	21
22	0.462425	0.282503	0.182765	0.122645	0.084616	0.059717	0.042965	0.031435	0.023345	0.017571	0.013387	0.010314	22
23	0.481587	0.301432	0.199007	0.135923	0.095257	0.068177	0.049678	0.036769	0.027595	0.020971	0.016120	0.012521	23
24	0.499486	0.319577	0.214919	0.149181	0.106063	0.076901	0.056697	0.042416	0.032148	0.024653	0.019108	0.014956	24
25	0.516238	0.336959	0.230467	0.162364	0.116978	0.085838	0.063980	0.048346	0.036980	0.028599	0.022341	0.017614	25
26	0.531942	0.353606	0.245631	0.175429	0.127951	0.094941	0.071488	0.054525	0.042067	0.032794	0.025807	0.020487	26
27	0.546689	0.369546	0.260395	0.188340	0.138940	0.104168	0.079183	0.060924	0.047385	0.037217	0.029493	0.023565	27
28	0.560561	0.384810	0.274752	0.201068	0.149909	0.113482	0.087032	0.067514	0.052911	0.041851	0.033384	0.026838	28
29	0.573629	0.399430	0.288701	0.213591	0.160826	0.122851	0.095005	0.074268	0.058622	0.046678	0.037467	0.030296	29
30	0.585561	0.413438	0.302243	0.225894	0.171667	0.132247	0.103073	0.081161	0.064496	0.051680	0.041727	0.033928	30
40	0.679228	0.525996	0.417050	0.335433	0.272668	0.223571	0.184671	0.153533	0.128393	0.107941	0.091192	0.077392	40
60	0.779306	0.659576	0.566032	0.489695	0.426135	0.372561	0.327012	0.288026	0.254476	0.225471	0.200293	0.178361	60
80	0.831906	0.735024	0.655779	0.588321	0.529875	0.478709	0.433602	0.393626	0.358051	0.326284	0.297833	0.272287	80
100	0.864288	0.783251	0.715144	0.655689	0.602930	0.555673	0.513081	0.474521	0.439488	0.407570	0.378421	0.351744	100
120	0.886219	0.816680	0.757179	0.704361	0.656738	0.613420	0.573796	0.537400	0.503866	0.472893	0.444226	0.417647	120
140	0.902052	0.841199	0.788462	0.741086	0.697881	0.658148	0.621410	0.587314	0.555578	0.525974	0.498306	0.472408	140
170	0.918970	0.867751	0.822764	0.781839	0.744063	0.708913	0.676042	0.645194	0.616167	0.588800	0.562955	0.538514	170
200	0.930906	0.886705	0.847518	0.811553	0.778074	0.746666	0.717058	0.689053	0.662499	0.637274	0.613274	0.590412	200
240	0.942249	0.904887	0.871471	0.840546	0.811527	0.784091	0.758031	0.733198	0.709478	0.686784	0.665038	0.644178	240
320	0.956525	0.928004	0.902213	0.878097	0.855239	0.833417	0.812491	0.792362	0.772959	0.754224	0.736112	0.718583	320
440	0.968286	0.947243	0.928043	0.909937	0.892635	0.875985	0.859892	0.844294	0.829142	0.814403	0.800046	0.786051	440
600	0.976693	0.961103	0.946788	0.933208	0.920155	0.907522	0.895244	0.883276	0.871588	0.860157	0.848964	0.837994	600
800	0.982494	0.970720	0.959861	0.949517	0.939535	0.929836	0.920373	0.911114	0.902038	0.893128	0.884371	0.875758	800
1000	0.985983	0.976524	0.967778	0.959426	0.951346	0.943478	0.935782	0.928236	0.920822	0.913527	0.906342	0.899259	1000
INF	1.000000	1.000000	1.000000	1.000000	1.000000	1.000000	1.000000	1.000000	1.000000	1.000000	1.000000	1.000000	INF

$U_{p,q,n}$

p=7 C=0.01

n	q=1	2	3	4	5	6	7	8	9	10	11	12	n
1	0.000000	0.000000	0.000000	0.000000	0.000000	0.000000	0.000000	0.000000	0.000000	0.000000	0.000000	0.000000	1
2	0.000000	0.000000	0.000000	0.000000	0.000000	0.000000	0.000000	0.000000	0.000000	0.000000	0.000C00	0.000000	2
3	0.000000	0.000000	0.000000	0.000000	0.000000	0.000000	0.000000	0.000000	0.000000	0.000000	0.000000	0.000000	3
4	0.000000	0.000000	0.00000C	0.000000	0.000000	0.000000	0.000000	0.000000	0.000000	0.000000	0.000000	0.000000	4
5	0.000000	0.000C00	0.000000	0.000000	0.000000	0.000000	0.000000	0.000000	0.000000	0.000000	0.000000	0.000000	5
6	0.000005	0.000001	0.000000	0.000000	0.000000	0.000000	0.000000	0.000000	0.000000	0.000000	0.000000	0.000000	6
7	0.000486	C.000068	0.000C19	0.C00007	0.000003	0.000002	0.000001	0.000001	0.000001	C.000000	C.000000	0.000000	7
8	0.004798	0.000782	0.000215	0.C00079	0.000035	0.000018	0.000010	0.000006	0.000004	0.000C03	0.000002	0.000002	8
9	0.017314	0.003481	0.001C47	0.C00398	0.000177	0.000089	0.000049	0.000029	0.000019	0.000012	0.000009	0.000006	9
10	0.038208	0.009312	0.003116	0.001259	0.000582	0.000297	0.000165	0.000098	0.000061	0.000040	0.000028	0.000020	10
11	0.064845	C.01856C	C.006864	0.002966	0.001432	0.000754	0.000426	0.000255	0.000160	0.000105	C.000C72	0.000051	11
12	0.094551	C.030855	0.012459	0.005732	0.002899	0.001578	0.000912	0.000555	0.000353	C.000233	0.000159	0.000112	12
13	0.125434	0.045575	0.019845	0.009662	0.005103	0.002873	0.001704	0.001057	0.000682	0.000455	C.000313	0.000221	13
14	0.156314	C.062081	0.028840	0.C14763	0.008113	0.004715	0.002870	0.001817	0.001191	0.000804	0.000558	0.000397	14
15	0.186495	C.079813	0.C392C3	0.02C973	0.011946	0.007151	0.004460	0.002881	0.0C1919	C.001314	0.00C522	0.000661	15
16	0.215591	0.098313	0.050682	0.028192	0.016584	0.010199	0.006508	0.004286	0.002902	0.002013	0.001428	0.001034	16
17	0.243398	C.117225	0.063C3S	0.C36299	0.021981	0.013856	0.009030	0.006056	0.004164	0.002928	0.002101	0.001535	17
18	0.269838	0.136276	0.076C63	0.045168	0.028076	0.018099	0.012027	0.008203	0.005725	0.004C77	0.002958	0.002182	18
19	0.294893	C.155259	0.089567	0.054675	0.034797	0.022896	0.015490	0.010733	0.007595	C.C05476	0.004016	0.002991	19
20	0.318593	C.174C26	0.103395	0.C64703	0.042071	0.028206	0.019400	0.013641	0.009779	0.007133	0.005285	0.C03972	20
21	0.340992	0.192467	0.117419	0.075148	0.049825	0.033983	0.023733	0.016917	0.012276	0.009053	0.006773	0.005135	21
22	0.362149	0.210506	C.131529	0.085915	0.057989	0.040181	0.028460	0.020544	0.015079	0.011235	C.CC8484	0.006487	22
23	0.382138	0.228C88	0.14564C	0.C96921	0.066497	0.046751	0.033549	0.024505	0.018179	0.013677	0.010420	0.008031	23
24	0.401C34	0.245182	0.159680	0.108094	0.075288	0.053649	0.038968	0.028778	0.021564	0.016372	0.012577	0.009768	24
25	0.418901	0.261767	0.173594	C.119371	0.084307	0.060830	0.044686	0.033340	0.025219	0.019311	0.014553	0.011698	25
26	0.435815	0.277834	0.187337	0.130700	0.093504	0.068253	0.050669	0.038170	0.029127	0.022486	0.017542	0.013817	26
27	0.451836	C.293382	0.200875	0.142036	0.102837	0.075880	0.056888	0.043242	0.033272	C.C25883	0.020335	0.016123	27
28	0.467026	0.308414	0.214182	0.153341	0.112264	0.083676	0.063314	0.048535	0.037637	0.029490	0.023325	0.C18609	28
29	0.481442	0.322941	0.227238	0.164581	0.121752	0.091609	0.069918	0.054026	0.042205	0.033296	0.026502	0.021269	29
30	0.495137	0.336973	0.240029	0.175732	0.131271	0.099650	0.076676	0.059694	0.046958	0.037285	0.029857	0.024096	30
40	0.601481	0.453452	0.352532	0.279077	0.223835	0.181405	0.148306	0.122167	0.101312	0.084528	0.070914	0.059797	40
60	0.720662	0.599212	0.507477	0.434307	0.374443	0.324703	0.282924	0.247545	0.217383	0.191527	0.169255	0.149991	60
80	0.785257	C.684670	0.604805	0.538151	0.481262	0.432066	0.389140	0.351441	0.318157	0.288649	0.262392	0.238958	80
100	0.825659	0.740370	0.670619	0.610810	0.558448	0.512056	0.470627	0.433418	0.399849	0.369460	0.341864	0.316744	100
120	0.853292	0.779445	0.717860	0.664096	0.616219	0.573107	0.533999	0.498338	0.465691	0.435710	0.408105	0.382634	120
140	0.873373	0.808339	0.753345	0.704715	0.660882	0.620947	0.584311	0.550538	0.519289	0.490293	0.463324	0.438191	140
170	0.894951	0.839849	0.792566	0.750184	0.711466	0.675794	0.642653	0.611744	0.582816	0.555676	0.530155	0.506117	170
200	0.910250	0.862477	0.821063	0.783589	0.749029	0.716923	0.686859	0.658572	0.631886	0.606648	0.582733	0.560038	200
240	0.924846	0.884289	0.848791	0.816382	0.786225	0.757986	0.731332	0.706070	0.682043	0.659149	0.637294	0.616401	240
320	0.943294	C.912166	0.884592	0.859136	0.835186	0.812531	0.790932	0.770258	0.750414	0.731325	0.712926	0.695173	320
440	0.958556	0.935488	0.914854	0.895630	0.877381	0.859974	0.843239	0.827091	0.811467	0.796319	0.781611	0.767310	440
600	0.969501	0.952359	0.936918	0.922439	0.908607	0.895334	0.882498	0.870040	0.857917	0.846097	0.834556	0.823272	600
800	0.977071	0.964096	^.952354	0.941294	0.930682	0.920458	0.910529	0.900854	0.891402	0.882150	0.873081	0.864181	800
1000	0.981630	0.971194	0.961722	0.952776	0.944171	0.935859	0.927767	0.919864	0.912124	0.904530	0.897C70	0.889731	1000
INF	1.000000	1.000000	1.000000	1.000000	1.000000	1.000000	1.000000	1.000000	1.000000	1.000000	1.000000	1.000000	INF

Tafel 1 (Forts.) Table 1 (cont.)

$U_{p,q,n}$

p=7 C=0.05

n	q=13	14	15	18	21	24	27	30	40	60	80	100	120	n
1	0.000000	0.000000	0.000000	0.000000	0.000000	0.000000	0.000000	0.000000	0.000000	0.000000	0.000000	0.000000	0.000000	1
2	0.000000	0.000000	0.000000	0.000000	0.000000	0.000000	0.000000	0.000000	0.000000	0.000000	0.000000	0.000000	0.000000	2
3	0.000000	0.000000	0.000000	0.000000	0.000000	0.000000	0.000000	0.000000	0.000000	0.000000	0.000000	0.000000	0.000000	3
4	0.000000	0.000000	0.000000	0.000000	0.000000	0.000000	0.000000	0.000000	0.000000	0.000000	0.000000	0.000000	0.000000	4
5	0.000000	0.000000	0.000000	0.000000	0.000000	0.000000	0.000000	0.000000	0.000000	0.000000	0.000000	0.000000	0.000000	5
6	0.000000	0.000000	0.000000	0.000000	0.000000	0.000000	0.000000	0.000000	0.000000	0.000000	0.000000	0.000000	0.000000	6
7	0.000001	0.000001	0.000001	0.000000	0.000000	0.000000	0.000000	0.000000	0.000000	0.000000	0.000000	0.000000	0.000000	7
8	0.000003	0.000003	0.000002	0.000001	0.000001	0.000001	0.000000	0.000000	0.000000	0.000000	0.000000	0.000000	0.000000	8
9	0.000012	0.000009	0.000007	0.000004	0.000002	0.000002	0.000001	0.000001	0.000001	0.000000	0.000000	0.000000	0.000000	9
10	0.000035	0.000026	0.000020	0.000010	0.000006	0.000004	0.000003	0.000002	0.000001	0.000000	0.000000	0.000000	0.000000	10
11	0.000085	0.000063	0.000047	0.000023	0.000013	0.000008	0.000005	0.000004	0.000002	0.000001	0.000000	0.000000	0.000000	11
12	0.000179	0.000131	0.000099	0.000046	0.000025	0.000015	0.000010	0.000007	0.000003	0.000001	0.000001	0.000001	0.000000	12
13	0.000338	0.000249	0.000187	0.000087	0.000046	0.000027	0.000017	0.000012	0.000005	0.000002	0.000001	0.000001	0.000001	13
14	0.000584	0.000432	0.000325	0.000152	0.000080	0.000047	0.000029	0.000020	0.000007	0.000002	0.000001	0.000001	0.000001	14
15	0.000941	0.000701	0.000530	0.000250	0.000132	0.000076	0.000047	0.000031	0.000011	0.000003	0.000002	0.000001	0.000001	15
16	0.001430	0.001073	0.000817	0.000390	0.000206	0.000119	0.000073	0.000048	0.000016	0.000005	0.000002	0.000001	0.000001	16
17	0.002072	0.001567	0.001201	0.000581	0.000309	0.000178	0.000109	0.000071	0.000024	0.000006	0.000003	0.000002	0.000001	17
18	0.002882	0.002198	0.001696	0.000833	0.000446	0.000257	0.000158	0.000103	0.000033	0.000008	0.000004	0.000002	0.000002	18
19	0.003875	0.002978	0.002315	0.001155	0.000625	0.000362	0.000223	0.000145	0.000046	0.000011	0.000005	0.000003	0.000002	19
20	0.005059	0.003919	0.003067	0.001555	0.000850	0.000496	0.000306	0.000198	0.000062	0.000014	0.000006	0.000003	0.000002	20
21	0.006443	0.005028	0.003961	0.002042	0.001128	0.000663	0.000410	0.000267	0.000083	0.000018	0.000007	0.000004	0.000003	21
22	0.008029	0.006311	0.005005	0.002621	0.001465	0.000867	0.000540	0.000351	0.000109	0.000024	0.000009	0.000005	0.000003	22
23	0.009820	0.007771	0.006201	0.003259	0.001866	0.001113	0.000697	0.000455	0.000141	0.000030	0.000011	0.000006	0.000004	23
24	0.011813	0.009409	0.007554	0.004081	0.002335	0.001405	0.000884	0.000579	0.000180	0.000038	0.000014	0.000007	0.000004	24
25	0.014006	0.011226	0.009065	0.004971	0.002877	0.001746	0.001106	0.000727	0.000227	0.000047	0.000017	0.000009	0.000005	25
26	0.016394	0.013219	0.010733	0.005970	0.003495	0.002139	0.001363	0.000901	0.000283	0.000057	0.000021	0.000010	0.000006	26
27	0.018972	0.015383	0.012557	0.007081	0.004191	0.002588	0.001660	0.001103	0.000348	0.000070	0.000025	0.000012	0.000007	27
28	0.021732	0.017716	0.014534	0.008305	0.004969	0.003095	0.001999	0.001334	0.000424	0.000085	0.000030	0.000014	0.000008	28
29	0.024666	0.020212	0.016661	0.009643	0.005830	0.003662	0.002381	0.001597	0.000513	0.000102	0.000035	0.000017	0.000010	29
30	0.027767	0.022864	0.018934	0.011093	0.006775	0.004292	0.002809	0.001894	0.000614	0.000122	0.000041	0.000019	0.000011	30
40	0.065960	0.056440	0.048475	0.031352	0.020844	0.014199	0.009886	0.007021	0.002543	0.000530	0.000171	0.000074	0.000039	40
60	0.159192	0.142390	0.127620	0.092926	0.068706	0.051496	0.039074	0.029984	0.013317	0.003431	0.001171	0.000494	0.000245	60
80	0.249296	0.228561	0.209825	0.163522	0.128706	0.102208	0.081826	0.065998	0.033799	0.010596	0.004017	0.001776	0.000891	80
100	0.327287	0.304828	0.284174	0.231422	0.189792	0.156640	0.130031	0.108521	0.061468	0.022462	0.009429	0.004438	0.002303	100
120	0.392966	0.370016	0.348651	0.292809	0.247216	0.209725	0.178700	0.152885	0.093298	0.038405	0.017595	0.008808	0.004762	120
140	0.448134	0.425355	0.403958	0.347068	0.299436	0.259316	0.225348	0.196457	0.126958	0.057431	0.028311	0.014983	0.008437	140
170	0.515373	0.493442	0.472638	0.416287	0.367787	0.325843	0.289420	0.257678	0.177657	0.089646	0.048268	0.027424	0.016328	170
200	0.568611	0.547802	0.527925	0.473356	0.425449	0.383218	0.345862	0.312721	0.226213	0.124011	0.071544	0.043067	0.026902	200
240	0.624149	0.604901	0.586391	0.534914	0.488846	0.447473	0.410212	0.376572	0.285673	0.170269	0.105494	0.067526	0.044468	240
320	0.701604	0.685146	0.669184	0.624048	0.582647	0.544565	0.509456	0.477030	0.385313	0.257045	0.175779	0.122769	0.087353	320
440	0.772395	0.759064	0.746041	0.708700	0.673732	0.640910	0.610046	0.580981	0.495439	0.365145	0.273164	0.206989	0.158644	440
600	0.827235	0.816676	0.806308	0.776279	0.747733	0.720540	0.694596	0.669812	0.594683	0.472618	0.379024	0.306337	0.249322	600
800	0.867279	0.858928	0.850700	0.826696	0.803638	0.781445	0.760054	0.739413	0.675516	0.566771	0.478245	0.405517	0.345351	800
1000	0.892271	0.885374	0.878563	0.858610	0.839320	0.820637	0.802518	0.784927	0.729765	0.633153	0.551529	0.482052	0.422602	1000
INF	1.000000	1.000000	1.000000	1.000000	1.000000	1.000000	1.000000	1.000000	1.000000	1.000000	1.000000	1.000000	1.000000	INF

Tafel 1 (Forts.) Table 1 (cont.)

$U_{p,q,n}$

p=7 C=0.01

n	q=13	14	15	18	21	24	27	30	40	60	80	100	120	n
1	0.000000	0.000000	0.000000	0.000000	0.000000	0.000000	0.000000	0.000000	0.000000	0.000000	0.000000	0.000000	0.000000	1
2	0.000000	0.000000	0.000000	0.000000	0.000000	0.000000	0.000000	0.000000	0.000000	0.000000	0.000000	0.000000	0.000000	2
3	0.000000	0.000000	0.000000	0.000000	0.000000	0.000000	0.000000	0.000000	0.000000	0.000000	0.000000	0.000000	0.000000	3
4	0.000000	0.000000	0.000000	0.000000	0.000000	0.000000	0.000000	0.000000	0.000000	0.000000	0.000000	0.000000	0.000000	4
5	0.000000	0.000000	0.000000	0.000000	0.000000	0.000000	0.000000	0.000000	0.000000	0.000000	0.000000	0.000000	0.000000	5
6	0.000000	0.000000	0.000000	0.000000	0.000000	0.000000	0.000000	0.000000	0.000000	0.000000	0.000000	0.000000	0.000000	6
7	0.000000	0.000000	0.000000	0.000000	0.000000	0.000000	0.000000	0.000000	0.000000	0.000000	0.000000	0.000000	0.000000	7
8	0.000001	0.000001	0.000001	0.000001	0.000000	0.000000	0.000000	0.000000	0.000000	0.000000	0.000000	0.000000	0.000000	8
9	0.000005	0.000004	0.000003	0.000002	0.000001	0.000001	0.000001	0.000000	0.000000	0.000000	0.000000	0.000000	0.000000	9
10	0.000014	0.000011	0.000008	0.000004	0.000003	0.000002	0.000001	0.000001	0.000001	0.000000	0.000000	0.000000	0.000000	10
11	0.000037	0.000027	0.000021	0.000010	0.000006	0.000004	0.000003	0.000002	0.000001	0.000000	0.000000	0.000000	0.000000	11
12	0.000081	0.000060	0.000046	0.000022	0.000012	0.000008	0.000005	0.000004	0.000002	0.000001	0.000000	0.000000	0.000000	12
13	0.000160	0.000119	0.000090	0.000043	0.000024	0.000014	0.000009	0.000006	0.000003	0.000001	0.000001	0.000000	0.000000	13
14	0.000289	0.000215	0.000162	0.000078	0.000042	0.000025	0.000016	0.000011	0.000004	0.000001	0.000001	0.000001	0.000001	14
15	0.000484	0.000361	0.000274	0.000131	0.000070	0.000041	0.000026	0.000018	0.000007	0.000002	0.000001	0.000001	0.000001	15
16	0.000762	0.000572	0.000436	0.000210	0.000112	0.000066	0.000041	0.000028	0.000010	0.000003	0.000002	0.000001	0.000001	16
17	0.001141	0.000861	0.000660	0.000321	0.000172	0.000100	0.000063	0.000042	0.000014	0.000004	0.000002	0.000001	0.000001	17
18	0.001635	0.001243	0.000958	0.000471	0.000254	0.000148	0.000092	0.000061	0.000021	0.000005	0.000003	0.000002	0.000001	18
19	0.002259	0.001730	0.001341	0.000667	0.000362	0.000212	0.000132	0.000087	0.000029	0.000007	0.000003	0.000002	0.000001	19
20	0.003025	0.002332	0.001819	0.000918	0.000503	0.000295	0.000184	0.000121	0.000039	0.000010	0.000004	0.000002	0.000002	20
21	0.003942	0.003060	0.002401	0.001229	0.000679	0.000401	0.000250	0.000164	0.000053	0.000013	0.000005	0.000003	0.000002	21
22	0.005017	0.003922	0.003096	0.001608	0.000897	0.000532	0.000333	0.000219	0.000070	0.000016	0.000007	0.000004	0.000002	22
23	0.006256	0.004923	0.003909	0.002060	0.001160	0.000693	0.000436	0.000286	0.000092	0.000021	0.000008	0.000004	0.000003	23
24	0.007663	0.006068	0.004846	0.002590	0.001474	0.000887	0.000560	0.000369	0.000118	0.000026	0.000010	0.000005	0.000003	24
25	0.009238	0.007359	0.005911	0.003203	0.001843	0.001116	0.000708	0.000468	0.000150	0.000032	0.000012	0.000006	0.000004	25
26	0.010981	0.008799	0.007106	0.003904	0.002269	0.001385	0.000883	0.000586	0.000188	0.000040	0.000015	0.000008	0.000005	26
27	0.012891	0.010388	0.008433	0.004695	0.002756	0.001696	0.001087	0.000724	0.000233	0.000049	0.000018	0.000009	0.000005	27
28	0.014965	0.012124	0.009891	0.005578	0.003308	0.002051	0.001323	0.000884	0.000286	0.000060	0.000021	0.000011	0.000006	28
29	0.017198	0.014005	0.011481	0.006555	0.003927	0.002454	0.001592	0.001068	0.000347	0.000072	0.000026	0.000012	0.000007	29
30	0.019587	0.016029	0.013200	0.007628	0.004615	0.002906	0.001896	0.001279	0.000419	0.000086	0.000030	0.000015	0.000008	30
40	0.050661	0.043110	0.036837	0.023507	0.015460	0.010441	0.007221	0.005104	0.001840	0.000390	0.000129	0.000057	0.000030	40
60	0.133263	0.118688	0.105949	0.076312	0.055899	0.041560	0.031315	0.023884	0.010457	0.002675	0.000920	0.000393	0.000197	60
80	0.217983	0.199164	0.182241	0.140766	0.109937	0.086704	0.068989	0.055339	0.027930	0.008623	0.003263	0.001451	0.000733	80
100	0.293825	0.272874	0.253688	0.205048	0.167052	0.137058	0.113169	0.093992	0.052531	0.018871	0.007867	0.003705	0.001930	100
120	0.359085	0.337278	0.317054	0.264549	0.222074	0.187422	0.158947	0.135402	0.081637	0.033048	0.015004	0.007487	0.004052	120
140	0.414729	0.392796	0.372264	0.318016	0.272978	0.235318	0.203636	0.176848	0.113043	0.050336	0.024569	0.012938	0.007275	140
170	0.483441	0.462024	0.441772	0.387223	0.340630	0.300598	0.266036	0.236072	0.161215	0.080201	0.042759	0.024140	0.014325	170
200	0.538467	0.517944	0.498396	0.445009	0.398466	0.357683	0.321797	0.290112	0.208094	0.112625	0.064372	0.038491	0.023938	200
240	0.596402	0.577239	0.558859	0.507987	0.462748	0.422342	0.386125	0.353570	0.266267	0.156922	0.096404	0.061307	0.040179	240
320	0.678024	0.661444	0.645401	0.600225	0.559017	0.521291	0.486656	0.454787	0.365234	0.241449	0.163958	0.113866	0.080645	320
440	0.753392	0.739830	0.726609	0.688842	0.653647	0.620747	0.589920	0.560982	0.476300	0.348593	0.259342	0.195617	0.149346	440
600	0.812231	0.801418	0.790820	0.760223	0.731260	0.703771	0.677626	0.652721	0.577599	0.456592	0.364600	0.293616	0.238223	600
800	0.855439	0.846846	0.838392	0.813811	0.790288	0.767721	0.746032	0.725156	0.660818	0.552153	0.464349	0.392610	0.333526	800
1000	0.882506	0.875387	0.868369	0.847871	0.828129	0.809065	0.790625	0.772764	0.716985	0.619977	0.538564	0.469612	0.410844	1000
INF	1.000000	1.000000	1.000000	1.000000	1.000000	1.000000	1.000000	1.000000	1.000000	1.000000	1.000000	1.000000	1.000000	INF

Tafel 1 (Forts.) Table 1 (cont.)

$U_{p,q,n}$

p=8 C=0.05

q

n	1	2	3	4	5	6	7	8	9	10	11	12	n
1	0.000000	0.000000	0.000000	0.000000	0.000000	0.000000	0.000000	0.000000	0.000000	0.000000	0.000000	0.000000	1
2	0.000000	0.000000	0.000000	0.000000	0.000000	0.000000	0.000000	0.000000	0.000000	0.000000	0.000000	0.000000	2
3	0.000000	0.000000	0.000000	0.000000	0.000000	0.000000	0.000000	0.000000	0.000000	0.000000	0.000000	0.000000	3
4	0.000000	0.000000	0.000000	0.000000	0.000000	0.000000	0.000000	0.000000	0.000000	0.000000	0.000000	0.000000	4
5	0.000000	0.000000	0.000000	0.000000	0.000000	0.000000	0.000000	0.000000	0.000000	0.000000	0.000000	0.000000	5
6	0.000000	0.000000	0.000000	0.000000	0.000000	0.000000	0.000000	0.000000	0.000000	0.000000	0.000000	0.000000	6
7	0.000138	0.000015	0.000004	0.000001	0.000001	0.000000	0.000000	0.000000	0.000000	0.000000	0.000000	0.000000	7
8	0.003295	0.000393	0.000090	0.000029	0.000012	0.000006	0.000003	0.000002	0.000001	0.000001	0.000001	0.000000	8
9	0.017079	0.002632	0.000659	0.000218	0.000087	0.000040	0.000020	0.000011	0.000007	0.000004	0.000003	0.000002	9
10	0.043574	0.008626	0.002458	0.000872	0.000361	0.000168	0.000086	0.000047	0.000028	0.000017	0.000011	0.000008	10
11	0.078039	0.019031	0.006148	0.002365	0.001032	0.000497	0.000259	0.000144	0.000085	0.000052	0.000034	0.000023	11
12	0.115676	0.033314	0.012011	0.004993	0.002304	0.001155	0.000619	0.000351	0.000209	0.000130	0.000084	0.000056	12
13	0.153630	0.050518	0.019990	0.008908	0.004335	0.002263	0.001252	0.000727	0.000441	0.000278	0.000181	0.000122	13
14	0.190453	0.069716	0.029839	0.014129	0.007216	0.003915	0.002234	0.001331	0.000824	0.000527	0.000347	0.000235	14
15	0.225477	0.090151	0.041241	0.020590	0.010980	0.006173	0.003628	0.002215	0.001399	0.000910	0.000608	0.000416	15
16	0.258443	0.111245	0.053875	0.028171	0.015610	0.009065	0.005476	0.003422	0.002203	0.001457	0.000987	0.000683	16
17	0.289300	0.132575	0.067447	0.036729	0.021061	0.012594	0.007801	0.004982	0.003269	0.002197	0.001509	0.001057	17
18	0.318105	0.153836	0.081699	0.046115	0.027265	0.016740	0.010611	0.006915	0.004617	0.003151	0.002194	0.001555	18
19	0.344966	0.174814	0.096415	0.056185	0.034144	0.021472	0.013900	0.009228	0.006265	0.004339	0.003060	0.002194	19
20	0.370015	0.195359	0.111416	0.066805	0.041616	0.026747	0.017653	0.011923	0.008219	0.005771	0.004120	0.002987	20
21	0.393387	0.215374	0.126559	0.077857	0.049601	0.032519	0.021845	0.014991	0.010483	0.007456	0.005386	0.003946	21
22	0.415217	0.234796	0.141726	0.089233	0.058021	0.038737	0.026450	0.018419	0.013053	0.009397	0.006863	0.005078	22
23	0.435632	0.253588	0.156826	0.100843	0.066804	0.045350	0.031435	0.022192	0.015923	0.011593	0.008555	0.006390	23
24	0.454749	0.271732	0.171785	0.112606	0.075884	0.052311	0.036769	0.026287	0.019081	0.014041	0.010462	0.007885	24
25	0.472677	0.289225	0.186549	0.124457	0.085199	0.059573	0.042416	0.030685	0.022515	0.016733	0.012583	0.009565	25
26	0.489514	0.306072	0.201075	0.136338	0.094698	0.067091	0.048346	0.035361	0.026210	0.019663	0.014914	0.011428	26
27	0.505352	0.322285	0.215331	0.148203	0.104332	0.074826	0.054525	0.040293	0.030150	0.022818	0.017449	0.013472	27
28	0.520271	0.337880	0.229293	0.160010	0.114060	0.082739	0.060924	0.045457	0.034319	0.026189	0.020182	0.015694	28
29	0.534345	0.352879	0.242945	0.171728	0.123844	0.090796	0.067514	0.050831	0.038700	0.029764	0.023104	0.018089	29
30	0.547639	0.367302	0.256277	0.183330	0.133653	0.098967	0.074268	0.056394	0.043276	0.033529	0.026207	0.020651	30
40	0.648630	0.484826	0.371902	0.289857	0.228618	0.182082	0.146235	0.118316	0.096365	0.078964	0.065068	0.053897	40
60	0.757690	0.627279	0.527185	0.447009	0.381482	0.327255	0.281978	0.243910	0.211718	0.184362	0.161015	0.141011	60
80	0.815243	0.708843	0.622840	0.550577	0.488795	0.435425	0.388992	0.348380	0.312704	0.281253	0.253441	0.228779	80
100	0.850742	0.761330	0.686819	0.622411	0.565838	0.515687	0.470954	0.430871	0.394827	0.362322	0.332935	0.306310	100
120	0.874811	0.797857	0.732425	0.674791	0.623251	0.576764	0.534599	0.496197	0.461114	0.428982	0.399491	0.372376	120
140	0.892201	0.824719	0.766516	0.714559	0.667497	0.624521	0.585067	0.548712	0.515117	0.484002	0.455129	0.428296	140
170	0.910793	0.853874	0.804039	0.758920	0.717494	0.679163	0.643522	0.610267	0.579158	0.549999	0.522621	0.496881	170
200	0.923918	0.874725	0.831204	0.791410	0.754525	0.720081	0.687764	0.657345	0.628642	0.601508	0.575820	0.551470	200
240	0.936396	0.894758	0.857556	0.823223	0.791114	0.760867	0.732246	0.705079	0.679234	0.654605	0.631100	0.608645	240
320	0.952108	0.920269	0.891472	0.864586	0.839159	0.814944	0.791784	0.769570	0.748216	0.727659	0.707843	0.688723	320
440	0.965057	0.941534	0.920045	0.899793	0.880463	0.861889	0.843968	0.826629	0.809821	0.793502	0.777641	0.762209	440
600	0.974316	0.956873	0.940825	0.925599	0.910972	0.896826	0.883093	0.869724	0.856684	0.843948	0.831494	0.819306	600
800	0.980707	0.967524	0.955338	0.943721	0.932512	0.921624	0.911008	0.900630	0.890464	0.880494	0.870704	0.861085	800
1000	0.984551	0.973956	0.964134	0.954746	0.945661	0.936815	0.928167	0.919691	0.911367	0.903183	0.895128	0.887192	1000
INF	1.000000	1.000000	1.000000	1.000000	1.000000	1.000000	1.000000	1.000000	1.000000	1.000000	1.000000	1.000000	INF

$U_{p,q,n}$

p=8 C=0.01

n	q = 1	2	3	4	5	6	7	8	9	10	11	12	n
1	0.000000	0.000000	0.000000	0.000000	0.000000	0.000000	0.000000	0.000000	0.000000	0.000000	0.000000	0.000000	1
2	0.000000	0.000000	0.000000	0.000000	0.000000	0.000000	0.000000	0.000000	0.000000	0.000000	0.000000	0.000000	2
3	0.000000	0.000000	0.000000	0.000000	0.000000	0.000000	0.000000	0.000000	0.000000	0.000000	0.000000	0.000000	3
4	0.000000	0.000000	0.000000	0.000000	0.000000	0.000000	0.000000	0.000000	0.000000	0.000000	0.000000	0.000000	4
5	0.000000	0.000000	0.000000	0.000000	0.000000	0.000000	0.000000	0.000000	0.000000	0.000000	0.000000	0.000000	5
6	0.000000	0.000000	0.000000	0.000000	0.000000	0.000000	0.000000	0.000000	0.000000	0.000000	0.000000	0.000000	6
7	0.000021	0.000002	0.000001	0.000000	0.000000	0.000000	0.000000	0.000000	0.000000	0.000000	0.000000	0.000000	7
8	0.000738	0.000088	0.000021	0.000007	0.000003	0.000001	0.000001	0.000000	0.000000	0.000000	0.000000	0.000000	8
9	0.005130	0.000759	0.000188	0.000063	0.000025	0.000012	0.000006	0.000004	0.000002	0.000001	0.000001	0.000001	9
10	0.016457	0.003031	0.000837	0.000293	0.000121	0.000057	0.000029	0.000016	0.000010	0.000006	0.000004	0.000003	10
11	0.034984	0.007819	0.002414	0.000906	0.000390	0.000187	0.000098	0.000055	0.000033	0.000020	0.000013	0.000009	11
12	0.058795	0.015460	0.005293	0.002131	0.000964	0.000477	0.000255	0.000144	0.000086	0.000054	0.000035	0.000024	12
13	0.085701	0.025773	0.009664	0.004155	0.001974	0.001014	0.000555	0.000321	0.000194	0.000123	0.000080	0.000054	13
14	0.114026	0.038323	0.015549	0.007095	0.003528	0.001878	0.001057	0.000624	0.000384	0.000245	0.000162	0.000110	14
15	0.142659	0.052612	0.022851	0.010993	0.005702	0.003140	0.001817	0.001097	0.000687	0.000445	0.000296	0.000203	15
16	0.170909	0.068175	0.031408	0.015836	0.008535	0.004851	0.002881	0.001778	0.001134	0.000744	0.000502	0.000347	16
17	0.198367	0.084613	0.041037	0.021570	0.012034	0.007041	0.004286	0.002700	0.001752	0.001168	0.000797	0.000556	17
18	0.224804	0.101603	0.051550	0.028118	0.016183	0.009723	0.006056	0.003890	0.002567	0.001736	0.001200	0.000846	18
19	0.250102	0.118888	0.062771	0.035392	0.020951	0.012897	0.008203	0.005367	0.003600	0.002468	0.001727	0.001231	19
20	0.274219	0.136267	0.074542	0.043297	0.026294	0.016549	0.010733	0.007144	0.004864	0.003380	0.002393	0.001723	20
21	0.297154	0.153589	0.086724	0.051742	0.032161	0.020658	0.013641	0.009225	0.006371	0.004484	0.003210	0.002335	21
22	0.318534	0.170736	0.099197	0.060641	0.038499	0.025196	0.016917	0.011612	0.008127	0.005789	0.004189	0.003076	22
23	0.339603	0.187624	0.111859	0.069912	0.045254	0.030131	0.020544	0.014298	0.010133	0.007299	0.005337	0.003955	23
24	0.359212	0.204189	0.124624	0.079482	0.052374	0.035431	0.024505	0.017276	0.012388	0.009019	0.006658	0.004978	24
25	0.377817	0.220388	0.137421	0.089287	0.059809	0.041060	0.028778	0.020535	0.014887	0.010948	0.008156	0.006149	25
26	0.395477	0.236190	0.150193	0.099266	0.067511	0.046985	0.033340	0.024060	0.017624	0.013084	0.009833	0.007472	26
27	0.412247	0.251575	0.162890	0.109371	0.075438	0.053173	0.038170	0.027838	0.020590	0.015423	0.011686	0.008948	27
28	0.428183	0.266532	0.175473	0.119555	0.083549	0.059592	0.043242	0.031852	0.023775	0.017959	0.013714	0.010578	28
29	0.443336	0.281057	0.187912	0.129781	0.091808	0.066212	0.048535	0.036085	0.027167	0.020686	0.015914	0.012359	29
30	0.457755	0.295150	0.200181	0.140015	0.100183	0.073004	0.054026	0.040521	0.030756	0.023596	0.018280	0.014290	30
40	0.570452	0.414141	0.310972	0.238227	0.185181	0.145635	0.115665	0.092650	0.074781	0.060775	0.049703	0.040884	40
60	0.697943	0.566845	0.469642	0.393584	0.332536	0.282759	0.241720	0.207592	0.179013	0.154941	0.134560	0.117227	60
80	0.767444	0.657821	0.571844	0.501042	0.441431	0.390577	0.346803	0.308868	0.275817	0.246896	0.221493	0.199107	80
100	0.811035	0.717584	0.641821	0.577506	0.521783	0.472932	0.429764	0.391400	0.357150	0.326462	0.298881	0.274029	100
120	0.840895	0.759705	0.692428	0.634154	0.582694	0.536747	0.495425	0.458070	0.424165	0.393293	0.365111	0.339324	120
140	0.862619	0.790948	0.730631	0.677631	0.630186	0.587267	0.548177	0.512402	0.479541	0.449270	0.421319	0.395458	140
170	0.885982	0.825107	0.773033	0.726555	0.684388	0.645690	0.609966	0.576841	0.546023	0.517277	0.490407	0.465247	170
200	0.902559	0.849693	0.803953	0.762678	0.724868	0.689844	0.657190	0.626631	0.597943	0.570945	0.545489	0.521449	200
240	0.918384	0.873433	0.834124	0.798284	0.765147	0.734170	0.705043	0.677532	0.651483	0.626761	0.603257	0.580879	240
320	0.938398	0.903831	0.873198	0.844911	0.818449	0.793424	0.769628	0.746913	0.725172	0.704312	0.684271	0.664991	320
440	0.954965	0.929310	0.906326	0.884882	0.864628	0.845292	0.826734	0.808858	0.791594	0.774890	0.758703	0.742995	440
600	0.966853	0.947766	0.930536	0.914342	0.898941	0.884139	0.869838	0.855973	0.842497	0.829376	0.816581	0.804090	600
800	0.975077	0.960620	0.947501	0.935108	0.923267	0.911833	0.900736	0.889929	0.879380	0.869062	0.858958	0.849052	800
1000	0.980031	0.968398	0.957807	0.947773	0.938158	0.928847	0.919787	0.910939	0.902280	0.893788	0.885451	0.877255	1000
INF	1.000000	1.000000	1.000000	1.000000	1.000000	1.000000	1.000000	1.000000	1.000000	1.000000	1.000000	1.000000	INF

Tafel 1 (Forts.) Table 1 (cont.)

$U_{p,q,n}$

p=8 C=0.05

n	q=13	14	15	18	21	24	27	30	40	60	80	100	120	n
1	0.000000	0.000000	0.000000	0.000000	0.000000	0.000000	0.000000	0.000000	0.000000	0.000000	0.000000	0.000000	0.000000	1
2	0.000000	0.000000	0.000000	0.000000	0.000000	0.000000	0.000000	0.000000	0.000000	0.000000	0.000000	0.000000	0.000000	2
3	0.000000	0.000000	0.000000	0.000000	0.000000	0.000000	0.000000	0.000000	0.000000	0.000000	0.000000	0.000000	0.000000	3
4	0.000000	0.000000	0.000000	0.000000	0.000000	0.000000	0.000000	0.000000	0.000000	0.000000	0.000000	0.000000	0.000000	4
5	0.000000	0.000000	0.000000	0.000000	0.000000	0.000000	0.000000	0.000000	0.000000	0.000000	0.000000	0.000000	0.000000	5
6	0.000000	0.000000	0.000000	0.000000	0.000000	0.000000	0.000000	0.000000	0.000000	0.000000	0.000000	0.000000	0.000000	6
7	0.000000	0.000000	0.000000	0.000000	0.000000	0.000000	0.000000	0.000000	0.000000	0.000000	0.000000	0.000000	0.000000	7
8	0.000000	0.000000	0.000000	0.000000	0.000000	0.000000	0.000000	0.000000	0.000000	0.000000	0.000000	0.000000	0.000000	8
9	0.000001	0.000001	0.000001	0.000000	0.000000	0.000000	0.000000	0.000000	0.000000	0.000000	0.000000	0.000000	0.000000	9
10	0.000005	0.000004	0.000003	0.000001	0.000001	0.000000	0.000000	0.000000	0.000000	0.000000	0.000000	0.000000	0.000000	10
11	0.000016	0.000011	0.000008	0.000004	0.000002	0.000001	0.000001	0.000001	0.000000	0.000000	0.000000	0.000000	0.000000	11
12	0.000039	0.000027	0.000020	0.000009	0.000004	0.000003	0.000002	0.000001	0.000000	0.000000	0.000000	0.000000	0.000000	12
13	0.000084	0.000059	0.000043	0.000018	0.000009	0.000005	0.000003	0.000002	0.000001	0.000000	0.000000	0.000000	0.000000	13
14	0.000163	0.000116	0.000084	0.000036	0.000018	0.000010	0.000006	0.000004	0.000001	0.000000	0.000000	0.000000	0.000000	14
15	0.000291	0.000208	0.000151	0.000065	0.000031	0.000017	0.000010	0.000006	0.000002	0.000001	0.000000	0.000000	0.000000	15
16	0.000483	0.000348	0.000255	0.000110	0.000053	0.000029	0.000017	0.000010	0.000003	0.000001	0.000000	0.000000	0.000000	16
17	0.000754	0.000547	0.000404	0.000176	0.000086	0.000046	0.000027	0.000017	0.000005	0.000001	0.000000	0.000000	0.000000	17
18	0.001121	0.000821	0.000610	0.000270	0.000133	0.000071	0.000041	0.000025	0.000007	0.000002	0.000001	0.000000	0.000000	18
19	0.001598	0.001180	0.000883	0.000398	0.000198	0.000106	0.000061	0.000038	0.000010	0.000002	0.000001	0.000000	0.000000	19
20	0.002196	0.001636	0.001234	0.000566	0.000284	0.000154	0.000089	0.000055	0.000015	0.000003	0.000001	0.000001	0.000000	20
21	0.002929	0.002201	0.001673	0.000781	0.000396	0.000216	0.000125	0.000077	0.000021	0.000004	0.000001	0.000001	0.000000	21
22	0.003804	0.002882	0.002206	0.001049	0.000539	0.000296	0.000173	0.000106	0.000028	0.000005	0.000002	0.000001	0.000001	22
23	0.004828	0.003686	0.002843	0.001376	0.000716	0.000397	0.000233	0.000143	0.000038	0.000007	0.000002	0.000001	0.000001	23
24	0.006007	0.004621	0.003589	0.001768	0.000932	0.000521	0.000307	0.000190	0.000051	0.000009	0.000003	0.000001	0.000001	24
25	0.007343	0.005690	0.004448	0.002228	0.001190	0.000672	0.000399	0.000248	0.000066	0.000011	0.000004	0.000002	0.000001	25
26	0.008839	0.006897	0.005426	0.002763	0.001494	0.000852	0.000509	0.000318	0.000085	0.000014	0.000004	0.000002	0.000001	26
27	0.010494	0.008242	0.006524	0.003375	0.001848	0.001064	0.000640	0.000401	0.000108	0.000018	0.000005	0.000002	0.000001	27
28	0.012307	0.009726	0.007744	0.004068	0.002255	0.001311	0.000795	0.000501	0.000136	0.000022	0.000007	0.000003	0.000002	28
29	0.014275	0.011350	0.009087	0.004845	0.002718	0.001595	0.000974	0.000618	0.000169	0.000027	0.000008	0.000004	0.000002	29
30	0.016396	0.013110	0.010553	0.005706	0.003239	0.001919	0.001181	0.000753	0.000208	0.000033	0.000010	0.000004	0.000002	30
40	0.044861	0.037511	0.031500	0.019102	0.011957	0.007698	0.005083	0.003434	0.001076	0.000180	0.000050	0.000019	0.000009	40
60	0.123813	0.108977	0.096142	0.066867	0.047324	0.034018	0.024800	0.018315	0.007235	0.001537	0.000452	0.000170	0.000077	60
80	0.206857	0.187326	0.169891	0.127776	0.097187	0.074671	0.057903	0.045283	0.021065	0.005594	0.001850	0.000731	0.000333	80
100	0.282141	0.260164	0.240149	0.189975	0.151482	0.121657	0.098349	0.079992	0.041775	0.013220	0.004906	0.002077	0.000984	100
120	0.347404	0.324375	0.303109	0.248396	0.204777	0.169730	0.141382	0.118311	0.067303	0.024412	0.010009	0.004543	0.002253	120
140	0.403323	0.380053	0.358346	0.301415	0.254718	0.216170	0.184171	0.157480	0.095665	0.038660	0.017234	0.008334	0.004330	140
170	0.472653	0.449823	0.428292	0.370625	0.321831	0.280335	0.244890	0.214500	0.140330	0.064270	0.031693	0.016618	0.009194	170
200	0.528363	0.506416	0.485554	0.428831	0.379751	0.337103	0.299909	0.267370	0.184797	0.093057	0.049659	0.027815	0.016253	200
240	0.587171	0.566619	0.546936	0.492630	0.444617	0.402012	0.364097	0.330267	0.241027	0.133557	0.077331	0.046465	0.028837	240
320	0.670257	0.652413	0.635157	0.586655	0.542569	0.502382	0.465665	0.432056	0.338734	0.213488	0.138378	0.091866	0.062290	320
440	0.747184	0.732548	0.718280	0.677545	0.639648	0.604307	0.571292	0.540405	0.450737	0.318314	0.228612	0.166585	0.122969	440
600	0.807371	0.795675	0.784208	0.751096	0.719768	0.690065	0.661861	0.635046	0.554586	0.426844	0.331883	0.260324	0.205812	600
800	0.851625	0.842319	0.833158	0.806495	0.780970	0.756487	0.732973	0.710364	0.640903	0.524739	0.432379	0.358229	0.298265	800
1000	0.879370	0.871655	0.864043	0.841781	0.820317	0.799584	0.779531	0.760119	0.699610	0.595113	0.508485	0.436115	0.375323	1000
INF	1.000000	1.000000	1.000000	1.000000	1.000000	1.000000	1.000000	1.000000	1.000000	1.000000	1.000000	1.000000	1.000000	INF

$U_{p,q,n}$

p=8 C=0.01

n	q=13	14	15	18	21	24	27	30	40	60	80	100	120	n
1	0.000000	0.000000	0.000000	0.000000	0.000000	0.000000	0.000000	0.000000	0.000000	0.000000	0.000000	0.000000	0.000000	1
2	0.000000	0.000000	0.000000	0.000000	0.000000	0.000000	0.000000	0.000000	0.000000	0.000000	0.000000	0.000000	0.000000	2
3	0.000000	0.000000	0.000000	0.000000	0.000000	0.000000	0.000000	0.000000	0.000000	0.000000	0.000000	0.000000	0.000000	3
4	0.000000	0.000000	0.000000	0.000000	0.000000	0.000000	0.000000	0.000000	0.000000	0.000000	0.000000	0.000000	0.000000	4
5	0.000000	0.000000	0.000000	0.000000	0.000000	0.000000	0.000000	0.000000	0.000000	0.000000	0.000000	0.000000	0.000000	5
6	0.000000	0.000000	0.000000	0.000000	0.000000	0.000000	0.000000	0.000000	0.000000	0.000000	0.000000	0.000000	0.000000	6
7	0.000000	0.000000	0.000000	0.000000	0.000000	0.000000	0.000000	0.000000	0.000000	0.000000	0.000000	0.000000	0.000000	7
8	0.000000	0.000000	0.000000	0.000000	0.000000	0.000000	0.000000	0.000000	0.000000	0.000000	0.000000	0.000000	0.000000	8
9	0.000001	0.000000	0.000000	0.000000	0.000000	0.000000	0.000000	0.000000	0.000000	0.000000	0.000000	0.000000	0.000000	9
10	0.000002	0.000002	0.000001	0.000001	0.000000	0.000000	0.000000	0.000000	0.000000	0.000000	0.000000	0.000000	0.000000	10
11	0.000006	0.000005	0.000003	0.000002	0.000001	0.000001	0.000000	0.000000	0.000000	0.000000	0.000000	0.000000	0.000000	11
12	0.000017	0.000012	0.000009	0.000004	0.000002	0.000001	0.000001	0.000001	0.000000	0.000000	0.000000	0.000000	0.000000	12
13	0.000038	0.000027	0.000020	0.000009	0.000004	0.000003	0.000002	0.000001	0.000000	0.000000	0.000000	0.000000	0.000000	13
14	0.000077	0.000055	0.000040	0.000017	0.000009	0.000005	0.000003	0.000002	0.000001	0.000000	0.000000	0.000000	0.000000	14
15	0.000142	0.000102	0.000075	0.000032	0.000016	0.000009	0.000005	0.000003	0.000001	0.000000	0.000000	0.000000	0.000000	15
16	0.000245	0.000176	0.000130	0.000057	0.000028	0.000015	0.000009	0.000006	0.000002	0.000000	0.000000	0.000000	0.000000	16
17	0.000396	0.000287	0.000212	0.000093	0.000046	0.000025	0.000015	0.000009	0.000003	0.000001	0.000000	0.000000	0.000000	17
18	0.000608	0.000444	0.000330	0.000147	0.000073	0.000040	0.000023	0.000015	0.000004	0.000001	0.000000	0.000000	0.000000	18
19	0.000892	0.000657	0.000491	0.000221	0.000111	0.000060	0.000035	0.000022	0.000006	0.000001	0.000001	0.000000	0.000000	19
20	0.001260	0.000935	0.000703	0.000322	0.000162	0.000089	0.000052	0.000032	0.000009	0.000002	0.000001	0.000000	0.000000	20
21	0.001722	0.001288	0.000975	0.000453	0.000230	0.000126	0.000074	0.000046	0.000013	0.000003	0.000001	0.000001	0.000000	21
22	0.002289	0.001725	0.001315	0.000621	0.000319	0.000176	0.000103	0.000064	0.000018	0.000003	0.000001	0.000001	0.000000	22
23	0.002968	0.002253	0.001729	0.000830	0.000431	0.000239	0.000141	0.000088	0.000024	0.000004	0.000002	0.000001	0.000000	23
24	0.003765	0.002879	0.002225	0.001085	0.000570	0.000319	0.000189	0.000118	0.000032	0.000006	0.000002	0.000001	0.000001	24
25	0.004687	0.003610	0.002807	0.001390	0.000738	0.000417	0.000248	0.000155	0.000042	0.000008	0.000003	0.000001	0.000001	25
26	0.005738	0.004449	0.003480	0.001751	0.000941	0.000535	0.000321	0.000201	0.000055	0.000010	0.000003	0.000001	0.000001	26
27	0.006920	0.005400	0.004250	0.002171	0.001180	0.000677	0.000408	0.000257	0.000071	0.000012	0.000004	0.000002	0.000001	27
28	0.008234	0.006466	0.005118	0.002653	0.001458	0.000844	0.000512	0.000323	0.000090	0.000015	0.000005	0.000002	0.000001	28
29	0.009682	0.007648	0.006087	0.003201	0.001780	0.001039	0.000634	0.000403	0.000112	0.000019	0.000006	0.000003	0.000001	29
30	0.011264	0.008948	0.007160	0.003817	0.002146	0.001264	0.000777	0.000495	0.000139	0.000023	0.000007	0.000003	0.000002	30
40	0.033812	0.028104	0.023471	0.014029	0.008682	0.005539	0.003633	0.002443	0.000763	0.000131	0.000037	0.000015	0.000007	40
60	0.102425	0.089736	0.078823	0.054184	0.037965	0.027055	0.019577	0.014364	0.005590	0.001181	0.000350	0.000133	0.000061	60
80	0.179321	0.161787	0.146213	0.108918	0.082153	0.062655	0.048267	0.037525	0.017187	0.004493	0.001485	0.000590	0.000272	80
100	0.251582	0.231266	0.212845	0.167022	0.132233	0.105525	0.084819	0.068628	0.035331	0.010982	0.004049	0.001717	0.000817	100
120	0.315683	0.293973	0.274004	0.222984	0.182692	0.150585	0.124801	0.103955	0.058376	0.020801	0.008449	0.003825	0.001900	120
140	0.371490	0.349242	0.328564	0.274674	0.230854	0.194952	0.165348	0.140800	0.084536	0.033591	0.014819	0.007130	0.003701	140
170	0.441652	0.419496	0.398669	0.343204	0.296637	0.257300	0.223900	0.195415	0.126543	0.057080	0.027853	0.014510	0.008003	170
200	0.498716	0.477192	0.456793	0.401619	0.354218	0.313280	0.277771	0.246858	0.169082	0.083982	0.044372	0.024680	0.014359	200
240	0.559547	0.539189	0.519744	0.466355	0.419455	0.378070	0.341418	0.308860	0.223644	0.122434	0.070251	0.041920	0.025886	240
320	0.646423	0.628525	0.611259	0.562930	0.519246	0.479615	0.443558	0.410679	0.319992	0.199724	0.128486	0.084784	0.057205	320
440	0.727736	0.712904	0.698476	0.657435	0.619437	0.584149	0.551300	0.520668	0.432247	0.302980	0.216312	0.156852	0.115304	440
600	0.791884	0.779948	0.768266	0.734641	0.702963	0.673036	0.644710	0.617854	0.537667	0.411459	0.318452	0.248824	0.196063	600
800	0.839330	0.829784	0.820403	0.793183	0.767221	0.742401	0.718629	0.695830	0.626094	0.510353	0.419016	0.346099	0.287396	800
1000	0.869192	0.861255	0.853435	0.830632	0.808726	0.787629	0.767278	0.747621	0.686600	0.581948	0.495772	0.424141	0.364209	1000
INF	1.000000	1.000000	1.000000	1.000000	1.000000	1.000000	1.000000	1.000000	1.000000	1.000000	1.000000	1.000000	1.000000	INF

TABLE 2

The Θ_{max}-Criterion of K. C. S. Pillai: A Version of the Maximum Root Criterion of S. N. Roy

(a) Content of the tables and definition of the test statistic :

The tables contain upper percentage points of the zero distribution of the largest root Θ_{max} of the determinant equation

$$| \underline{S}_h - \Theta(\underline{S}_e + \underline{S}_h)| = 0 .$$

For the relationship with other test statistics, use

$$\Theta_{max} = \frac{\lambda_{max}}{1 + \lambda_{max}} ,$$

where λ_{max} is determined by

$$| \underline{S}_h - \lambda\underline{S}_e | = 0 .$$

In addition

$$\Theta_{max} = 1 - \mu_{min} ,$$

where μ_{min} is determined by

$$| \underline{S}_e + \mu(\underline{S}_e - \underline{S}_h) | = 0 .$$

(b) Extent of the table and definition of the parameters :

(1) The parameter α :

α = error probability

α = 5% and 1% .

(2) The parameter s :

$s = \min(n_h, p)$ where p = dimension of the variate

n_h = degrees of freedom of the hypothesis.

$s = 2(1)20$.

(3) The parameter m :

$$m = \frac{|p - n_h| - 1}{2}$$

$m = 0(1)5;7;10;15.$

(4) The parameter n :

$$n = \frac{n_e - p - 1}{2}$$, where n_e = degrees of freedom of the error

$n = 5(5)50;48;60;80;120;240;\infty$.

(c) Suggestions for applications :

(1) Multivariate general linear hypothesis, especially MANOVA-problems, multivariate regression and covariance analyses. Multivariate confidence regions.

(2) To test the independence of p (dependent) from q (independent) variates (multivariate regression analysis) where

$$\underline{S}_h = \underline{R}_{yx} \cdot \underline{R}_x^{-1} \cdot \underline{R}_{xy} \qquad \text{and}$$

$$\underline{S}_e = \underline{R}_y - \underline{R}_{yx} \cdot \underline{R}_x^{-1} \cdot \underline{R}_{xy}$$

(Standardized Samples ; $\underline{R}_x$, $\underline{R}_y$, $\underline{R}_{xy}$, $\underline{R}_{yx}$ = correlation matrices)

and

p = dimension of the (dependent) vector $\underline{y}$

q = dimension of the (independent) vector $\underline{x}$,

using the initial parameters

$$s = \min (p , q)$$

$$m = \frac{| p - q | - 1}{2}$$

$$n = \frac{N - p - q - 2}{2} \quad \text{where } N = \text{sample size.}$$

(d) References :

(1) Original tables due to

PILLAI, K. C. S. : Statistical Tables for Tests of Multivariate Hypotheses.
Manila/Philippines : The Statistical Center. The University of the Philippines, 1960.

(2) Copies of the present tables in modified version due to

PEARSON, E. S. + HARTLEY, H. O. : Biometrika Tables for Statisticians. Volume II. (Table 48).
Cambridge : Cambridge University Press, 1972.
(Published for the Biometrika Trusties).

(3) For the Θ_{max}-criterion :

ROY, S. N. : Some Aspects of Multivariate Analysis.
New York : Wiley, 1957. + Calcutta : Indian Statistical
Institute, 1957.

(e) Further remarks :

The tables may also be used to determin the lower 5% and 1% points of the smallest root Θ_{min} .

For this , use :

$$P (\Theta_{max} \leqq x ; m , n) = 1 - P (\Theta_{min} < 1 - x ; n,m) .$$

Note that in this case the roles of the parameters m and n have been reversed.

Obere 5-Prozent-Punkte

Upper 5 per cent points

n \ m	0	1	2	3	4	5	7	10	15
					s = 2				
5	0·5646	0·6507	0·7063	0·7459	0·7758	0·7992	0·8337	0·8676	0·9011
10	·3737	·4550	·5143	·5605	·5981	·6293	·6786	·7316	·7889
15	·2780	·3477	·4015	·4455	·4826	·5145	·5670	·6266	·6955
20	·2211	·2809	·3287	·3688	·4034	·4339	·4855	·5462	·6198
25	·1835	·2355	·2780	·3143	·3463	·3748	·4239	·4835	·5580
30	0·1568	0·2027	0·2408	0·2738	0·3031	0·3296	0·3760	0·4333	0·5071
35	·1369	·1780	·2124	·2425	·2696	·2942	·3377	·3924	·4644
40	·1214	·1585	·1898	·2175	·2425	·2655	·3064	·3585	·4282
45	·1093	·1431	·1718	·1974	·2206	·2420	·2805	·3300	·3973
50	·0993	·1304	·1569	·1807	·2023	·2224	·2586	·3057	·3704
48	0·1031	0·1352	0·1626	0·1870	0·2093	0·2299	0·2670	0·3150	0·3807
60	·0836	·1103	·1333	·1540	·1731	·1909	·2233	·2661	·3260
80	·0638	·0846	·1027	·1192	·1346	·1490	·1756	·2114	·2630
120	·0433	·0577	·0704	·0821	·0931	·1035	·1230	·1498	·1896
240	·0220	·0295	·0362	·0424	·0483	·0540	·0647	·0798	·1030
∞	·0000	·0000	·0000	·0000	·0000	·0000	·0000	·0000	·0000
					s = 3				
5	0·6689	0·7292	0·7698	0·7994	0·8221	0·8400	0·8668	0·8933	0·9199
10	·4718	·5373	·5862	·6249	·6564	·6828	·7246	·7696	·8185
15	·3620	·4219	·4690	·5079	·5407	·5691	·6157	·6687	·7298
20	·2931	·3465	·3898	·4265	·4582	·4861	·5334	·5889	·6559
25	·2461	·2937	·3332	·3671	·3970	·4237	·4697	·5252	·5944
30	0·2120	0·2548	0·2907	0·3221	0·3500	0·3752	0·4192	0·4734	0·5429
35	·1863	·2250	·2579	·2869	·3129	·3366	·3784	·4308	·4993
40	·1660	·2013	·2316	·2584	·2828	·3050	·3447	·3950	·4620
45	·1499	·1823	·2103	·2353	·2581	·2790	·3165	·3647	·4298
50	·1367	·1666	·1926	·2160	·2373	·2570	·2926	·3387	·4017
48	0·1417	0·1726	0·1994	0·2234	0·2452	0·2654	0·3018	0·3486	0·4125
60	·1157	·1417	·1644	·1850	·2040	·2217	·2538	·2961	·3550
80	·0888	·1093	·1274	·1441	·1595	·1740	·2008	·2366	·2880
120	·0606	·0750	·0879	·0999	·1111	·1217	·1415	·1687	·2089
240	·0310	·0386	·0455	·0519	·0580	·0639	·0750	·0905	·1143
∞	·0000	·0000	·0000	·0000	·0000	·0000	·0000	·0000	·0000
					s = 4				
5	0·7387	0·7825	0·8131	0·8360	0·8537	0·8679	0·8892	0·9108	0·9326
10	·5472	·6004	·6412	·6737	·7004	·7229	·7588	·7976	·8401
15	·4307	·4822	·5235	·5578	·5869	·6121	·6538	·7012	·7561
20	·3543	·4017	·4409	·4742	·5031	·5286	·5719	·6228	·6843
25	·3006	·3439	·3802	·4117	·4395	·4644	·5072	·5590	·6235
30	0·2609	0·3004	0·3341	0·3636	0·3899	0·4137	0·4552	0·5064	0·5720
35	·2306	·2667	·2978	·3254	·3502	·3728	·4127	·4626	·5279
40	·2063	·2396	·2685	·2943	·3177	·3391	·3773	·4256	·4899
45	·1870	·2178	·2447	·2688	·2908	·3111	·3475	·3941	·4569
50	·1709	·1995	·2247	·2473	·2681	·2873	·3220	·3668	·4279
48	0·1770	0·2065	0·2323	0·2555	0·2768	0·2964	0·3317	0·3772	0·4391
60	1454	·1704	·1927	·2129	·2315	·2488	·2805	·3219	·3796
80	·1122	·1322	·1501	·1666	·1820	·1964	·2230	·2586	·3094
120	·0770	·0913	·1042	·1162	·1274	·1381	·1581	·1854	·2257
240	·0397	·0473	·0542	·0608	·0670	·0730	·0843	·1002	·1243
∞	·0000	·0000	·0000	·0000	·0000	·0000	·0000	·0000	·0000

Obere 1-Prozent-Punkte

Upper 1 per cent points

n \ m	0	1	2	3	4	5	7	10	15
					s = 2				
5	0·6770	0·7446	0·7872	0·8171	0·8394	0·8568	0·8820	0·9066	0·9306
10	·4701	·5443	·5971	·6377	·6703	·6971	·7391	·7834	·8309
15	·3573	·4247	·4757	·5168	·5511	·5803	·6279	·6812	·7418
20	·2875	·3473	·3941	·4329	·4661	·4951	·5435	·5998	·6670
25	·2404	·2935	·3360	·3719	·4032	·4309	·4782	·5347	·6045
30	0·2065	0·2540	0·2926	0·3258	0·3550	0·3812	0·4265	0·4819	0·5521
35	·1811	·2239	·2592	·2898	·3171	·3417	·3847	·4383	·5077
40	·1610	·2000	·2325	·2608	·2863	·3094	·3503	·4017	·4697
45	·1452	·1810	·2110	·2373	·2611	·2828	·3215	·3708	·4369
50	·1322	·1652	·1931	·2177	·2399	·2604	·2971	·3443	·4083
48	0·1372	0·1712	0·1999	0·2251	0·2480	0·2689	0·3064	0·3544	0·4193
60	·1117	·1403	·1646	·1863	·2061	·2244	·2576	·3008	·3607
80	·0855	·1080	·1273	·1448	·1609	·1759	·2035	·2402	·2925
120	·0582	·0740	·0877	·1002	·1118	·1228	·1433	·1711	·2120
240	·0297	·0380	·0453	·0520	·0583	·0644	·0758	·0917	·1160
∞	·0000	·0000	·0000	·0000	·0000	·0000	·0000	·0000	·0000
					s = 3				
5	0·7582	0·8040	0·8344	0·8564	0·8730	0·8862	0·9056	0·9247	0·9437
10	·5586	·6164	·6590	·6923	·7192	·7416	·7767	·8141	·8544
15	·4375	·4937	·5374	·5730	·6029	·6285	·6703	·7172	·7708
20	·3586	·4104	·4519	·4867	·5166	·5428	·5866	·6376	·6985
25	·3034	·3506	·3893	·4223	·4511	·4767	·5203	·5726	·6370
30	0·2629	0·3058	0·3416	0·3726	0·3999	0·4245	0·4670	0·5189	0·5846
35	·2319	·2712	·3043	·3332	·3591	·3824	·4233	·4741	·5397
40	·2073	·2434	·2742	·3012	·3256	·3477	·3869	·4361	·5010
45	·1876	·2211	·2497	·2750	·2979	·3189	·3563	·4038	·4673
50	·1714	·2024	·2291	·2529	·2746	·2944	·3301	·3758	·4378
48	0·1776	0·2095	0·2369	0·2613	0·2835	0·3038	0·3401	0·3865	0·4491
60	·1456	·1727	·1963	·2175	·2369	·2549	·2874	·3298	·3883
80	·1121	·1338	·1528	·1701	·1861	·2010	·2284	·2648	·3166
120	·0769	·0922	·1059	·1185	·1302	·1413	·1619	·1899	·2309
240	·0395	·0477	·0551	·0619	·0684	·0746	·0863	·1025	·1272
∞	·0000	·0000	·0000	·0000	·0000	·0000	·0000	·0000	·0000
					s = 4				
5	0·8110	0·8436	0·8662	0·8830	0·8959	0·9062	0·9216	0·9370	0·9526
10	·6247	·6708	·7057	·7334	·7560	·7748	·8047	·8369	·8717
15	·5016	·5490	·5867	·6177	·6439	·6664	·7034	·7452	·7930
20	·4175	·4627	·4997	·5309	·5579	·5815	·6213	·6678	·7233
25	·3570	·3992	·4343	·4645	·4910	·5146	·5550	·6033	·6631
30	0·3117	0·3507	0·3837	0·4125	0·4380	0·4609	0·5007	0·5494	0·6111
35	·2765	·3126	·3435	·3707	·3951	·4171	·4558	·5039	·5661
40	·2483	·2819	·3108	·3365	·3596	·3807	·4181	·4651	·5270
45	·2255	·2568	·2839	·3081	·3301	·3502	·3861	·4318	·4928
50	·2066	·2358	·2612	·2841	·3049	·3241	·3586	·4028	·4626
48	0·2138	0·2438	0·2699	0·2933	0·3145	0·3341	0·3691	0·4139	0·4742
60	·1763	·2021	·2249	·2454	·2643	·2818	·3135	·3548	·4118
80	·1367	·1575	·1760	·1930	·2087	·2234	·2505	·2864	·3373
120	·0943	·1092	·1227	·1352	·1469	·1579	·1785	·2065	·2474
240	·0488	·0568	·0642	·0711	·0777	·0839	·0957	·1122	·1371
∞	·0000	·0000	·0000	·0000	·0000	·0000	·0000	·0000	·0000

Obere 5-Prozent-Punkte

Upper 5 per cent points

n \ m	0	1	2	3	4	5	7	10	15
					s = 5				
5	0·7882	0·8210	0·8447	0·8627	0·8768	0·8883	0·9058	0·9236	0·9419
10	·6069	·6507	·6849	·7125	·7354	·7547	·7858	·8197	·8570
15	·4883	·5328	·5690	·5993	·6252	·6477	·6850	·7277	·7773
20	·4072	·4495	·4847	·5150	·5414	·5647	·6043	·6511	·7077
25	·3488	·3881	·4215	·4507	·4764	·4995	·5394	·5877	·6480
30	0·3049	0·3413	0·3726	0·4003	0·4250	0·4474	0·4865	0·5349	0·5967
35	·2708	·3045	·3338	·3599	·3834	·4049	·4428	·4904	·5525
40	·2434	·2746	·3021	·3267	·3490	·3696	·4061	·4525	·5141
45	·2212	·2503	·2761	·2992	·3204	·3400	·3750	·4200	·4806
50	·2027	·2299	·2541	·2760	·2961	·3147	·3483	·3918	·4510
48	0·2097	0·2377	0·2625	0·2849	0·3054	0·3244	0·3585	0·4026	0·4624
60	·1732	·1973	·2188	·2385	·2567	·2736	·3045	·3450	·4013
80	·1344	·1539	·1714	·1877	·2028	·2171	·2433	·2785	·3286
120	·0928	·1068	·1196	·1316	·1428	·1535	·1735	·2008	·2409
240	·0481	·0557	·0627	·0693	·0756	·0816	·0931	·1091	·1335
∞	·0000	·0000	·0000	·0000	·0000	·0000	·0000	·0000	·0000
					s = 6				
5	0·8247	0·8499	0·8686	0·8830	0·8945	0·9039	0·9185	0·9335	0·9491
10	·6552	·6917	·7206	·7442	·7640	·7808	·8079	·8377	·8708
15	·5372	·5759	·6077	·6346	·6577	·6779	·7115	·7500	·7951
20	·4535	·4913	·5231	·5506	·5747	·5960	·6324	·6754	·7278
25	·3919	·4276	·4583	·4852	·5091	·5306	·5677	·6128	·6692
30	0·3447	0·3782	0·4074	0·4333	0·4565	0·4775	0·5144	0·5600	0·6184
35	·3076	·3390	·3665	·3912	·4135	·4338	·4699	·5151	·5743
40	·2775	·3069	·3329	·3563	·3777	·3973	·4322	·4767	·5357
45	·2530	·2806	·3051	·3273	·3476	·3664	·4001	·4434	·5017
50	·2324	·2583	·2815	·3025	·3219	·3399	·3724	·4144	·4717
48	0·2403	0·2668	0·2905	0·3120	0·3317	0·3500	0·3830	0·4256	0·4833
60	·1995	·2226	·2434	·2624	·2801	·2966	·3267	·3662	·4210
80	·1556	·1745	·1916	·2075	·2224	·2364	·2623	·2969	·3462
120	·1081	·1218	·1344	·1463	·1574	·1681	·1880	·2151	·2551
240	·0563	·0638	·0708	·0775	·0838	·0899	·1014	·1176	·1422
∞	·0000	·0000	·0000	·0000	·0000	·0000	·0000	·0000	·0000
					s = 7				
5	0·8523	0·8722	0·8872	0·8990	0·9085	0·9163	0·9286	0·9413	0·9548
10	·6950	·7257	·7503	·7707	·7878	·8025	·8263	·8527	·8823
15	·5792	·6130	·6412	·6651	·6858	·7039	·7342	·7692	·8104
20	·4944	·5282	·5570	·5820	·6040	·6236	·6570	·6968	·7453
25	·4305	·4631	·4914	·5162	·5384	·5583	·5930	·6351	·6879
30	0·3809	0·4119	0·4390	0·4632	0·4850	0·5048	0·5395	0·5825	0·6378
35	·3415	·3707	·3965	·4198	·4409	·4603	·4944	·5374	·5939
40	·3093	·3369	·3615	·3837	·4040	·4227	·4561	·4986	·5552
45	·2828	·3088	·3322	·3533	·3728	·3908	·4232	·4648	·5210
50	·2604	·2850	·3072	·3273	·3459	·3632	·3946	·4352	·4907
48	0·2690	0·2941	0·3167	0·3373	0·3562	0·3738	0·4056	0·4466	0·5024
60	·2244	·2465	·2665	·2850	·3021	·3181	·3474	·3858	·4392
80	·1759	·1942	·2109	·2264	·2410	·2547	·2801	·3141	·3626
120	·1229	·1363	·1487	·1605	·1715	·1820	·2018	·2287	·2684
240	·0644	·0718	·0788	·0854	·0918	·0979	·1095	·1257	·1504
∞	·0000	·0000	·0000	·0000	·0000	·0000	·0000	·0000	·0000

Obere 1-Prozent-Punkte

Upper 1 per cent points

n \ m	0	1	2	3	4	5	7	10	15
					S = 5				
5	0·8477	0·8719	0·8892	0·9023	0·9125	0·9208	0·9334	0·9461	0·9591
10	·6762	·7136	·7425	·7658	·7850	·8011	·8268	·8548	·8853
15	·5544	·5948	·6274	·6546	·6777	·6977	·7306	·7680	·8111
20	·4677	·5074	·5404	·5684	·5928	·6143	·6505	·6930	·7440
25	·4038	·4415	·4735	·5011	·5255	·5473	·5846	·6295	·6851
30	0·3549	0·3904	0·4208	0·4475	0·4713	0·4927	0·5301	0·5757	0·6337
35	·3165	·3498	·3786	·4041	·4270	·4478	·4844	·5299	·5889
40	·2854	·3166	·3438	·3681	·3900	·4101	·4457	·4906	·5497
45	·2601	·2893	·3150	·3380	·3590	·3783	·4127	·4565	·5151
50	·2388	·2663	·2906	·3124	·3324	·3509	·3841	·4267	·4844
48	0·2469	0·2751	0·2999	0·3222	0·3426	0·3614	0·3950	0·4382	0·4962
60	·2048	·2293	·2512	·2710	·2893	·3063	·3371	·3772	·4326
80	·1596	·1796	·1977	·2142	·2296	·2441	·2706	·3059	·3559
120	·1108	·1253	·1386	·1509	·1625	·1735	·1940	·2217	·2624
240	·0577	·0656	·0730	·0799	·0865	·0927	·1047	·1212	·1464
∞	·0000	·0000	·0000	·0000	·0000	·0000	·0000	·0000	·0000
					S = 6				
5	0·8745	0·8929	0·9065	0·9169	0·9252	0·9320	0·9424	0·9531	0·9642
10	·7173	·7482	·7724	·7922	·8086	·8225	·8449	·8694	·8964
15	·5986	·6334	·6619	·6858	·7063	·7240	·7535	·7872	·8262
20	·5111	·5462	·5757	·6010	·6231	·6426	·6757	·7147	·7616
25	·4450	·4790	·5081	·5335	·5559	·5760	·6106	·6524	·7042
30	0·3936	0·4261	0·4542	0·4789	0·5011	0·5211	0·5561	0·5990	0·6536
35	·3527	·3835	·4103	·4342	·4557	·4754	·5100	·5531	·6090
40	·3194	·3484	·3740	·3969	·4177	·4367	·4706	·5134	·5698
45	·2919	·3193	·3436	·3655	·3854	·4038	·4368	·4788	·5350
50	·2687	·2946	·3177	·3386	·3577	·3755	·4074	·4485	·5041
48	0·2775	0·3040	0·3276	0·3488	0·3683	0·3863	0·4187	0·4602	0·5160
60	·2315	·2548	·2757	·2948	·3125	·3289	·3588	·3977	·4515
80	·1814	·2006	·2181	·2342	·2493	·2634	·2894	·3240	·3730
120	·1266	·1407	·1538	·1659	·1774	·1882	·2085	·2360	·2763
240	·0663	·0741	·0814	·0883	·0949	·1012	·1132	·1298	·1550
∞	·0000	·0000	·0000	·0000	·0000	·0000	·0000	·0000	·0000
					S = 7				
5	0·8947	0·9091	0·9199	0·9284	0·9352	0·9408	0·9496	0·9587	0·9682
10	·7508	·7766	·7971	·8140	·8282	·8403	·8599	·8815	·9056
15	·6363	·6665	·6914	·7126	·7308	·7467	·7732	·8037	·8392
20	·5490	·5803	·6068	·6297	·6497	·6675	·6978	·7336	·7770
25	·4817	·5125	·5391	·5624	·5831	·6016	·6338	·6726	·7210
30	0·4286	0·4583	0·4843	0·5073	0·5280	0·5467	0·5795	0·6198	0·6713
35	·3858	·4142	·4393	·4617	·4820	·5005	·5332	·5740	·6272
40	·3506	·3777	·4017	·4233	·4431	·4612	·4934	·5342	·5881
45	·3213	·3471	·3700	·3908	·4099	·4275	·4590	·4993	·5532
50	·2965	·3210	·3429	·3629	·3812	·3982	·4289	·4685	·5221
48	0·3060	0·3309	0·3533	0·3736	0·3922	0·4094	0·4405	0·4803	0·5341
60	·2565	·2787	·2987	·3171	·3341	·3500	·3789	·4167	·4689
80	·2020	·2205	·2375	·2532	·2678	·2816	·3070	·3409	·3889
120	·1418	·1556	·1683	·1803	·1916	·2023	·2223	·2496	·2894
240	·0747	·0824	·0897	·0966	·1031	·1094	·1214	·1380	·1633
∞	·0000	·0000	·0000	·0000	·0000	·0000	·0000	·0000	·0000

Tafel 2 (Forts.) Table 2 (cont.)

Obere 5-Prozent-Punkte
Upper 5 per cent points

n \ m	0	1	2	3	4	5	7	10	15
					s = 8				
5	0·8739	0·8898	0·9020	0·9118	0·9197	0·9263	0·9367	0·9478	0·9595
10	·7281	·7542	·7754	·7931	·8080	·8209	·8419	·8655	·8921
15	·6156	·6453	·6703	·6917	·7103	·7266	·7541	·7859	·8236
20	·5307	·5611	·5872	·6100	·6302	·6481	·6789	·7157	·7607
25	·4655	·4953	·5213	·5443	·5648	·5834	·6157	·6551	·7047
30	0·4141	0·4428	0·4680	0·4906	0·5110	0·5296	0·5623	0·6029	0·6552
35	·3728	·4001	·4243	·4462	·4662	·4845	·5170	·5578	·6115
40	·3388	·3648	·3880	·4091	·4284	·4463	·4782	·5188	·5730
45	·3106	·3352	·3574	·3776	·3962	·4136	·4447	·4846	·5387
50	·2867	·3101	·3312	·3506	·3685	·3852	·4154	·4546	·5081
48	0·2958	0·3197	0·3412	0·3609	0·3791	0·3961	0·4267	0·4662	0·5200
60	·2480	·2692	·2885	·3063	·3229	·3384	·3668	·4041	·4560
80	·1954	·2131	·2293	·2445	·2587	·2722	·2971	·3304	·3779
120	·1372	·1503	·1626	·1741	·1850	·1954	·2149	·2417	·2810
240	·0723	·0797	·0866	·0932	·0996	·1057	·1173	·1335	·1583
∞	·0000	·0000	·0000	·0000	·0000	·0000	·0000	·0000	·0000
					s = 9				
5	0·8910	0·9039	0·9141	0·9222	0·9289	0·9346	0·9435	0·9531	0·9635
10	·7560	·7784	·7968	·8122	·8253	·8367	·8554	·8765	·9006
15	·6473	·6736	·6959	·7151	·7318	·7466	·7716	·8007	·8353
20	·5631	·5906	·6143	·6351	·6536	·6701	·6985	·7326	·7745
25	·4972	·5245	·5485	·5697	·5888	·6061	·6363	·6732	·7198
30	0·4446	0·4712	0·4947	0·5158	0·5350	0·5524	0·5832	0·6216	0·6711
35	·4018	·4274	·4502	·4709	·4897	·5070	·5378	·5767	·6278
40	·3664	·3908	·4128	·4329	·4512	·4682	·4987	·5376	·5894
45	·3367	·3600	·3811	·4005	·4182	·4348	·4647	·5032	·5551
50	·3115	·3336	·3539	·3725	·3896	·4057	·4349	·4727	·5244
48	0·3211	0·3437	0·3643	0·3833	0·4006	0·4169	0·4464	0·4845	0·5362
60	·2704	·2907	·3093	·3265	·3426	·3576	·3852	·4214	·4718
80	·2141	·2312	·2470	·2618	·2757	·2888	·3132	·3458	·3924
120	·1511	·1639	·1760	·1873	·1981	·2083	·2277	·2542	·2931
240	·0801	·0874	·0943	·1009	·1072	·1133	·1249	·1412	·1660
∞	·0000	·0000	·0000	·0000	·0000	·0000	·0000	·0000	·0000
					s = 10				
5	0·9049	0·9155	0·9240	0·9309	0·9366	0·9414	0·9492	0·9576	0·9672
10	·7798	·7991	·8151	·8287	·8403	·8504	·8671	·8861	·9079
15	·6752	·6986	·7185	·7358	·7510	·7644	·7871	·8138	·8457
20	·5922	·6171	·6387	·6578	·6747	·6900	·7162	·7479	·7870
25	·5261	·5512	·5733	·5930	·6108	·6269	·6551	·6897	·7336
30	0·4726	0·4973	0·5193	0·5391	0·5570	0·5734	0·6025	0·6388	0·6857
35	·4287	·4527	·4742	·4937	·5114	·5278	·5571	·5942	·6429
40	·3922	·4152	·4360	·4550	·4725	·4887	·5178	·5550	·6047
45	·3614	·3834	·4035	·4218	·4389	·4548	·4834	·5204	·5704
50	·3350	·3561	·3753	·3931	·4097	·4252	·4532	·4897	·5396
48	0·3451	0·3665	0·3861	0·4041	0·4209	0·4366	0·4648	0·5015	0·5516
60	·2918	·3113	·3292	·3458	·3613	·3758	·4025	·4377	·4867
80	·2321	·2487	·2640	·2784	·2920	·3048	·3286	·3605	·4062
120	·1646	·1772	·1890	·2002	·2108	·2209	·2400	·2662	·3045
240	·0878	·0950	·1019	·1084	·1147	·1208	·1323	·1486	·1734
∞	·0000	:0000	·0000	·0000	·0000	·0000	·0000	·0000	0000

Obere 1-Prozent-Punkte

Upper 1 per cent points

n \ m	0	1	2	3	4	5	7	10	15
					s = **8**				
5	0·9103	0·9218	0·9305	0·9375	0·9432	0·9480	0·9554	0·9632	0·9715
10	·7785	·8003	·8179	·8325	·8448	·8554	·8727	·8919	·9135
15	·6687	·6950	·7171	·7359	·7522	·7665	·7904	·8180	·8505
20	·5825	·6104	·6343	·6550	·6733	·6896	·7174	·7504	·7907
25	·5147	·5427	·5670	·5884	·6075	·6247	·6546	·6908	·7361
30	0·4604	0·4877	0·5118	0·5332	0·5524	0·5700	0·6007	0·6387	0·6873
35	·4162	·4425	·4659	·4870	·5060	·5235	·5544	·5932	·6436
40	·3795	·4048	·4274	·4479	·4665	·4837	·5144	·5533	·6047
45	·3488	·3730	·3947	·4145	·4326	·4494	·4795	·5181	·5699
50	·3226	·3457	·3665	·3856	·4032	·4195	·4489	·4870	·5387
48	0·3326	0·3561	0·3773	0·3967	0·4145	0·4310	0·4607	0·4990	0·5508
60	·2801	·3012	·3205	·3381	·3546	·3699	·3979	·4345	·4851
80	·2217	·2396	·2560	·2712	·2855	·2989	·3237	·3569	·4039
120	·1564	·1699	·1824	·1941	·2052	·2158	·2355	·2625	·3019
240	·0829	·0906	·0977	·1046	·1111	·1174	·1293	·1459	·1712
∞	·0000	·0000	·0000	·0000	·0000	·0000	·0000	·0000	·0000
					s = **9**				
5	0·9226	0·9319	0·9392	0·9450	0·9498	0·9538	0·9602	0·9670	0·9743
10	·8018	·8203	·8355	·8482	·8590	·8683	·8836	·9008	·9203
15	·6968	·7199	·7395	·7563	·7709	·7838	·8055	·8307	·8605
20	·6122	·6373	·6589	·6777	·6944	·7094	·7349	·7655	·8028
25	·5444	·5699	·5922	·6120	·6296	·6456	·6734	·7072	·7498
30	0·4894	0·5146	0·5369	0·5568	0·5749	0·5913	0·6201	0·6559	0·7018
35	·4441	·4687	·4906	·5103	·5283	·5447	·5740	·6107	·6587
40	·4063	·4300	·4513	·4706	·4883	·5046	·5338	·5709	·6201
45	·3744	·3972	·4178	·4365	·4538	·4698	·4987	·5357	·5854
50	·3471	·3689	·3887	·4069	·4237	·4394	·4677	·5043	·5541
48	0·3575	0·3797	0·3999	0·4183	0·4353	0·4511	0·4796	0·5164	0·5662
60	·3025	·3226	·3410	·3580	·3739	·3887	·4157	·4511	·5002
80	·2406	·2578	·2736	·2884	·3023	·3154	·3396	·3720	·4180
120	·1706	·1837	·1959	·2075	·2184	·2288	·2482	·2748	·3138
240	·0910	·0985	·1056	·1124	·1189	·1251	·1370	·1536	·1789
∞	·0000	·0000	·0000	·0000	·0000	·0000	·0000	·0000	·0000
					s = **10**				
5	0·9326	0·9402	0·9462	0·9512	0·9552	0·9587	0·9642	0·9702	0·9768
10	·8215	·8374	·8506	·8617	·8712	·8795	·8931	·9085	·9262
15	·7213	·7418	·7593	·7743	·7875	·7992	·8189	·8419	·8693
20	·6387	·6613	·6809	·6981	·7134	·7271	·7507	·7790	·8138
25	·5714	·5947	·6152	·6334	·6498	·6646	·6905	·7222	·7622
30	0·5160	0·5393	0·5600	0·5786	0·5955	0·6108	0·6380	0·6717	0·7152
35	·4700	·4929	·5134	·5319	·5489	·5644	·5921	·6270	·6727
40	·4313	·4536	·4736	·4919	·5087	·5242	·5519	·5873	·6344
45	·3984	·4199	·4394	·4573	·4737	·4890	·5166	·5520	·5998
50	·3701	·3908	·4097	·4270	·4431	·4581	·4853	·5205	·5685
48	0·3809	0·4020	0·4211	0·4387	0·4549	0·4700	0·4974	0·5327	0·5807
60	·3237	·3429	·3606	·3769	·3922	·4064	·4326	·4669	·5144
80	·2587	·2752	·2906	·3049	·3184	·3312	·3548	·3864	·4313
120	·1844	·1971	·2091	·2204	·2311	·2413	·2605	·2867	·3251
240	·0989	·1063	·1133	·1200	·1265	·1327	·1446	·1611	·1863
∞	·0000	·0000	·0000	·0000	·0000	·0000	·0000	·0000	·0000

TABLE 3

The Generalized F-Criterion of R. D. Bock: A Version of the Maximal Root Criterion of S. N. Roy

(a) Content of the tables and definition of the test statistic :

The tables contain the upper percentage points of the generalized F-statistic

$$F_o = \frac{t}{r} \cdot \lambda_1 .$$

Here λ_1 is by definition the largest root of the determinant equation

$$| \underline{S}_h - \lambda \underline{S}_e | = 0 ,$$

or eqivalently, the largest eigenvalue of the matrix $\underline{S}_h \cdot \underline{S}_e^{-1}$.
The parameters t and r will be defined in (b).

Because of the relationship $\lambda = \Theta/(1 - \Theta)$ we have in this criterion a modification of the maximal root criterion of S. N. Roy .
We also have

$$\lambda_1 = (1 - \mu_{min}) / \mu_{min}$$, where μ_{min} is determined

by

$$| \underline{S}_e - \mu(\underline{S}_e + \underline{S}_h) | = 0 .$$

(b) Extent of the table and definition of the parameters :

(1) The parameter α :

α = error probability

α = 5% and 1% .

(2) The parameter s :

$s = \min(n_h, p)$, where p = dimension of the variate
n_h = degrees of freedom of the hypothesis

$s = 1(1)10, 12(2)20$.

(3) The parameter r :

$r = |n_h - p| + 1$

$r = 1, 2(2)12, 16, 22, 32$.

(4) The parameter t :

$t = n_e - p + 1$ where n_e = degrees of freedom of the error

$t = 12(2)20, 25(5)50, 60(10)100, 150, 200, 300, 500, 1000, 2000.$
(For $s = 1$, several additional values of t have been taken into consideration).

(c) Suggestions for applications :

(1) Multivariate general linear hypotheses, especially MANOVA-problems. Multivariate regression and covariance analyses. Generalized confidence regions of S. N. ROY.

(2) For testing the independence of p (dependent) from q (independent) variates (multivariate regression analysis),

where

$$\underline{S}_h = \underline{R}_{yx} \cdot \underline{R}_x^{-1} \cdot \underline{R}_{xy} \qquad \text{and}$$

$$\underline{S}_e = \underline{R}_y - \underline{R}_{yx} \cdot \underline{R}_x^{-1} - \underline{R}_{xy}$$

(Standardized samples ; $\underline{R}_x$, $\underline{R}_y$, $\underline{R}_{xy}$, $\underline{R}_{yx}$ = correlation matrices)

and

p = dimension of the (dependent) vector $\underline{y}$

q = dimension of the (independent) vector $\underline{x}$,

using the initial parameters

$$s = \min (p , q)$$

$$r = | p - q | + 1$$

$$t = N - p - q \quad \text{where } N = \text{sample size} .$$

(d) References :

(1) For the source of the tables :

Bock, R. D. : Multivariate Statistical Methods in Behavioral Research.
New York : McGraw-Hill, 1975 (+ Preprints 1969/1971).

(2) For the Θ_{max}-criterion :

ROY, S. N. : Some Aspects of Multivariate Analysis.
New York : Wiley, 1957. + Calcutta : Indian Statistical Institute, 1957.

(e) <u>Further remarks</u> :

When $s = 1$, i. e. in the univariate case $p = 1$, the tables give the percentage points of the well-known univariate F-distribution.

In this case we have :

$$f_1 = n_h = r$$

$$f_2 = n_e = t \quad .$$

Tafel 3 Table 3

$s = 1$ *Obere 5% Punkte* / **Upper 5% points**

t \ r:	1	2	4	6	8	10	12	16	22	32
1	161.00	200.00	225.00	234.00	239.00	242.00	244.00	246.00	249.00	250.00
2	18.50	19.00	19.20	19.30	19.40	19.40	19.40	19.40	19.50	19.50
3	10.10	9.55	9.12	8.94	8.85	8.79	8.74	8.69	8.65	8.61
4	7.71	6.94	6.39	6.16	6.04	5.96	5.91	5.84	5.79	5.74
5	6.61	5.79	5.19	4.95	4.82	4.74	4.68	4.60	4.54	4.49
6	5.99	5.14	4.53	4.28	4.15	4.06	4.00	3.92	3.86	3.80
7	5.59	4.74	4.12	3.87	3.73	3.64	3.57	3.49	3.43	3.37
8	5.32	4.46	3.84	3.58	3.44	3.35	3.28	3.20	3.13	3.07
9	5.12	4.26	3.63	3.37	3.23	3.14	3.07	2.99	2.92	2.85
10	4.96	4.10	3.48	3.22	3.07	2.98	2.91	2.83	2.75	2.69
11	4.84	3.98	3.36	3.09	2.95	2.85	2.79	2.70	2.63	2.56
12	4.75	3.89	3.26	3.00	2.85	2.75	2.69	2.60	2.52	2.46
13	4.67	3.81	3.18	2.92	2.77	2.67	2.60	2.51	2.44	2.37
14	4.60	3.74	3.11	2.85	2.70	2.60	2.53	2.44	2.37	2.30
15	4.54	3.68	3.06	2.79	2.64	2.54	2.48	2.38	2.31	2.24
16	4.49	3.63	3.01	2.74	2.59	2.49	2.42	2.33	2.25	2.18
17	4.45	3.59	2.96	2.70	2.55	2.45	2.38	2.29	2.21	2.14
18	4.41	3.55	2.93	2.66	2.51	2.41	2.34	2.25	2.17	2.10
19	4.38	3.52	2.90	2.63	2.48	2.38	2.31	2.21	2.13	2.06
20	4.35	3.49	2.87	2.60	2.45	2.35	2.28	2.18	2.10	2.03
22	4.30	3.44	2.82	2.55	2.40	2.30	2.23	2.13	2.05	1.97
24	4.26	3.40	2.78	2.51	2.36	2.25	2.18	2.09	2.00	1.93
26	4.23	3.37	2.74	2.47	2.32	2.22	2.15	2.05	1.97	1.89
28	4.20	3.34	2.71	2.45	2.29	2.19	2.12	2.02	1.93	1.86
30	4.17	3.32	2.69	2.42	2.27	2.16	2.09	1.99	1.91	1.83
32	4.15	3.29	2.67	2.40	2.24	2.14	2.07	1.97	1.88	1.81
34	4.13	3.28	2.65	2.38	2.23	2.12	2.05	1.95	1.86	1.79
36	4.11	3.26	2.63	2.36	2.21	2.11	2.03	1.93	1.85	1.77
38	4.10	3.24	2.62	2.35	2.19	2.09	2.02	1.92	1.83	1.75
40	4.08	3.23	2.61	2.34	2.18	2.08	2.00	1.90	1.81	1.73
42	4.07	3.22	2.59	2.32	2.17	2.06	1.99	1.89	1.80	1.72
44	4.06	3.21	2.58	2.31	2.16	2.05	1.98	1.88	1.79	1.71
46	4.05	3.20	2.57	2.30	2.15	2.04	1.97	1.87	1.78	1.70
48	4.04	3.19	2.57	2.29	2.14	2.03	1.96	1.86	1.77	1.69
50	4.03	3.18	2.56	2.29	2.13	2.03	1.95	1.85	1.76	1.68
55	4.02	3.16	2.54	2.27	2.11	2.01	1.93	1.83	1.74	1.66
60	4.00	3.15	2.53	2.25	2.10	1.99	1.92	1.82	1.72	1.64
65	3.99	3.14	2.51	2.24	2.08	1.98	1.90	1.80	1.71	1.62
70	3.98	3.13	2.50	2.23	2.07	1.97	1.89	1.79	1.70	1.61
80	3.96	3.11	2.49	2.21	2.06	1.95	1.88	1.77	1.68	1.59
90	3.95	3.10	2.47	2.20	2.04	1.94	1.86	1.76	1.66	1.57
100	3.94	3.09	2.46	2.19	2.03	1.93	1.85	1.75	1.65	1.56
125	3.92	3.07	2.44	2.17	2.01	1.91	1.83	1.72	1.63	1.54
150	3.90	3.06	2.43	2.16	2.00	1.89	1.82	1.71	1.61	1.52
200	3.89	3.04	2.42	2.14	1.98	1.88	1.80	1.69	1.60	1.50
300	3.87	3.03	2.40	2.13	1.97	1.86	1.78	1.68	1.58	1.48
500	3.86	3.01	2.39	2.12	1.96	1.85	1.77	1.66	1.56	1.47
1,000	3.85	3.00	2.38	2.11	1.95	1.84	1.76	1.65	1.55	1.46
2,000	3.84	3.00	2.37	2.10	1.94	1.83	1.75	1.64	1.54	1.44

$s = 1$ *Obere 1% Punkte* / **Upper 1% points**

t \ r:	1	2	4	6	8	10	12	16	22	32
1†	405.00	500.00	563.00	586.00	598.00	606.00	611.00	617.00	622.00	627.00
2	98.50	99.00	99.20	99.30	99.40	99.40	99.40	99.40	99.50	99.50
3	34.10	30.80	28.70	27.90	27.50	27.20	27.10	26.80	26.60	26.50
4	21.20	18.00	16.00	15.20	14.80	14.50	14.40	14.20	14.00	13.80
5	16.30	13.30	11.40	10.70	10.30	10.10	9.89	9.68	9.51	9.36
6	13.70	10.90	9.15	8.47	8.10	7.87	7.72	7.52	7.35	7.21
7	12.20	9.55	7.85	7.19	6.84	6.62	6.47	6.27	6.11	5.97
8	11.30	8.65	7.01	6.37	6.03	5.81	5.67	5.48	5.32	5.18
9	10.60	8.02	6.42	5.80	5.47	5.26	5.11	4.92	4.77	4.63
10	10.00	7.56	5.99	5.39	5.06	4.85	4.71	4.52	4.36	4.23
11	9.65	7.21	5.67	5.07	4.74	4.54	4.40	4.21	4.06	3.92
12	9.33	6.93	5.41	4.82	4.50	4.30	4.16	3.97	3.82	3.68
13	9.07	6.70	5.21	4.62	4.30	4.10	3.96	3.78	3.62	3.49
14	8.86	6.51	5.04	4.46	4.14	3.94	3.80	3.62	3.46	3.33
15	8.68	6.36	4.89	4.32	4.00	3.80	3.67	3.49	3.33	3.19
16	8.53	6.23	4.77	4.20	3.89	3.69	3.55	3.37	3.22	3.08
17	8.40	6.11	4.67	4.10	3.79	3.59	3.46	3.27	3.12	2.98
18	8.29	6.01	4.58	4.01	3.71	3.51	3.37	3.19	3.03	2.90
19	8.18	5.93	4.50	3.94	3.63	3.43	3.30	3.12	2.96	2.82
20	8.10	5.85	4.43	3.87	3.56	3.37	3.23	3.05	2.90	2.76
22	7.95	5.72	4.31	3.76	3.45	3.26	3.12	2.94	2.78	2.65
24	7.82	5.61	4.22	3.67	3.36	3.17	3.03	2.85	2.70	2.56
26	7.72	5.53	4.14	3.59	3.29	3.09	2.96	2.78	2.62	2.48
28	7.64	5.45	4.07	3.53	3.23	3.03	2.90	2.72	2.56	2.42
30	7.56	5.39	4.02	3.47	3.17	2.98	2.84	2.66	2.51	2.37
32	7.50	5.34	3.97	3.43	3.13	2.93	2.80	2.62	2.46	2.32
34	7.44	5.29	3.93	3.39	3.09	2.89	2.76	2.58	2.42	2.28
36	7.40	5.25	3.89	3.35	3.05	2.86	2.72	2.54	2.38	2.24
38	7.35	5.21	3.86	3.32	3.02	2.83	2.69	2.51	2.35	2.21
40	7.31	5.18	3.83	3.29	2.99	2.80	2.66	2.48	2.33	2.18
42	7.28	5.15	3.80	3.27	2.97	2.78	2.64	2.46	2.30	2.16
44	7.25	5.12	3.78	3.24	2.95	2.75	2.62	2.44	2.28	2.13
46	7.22	5.10	3.76	3.22	2.93	2.73	2.60	2.42	2.26	2.11
48	7.19	5.08	3.74	3.20	2.91	2.72	2.58	2.40	2.24	2.10
50	7.17	5.06	3.72	3.19	2.89	2.70	2.56	2.38	2.22	2.08
55	7.12	5.01	3.68	3.15	2.85	2.66	2.53	2.34	2.18	2.04
60	7.08	4.98	3.65	3.12	2.82	2.63	2.50	2.31	2.15	2.01
65	7.04	4.95	3.62	3.09	2.80	2.61	2.47	2.29	2.13	1.98
70	7.01	4.92	3.60	3.07	2.78	2.59	2.45	2.27	2.11	1.97
80	6.96	4.88	3.56	3.04	2.74	2.55	2.42	2.23	2.07	1.92
90	6.93	4.85	3.54	3.01	2.72	2.52	2.39	2.21	2.04	1.90
100	6.90	4.82	3.51	2.99	2.69	2.50	2.37	2.19	2.02	1.87
125	6.84	4.78	3.47	2.95	2.66	2.47	2.33	2.15	1.98	1.83
150	6.81	4.75	3.45	2.92	2.63	2.44	2.31	2.12	1.96	1.81
200	6.76	4.71	3.41	2.89	2.60	2.41	2.27	2.09	1.93	1.77
300	6.72	4.68	3.38	2.86	2.57	2.38	2.24	2.06	1.89	1.74
500	6.69	4.65	3.36	2.84	2.55	2.36	2.22	2.04	1.87	1.72
1,000	6.66	4.63	3.34	2.82	2.53	2.34	2.20	2.02	1.85	1.70
2,000	6.63	4.61	3.32	2.80	2.51	2.32	2.18	2.00	1.83	1.68

† Entries in this row should be multiplied by 10. / *Die Eintragungen in dieser Zeile sind mit 10 zu multiplizieren.*

$s = 2$ *Obere 5% Punkte* / **Upper 5% points**

t \ r:	1	2	4	6	8	10	12	16	22	32
12	12.23	7.78	5.59	4.81	4.40	4.15	3.98	3.76	3.58	3.42
14	11.87	7.54	5.39	4.62	4.22	3.98	3.81	3.59	3.41	3.25
16	11.50	7.29	5.19	4.44	4.05	3.80	3.63	3.42	3.24	3.08
18	11.14	7.05	4.99	4.25	3.87	3.62	3.46	3.25	3.07	2.91
20	10.78	6.81	4.79	4.07	3,69	3.45	3.29	3.08	2.90	2.74
25	10.24	6.44	4.49	3.79	3.42	3.19	3.03	2.82	2.64	2.48
30	9.95	6.24	4.33	3.64	3.27	3.04	2.88	2.68	2.50	2.34
35	9.74	6.10	4.22	3.53	3.17	2.94	2.78	2.58	2.40	2.24
40	9.59	6.00	4.14	3.46	3.10	2.87	2.71	2.51	2.33	2.17
45	9.48	5.93	4.07	3.40	3.04	2.81	2.66	2.45	2.27	2.11
50	9.39	5.87	4.03	3.35	3.00	2.77	2.61	2.41	2.23	2.07
60	9.27	5.78	3.95	3.29	2.93	2.71	2.55	2.35	2.17	2.00
70	9.19	5.72	3.91	3.25	2.89	2.67	2.51	2.31	2.13	1.96
80	9.12	5.68	3.87	3.21	2.86	2.63	2.48	2.27	2.09	1.93
90	9.07	5.65	3.84	3.19	2.83	2.61	2.45	2.25	2.07	1.90
100	9.04	5.62	3.83	3.17	2.82	2.59	2.44	2.23	2.05	1.89
150	8.91	5.53	3.75	3.10	2.75	2.53	2.37	2.17	1.98	1.82
200	8.84	5.49	3.72	3.07	2.72	2.50	2.34	2.14	1.95	1.78
300	8.79	5.45	3.69	3.04	2.69	2.47	2.31	2.11	1.92	1.75
500	8.74	5.42	3.66	3.02	2.67	2.45	2.29	2.09	1.90	1.73
1,000	8.71	5.39	3.64	3.00	2.65	2.43	2.27	2.07	1.88	1.71
2,000	8.69	5.38	3.63	2.99	2.64	2.42	2.27	2.06	1.87	1.70

$s = 2$ *Obere 1% Punkte* / **Upper 1% points**

t \ r:	1	2	4	6	8	10	12	16	22	32
12	20.36	12.57	8.75	7.40	6.70	6.27	5.98	5.61	5.30	5.03
14	19.50	12.01	8.31	7.01	6.33	5.91	5.63	5.27	4.96	4.70
16	18.63	11.45	7.88	6.61	5.96	5.55	5.28	4.92	4.62	4.37
18	17.77	10.89	7.44	6.22	5.59	5.19	4.92	4.58	4.29	4.04
20	16.90	10.32	7.00	5.83	5.21	4.83	4.57	4.24	3.95	3.71
25	15.64	9.50	6.37	5.26	4.67	4.31	4.06	3.74	3.46	3.23
30	14.98	9.07	6.04	4.96	4.39	4.04	3.79	3.48	3.21	2.97
35	14.52	8.77	5.81	4.75	4.20	3.85	3.61	3.30	3.03	2.80
40	14.20	8.56	5.65	4.61	4.06	3.72	3.48	3.17	2.91	2.68
45	13.96	8.40	5.53	4.50	3.96	3.62	3.39	3.08	2.82	2.59
50	13.77	8.28	5.44	4.42	3.88	3.54	3.31	3.01	2.75	2.51
60	13.50	8.10	5.30	4.30	3.77	3.43	3.20	2.90	2.64	2.41
70	13.33	7.99	5.22	4.22	3.69	3.36	3.13	2.83	2.57	2.34
80	13.17	7.89	5.14	4.15	3.63	3.30	3.07	2.77	2.51	2.28
90	13.08	7.83	5.10	4.11	3.59	3.26	3.04	2.74	2.48	2.25
100	13.01	7.78	5.06	4.08	3.56	3.23	3.01	2.71	2.45	2.22
150	12.74	7.60	4.92	3.96	3.45	3.12	2.90	2.60	2.34	2.11
200	12.61	7.52	4.86	3.90	3.39	3.07	2.85	2.55	2.30	2.06
300	12.49	7.44	4.80	3.85	3.34	3.02	2.80	2.51	2.25	2.01
500	12.40	7.38	4.76	3.81	3.31	2.99	2.76	2.47	2.21	1.98
1,000	12.32	7.33	4.72	3.78	3.27	2.95	2.73	2.44	2.18	1.95
2,000	12.29	7.31	4.70	3.76	3.26	2.94	2.72	2.43	2.17	1.93

$s = 3$ *Obere 5% Punkte* / **Upper 5% points**

t	r: 1	2	4	6	8	10	12	16	22	32
12	20.28	12.12	8.08	6.69	5.98	5.54	5.25	4.88	4.57	4.31
14	19.55	11.66	7.74	6.39	5.70	5.28	4.99	4.63	4.32	4.06
16	18.82	11.20	7.40	6.09	5.42	5.01	4.73	4.37	4.08	3.82
18	18.09	10.74	7.06	5.79	5.14	4.74	4.47	4.12	3.83	3.58
20	17.36	10.29	6.72	5.49	4.86	4.47	4.21	3.87	3.59	3.34
25	16.28	9.60	6.22	5.05	4.44	4.07	3.82	3.49	3.22	2.98
30	15.69	9.23	5.95	4.81	4.22	3.85	3.61	3.29	3.02	2.78
35	15.28	8.97	5.76	4.64	4.06	3.70	3.46	3.14	2.88	2.64
40	14.98	8.78	5.62	4.52	3.95	3.59	3.35	3.04	2.77	2.54
45	14.76	8.64	5.52	4.43	3.86	3.51	3.27	2.96	2.70	2.47
50	14.59	8.53	5.44	4.36	3.80	3.45	3.21	2.90	2.64	2.41
60	14.34	8.37	5.32	4.25	3.70	3.35	3.12	2.81	2.55	2.32
70	14.17	8.27	5.25	4.19	3.64	3.30	3.06	2.76	2.49	2.26
80	14.03	8.18	5.18	4.13	3.58	3.24	3.01	2.70	2.44	2.21
90	13.94	8.12	5.14	4.10	3.55	3.21	2.98	2.67	2.41	2.18
100	13.88	8.08	5.11	4.07	3.52	3.19	2.95	2.65	2.39	2.15
150	13.62	7.92	4.99	3.96	3.42	3.09	2.86	2.56	2.30	2.06
200	13.50	7.84	4.93	3.91	3.38	3.04	2.81	2.51	2.25	2.02
300	13.38	7.77	4.88	3.86	3.33	3.00	2.77	2.47	2.21	1.98
500	13.29	7.71	4.84	3.83	3.30	2.97	2.74	2.44	2.18	1.94
1,000	13.22	7.66	4.80	3.80	3.27	2.94	2.71	2.42	2.15	1.92
2,000	13.19	7.64	4.79	3.78	3.26	2.93	2.70	2.40	2.14	1.90

$s = 3$ *Obere 1% Punkte* / **Upper 1% points**

t	r: 1	2	4	6	8	10	12	16	22	32
12	31.94	18.81	12.30	10.08	8.94	8.25	7.79	7.19	6.70	6.28
14	30.35	17.83	11.61	9.48	8.39	7.73	7.28	6.71	6.24	5.83
16	28.76	16.85	10.92	8.88	7.84	7.21	6.78	6.23	5.77	5.38
18	27.17	15.88	10.22	8.28	7.29	6.68	6.27	5.75	5.31	4.93
20	25.59	14.90	9.53	7.69	6.74	6.16	5.77	5.26	4.84	4.48
25	23.28	13.48	8.53	6.82	5.94	5.40	5.04	4.57	4.17	3.83
30	22.08	12.74	8.01	6.37	5.53	5.01	4.66	4.21	3.83	3.50
35	21.26	12.23	7.65	6.07	5.25	4.75	4.40	3.96	3.59	3.27
40	20.69	11.88	7.41	5.86	5.06	4.56	4.23	3.79	3.42	3.10
45	20.26	11.62	7.22	5.70	4.91	4.42	4.09	3.66	3.30	2.98
50	19.92	11.41	7.08	5.57	4.80	4.32	3.99	3.57	3.20	2.89
60	19.43	11.11	6.87	5.39	4.63	4.16	3.84	3.42	3.06	2.75
70	19.12	10.92	6.74	5.28	4.53	4.06	3.74	3.33	2.98	2.66
80	18.85	10.75	6.62	5.18	4.44	3.98	3.66	3.25	2.90	2.59
90	18.69	10.65	6.55	5.12	4.38	3.92	3.61	3.20	2.85	2.54
100	18.56	10.57	6.49	5.07	4.34	3.88	3.57	3.16	2.81	2.50
150	18.08	10.28	6.29	4.90	4.18	3.73	3.42	3.02	2.67	2.37
200	17.85	10.14	6.19	4.82	4.10	3.66	3.35	2.95	2.61	2.30
300	17.64	10.01	6.10	4.74	4.03	3.59	3.29	2.89	2.55	2.24
500	17.48	9.91	6.03	4.68	3.98	3.54	3.24	2.84	2.50	2.19
1,000	17.34	9.83	5.98	4.63	3.93	3.50	3.20	2.80	2.46	2.16
2,000	17.28	9.79	5.95	4.61	3.91	3.48	3.18	2.78	2.45	2.14

$s = 4$ *Obere 5% Punkte* / **Upper 5% points**

t \ r:	1	2	4	6	8	10	12	16	22	32
12	29.37	16.97	10.80	8.70	7.64	7.00	6.57	6.02	5.57	5.19
14	28.15	16.23	10.29	8.27	7.25	6.63	6.21	5.68	5.24	4.87
16	26.94	15.50	9.78	7.84	6.86	6.26	5.85	5.34	4.92	4.56
18	25.72	14.76	9.28	7.41	6.46	5.89	5.50	5.00	4.59	4.24
20	24.51	14.03	8.77	6.98	6.07	5.51	5.14	4.66	4.27	3.93
25	22.70	12.94	8.02	6.34	5.49	4.96	4.61	4.16	3.78	3.46
30	21.72	12.34	7.61	6.00	5.17	4.67	4.32	3.89	3.52	3.20
35	21.04	11.93	7.33	5.76	4.95	4.46	4.12	3.70	3.33	3.02
40	20.56	11.64	7.13	5.59	4.80	4.31	3.98	3.56	3.20	2.90
45	20.19	11.42	6.98	5.46	4.68	4.20	3.87	3.46	3.10	2.80
50	19.90	11.25	6.86	5.36	4.59	4.11	3.79	3.38	3.03	2.72
60	19.49	10.99	6.69	5.21	4.45	3.99	3.67	3.26	2.91	2.61
70	19.22	10.83	6.58	5.12	4.37	3.90	3.59	3.18	2.84	2.54
80	18.98	10.69	6.48	5.03	4.29	3.83	3.52	3.12	2.78	2.47
90	18.84	10.60	6.42	4.98	4.24	3.79	3.48	3.08	2.74	2.44
100	18.73	10.53	6.37	4.94	4.21	3.75	3.44	3.05	2.70	2.40
150	18.31	10.28	6.20	4.79	4.07	3.63	3.32	2.92	2.59	2.29
200	18.10	10.15	6.11	4.72	4.01	3.56	3.26	2.87	2.53	2.23
300	17.92	10.04	6.04	4.66	3.95	3.51	3.20	2.82	2.48	2.18
500	17.77	9.96	5.98	4.61	3.90	3.46	3.16	2.77	2.44	2.14
1,000	17.65	9.88	5.93	4.57	3.86	3.42	3.13	2.74	2.40	2.10
2,000	17.60	9.85	5.91	4.55	3.84	3.41	3.11	2.72	2.39	2.09

$s = 4$ *Obere 1% Punkte* / **Upper 1% points**

t \ r:	1	2	4	6	8	10	12	16	22	32
12	44.96	25.74	16.19	12.95	11.32	10.33	9.66	8.82	8.12	7.53
14	42.46	24.26	15.19	12.12	10.57	9.62	8.99	8.19	7.52	6.96
16	39.96	22.77	14.19	11.29	9.82	8.92	8.32	7.56	6.92	6.39
18	37.46	21.28	13.20	10.46	9.07	8.22	7.65	6.93	6.33	5.82
20	34.95	19.80	12.20	9.62	8.31	7.52	6.98	6.30	5.73	5.24
25	31.33	17.65	10.77	8.42	7.23	6.51	6.01	5.39	4.87	4.42
30	29.47	16.54	10.03	7.81	6.68	5.99	5.52	4.93	4.43	4.00
35	28.19	15.79	9.53	7.40	6.31	5.64	5.19	4.61	4.13	3.71
40	27.30	15.26	9.18	7.11	6.05	5.40	4.96	4.39	3.92	3.51
45	26.64	14.87	8.92	6.89	5.85	5.21	4.78	4.23	3.76	3.36
50	26.12	14.56	8.72	6.72	5.70	5.07	4.65	4.10	3.64	3.24
60	25.37	14.12	8.42	6.48	5.48	4.87	4.45	3.92	3.47	3.08
70	24.90	13.84	8.24	6.33	5.34	4.74	4.33	3.80	3.36	2.97
80	24.48	13.59	8.08	6.19	5.22	4.63	4.22	3.70	3.26	2.87
90	24.23	13.45	7.98	6.11	5.15	4.56	4.16	3.64	3.20	2.82
100	24.03	13.33	7.90	6.04	5.09	4.50	4.10	3.59	3.16	2.77
150	23.30	12.89	7.62	5.81	4.88	4.31	3.91	3.41	2.98	2.60
200	22.95	12.69	7.48	5.69	4.78	4.21	3.83	3.33	2.90	2.52
300	22.63	12.50	7.36	5.59	4.68	4.13	3.74	3.25	2.83	2.45
500	22.38	12.35	7.26	5.51	4.61	4.06	3.68	3.19	2.77	2.39
1,000	22.18	12.23	7.18	5.45	4.55	4.00	3.63	3.14	2.72	2.35
2,000	22.08	12.17	7.14	5.42	4.53	3.98	3.60	3.12	2.70	2.33

Tafel 3 (Forts.) Table 3 (cont.)

$s = 5$ *Obere 5% Punkte* / **Upper 5% points**

t \ r:	1	2	4	6	8	10	12	16	22	32
12	39.52	22.33	13.76	10.88	9.42	8.54	7.95	7.21	6.59	6.07
14	37.70	21.26	13.06	10.30	8.90	8.06	7.49	6.78	6.18	5.68
16	35.87	20.19	12.36	9.71	8.38	7.57	7.03	6.34	5.77	5.29
18	34.05	19.12	11.65	9.13	7.86	7.08	6.57	5.91	5.36	4.90
20	32.23	18.05	10.95	8.55	7.34	6.60	6.10	5.48	4.96	4.51
25	29.53	16.47	9.91	7.69	6.57	5.88	5.42	4.83	4.35	3.93
30	28.06	15.61	9.35	7.23	6.15	5.49	5.05	4.49	4.02	3.62
35	27.05	15.01	8.96	6.90	5.86	5.22	4.79	4.25	3.79	3.40
40	26.33	14.60	8.68	6.68	5.66	5.03	4.61	4.08	3.63	3.24
45	25.79	14.28	8.47	6.50	5.50	4.89	4.48	3.95	3.50	3.12
50	25.37	14.03	8.31	6.37	5.38	4.78	4.37	3.85	3.41	3.03
60	24.75	13.66	8.07	6.17	5.20	4.61	4.21	3.70	3.27	2.89
70	24.35	13.43	7.92	6.05	5.09	4.51	4.11	3.60	3.18	2.80
80	24.00	13.23	7.79	5.94	4.99	4.42	4.02	3.52	3.10	2.73
90	23.79	13.11	7.71	5.87	4.93	4.36	3.97	3.47	3.05	2.68
100	23.62	13.00	7.64	5.82	4.88	4.31	3.93	3.43	3.01	2.64
150	23.00	12.64	7.40	5.62	4.71	4.15	3.77	3.28	2.86	2.50
200	22.70	12.46	7.29	5.52	4.62	4.07	3.69	3.21	2.80	2.43
300	22.42	12.30	7.18	5.44	4.54	4.00	3.62	3.14	2.73	2.37
500	22.21	12.18	7.10	5.37	4.48	3.94	3.57	3.09	2.68	2.32
1,000	22.03	12.07	7.03	5.31	4.43	3.89	3.52	3.05	2.64	2.28
2,000	21.95	12.03	7.00	5.29	4.41	3.87	3.50	3.03	2.62	2.26

$s = 5$ *Obere 1% Punkte* / **Upper 1% points**

t \ r:	1	2	4	6	8	10	12	16	22	32
12	59.47	33.41	20.42	16.05	13.85	12.52	11.63	10.50	9.58	8.80
14	55.86	31.32	19.08	14.96	12.88	11.62	10.78	9.72	8.84	8.10
16	52.26	29.23	17.74	13.86	11.91	10.72	9.93	8.93	8.10	7.40
18	48.66	27.15	16.39	12.77	10.93	9.83	9.08	8.14	7.36	6.70
20	45.05	25.06	15.05	11.67	9.96	8.93	8.23	7.35	6.62	6.01
25	39.86	22.05	13.12	10.10	8.57	7.64	7.02	6.22	5.56	5.00
30	37.20	20.52	12.14	9.30	7.86	6.99	6.40	5.65	5.03	4.50
35	35.38	19.47	11.46	8.76	7.38	6.54	5.98	5.26	4.66	4.15
40	34.12	18.74	11.00	8.38	7.05	6.24	5.69	4.99	4.41	3.91
45	33.18	18.20	10.65	8.10	6.80	6.01	5.47	4.79	4.22	3.73
50	32.45	17.78	10.39	7.88	6.61	5.83	5.31	4.64	4.07	3.59
60	31.39	17.16	10.00	7.57	6.33	5.57	5.06	4.41	3.86	3.39
70	30.72	16.78	9.76	7.37	6.15	5.41	4.91	4.27	3.73	3.26
80	30.14	16.44	9.54	7.19	6.00	5.27	4.78	4.15	3.61	3.15
90	29.79	16.24	9.41	7.09	5.91	5.19	4.70	4.07	3.54	3.08
100	29.50	16.08	9.31	7.01	5.83	5.12	4.63	4.01	3.49	3.03
150	28.47	15.49	8.93	6.70	5.57	4.87	4.40	3.79	3.28	2.83
200	27.98	15.20	8.75	6.56	5.44	4.75	4.29	3.69	3.18	2.74
300	27.53	14.94	8.59	6.43	5.32	4.64	4.18	3.59	3.09	2.65
500	27.18	14.75	8.46	6.32	5.23	4.56	4.11	3.52	3.02	2.58
1,000	26.90	14.58	8.36	6.24	5.16	4.49	4.04	3.46	2.97	2.53
2,000	26.77	14.51	8.31	6.20	5.12	4.46	4.01	3.43	2.94	2.50

$s = 6$ *Obere 5% Punkte* / **Upper 5% points**

t \ r:	1	2	4	6	8	10	12	16	22	32
12	50.73	28.22	16.99	13.22	11.32	10.18	9.41	8.45	7.65	6.99
14	48.18	26.76	16.06	12.46	10.66	9.57	8.83	7.91	7.15	6.52
16	45.63	25.29	15.13	11.71	9.99	8.95	8.26	7.38	6.66	6.05
18	43.09	23.83	14.20	10.96	9.33	8.34	7.68	6.85	6.16	5.58
20	40.54	22.37	13.27	10.21	8.67	7.73	7.10	6.31	5.66	5.10
25	36.77	20.20	11.90	9.10	7.69	6.83	6.25	5.53	4.92	4.41
30	34.73	19.04	11.16	8.50	7.16	6.34	5.80	5.10	4.52	4.03
35	33.32	18.23	10.65	8.09	6.79	6.01	5.48	4.81	4.25	3.77
40	32.33	17.66	10.28	7.80	6.54	5.77	5.25	4.60	4.05	3.58
45	31.58	17.23	10.01	7.58	6.34	5.59	5.08	4.44	3.90	3.44
50	30.99	16.89	9.80	7.40	6.19	5.45	4.95	4.32	3.79	3.33
60	30.13	16.40	9.49	7.15	5.97	5.24	4.76	4.14	3.62	3.17
70	29.58	16.08	9.29	6.99	5.82	5.11	4.63	4.02	3.51	3.07
80	29.10	15.81	9.11	6.85	5.70	5.00	4.53	3.92	3.41	2.97
90	28.81	15.64	9.01	6.76	5.62	4.93	4.46	3.86	3.36	2.92
100	28.57	15.50	8.92	6.69	5.56	4.87	4.41	3.81	3.31	2.87
150	27.71	15.01	8.61	6.44	5.34	4.67	4.21	3.63	3.14	2.71
200	27.30	14.77	8.46	6.32	5.23	4.57	4.12	3.54	3.05	2.63
300	26.92	14.56	8.32	6.21	5.14	4.48	4.03	3.46	2.98	2.55
500	26.63	14.39	8.22	6.13	5.06	4.41	3.97	3.40	2.92	2.50
1,000	26.38	14.25	8.13	6.05	5.00	4.35	3.91	3.35	2.87	2.45
2,000	26.27	14.18	8.09	6.02	4.97	4.32	3.89	3.32	2.85	2.43

$s = 6$ *Obere 1% Punkte* / **Upper 1% points**

t \ r:	1	2	4	6	8	10	12	16	22	32
12	75.49	41.82	25.02	19.39	16.56	14.85	13.71	12.27	11.08	10.09
14	70.59	39.04	23.29	18.00	15.34	13.74	12.67	11.31	10.20	9.26
16	65.70	36.26	21.55	16.61	14.13	12.63	11.62	10.36	9.31	8.43
18	60.80	33.48	19.81	15.22	12.91	11.52	10.58	9.40	8.43	7.60
20	55.91	30.69	18.08	13.83	11.70	10.40	9.54	8.44	7.54	6.78
25	48.87	26.70	15.58	11.84	9.96	8.81	8.05	7.08	6.27	5.59
30	45.29	24.67	14.33	10.84	9.08	8.01	7.30	6.39	5.64	4.99
35	42.85	23.29	13.47	10.16	8.48	7.47	6.79	5.92	5.20	4.59
40	41.17	22.34	12.88	9.69	8.07	7.09	6.43	5.60	4.90	4.31
45	39.91	21.62	12.43	9.33	7.77	6.81	6.17	5.36	4.68	4.10
50	38.93	21.07	12.09	9.06	7.53	6.60	5.97	5.17	4.50	3.93
60	37.51	20.27	11.60	8.67	7.19	6.28	5.68	4.90	4.25	3.70
70	36.63	19.77	11.29	8.42	6.97	6.09	5.49	4.73	4.10	3.55
80	35.85	19.33	11.01	8.21	6.78	5.92	5.33	4.59	3.96	3.42
90	35.38	19.06	10.85	8.08	6.67	5.81	5.24	4.50	3.88	3.34
100	35.00	18.85	10.72	7.97	6.58	5.73	5.16	4.43	3.81	3.28
150	33.63	18.07	10.24	7.59	6.25	5.43	4.87	4.17	3.57	3.05
200	32.97	17.71	10.02	7.41	6.09	5.29	4.74	4.04	3.45	2.94
300	32.38	17.37	9.81	7.25	5.95	5.16	4.62	3.93	3.35	2.84
500	31.93	17.11	9.65	7.12	5.84	5.06	4.52	3.84	3.27	2.76
1,000	31.55	16.90	9.52	7.02	5.75	4.97	4.44	3.77	3.20	2.70
2,000	31.37	16.80	9.46	6.97	5.71	4.93	4.41	3.74	3.17	2.67

$s = 7$ *Obere 5% Punkte* / **Upper 5% points**

t \ r:	1	2	4	6	8	10	12	16	22	32
12	62.95	34.62	20.47	15.73	13.35	11.91	10.95	9.75	8.75	7.92
14	59.57	32.71	19.29	14.79	12.53	11.17	10.25	9.11	8.15	7.37
16	56.19	30.80	18.10	13.85	11.71	10.42	9.55	8.47	7.56	6.81
18	52.81	28.89	16.92	12.90	10.89	9.67	8.85	7.83	6.97	6.26
20	49.44	26.98	15.74	11.96	10.06	8.92	8.15	7.18	6.38	5.71
25	44.44	24.15	13.99	10.57	8.85	7.81	7.12	6.24	5.51	4.89
30	41.75	22.63	13.05	9.83	8.20	7.22	6.56	5.73	5.04	4.45
35	39.88	21.58	12.40	9.31	7.75	6.81	6.18	5.38	4.71	4.14
40	38.56	20.83	11.94	8.95	7.44	6.52	5.91	5.13	4.48	3.93
45	37.56	20.27	11.59	8.67	7.20	6.30	5.70	4.94	4.31	3.76
50	36.78	19.83	11.32	8.46	7.01	6.13	5.54	4.79	4.17	3.63
60	35.63	19.19	10.93	8.14	6.74	5.88	5.31	4.58	3.97	3.45
70	34.92	18.79	10.68	7.95	6.57	5.73	5.16	4.44	3.84	3.33
80	34.29	18.43	10.46	7.77	6.41	5.59	5.03	4.32	3.73	3.22
90	33.92	18.22	10.33	7.67	6.32	5.50	4.95	4.25	3.66	3.16
100	33.60	18.04	10.22	7.58	6.25	5.43	4.89	4.19	3.61	3.10
150	32.43	17.40	9.83	7.27	5.97	5.19	4.65	3.98	3.41	2.91
200	31.95	17.10	9.63	7.12	5.84	5.07	4.54	3.87	3.31	2.82
300	31.42	16.81	9.46	6.98	5.72	4.96	4.44	3.78	3.22	2.73
500	31.03	16.59	9.33	6.87	5.63	4.87	4.36	3.70	3.15	2.67
1,000	30.71	16.41	9.21	6.78	5.55	4.80	4.29	3.64	3.09	2.61
2,000	30.58	16.33	9.16	6.74	5.52	4.77	4.26	3.61	3.07	2.59

$s = 7$ *Obere 1% Punkte* / **Upper 1% points**

t \ r:	1	2	4	6	8	10	12	16	22	32
12	93.02	50.98	30.00	22.97	19.45	17.32	15.90	14.12	12.65	11.42
14	86.65	47.41	27.82	21.26	17.97	15.98	14.65	12.98	11.61	10.46
16	80.28	43.85	25.64	19.55	16.48	14.64	13.40	11.85	10.57	9.49
18	73.90	40.28	23.47	17.83	15.00	13.29	12.15	10.71	9.53	8.53
20	67.53	36.71	21.29	16.12	13.52	11.95	10.90	9.58	8.48	7.56
25	58.39	31.60	18.18	13.67	11.40	10.03	9.11	7.95	7.00	6.18
30	53.78	29.02	16.61	12.44	10.34	9.07	8.22	7.14	6.25	5.50
35	50.63	27.26	15.54	11.61	9.62	8.42	7.61	6.59	5.75	5.03
40	48.46	26.05	14.81	11.03	9.12	7.97	7.19	6.21	5.40	4.70
45	46.84	25.15	14.26	10.60	8.75	7.64	6.88	5.93	5.14	4.46
50	45.58	24.45	13.84	10.27	8.47	7.38	6.64	5.71	4.94	4.28
60	43.76	23.43	13.23	9.79	8.05	7.00	6.29	5.40	4.65	4.01
70	42.63	22.80	12.85	9.49	7.80	6.77	6.08	5.20	4.47	3.84
80	41.63	22.25	12.51	9.23	7.57	6.56	5.89	5.03	4.31	3.69
90	41.04	21.92	12.31	9.07	7.44	6.44	5.77	4.92	4.21	3.60
100	40.55	21.64	12.14	8.94	7.33	6.34	5.68	4.84	4.13	3.53
150	38.80	20.67	11.56	8.48	6.93	5.98	5.35	4.54	3.85	3.27
200	37.96	20.21	11.28	8.27	6.74	5.82	5.19	4.39	3.71	3.14
300	37.21	19.78	11.02	8.07	6.57	5.66	5.05	4.26	3.60	3.03
500	36.63	19.46	10.83	7.92	6.44	5.54	4.93	4.16	3.51	2.94
1,000	36.15	19.19	10.67	7.79	6.33	5.44	4.84	4.08	3.43	2.87
2,000	35.92	19.07	10.59	7.73	6.28	5.40	4.80	4.04	3.40	2.83

Tafel 3 (Forts.) Table 3 (cont.)

$s = 8$ *Obere 5% Punkte* / **Upper 5% points**

t \ r:	1	2	4	6	8	10	12	16	22	32
12	76.34	41.58	24.22	18.41	15.51	13.74	12.57	11.10	9.90	8.88
14	71.99	39.16	22.75	17.26	14.51	12.85	11.74	10.34	9.21	8.24
16	67.65	36.73	21.28	16.11	13.52	11.95	10.90	9.59	8.52	7.60
18	63.30	34.31	19.81	14.96	12.53	11.05	10.07	8.83	7.82	6.96
20	58.95	31.88	18.35	13.81	11.53	10.15	9.24	8.08	7.13	6.32
25	52.53	28.31	16.18	12.11	10.07	8.83	8.01	6.97	6.11	5.38
30	49.09	26.39	15.02	11.21	9.29	8.13	7.35	6.37	5.56	4.87
35	46.70	25.06	14.21	10.58	8.75	7.64	6.89	5.96	5.18	4.52
40	45.00	24.12	13.65	10.13	8.36	7.30	6.57	5.67	4.91	4.27
45	43.74	23.42	13.22	9.80	8.08	7.03	6.33	5.45	4.71	4.08
50	42.77	22.86	12.89	9.54	7.85	6.83	6.14	5.28	4.55	3.94
60	41.29	22.60	12.40	9.16	7.52	6.53	5.87	5.02	4.32	3.72
70	40.37	21.55	12.10	8.92	7.32	6.35	5.69	4.87	4.17	3.58
80	39.56	21.10	11.83	8.71	7.13	6.18	5.54	4.72	4.04	3.46
90	39.08	20.83	11.67	8.58	7.02	6.08	5.45	4.64	3.97	3.39
100	38.68	20.61	11.53	8.47	6.93	6.00	5.37	4.57	3.90	3.33
150	37.23	19.80	11.50	8.10	6.61	5.70	5.09	4.32	3.67	3.11
200	36.56	19.42	10.82	7.91	6.45	5.56	4.96	4.20	3.56	3.01
300	35.94	19.06	10.60	7.75	6.31	5.43	4.84	4.09	3.46	2.91
500	35.42	18.79	10.43	7.62	6.19	5.33	4.75	4.00	3.38	2.83
1,000	35.01	18.56	10.30	7.51	6.10	5.25	4.67	3.93	3.31	2.77
2,000	34.84	18.46	10.24	7.46	6.06	5.21	4.63	3.90	3.28	2.74

$s = 8$ *Obere 1% Punkte* / **Upper 1% points**

t \ r:	1	2	4	6	8	10	12	16	22	32
12	112.06	60.90	35.34	26.80	22.52	19.94	18.22	16.05	14.28	12.79
14	104.03	56.45	32.68	24.73	20.75	18.35	16.74	14.73	13.07	11.68
16	96.00	52.01	30.02	22.66	18.98	16.75	15.27	13.40	11.87	10.58
18	87.97	47.56	27.36	20.60	17.21	15.16	13.79	12.08	10.66	9.47
20	79.94	43.12	24.70	18.53	15.43	13.57	12.32	10.75	9.46	8.37
25	68.44	36.76	20.90	15.58	12.91	11.30	10.22	8.86	7.74	6.79
30	62.66	33.57	18.99	14.11	11.65	10.17	9.17	7.92	6.88	6.00
35	58.72	31.39	17.70	13.10	10.79	9.40	8.46	7.28	6.30	5.47
40	56.01	29.90	16.81	12.42	10.20	8.87	7.97	6.84	5.90	5.10
45	53.98	28.78	16.14	11.90	9.77	8.47	7.61	6.51	5.60	4.83
50	52.42	27.92	15.63	11.51	9.43	8.17	7.33	6.26	5.37	4.62
60	50.16	26.67	14.89	10.93	8.94	7.73	6.92	5.90	5.04	4.31
70	48.75	25.90	14.43	10.58	8.64	7.46	6.67	5.67	4.83	4.12
80	47.51	25.21	14.02	10.26	8.37	7.22	6.45	5.47	4.65	3.95
90	46.77	24.81	13.78	10.08	8.21	7.08	6.32	5.35	4.55	3.85
100	46.16	24.47	13.58	9.92	8.08	6.96	6.21	5.25	4.46	3.77
150	43.99	23.28	12.88	9.38	7.61	6.54	5.82	4.90	4.14	3.48
200	42.96	22.71	12.54	9.12	7.39	6.34	5.64	4.74	3.99	3.34
300	42.02	22.19	12.24	8.88	7.19	6.16	5.47	4.59	3.85	3.21
500	41.30	21.80	12.00	8.70	7.04	6.02	5.34	4.47	3.74	3.11
1,000	40.70	21.47	11.81	8.55	6.91	5.91	5.24	4.38	3.66	3.03
2,000	40.43	21.32	11.71	8.48	6.85	5.86	5.19	4.33	3.62	2.99

$s = 9$ *Obere 5% Punkte* / **Upper 5% points**

t \ r:	1	2	4	6	8	10	12	16	22	32
12	90.90	49.05	28.22	21.28	17.78	15.68	14.29	12.52	11.08	9.90
14	85.44	46.05	26.44	19.90	16.60	14.62	13.31	11.65	10.29	9.16
16	79.97	43.06	24.66	18.52	15.42	13.56	12.33	10.77	9.49	8.43
18	74.51	40.07	22.88	17.14	14.25	12.51	11.35	9.89	8.69	7.70
20	69.05	37.07	21.10	15.76	13.07	11.45	10.37	9.01	7.89	6.96
25	61.02	32.67	18.48	13.72	11.34	9.89	8.93	7.72	6.72	5.88
30	56.75	30.31	17.07	12.64	10.41	9.06	8.16	7.03	6.09	5.30
35	53.78	28.67	16.10	11.89	9.77	8.49	7.63	6.55	5.66	4.90
40	51.67	27.53	15.42	11.36	9.32	8.09	7.26	6.22	5.35	4.62
45	50.10	26.66	14.90	10.96	8.98	7.78	6.98	5.96	5.12	4.41
50	48.88	25.98	14.50	10.65	8.71	7.54	6.76	5.77	4.94	4.24
60	47.07	24.99	13.92	10.20	8.33	7.19	6.43	5.47	4.68	4.00
70	45.96	24.37	13.55	9.91	8.08	6.98	6.23	5.29	4.51	3.84
80	44.98	23.82	13.22	9.66	7.87	6.78	6.05	5.13	4.36	3.70
90	44.38	23.49	13.02	9.51	7.74	6.66	5.94	5.03	4.27	3.62
100	43.87	23.21	12.86	9.38	7.63	6.57	5.85	4.95	4.20	3.56
150	42.06	22.23	12.28	8.93	7.24	6.22	5.54	4.67	3.94	3.31
200	41.26	21.77	11.99	8.71	7.06	6.06	5.38	4.53	3.81	3.19
300	40.39	21.32	11.74	8.52	6.89	5.91	5.24	4.40	3.69	3.08
500	39.82	20.99	11.54	8.37	6.76	5.79	5.13	4.30	3.60	3.00
1,000	39.29	20.71	11.38	8.24	6.65	5.69	5.04	4.22	3.53	2.93
2,000	39.09	20.59	11.30	8.18	6.60	5.65	5.00	4.18	3.49	2.89

$s = 9$ *Obere 1% Punkte* / **Upper 1% points**

t \ r:	1	2	4	6	8	10	12	16	22	32
12	132.59	71.56	41.07	30.88	25.77	22.70	20.65	18.08	15.96	14.22
14	122.72	66.15	37.87	28.43	23.69	20.84	18.94	16.55	14.59	12.96
16	112.85	60.73	34.68	25.98	21.61	18.98	17.23	15.02	13.21	11.70
18	102.98	55.32	31.49	23.52	19.53	17.12	15.51	13.50	11.83	10.45
20	93.12	49.91	28.30	21.07	17.44	15.26	13.80	11.97	10.46	9.19
25	79.01	42.18	23.74	17.58	14.48	12.61	11.36	9.79	8.50	7.41
30	71.94	38.31	21.47	15.84	13.00	11.30	10.15	8.71	7.52	6.52
35	67.13	35.68	19.91	14.65	12.00	10.40	9.33	7.98	6.86	5.92
40	63.83	33.88	18.87	13.84	11.32	9.79	8.77	7.48	6.41	5.51
45	61.36	32.53	18.08	13.24	10.80	9.33	8.35	7.10	6.07	5.20
50	59.45	31.48	17.47	12.77	10.41	8.98	8.02	6.82	5.81	4.96
60	56.70	29.98	16.59	12.10	9.84	8.47	7.56	6.40	5.44	4.62
70	54.99	29.05	16.05	11.68	9.49	8.16	7.27	6.14	5.21	4.41
80	53.48	28.23	15.56	11.32	9.17	7.88	7.01	5.91	5.00	4.22
90	52.59	27.74	15.28	11.10	8.99	7.72	6.86	5.78	4.88	4.11
100	51.85	27.34	15.04	10.92	8.84	7.58	6.74	5.67	4.78	4.02
150	49.21	25.90	14.20	10.28	8.30	7.10	6.29	5.27	4.42	3.69
200	47.97	25.22	13.81	9.97	8.04	6.87	6.08	5.08	4.25	3.53
300	46.83	24.60	13.44	9.70	7.81	6.66	5.89	4.91	4.10	3.39
500	45.96	24.12	13.17	9.49	7.63	6.50	5.75	4.78	3.98	3.28
1,000	45.23	23.73	12.94	9.31	7.48	6.37	5.62	4.67	3.88	3.19
2,000	44.91	23.55	12.83	9.23	7.41	6.31	5.57	4.62	3.83	3.15

Tafel 3 (Forts.) Table 3 (**cont.**)

$s = 10$ *Obere 5% Punkte* / **Upper 5% points**

t \ r:	1	2	4	6	8	10	12	16	22	32
12	106.38	57.09	32.50	24.32	20.21	17.73	16.06	14.01	12.32	11.06
14	99.72	53.46	30.38	22.68	18.83	16.50	14.94	13.01	11.41	10.20
16	93.06	49.84	28.25	21.06	17.45	15.27	13.81	12.00	10.50	9.35
18	86.41	46.21	26.13	19.42	16.07	14.04	12.68	10.99	9.60	8.49
20	79.75	42.58	24.00	17.79	14.68	12.81	11.55	9.98	8.69	7.63
25	69.95	37.25	20.88	15.40	12.65	11.00	9.89	8.50	7.35	6.39
30	64.72	34.40	19.21	14.12	11.57	10.04	9.01	7.71	6.64	5.74
35	61.12	32.43	18.06	13.24	10.83	9.37	8.39	7.16	6.15	5.29
40	58.61	31.05	17.25	12.62	10.30	8.90	7.96	6.78	5.80	4.98
45	56.70	30.01	16.64	12.15	9.90	8.55	7.64	6.49	5.54	4.74
50	55.19	29.19	16.16	11.79	9.59	8.27	7.38	6.26	5.33	4.55
60	53.02	28.00	15.46	11.26	9.15	7.87	7.01	5.93	5.04	4.28
70	51.65	27.25	15.02	10.92	8.86	7.62	6.78	5.73	4.85	4.10
80	50.44	26.59	14.63	10.62	8.61	7.39	6.57	5.54	4.68	3.95
90	49.71	26.19	14.40	10.45	8.46	7.26	6.45	5.43	4.58	3.86
100	49.10	25.86	14.21	10.30	8.33	7.15	6.35	5.34	4.50	3.78
150	46.90	24.68	13.52	9.77	7.89	6.75	5.98	5.01	4.20	3.51
200	45.92	24.11	13.18	9.52	7.67	6.55	5.80	4.85	4.06	3.38
300	44.94	23.59	12.88	9.29	7.48	6.38	5.64	4.71	3.93	3.25
500	44.25	23.20	12.65	9.11	7.33	6.24	5.52	4.60	3.83	3.16
1,000	43.62	22.86	12.45	8.96	7.20	6.13	5.41	4.50	3.74	3.08
2,000	43.32	22.71	12.37	8.89	7.14	6.08	5.37	4.46	3.70	3.04

$s = 10$ *Obere 1% Punkte* / **Upper 1% points**

t \ r:	1	2	4	6	8	10	12	16	22	32
12	155.00	82.99	47.17	35.20	29.21	25.61	23.21	20.19	17.73	15.79
14	143.04	76.52	43.40	32.34	26.80	23.46	21.24	18.45	16.17	14.36
16	131.07	70.05	39.63	29.47	24.38	21.32	19.28	16.71	14.61	12.92
18	119.11	63.57	35.86	26.60	21.97	19.17	17.31	14.97	13.05	11.49
20	107.15	57.10	32.09	23.74	19.55	17.03	15.34	13.23	11.49	10.06
25	90.10	47.86	26.72	19.66	16.11	13.98	12.55	10.75	9.28	8.04
30	81.62	43.26	24.05	17.63	14.41	12.46	11.16	9.53	8.18	7.05
35	75.86	40.13	22.24	16.26	13.25	11.44	10.22	8.70	7.44	6.38
40	71.90	37.99	21.00	15.32	12.46	10.74	9.58	8.13	6.93	5.92
45	68.96	36.39	20.07	14.61	11.87	10.21	9.11	7.71	6.55	5.58
50	66.68	35.15	19.36	14.07	11.41	9.81	8.74	7.38	6.26	5.31
60	63.40	33.37	18.33	13.29	10.76	9.23	8.21	6.91	5.84	4.93
70	61.36	32.27	17.69	12.81	10.35	8.87	7.88	6.62	5.58	4.70
80	59.55	31.29	17.13	12.38	9.99	8.55	7.58	6.36	5.35	4.49
90	58.49	30.72	16.79	12.13	9.78	8.37	7.41	6.21	5.22	4.37
100	57.61	30.24	16.52	11.92	9.61	8.21	7.27	6.09	5.10	4.26
150	54.47	28.54	15.54	11.18	8.99	7.66	6.77	5.64	4.70	3.90
200	52.99	27.74	15.08	10.83	8.69	7.40	6.53	5.43	4.51	3.73
300	51.64	27.01	14.66	10.51	8.42	7.16	6.31	5.24	4.34	3.57
500	50.61	26.45	14.33	10.27	8.22	6.98	6.15	5.09	4.21	3.45
1,000	49.75	25.98	14.07	10.07	8.05	6.83	6.01	4.97	4.10	3.35
2,000	49.36	25.77	13.94	9.97	7.97	6.76	5.94	4.91	4.05	3.30

Tafel 3 (Forts.) Table 3 (cont.)

$s = 11$ *Obere 5% Punkte* / **Upper 5% points**

t \ r:	1	2	4	6	8	10	12	16	22	32
12	122.70	65.58	37.03	27.50	22.73	19.86	17.95	15.55	13.58	11.94
14	114.77	61.28	34.54	25.60	21.14	18.45	16.65	14.40	12.56	11.02
16	106.84	56.97	32.04	23.71	19.54	17.04	15.36	13.26	11.54	10.10
18	98.92	52.67	29.54	21.82	17.95	15.62	14.07	12.12	10.52	9.18
20	90.99	48.37	27.05	19.93	16.36	14.21	12.78	10.98	9.50	8.26
25	79.32	42.04	23.38	17.15	14.02	12.14	10.88	9.30	8.00	6.90
30	73.10	38.66	21.43	15.66	12.77	11.03	9.87	8.40	7.20	6.18
35	68.79	36.33	20.08	14.64	11.91	10.27	9.17	7.79	6.65	5.68
40	65.76	34.69	19.13	13.92	11.31	9.73	8.68	7.35	6.26	5.33
45	63.48	33.45	18.41	13.38	10.85	9.33	8.31	7.03	5.96	5.07
50	61.70	32.48	17.85	12.96	10.50	9.01	8.02	6.77	5.73	4.86
60	59.12	31.08	17.04	12.34	9.98	8.55	7.60	6.40	5.40	4.56
70	57.49	30.19	16.53	11.95	9.65	8.26	7.33	6.16	5.19	4.37
80	56.02	29.41	16.07	11.61	9.36	8.01	7.10	5.95	5.00	4.19
90	55.15	28.94	15.80	11.40	9.19	7.85	6.96	5.83	4.89	4.09
100	54.43	28.55	15.58	11.23	9.05	7.73	6.84	5.73	4.80	4.01
150	51.86	27.15	14.77	10.62	8.53	7.27	6.42	5.35	4.47	3.70
200	50.63	26.48	14.38	10.33	8.29	7.05	6.23	5.18	4.31	3.56
300	49.49	25.87	14.03	10.06	8.06	6.85	6.04	5.02	4.16	3.42
500	48.62	25.40	13.76	9.85	7.89	6.70	5.90	4.89	4.05	3.32
1,000	47.90	25.00	13.53	9.68	7.74	6.57	5.78	4.78	3.95	3.23
2,000	47.56	24.82	13.42	9.60	7.68	6.51	5.73	4.74	3.91	3.19

$s = 11$ *Obere 1% Punkte* / **Upper 1% points**

t \ r:	1	2	4	6	8	10	12	16	22	32
12	178.29	95.18	53.65	39.78	32.84	28.67	25.88	22.40	19.54	17.15
14	164.18	87.55	49.26	36.47	30.07	26.22	23.65	20.43	17.79	15.58
16	150.07	79.93	44.87	33.16	27.29	23.77	21.42	18.47	16.05	14.02
18	135.96	72.30	40.48	29.85	24.52	21.32	19.18	16.50	14.30	12.46
20	121.85	64.67	36.09	26.54	21.75	18.87	16.95	14.54	12.56	10.90
25	101.72	53.81	29.84	21.83	17.81	15.39	13.78	11.75	10.08	8.68
30	91.71	48.41	26.74	19.49	15.86	13.67	12.21	10.37	8.86	7.58
35	84.91	44.74	24.64	17.91	14.54	12.51	11.15	9.44	8.03	6.84
40	80.26	42.23	23.20	16.83	13.63	11.71	10.42	8.80	7.46	6.33
45	76.78	40.36	22.13	16.03	12.96	11.12	9.88	8.33	7.04	5.96
50	74.10	38.92	21.30	15.40	12.44	10.66	9.46	7.96	6.72	5.67
60	70.25	36.84	20.11	14.51	11.70	10.00	8.87	7.43	6.25	5.25
70	67.85	35.55	19.37	13.96	11.24	9.59	8.49	7.11	5.96	4.99
80	65.73	34.41	18.72	13.47	10.83	9.23	8.17	6.82	5.70	4.76
90	64.48	33.74	18.34	13.18	10.59	9.02	7.97	6.65	5.55	4.62
100	63.45	33.18	18.02	12.94	10.39	8.85	7.81	6.51	5.43	4.51
150	59.78	31.21	16.89	12.09	9.68	8.22	7.25	6.01	4.99	4.11
200	58.04	30.27	16.35	11.69	9.35	7.93	6.98	5.78	4.78	3.92
300	56.45	29.42	15.87	11.33	9.04	7.66	6.73	5.56	4.58	3.75
500	55.25	28.77	15.50	11.05	8.81	7.46	6.55	5.40	4.44	3.61
1,000	54.25	28.23	15.19	10.82	8.62	7.29	6.39	5.26	4.31	3.50
2,000	53.77	27.98	15.05	10.71	8.53	7.21	6.32	5.20	4.26	3.44

Tafel 3 (Forts.) Table 3 (cont.)

$s = 12$ *Obere 5% Punkte* / **Upper 5% points**

t \ r:	1	2	4	6	8	10	12	16	22	32
12	140.51	74.66	41.84	30.88	25.40	22.11	19.91	17.16	14.90	13.01
14	131.11	69.61	38.94	28.70	23.58	20.50	18.45	15.87	13.76	11.99
16	121.72	64.56	36.04	26.53	21.76	18.90	16.98	14.59	12.62	10.97
18	112.32	59.51	33.15	24.35	19.94	17.29	15.52	13.30	11.48	9.95
20	102.93	54.47	30.25	22.17	18.12	15.68	14.05	12.01	10.34	8.93
25	89.13	47.05	26.00	18.97	15.44	13.32	11.90	10.13	8.66	7.44
30	81.80	43.10	23.73	17.26	14.02	12.07	10.76	9.12	7.77	6.64
35	76.73	40.37	22.17	16.09	13.04	11.20	9.97	8.43	7.16	6.08
40	73.17	38.45	21.07	15.26	12.35	10.59	9.42	7.94	6.72	5.70
45	70.48	37.00	20.24	14.64	11.83	10.13	9.00	7.57	6.40	5.40
50	68.39	35.88	19.60	14.15	11.42	9.77	8.67	7.29	6.14	5.17
60	65.34	34.23	18.66	13.45	10.83	9.25	8.20	6.87	5.77	4.84
70	63.42	33.20	18.07	13.00	10.46	8.93	7.90	6.61	5.54	4.63
80	61.71	32.28	17.54	12.61	10.13	8.64	7.63	6.37	5.33	4.44
90	60.69	31.73	17.23	12.37	9.93	8.46	7.48	6.23	5.20	4.33
100	59.85	31.28	16.97	12.18	9.77	8.32	7.34	6.12	5.10	4.24
150	56.83	29.65	16.04	11.48	9.18	7.80	6.87	5.70	4.73	3.90
200	55.38	28.87	15.59	11.14	8.90	7.56	6.65	5.51	4.55	3.74
300	54.05	28.16	15.18	10.83	8.65	7.33	6.44	5.32	4.39	3.59
500	53.03	27.61	14.87	10.60	8.45	7.15	6.28	5.18	4.27	3.48
1,000	52.18	27.15	14.60	10.40	8.29	7.01	6.15	5.06	4.16	3.37
2,000	51.80	26.93	14.48	10.31	8.21	6.94	6.09	5.00	4.10	3.32

$s = 12$ *Obere 1% Punkte* / **Upper 1% points**

t \ r:	1	2	4	6	8	10	12	16	22	32
12	203.70	108.13	60.51	44.61	36.66	31.88	28.69	24.70	21.42	18.69
14	187.13	99.26	55.45	40.82	33.51	29.10	26.17	22.50	19.48	16.96
16	170.57	90.39	50.39	37.04	30.35	26.33	23.66	20.30	17.54	15.24
18	154.01	81.52	45.33	33.25	27.20	23.56	21.14	18.10	15.60	13.51
20	137.44	72.65	40.27	29.46	24.05	20.79	18.62	15.90	13.66	11.78
25	113.90	60.02	33.08	24.08	19.57	16.86	15.05	12.77	10.90	9.33
30	102.26	53.77	29.52	21.42	17.36	14.92	13.29	11.24	9.55	8.13
35	94.34	49.52	27.11	19.62	15.87	13.60	12.09	10.19	8.63	7.31
40	88.90	46.62	25.46	18.39	14.84	12.71	11.28	9.48	8.00	6.76
45	84.86	44.46	24.23	17.47	14.08	12.04	10.67	8.96	7.54	6.34
50	81.74	42.79	23.29	16.77	13.50	11.53	10.21	8.55	7.18	6.02
60	77.26	40.39	21.93	15.75	12.65	10.79	9.54	7.96	6.66	5.56
70	74.49	38.90	21.09	15.13	12.13	10.33	9.12	7.60	6.34	5.28
80	72.04	37.59	20.34	14.57	11.67	9.92	8.75	7.28	6.06	5.03
90	70.59	36.81	19.90	14.24	11.40	9.69	8.54	7.09	5.90	4.88
100	69.39	36.17	19.54	13.97	11.18	9.49	8.36	6.94	5.76	4.76
150	65.13	33.89	18.25	13.01	10.38	8.79	7.73	6.38	5.27	4.32
200	63.12	32.82	17.64	12.56	10.00	8.46	7.43	6.12	5.04	4.11
300	61.29	31.84	17.09	12.14	9.66	8.16	7.16	5.88	4.83	3.92
500	59.89	31.09	16.66	11.83	9.40	7.93	6.95	5.70	4.67	3.78
1,000	58.73	30.47	16.31	11.57	9.18	7.74	6.77	5.55	4.53	3.65
2,000	58.19	30.18	16.15	11.45	9.08	7.65	6.69	5.48	4.47	3.59

$s = 13$ *Obere 5% Punkte* / **Upper 5% points**

t \ r:	1	2	4	6	8	10	12	16	22	32
12	159.15	84.25	46.91	34.44	28.21	24.46	21.96	18.84	16.27	14.13
14	148.21	78.41	43.58	31.96	26.14	22.65	20.32	17.40	15.00	13.00
16	137.28	72.56	40.26	29.47	24.08	20.84	18.67	15.96	13.74	11.88
18	126.34	66.71	36.93	26.99	22.01	19.02	17.02	14.53	12.47	10.75
20	115.41	60.86	33.60	24.50	19.95	17.21	15.38	13.09	11.21	9.63
25	99.36	52.27	28.72	20.86	16.92	14.55	12.96	10.98	9.35	7.98
30	90.83	47.70	26.12	18.92	15.31	13.13	11.68	9.86	8.36	7.10
35	84.94	44.55	24.33	17.58	14.19	12.16	10.79	9.08	7.68	6.49
40	80.81	42.33	23.07	16.64	13.41	11.47	10.17	8.54	7.20	6.07
45	77.69	40.66	22.13	15.93	12.83	10.96	9.71	8.13	6.84	5.75
50	75.26	39.36	21.39	15.38	12.37	10.55	9.34	7.81	6.56	5.49
60	71.71	37.46	20.31	14.58	11.70	9.97	8.81	7.35	6.15	5.13
70	69.49	36.27	19.63	14.07	11.28	9.60	8.47	7.06	5.89	4.90
80	67.52	35.21	19.03	13.62	10.91	9.27	8.18	6.80	5.66	4.69
90	66.34	34.58	18.68	13.36	10.69	9.08	8.00	6.64	5.52	4.57
100	65.36	34.05	18.38	13.14	10.50	8.92	7.85	6.51	5.41	4.47
150	61.84	32.17	17.31	12.34	9.84	8.34	7.33	6.05	5.00	4.10
200	60.17	31.28	16.81	11.96	9.53	8.06	7.07	5.83	4.80	3.92
300	58.64	30.45	16.34	11.61	9.24	7.81	6.84	5.63	4.62	3.76
500	57.45	29.82	15.98	11.35	9.01	7.61	6.67	5.48	4.48	3.64
1,000	56.46	29.29	15.68	11.12	8.83	7.44	6.52	5.34	4.37	3.25
2,000	56.01	29.04	15.54	11.02	8.74	7.37	6.45	5.28	4.32	3.48

$s = 13$ *Obere 1% Punkte* / **Upper 1% points**

t \ r:	1	2	4	6	8	10	12	16	22	32
12	230.29	121.82	67.74	49.69	40.66	35.24	31.62	27.09	23.38	20.27
14	211.17	111.62	61.97	45.40	37.10	32.13	28.80	24.64	21.23	18.38
16	192.06	101.41	56.20	41.11	33.55	29.02	25.98	22.19	19.09	16.48
18	172.94	91.21	50.43	36.81	30.00	25.91	23.17	19.75	16.94	14.59
20	153.83	81.01	44.65	32.52	26.44	22.79	20.35	17.30	14.79	12.69
25	126.62	66.50	36.46	26.43	21.40	18.38	16.36	13.83	11.75	10.01
30	113.14	59.33	32.41	23.42	18.92	16.21	14.39	12.13	10.26	8.69
35	104.03	54.46	29.67	21.39	17.23	14.74	13.07	10.97	9.25	7.80
40	97.82	51.14	27.79	20.00	16.09	13.73	12.16	10.18	8.56	7.19
45	93.18	48.67	26.40	18.96	15.23	12.99	11.49	9.60	8.04	6.73
50	89.58	46.76	25.33	18.17	14.58	12.41	10.96	9.15	7.65	6.39
60	84.45	44.02	23.79	17.02	13.63	11.59	10.22	8.50	7.08	5.89
70	81.26	42.32	22.83	16.32	13.05	11.08	9.76	8.10	6.73	5.58
80	78.43	40.82	21.99	15.69	12.53	10.63	9.35	7.75	6.42	5.30
90	76.77	39.93	21.49	15.32	12.23	10.36	9.11	7.54	6.24	5.14
100	75.40	39.20	21.08	15.02	11.98	10.14	8.91	7.37	6.09	5.01
150	70.53	36.61	19.62	13.94	11.08	9.36	8.21	6.76	5.55	4.53
200	68.22	35.38	18.93	13.43	10.66	9.00	7.88	6.47	5.30	4.30
300	66.13	34.26	18.31	12.96	10.28	8.66	7.58	6.21	5.07	4.10
500	64.53	33.41	17.83	12.61	9.99	8.41	7.35	6.01	4.89	3.94
1,000	63.19	32.70	17.43	12.32	9.75	8.19	7.15	5.84	4.75	3.45
2,000	62.59	32.38	17.25	12.18	9.63	8.10	7.07	5.76	4.68	3.75

Tafel 3 (Forts.) Table 3 (cont.)

$s = 14$ *Obere 5% Punkte* / **Upper 5% points**

t \ r:	1	2	4	6	8	10	12	16	22	32
12	178.72	94.38	52.24	38.18	31.15	26.92	24.11	20.57	17.68	15.27
14	166.15	87.68	48.45	35.37	28.83	24.89	22.27	18.98	16.29	14.04
16	153.59	80.97	44.67	32.56	26.50	22.86	20.43	17.39	14.89	12.81
18	141.03	74.27	40.89	29.75	24.18	20.83	18.59	15.79	13.49	11.57
20	128.47	67.56	37.10	26.94	21.85	18.79	16.75	14.20	12.10	10.34
25	110.03	57.71	31.55	22.82	18.44	15.81	14.06	11.86	10.05	8.53
30	100.24	52.48	28.60	20.63	16.63	14.23	12.63	10.62	8.96	7.57
35	93.47	48.87	26.57	19.12	15.39	13.14	11.64	9.76	8.21	6.91
40	88.72	46.33	25.14	18.06	14.51	12.38	10.95	9.16	7.68	6.44
45	85.14	44.42	24.06	17.26	13.85	11.80	10.43	8.71	7.29	6.09
50	82.33	42.93	23.22	16.64	13.34	11.35	10.02	8.35	6.98	5.82
60	78.25	40.76	22.00	15.73	12.59	10.70	9.43	7.84	6.53	5.42
70	75.69	39.40	21.24	15.16	12.12	10.29	9.06	7.52	6.24	5.17
80	73.42	38.19	20.55	14.66	11.70	9.92	8.73	7.23	5.99	4.94
90	72.07	37.47	20.15	14.36	11.45	9.70	8.53	7.06	5.84	4.81
100	70.95	36.87	19.81	14.11	11.25	9.52	8.37	6.91	5.71	4.70
150	66.92	34.73	18.60	13.21	10.51	8.88	7.78	6.41	5.27	4.30
200	65.00	33.70	18.03	12.79	10.15	8.57	7.50	6.16	5.05	4.11
300	63.24	32.76	17.50	12.39	9.83	8.28	7.25	5.94	4.85	3.93
500	61.88	32.04	17.09	12.09	9.58	8.07	7.05	5.77	4.70	3.79
1,000	60.75	31.43	16.75	11.84	9.37	7.88	6.88	5.62	4.58	3.68
2,000	60.22	31.16	16.60	11.72	9.28	7.80	6.81	5.56	4.52	3.63

$s = 14$ *Obere 1% Punkte* / **Upper 1% points**

t \ r:	1	2	4	6	8	10	12	16	22	32
12	258.50	136.31	75.35	55.03	44.85	38.75	34.68	29.58	25.39	21.91
14	236.62	124.68	68.82	50.20	40.87	35.28	31.54	26.87	23.03	19.84
16	214.74	113.05	62.29	45.37	36.90	31.81	28.42	24.16	20.68	17.77
18	192.85	101.41	55.76	40.54	32.92	28.34	25.28	21.46	18.32	15.70
20	170.97	89.78	49.23	35.71	28.94	24.87	22.15	18.75	15.96	13.63
25	139.86	73.25	39.96	28.86	23.29	19.95	17.72	14.92	12.62	10.69
30	124.49	65.10	35.40	25.49	20.52	17.53	15.54	13.04	10.98	9.26
35	114.08	59.57	32.31	23.20	18.64	15.90	14.07	11.77	9.88	8.29
40	107.00	55.80	30.20	21.65	17.36	14.79	13.06	10.90	9.12	7.62
45	101.71	53.00	28.63	20.49	16.41	13.96	12.32	10.26	8.56	7.13
50	97.62	50.83	27.42	19.60	15.68	13.32	11.74	9.76	8.13	6.75
60	91.77	47.73	25.69	18.32	14.63	12.40	10.92	9.05	7.51	6.21
70	88.16	45.81	24.61	17.53	13.98	11.84	10.41	8.61	7.13	5.87
80	84.96	44.11	23.66	16.83	13.40	11.34	9.96	8.22	6.79	5.58
90	83.08	43.11	23.10	16.42	13.06	11.04	9.69	7.99	6.59	5.40
100	81.52	42.28	22.64	16.08	12.79	10.80	9.47	7.80	6.42	5.26
150	75.98	39.34	21.00	14.87	11.79	9.94	8.70	7.13	5.84	4.74
200	73.36	37.96	20.23	14.30	11.33	9.53	8.33	6.82	5.56	4.50
300	70.99	36.70	19.53	13.79	10.90	9.16	8.00	6.53	5.31	4.28
500	69.17	35.74	18.99	13.39	10.58	8.88	7.74	6.31	5.12	4.11
1,000	67.66	34.93	18.55	13.06	10.31	8.65	7.53	6.13	4.96	3.97
2,000	66.97	34.57	18.34	12.91	10.18	8.54	7.44	6.04	4.89	3.90

$s = 15$ *Obere 5% Punkte* / **Upper 5% points**

t \ r:	1	2	4	6	8	10	12	16	22	32
12	199.51	105.05	57.84	42.10	34.22	29.49	26.33	22.38	19.15	16.45
14	185.18	97.43	53.57	38.95	31.63	27.23	24.29	20.62	17.61	15.10
16	170.85	89.80	49.30	35.79	29.03	24.97	22.25	18.86	16.08	13.76
18	156.53	82.18	45.03	32.64	26.43	22.70	20.22	17.10	14.55	12.42
20	142.20	74.55	40.76	29.48	23.83	20.44	18.18	15.34	13.02	11.07
25	121.15	63.37	34.49	24.85	20.02	17.12	15.19	12.76	10.77	9.10
30	109.95	57.43	31.17	22.40	18.01	15.37	13.60	11.39	9.58	8.05
35	102.23	53.33	28.87	20.70	16.61	14.15	12.51	10.45	8.76	7.33
40	96.83	50.45	27.26	19.52	15.64	13.30	11.74	9.79	8.18	6.83
45	92.75	48.29	26.05	18.62	14.90	12.66	11.17	9.29	7.74	6.44
50	89.56	46.60	25.10	17.93	14.33	12.16	10.72	8.90	7.41	6.15
60	84.94	44.14	23.73	16.91	13.50	11.44	10.06	8.33	6.91	5.71
70	82.03	42.60	22.87	16.28	12.97	10.98	9.65	7.98	6.60	5.44
80	79.44	41.22	22.10	15.71	12.51	10.58	9.29	7.66	6.33	5.19
90	77.90	40.41	21.64	15.37	12.23	10.33	9.07	7.48	6.16	5.05
100	76.62	39.73	21.27	15.10	12.00	10.14	8.89	7.32	6.02	4.93
150	72.05	37.30	19.91	14.10	11.18	9.42	8.24	6.76	5.54	4.50
200	69.87	36.14	19.26	13.62	10.78	9.08	7.93	6.49	5.30	4.29
300	67.87	35.08	18.67	13.18	10.42	8.76	7.65	6.25	5.09	4.10
500	66.32	34.26	18.21	12.84	10.15	8.52	7.43	6.06	4.92	3.95
1,000	65.03	33.58	17.83	12.56	9.91	8.32	7.25	5.90	4.78	3.83
2,000	64.43	33.26	17.65	12.43	9.81	8.23	7.17	5.83	4.72	3.77

$s = 15$ *Obere 1% Punkte* / **Upper 1% points**

t \ r:	1	2	4	6	8	10	12	16	22	32
12	288.10	151.52	83.36	60.62	49.24	42.40	37.85	32.14	27.48	23.59
14	263.26	138.37	76.02	55.22	44.82	38.56	34.39	29.17	24.90	21.34
16	238.43	125.22	68.68	49.82	40.39	34.71	30.93	26.20	22.32	19.08
18	213.59	112.07	61.35	44.43	35.96	30.87	27.47	23.22	19.74	16.83
20	188.76	98.92	54.01	39.03	31.53	27.02	24.02	20.25	17.16	14.58
25	153.52	80.26	43.61	31.38	25.25	21.57	19.12	16.04	13.51	11.[illegible]0
30	136.20	71.07	38.49	27.62	22.17	18.90	16.72	13.98	11.72	9.84
35	124.46	64.85	35.03	25.08	20.09	17.09	15.09	12.58	10.52	8.79
40	116.45	60.60	32.67	23.35	18.67	15.86	13.99	11.63	9.70	8.07
45	110.48	57.44	30.92	22.06	17.62	14.95	13.17	10.93	9.08	7.54
50	105.87	55.01	29.57	21.07	16.81	14.25	12.53	10.38	8.62	7.13
60	99.27	51.52	27.63	19.64	15.64	13.24	11.63	9.61	7.94	6.54
70	95.20	49.36	26.43	18.77	14.92	12.61	11.07	9.13	7.52	6.17
80	91.59	47.45	25.37	17.99	14.29	12.06	10.57	8.70	7.16	5.85
90	89.46	46.33	24.74	17.53	13.91	11.74	10.28	8.45	6.94	5.66
100	87.71	45.40	24.23	17.16	13.61	11.47	10.04	8.24	6.76	5.51
150	81.47	42.11	22.40	15.82	12.51	10.52	9.19	7.51	6.13	4.95
200	78.53	40.55	21.54	15.18	11.99	10.07	8.79	7.17	5.83	4.69
300	75.87	39.14	20.76	14.61	11.53	9.67	8.42	6.85	5.56	4.45
500	73.82	38.06	20.16	14.17	11.17	9.36	8.14	6.61	5.35	4.27
1,000	72.12	37.16	19.66	13.81	10.87	9.10	7.91	6.42	5.17	4.12
2,000	71.34	36.75	19.43	13.64	10.73	8.98	7.81	6.32	5.10	4.05

Tafel 3 (Forts.) Table 3 (cont.)

$s = 16$ *Obere 5% Punkte* / **Upper 5% points**

t \ r:	1	2	4	6	8	10	12	16	22	32
12	221.49	116.25	63.71	46.19	37.42	32.16	28.65	24.26	20.66	17.65
14	205.24	107.65	58.92	42.67	34.54	29.66	26.40	22.32	18.99	16.20
16	189.00	99.05	54.13	39.15	31.65	27.15	24.15	20.39	17.31	14.74
18	172.75	90.45	49.34	35.63	28.77	24.65	21.90	18.46	15.64	13.28
20	156.51	81.85	44.56	32.12	25.89	22.15	19.65	16.52	13.96	11.82
25	132.67	69.24	37.53	26.96	21.66	18.48	16.35	13.69	11.51	9.68
30	120.01	62.55	33.81	24.22	19.42	16.53	14.61	12.19	10.21	8.55
35	111.28	57.93	31.25	22.34	17.88	15.19	13.40	11.16	9.31	7.76
40	105.17	54.70	29.45	21.02	16.79	14.25	12.56	10.43	8.68	7.22
45	100.57	52.26	28.09	20.02	15.98	13.55	11.92	9.89	8.21	6.80
50	96.97	50.36	27.03	19.25	15.34	13.00	11.43	9.46	7.84	6.48
60	91.76	47.59	25.50	18.12	14.42	12.19	10.71	8.84	7.30	6.01
70	88.48	45.86	24.53	17.41	13.84	11.69	10.26	8.45	6.97	5.71
80	85.56	44.31	23.67	16.78	13.32	11.24	9.85	8.11	6.67	5.45
90	83.83	43.40	23.16	16.41	13.02	10.98	9.61	7.90	6.49	5.29
100	82.39	42.63	22.74	16.10	12.76	10.76	9.42	7.73	6.34	5.16
150	77.23	39.91	21.23	14.98	11.85	9.97	8.71	7.12	5.81	4.70
200	74.76	38.60	20.50	14.45	11.42	9.59	8.37	6.83	5.55	4.47
300	72.51	37.41	19.84	13.97	11.02	9.24	8.06	6.56	5.32	4.27
500	70.77	36.49	19.33	13.59	10.71	8.98	7.82	6.35	5.14	4.11
1,000	69.32	35.72	18.90	13.28	10.46	8.75	7.62	6.18	4.99	3.97
2,000	68.65	35.37	18.71	13.14	10.34	8.65	7.52	6.10	4.92	3.91

$s = 16$ *Obere 1% Punkte* / **Upper 1% points**

t \ r:	1	2	4	6	8	10	12	16	22	32
12	319.31	167.51	91.73	66.45	53.81	46.21	41.16	34.83	29.64	25.31
14	291.37	152.75	83.54	60.45	48.91	41.97	37.36	31.57	26.83	22.87
16	263.43	137.99	75.36	54.46	44.01	37.73	33.55	28.31	24.02	20.43
18	235.48	123.23	67.17	48.47	39.12	33.49	29.75	25.06	21.21	18.00
20	207.54	108.47	58.99	42.48	34.22	29.25	25.94	21.80	18.40	15.56
25	167.90	87.54	47.38	33.99	27.28	23.25	20.56	17.19	14.43	12.11
30	148.40	77.25	41.69	29.82	23.88	20.31	17.93	14.94	12.49	10.43
35	135.19	70.29	37.84	27.01	21.58	18.32	16.15	13.42	11.17	9.29
40	126.17	65.54	35.21	25.09	20.02	16.97	14.93	12.38	10.28	8.52
45	119.48	62.01	33.26	23.66	18.86	15.96	14.03	11.61	9.62	7.95
50	114.34	59.29	31.76	22.57	17.96	15.19	13.34	11.02	9.11	7.50
60	106.94	55.40	29.61	21.00	16.68	14.08	12.35	10.17	8.38	6.87
70	102.37	52.99	28.28	20.02	15.89	13.40	11.74	9.65	7.93	6.48
80	98.32	50.85	27.10	19.17	15.19	12.80	11.19	9.19	7.53	6.13
90	95.95	49.60	26.41	18.66	14.78	12.44	10.88	8.91	7.30	5.93
100	93.99	48.57	25.84	18.25	14.44	12.15	10.62	8.69	7.10	5.76
150	87.02	44.89	23.81	16.77	13.23	11.11	9.68	7.89	6.42	5.16
200	83.73	43.16	22.86	16.07	12.67	10.62	9.24	7.52	6.09	4.88
300	80.75	41.59	21.99	15.44	12.15	10.17	8.85	7.18	5.80	4.63
500	78.47	40.39	21.33	14.96	11.76	9.83	8.54	6.92	5.57	4.43
1,000	76.58	39.39	20.78	14.55	11.43	9.55	8.29	6.70	5.39	4.27
2,000	75.71	38.94	20.53	14.37	11.28	9.42	8.17	6.60	5.30	4.19

Tafel 3 (Forts.) Table 3 (cont.)

s = 17 *Obere 5% Punkte* / **Upper 5% points**

t \ r:	1	2	4	6	8	10	12	16	22	32
12	244.40	127.99	69.85	50.46	40.77	34.93	31.05	26.20	22.22	18.90
14	226.15	118.35	64.52	46.56	37.58	32.18	28.58	24.09	20.40	17.32
16	207.90	108.72	59.18	42.66	34.39	29.42	26.11	21.97	18.58	15.74
18	189.65	99.09	53.84	38.75	31.20	26.67	23.64	19.86	16.76	14.16
20	171.41	89.45	48.51	34.85	28.02	23.91	21.17	17.75	14.93	12.58
25	144.64	75.33	40.69	29.13	23.35	19.87	17.55	14.65	12.26	10.27
30	130.45	67.84	36.55	26.10	20.88	17.74	15.64	13.01	10.85	9.05
35	120.66	62.68	33.69	24.02	19.17	16.26	14.32	11.88	9.88	8.20
40	113.80	59.06	31.69	22.55	17.98	15.23	13.39	11.09	9.20	7.61
45	108.64	56.34	30.18	21.45	17.08	14.45	12.70	10.50	8.68	7.17
50	104.60	54.21	29.01	20.60	16.38	13.85	12.15	10.03	8.28	6.82
60	98.74	51.12	27.30	19.35	15.36	12.97	11.37	9.36	7.70	6.31
70	95.06	49.18	26.23	18.56	14.72	12.41	10.87	8.93	7.33	5.99
80	91.78	47.46	25.28	17.87	14.16	11.92	10.43	8.56	7.01	5.71
90	89.84	46.43	24.71	17.45	13.82	11.63	10.17	8.33	6.82	5.54
100	88.23	45.58	24.24	17.11	13.54	11.39	9.95	8.14	6.66	5.40
150	82.46	42.54	22.56	15.88	12.54	10.52	9.17	7.48	6.08	4.90
200	79.71	41.08	21.76	15.30	12.06	10.11	8.80	7.16	5.80	4.66
300	77.18	39.75	21.02	14.76	11.62	9.73	8.46	6.87	5.55	4.43
500	75.24	38.73	20.45	14.34	11.28	9.43	8.20	6.64	5.36	4.26
1,000	73.61	37.87	19.98	14.00	11.00	9.19	7.98	6.46	5.19	4.12
2,000	72.86	37.48	19.76	13.84	10.87	9.08	7.88	6.37	5.12	4.05

s = 17 *Obere 1% Punkte* / **Upper 1% points**

t \ r:	1	2	4	6	8	10	12	16	22	32
12	351.98	184.23	100.48	72.54	58.57	50.19	44.60	37.59	31.86	27.08
14	320.74	167.78	91.40	65.92	53.18	45.54	40.43	34.04	28.82	24.45
16	289.50	151.32	82.32	59.30	47.79	40.88	36.27	30.50	25.77	21.82
18	258.26	134.87	73.24	52.68	42.40	36.22	32.11	26.95	22.72	19.19
20	227.02	118.41	64.16	46.06	37.00	31.57	27.94	23.40	19.67	16.56
25	182.72	95.09	51.30	36.69	29.37	24.98	22.05	18.38	15.37	12.85
30	160.97	83.64	44.99	32.09	25.64	21.76	19.17	15.93	13.26	11.04
35	146.25	75.90	40.73	28.99	23.12	19.58	17.23	14.27	11.85	9.81
40	136.22	70.62	37.82	26.88	21.40	18.10	15.90	13.15	10.88	8.98
45	128.76	66.70	35.67	25.31	20.12	17.01	14.92	12.31	10.17	8.36
50	123.01	63.68	34.01	24.10	19.14	16.16	14.17	11.67	9.61	7.89
60	114.78	59.36	31.63	22.38	17.74	14.95	13.09	10.75	8.83	7.21
70	109.69	56.68	30.16	21.31	16.87	14.20	12.42	10.18	8.34	6.79
80	105.19	54.32	28.86	20.36	16.10	13.54	11.83	9.68	7.91	6.42
90	102.55	52.93	28.10	19.81	15.65	13.15	11.48	9.38	7.66	6.20
100	100.38	51.79	27.47	19.35	15.28	12.83	11.20	9.14	7.45	6.02
150	92.63	47.71	25.24	17.73	13.96	11.70	10.18	8.28	6.71	5.38
200	88.98	45.79	24.18	16.96	13.34	11.16	9.70	7.87	6.36	5.08
300	85.67	44.05	23.23	16.27	12.78	10.68	9.27	7.50	6.04	4.80
500	83.13	42.72	22.50	15.74	12.35	10.31	8.94	7.22	5.80	4.59
1,000	81.03	41.62	21.89	15.30	11.99	10.00	8.67	6.99	5.60	4.42
2,000	80.06	41.11	21.62	15.10	11.83	9.86	8.54	6.88	5.51	4.34

$s = 18$ *Obere 5% Punkte* / **Upper 5% points**

t \ r:	1	2	4	6	8	10	12	16	22	32
12	268.48	140.27	76.26	54.90	44.23	37.82	33.54	28.20	23.83	20.30
14	248.09	129.54	70.35	50.60	40.73	34.80	30.84	25.90	21.85	18.57
16	227.70	118.81	64.44	46.29	37.23	31.78	28.14	23.60	19.88	16.85
18	207.31	108.09	58.53	41.99	33.72	28.76	25.44	21.30	17.91	15.12
20	186.92	97.36	52.61	37.69	30.22	25.74	22.74	19.00	15.93	13.39
25	157.04	81.64	43.95	31.38	25.09	21.31	18.79	15.63	13.04	10.88
30	141.21	73.31	39.37	28.04	22.38	18.97	16.70	13.85	11.51	9.56
35	130.29	67.57	36.21	25.74	20.51	17.36	15.26	12.62	10.46	8.65
40	122.65	63.55	34.00	24.13	19.20	16.23	14.25	11.76	9.72	8.01
45	116.90	60.52	32.33	22.92	18.21	15.38	13.49	11.12	9.17	7.53
50	112.40	58.16	31.03	21.98	17.44	14.72	12.89	10.61	8.74	7.16
60	105.88	54.73	29.14	20.60	16.33	13.75	12.03	9.88	8.11	6.62
70	101.78	52.57	27.96	19.74	15.62	13.15	11.49	9.42	7.71	6.27
80	98.14	50.66	26.91	18.97	15.00	12.61	11.01	9.01	7.36	5.97
90	95.98	49.52	26.28	18.52	14.63	12.29	10.73	8.77	7.15	5.79
100	94.19	48.58	25.76	18.14	14.32	12.03	10.49	8.57	6.98	5.64
150	87.76	45.20	23.90	16.79	13.22	11.08	9.64	7.84	6.35	5.10
200	84.68	43.58	23.02	16.15	12.70	10.63	9.24	7.50	6.06	4.84
300	81.87	42.11	22.20	15.55	12.22	10.21	8.87	7.18	5.79	4.60
500	79.71	40.97	21.58	15.10	11.85	9.89	8.58	6.94	5.58	4.42
1,000	77.90	40.02	21.06	14.72	11.54	9.63	8.35	6.73	5.40	4.27
2,000	77.07	39.58	20.82	14.54	11.40	9.50	8.24	6.64	5.32	4.19

$s = 18$ *Obere 1% Punkte* / **Upper 1% points**

t \ r:	1	2	4	6	8	10	12	16	22	32
12	386.24	201.76	109.61	78.91	63.52	54.30	48.14	40.46	34.15	28.99
14	351.51	183.51	99.59	71.62	57.61	49.22	43.60	36.60	30.86	26.14
16	316.77	165.26	89.57	64.34	51.70	44.13	39.07	32.75	27.57	23.30
18	282.03	147.01	79.55	57.06	45.80	39.04	34.53	28.90	24.27	20.45
20	247.29	128.76	69.53	49.78	39.89	33.96	30.00	25.04	20.98	17.61
25	198.08	102.91	55.34	39.47	31.53	26.76	23.58	19.60	16.33	13.60
30	173.94	90.24	48.39	34.43	27.45	23.25	20.45	16.94	14.06	11.65
35	157.62	81.67	43.70	31.03	24.69	20.88	18.33	15.15	12.53	10.34
40	146.51	75.84	40.51	28.72	22.81	19.27	16.90	13.93	11.49	9.45
45	138.25	71.51	38.14	27.00	21.42	18.07	15.83	13.03	10.72	8.79
50	131.90	68.18	36.31	25.68	20.35	17.15	15.01	12.33	10.13	8.28
60	122.78	63.40	33.70	23.78	18.82	15.83	13.84	11.33	9.28	7.55
70	117.15	60.45	32.08	22.61	17.87	15.02	13.11	10.72	8.75	7.10
80	112.17	57.84	30.66	21.58	17.03	14.30	12.47	10.18	8.29	6.70
90	109.25	56.31	29.82	20.97	16.54	13.88	12.10	9.86	8.02	6.47
100	106.85	55.05	29.13	20.47	16.14	13.53	11.79	9.60	7.80	6.28
150	98.29	50.56	26.67	18.70	14.70	12.29	10.68	8.67	7.00	5.59
200	94.26	48.44	25.52	17.86	14.02	11.71	10.17	8.23	6.62	5.27
300	90.59	46.53	24.47	17.10	13.41	11.19	9.70	7.83	6.28	4.98
500	87.80	45.06	23.67	16.53	12.94	10.78	9.34	7.53	6.02	4.75
1,000	85.48	43.85	23.01	16.04	12.55	10.45	9.04	7.27	5.81	4.57
2,000	84.42	43.29	22.70	15.83	12.38	10.30	8.91	7.16	5.71	4.48

$s = 19$ Obere 5% Punkte / **Upper 5% points**

t \ r:	1	2	4	6	8	10	12	16	22	32
12	293.50	153.07	82.91	59.54	47.83	40.82	36.13	30.27	25.48	21.48
14	270.88	141.19	76.40	54.81	43.99	37.52	33.19	27.77	23.35	19.65
16	248.26	129.31	69.89	50.08	40.16	34.22	30.25	25.28	21.22	17.82
18	225.63	117.44	63.37	45.35	36.33	30.92	27.31	22.79	19.09	16.00
20	203.01	105.56	56.86	40.62	32.50	27.62	24.37	20.29	16.96	14.17
25	169.86	88.16	47.32	33.70	26.88	22.79	20.06	16.64	13.84	11.49
30	152.30	78.95	42.28	30.04	23.92	20.24	17.79	14.71	12.19	10.08
35	140.20	72.59	38.80	27.52	21.87	18.48	16.22	13.38	11.05	9.11
40	131.71	68.15	36.36	25.75	20.44	17.25	15.12	12.45	10.26	8.42
45	125.34	64.81	34.53	24.42	19.37	16.33	14.30	11.75	9.66	7.91
50	120.38	62.20	33.10	23.39	18.52	15.60	13.65	11.21	9.19	7.51
60	113.15	58.41	31.02	21.88	17.31	14.55	12.72	10.41	8.52	6.93
70	108.62	56.03	29.72	20.94	16.54	13.90	12.13	9.91	8.09	6.56
80	104.59	53.91	28.56	20.10	15.86	13.31	11.61	9.47	7.71	6.23
90	102.20	52.66	27.87	19.60	15.46	12.96	11.30	9.21	7.49	6.04
100	100.21	51.61	27.30	19.19	15.12	12.67	11.04	8.99	7.30	5.88
150	93.09	47.88	25.26	17.71	13.92	11.64	10.12	8.21	6.63	5.30
200	89.71	46.10	24.29	17.00	13.35	11.15	9.68	7.83	6.31	5.02
300	86.60	44.47	23.39	16.35	12.82	10.70	9.28	7.49	6.02	4.77
500	84.20	43.22	22.71	15.86	12.42	10.35	8.97	7.23	5.79	4.58
1,000	82.20	42.17	22.13	15.44	12.08	10.06	8.71	7.01	5.60	4.41
2,000	81.28	41.69	21.87	15.25	11.93	9.93	8.59	6.91	5.52	4.34

$s = 19$ Obere 1% Punkte / **Upper 1% points**

t \ r:	1	2	4	6	8	10	12	16	22	32
12	422.14	220.07	119.15	85.49	68.66	58.56	51.83	43.42	36.51	30.77
14	383.69	199.93	108.14	77.52	62.21	53.03	46.90	39.25	32.96	27.74
16	345.25	179.78	97.12	69.55	55.76	47.49	41.97	35.08	29.42	24.71
18	306.81	159.64	86.11	61.59	49.32	41.96	37.05	30.91	25.87	21.68
20	268.36	139.49	75.10	53.62	42.87	36.42	32.12	26.74	22.32	18.64
25	213.94	110.98	59.52	42.35	33.76	28.60	25.16	20.85	17.32	14.36
30	187.31	97.04	51.90	36.84	29.31	24.78	21.76	17.98	14.88	12.28
35	169.31	87.61	46.76	33.13	26.30	22.20	19.47	16.04	13.23	10.88
40	157.07	81.20	43.26	30.60	24.26	20.45	17.91	14.73	12.12	9.92
45	147.98	76.44	40.66	28.72	22.75	19.16	16.76	13.75	11.29	9.22
50	140.98	72.77	38.66	27.28	21.58	18.16	15.87	13.00	10.65	8.67
60	130.95	67.53	35.80	25.22	19.91	16.73	14.60	11.93	9.74	7.90
70	124.77	64.29	34.04	23.94	18.89	15.85	13.82	11.27	9.18	7.42
80	119.29	61.42	32.48	22.82	17.98	15.07	13.12	10.68	8.68	6.99
90	116.08	59.74	31.56	22.16	17.44	14.61	12.72	10.34	8.39	6.74
100	113.43	58.36	30.81	21.61	17.00	14.23	12.38	10.06	8.15	6.54
150	104.00	53.43	28.13	19.68	15.44	12.90	11.19	9.06	7.29	5.81
200	99.57	51.11	26.86	18.77	14.71	12.27	10.63	8.58	6.89	5.46
300	95.56	49.01	25.72	17.94	14.04	11.70	10.13	8.16	6.53	5.15
500	92.49	47.41	24.85	17.31	13.54	11.26	9.74	7.83	6.25	4.91
1,000	89.93	46.07	24.12	16.79	13.11	10.90	9.42	7.56	6.02	4.72
2,000	88.77	45.46	23.79	16.55	12.92	10.74	9.27	7.43	5.91	4.62

Tafel 3 (Forts.) Table 3 (cont.)

$s = 20$ Obere 5% Punkte / **Upper 5% points**

t \ r:	1	2	4	6	8	10	12	16	22	32
12	319.69	166.41	89.85	64.33	51.56	43.91	38.79	32.41	27.17	22.97
14	294.70	153.33	82.70	59.17	47.38	40.33	35.60	29.71	24.88	20.99
16	269.71	140.24	75.56	54.00	43.20	36.74	32.42	27.02	22.59	19.00
18	244.71	127.15	68.41	48.83	39.03	33.15	29.23	24.32	20.30	17.01
20	219.72	114.06	61.26	43.66	34.85	29.57	26.04	21.62	18.01	15.02
25	183.11	94.89	50.80	36.09	28.73	24.32	21.37	17.68	14.65	12.13
30	163.75	84.75	45.27	32.09	25.50	21.54	18.90	15.60	12.88	10.61
35	150.40	77.77	41.45	29.34	23.27	19.63	17.20	14.16	11.66	9.57
40	141.07	72.88	38.79	27.41	21.72	18.30	16.01	13.15	10.81	8.84
45	134.05	69.20	36.78	25.96	20.55	17.29	15.12	12.40	10.16	8.29
50	128.56	66.33	35.21	24.83	19.63	16.51	14.42	11.81	9.66	7.86
60	120.59	62.16	32.94	23.19	18.30	15.37	13.41	10.95	8.93	7.24
70	115.60	59.55	31.51	22.16	17.47	14.65	12.77	10.42	8.48	6.85
80	111.15	57.22	30.25	21.24	16.73	14.02	12.21	9.94	8.07	6.50
90	108.52	55.84	29.49	20.70	16.29	13.64	11.87	9.65	7.83	6.29
100	106.33	54.70	28.87	20.25	15.93	13.33	11.60	9.42	7.63	6.12
150	98.49	50.60	26.63	18.63	14.62	12.21	10.60	8.58	6.91	5.51
200	94.76	48.64	25.57	17.86	14.00	11.68	10.13	8.18	6.57	5.21
300	91.34	46.85	24.59	17.16	13.43	11.19	9.69	7.81	6.25	4.94
500	88.70	45.47	23.84	16.61	12.99	10.81	9.36	7.52	6.01	4.73
1,000	86.50	44.32	23.21	16.16	12.63	10.50	9.08	7.29	5.81	4.56
2,000	85.49	43.79	22.92	15.95	12.46	10.35	8.95	7.18	5.72	4.48

$s = 20$ Obere 1% Punkte / **Upper 1% points**

t \ r:	1	2	4	6	8	10	12	16	22	32
12	459.41	239.10	129.04	92.34	73.99	62.97	55.63	46.45	38.92	32.81
14	417.11	216.98	117.00	83.65	66.98	56.97	50.30	41.96	35.12	29.54
16	374.82	194.86	104.95	74.97	59.97	50.97	44.97	37.46	31.31	26.28
18	332.52	172.74	92.91	66.28	52.96	44.97	39.64	32.97	27.51	23.01
20	290.22	150.62	80.86	57.60	45.96	38.97	34.31	28.48	23.70	19.75
25	230.38	119.33	63.83	45.31	36.05	30.49	26.78	22.13	18.33	15.14
30	201.14	104.05	55.51	39.32	31.22	26.35	23.11	19.04	15.71	12.92
35	181.39	93.72	49.90	35.28	27.96	23.57	20.63	16.96	13.95	11.43
40	167.95	86.70	46.08	32.53	25.75	21.67	18.95	15.55	12.75	10.41
45	157.97	81.49	43.25	30.49	24.10	20.27	17.71	14.50	11.86	9.66
50	150.29	77.48	41.07	28.92	22.84	19.19	16.75	13.69	11.18	9.08
60	139.29	71.74	37.95	26.68	21.03	17.64	15.38	12.54	10.21	8.25
70	132.51	68.20	36.03	25.29	19.92	16.69	14.53	11.83	9.60	7.74
80	126.50	65.06	34.33	24.07	18.93	15.85	13.78	11.20	9.07	7.28
90	122.99	63.22	33.33	23.35	18.36	15.35	13.35	10.83	8.76	7.02
100	120.09	61.71	32.51	22.76	17.88	14.95	12.99	10.53	8.50	6.80
150	109.78	56.33	29.59	20.66	16.19	13.50	11.71	9.45	7.59	6.02
200	104.93	53.80	28.22	19.68	15.40	12.82	11.10	8.94	7.16	5.66
300	100.53	51.51	26.98	18.78	14.68	12.21	10.56	8.49	6.77	5.33
500	97.17	49.75	26.03	18.10	14.13	11.74	10.14	8.13	6.48	5.07
1,000	94.39	48.30	25.24	17.54	13.67	11.35	9.80	7.84	6.23	4.86
2,000	93.11	47.63	24.88	17.28	13.46	11.17	9.64	7.71	6.12	4.77

TABLE 4

The Nomograms of D. L. Heck for the Distribution of the Θ_{max}-Criterion of S. N. Roy

(a) Content of the tables and definition of the test statistic :

The nomograms contain the upper percentage points of the zero distribution of the test statistic

$$x_\alpha = x_\alpha(s, m, n) = \Theta_{max} ,$$

where Θ_{max} is the largest root of the determinant equation

$$| \underline{S}_h - \Theta(\underline{S}_e - \underline{S}_h) | = 0 ,$$

or equivalently, the largest eigenvalue of the matrix $(\underline{S}_e+\underline{S}_h)^{-1}\cdot\underline{S}_h$.

The relationship with other test statistics is given by the expression

$$\Theta_{max} = \lambda_{max}/(1 + \lambda_{max}) , \text{ where } \lambda_{max} \text{ is a root}$$

of the equation

$$| \underline{S}_h - \lambda\underline{S}_e| = 0 , \text{ or equivalently, an eigenvalue}$$

of $\underline{S}_e^{-1}\cdot\underline{S}_h$.

In addition

$$\Theta_{max} = 1 - \mu_{min} , \text{ where } \mu_{min} \text{ is determi-}$$

mined by $| \underline{S}_e - \mu(\underline{S}_e + \underline{S}_h) | = 0$.

(b) Extent of the tables and definition of the parameters :

(1) The parameter α :

α = error probability

α = 0.01 ; 0.025 ; 0.05 ;

(2) The parameter s :

$s = \min(n_h, p)$ with p = dimension of the variates.
n_h = degrees of freedom of the hypothesis.

$s = 2(1)5$.

(3) The parameter m :

$m = (|n_h - p| - 1)/2$

$m = -\frac{1}{2}, 0(1)10$.

(4) The parameter n :

$n = (n_e - p - 1)/2$ with n_e = degree of freedom of the error

$n = 5, \ldots, 1000$.

(c) Suggestions for applications :

(1) Using the nomograms in practice :

The twelve nomogrmas (Chart I through Chart XII) allow one to find the critical value $x_\alpha(s,m,n)$ for

$$P(\Theta_{max} > x_\alpha(s,m,n)) = \alpha .$$

For each combination of α and s there is a seperate nomogram, containing a family of 12 curves, one for each of the twelve values of the parameter $m = -1/2 ; 0(1)10$.

In order facilitate the representation, each family of curves has been represented in two sections: The lower family (resp. lower section) is the continuation of the upper, and with an overlap for the x_α from 0.50 to 0.55 . Correspondingly, there are two scales for x_α at the bottom of the nomogram : The upper scale corresponds to the upper family and the lower scale to the lower family.
The lowermost curve in each family (with the exception of Chart III) corresponds to the parameter value $m=-1/2$, the second lowest to the value $m=0$, the third lowest to $m=1$, and so on to the uppermost curve, which corresponds to $m=10$.
The scale for n is logarithmic along the vertical axis.

In order to determine the percentage point $x_\alpha = x_\alpha(s,m,n)$, take the nomogram corresponding to the given value of n . Then follow the horizontal line emanating from this point to the right until the intersection with the curve corresponding to the given value of the parameter m is reached.If this curve belongs to the upper family , descend vertically from the intersection point to the upper scale on the horizontal axis, and read off the value of the percentage point x_α . If the curve belongs to the lower family, the value of x_α is then read off the lower scale.

The values of $x_\alpha(s,m,n)$ may be read off the nomograms to an accuracy of two decimal places. To increase the precision, observe the considerations regarding Table 4a in section (e) .

(2) Several possibilities for application :

(aa) Multivariate general linear hypotheses, especially MANOVA-problems; multivariate regression and covariance analyses; confidence regions; Canonical discriminance analyses.

(bb) To test the independence of p (dependent) from q (independent) variates (multivariate regression analysis, standardized samples), use the initial parameters

$$s = \min(p, q)$$

$$m = (|p - q| - 1)/2$$

$$n = (N - p - q - 2)/2$$

with N = sample size.

(d) References :

(1) For the Θ_{max}-criterion :

ROY, S. N. : Some Aspects of Multivariate Analysis. New York: Wiley, 1957. + Calcutta: Indian Statistical Institute, 1957.

(2) For the source of the nomograms :

HECK, D. L. : Charts of some upper percentage points of the distribution of the largest characteristic root. The Annals of Mathematical Statistics 31,625-642(1960).

(e) Further remarks :

(1) Increasing accuracy with the asymptotic statistic $z_\alpha(s,m)$:

For purposes of greater accuracy, Table 4a gives the values of the asymptotic statistic $z_\alpha(s,m)$ for $s = 2(1)5$, as well as $m = -1/2;0(1)10$ and $\alpha = 0.01$; 0.025 ; 0.05
For $n < 100$ these z_α can be used to determine $x_\alpha(s.m,n)$ to within an error of at most five units in the fourth decimal place. To this end, take the given parameter combination (α, s, m, n), find the corresponding value of $z_\alpha = z_\alpha(s, m)$ in Table 4a, and substitute this in the expression

$$z = -(m + 2n + s + 1) \cdot \ln(1 - x) \quad .$$

Then use

$$y = z(s, m) / (m + 2n + s + 1)$$

to solve for $x = x_\alpha(s, m, n) = 1 - e^{-y}$.

(2) In some test situations one obtains the parameter value $s = 1$. In this case, the only positive eigenvalue Θ of $\underline{S}_h(\underline{S}_h + \underline{S}_e)^{-1}$ belongs to a beta distribution with density function

$$f(\Theta) = \frac{1}{B(m+1, n+1)} \cdot \Theta^m \cdot (1-\Theta)^n$$

for $0 \leqq \Theta \leqq 1$.

The critical value of Θ can be found either in the tables for the <u>incomplete beta function</u> or by means of the substitution

$$F = \frac{n+1}{m+1} \cdot \frac{\Theta}{1-\Theta}$$

in tables of the <u>F-distribution</u> with degrees of freedom $f_1 = 2m + 2$ and $f_2 = 2n + 2$.

Tafel 4 Table 4

CHART I

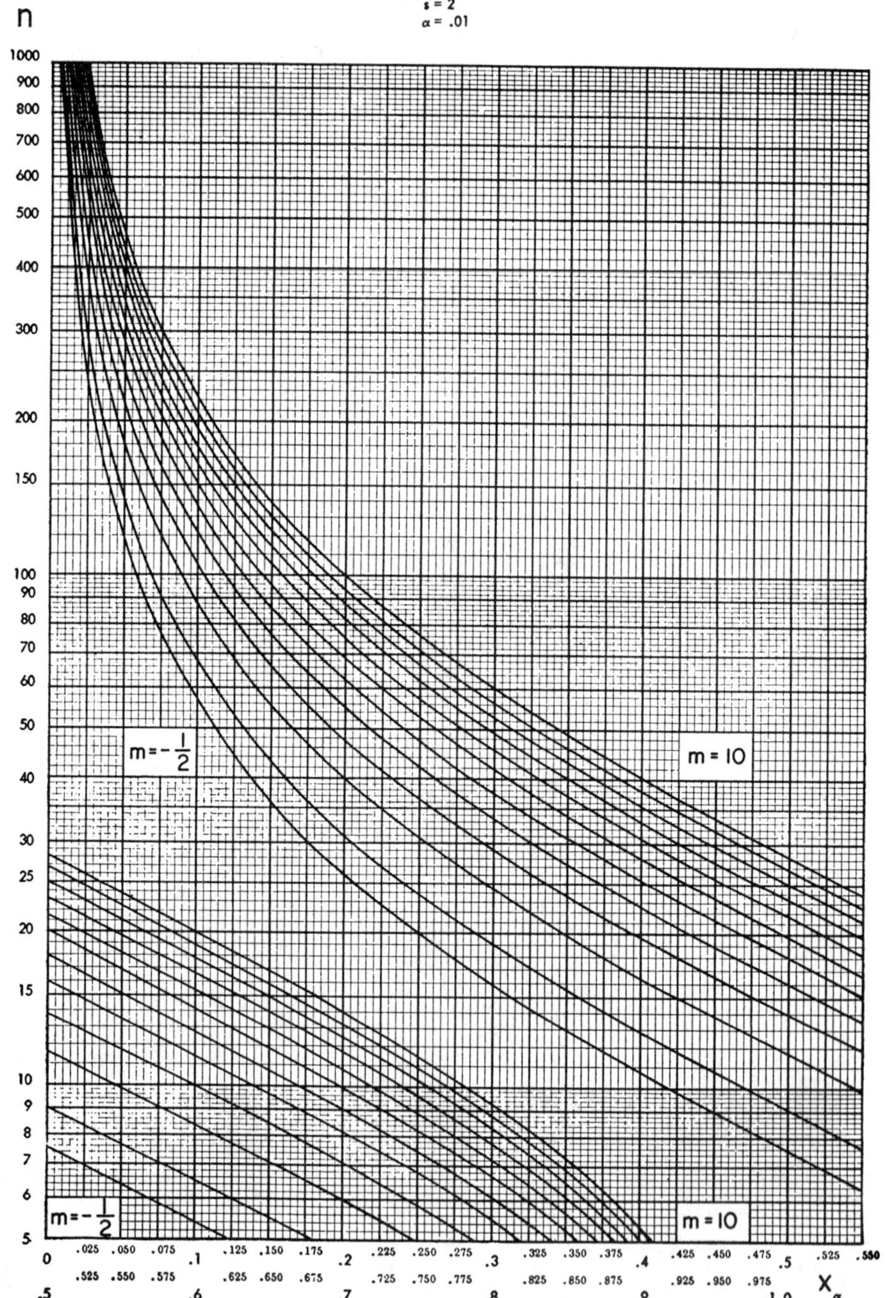

Tafel 4 (Forts.)

Table 4 (cont.)

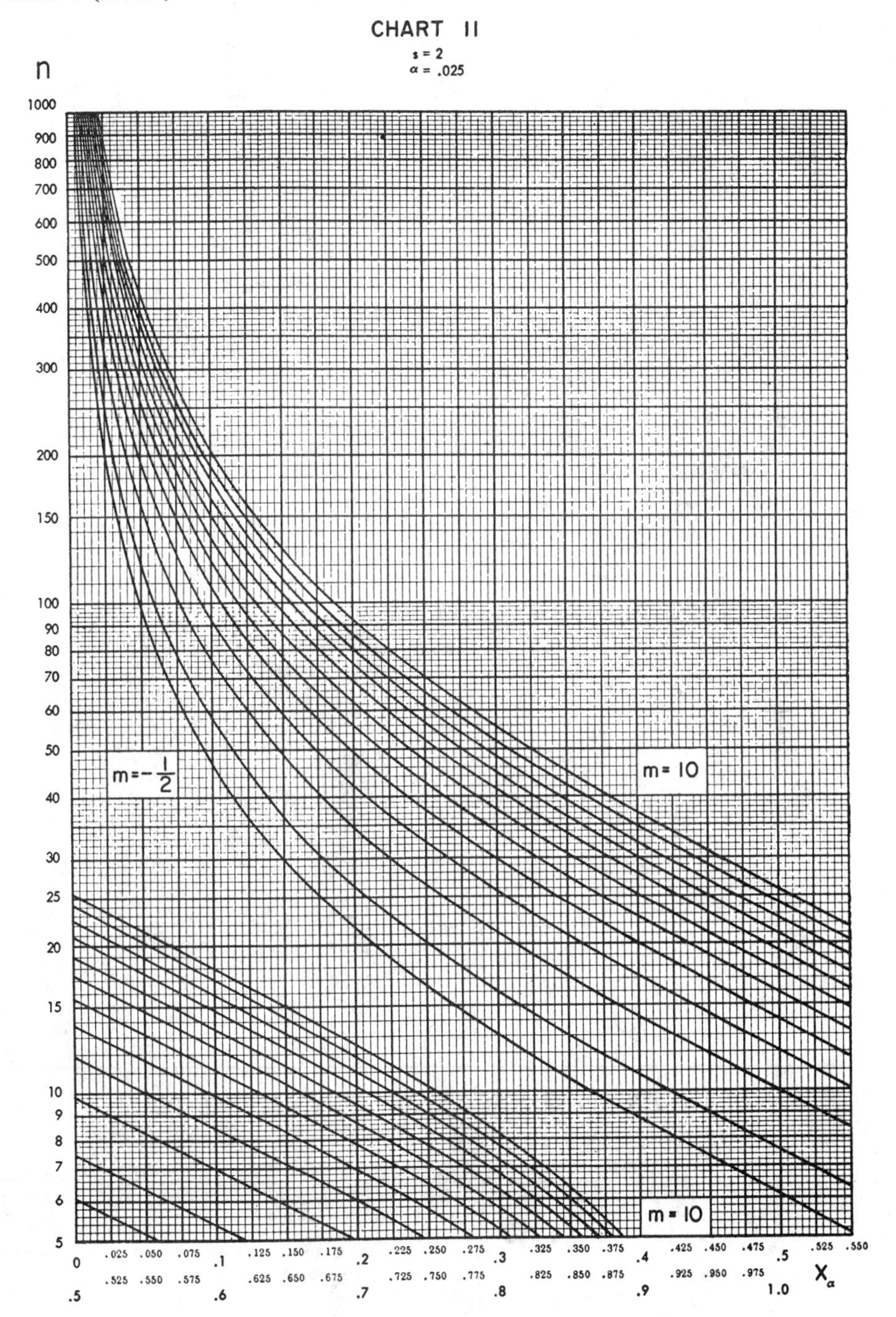

Tafel 4 (Forts.) Table 4 (cont.)

CHART III

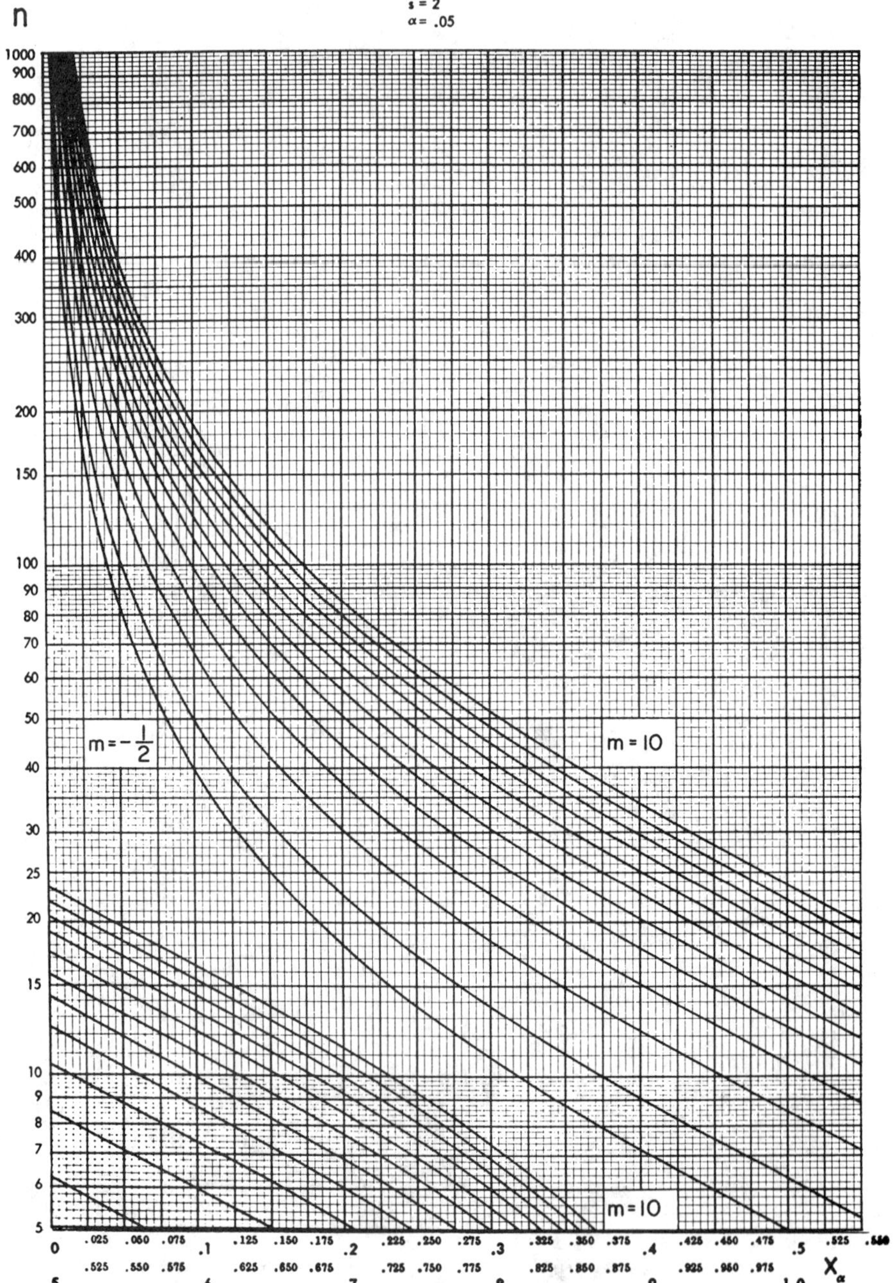

Tafel 4 (Forts.) Table 4 (cont.)

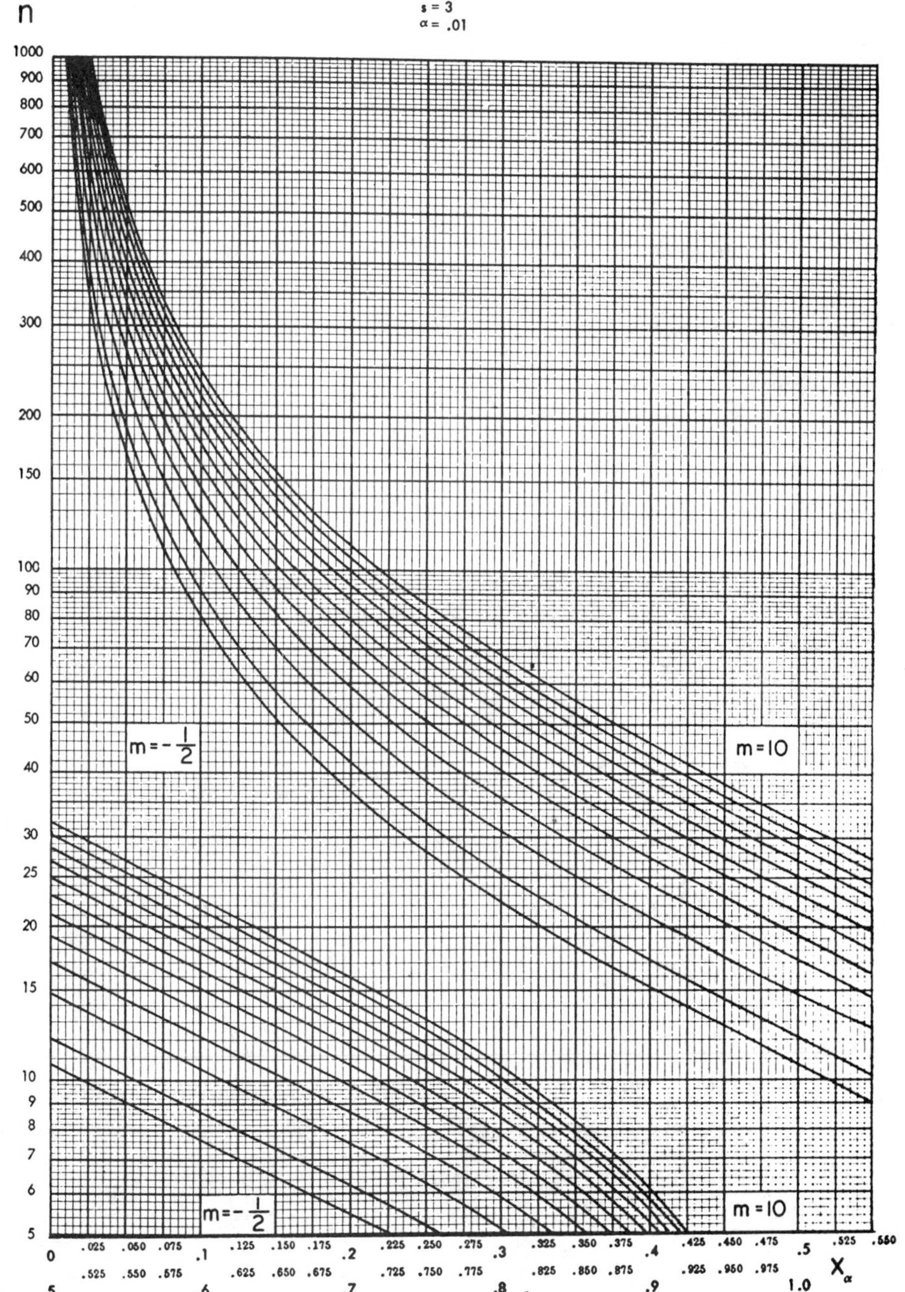

Tafel 4 (Forts.) Table 4 (cont.)

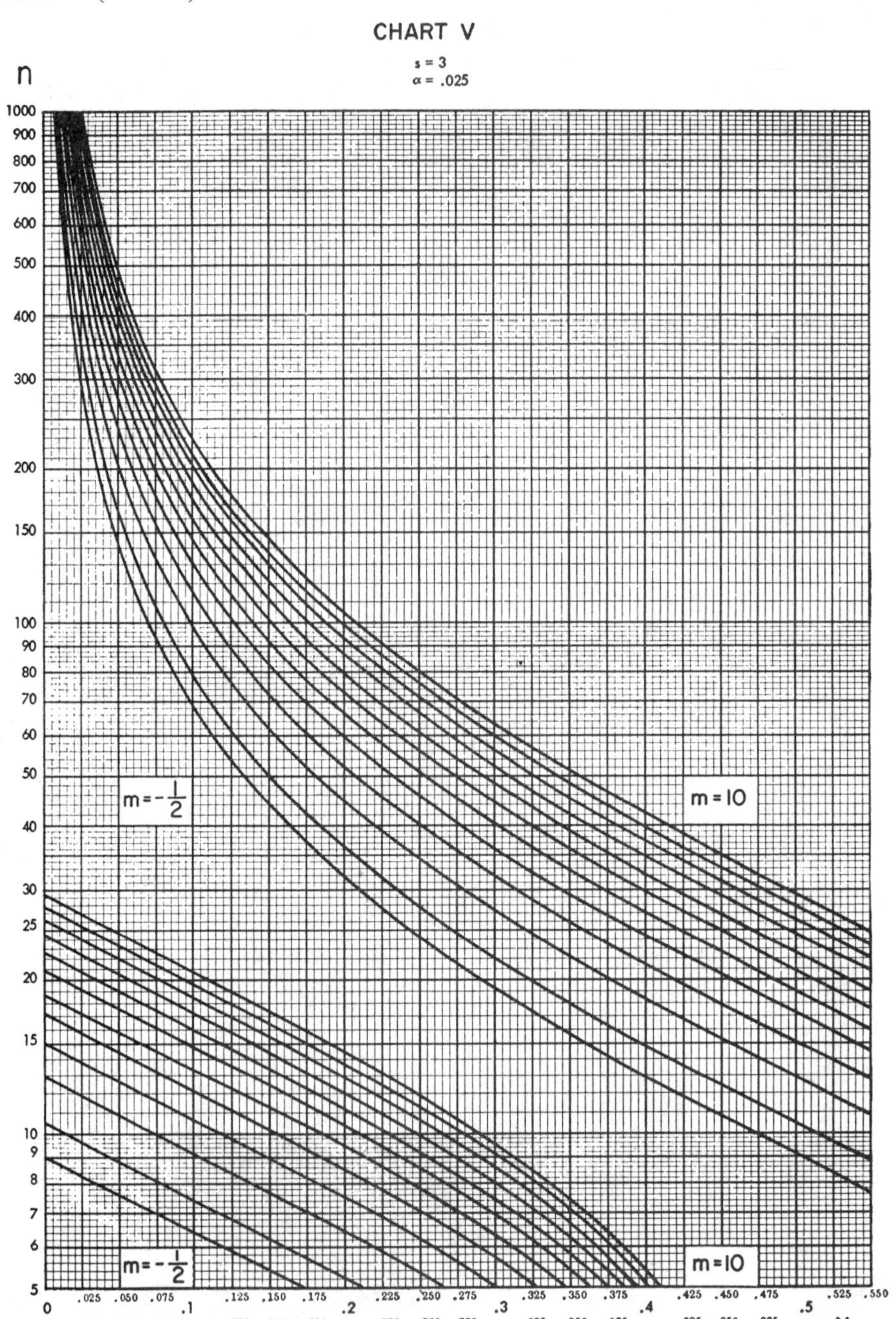

Tafel 4 (Forts.) Table 4 (cont.)

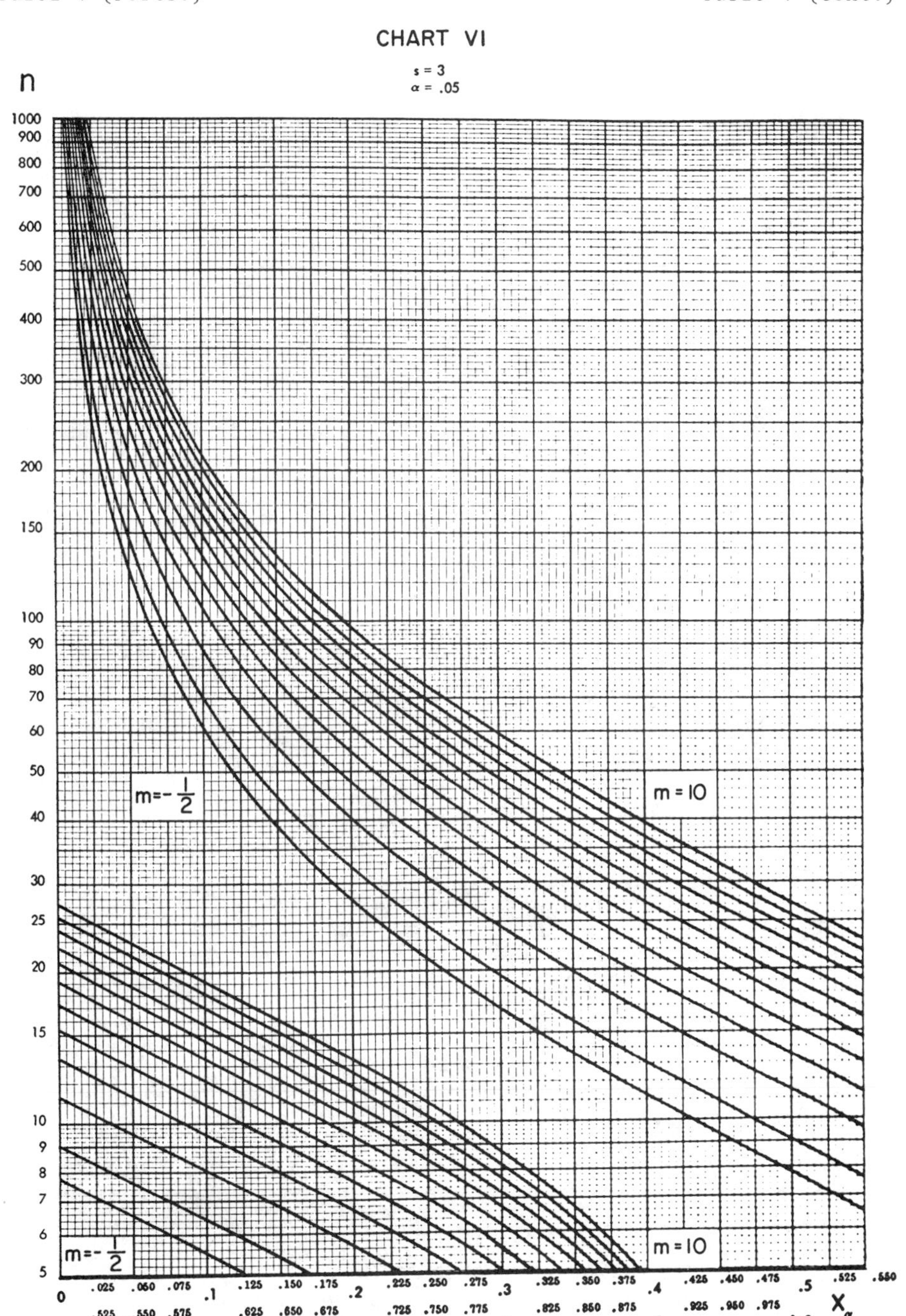

Tafel 4 (Forts.) Table 4 (cont.)

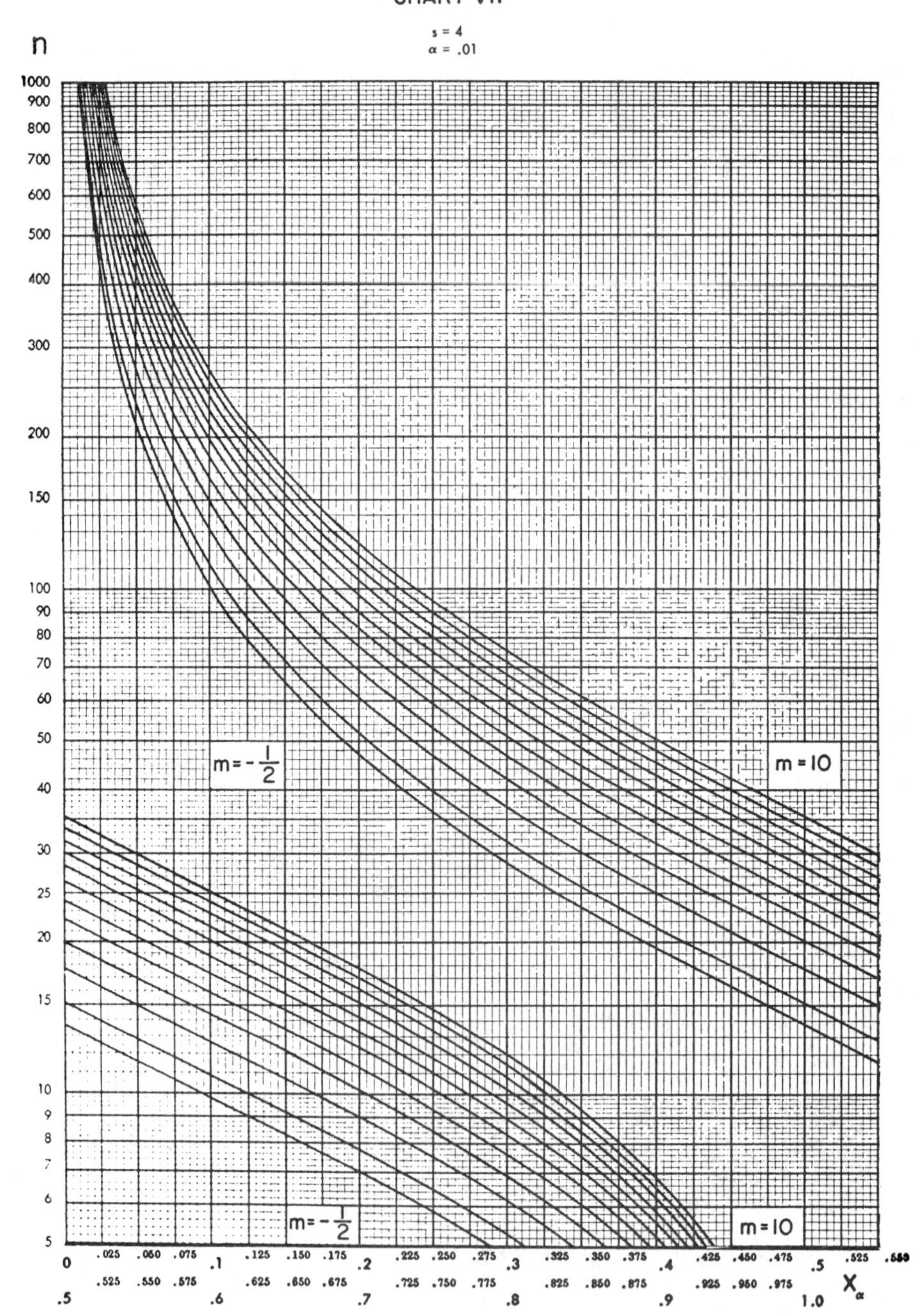

Tafel 4 (Forts.) Table 4 (cont.)

CHART VIII

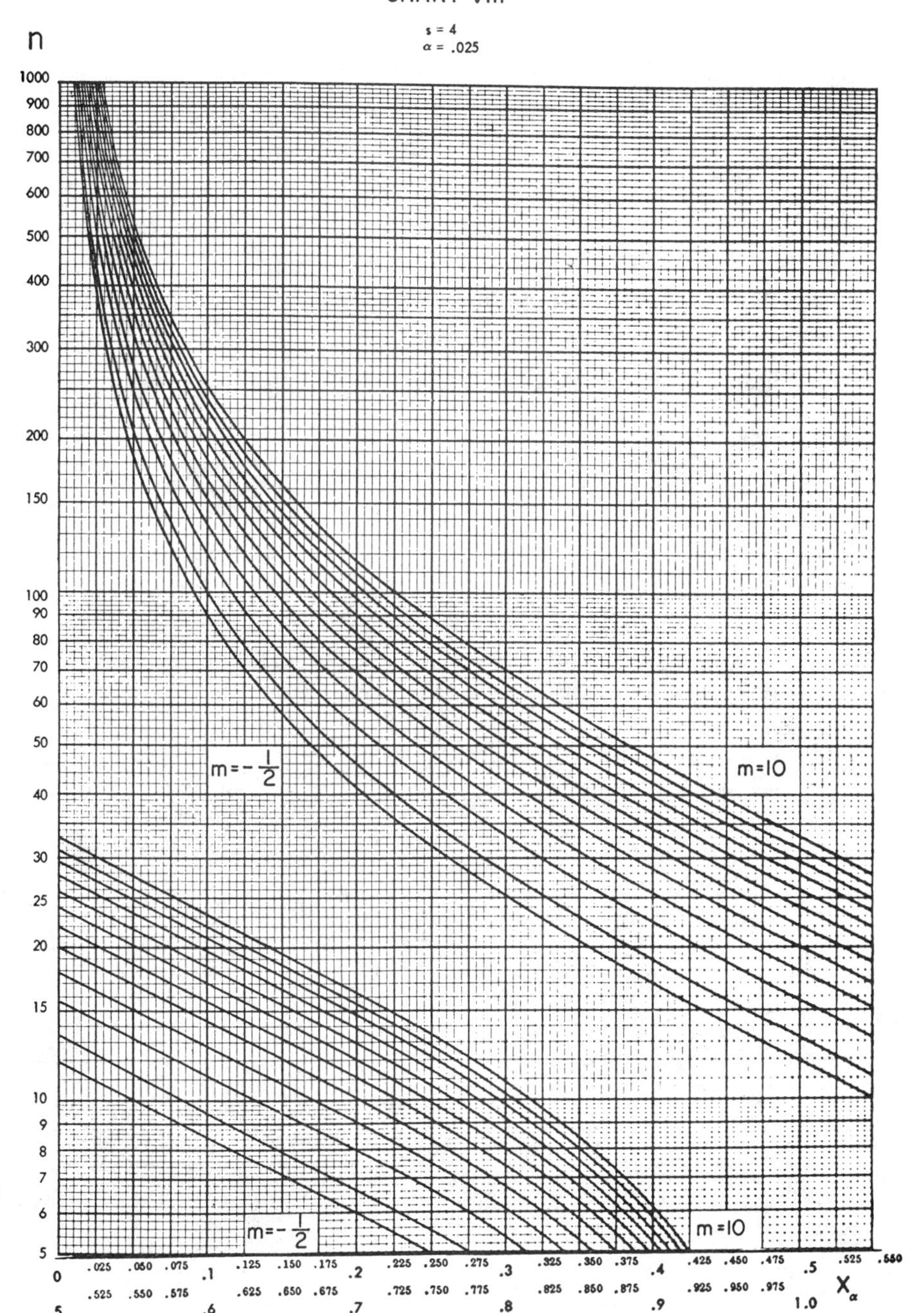

Tafel 4 (Forts.) Table 4 (cont.)

CHART IX

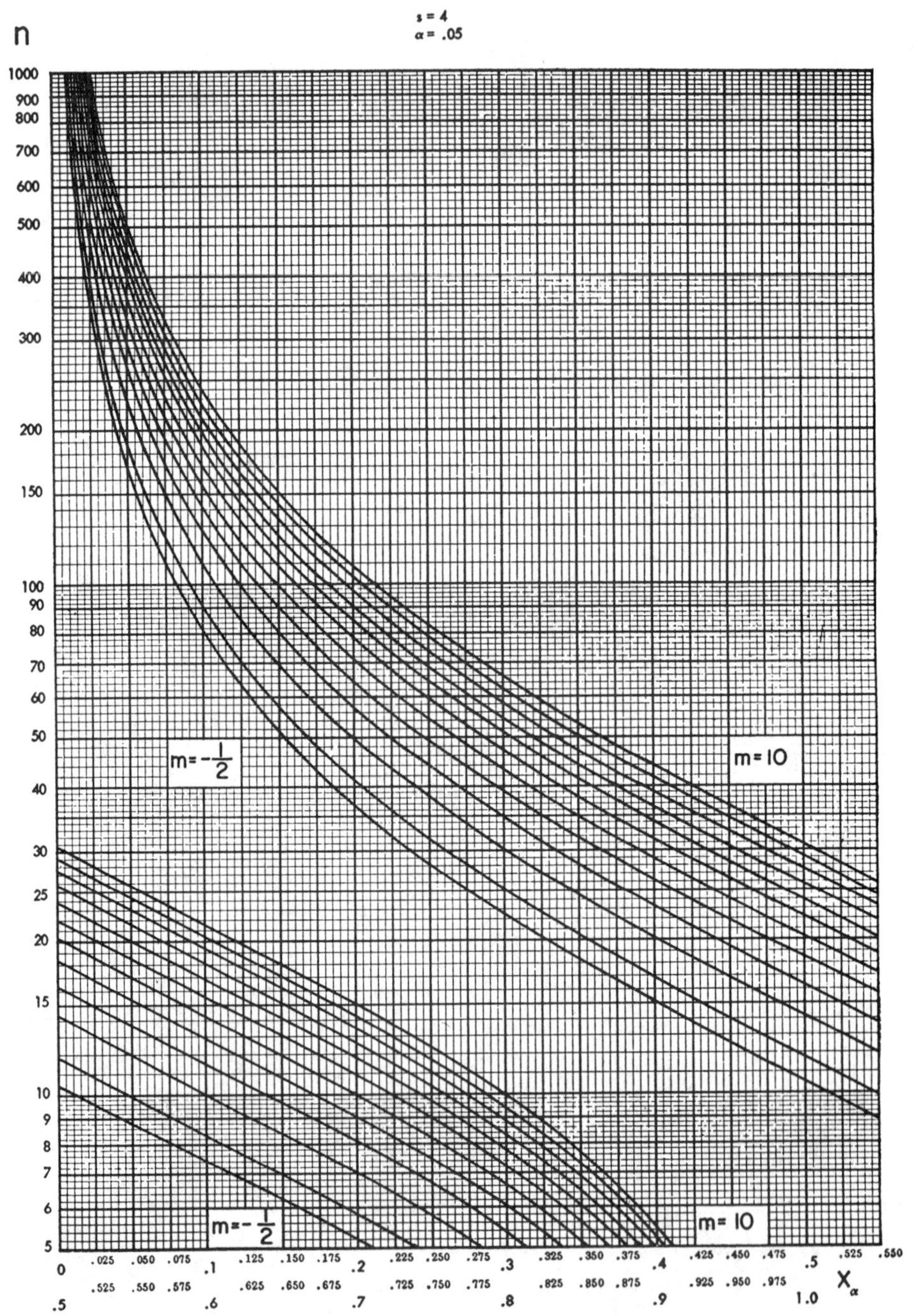

Tafel 4 (Forts.)

Table 4 (cont.)

CHART X

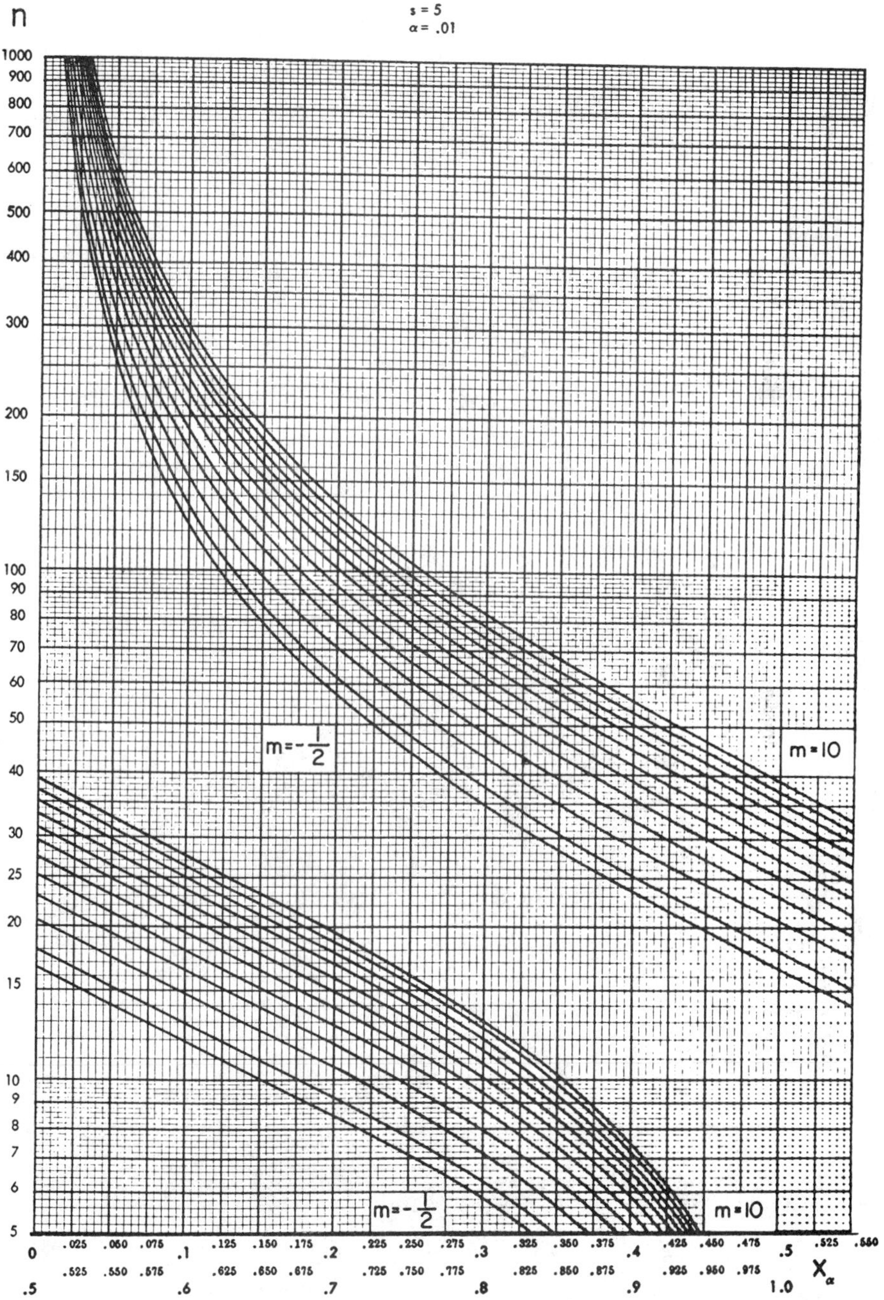

$s = 5$
$\alpha = .01$

Tafel 4 (Forts.) Table 4 (cont.)

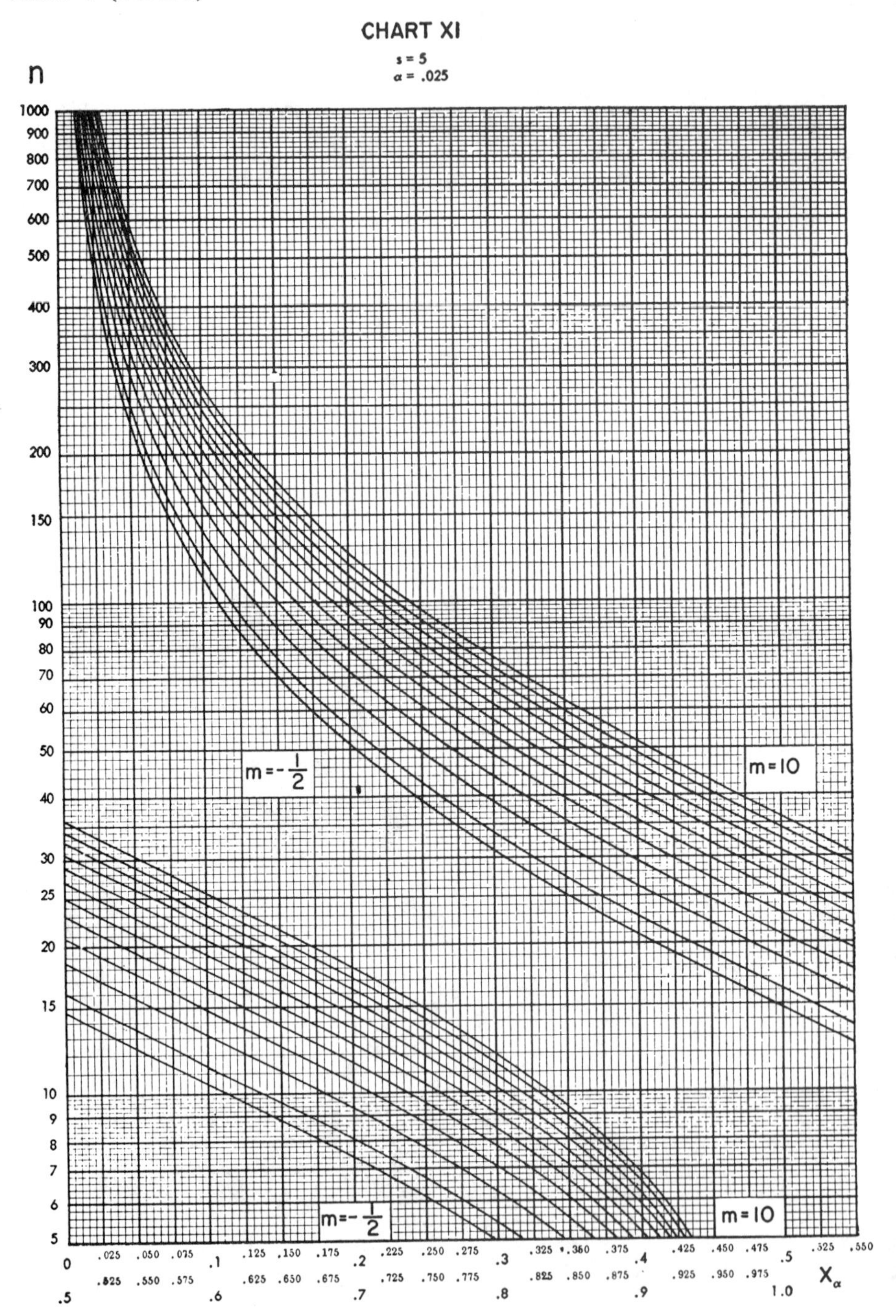

Tafel 4 (Forts.) Table 4 (cont.)

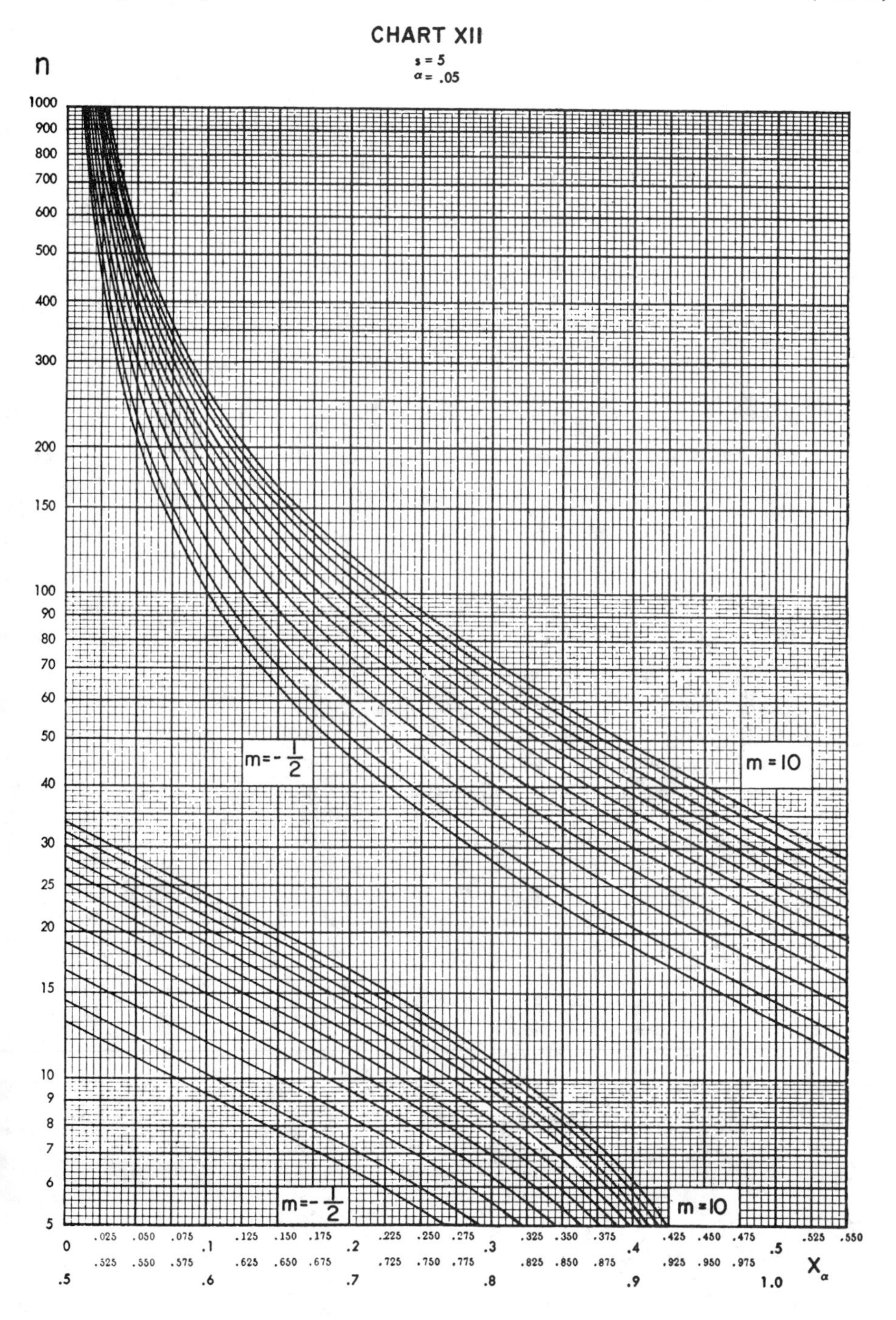

Tafel 4a | Table 4a

α / m	s = 2			s = 3		
	.01	.025	.05	.01	.025	.05
$-\frac{1}{2}$	12.1601	10.1465	8.5941	17.1762	14.9006	13.1141
0	14.5680	12.4157	10.7393	19.5012	17.1192	15.2389
1	18.7346	16.3599	14.4873	23.6906	21.1262	19.0866
2	22.4664	19.9086	17.8762	27.5181	24.7971	22.6216
3	25.9526	23.2352	21.0641	31.1203	28.2597	25.9635
4	29.2755	26.4145	24.1192	34.5647	31.5768	29.1708
5	32.4795	29.4870	27.0779	37.8905	34.7848	32.2774
6	35.5920	32.4773	29.9628	41.1230	37.9071	35.3050
7	38.6311	35.4018	32.7886	44.2795	40.9597	38.2685
8	41.6098	38.2722	35.5658	47.3726	43.9542	41.1785
9	44.5375	41.0970	38.3021	50.4118	46.8993	44.0430
10	47.4215	43.8827	41.0033	53.4042	49.8017	46.8684

α / m	s = 4			s = 5		
	.01	.025	.05	.01	.025	.05
$-\frac{1}{2}$	21.9646	19.4847	17.5183	26.6206	23.9697	21.8538
0	24.2395	21.6713	19.6277	28.8613	26.1339	23.9515
1	28.4328	25.7078	23.5278	33.0524	30.1861	27.8835
2	32.3175	29.4540	27.1543	36.9748	33.9834	31.5731
3	35.9964	33.0074	30.5996	40.7087	37.6027	35.0938
4	39.5253	36.4207	33.9135	44.3009	41.0883	38.4880
5	42.9387	39.7262	37.1265	47.7814	44.4688	41.7829
6	46.2593	42.9454	40.2588	51.1710	47.7639	44.9971
7	49.5034	46.0934	43.3246	54.4847	50.9876	48.1441
8	52.6831	49.1815	46.3345	57.7338	54.1508	51.2340
9	55.8073	52.2182	49.2964	60.9269	57.2615	54.2745
10	58.8833	55.2102	52.2166	64.0709	60.3264	57.2717

TABLE 5

The Θ_{max}-Criterion of F. G. Foster and D. H. Rees: A Version of the Maximum Root Criterion of S. N. Roy

(a) Content of the tables and definition of the parameters :

The tables contain the upper percentage points of the zero distribution of the largest root Θ_{max} of the determinant equation

$$| \underline{S}_h - \Theta(\underline{S}_h + \underline{S}_e) | = 0 \qquad \text{or equivalently ,}$$

the largest eigenvalue of the matrix $(\underline{S}_h + \underline{S}_e)^{-1} \cdot \underline{S}_h$.

The relationship with other test statistics , as well as a computational formula, is given by the expression

$$\Theta_{max} = \frac{\lambda_{max}}{1 + \lambda_{max}}$$ where λ_{max} is the largest

root of the determinant equation $| \underline{S}_h - \lambda \underline{S}_e | = 0$, or equivalently, the largest eigenvalue of the matrix $\underline{S}_e^{-1} \cdot \underline{S}_h$.

In addition

$$\Theta_{max} = 1 - \mu_{min}$$, where μ_{min} is determined by

$$| \underline{S}_e - \mu(\underline{S}_e + \underline{S}_h) | = 0 \quad .$$

(b) Extent of the tables and and definition of the parameters :

(1) The parameter P :

P = confidence probability

$P = 1 - \alpha = 0.80 \ ; \ 0.85 \ ; \ 0.90 \ ; \ 0.95 \ ; \ 0.99$.

(2) The parameter p :

p = dimension of the variate

$p = 2 \ ; \ 3 \ ; \ 4$.

(3) The parameter ν_2 :

$\nu_2 = n_h$ = degrees of freedom of the hypothesis

$\nu_2 = 2 \ ; \ 3 \ ; \ 5(2)27$ when $p = 2$

$\nu_2 = 3(1)10$ when $p = 3$

$\nu_2 = 4(1)11$ when $p = 4$.

(4) The parameter ν_1 :

$\nu_1 = n_e$ = degrees of freedom of the error

$\nu_1 = 5(2)41(10)101;121;161$ when $p = 2$

$\nu_1 = 4(2)46(4)70; \ 99 \ ; \ 194$ when $p = 3$

$\nu_1 = 5(2)47(4)71; \ 99 \ ; \ 195$ when $p = 4$.

(c) Suggestions for applications :

(1) Multivariate general linear hypotheses, especially MANOVA-problems; multivariate regression and covariance analyses. Multivariate confidence regions. Canonical discriminant analyses.

(2) Testing the independence of p (dependent) from q (independent) variates (multivariate regression analyses, canonical correlation analysis).

(d) References :

(1) Original tables due to

FOSTER, F. G. + REES, D. H. : Upper percentage points of the generalized beta distribution. I. Biometrika 44, 237 - 247 (1957).

FOSTER, F. G. : Upper percentage points of the generalized beta distribution. II. Biometrika 44, 441 - 453 (1957).

FOSTER, F. G. : Upper percentage points of the generalized beta distribution. III. Biometrika 45, 492 - 503 (1958).

(2) Copies of the present tables in modified version due to

PEARSON, E. S. + HARTLEY, H. O.(eds.): Biometrika Tables for Statisticians. Volume II. (Table 49). Cambridge : Cambridge University Press, 1972. (Published for the Biometrika Trusties).

(3) For the Θ_{max}-criterion :

ROY, S. N. : Some Aspects of Multivariate Analysis. New York : Wiley, 1957 . + Calcutta : Indian Statistical Institute, 1957 .

(e) Further remarks :

(1) Because of the mathematical form of the distribution of the maximal root Θ_{max} , FOSTER & REES (1957) introduce

the term generalized beta distribution.

(2) For $p = 3$ and $p = 4$ the (original) tables of FOSTER include additional values of the parameter $\nu_1 = n_e$. (For interpolation, see PEARSON & HARTLEY (eds.) (1972) p. 103).

$p = 2$

ν_1	P \ ν_2	2	3	5	7	9	11	13	15	17	19	21
5	0·80	0·7011	0·7728	0·8454	0·8825	0·9052	0·9205	0·9315	0·9398	0·9464	0·9516	0·9559
	·85	·7449	·8075	·8696	·9013	·9205	·9334	·9427	·9497	·9552	·9596	·9632
	·90	·7950	·8463	·8968	·9221	·9374	·9476	·9550	·9605	·9649	·9683	·9712
	·95	·8577	·8943	·9296	·9471	·9576	·9645	·9696	·9733	·9763	·9787	·9806
	·99	·9377	·9542	·9698	·9774	·9819	·9850	·9872	·9888	·9900	·9910	·9918
7	0·80	0·5638	0·6469	0·7416	0·7954	0·8303	0·8550	0·8734	0·8876	0·8989	0·9082	0·9158
	·85	·6085	·6851	·7712	·8194	·8507	·8726	·8889	·9014	·9114	·9196	·9264
	·90	·6628	·7307	·8058	·8474	·8741	·8928	·9066	·9173	·9257	·9326	·9383
	·95	·7370	·7919	·8514	·8839	·9045	·9189	·9295	·9376	·9440	·9493	·9536
	·99	·8498	·8826	·9173	·9358	·9475	·9556	·9615	·9660	·9695	·9724	·9748
9	0·80	0·4688	0·5526	0·6558	0·7189	0·7619	0·7933	0·8173	0·8363	0·8517	0·8644	0·8751
	·85	·5108	·5903	·6869	·7452	·7848	·8136	·8354	·8527	·8666	·8782	·8878
	·90	·5632	·6366	·7244	·7768	·8120	·8374	·8567	·8720	·8842	·8943	·9027
	·95	·6383	·7017	·7761	·8197	·8487	·8696	·8853	·8976	·9076	·9157	·9225
	·99	·7635	·8074	·8575	·8862	·9051	·9185	·9286	·9364	·9427	·9478	·9521
11	0·80	0·4003	0·4810	0·5859	0·6536	0·7016	0·7376	0·7657	0·7884	0·8069	0·8225	0·8357
	·85	·4389	·5169	·6169	·6808	·7257	·7592	·7854	·8063	·8235	·8378	·8500
	·90	·4880	·5617	·6551	·7138	·7548	·7854	·8089	·8278	·8433	·8561	·8670
	·95	·5603	·6267	·7091	·7600	·7952	·8212	·8413	·8573	·8702	·8810	·8902
	·99	·6878	·7381	·7989	·8357	·8607	·8790	·8929	·9039	·9128	·9202	·9265
13	0·80	0·3490	0·4253	0·5286	0·5982	0·6489	0·6880	0·7190	0·7443	0·7654	0·7832	0·7984
	·85	·3843	·4590	·5589	·6252	·6735	·7103	·7395	·7632	·7829	·7996	·8138
	·90	·4298	·5016	·5965	·6587	·7035	·7375	·7644	·7862	·8042	·8194	·8324
	·95	·4981	·5646	·6507	·7063	·7459	·7757	·7992	·8181	·8337	·8468	·8580
	·99	·6233	·6770	·7446	·7872	·8171	·8394	·8568	·8706	·8821	·8915	·8997
15	0·80	0·3091	0·3809	0·4812	0·5508	0·6030	0·6439	0·6769	0·7042	0·7271	0·7467	0·7636
	·85	·3415	·4124	·5102	·5775	·6275	·6664	·6978	·7236	·7453	·7638	·7798
	·90	·3837	·4527	·5468	·6106	·6577	·6942	·7235	·7475	·7675	·7847	·7993
	·95	·4478	·5130	·6003	·6584	·7011	·7338	·7598	·7810	·7989	·8138	·8268
	·99	·5687	·6237	·6954	·7422	·7758	·8013	·8216	·8378	·8512	·8629	·8726
17	0·80	0·2774	0·3447	0·4412	0·5101	0·5628	0·6047	0·6389	0·6676	0·6919	0·7129	0·7312
	·85	·3072	·3741	·4690	·5360	·5869	·6272	·6600	·6874	·7105	·7304	·7478
	·90	·3463	·4122	·5043	·5685	·6169	·6550	·6860	·7116	·7334	·7519	·7681
	·95	·4065	·4697	·5564	·6160	·6605	·6951	·7232	·7464	·7659	·7825	·7969
	·99	·5222	·5773	·6512	·7008	·7373	·7652	·7875	·8061	·8212	·8346	·8458
19	0·80	0·2515	0·3148	0·4073	0·4748	0·5273	0·5697	0·6047	0·6344	0·6597	0·6817	0·7010
	·85	·2791	·3424	·4338	·4999	·5509	·5919	·6257	·6541	·6785	·6995	·7179
	·90	·3155	·3782	·4677	·5315	·5805	·6196	·6517	·6786	·7016	·7214	·7387
	·95	·3719	·4327	·5182	·5782	·6238	·6599	·6894	·7139	·7349	·7528	·7684
	·99	·4823	·5369	·6116	·6630	·7014	·7313	·7555	·7756	·7926	·8071	·8198
21	0·80	0·2300	0·2895	0·3782	0·4439	0·4959	0·5383	0·5738	0·6040	0·6301	0·6529	0·6729
	·85	·2557	·3155	·4034	·4682	·5189	·5602	·5946	·6237	·6489	·6707	·6900
	·90	·2897	·3493	·4358	·4988	·5479	·5875	·6204	·6482	·6721	·6929	·7112
	·95	·3427	·4012	·4847	·5445	·5906	·6277	·6581	·6838	·7058	·7248	·7415
	·99	·4479	·5014	·5762	·6285	·6685	·6997	·7254	·7469	·7652	·7810	·7946

Tafel 5 (Forts.) Table 5 (cont.)

$p = 2$

ν_1	P \ ν_2	2	3	5	7	9	11	13	15	17	19	21
23	0·80	0·2119	0·2680	0·3528	0·4167	0·4678	0·5101	0·5457	0·5763	0·6028	0·6262	0·6469
	·85	·2359	·2924	·3769	·4401	·4903	·5315	·5661	·5958	·6215	·6441	·6640
	·90	·2677	·3244	·4080	·4699	·5185	·5584	·5918	·6202	·6448	·6663	·6853
	·95	·3177	·3737	·4551	·5143	·5606	·5981	·6294	·6558	·6787	·6986	·7160
	·99	·4179	·4701	·5443	·5971	·6380	·6703	·6970	·7197	·7391	·7558	·7705
25	0·80	0·1964	0·2494	0·3306	0·3926	0·4428	0·4846	0·5202	0·5509	0·5777	0·6015	0·6226
	·85	·2189	·2725	·3536	·4151	·4645	·5055	·5402	·5702	·5963	·6193	·6397
	·90	·2488	·3027	·3834	·4439	·4921	·5319	·5655	·5944	·6194	·6415	·6610
	·95	·2960	·3498	·4287	·4872	·5333	·5710	·6027	·6298	·6533	·6738	·6920
	·99	·3915	·4424	·5155	·5685	·6096	·6429	·6708	·6941	·7143	·7319	·7474
27	0·80	0·1830	0·2333	0·3110	0·3711	0·4202	0·4615	0·4968	0·5275	0·5546	0·5785	0·6000
	·85	·2042	·2551	·3331	·3928	·4413	·4818	·5165	·5465	·5728	·5962	·6170
	·90	·2324	·2839	·3616	·4206	·4682	·5077	·5413	·5704	·5958	·6183	·6383
	·95	·2771	·3286	·4052	·4626	·5084	·5462	·5781	·6056	·6296	·6506	·6693
	·99	·3682	·4176	·4895	·5422	·5837	·6175	·6458	·6700	·6909	·7092	·7254
29	0·80	0·1713	0·2191	0·2936	0·3518	0·3998	0·4404	0·4754	0·5060	0·5331	0·5572	0·5789
	·85	·1913	·2398	·3147	·3727	·4202	·4603	·4947	·5247	·5512	·5747	·5958
	·90	·2180	·2672	·3420	·3996	·4463	·4855	·5191	·5482	·5738	·5966	·6170
	·95	·2604	·3099	·3840	·4404	·4855	·5232	·5553	·5830	·6074	·6288	·6480
	·99	·3475	·3954	·4659	·5182	·5597	·5938	·6225	·6472	·6687	·6876	·7043
31	0·80	0·1611	0·2065	0·2780	0·3344	0·3812	0·4211	0·4557	0·4861	0·5131	0·5373	0·5591
	·85	·1800	·2262	·2982	·3546	·4010	·4405	·4746	·5045	·5309	·5546	·5759
	·90	·2053	·2523	·3246	·3805	·4264	·4651	·4985	·5277	·5534	·5763	·5969
	·95	·2457	·2931	·3650	·4200	·4647	·5022	·5342	·5620	·5866	·6082	·6278
	·99	·3290	·3754	·4444	·4961	·5374	·5717	·6008	·6258	·6477	·6671	·6843
33	0·80	0·1519	0·1953	0·2639	0·3186	0·3643	0·4034	0·4375	0·4677	0·4946	0·5187	0·5406
	·85	·1699	·2141	·2834	·3381	·3835	·4223	·4560	·4857	·5121	·5358	·5572
	·90	·1940	·2389	·3087	·3632	·4081	·4464	·4794	·5084	·5342	·5572	·5781
	·95	·2324	·2781	·3478	·4015	·4455	·4825	·5145	·5424	·5671	·5890	·6088
	·99	·3123	·3573	·4247	·4757	·5168	·5511	·5803	·6057	·6279	·6476	·6653
35	0·80	0·1438	0·1852	0·2512	0·3042	0·3487	0·3871	0·4208	0·4506	0·4773	0·5014	0·5233
	·85	·1609	·2032	·2699	·3230	·3674	·4055	·4388	·4682	·4945	·5181	·5396
	·90	·1838	·2270	·2943	·3473	·3913	·4290	·4617	·4906	·5163	·5394	·5603
	·95	·2206	·2645	·3320	·3845	·4277	·4644	·4962	·5240	·5487	·5708	·5907
	·99	·2972	·3408	·4066	·4569	·4974	·5318	·5612	·5867	·6091	·6292	·6471
37	0·80	0·1365	0·1762	0·2397	0·2910	0·3344	0·3721	0·4052	0·4347	0·4612	0·4852	0·5070
	·85	·1528	·1933	·2577	·3092	·3526	·3900	·4228	·4519	·4781	·5017	·5231
	·90	·1747	·2161	·2812	·3328	·3759	·4129	·4453	·4739	·4995	·5226	·5434
	·95	·2098	·2521	·3175	·3689	·4113	·4475	·4791	·5068	·5315	·5536	·5737
	·99	·2834	·3257	·3900	·4394	·4797	·5138	·5430	·5688	·5915	·6117	·6299
39	0·80	0·1299	0·1679	0·2292	0·2790	0·3213	0·3582	0·3907	0·4198	0·4460	0·4699	0·4916
	·85	·1455	·1844	·2465	·2965	·3389	·3756	·4079	·4367	·4627	·4861	·5076
	·90	·1664	·2062	·2691	·3194	·3616	·3980	·4299	·4583	·4837	·5067	·5277
	·95	·2001	·2408	·3044	·3544	·3961	·4319	·4631	·4906	·5152	·5375	·5576
	·99	·2709	·3119	·3747	·4232	·4631	·4969	·5260	·5517	·5747	·5950	·6134

Tafel 5 (Forts.) Table 5 (cont.)

$p = 2$

ν_1	P \ ν_2	2	3	5	7	9	11	13	15	17	19	21
41	0·80	0·1239	0·1604	0·2195	0·2679	0·3091	0·3452	0·3772	0·4059	0·4319	0·4555	0·4771
	·85	·1388	·1762	·2362	·2849	·3262	·3622	·3940	·4225	·4482	·4715	·4928
	·90	·1589	·1972	·2582	·3070	·3483	·3840	·4155	·4436	·4689	·4918	·5127
	·95	·1912	·2306	·2922	·3411	·3819	·4172	·4480	·4754	·4999	·5221	·5423
	·99	·2594	·2993	·3605	·4079	·4472	·4812	·5100	·5358	·5585	·5792	·5977
51	0·80	0·1006	0·1311	0·1814	0·2233	0·2598	0·2923	0·3216	0·3482	0·3725	0·3950	0·4157
	·85	·1129	·1443	·1955	·2380	·2748	·3073	·3366	·3631	·3874	·4097	·4303
	·90	·1295	·1619	·2142	·2571	·2941	·3267	·3559	·3823	·4063	·4284	·4487
	·95	·1565	·1900	·2434	·2868	·3239	·3564	·3853	·4114	·4350	·4566	·4765
	·99	·2140	·2486	·3029	·3462	·3828	·4144	·4424	·4675	·4900	·5106	·5295
61	0·80	0·0846	0·1109	0·1545	0·1914	0·2240	0·2534	0·2801	0·3047	0·3274	0·3485	0·3682
	·85	·0951	·1221	·1667	·2043	·2372	·2668	·2936	·3182	·3409	·3620	·3816
	·90	·1093	·1373	·1830	·2211	·2544	·2842	·3111	·3357	·3583	·3792	·3987
	·95	·1324	·1615	·2086	·2474	·2810	·3109	·3378	·3622	·3847	·4054	·4246
	·99	·1821	·2126	·2610	·3004	·3341	·3637	·3903	·4142	·4361	·4561	·4743
71	0·80	0·0730	0·0960	0·1345	0·1675	0·1969	0·2236	0·2481	0·2708	0·2919	0·3117	0·3303
	·85	·0822	·1059	·1454	·1789	·2087	·2357	·2604	·2832	·3043	·3241	·3427
	·90	·0946	·1191	·1597	·1939	·2242	·2514	·2762	·2991	·3203	·3401	·3586
	·95	·1148	·1404	·1824	·2175	·2481	·2756	·3006	·3236	·3447	·3644	·3828
	·99	·1584	·1855	·2293.	·2651	·2963	·3239	·3488	·3715	·3924	·4118	·4298
81	0·80	0·0643	0·0847	0·1191	0·1489	0·1756	0·2000	0·2226	0·2437	0·2634	0·2819	0·2994
	·85	·0723	·0934	·1288	·1592	·1863	·2110	·2338	·2550	·2748	·2934	·3109
	·90	·0833	·1052	·1417	·1727	·2003	·2253	·2483	·2697	·2896	·3082	·3257
	·95	·1013	·1242	·1620	·1939	·2221	·2475	·2707	·2922	·3122	·3308	·3483
	·99	·1402	·1646	·2044	·2374	·2662	·2919	·3153	·3368	·3567	·3751	·3924
91	0·80	0·0574	0·0757	0·1069	0·1340	0·1585	0·1810	0·2019	0·2215	0·2399	0·2573	0·2738
	·85	·0646	·0836	·1156	·1433	·1682	·1910	·2122	·2319	·2505	·2680	·2845
	·90	·0745	·0942	·1273	·1556	·1810	·2042	·2255	·2455	·2642	·2818	·2984
	·95	·0906	·1114	·1457	·1750	·2010	·2245	·2462	·2664	·2852	·3029	·3195
	·99	·1257	·1479	·1844	·2149	·2416	·2657	·2874	·3081	·3270	·3444	·3608
101	0·80	0·0518	0·0685	0·0969	0·1218	0·1444	0·1652	0·1847	0·2029	0·2202	0·2366	0·2522
	·85	·0584	·0756	·1049	·1304	·1533	·1745	·1942	·2127	·2301	·2466	·2622
	·90	·0673	·0853	·1155	·1416	·1651	·1866	·2066	·2253	·2428	·2595	·2752
	·95	·0820	·1009	·1325	·1594	·1835	·2055	·2258	·2447	·2625	·2792	·2951
	·99	·1140	·1343	·1678	·1961	·2211	·2436	·2644	·2836	·3015	·3184	·3342
121	0·80	0·0434	0·0575	0·0817	0·1030	0·1225	0·1407	0·1577	0·1739	0·1892	0·2038	0·2178
	·85	·0489	·0635	·0885	·1104	·1303	·1487	·1660	·1824	·1978	·2126	·2267
	·90	·0565	·0717	·0976	·1200	·1404	·1593	·1768	·1934	·2091	·2240	·2382
	·95	·0688	·0849	·1120	·1353	·1563	·1756	·1936	·2105	·2264	·2415	·2559
	·99	·0960	·1133	·1423	·1670	·1891	·2090	·2274	·2447	·2610	·2763	·2907
161	0·80	0·0328	0·0435	0·0622	0·0788	0·0941	0·1085	0·1221	0·1351	0·1476	0·1596	0·1711
	·85	·0370	·0481	·0674	·0844	·1001	·1148	·1287	·1419	·1545	·1666	·1783
	·90	·0427	·0544	·0744	·0920	·1081	·1231	·1372	·1507	·1635	·1758	·1876
	·95	·0521	·0645	·0856	·1039	·1206	·1361	·1506	·1644	·1775	·1900	·2020
	·99	·0730	·0864	·1092	·1288	·1464	·1627	·1778	·1921	·2056	·2185	·2309

$p = 3$

ν_1	ν_2 / P	3	4	5	6	7	8	9	10
4	0·80	0·9516	0·9635	0·9707	0·9755	0·9790	0·9816	0·9836	0·9852
	·85	·9645	·9733	·9786	·9821	·9846	·9865	·9880	·9892
	·90	·9769	·9826	·9861	·9884	·9900	·9913	·9922	·9930
	·95	·9887	·9915	·9932	·9943	·9951	·9957	·9962	·9966
	·99	·9978	·9983	·9987	·9989	·9990	·9992	·9993	·9993
6	0·80	0·8294	0·8629	0·8852	0·9012	0·9132	0·9226	0·9302	0·9364
	·85	·8561	·8846	·9036	·9171	·9273	·9352	·9416	·9468
	·90	·8857	·9087	·9238	·9346	·9427	·9490	·9540	·9582
	·95	·9218	·9378	·9482	·9556	·9612	·9655	·9689	·9717
	·99	·9664	·9734	·9779	·9811	·9835	·9853	·9868	·9880
8	0·80	0·7180	0·7639	0·7965	0·8210	0·8400	0·8554	0·8680	0·8786
	·85	·7497	·7911	·8203	·8421	·8591	·8728	·8840	·8933
	·90	·7871	·8229	·8480	·8667	·8812	·8929	·9024	·9103
	·95	·8365	·8646	·8842	·8986	·9098	·9188	·9261	·9322
	·99	·9086	·9247	·9359	·9441	·9504	·9554	·9595	·9628
10	0·80	0·6281	0·6798	0·7181	0·7478	0·7717	0·7913	0·8077	0·8217
	·85	·6610	·7090	·7443	·7716	·7935	·8114	·8264	·8392
	·90	·7008	·7440	·7757	·8000	·8195	·8353	·8486	·8598
	·95	·7560	·7922	·8185	·8386	·8546	·8676	·8784	·8876
	·99	·8439	·8679	·8852	·8983	·9086	·9170	·9239	·9298
12	0·80	0·5564	0·6102	0·6513	0·6840	0·7108	0·7332	0·7523	0·7688
	·85	·5887	·6396	·6783	·7089	·7340	·7549	·7726	·7879
	·90	·6287	·6757	·7112	·7392	·7620	·7810	·7971	·8109
	·95	·6857	·7266	·7574	·7815	·8010	·8172	·8309	·8426
	·99	·7816	·8113	·8334	·8505	·8643	·8757	·8853	·8934
14	0·80	0·4986	0·5525	0·5946	0·6289	0·6574	0·6816	0·7025	0·7207
	·85	·5297	·5813	·6215	·6541	·6811	·7039	·7236	·7407
	·90	·5687	·6172	·6548	·6851	·7101	·7313	·7494	·7652
	·95	·6254	·6689	·7024	·7292	·7512	·7698	·7857	·7994
	·99	·7245	·7582	·7837	·8040	·8206	·8344	·8462	·8564
16	0·80	0·4512	0·5042	0·5464	0·5812	0·6106	0·6359	0·6579	0·6773
	·85	·4809	·5321	·5728	·6062	·6343	·6584	·6794	·6978
	·90	·5185	·5672	·6057	·6372	·6637	·6863	·7058	·7230
	·95	·5739	·6185	·6535	·6820	·7058	·7261	·7436	·7589
	·99	·6735	·7096	·7376	·7601	·7788	·7947	·8083	·8201
18	0·80	0·4118	0·4633	0·5050	0·5398	0·5696	0·5954	0·6181	0·6383
	·85	·4401	·4902	·5307	·5643	·5930	·6179	·6396	·6589
	·90	·4760	·5242	·5629	·5950	·6222	·6457	·6663	·6845
	·95	·5296	·5745	·6103	·6398	·6647	·6861	·7048	·7212
	·99	·6282	·6657	·6953	·7195	·7398	·7571	·7721	·7853
20	0·80	0·3787	0·4284	0·4692	0·5037	0·5334	0·5595	0·5825	0·6031
	·85	·4054	·4542	·4940	·5276	·5564	·5816	·6038	·6237
	·90	·4397	·4870	·5254	·5576	·5852	·6093	·6304	·6493
	·95	·4914	·5359	·5719	·6019	·6275	·6497	·6692	·6864
	·99	·5879	·6262	·6568	·6821	·7035	·7220	·7381	·7523
22	0·80	0·3504	0·3983	0·4380	0·4719	0·5014	0·5274	0·5505	0·5714
	·85	·3758	·4230	·4619	·4951	·5238	·5491	·5715	·5917
	·90	·4085	·4545	·4924	·5244	·5521	·5764	·5979	·6172
	·95	·4580	·5019	·5377	·5679	·5938	·6165	·6365	·6544
	·99	·5521	·5906	·6218	·6478	·6700	·6893	·7063	·7214

$p = 3$

ν_1	ν_2 / P	3	4	5	6	7	8	9	10
24	0·80	0·3259	0·3720	0·4106	0·4438	0·4728	0·4986	0·5217	0·5426
	·85	·3501	·3957	·4337	·4662	·4947	·5198	·5423	·5626
	·90	·3813	·4260	·4631	·4947	·5223	·5466	·5683	·5878
	·95	·4288	·4718	·5072	·5373	·5633	·5862	·6066	·6249
	·99	·5201	·5586	·5900	·6164	·6392	·6590	·6766	·6923
26	0·80	0·3046	0·3490	0·3864	0·4187	0·4473	0·4727	0·4957	0·5165
	·85	·3276	·3716	·4086	·4404	·4685	·4934	·5158	·5361
	·90	·3574	·4007	·4370	·4681	·4954	·5196	·5413	·5610
	·95	·4030	·4450	·4798	·5096	·5356	·5586	·5792	·5977
	·99	·4914	·5296	·5610	·5876	·6107	·6309	·6490	·6651
28	0·80	0·2859	0·3286	0·3648	0·3963	0·4243	0·4493	0·4720	0·4926
	·85	·3078	·3503	·3861	·4173	·4448	·4694	·4917	·5119
	·90	·3363	·3783	·4136	·4441	·4710	·4951	·5167	·5363
	·95	·3801	·4210	·4552	·4846	·5104	·5333	·5539	·5726
	·99	·4656	·5033	·5345	·5612	·5844	·6049	·6232	·6398
30	0·80	0·2694	0·3104	0·3455	0·3761	0·4035	0·4280	0·4504	0·4708
	·85	·2903	·3312	·3660	·3964	·4234	·4476	·4696	·4897
	·90	·3175	·3581	·3925	·4224	·4489	·4726	·4941	·5136
	·95	·3595	·3993	·4328	·4618	·4873	·5101	·5307	·5493
	·99	·4423	·4794	·5103	·5369	·5601	·5808	·5993	·6160
32	0·80	0·2546	0·2941	0·3280	0·3579	0·3845	0·4087	0·4307	0·4508
	·85	·2746	·3141	·3479	·3775	·4039	·4277	·4494	·4693
	·90	·3007	·3400	·3734	·4027	·4287	·4521	·4733	·4927
	·95	·3411	·3798	·4125	·4410	·4661	·4887	·5092	·5278
	·99	·4211	·4575	·4881	·5144	·5376	·5583	·5769	·5938
34	0·80	0·2414	0·2795	0·3123	0·3413	0·3673	0·3909	0·4125	0·4324
	·85	·2605	·2986	·3314	·3602	·3861	·4095	·4308	·4504
	·90	·2855	·3236	·3561	·3847	·4102	·4332	·4542	·4734
	·95	·3244	·3620	·3940	·4219	·4467	·4690	·4893	·5079
	·99	·4018	·4375	·4676	·4937	·5168	·5374	·5561	·5730
36	0·80	0·2295	0·2662	0·2980	0·3261	0·3515	0·3746	0·3958	0·4154
	·85	·2478	·2846	·3164	·3445	·3697	·3927	·4137	·4330
	·90	·2718	·3087	·3403	·3682	·3932	·4158	·4364	·4555
	·95	·3093	·3458	·3770	·4043	·4287	·4508	·4709	·4893
	·99	·3842	·4191	·4488	·4746	·4974	·5180	·5366	·5536
38	0·80	0·2187	0·2541	0·2849	0·3123	0·3370	0·3596	0·3804	0·3997
	·85	·2363	·2719	·3027	·3300	·3547	·3772	·3978	·4169
	·90	·2594	·2951	·3258	·3530	·3775	·3997	·4200	·4388
	·95	·2955	·3309	·3614	·3882	·4122	·4339	·4537	·4720
	·99	·3680	·4022	·4313	·4568	·4794	·4998	·5183	·5353
40	0·80	0·2088	0·2430	0·2729	0·2995	0·3237	0·3458	0·3662	0·3851
	·85	·2258	·2602	·2901	·3167	·3408	·3628	·3831	·4018
	·90	·2480	·2826	·3125	·3390	·3630	·3848	·4048	·4233
	·95	·2828	·3173	·3470	·3732	·3968	·4182	·4377	·4558
	·99	·3531	·3865	·4151	·4402	·4626	·4828	·5012	·5181
42	0·80	0·1998	0·2329	0·2619	0·2878	0·3113	0·3329	0·3529	0·3715
	·85	·2161	·2495	·2785	·3045	·3280	·3495	·3694	·3878
	·90	·2376	·2711	·3002	·3261	·3495	·3709	·3906	·4088
	·95	·2712	·3048	·3337	·3594	·3825	·4035	·4228	·4407
	·99	·3393	·3720	·4001	·4248	·4469	·4669	·4851	·5020

Tafel 5 (Forts.) Table 5 (cont.)

$p = 3$

ν_1	ν_2 / P	3	4	5	6	7	8	9	10
44	0·80	0·1916	0·2236	0·2517	0·2769	0·2999	0·3210	0·3406	0·3588
	·85	·2073	·2396	·2678	·2931	·3161	·3371	·3566	·3748
	·90	·2280	·2605	·2888	·3141	·3370	·3580	·3773	·3953
	·95	·2605	·2931	·3241	·3465	·3692	·3898	·4089	·4265
	·99	·3266	·3586	·3861	·4104	·4321	·4519	·4700	·4867
46	0·80	0·1840	0·2150	0·2423	0·2668	0·2892	0·3099	0·3290	0·3469
	·85	·1991	·2304	·2579	·2825	·3050	·3256	·3447	·3625
	·90	·2192	·2507	·2783	·3030	·3254	·3459	·3649	·3826
	·95	·2506	·2824	·3099	·3345	·3567	·3770	·3958	·4132
	·99	·3147	·3460	·3730	·3969	·4183	·4379	·4558	·4724
50	0·80	0·1704	0·1996	0·2254	0·2487	0·2700	0·2898	0·3082	0·3254
(4)*	·85	·1846	·2141	·2401	·2635	·2849	·3047	·3231	·3403
	·90	·2034	·2332	·2594	·2829	·3043	·3241	·3424	·3595
	·95	·2329	·2630	·2893	·3128	·3342	·3538	·3719	·3888
	·99	·2934	·3233	·3493	·3723	·3932	·4122	·4297	·4460
54	0·80	0·1588	0·1863	0·2107	0·2328	0·2532	0·2721	0·2898	0·3064
	·85	·1721	·1999	·2246	·2469	·2674	·2863	·3040	·3206
	·90	·1897	·2179	·2428	·2653	·2858	·3048	·3225	·3390
	·95	·2175	·2461	·2712	·2937	·3142	·3332	·3507	·3671
	·99	·2748	·3034	·3284	·3506	·3708	·3893	·4064	·4223
58	0·80	0·1486	0·1746	0·1978	0·2189	0·2383	0·2565	0·2734	0·2894
	·85	·1611	·1875	·2110	·2322	·2518	·2700	·2870	·3030
	·90	·1777	·2045	·2282	·2497	·2694	·2876	·3047	·3207
	·95	·2041	·2313	·2552	·2768	·2966	·3148	·3318	·3477
	·99	·2584	·2858	·3098	·3312	·3508	·3687	·3854	·4009
62	0·80	0·1396	0·1643	0·1864	0·2065	0·2251	0·2425	0·2588	0·2742
	·85	·1515	·1765	·1989	·2192	·2380	·2555	·2718	·2873
	·90	·1672	·1927	·2153	·2358	·2547	·2723	·2887	·3042
	·95	·1921	·2181	·2410	·2617	·2807	·2983	·3148	·3302
	·99	·2438	·2701	·2932	·3139	·3328	·3502	·3664	·3816
66	0·80	0·1317	0·1552	0·1762	0·1955	0·2133	0·2300	0·2457	0·2605
(3)*	·85	·1429	·1668	·1881	·2076	·2256	·2424	·2582	·2731
	·90	·1578	·1821	·2038	·2234	·2416	·2585	·2744	·2893
	·95	·1815	·2063	·2283	·2482	·2665	·2835	·2994	·3144
	·99	·2308	·2560	·2782	·2982	·3165	·3335	·3492	·3640
70	0·80	0·1246	0·1470	0·1671	0·1855	0·2026	0·2187	0·2338	0·2482
	·85	·1353	·1580	·1784	·1971	·2144	·2306	·2458	·2602
	·90	·1495	·1727	·1934	·2123	·2297	·2461	·2614	·2758
	·95	·1720	·1958	·2168	·2360	·2536	·2701	·2855	·3000
	·99	·2191	·2433	·2647	·2840	·3018	·3182	·3335	·3479
98	0·80	0·0905	0·1073	0·1226	0·1368	0·1501	0·1627	0·1747	0·1861
(2)*	·85	·0984	·1156	·1312	·1456	·1591	·1719	·1840	·1955
	·90	·1090	·1266	·1425	·1572	·1709	·1839	·1962	·2079
	·95	·1259	·1441	·1604	·1755	·1895	·2026	·2151	·2270
	·99	·1616	·1805	·1975	·2130	·2273	·2408	·2535	·2655
194	0·80	0·0467	0·0557	0·0641	0·0719	0·0794	0·0866	0·0935	0·1001
(1)*	·85	·0508	·0602	·0687	·0768	·0844	·0917	·0987	·1055
	·90	·0565	·0661	·0749	·0831	·0910	·0984	·1056	·1125
	·95	·0655	·0756	·0847	·0933	·1013	·1090	·1164	·1235
	·99	·0849	·0956	·1054	·1144	·1229	·1310	·1388	·1462

Tafel 5 (Forts.) Table 5 (cont.)

$p = 4$

ν_1 \ ν_2	P	4	5	6	7	8	9	10	11
5	0·80	0·9725	0·9779	0·9816	0·9842	0·9862	0·9877	0·9889	0·9899
	·85	·9799	·9839	·9865	·9885	·9899	·9910	·9919	·9926
	·90	·9869	·9895	·9913	·9925	·9934	·9942	·9947	·9952
	·95	·9936	·9949	·9957	·9963	·9968	·9972	·9974	·9977
	·99	·9987	·9990	·9992	·9993	·9994	·9994	·9995	·9995
7	0·80	0·8902	0·9083	0·9212	0·9309	0·9385	0·9445	0·9495	0·9536
	·85	·9078	·9231	·9340	·9422	·9485	·9536	·9578	·9613
	·90	·9272	·9394	·9480	·9545	·9595	·9636	·9668	·9696
	·95	·9505	·9589	·9648	·9692	·9726	·9754	·9776	·9795
	·99	·9789	·9825	·9850	·9869	·9884	·9895	·9905	·9913
9	0·80	0·8034	0·8311	0·8518	0·8678	0·8807	0·8913	0·9001	0·9075
	·85	·8264	·8512	·8695	·8838	·8952	·9045	·9123	·9189
	·90	·8532	·8744	·8900	·9022	·9118	·9197	·9263	·9319
	·95	·8882	·9045	·9166	·9259	·9333	·9393	·9443	·9486
	·99	·9381	·9473	·9541	·9593	·9634	·9668	·9696	·9719
11	0·80	0·7259	0·7596	0·7856	0·8063	0·8233	0·8374	0·8494	0·8598
	·85	·7514	·7824	·8062	·8251	·8405	·8534	·8643	·8737
	·90	·7820	·8095	·8306	·8473	·8610	·8723	·8819	·8901
	·95	·8236	·8463	·8636	·8773	·8884	·8976	·9054	·9121
	·99	·8885	·9032	·9144	·9231	·9302	·9361	·9410	·9453
13	0·80	0·6594	0·6965	0·7259	0·7498	0·7697	0·7865	0·8010	0·8136
	·85	·6858	·7206	·7479	·7701	·7886	·8042	·8176	·8293
	·90	·7180	·7497	·7746	·7947	·8114	·8255	·8376	·8481
	·95	·7632	·7904	·8116	·8287	·8429	·8548	·8650	·8738
	·99	·8374	·8567	·8716	·8836	·8934	·9017	·9088	·9149
15	0·80	0·6028	0·6417	0·6730	0·6989	0·7208	0·7396	0·7559	0·7703
	·85	·6292	·6661	·6957	·7201	·7407	·7584	·7737	·7871
	·90	·6619	·6961	·7235	·7460	·7650	·7812	·7952	·8075
	·95	·7085	·7387	·7628	·7825	·7991	·8131	·8253	·8360
	·99	·7882	·8110	·8290	·8436	·8559	·8663	·8752	·8830
17	0·80	0·5544	0·5940	0·6264	0·6536	0·6768	0·6969	0·7146	0·7302
	·85	·5804	·6183	·6492	·6751	·6971	·7162	·7329	·7477
	·90	·6128	·6485	·6774	·7016	·7222	·7399	·7554	·7691
	·95	·6599	·6920	·7180	·7396	·7579	·7737	·7874	·7995
	·99	·7424	·7677	·7881	·8049	·8190	·8312	·8417	·8510
19	0·80	0·5128	0·5524	0·5853	0·6131	0·6372	0·6582	0·6768	0·6933
	·85	·5381	·5763	·6079	·6346	·6576	·6777	·6955	·7112
	·90	·5700	·6063	·6361	·6614	·6830	·7019	·7185	·7333
	·95	·6166	·6499	·6771	·7000	·7197	·7367	·7517	·7649
	·99	·7003	·7275	·7495	·7680	·7837	·7973	·8092	·8197
21	0·80	0·4768	0·5160	0·5489	0·5770	0·6015	0·6231	0·6423	0·6595
	·85	·5013	·5394	·5712	·5983	·6219	·6427	·6611	·6776
	·90	·5323	·5688	·5991	·6250	·6474	·6670	·6845	·7000
	·95	·5782	·6120	·6400	·6638	·6844	·7024	·7183	·7325
	·99	·6619	·6903	·7136	·7333	·7502	·7650	·7780	·7895
23	0·80	0·4453	0·4839	0·5165	0·5447	0·5693	0·5912	0·6108	0·6285
	·85	·4690	·5066	·5383	·5656	·5895	·6107	·6296	·6466
	·90	·4991	·5353	·5658	·5920	·6148	·6350	·6530	·6692
	·95	·5439	·5779	·6064	·6307	·6519	·6706	·6872	·7021
	·99	·6270	·6561	·6803	·7010	·7188	·7345	·7483	·7607

Tafel 5 (Forts.) Table 5 (cont.)

$p = 4$

ν_1 \ ν_2	P	4	5	6	7	8	9	10	11
25	0·80	0·4177	0·4554	0·4876	0·5155	0·5402	0·5622	0·5820	0·6000
	·85	·4405	·4774	·5089	·5361	·5601	·5815	·6007	·6180
	·90	·4696	·5054	·5358	·5621	·5851	·6056	·6240	·6406
	·95	·5133	·5472	·5758	·6004	·6220	·6412	·6583	·6737
	·99	·5951	·6247	·6495	·6708	·6894	·7057	·7203	·7334
27	0·80	0·3932	0·4300	0·4616	0·4893	0·5138	0·5358	0·5557	0·5738
	·85	·4152	·4513	·4823	·5094	·5333	·5548	·5741	·5917
	·90	·4433	·4786	·5086	·5348	·5580	·5786	·5973	·6141
	·95	·4858	·5194	·5479	·5727	·5945	·6139	·6314	·6472
	·99	·5661	·5959	·6211	·6428	·6619	·6787	·6938	·7075
29	0·80	0·3713	0·4072	0·4382	0·4655	0·4898	0·5116	0·5315	0·5496
	·85	·3925	·4279	·4583	·4851	·5089	·5303	·5496	·5673
	·90	·4197	·4543	·4840	·5100	·5331	·5538	·5725	·5896
	·95	·4609	·4941	·5225	·5472	·5691	·5887	·6064	·6225
	·99	·5395	·5694	·5948	·6168	·6362	·6534	·6690	·6830
31	0·80	0·3518	0·3867	0·4170	0·4438	0·4678	0·4895	0·5092	0·5274
	·85	·3722	·4067	·4365	·4629	·4865	·5077	·5271	·5448
	·90	·3985	·4323	·4615	·4873	·5102	·5309	·5496	·5668
	·95	·4384	·4711	·4992	·5238	·5457	·5654	·5832	·5995
	·99	·5152	·5450	·5704	·5926	·6122	·6297	·6456	·6600
33	0·80	0·3341	0·3681	0·3977	0·4240	0·4476	0·4691	0·4887	0·5067
	·85	·3538	·3874	·4167	·4426	·4659	·4870	·5062	·5239
	·90	·3792	·4123	·4410	·4664	·4891	·5097	·5284	·5456
	·95	·4180	·4500	·4778	·5022	·5240	·5437	·5616	·5779
	·99	·4929	·5224	·5478	·5700	·5898	·6075	·6236	·6382
35	0·80	0·3181	0·3511	0·3801	0·4058	0·4291	0·4503	0·4697	0·4876
	·85	·3371	·3699	·3985	·4240	·4469	·4678	·4869	·5044
	·90	·3617	·3940	·4222	·4472	·4697	·4901	·5087	·5258
	·95	·3993	·4307	·4581	·4822	·5039	·5235	·5414	·5578
	·99	·4724	·5016	·5268	·5490	·5688	·5867	·6029	·6177
37	0·80	0·3036	0·3357	0·3639	0·3891	0·4120	0·4329	0·4521	0·4698
	·85	·3220	·3538	·3818	·4068	·4294	·4500	·4689	·4863
	·90	·3457	·3773	·4049	·4295	·4516	·4718	·4903	·5074
	·95	·3822	·4130	·4399	·4637	·4852	·5047	·5225	·5389
	·99	·4534	·4823	·5073	·5294	·5492	·5671	·5834	·5983
39	0·80	0·2903	0·3215	0·3490	0·3738	0·3962	0·4167	0·4357	0·4532
	·85	·3081	·3391	·3665	·3910	·4132	·4335	·4522	·4695
	·90	·3311	·3618	·3889	·4131	·4349	·4549	·4732	·4901
	·95	·3664	·3966	·4230	·4466	·4678	·4871	·5048	·5212
	·99	·4359	·4644	·4891	·5111	·5308	·5487	·5650	·5800
41	0·80	0·2782	0·3085	0·3353	0·3595	0·3815	0·4017	0·4204	0·4377
	·85	·2953	·3255	·3523	·3763	·3981	·4181	·4365	·4537
	·90	·3176	·3476	·3741	·3978	·4193	·4390	·4572	·4740
	·95	·3519	·3814	·4074	·4306	·4516	·4707	·4883	·5046
	·99	·4196	·4476	·4722	·4939	·5135	·5314	·5477	·5627
43	0·80	0·2670	0·2965	0·3227	0·3463	0·3679	0·3878	0·4061	0·4233
	·85	·2836	·3130	·3391	·3626	·3841	·4037	·4219	·4389
	·90	·3051	·3345	·3604	·3837	·4048	·4242	·4422	·4588
	·95	·3385	·3674	·3929	·4157	·4364	·4553	·4727	·4889
	·99	·4044	·4321	·4563	·4778	·4973	·5151	·5313	·5464

Tafel 5 (Forts.) Table 5 (cont.)

$p = 4$

ν_1	P \ ν_2	4	5	6	7	8	9	10	11
45	0·80	0·2566	0·2853	0·3109	0·3340	0·3552	0·3747	0·3928	0·4097
	·85	·2727	·3014	·3269	·3499	·3710	·3903	·4083	·4250
	·90	·2936	·3222	·3476	·3704	·3913	·4104	·4281	·4446
	·95	·3260	·3543	·3793	·4017	·4221	·4409	·4581	·4742
	·99	·3903	·4175	·4414	·4627	·4820	·4997	·5159	·5309
47	0·80	0·2471	0·2750	0·2999	0·3226	0·3433	0·3625	0·3803	0·3969
	·85	·2626	·2906	·3155	·3381	·3587	·3778	·3954	·4119
	·90	·2830	·3109	·3357	·3581	·3786	·3974	·4149	·4312
	·95	·3144	·3421	·3666	·3887	·4088	·4273	·4444	·4603
	·99	·3772	·4039	·4274	·4485	·4676	·4851	·5013	·5162
51	0·80	0·2299	0·2564	0·2802	0·3018	0·3218	0·3403	0·3575	0·3737
(4)*	·85	·2446	·2712	·2950	·3166	·3365	·3549	·3720	·3881
	·90	·2638	·2904	·3141	·3357	·3554	·3737	·3907	·4066
	·95	·2936	·3200	·3436	·3649	·3844	·4024	·4191	·4347
	·99	·3533	·3791	·4019	·4224	·4412	·4584	·4743	·4891
55	0·80	0·2149	0·2402	0·2629	0·2836	0·3028	0·3206	0·3372	0·3529
	·85	·2288	·2542	·2769	·2977	·3168	·3346	·3512	·3668
	·90	·2470	·2724	·2951	·3159	·3349	·3526	·3691	·3846
	·95	·2752	·3006	·3233	·3439	·3628	·3802	·3965	·4117
	·99	·3322	·3571	·3792	·3992	·4175	·4343	·4499	·4645
59	0·80	0·2018	0·2258	0·2475	0·2674	0·2858	0·3030	0·3191	0·3343
	·85	·2150	·2391	·2609	·2808	·2993	·3164	·3325	·3476
	·90	·2322	·2565	·2783	·2983	·3167	·3338	·3498	·3648
	·95	·2591	·2834	·3052	·3251	·3434	·3603	·3762	·3910
	·99	·3134	·3374	·3589	·3783	·3961	·4126	·4279	·4422
63	0·80	0·1902	0·2131	0·2339	0·2530	0·2707	0·2873	0·3029	0·3176
	·85	·2027	·2258	·2467	·2658	·2836	·3001	·3157	·3304
	·90	·2191	·2423	·2633	·2825	·3002	·3168	·3323	·3470
	·95	·2447	·2680	·2890	·3082	·3259	·3424	·3578	·3723
	·99	·2966	·3198	·3406	·3594	·3768	·3929	·4079	·4219
67	0·80	0·1798	0·2018	0·2217	0·2400	0·2571	0·2731	0·2881	0·3024
(3)*	·85	·1917	·2138	·2339	·2523	·2694	·2854	·3005	·3148
	·90	·2074	·2296	·2498	·2683	·2854	·3015	·3165	·3307
	·95	·2318	·2542	·2745	·2930	·3101	·3261	·3411	·3552
	·99	·2816	·3039	·3240	·3424	·3592	·3749	·3896	·4034
71	0·80	0·1706	0·1915	0·2107	0·2283	0·2448	0·2602	0·2748	0·2886
	·85	·1819	·2031	·2223	·2401	·2566	·2721	·2867	·3005
	·90	·1968	·2182	·2376	·2554	·2720	·2875	·3021	·3160
	·95	·2202	·2418	·2613	·2792	·2958	·3113	·3259	·3396
	·99	·2679	·2896	·3090	·3268	·3432	·3585	·3729	·3864
99	0·80	0·1252	0·1414	0·1562	0·1701	0·1832	0·1956	0·2074	0·2186
(2)*	·85	·1338	·1502	·1652	·1792	·1924	·2049	·2168	·2281
	·90	·1451	·1618	·1770	·1912	·2045	·2171	·2291	·2405
	·95	·1630	·1800	·1955	·2098	·2233	·2360	·2481	·2596
	·99	·2000	·2173	·2331	·2477	·2614	·2742	·2863	·2979
195	0·80	0·0655	0·0744	0·0828	0·0907	0·0982	0·1055	0·1125	0·1192
(1)*	·85	·0701	·0793	·0878	·0958	·1035	·1108	·1179	·1247
	·90	·0763	·0857	·0943	·1025	·1104	·1178	·1250	·1320
	·95	·0862	·0958	·1047	·1132	·1212	·1288	·1362	·1433
	·99	·1068	·1169	·1263	·1351	·1434	·1514	·1590	·1663

TABLE 6

The Trace Criterion T_o^2 of H. Hotelling and D. N. Lawley in the Version of K. C. S. Pillai

(a) Content of the tables and definition of the test statistic :

The tables contain the upper percentage points of the zero distribution of the test statistic

$$\tau \equiv U^{(s)} \equiv T_o^2/n_e = \text{trace} (\underline{S}_h \cdot \underline{S}_e^{-1})$$

$$= \sum_{i=1}^{s} \lambda_i \quad .$$

Here $U^{(s)}$ is as defined in K. C. S. PILLAI (1960) ,

T_o^2 is as defined in H. HOTELLING (1951) .

The relationship with other test statistics is given by the expression

$$\lambda_i = \frac{\Theta_i}{1 - \Theta_i} \quad , \text{ where the } \Theta_i \text{ are the eigenvalues}$$

of the matrix $(\underline{S}_h + \underline{S}_e)^{-1} \cdot \underline{S}_h$.

In addition $\lambda_i = (1 - \mu_k) / \mu_k$, where μ_k is determined by

$$| \underline{S}_e - \mu(\underline{S}_e + \underline{S}_h) | = 0 \quad .$$

(b) Extent of the tables and definition of the parameters :

(1) The parameter α :

α = error probability

α = 5% and 1% .

(2) The parameter s :

$s = \min(p, n_h)$ where p = dimension of the variates
n_h = degrees of freedom of the hypothesis

= 2(1)8 .

(3) The parameter m :

$m = 0.5(|p - n_h| - 1)$

for different values of m between -0.5 and 1000 .

(4) The parameter n :

$n = 0.5(n_e - p - 1)$ where n_e = degrees of freedom of the error

= 15(5)50,60(20)100,130,160,200 with $s = 2$; 3

= 10(5)50,60(20)100,130,160,200 with $s = 4(1)8$.

(c) Suggestions for applications :

(1) Multivariate general linear hypotheses, especially MANOVA-problems; multivariate regression and covariance analyses. Multivariate confidence regions. Canonical discriminant analyses.

(2) To test the independence of p (dependent) from q (independent) variates (multivariate regression analysis) where

$$\underline{S}_h = \underline{R}_{yx} \cdot \underline{R}_x^{-1} \cdot \underline{R}_{xy} \quad \text{and} \quad \underline{S}_e = \underline{R}_y - \underline{R}_{yx} \cdot \underline{R}_x^{-1} \cdot \underline{R}_{xy}$$

(Standardized samples ; $\underline{R}_x$, $\underline{R}_y$, $\underline{R}_{xy}$, $\underline{R}_{yx}$ = correlation matrices)

and p = dimension of the (dependent) vector $\underline{y}$

q = dimension of the (independent) vector $\underline{x}$

using the initial parameters

$$s = \min(p, q)$$

$$m = \frac{p - q - 1}{2}$$

$$n = \frac{N - p - q - 2}{2}$$

where N = sample size .

(d) References :

(1) For the test criterion :

HOTELLING, H.: A generalized T-test and measure of multivariate dispersion. Proceedings of the Second Berkeley Symposium on Mathematical Statistics and Probability Vol. 2, 23 - 41 (1951)

LAWLEY, D. N. : A generalization of Fisher's Z test.
Biometrika 30, 180 - 187 (1938).

(2) For the source of the tables :

PILLAI, K. C. S. : Statistical Tables for Tests of
Multivariate Hypotheses. (Table 3).
Manila/Philippines : The Statistical Center,
The University of the Philippines, 1960 .

(e) Further remarks :

For values outside the tabulated region, a suitable approximation is given by the F-distribution with

$$F = \frac{f_2}{f_1} \cdot \frac{U^{(s)}}{s}$$

and the degrees of freedom

$$f_1 = s(2m + s + 1) \quad \text{and}$$

$$f_2 = 2(s.n + 1) \quad .$$

Tafel 6

Table 6

$s = 2$

Upper 5% Points

Obere 5% Punkte

m \ n	15	20	25	30	35	40	45	50	60	80	100	130	160	200
1	--	--	--	--	--	--	--	--	--	0.116	0.092	0.071	0.058	0.046
1.5	--	--	--	--	0.310	0.270	0.239	0.215	0.179	0.133	0.106	0.081	0.066	0.053
2	--	--	0.496	0.410	0.350	0.305	0.270	0.242	0.202	0.150	0.120	0.092	0.075	0.060
2.5	--	--	0.551	0.456	0.389	0.339	0.300	0.269	0.224	0.167	0.133	0.102	0.083	0.066
3	--	0.768	0.607	0.502	0.428	0.373	0.330	0.296	0.246	0.183	0.146	0.112	0.091	0.073
3.5	--	0.838	0.662	0.547	0.466	0.406	0.360	0.323	0.268	0.200	0.159	0.122	0.099	0.079
4	--	0.907	0.717	0.592	0.505	0.439	0.389	0.349	0.290	0.216	0.172	0.132	0.107	0.085
4.5	--	0.976	0.771	0.637	0.542	0.472	0.418	0.375	0.311	0.232	0.185	0.142	0.115	0.092
5	--	1.045	0.824	0.681	0.580	0.505	0.447	0.401	0.332	0.248	0.197	0.151	0.123	0.098
10	--	1.718	1.352	1.115	0.947	0.823	0.728	0.652	0.541	0.402	0.320	0.245	0.199	0.159
15	3.263	2.381	1.870	1.539	1.306	1.134	1.003	0.899	0.744	0.553	0.440	0.337	0.273	0.217
20	4.171	3.039	2.383	1.958	1.661	1.442	1.275	1.142	0.945	0.701	0.558	0.426	0.345	0.275
25	5.077	3.692	2.894	2.374	2.014	1.749	1.545	1.383	1.143	0.848	0.674	0.515	0.417	0.332
30	5.981	4.345	3.403	2.792	2.366	2.053	1.813	1.622	1.341	0.994	0.790	0.603	0.488	0.389
50	9.590	6.947	5.430	4.449	3.767	3.266	2.881	2.577	2.127	1.573	1.248	0.953	0.770	0.613
60	11.39	8.246	6.442	5.275	4.466	3.870	3.413	3.051	2.516	1.861	1.476	1.126	0.910	0.724
80	14.99	10.84	8.463	6.926	5.860	5.074	4.475	3.998	3.295	2.435	1.930	1.580	1.188	0.945
100	18.59	13.43	10.48	8.575	7.253	6.280	5.534	4.944	4.072	3.007	2.382	1.814	1.465	1.165
130	23.99	17.32	13.51	11.05	9.340	8.084	7.122	6.361	5.237	3.864	3.059	2.328	1.879	1.494
160	29.39	21.21	16.54	13.52	11.43	9.888	8.709	7.777	6.400	4.721	3.735	2.810	2.292	1.822
200	36.58	26.40	20.57	16.81	14.21	12.29	10.82	9.663	7.951	5.861	4.635	3.525	2.842	2.258
500	90.54	65.27	50.83	41.51	35.06	30.31	26.68	23.80	19.57	14.40	11.38	8.640	6.959	5.521
1000	180	130	101	82.67	69.79	60.34	53.09	47.36	38.92	28.63	22.61	17.16	13.81	10.95

Tafel 6 (Forts.) Table 6 (cont.)

$s = 2$

Upper 1% Points

Obere 1% Punkte

m \ n	15	20	25	30	35	40	45	50	60	80	100	130	160	200
1	--	--	--	--	--	--	--	--	--	0.148	0.118	0.091	0.073	0.059
1.5	--	--	--	--	0.396	0.344	0.304	0.273	0.226	0.168	0.134	0.102	0.083	0.066
2	--	--	0.632	0.520	0.442	0.384	0.339	0.303	0.251	0.187	0.149	0.114	0.092	0.074
2.5	--	--	0.697	0.572	0.485	0.422	0.372	0.334	0.276	0.205	0.163	0.125	0.101	0.081
3	--	0.970	0.760	0.624	0.529	0.458	0.406	0.364	0.301	0.224	0.178	0.136	0.110	0.088
3.5	--	1.051	0.822	0.675	0.572	0.497	0.439	0.393	0.325	0.242	0.192	0.147	0.119	0.095
4	--	1.130	0.884	0.725	0.616	0.534	0.472	0.422	0.349	0.259	0.206	0.158	0.128	0.102
4.5	--	1.210	0.945	0.776	0.658	0.571	0.504	0.451	0.373	0.277	0.220	0.168	0.136	0.109
5	--	1.288	1.005	0.825	0.700	0.607	0.536	0.480	0.396	0.294	0.234	0.179	0.145	0.116
10	--	2.057	1.602	1.310	1.108	0.960	0.846	0.757	0.625	0.462	0.367	0.280	0.227	0.181
15	3.934	2.812	2.183	1.782	1.505	1.302	1.147	1.024	0.845	0.624	0.495	0.378	0.305	0.243
20	4.974	3.551	2.758	2.243	1.895	1.640	1.443	1.288	1.061	0.783	0.621	0.473	0.382	0.304
25	6.043	4.302	3.330	2.712	2.286	1.974	1.736	1.549	1.275	0.940	0.745	0.567	0.458	0.365
30	7.092	5.043	3.900	3.173	2.673	2.306	2.028	1.807	1.488	1.097	0.868	0.661	0.533	0.424
50	11.28	7.996	6.167	5.007	4.202	3.626	3.185	2.837	2.328	1.711	1.353	1.029	0.829	0.659
60	13.37	9.468	7.297	5.921	4.973	4.284	3.761	3.350	2.748	2.019	1.594	1.211	0.976	0.775
80	17.55	12.41	9.553	7.745	6.500	5.595	4.910	4.371	3.583	2.628	2.073	1.733	1.267	1.005
100	21.73	15.35	11.81	9.567	8.024	6.905	6.056	5.390	4.415	3.235	2.550	1.933	1.556	1.234
130	27.99	19.76	15.19	12.30	10.31	8.866	7.774	6.916	5.660	4.143	3.263	2.472	1.988	1.576
160	34.25	24.17	18.56	15.03	12.59	10.83	9.489	8.440	6.904	5.050	4.782	2.962	2.419	1.917
200	42.60	30.04	23.07	18.67	15.63	13.44	11.78	10.47	8.562	6.258	4.923	3.724	2.993	2.370
500	105	74.09	56.83	45.94	38.44	33.02	28.91	25.69	20.98	15.30	12.02	9.072	7.279	5.753
1000	210	147	113	91.38	76.45	65.64	57.46	51.04	41.66	30.36	23.83	17.97	14.41	11.38

$s = 3$

Upper 5% Points

Obere 5% Punkte

m \ n	15	20	25	30	35	40	45	50	60	80	100	130	160	200
-0.5	--	--	--	--	--	--	--	--	--	--	0.085	0.066	0.053	0.042
0	--	--	--	0.362	0.309	0.269	0.239	0.214	0.178	0.133	0.106	0.081	0.066	0.053
.5	--	0.658	0.520	0.431	0.368	0.321	0.284	0.255	0.212	0.158	0.126	0.097	0.079	0.063
1	--	0.761	0.603	0.499	0.425	0.371	0.329	0.295	0.245	0.183	0.146	0.112	0.091	0.073
1.5	1.172	0.864	0.683	0.566	0.482	0.420	0.372	0.334	0.278	0.207	0.165	0.127	0.103	0.082
2	1.310	0.964	0.763	0.632	0.538	0.469	0.416	0.373	0.310	0.231	0.184	0.141	0.115	0.092
2.5	1.446	1.065	0.843	0.696	0.594	0.517	0.458	0.411	0.341	0.254	0.203	0.156	0.126	0.101
3	1.581	1.164	0.921	0.761	0.649	0.565	0.500	0.449	0.372	0.278	0.222	0.170	0.138	0.110
3.5	1.717	1.263	0.998	0.825	0.703	0.612	0.542	0.486	0.403	0.293	0.240	0.184	0.149	0.119
4	1.852	1.362	1.076	0.889	0.757	0.659	0.583	0.524	0.434	0.324	0.258	0.198	0.161	0.128
4.5	1.986	1.459	1.153	0.952	0.810	0.706	0.625	0.561	0.465	0.347	0.277	0.212	0.172	0.137
5	2.120	1.557	1.229	1.015	0.864	0.752	0.666	0.598	0.496	0.369	0.295	0.226	0.183	0.146
10	3.438	2.517	1.982	1.634	1.391	1.210	1.071	0.961	0.796	0.593	0.472	0.362	0.293	0.234
15	4.742	3.494	2.725	2.245	1.933	1.661	1.469	1.316	1.090	0.811	0.646	0.495	0.401	0.320
20	6.039	4.407	3.478	2.852	2.424	2.107	1.862	1.668	1.381	1.026	0.817	0.626	0.507	0.404
25	7.336	5.345	4.197	3.455	2.935	2.550	2.253	2.018	1.670	1.241	0.987	0.755	0.612	0.488
30	8.628	6.282	4.930	4.057	3.445	2.991	2.643	2.367	1.958	1.454	1.156	0.884	0.716	0.571
50	13.79	10.02	7.853	6.455	5.475	4.750	4.194	3.754	3.102	2.300	1.827	1.396	1.129	0.900
60	16.37	11.89	9.312	7.651	6.487	5.627	4.967	4.445	3.671	2.721	2.161	1.650	1.335	1.063
80	21.52	15.62	12.23	10.04	8.509	7.378	6.510	5.824	4.808	3.561	2.826	2.157	1.744	1.389
100	26.67	19.34	15.14	12.43	10.53	9.127	8.052	7.201	5.943	4.399	3.490	2.663	2.152	1.713
130	34.39	24.93	19.51	16.01	13.56	11.75	10.36	9.266	7.644	5.654	4.483	3.419	2.762	2.198
160	42.12	30.52	23.87	19.59	16.58	14.37	12.67	11.33	9.344	6.909	5.476	4.175	3.371	2.682
200	52.41	37.98	29.70	24.36	20.62	17.86	15.75	14.08	11.61	8.580	6.799	5.181	4.183	3.327
500	130	93.85	73.34	60.12	50.87	44.05	38.82	34.69	28.58	21.10	16.70	12.71	10.25	8.147
1000	258	187	146	120	101	87.68	77.26	69.03	56.86	41.96	33.20	25.26	20.36	16.17

Tafel 6 (Forts.) Table 6 (cont.)

s = 3

Upper 1% Points

Obere 1% Punkte

m \ n	15	20	25	30	35	40	45	50	60	80	100	130	160	200
-0.5	--	--	--	--	--	--	--	--	--	--	--	0.084	0.068	0.055
0	--	--	--	0.464	0.394	0.341	0.303	0.271	0.225	0.167	0.133	0.102	0.083	0.066
.5	--	0.837	0.658	0.541	0.459	0.400	0.353	0.317	0.262	0.195	0.156	0.119	0.097	0.077
1	--	0.956	0.750	0.617	0.524	0.456	0.403	0.361	0.299	0.223	0.177	0.136	0.110	0.088
1.5	1.475	1.071	0.840	0.691	0.587	0.511	0.451	0.404	0.335	0.249	0.198	0.152	0.123	0.098
2	1.634	1.186	0.930	0.765	0.649	0.564	0.499	0.447	0.370	0.275	0.219	0.168	0.136	0.109
2.5	1.791	1.298	1.018	0.837	0.711	0.617	0.546	0.489	0.405	0.301	0.240	0.183	0.149	0.119
3	1.947	1.410	1.106	0.909	0.771	0.670	0.592	0.531	0.439	0.326	0.260	0.199	0.161	0.129
3.5	2.102	1.521	1.193	0.980	0.831	0.722	0.638	0.572	0.473	0.339	0.279	0.214	0.173	0.138
4	2.252	1.632	1.278	1.050	0.891	0.774	0.684	0.612	0.506	0.376	0.299	0.229	0.186	0.148
4.5	2.408	1.742	1.364	1.120	0.950	0.825	0.729	0.653	0.540	0.401	0.319	0.244	0.198	0.158
5	2.560	1.852	1.449	1.190	1.009	0.876	0.774	0.693	0.573	0.424	0.338	0.259	0.210	0.167
10	4.060	2.925	2.283	1.871	1.583	1.372	1.211	1.084	0.895	0.664	0.528	0.404	0.327	0.260
15	5.541	4.026	3.101	2.536	2.180	1.858	1.639	1.465	1.210	0.896	0.712	0.544	0.440	0.351
20	7.009	5.030	3.938	3.196	2.701	2.338	2.061	1.842	1.520	1.124	0.893	0.681	0.551	0.439
25	8.479	6.072	4.719	3.849	3.253	2.815	2.480	2.216	1.825	1.351	1.072	0.818	0.661	0.526
30	9.943	7.108	5.522	4.505	3.804	3.290	2.897	2.588	2.132	1.576	1.249	0.953	0.770	0.613
50	15.79	11.26	8.723	7.113	5.992	5.176	4.555	4.065	3.344	2.466	1.953	1.488	1.201	0.955
60	18.70	13.33	10.32	8.403	7.082	6.116	5.379	4.799	3.947	2.909	2.302	1.752	1.414	1.125
80	24.53	17.46	13.51	10.99	9.259	7.991	7.026	6.266	5.149	3.791	2.997	2.280	1.839	1.461
100	30.36	21.53	16.70	13.58	11.43	9.864	8.669	7.729	6.348	4.670	3.691	2.806	2.261	1.796
130	39.10	27.79	21.48	17.46	14.69	12,67	11.13	9.922	8.145	5.986	4.728	3.591	2.893	2.297
160	47.83	33.99	26.25	21.33	17.95	15.48	13.59	12.11	9.939	7.301	5.764	4.376	3.523	2.796
200	59.48	41.95	32.62	26.50	22.29	19.21	16.87	15.03	12.33	9.052	7.143	5.419	4.362	3.460
500	147	104	80.36	65.24	54.83	47.24	41.45	36.91	30.25	22.17	17.47	13.23	10.64	8.424
1000	292	207	160	130	109	93.93	82.41	73.36	60.09	44.01	34.66	26.23	21.08	16.68

Tafel 6 (Forts.) Table 6 (cont.)

$s = 4$

Upper 5% Points
Obere 5% Punkte

m \ n	10	15	20	25	30	35	40	45	50	60	80	100	130	160	200
-0.5	--	--	0.691	0.547	0.453	0.387	0.337	0.299	0.268	0.223	0.167	0.133	0.102	0.083	0.066
0	--	1.122	0.827	0.655	0.542	0.463	0.403	0.357	0.321	0.267	0.199	0.159	0.122	0.099	0.079
.5	--	1.302	0.961	0.761	0.630	0.537	0.468	0.415	0.372	0.309	0.231	0.184	0.141	0.114	0.091
1	--	1.482	1.092	0.865	0.716	0.610	0.531	0.471	0.426	0.351	0.262	0.209	0.160	0.130	0.104
1.5	--	1.660	1.223	0.968	0.800	0.682	0.594	0.526	0.472	0.392	0.293	0.234	0.179	0.145	0.116
2	--	1.837	1.352	1.070	0.884	0.753	0.656	0.581	0.522	0.433	0.324	0.258	0.198	0.161	0.128
2.5	--	2.012	1.480	1.171	0.967	0.824	0.718	0.636	0.571	0.474	0.354	0.282	0.216	0.176	0.140
3	--	2.186	1.608	1.271	1.050	0.895	0.780	0.691	0.620	0.515	0.384	0.306	0.235	0.191	0.152
3.5	--	2.359	1.735	1.371	1.132	0.965	0.841	0.745	0.669	0.555	0.414	0.330	0.253	0.205	0.164
4	--	2.532	1.862	1.470	1.214	1.035	0.902	0.799	0.717	0.595	0.444	0.354	0.271	0.220	0.176
4.5	--	2.705	1.988	1.569	1.296	1.104	0.962	0.852	0.765	0.636	0.473	0.378	0.289	0.234	0.187
5	--	2.877	2.113	1.668	1.378	1.174	1.023	0.906	0.813	0.674	0.503	0.401	0.307	0.249	0.199
10	7.174	4.578	3.353	2.644	2.183	1.855	1.618	1.430	1.284	1.064	0.793	0.632	0.484	0.393	0.313
15	9.839	6.266	4.581	3.610	2.978	2.533	2.203	1.949	1.748	1.448	1.078	0.859	0.658	0.533	0.425
20	12.15	7.404	5.803	4.570	3.768	3.204	2.784	2.463	2.207	1.828	1.361	1.083	0.829	0.672	0.536
25	15.15	9.622	7.021	5.526	4.553	3.869	3.363	2.974	2.665	2.207	1.641	1.306	1.000	0.810	0.646
30	17.80	11.30	8.238	6.481	5.337	4.533	4.050	3.464	3.121	2.583	1.920	1.528	1.169	0.947	0.755
50	28.40	17.98	13.09	10.29	8.464	7.184	6.238	5.505	4.938	4.082	3.031	2.410	1.843	1.491	1.189
60	33.69	21.33	15.52	12.19	10.02	8.505	7.385	6.524	5.841	4.829	3.584	2.848	2.177	1.762	1.404
80	44.28	28.00	20.36	15.99	13.14	11.15	9.674	8.544	7.648	6.320	4.687	3.724	2.806	2.301	1.834
100	54.86	34.67	25.21	19.78	16.26	13.78	11.96	10.56	9.453	7.809	5.789	4.597	3.511	2.839	2.261
130	70.74	44.68	32.47	25.48	20.38	17.74	15.39	13.59	12.16	10.04	7.439	5.905	4.508	3.644	2.902
160	86.61	54.69	39.73	31.17	25.59	21.70	18.82	16.61	14.86	12.27	9.088	7.212	5.503	4.448	3.541
200	107	68.03	49.42	38.75	31.82	26.97	23.39	20.64	18.46	15.24	11.28	8.953	6.829	5.514	4.392
500	267	168	122	95.64	78.49	66.49	57.64	50.85	45.48	37.51	27.75	21.99	16.76	13.53	10.76
1000	531	335	243	190	156	132	115	101	90.48	74.62	55.18	43.72	33.30	26.88	21.36

$s = 4$

Upper 1% Points

Obere 1% Punkte

m \ n	10	15	20	25	30	35	40	45	50	60	80	100	130	160	200
-0.5	--	--	0.875	0.687	0.565	0.480	0.418	0.370	0.331	0.274	0.204	0.163	0.125	0.101	0.081
0	--	1.413	1.028	0.807	0.664	0.565	0.491	0.434	0.389	0.322	0.240	0.191	0.146	0.119	0.095
.5	--	1.620	1.177	0.925	0.761	0.647	0.562	0.497	0.446	0.369	0.275	0.219	0.168	0.136	0.108
1	--	1.824	1.325	1.040	0.856	0.727	0.632	0.559	0.501	0.415	0.309	0.246	0.188	0.153	0.122
1.5	--	2.025	1.470	1.154	0.949	0.807	0.701	0.620	0.556	0.460	0.342	0.272	0.209	0.169	0.135
2	--	2.224	1.615	1.266	1.042	0.885	0.769	0.680	0.609	0.504	0.375	0.299	0.229	0.185	0.148
2.5	--	2.421	1.758	1.378	1.134	0.963	0.836	0.739	0.662	0.548	0.408	0.325	0.247	0.201	0.161
3	--	2.618	1.899	1.489	1.225	1.039	0.902	0.798	0.715	0.592	0.440	0.350	0.268	0.217	0.173
3.5	--	2.813	2.039	1.599	1.314	1.116	0.969	0.856	0.767	0.635	0.472	0.376	0.288	0.233	0.186
4	--	3.008	2.180	1.708	1.402	1.191	1.034	0.914	0.819	0.678	0.504	0.401	0.307	0.249	0.198
4.5	--	3.202	2.319	1.817	1.493	1.266	1.100	0.972	0.870	0.722	0.535	0.426	0.326	0.264	0.211
5	--	3.395	2.458	1.925	1.581	1.341	1.165	1.029	0.922	0.763	0.567	0.451	0.345	0.280	0.223
10	8.574	5.305	3.827	2.991	2.452	2.075	1.803	1.591	1.425	1.179	0.875	0.696	0.532	0.431	0.344
15	11.66	7.189	5.177	4.038	3.309	2.802	2.430	2.144	1.919	1.585	1.175	0.934	0.714	0.578	0.461
20	14.74	9.065	6.520	5.079	4.159	3.521	3.050	2.690	2.406	1.986	1.472	1.169	0.893	0.723	0.576
25	17.81	10.94	7.857	6.115	5.004	4.233	3.666	3.233	2.890	2.385	1.767	1.403	1.071	0.867	0.691
30	20.88	12.80	9.191	7.148	5.847	4.942	4.396	3.744	3.373	2.782	2.059	1.634	1.248	1.009	0.804
50	33.13	20.26	14.51	11.27	9.203	7.772	6.722	5.915	5.290	4.357	3.221	2.553	1.947	1.573	1.252
60	39.25	23.99	17.17	13.32	10.88	9.182	7.940	6.991	6.245	5.142	3.798	3.010	2.294	1.852	1.474
80	51.49	31.42	22.47	17.43	14.22	12.00	10.37	9.128	8.150	6.707	4.950	3.920	2.929	2.409	1.916
100	63.73	38.86	27.77	21.53	17.56	14.81	12.80	11.26	10.05	8.269	6.098	4.827	3.673	2.964	2.356
130	82.08	50.02	35.72	27.68	21.74	19.03	16.43	14.46	12.90	10.61	7.818	6.184	4.703	3.793	3.014
160	100	61.17	43.67	33.83	27.57	23.24	20.07	17.65	15.75	12.95	9.536	7.539	5.732	4.621	3.671
200	125	76.04	54.27	42.03	34.24	28.86	24.92	21.91	19.54	16.06	11.82	9.245	7.101	5.717	4.545
500	308	188	134	103	84.27	70.97	61.24	53.83	47.99	39.40	28.97	22.86	17.35	13.97	11.08
1000	614	373	266	206	168	141	122	107	95.39	78.29	57.52	45.38	34.41	27.69	21.95

Tafel 6 (Forts.) Table 6 (cont.)

$s = 5$

Upper 5% Points

Obere 5% Punkte

m \ n	10	15	20	25	30	35	40	45	50	60	80	100	130	160	200
-0.5	--	1.345	0.992	0.786	0.651	0.555	0.484	0.429	0.385	0.319	0.238	0.190	0.146	0.118	0.095
0	--	1.568	1.156	0.915	0.757	0.645	0.562	0.498	0.447	0.371	0.277	0.221	0.170	0.138	0.110
.5	--	1.787	1.316	1.042	0.861	0.734	0.640	0.567	0.509	0.423	0.316	0.252	0.193	0.157	0.125
1	--	2.003	1.476	1.167	0.965	0.822	0.717	0.635	0.570	0.474	0.354	0.282	0.216	0.176	0.140
1.5	--	2.218	1.635	1.291	1.067	0.910	0.793	0.703	0.631	0.524	0.391	0.312	0.239	0.194	0.155
2	--	2.433	1.790	1.415	1.170	0.997	0.869	0.770	0.691	0.574	0.428	0.342	0.262	0.213	0.170
2.5	4.119	2.646	1.946	1.538	1.271	1.084	0.945	0.837	0.751	0.623	0.465	0.371	0.285	0.231	0.184
3	4.451	2.858	2.101	1.660	1.373	1.170	1.020	0.903	0.811	0.673	0.502	0.400	0.307	0.249	0.199
3.5	4.782	3.069	2.256	1.782	1.473	1.256	1.094	0.970	0.870	0.722	0.539	0.430	0.329	0.267	0.213
4	5.112	3.279	2.410	1.903	1.574	1.341	1.169	1.035	0.929	0.771	0.575	0.459	0.352	0.285	0.228
4.5	5.442	3.489	2.563	2.025	1.674	1.427	1.244	1.101	0.988	0.820	0.611	0.488	0.374	0.303	0.242
5	5.771	3.699	2.716	2.146	1.774	1.512	1.317	1.166	1.046	0.868	0.647	0.516	0.396	0.321	0.256
10	9.043	5.781	4.235	3.344	2.761	2.350	2.046	1.811	1.624	1.347	1.004	0.800	0.613	0.497	0.397
15	12.30	7.845	5.742	4.530	3.737	3.179	2.766	2.448	2.195	1.820	1.355	1.080	0.827	0.671	0.535
20	15.54	9.902	7.242	5.709	4.707	4.003	3.482	3.081	2.762	2.289	1.704	1.357	1.039	0.842	0.672
25	18.78	11.96	8.739	6.885	5.673	4.824	4.195	3.711	3.327	2.755	2.050	1.632	1.250	1.013	0.808
30	22.02	14.01	10.23	8.058	6.638	5.643	4.906	4.339	3.889	3.220	2.395	1.907	1.460	1.183	0.944
35	25.26	16.06	11.73	9.230	7.602	6.460	5.616	4.966	4.450	3.684	2.740	2.180	1.669	1.352	1.078
40	28.50	18.10	13.22	10.40	8.564	7.277	6.325	5.592	5.010	4.147	3.083	2.453	1.877	1.520	1.213
45	31.73	20.15	14.71	11.57	9.526	8.092	7.033	6.217	5.570	4.609	3.426	2.726	2.086	1.689	1.347
50	34.97	22.20	16.20	12.74	10.49	8.908	7.740	6.842	6.129	5.071	3.769	2.998	2.293	1.857	1.481
60	41.44	26.29	19.18	15.08	12.41	10.54	9.154	8.090	7.246	5.994	4.453	3.541	2.708	2.192	1.748
80	54.38	34.47	25.13	19.75	16.25	13.79	11.98	10.58	9.477	7.837	5.819	4.625	3.536	2.861	2.281
100	67.32	42.65	31.08	24.42	20.08	17.05	14.80	13.07	11.71	9.678	7.182	5.707	4.361	3.528	2.812
130	86.72	54.91	40.01	31.42	25.83	21.92	19.03	16.81	15.05	12.44	9.225	7.328	5.598	4.527	3.607
160	106	67.18	48.93	38.42	31.58	26.80	23.26	20.54	18.39	15.19	11.27	8.947	6.832	5.525	4.400
200	132	83.53	60.83	47.75	39.24	33.30	28.90	25.52	22.84	18.87	13.99	11.10	8.477	6.853	5.457
500	326	206	150	118	96.72	82.02	71.17	62.81	56.20	46.41	34.37	27.27	20.80	16.80	13.37
1000	649	411	299	234	193	163	142	125	112	92.30	68.33	54.19	41.32	33.37	26.54

$s = 5$

Upper 1% Points

Obere 1% Punkte

m \ n	10	15	20	25	30	35	40	45	50	60	80	100	130	160	200
-0.5	--	1.668	1.212	0.952	0.784	0.666	0.579	0.512	0.459	0.380	0.283	0.226	0.173	0.140	0.112
0	--	1.917	1.393	1.094	0.901	0.766	0.666	0.589	0.528	0.437	0.325	0.259	0.198	0.161	0.128
.5	--	2.162	1.572	1.234	1.016	0.863	0.750	0.664	0.595	0.492	0.366	0.292	0.223	0.181	0.144
1	--	2.405	1.748	1.372	1.129	0.959	0.834	0.737	0.661	0.547	0.407	0.324	0.248	0.201	0.161
1.5	--	2.646	1.922	1.509	1.241	1.054	0.916	0.810	0.726	0.601	0.447	0.356	0.273	0.221	0.176
2	--	2.885	2.095	1.644	1.352	1.148	0.998	0.882	0.790	0.654	0.487	0.388	0.297	0.241	0.192
2.5	5.005	3.123	2.267	1.778	1.463	1.241	1.078	0.953	0.854	0.708	0.526	0.419	0.321	0.260	0.208
3	5.387	3.360	2.438	1.912	1.572	1.334	1.159	1.025	0.918	0.760	0.566	0.450	0.345	0.279	0.223
3.5	5.768	3.596	2.607	2.045	1.680	1.426	1.239	1.095	0.982	0.813	0.604	0.481	0.369	0.299	0.238
4	6.148	3.831	2.776	2.177	1.788	1.520	1.319	1.166	1.045	0.865	0.643	0.512	0.392	0.318	0.253
4.5	6.527	4.064	2.945	2.308	1.896	1.609	1.400	1.236	1.107	0.916	0.682	0.543	0.415	0.337	0.268
5	6.906	4.297	3.113	2.439	2.003	1.701	1.477	1.305	1.170	0.968	0.720	0.573	0.439	0.355	0.284
10	10.66	6.604	4.774	3.731	3.064	2.598	2.255	1.992	1.783	1.475	1.096	0.872	0.667	0.540	0.431
15	14.39	8.891	6.412	5.008	4.108	3.481	3.020	2.666	2.386	1.973	1.464	1.164	0.890	0.721	0.575
20	18.11	11.17	8.042	6.278	5.145	4.358	3.778	3.334	2.984	2.465	1.829	1.453	1.111	0.899	0.717
25	21.81	13.44	9.667	7.541	6.178	5.229	4.532	3.999	3.578	2.954	2.191	1.740	1.330	1.076	0.858
30	25.52	15.70	11.29	8.802	7.207	6.098	5.284	4.662	4.169	3.442	2.551	2.026	1.548	1.252	0.997
35	29.22	17.96	12.91	10.06	8.235	6.966	6.034	5.322	4.759	3.927	2.909	2.310	1.764	1.426	1.136
40	32.93	20.23	14.53	11.32	9.261	7.832	6.783	5.981	5.348	4.412	3.267	2.593	1.980	1.601	1.275
45	36.63	22.49	16.14	12.57	10.29	8.696	7.531	6.639	5.935	4.896	3.624	2.876	2.195	1.774	1.413
50	40.32	24.75	17.76	13.83	11.31	9.560	8.278	7.297	6.522	5.379	3.981	3.158	2.409	1.947	1.551
60	47.72	29.26	20.99	16.34	13.36	11.29	9.770	8.610	7.694	6.343	4.692	3.720	2.836	2.293	1.825
80	62.51	38.29	27.45	21.35	17.44	14.74	12.75	11.23	10.03	8.268	6.111	4.842	3.691	2.981	2.372
100	77.30	47.32	33.90	26.36	21.53	18.18	15.73	13.85	12.37	10.19	7.526	5.961	4.542	3.667	2.917
130	99.47	60.86	43.58	33.87	27.64	23.34	20.19	17.77	15.87	13.07	9.646	7.636	5.815	4.693	3.732
160	122	74.40	53.25	41.37	33.77	28.50	24.65	21.70	19.37	15.95	11.76	9.309	7.086	5.717	4.545
200	151	92.44	66.15	51.38	41.93	35.38	30.59	26.92	24.03	19.78	14.58	11.54	8.778	7.081	5.627
500	373	228	163	126	103	86.96	75.14	66.10	58.99	48.50	35.72	28.23	21.45	17.29	13.72
1000	742	453	324	251	205	173	149	131	117	96.36	70.93	56.03	42.55	34.27	27.19

Tafel 6 (Forts.) Table 6 (cont.)

s = 6

Upper 5% Points

Obere 5% Punkte

m \ n	10	15	20	25	30	35	40	45	50	60	80	100	130	160	200
-0.5	--	1.829	1.347	1.066	0.882	0.752	0.655	0.580	0.521	0.433	0.323	0.258	0.198	0.161	0.128
0	--	2.086	1.537	1.216	1.005	0.857	0.747	0.662	0.594	0.494	0.369	0.294	0.226	0.183	0.146
.5	3.641	2.342	1.725	1.363	1.127	0.961	0.838	0.743	0.667	0.554	0.413	0.330	0.253	0.205	0.164
1	4.037	2.594	1.910	1.510	1.249	1.065	0.928	0.823	0,739	0.613	0.458	0.365	0.280	0.227	0.181
1.5	4.431	2.848	2.095	1.656	1.370	1.168	1.018	0.902	0.810	0.672	0.502	0.400	0.307	0.249	0.199
2	4.823	3.098	2.279	1.802	1.490	1.271	1.107	0.981	0.881	0.731	0.545	0.435	0.334	0.271	0.216
2.5	5.214	3.349	2.462	1.947	1.610	1.373	1.196	1.060	0.951	0.789	0.589	0.470	0.360	0.292	0.233
3	5.604	3.599	2.645	2.091	1.729	1.474	1.284	1.138	1.021	0.847	0.632	0.504	0.387	0.314	0.251
3.5	5.994	3.848	2.827	2.235	1.848	1.575	1.372	1.216	1.091	0.905	0.676	0.539	0.413	0.335	0.268
4	6.382	4.096	3.008	2.379	1.967	1.676	1.460	1.293	1.161	0.963	0.719	0.573	0.440	0.357	0.285
4.5	6.770	4.344	3.189	2.522	2.085	1.777	1.547	1.370	1.230	1.021	0.761	0.607	0.466	0.378	0.302
5	7.158	4.591	3.370	2.665	2.203	1.877	1.634	1.448	1.299	1.078	0.804	0.641	0.492	0.399	0.319
10	11.02	7.045	5.168	4.081	3.370	2.869	2.498	2.212	1.984	1.646	1.227	0.978	0.750	0.608	0.485
15	14.86	9.483	6.951	5.484	4.525	3.851	3.352	2.967	2.661	2.206	1.644	1.310	1.004	0.814	0.650
20	18.70	11.92	8.727	6.881	5.675	4.829	4.201	3.718	3.334	2.763	2.058	1.639	1.256	1.018	0.813
25	22.53	14.34	10.50	8.273	6.822	5.803	5.048	4.466	4.004	3.317	2.470	1.967	1.507	1.221	0.974
30	26.35	16.77	12.27	9.664	7.966	6.775	5.892	5.212	4.672	3.870	2.881	2.294	1.757	1.423	1.136
35	30.18	19.19	14.04	11.05	9.109	7.745	6.735	5.957	5.339	4.422	3.291	2.620	2.006	1.625	1.296
40	34.00	21.61	15.81	12.44	10.25	8.715	7.577	6.701	6.005	4.973	3.700	2.945	2.254	1.826	1.457
45	37.83	24.03	17.57	13.82	11.39	9.684	8.418	7.444	6.671	5.523	4.108	3.269	2.502	2.027	1.617
50	41.65	26.45	19.34	15.21	12.53	10.65	9.253	8.182	7.333	6.072	4.516	3.593	2.750	2.227	1.776
60	49.29	31.29	22.86	17.99	14.81	12.59	10.93	9.666	8.662	7.170	5.331	4.241	3.244	2.627	2.095
80	64.58	40.97	29.92	23.53	19.37	16.45	14.29	12.63	11.32	9.364	6.957	5.532	4.231	3.425	2.731
100	79.86	50.64	36.97	29.06	23.92	20.32	17.64	15.59	13.97	11.55	8.581	6.822	5.216	4.221	3.365
130	103	65.14	47.55	37.37	30.75	26.11	22.64	20.03	17.94	14.84	11.02	8.754	6.691	5.413	4.314
160	126	79.65	58.12	45.67	37.57	31.90	27.69	24.47	21.91	18.12	13.45	10.68	8.164	6.604	5.262
200	156	98.99	72.22	56.73	46.67	39.62	34.38	30.38	27.20	22.49	16.69	13.26	10.13	8.190	6.524
500	385	244	178	140	115	97.50	84.63	74.74	66.90	55.28	40.98	32.53	24.83	20.07	15.98
1000	767	486	354	278	229	194	168	149	133	110	81.45	64.64	49.32	39.85	31.71

s = 6

Upper 1% Points
Obere 1% Punkte

m \ n	10	15	20	25	30	35	40	45	50	60	80	100	130	160	200
-0.5	--	2.208	1.602	1.261	1.038	0.882	0.767	0.678	0.608	0.503	0.374	0.298	0.228	0.185	0.148
0	--	2.495	1.814	1.425	1.172	0.996	0.866	0.766	0.686	0.568	0.423	0.337	0.258	0.209	0.167
.5	4.445	2.780	2.020	1.586	1.305	1.109	0.963	0.852	0.764	0.632	0.471	0.375	0.287	0.233	0.186
1	4.899	3.062	2.225	1.746	1.437	1.220	1.060	0.937	0.840	0.696	0.518	0.413	0.316	0.256	0.204
1.5	5.350	3.343	2.427	1.905	1.567	1.330	1.156	1.022	0.916	0.759	0.565	0.450	0.345	0.279	0.223
2	5.799	3.621	2.628	2.063	1.696	1.440	1.252	1.107	0.992	0.821	0.611	0.487	0.373	0.302	0.241
2.5	6.246	3.897	2.828	2.219	1.824	1.549	1.347	1.191	1.067	0.884	0.658	0.524	0.401	0.325	0.259
3	6.692	4.173	3.028	2.374	1.952	1.658	1.441	1.274	1.141	0.945	0.703	0.560	0.429	0.347	0.277
3.5	7.136	4.447	3.226	2.529	2.080	1.766	1.535	1.357	1.216	1.007	0.749	0.596	0.457	0.370	0.295
4	7.580	4.721	3.422	2.683	2.209	1.873	1.628	1.439	1.290	1.068	0.794	0.632	0.484	0.392	0.313
4.5	8.022	4.995	3.621	2.837	2.333	1.981	1.721	1.521	1.363	1.129	0.839	0.668	0.512	0.415	0.331
5	8.464	5.268	3.819	2.991	2.459	2.088	1.814	1.603	1.437	1.189	0.884	0.704	0.539	0.437	0.349
10	12.85	7.971	5.763	4.512	3.706	3.144	2.729	2.411	2.160	1.789	1.328	1.057	0.809	0.655	0.523
15	17.20	10.65	7.689	6.014	4.936	4.185	3.631	3.207	2.872	2.375	1.764	1.403	1.073	0.869	0.693
20	21.55	13.32	9.607	7.508	6.159	5.219	4.527	3.997	3.578	2.957	2.195	1.746	1.335	1.081	0.862
25	25.88	15.98	11.52	8.997	7.376	6.248	5.417	4.783	4.280	3.537	2.624	2.086	1.594	1.290	1.029
30	30.22	18.64	13.43	10.48	8.591	7.275	6.308	5.566	4.980	4.114	3.051	2.424	1.850	1.499	1.195
35	34.55	21.30	15.33	11.97	9.803	8.299	7.194	6.347	5.678	4.689	3.477	2.761	2.109	1.706	1.360
40	38.88	23.95	17.24	13.45	11.02	9.322	8.079	7.127	6.380	5.265	3.901	3.097	2.366	1.914	1.525
45	43.21	26.61	19.14	14.93	12.23	10.34	8.964	7.906	7.071	5.837	4.325	3.433	2.621	2.120	1.689
50	47.54	29.26	21.04	16.41	13.43	11.37	9.847	8.684	7.766	6.410	4.748	3.768	2.877	2.326	1.853
60	56.19	34.56	24.85	19.37	15.85	13.41	11.61	10.24	9.155	7.553	5.591	4.436	3.386	2.737	2.180
80	73.49	45.17	32.45	25.28	20.68	17.48	15.14	13.34	11.93	9.836	7.270	5.769	4.401	3.556	2.831
100	90.79	55.77	40.04	31.18	25.50	21.55	18.66	16.44	14.70	12.12	8.956	7.099	5.413	4.373	3.480
130	117	71.67	51.43	40.04	32.73	27.66	23.94	21.09	18.84	15.53	11.47	9.091	6.928	5.595	4.451
160	143	87.56	62.83	48.89	39.96	33.76	29.21	25.73	22.99	18.94	13.99	11.08	8.440	6.814	5.420
200	177	109	78.01	60.69	49.59	41.89	36.24	31.92	28.52	23.49	17.34	13.73	10.46	8.439	6.710
500	437	268	192	149	122	103	88.96	78.32	69.94	57.57	42.44	33.58	25.54	20.60	16.36
1000	869	533	382	297	242	204	177	156	139	114	84.27	66.64	50.66	40.83	32.42

$s = 7$

Upper 5% Points

Obere 5% Punkte

m \ n	10	15	20	25	30	35	40	45	50	60	80	100	130	160	200
-0.5	3.704	2.383	1.755	1.387	1.147	0.978	0.853	0.756	0.679	0.564	0.421	0.336	0.257	0.209	0.167
0	4.161	2.677	1.970	1.557	1.289	1.099	0.958	0.849	0.762	0.633	0.472	0.377	0.289	0.234	0.187
.5	4.616	2.968	2.184	1.727	1.429	1.219	1.062	0.941	0.845	0.701	0.523	0.417	0.320	0.260	0.208
1	5.069	3.259	2.397	1.896	1.568	1.337	1.166	1.033	0.927	0.769	0.574	0.458	0.351	0.285	0.228
1.5	5.521	3.549	2.609	2.064	1.707	1.456	1.268	1.124	1.009	0.837	0.625	0.498	0.382	0.310	0.248
2	5.971	3.837	2.820	2.231	1.845	1.573	1.370	1.214	1.090	0.905	0.675	0.538	0.413	0.335	0.268
2.5	6.420	4.125	3.031	2.398	1.983	1.690	1.472	1.304	1.171	0.972	0.725	0.578	0.444	0.360	0.287
3	6.868	4.411	3.241	2.564	2.120	1.807	1.574	1.394	1.251	1.039	0.775	0.618	0.474	0.385	0.307
3.5	7.317	4.697	3.451	2.729	2.257	1.923	1.675	1.484	1.332	1.105	0.825	0.658	0.505	0.409	0.327
4	7.765	4.982	3.660	2.895	2.393	2.039	1.776	1.573	1.412	1.17[illegible]	0.874	0.697	0.535	0.434	0.346
4.5	8.212	5.266	3.869	3.060	2.529	2.154	1.877	1.662	1.492	1.238	0.924	0.737	0.565	0.458	0.366
5	8.659	5.550	4.078	3.224	2.664	2.270	1.977	1.751	1.572	1.304	0.973	0.776	0.595	0.483	0.385
10	13.11	8.377	6.150	4.856	4.010	[illegible].415	2.974	2.633	2.363	1.960	1.461	1.165	0.893	0.724	0.578
15	17.53	11.19	8.207	6.476	5.345	4.550	3.961	3.507	3.145	2.608	1.944	1.549	1.187	0.963	0.769
20	21.95	13.99	10.26	8.089	6.674	5.680	4.943	4.375	3.924	3.253	2.423	1.931	1.480	1.199	0.957
25	26.36	16.79	12.31	9.698	8.000	6.807	5.922	5.241	4.699	3.895	2.901	2.311	1.770	1.435	1.145
30	30.77	19.59	14.35	11.31	9.323	7.931	6.899	6.104	5.473	4.535	3.377	2.690	2.060	1.669	1.332
35	35.18	22.38	16.39	12.91	10.65	9.054	7.875	6.966	6.246	5.174	3.852	3.067	2.349	1.903	1.519
40	39.58	25.18	18.43	14.52	11.97	10.18	8.849	7.827	7.017	5.813	4.326	3.444	2.637	2.136	1.705
45	43.99	27.97	20.47	16.12	13.29	11.30	9.823	8.688	7.788	6.450	4.799	3.821	2.925	2.369	1.891
50	48.39	30.76	22.51	17.72	14.61	12.42	10.80	9.548	8.558	7.088	5.272	4.197	3.213	2.602	2.076
60	57.20	36.34	26.59	20.92	17.24	14.66	12.74	11.27	10.10	18.360	6.217	4.948	3.787	3.067	2.446
80	74.81	47.50	34.74	27.33	22.51	19.13	16.62	14.70	13.17	10.90	8.104	6.447	4.932	3.993	3.185
100	92.42	58.66	42.88	33.73	27.78	23.60	20.51	18.13	16.24	13.44	9.988	7.944	6.075	4.918	3.922
130	119	75.40	55.10	43.33	35.68	30.31	26.33	23.27	20.85	17.25	12.81	10.19	7.789	6.303	5.025
160	145	92.13	67.32	52.92	43.57	37.01	32.15	28.41	25.45	21.05	15.63	12.43	9.500	7.687	6.126
200	180	114	83.60	65.72	54.10	45.94	39.91	35.27	31.59	26.12	19.40	15.41	11.78	9.530	7.594
500	444	282	206	162	133	113	98.07	86.64	77.59	64.14	47.58	37.79	28.86	23.33	18.58
1000	885	561	410	322	265	225	195	172	154	127	95.54	75.07	57.31	46.32	36.88

$s = 7$

Upper 1% Points

Obere 1% Punkte

m \ n	10	15	20	25	30	35	40	45	50	60	80	100	130	160	200
-0.5	4.515	2.824	2.053	1.612	1.327	1.125	0.978	0.865	0.776	0.643	0.478	0.381	0.292	0.237	0.189
0	5.037	3.150	2.289	1.798	1.476	1.255	1.091	0.964	0.865	0.717	0.533	0.425	0.325	0.264	0.211
.5	5.555	3.473	2.522	1.980	1.627	1.383	1.203	1.064	0.953	0.790	0.588	0.468	0.359	0.291	0.232
1	6.071	3.792	2.754	2.158	1.777	1.510	1.313	1.161	1.041	0.862	0.642	0.511	0.391	0.317	0.253
1.5	6.585	4.111	2.985	2.339	1.926	1.637	1.423	1.258	1.128	0.934	0.695	0.553	0.424	0.344	0.274
2	7.096	4.428	3.215	2.520	2.074	1.762	1.531	1.354	1.214	1.005	0.748	0.596	0.456	0.370	0.295
2.5	7.606	4.744	3.444	2.699	2.222	1.887	1.639	1.450	1.300	1.076	0.801	0.638	0.488	0.396	0.316
3	8.115	5.059	3.664	2.877	2.368	2.011	1.748	1.545	1.385	1.147	0.855	0.680	0.520	0.422	0.337
3.5	8.621	5.374	3.891	3.055	2.514	2.135	1.855	1.640	1.470	1.217	0.906	0.721	0.552	0.448	0.357
4	9.126	5.687	4.117	3.232	2.659	2.258	1.962	1.735	1.555	1.287	0.958	0.763	0.584	0.473	0.378
4.5	9.629	5.999	4.343	3.409	2.804	2.381	2.069	1.829	1.639	1.357	1.010	0.804	0.616	0.499	0.398
5	10.13	6.310	4.569	3.586	2.949	2.504	2.175	1.923	1.723	1.427	1.061	0.845	0.647	0.524	0.418
10	15.14	9.403	6.802	5.330	4.380	3.716	3.227	2.852	2.555	2.114	1.572	1.251	0.958	0.776	0.619
15	20.13	12.48	9.015	7.057	5.794	4.914	4.266	3.769	3.375	2.792	2.075	1.651	1.263	1.023	0.816
20	25.10	15.54	11.22	8.775	7.201	6.105	5.298	4.679	4.189	3.464	2.573	2.046	1.565	1.267	1.010
25	30.06	18.59	13.42	10.49	8.603	7.291	6.326	5.585	5.000	4.133	3.069	2.439	1.865	1.510	1.204
30	35.02	21.64	15.61	12.20	10.00	8.475	7.350	6.489	5.808	4.800	3.562	2.830	2.161	1.751	1.396
35	39.98	24.69	17.80	13.90	11.40	9.656	8.373	7.390	6.614	5.465	4.053	3.220	2.461	1.991	1.587
40	44.93	27.74	19.99	15.61	12.79	10.84	9.394	8.290	7.418	6.128	4.544	3.609	2.758	2.231	1.778
45	49.89	30.78	22.18	17.31	14.19	12.01	10.41	9.190	8.222	6.791	5.033	3.997	3.054	2.470	1.969
50	54.84	33.83	24.36	19.02	15.58	13.19	11.43	10.09	9.025	7.453	5.522	4.385	3.349	2.709	2.159
60	64.75	39.92	28.74	22.42	18.36	15.54	13.47	11.88	10.63	8.774	6.499	5.159	3.939	3.185	2.538
80	84.55	52.08	37.47	29.22	23.92	20.24	17.53	15.47	13.83	11.41	8.447	6.702	5.115	4.135	3.293
100	104	64.25	46.21	36.02	29.48	24.94	21.60	19.05	17.03	14.05	10.39	8.242	6.288	5.082	4.046
130	134	82.49	59.30	46.21	37.81	31.98	27.69	24.41	21.82	17.99	13.31	10.55	8.044	6.499	5.173
160	164	100	72.40	56.40	46.14	39.01	33.78	29.77	26.61	21.94	16.21	12.85	9.798	7.911	6.297
200	203	125	89.86	69.99	57.24	48.39	41.89	36.92	33.00	27.20	20.09	15.92	12.13	9.798	7.794
500	500	307	220	172	140	119	103	90.51	80.86	66.60	49.16	38.92	29.62	23.90	19.00
1000	995	611	438	342	279	236	204	180	161	132	97.57	77.21	58.75	47.38	37.64

Tafel 6 (Forts.)

Table 6 (cont.)

$s = 8$

Upper 5% Points

Obere 5% Punkte

m \ n	10	15	20	25	30	35	40	45	50	60	80	100	130	160	200
-0.5	4.678	3.008	2.214	1.751	1.448	1.235	1.077	0.954	0.856	0.711	0.531	0.423	0.325	0.263	0.210
0	5.192	3.339	2.456	1.943	1.607	1.371	1.194	1.058	0.950	0.788	0.589	0.469	0.360	0.292	0.233
.5	5.704	3.668	2.697	2.134	1.765	1.505	1.311	1.162	1.043	0.866	0.646	0.515	0.395	0.321	0.256
1	6.215	3.995	2.937	2.324	1.922	1.639	1.428	1.265	1.135	0.942	0.703	0.561	0.430	0.349	0.279
1.5	6.725	4.321	3.176	2.513	2.079	1.772	1.543	1.367	1.227	1.018	0.760	0.607	0.465	0.377	0.301
2	7.234	4.645	3.415	2.702	2.234	1.904	1.659	1.470	1.319	1.095	0.817	0.652	0.500	0.405	0.324
2.5	7.743	4.969	3.653	2.890	2.389	2.036	1.774	1.572	1.411	1.171	0.874	0.697	0.535	0.434	0.346
3	8.250	5.292	3.891	3.078	2.544	2.168	1.889	1.673	1.502	1.247	0.930	0.742	0.569	0.462	0.369
3.5	8.757	5.615	4.128	3.265	2.698	2.299	2.003	1.775	1.593	1.322	0.986	0.787	0.603	0.489	0.391
4	9.263	5.937	4.364	3.451	2.852	2.430	2.117	1.876	1.684	1.397	1.043	0.831	0.638	0.517	0.413
4.5	9.768	6.259	4.601	3.638	3.006	2.561	2.231	1.977	1.774	1.472	1.098	0.876	0.672	0.545	0.435
5	10.27	6.580	4.837	3.824	3.159	2.692	2.345	2.077	1.864	1.547	1.154	0.920	0.706	0.573	0.457
10	15.28	9.777	7.182	5.671	4.684	3.990	3.474	3.077	2.760	2.290	1.708	1.361	1.044	0.846	0.676
15	20.30	12.96	9.513	7.506	6.198	5.277	4.594	4.067	3.649	3.026	2.256	1.798	1.378	1.117	0.892
20	25.30	16.13	11.84	9.335	7.705	6.559	5.708	5.053	4.532	3.758	2.801	2.231	1.710	1.386	1.107
25	30.29	19.30	14.16	11.16	9.209	7.837	6.819	6.036	5.413	4.487	3.343	2.663	2.041	1.654	1.320
30	35.27	22.47	16.47	12.98	10.71	9.113	7.929	7.017	6.292	5.215	3.884	3.094	2.370	1.921	1.533
35	40.26	25.64	18.79	14.80	12.21	10.39	9.037	7.996	7.170	5.942	4.424	3.523	2.699	2.187	1.745
40	45.24	28.80	21.10	16.62	13.71	11.66	10.14	8.974	8.047	6.667	4.963	3.952	3.027	2.453	1.957
45	50.23	31.96	23.41	18.44	15.21	12.93	11.25	9.952	8.922	7.392	5.502	4.381	3.354	2.718	2.169
50	55.21	35.13	25.72	20.26	16.71	14.21	12.35	10.93	9.798	8.116	6.040	4.808	3.682	2.983	2.380
60	65.17	41.45	30.34	23.89	19.70	16.75	14.56	12.88	11.55	9.563	7.114	5.663	4.335	3.511	2.801
80	85.10	54.09	39.58	31.16	25.68	21.83	18.98	16.78	15.04	12.45	9.261	7.369	5.639	4.567	3.643
100	105	66.73	48.82	38.42	31.66	26.91	23.39	20.68	18.53	15.34	11.40	9.072	6.941	5.620	4.482
130	135	85.69	62.67	49.31	40.62	34.50	30.00	26.52	23.77	19.67	14.62	11.63	8.892	7.198	5.739
160	165	105	76.52	60.20	49.59	42.13	36.61	32.37	29.00	24.00	17.83	14.18	10.84	8.773	6.994
200	205	130	94.98	74.71	61.54	52.28	45.42	40.15	35.98	29.76	22.11	17.58	13.44	10.87	9.666
500	503	319	233	184	151	128	112	98.54	88.26	72.99	54.18	43.05	32.89	26.60	21.19
1000	1001	635	464	365	300	255	222	196	175	145	108	85.50	65.29	52.79	42.04

Tafel 6 (Forts.)

Table 6 (cont.)

$s = 8$

Upper 1% Points
Obere 1% Punkte

m \ n	10	15	20	25	30	35	40	45	50	60	80	100	130	160	200
-0.5	5.624	3.516	2.554	2.005	1.649	1.401	1.218	1.077	0.966	0.800	0.595	0.474	0.363	0.294	0.235
0	6.207	3.878	2.817	2.211	1.819	1.546	1.343	1.188	1.065	0.882	0.657	0.523	0.401	0.325	0.259
.5	6.787	4.238	3.079	2.415	1.988	1.689	1.468	1.298	1.164	0.964	0.718	0.571	0.438	0.355	0.283
1	7.364	4.598	3.340	2.619	2.155	1.831	1.591	1.408	1.262	1.045	0.778	0.619	0.475	0.385	0.307
1.5	7.939	4.955	3.598	2.822	2.322	1.973	1.715	1.516	1.359	1.126	0.838	0.667	0.511	0.414	0.331
2	8.510	5.311	3.854	3.023	2.488	2.113	1.837	1.625	1.456	1.206	0.897	0.715	0.548	0.444	0.354
2.5	9.080	5.665	4.110	3.225	2.653	2.254	1.959	1.732	1.552	1.286	0.957	0.762	0.584	0.473	0.378
3	9.649	6.018	4.366	3.425	2.818	2.394	2.080	1.839	1.648	1.365	1.016	0.809	0.620	0.502	0.401
3.5	10.22	6.370	4.620	3.625	2.982	2.533	2.201	1.946	1.744	1.444	1.075	0.856	0.656	0.531	0.424
4	10.78	6.722	4.875	3.824	3.146	2.672	2.322	2.053	1.840	1.523	1.133	0.903	0.691	0.560	0.447
4.5	11.35	7.073	5.128	4.023	3.309	2.810	2.442	2.159	1.935	1.602	1.192	0.949	0.727	0.589	0.470
5	11.92	7.423	5.382	4.221	3.472	2.948	2.562	2.265	2.030	1.681	1.250	0.996	0.763	0.618	0.493
10	17.55	10.91	7.897	6.188	5.086	4.317	3.750	3.314	2.969	2.457	1.828	1.455	1.114	0.902	0.720
15	23.15	14.37	10.39	8.139	6.684	5.671	4.924	4.351	3.897	3.225	2.397	1.907	1.459	1.182	0.943
20	28.75	17.82	12.88	10.08	8.275	7.018	6.092	5.381	4.819	3.986	2.964	2.356	1.802	1.459	1.164
25	34.33	21.27	15.36	12.01	9.860	8.360	7.255	6.408	5.737	4.744	3.523	2.801	2.142	1.734	1.383
30	39.91	24.71	17.84	13.95	11.44	9.699	8.415	7.431	6.653	5.500	4.082	3.245	2.481	2.009	1.602
35	45.49	28.14	20.31	15.87	13.02	11.04	9.573	8.452	7.566	6.254	4.640	3.688	2.819	2.282	1.819
40	51.07	31.58	22.78	17.80	14.60	12.37	10.73	9.472	8.478	7.006	5.197	4.129	3.156	2.554	2.036
45	56.64	35.01	25.25	19.73	16.18	13.70	11.89	10.49	9.389	7.757	5.752	4.570	3.492	2.826	2.253
50	62.22	38.45	27.72	21.65	17.75	15.04	13.04	11.51	10.30	8.508	6.307	5.010	3.828	3.097	2.469
60	73.36	45.31	32.66	25.50	20.90	17.70	15.35	13.54	12.12	10.01	7.415	5.889	4.498	3.639	2.900
80	95.59	59.03	42.53	33.19	27.19	23.02	19.95	17.60	15.75	13.00	9.627	7.642	5.835	4.718	3.759
100	118	72.75	52.39	40.87	33.48	28.33	24.56	21.66	19.37	15.99	11.84	9.392	7.168	5.795	4.615
130	151	93.32	67.18	52.40	42.91	36.29	31.46	27.74	24.81	20.47	15.14	12.01	9.165	7.407	5.897
160	185	114	81.97	63.92	52.33	44.28	38.35	33.82	30.24	24.94	18.45	14.63	11.16	9.016	7.177
200	229	141	102	79.27	64.90	54.90	47.55	41.91	37.48	30.90	22.85	18.12	13.82	11.16	8.880
500	563	347	250	194	159	135	116	103	91.76	75.61	55.86	44.25	33.71	27.21	21.63
1000	1120	690	496	386	316	267	231	204	182	150	111	87.79	66.83	53.92	42.86

TABLE 7

The Trace Criterion of H. Hotelling and D. N. Lawley in the Version $V^{(s)}$ of K. C. S. Pillai

(a) Content of the tables and definition of the test statistic :

The tables contain the upper percentage points of the zero distribution of the test statistic

$$V \equiv V^{(s)} = \operatorname{trace}\left(\underline{S}_h \cdot (\underline{S}_e + \underline{S}_h)^{-1} \right)$$

$$= \sum_{i=1}^{s} \Theta_i$$

The relationship with other test statistics is given by the expression

$$\Theta_i = \frac{\lambda_i}{1 + \lambda_i} \quad ,$$

where the λ_i are the eigenvalues of the matrix $\underline{S}_h \cdot \underline{S}_e^{-1}$.

In addition $\sum_{i=1}^{s} \Theta_i = \sum_{i=1}^{s} (1 - \mu_i)$, where μ_i is determined by $| \underline{S}_e - \mu(\underline{S}_e + \underline{S}_h) | = 0$.

(b) Extent of the tables and definition of the parameters :

(1) The parameter α :

α = error probability

α = 5% and 1% .

(2) The parameter s :

$s = \min(p, n_h)$ where p = dimension of the variate

n_h = degree of freedom of the hypothesis.

= 2(1)8 .

(3) The parameter m :

$m = 0.5(|p - n_h| - 1)$

for different values of m between -0.5 and 200.

(4) The parameter n :

$n = 0.5(n_e - p - 1)$ where n_e = degrees of freedom of the error

n = 5(5)50 $s = 2$

n = 5(5)50;60 $s = 3$

n = 5(5)50;60;80 $s = 4$

n = 5(5)50;60;80;100;130;160;200 with s = 5(1)8 .

(c) Suggestions for applications :

(1) Multivariate general linear hypotheses, especially MANOVA-problems; multivariate regression and covariance analyses. Multivariate confidence regions. Canonical discriminant analyses.

(2) To test the independence of p (independent) from q (independent) variates (multivariate regression analysis) where

$\underline{S}_h = \underline{R}_{yx} \cdot \underline{R}_x^{-1} \cdot \underline{R}_{xy}$ and $\underline{S}_e = \underline{R}_y - \underline{R}_{yx} \cdot \underline{R}_x^{-1} \cdot \underline{R}_{xy}$

(standardized samples ; $\underline{R}_x$, $\underline{R}_y$, $\underline{R}_{xy}$, $\underline{R}_{yx}$ = correlation matrices)

and p = dimension of the (dependent) vecotr $\underline{y}$

q = dimension of the (independent) vecotr $\underline{x}$

using the initial parameters

$$s = \min (p , q)$$

$$m = \frac{p - q \quad 1}{2}$$

$$n = \frac{N - p - q - 2}{2}$$

where N = sample size .

(d) References :

(1) For the tets criterion :

HOTELLING, H.: A generalized T-test and measure of multivariate dispersion. Proceedings of the Second Berkeley

Symposium on Mathematical Statistics and Probability Vol. 2 , 23 - 41 (1951) .

LAWLEY, D. N. : A generalization of Fisher's Z test. Biometrika 30 , 180 - 187 (1938) .

(2) For the source of the tables :

PILLAI, K. C. S. : Statistical Tables for Tests of Multivariate Hypotheses. (Table 2). Manila/Philippines : The Statistical Center, The University of the Philippines, 1960 .

(e) Further remarks :

For values outside the tabulated region , a suitable approximation is given by the F-distribution with

$$F = \frac{(2n + s + 1)}{(2m + s + 1)} \cdot \frac{V^{(s)}}{(s - V^{(s)})}$$

and the degrees of freedom

$$f_1 = s(2m + s + 1) \quad \text{and}$$

$$f_2 = s(2n + s + 1) \quad .$$

Tafel 7 Table 7

$s = 2$

Upper 5% Points

Obere 5% Punkte

m \ n	5	10	15	20	25	30	35	40	45	50
-.5	0.567	0.357	0.259	0.204	0.168	0.143	0.124	0.110	0.098	0.089
0	0.699	0.451	0.333	0.263	0.218	0.186	0.162	0.144	0.129	0.117
.5	0.807	0.532	0.397	0.316	0.263	0.225	0.196	0.174	0.157	0.142
1	0.897	0.604	0.455	0.364	0.304	0.261	0.228	0.203	0.183	0.166
1.5	0.974	0.668	0.507	0.409	0.342	0.294	0.258	0.230	0.208	0.189
2	1.040	0.725	0.556	0.451	0.379	0.327	0.287	0.256	0.231	0.210
2.5	1.097	0.778	0.601	0.490	0.413	0.357	0.314	0.281	0.254	0.231
3	1.148	0.826	0.643	0.526	0.445	0.386	0.341	0.305	0.276	0.252
3.5	1.194	0.870	0.682	0.561	0.476	0.414	0.366	0.328	0.297	0.271
4	1.234	0.910	0.719	0.594	0.506	0.440	0.390	0.350	0.317	0.290
4.5	1.271	0.948	0.754	0.625	0.534	0.466	0.413	0.371	0.337	0.308
5	1.305	0.982	0.786	0.655	0.561	0.490	0.436	0.392	0.356	0.326
10	1.522	1.235	1.035	0.890	0.781	0.695	0.626	0.570	0.522	0.482
15	1.636	1.387	1.198	1.053	0.940	0.848	0.772	0.709	0.655	0.609
20	1.706	1.487	1.314	1.174	1.061	0.967	0.889	0.822	0.765	0.715
25	1.753	1.559	1.399	1.268	1.157	1.064	0.985	0.917	0.857	0.806
30	1.788	1.613	1.466	1.342	1.236	1.145	1.066	0.997	0.936	0.883
35	1.813	1.656	1.519	1.402	1.301	1.212	1.134	1.066	1.005	0.951
40	1.833	1.690	1.563	1.452	1.355	1.270	1.194	1.126	1.066	1.011
45	1.850	1.718	1.599	1.495	1.402	1.319	1.246	1.179	1.119	1.066
50	1.863	1.741	1.630	1.531	1.442	1.363	1.291	1.226	1.167	1.114

$s = 2$

Upper 1% Points
Obere 1% Punkte

m \ n	5	10	15	20	25	30	35	40	45	50
.5	0.962	0.653	0.494	0.397	0.331	0.284	0.249	0.222	0.200	0.182
1	1.049	0.725	0.553	0.446	0.375	0.323	0.284	0.253	0.228	0.208
1.5	1.122	0.788	0.607	0.493	0.416	0.359	0.316	0.282	0.255	0.232
2	1.185	0.845	0.656	0.536	0.453	0.392	0.346	0.309	0.280	0.255
2.5	1.236	0.897	0.702	0.576	0.488	0.424	0.374	0.335	0.304	0.277
3	1.282	0.944	0.744	0.613	0.521	0.454	0.402	0.360	0.326	0.298
3.5	1.323	0.987	0.783	0.648	0.553	0.482	0.428	0.384	0.348	0.319
4	1.358	1.026	0.820	0.681	0.583	0.510	0.453	0.407	0.370	0.339
4.5	1.390	1.061	0.854	0.713	0.612	0.536	0.477	0.429	0.390	0.358
5	1.422	1.093	0.886	0.743	0.639	0.561	0.500	0.450	0.410	0.376
10	1.610	1.328	1.125	0.973	0.858	0.767	0.693	0.632	0.580	0.536
15	1.705	1.465	1.278	1.129	1.013	0.917	0.837	0.771	0.714	0.664
20	1.763	1.555	1.386	1.245	1.128	1.032	0.953	0.883	0.822	0.770
25	1.802	1.619	1.463	1.334	1.222	1.126	1.045	0.975	0.914	0.860
30	1.829	1.661	1.524	1.402	1.297	1.204	1.123	1.053	0.991	0.936
35	1.850	1.704	1.573	1.457	1.358	1.269	1.190	1.120	1.058	1.002
40	1.866	1.734	1.612	1.504	1.408	1.324	1.247	1.178	1.117	1.061
45	1.880	1.758	1.645	1.543	1.452	1.370	1.297	1.230	1.169	1.113
50	1.890	1.778	1.675	1.576	1.489	1.411	1.340	1.275	1.215	1.161

$s = 3$

Upper 5% Points

Obere 5% Punkte

m \ n	5	10	15	20	25	30	35	40	45	50	60
-.5	0.886	0.583	0.434	0.345	0.287	0.245	0.214	0.190	0.171	0.155	0.131
0	1.041	0.697	0.524	0.420	0.350	0.300	0.263	0.234	0.210	0.191	0.162
.5	1.173	0.800	0.606	0.488	0.409	0.351	0.308	0.274	0.247	0.225	0.191
1	1.288	0.892	0.682	0.552	0.464	0.399	0.351	0.313	0.282	0.257	0.218
1.5	1.388	0.977	0.752	0.612	0.516	0.445	0.392	0.350	0.316	0.288	0.245
2	1.476	1.053	0.818	0.668	0.565	0.489	0.431	0.386	0.349	0.318	0.271
2.5	1.556	1.124	0.879	0.722	0.612	0.531	0.469	0.420	0.380	0.347	0.296
3	1.628	1.190	0.937	0.772	0.657	0.571	0.505	0.453	0.411	0.375	0.321
3.5	1.691	1.250	0.991	0.820	0.700	0.610	0.540	0.485	0.440	0.403	0.344
4	1.750	1.307	1.042	0.866	0.741	0.647	0.574	0.516	0.469	0.429	0.368
4.5	1.804	1.360	1.091	0.910	0.781	0.683	0.607	0.546	0.497	0.455	0.390
5	1.854	1.410	1.136	0.952	0.818	0.717	0.639	0.576	0.524	0.481	0.412
10	2.188	1.778	1.495	1.290	1.133	1.010	0.911	0.830	0.762	0.704	0.612
15	2.369	2.008	1.737	1.529	1.365	1.234	1.124	1.033	0.955	0.888	0.779
20	2.486	2.164	1.912	1.710	1.545	1.410	1.297	1.200	1.117	1.044	0.923
25	2.565	2.277	2.043	1.851	1.690	1.554	1.439	1.340	1.253	1.177	1.050
30	2.623	2.363	2.146	1.964	1.808	1.675	1.560	1.459	1.371	1.293	1.160
35	2.668	2.431	2.229	2.056	1.907	1.777	1.663	1.563	1.474	1.395	1.259
40	2.703	2.486	2.297	2.133	1.990	1.864	1.752	1.653	1.565	1.485	1.348
45	2.731	2.531	2.354	2.199	2.061	1.939	1.831	1.733	1.645	1.566	1.427
50	2.755	2.569	2.402	2.255	2.123	2.005	1.899	1.804	1.717	1.638	1.500
60	2.791	2.629	2.480	2.346	2.225	2.115	2.015	1.924	1.840	1.763	1.690

$s = 3$

Upper 1% Points
Obere 1% Punkte

m \ n	5	10	15	20	25	30	35	40	45	50	60
-.5	1.052	0.710	0.535	0.429	0.359	0.308	0.269	0.240	0.216	0.196	0.166
0	1.212	0.829	0.630	0.508	0.426	0.366	0.321	0.286	0.258	0.235	0.199
.5	1.345	0.934	0.716	0.580	0.488	0.421	0.370	0.330	0.298	0.272	0.231
1	1.459	1.028	0.794	0.647	0.545	0.472	0.415	0.371	0.335	0.306	0.260
1.5	1.554	1.114	0.866	0.709	0.600	0.520	0.458	0.410	0.371	0.339	0.289
2	1.639	1.191	0.933	0.767	0.651	0.565	0.500	0.447	0.405	0.370	0.316
2.5	1.717	1.262	0.995	0.821	0.699	0.609	0.539	0.483	0.438	0.401	0.342
3	1.784	1.326	1.053	0.873	0.745	0.650	0.577	0.518	0.470	0.430	0.368
3.5	1.842	1.385	1.108	0.922	0.789	0.689	0.613	0.551	0.501	0.459	0.393
4	1.897	1.440	1.160	0.969	0.832	0.729	0.648	0.584	0.531	0.487	0.418
4.5	1.948	1.492	1.207	1.012	0.872	0.765	0.681	0.614	0.559	0.513	0.441
5	1.995	1.540	1.252	1.055	0.910	0.800	0.714	0.645	0.587	0.539	0.464
10	2.298	1.892	1.601	1.389	1.224	1.095	0.990	0.904	0.831	0.769	0.669
15	2.459	2.106	1.834	1.621	1.453	1.317	1.203	1.107	1.025	0.954	0.839
20	2.560	2.249	2.001	1.796	1.628	1.489	1.372	1.272	1.186	1.110	0.984
25	2.629	2.353	2.123	1.932	1.768	1.630	1.512	1.410	1.321	1.242	1.110
30	2.680	2.432	2.218	2.038	1.883	1.747	1.630	1.527	1.437	1.356	1.220
35	2.718	2.493	2.296	2.125	1.977	1.846	1.731	1.628	1.538	1.456	1.317
40	2.748	2.542	2.359	2.197	2.056	1.930	1.818	1.717	1.623	1.545	1.405
45	2.772	2.583	2.412	2.259	2.123	2.002	1.893	1.795	1.705	1.624	1.483
50	2.792	2.617	2.456	2.311	2.182	2.065	1.959	1.863	1.776	1.696	1.555
60	2.823	2.670	2.528	2.397	2.277	2.169	2.070	1.979	1.895	1.818	1.742

$s = 4$

Upper 5% Points
Obere 5% Punkte

m \ n	5	10	15	20	25	30	35	40	45	50	60	80
-.5	1.239	0.843	0.638	0.514	0.430	0.370	0.324	0.289	0.260	0.237	0.200	0.154
0	1.411	0.974	0.744	0.602	0.505	0.435	0.382	0.341	0.308	0.280	0.238	0.182
.5	1.560	1.094	0.842	0.684	0.576	0.497	0.438	0.391	0.353	0.322	0.273	0.210
1	1.693	1.203	0.932	0.761	0.643	0.556	0.490	0.438	0.396	0.362	0.308	0.237
1.5	1.812	1.304	1.017	0.834	0.707	0.613	0.541	0.484	0.438	0.401	0.341	0.264
2	1.919	1.396	1.097	0.903	0.768	0.667	0.590	0.529	0.479	0.438	0.374	0.289
2.5	2.015	1.482	1.172	0.969	0.826	0.720	0.637	0.572	0.519	0.475	0.406	0.314
3	2.103	1.563	1.243	1.032	0.882	0.770	0.683	0.614	0.557	0.510	0.437	0.339
3.5	2.183	1.638	1.311	1.092	0.935	0.818	0.727	0.654	0.595	0.545	0.467	0.363
4	2.257	1.709	1.374	1.149	0.987	0.865	0.770	0.694	0.631	0.579	0.497	0.387
4.5	2.325	1.776	1.435	1.204	1.037	0.910	0.811	0.732	0.666	0.612	0.526	0.410
5	2.389	1.839	1.493	1.257	1.085	0.954	0.852	0.769	0.701	0.644	0.554	0.433
10	2.829	2.312	1.952	1.689	1.488	1.329	1.201	1.095	1.007	0.932	0.810	0.643
15	3.079	2.615	2.268	2.000	1.789	1.619	1.479	1.359	1.258	1.171	1.028	0.827
20	3.241	2.825	2.499	2.238	2.026	1.850	1.702	1.577	1.468	1.374	1.216	0.990
25	3.355	2.979	2.675	2.425	2.216	2.040	1.890	1.761	1.648	1.548	1.382	1.136
30	3.439	3.098	2.814	2.576	2.373	2.200	2.049	1.918	1.803	1.701	1.527	1.269
35	3.503	3.191	2.926	2.700	2.505	2.335	2.186	2.055	1.939	1.835	1.658	1.389
40	3.555	3.267	3.019	2.804	2.616	2.451	2.305	2.175	2.059	1.955	1.775	1.499
45	3.610	3.330	3.097	2.892	2.712	2.552	2.409	2.281	2.166	2.062	1.881	1.599
50	3.631	3.383	3.164	2.969	2.795	2.640	2.501	2.376	2.262	2.158	1.977	1.692
60	3.685	3.468	3.271	3.094	2.933	2.788	2.656	2.536	2.426	2.325	2.145	1.858
80	3.756	3.582	3.420	3.270	3.132	3.005	2.886	2.777	2.676	2.581	2.410	2.127

Tafel 7 (Forts.) Table 7 (cont.)

$s = 4$

Upper 1% Points

Obere 1% Punkte

m \ n	5	10	15	20	25	30	35	40	45	50	60	80
.5	1.419	0.982	0.750	0.608	0.511	0.441	0.388	0.346	0.312	0.284	0.242	0.186
0	1.594	1.118	0.861	0.701	0.590	0.510	0.449	0.401	0.363	0.331	0.281	0.216
.5	1.744	1.241	0.962	0.786	0.665	0.576	0.508	0.454	0.411	0.372	0.319	0.246
1	1.875	1.352	1.056	0.866	0.735	0.638	0.563	0.504	0.457	0.417	0.356	0.275
1.5	1.992	1.454	1.143	0.941	0.801	0.696	0.616	0.553	0.501	0.458	0.391	0.302
2	2.098	1.547	1.224	1.013	0.864	0.753	0.667	0.599	0.543	0.497	0.425	0.330
2.5	2.189	1.632	1.301	1.080	0.924	0.807	0.716	0.643	0.584	0.535	0.458	0.356
3	2.273	1.712	1.373	1.144	0.981	0.858	0.763	0.687	0.624	0.572	0.490	0.382
3.5	2.350	1.787	1.440	1.206	1.036	0.908	0.808	0.729	0.663	0.608	0.522	0.407
4	2.420	1.857	1.504	1.264	1.089	0.956	0.853	0.769	0.701	0.643	0.553	0.431
4.5	2.486	1.922	1.564	1.319	1.140	1.003	0.895	0.808	0.737	0.677	0.583	0.456
5	2.547	1.984	1.622	1.371	1.188	1.048	0.937	0.847	0.773	0.711	0.612	0.479
10	2.956	2.440	2.071	1.801	1.591	1.425	1.290	1.173	1.084	1.004	0.875	0.696
15	3.185	2.729	2.378	2.104	1.888	1.713	1.567	1.443	1.336	1.246	1.096	0.883
20	3.331	2.924	2.601	2.336	2.120	1.940	1.788	1.659	1.547	1.449	1.285	1.048
25	3.432	3.068	2.768	2.518	2.306	2.126	1.973	1.841	1.724	1.622	1.450	1.196
30	3.507	3.178	2.898	2.662	2.459	2.283	2.130	1.996	1.878	1.773	1.595	1.329
35	3.564	3.264	3.003	2.780	2.586	2.415	2.264	2.131	2.012	1.906	1.724	1.448
40	3.610	3.334	3.091	2.879	2.693	2.527	2.380	2.248	2.130	2.024	1.840	1.558
45	3.666	3.392	3.164	2.963	2.784	2.625	2.481	2.352	2.236	2.129	1.945	1.658
50	3.677	3.441	3.227	3.035	2.864	2.710	2.571	2.444	2.329	2.224	2.040	1.750
60	3.725	3.518	3.327	3.153	2.995	2.851	2.720	2.600	2.489	2.387	2.190	1.914
80	3.787	3.622	3.465	3.319	3.183	3.058	2.941	2.833	2.732	2.637	2.465	2.180

Tafel 7 (Forts.) Table 7 (cont.)

$s = 5$

Upper 5% Points

Obere 5% Punkte

m \ n	5	10	15	20	25	30	35	40	45	50	60	80	100	130	160	200
-0.5	1.617	1.128	0.871	0.708	0.595	0.514	0.453	0.404	0.365	0.333	0.283	0.217	0.177	0.140	0.114	0.092
0	1.800	1.277	0.989	0.807	0.681	0.590	0.520	0.465	0.420	0.383	0.326	0.251	0.204	0.160	0.131	0.106
0.5	1.964	1.410	1.099	0.901	0.763	0.662	0.584	0.523	0.473	0.432	0.368	0.284	0.231	0.181	0.149	0.120
1	2.112	1.533	1.203	0.990	0.841	0.731	0.646	0.579	0.525	0.479	0.409	0.316	0.258	0.202	0.166	0.134
1.5	2.245	1.647	1.301	1.074	0.915	0.797	0.706	0.633	0.574	0.525	0.449	0.348	0.284	0.222	0.183	0.148
2	2.366	1.753	1.393	1.155	0.987	0.861	0.764	0.686	0.623	0.570	0.488	0.378	0.309	0.242	0.200	0.161
2.5	2.476	1.853	1.480	1.232	1.055	0.922	0.819	0.737	0.670	0.614	0.526	0.409	0.334	0.262	0.216	0.175
3	2.578	1.947	1.563	1.306	1.121	0.982	0.874	0.787	0.716	0.656	0.563	0.438	0.359	0.284	0.232	0.188
3.5	2.672	2.035	1.642	1.376	1.184	1.040	0.926	0.835	0.760	0.698	0.599	0.467	0.383	0.301	0.248	0.201
4	2.759	2.116	1.717	1.444	1.246	1.095	0.977	0.882	0.804	0.738	0.635	0.496	0.407	0.320	0.264	0.214
4.5	2.840	2.196	1.789	1.509	1.305	1.149	1.020	0.922	0.846	0.778	0.670	0.524	0.430	0.339	0.280	0.227
5	2.916	2.271	1.858	1.572	1.362	1.202	1.075	0.973	0.888	0.817	0.704	0.552	0.453	0.358	0.296	0.240
10	3.454	2.838	2.408	2.090	1.847	1.653	1.497	1.367	1.258	1.165	1.015	0.807	0.670	0.534	0.444	0.362
15	3.769	3.212	2.793	2.470	2.213	2.005	1.833	1.688	1.564	1.450	1.281	1.032	0.864	0.695	0.580	0.476
20	3.977	3.474	3.078	2.762	2.503	2.289	2.109	1.955	1.821	1.705	1.512	1.233	1.040	0.843	0.708	0.584
25	4.125	3.668	3.298	2.993	2.739	2.524	2.340	2.181	2.042	1.920	1.715	1.413	1.201	0.980	0.828	0.686
30	4.236	3.819	3.473	3.181	2.933	2.721	2.537	2.376	2.234	2.108	1.895	1.564	1.347	1.109	0.911	0.783
35	4.322	3.939	3.614	3.336	3.097	2.889	2.706	2.545	2.403	2.275	2.056	1.724	1.485	1.228	1.048	0.876
40	4.390	4.036	3.730	3.467	3.236	3.033	2.854	2.694	2.551	2.423	2.201	1.860	1.611	1.341	1.148	0.964
45	4.446	4.118	3.830	3.578	3.356	3.159	2.984	2.826	2.684	2.556	2.332	1.985	1.728	1.446	1.244	1.048
50	4.493	4.186	3.915	3.675	3.461	3.270	3.098	2.944	2.803	2.676	2.452	2.101	1.837	1.546	1.335	1.129
60	4.566	4.295	4.052	3.833	3.635	3.456	3.293	3.144	3.008	2.883	2.661	2.306	2.035	1.729	1.503	1.280
80	4.663	4.445	4.243	4.058	3.886	3.728	3.582	3.446	3.320	3.203	2.991	1.641	2.363	2.042	1.797	1.549
100	4.724	4.542	4.370	4.210	4.059	3.919	3.787	3.664	3.548	3.439	3.239	2.901	2.627	2.300	2.045	1.782
130	4.784	4.637	4.497	4.364	4.237	4.117	4.003	3.895	3.793	3.695	3.514	3.199	2.935	2.612	2.352	2.077
160	4.822	4.700	4.581	4.467	4.358	4.254	4.154	4.058	3.967	3.879	3.714	3.423	3.173	2.859	2.601	2.322
200	4.856	4.756	4.658	4.563	4.471	4.382	4.296	4.213	4.133	4.056	3.910	3.647	3.415	3.118	2.868	2.591

$s = 5$

Upper 1% Points

Obere 1% Punkte

m \ n	5	10	15	20	25	30	35	40	45	50	60	80	100	130	160	200
-0.5	1.806	1.276	0.994	0.812	0.686	0.594	0.524	0.468	0.424	0.387	0.329	0.254	0.206	0.164	0.135	0.108
0	1.991	1.431	1.117	0.915	0.776	0.673	0.595	0.532	0.482	0.440	0.375	0.290	0.236	0.185	0.152	0.122
0.5	2.156	1.568	1.231	1.013	0.862	0.748	0.662	0.593	0.538	0.492	0.420	0.324	0.265	0.207	0.170	0.138
1	2.304	1.692	1.337	1.104	0.941	0.820	0.726	0.652	0.591	0.541	0.463	0.358	0.293	0.229	0.189	0.152
1.5	2.436	1.807	1.437	1.191	1.017	0.888	0.788	0.709	0.643	0.589	0.504	0.391	0.320	0.251	0.206	0.167
2	2.553	1.917	1.530	1.275	1.093	0.955	0.848	0.763	0.693	0.636	0.545	0.423	0.346	0.272	0.224	0.181
2.5	2.660	2.013	1.618	1.353	1.162	1.018	0.906	0.816	0.742	0.681	0.584	0.455	0.373	0.293	0.241	0.195
3	2.759	2.107	1.702	1.428	1.230	1.079	0.962	0.867	0.790	0.725	0.623	0.486	0.398	0.316	0.258	0.209
3.5	2.850	2.195	1.781	1.499	1.294	1.139	1.016	0.917	0.836	0.768	0.660	0.516	0.423	0.334	0.275	0.223
4	2.935	2.271	1.857	1.567	1.356	1.195	1.068	0.966	0.881	0.810	0.697	0.546	0.448	0.353	0.292	0.237
4.5	3.013	2.353	1.928	1.633	1.416	1.250	1.109	1.004	0.925	0.851	0.733	0.575	0.473	0.373	0.308	0.250
5	3.087	2.426	1.997	1.696	1.474	1.303	1.167	1.058	0.967	0.891	0.769	0.603	0.497	0.392	0.324	0.263
10	3.595	2.978	2.538	2.212	1.960	1.758	1.594	1.458	1.343	1.245	1.086	0.866	0.720	0.574	0.478	0.390
15	3.887	3.338	2.915	2.584	2.322	2.108	1.930	1.780	1.651	1.531	1.356	1.094	0.918	0.739	0.618	0.507
20	4.079	3.585	3.191	2.871	2.607	2.388	2.203	2.045	1.908	1.788	1.588	1.297	1.096	0.889	0.748	0.618
25	4.214	3.767	3.401	3.096	2.838	2.619	2.431	2.269	2.127	2.001	1.791	1.479	1.259	1.029	0.870	0.721
30	4.315	3.909	3.567	3.277	3.028	2.813	2.625	2.462	2.317	2.189	1.970	1.625	1.405	1.159	0.984	0.820
35	4.393	4.021	3.701	3.426	3.187	2.977	2.792	2.629	2.483	2.353	2.130	1.790	1.544	1.279	1.092	0.914
40	4.454	4.112	3.812	3.551	3.321	3.118	2.937	2.776	2.630	2.500	2.274	1.926	1.670	1.392	1.194	1.003
45	4.505	4.188	3.906	3.657	3.437	3.240	3.064	2.905	2.761	2.631	2.404	2.050	1.787	1.498	1.290	1.088
50	4.547	4.252	3.986	3.749	3.538	3.347	3.176	3.020	2.878	2.749	2.522	2.165	1.896	1.598	1.381	1.169
60	4.612	4.353	4.115	3.899	3.704	3.526	3.364	3.215	3.079	2.953	2.729	2.369	2.093	1.781	1.550	1.321
80	4.700	4.491	4.297	4.113	3.944	3.788	3.643	3.509	3.383	3.266	3.053	2.700	2.420	2.093	1.844	1.591
100	4.755	4.581	4.414	4.257	4.109	3.971	3.841	3.719	3.604	3.495	3.296	2.957	2.680	2.349	2.091	1.823
130	4.808	4.668	4.533	4.403	4.279	4.160	4.048	3.942	3.840	3.744	3.564	3.249	2.985	2.659	2.396	2.118
160	4.842	4.725	4.611	4.501	4.394	4.291	4.193	4.099	4.008	3.922	3.759	3.468	3.218	2.903	2.644	2.361
200	4.872	4.777	4.682	4.590	4.500	4.413	4.329	4.247	4.169	4.093	3.949	3.687	3.456	3.159	2.908	2.629

Tafel 7 (Forts.) Table 7 (cont.)

$s = 6$

Upper 5% Points / *Obere 5% Punkte*

m \ n	5	10	15	20	25	30	35	40	45	50	60	80	100	130	160	200
-0.5	2.015	1.446	1.127	0.923	0.782	0.678	0.598	0.536	0.485	0.443	0.377	0.291	0.237	0.185	0.152	0.123
0	2.208	1.601	1.256	1.033	0.877	0.762	0.674	0.604	0.547	0.500	0.427	0.330	0.269	0.210	0.173	0.140
0.5	2.382	1.745	1.377	1.137	0.968	0.843	0.747	0.670	0.608	0.556	0.475	0.368	0.300	0.235	0.193	0.156
1	2.541	1.880	1.491	1.237	1.056	0.921	0.817	0.734	0.666	0.610	0.522	0.405	0.331	0.259	0.213	0.173
1.5	2.684	2.006	1.600	1.332	1.140	0.996	0.885	0.796	0.723	0.663	0.568	0.441	0.361	0.283	0.233	0.189
2	2.818	2.124	1.704	1.422	1.221	1.069	0.951	0.856	0.779	0.714	0.612	0.477	0.390	0.307	0.253	0.205
2.5	2.941	2.236	1.802	1.509	1.298	1.139	1.015	0.915	0.833	0.764	0.656	0.512	0.419	0.330	0.272	0.220
3	3.054	2.341	1.895	1.593	1.374	1.207	1.077	0.972	0.886	0.813	0.699	0.546	0.447	0.353	0.291	0.236
3.5	3.160	2.439	1.985	1.673	1.446	1.273	1.137	1.028	0.937	0.861	0.741	0.580	0.476	0.375	0.310	0.251
4	3.258	2.532	2.070	1.750	1.516	1.337	1.196	1.082	0.987	0.908	0.783	0.613	0.504	0.398	0.328	0.267
4.5	3.350	2.621	2.153	1.825	1.584	1.399	1.253	1.135	1.037	0.954	0.823	0.646	0.531	0.420	0.347	0.282
5	3.436	2.705	2.231	1.897	1.650	1.460	1.309	1.186	1.084	0.999	0.863	0.678	0.559	0.442	0.365	0.297
10	4.063	3.362	2.864	2.495	2.210	1.983	1.798	1.645	1.516	1.405	1.226	0.978	0.813	0.649	0.540	0.441
15	4.443	3.800	3.315	2.938	2.639	2.394	2.191	2.020	1.874	1.749	1.539	1.242	1.041	0.838	0.701	0.576
20	4.697	4.112	3.652	3.282	2.982	2.728	2.516	2.334	2.177	2.039	1.811	1.479	1.249	1.014	0.853	0.704
25	4.882	4.346	3.913	3.556	3.258	3.005	2.789	2,602	2.437	2.293	2.050	1.692	1.440	1.177	0.995	0.825
30	5.020	4.529	4.122	3.780	3.489	3.239	3.022	2.832	2.665	2.516	2.264	1.885	1.615	1.329	1.129	0.940
35	5.125	4.675	4.292	3.966	3.684	3.439	3.224	3.034	2.865	2.714	2.455	2.061	1.776	1.471	1.255	1.051
40	5.212	4.795	4.435	4.123	3.851	3.611	3.400	3.211	3.042	2.890	2.627	2.223	1.926	1.605	1.375	1.155
45	5.283	4.894	4.555	4.257	3.995	3.762	3.554	3.368	3.200	3.048	2.784	2.371	2.065	1.730	1.489	1.255
50	5.342	4.979	4.656	4.373	4.121	3.895	3.692	3.508	3.342	3.191	2.926	2.509	2.195	1.849	1.597	1.351
60	5.435	5.114	4.825	4.565	4.330	4.118	3.924	3.748	3.587	3.439	3.176	2.754	2.431	2.067	1.798	1.532
80	5.560	5.300	5.060	4.838	4.635	4.446	4.272	4.111	3.961	3.822	3.570	3.154	2.824	2.441	2.149	1.853
100	5.640	5.421	5.216	5.024	4.845	4.677	4.520	4.373	4.235	4.105	3.868	3.466	3.138	2.749	2.445	2.131
130	5.717	5.541	5.373	5.213	5.062	4.918	4.782	4.654	4.531	4.414	4.198	3.823	3.508	3.122	2.724	2.483
160	5.766	5.619	5.477	5.341	5.210	5.085	4.965	4.851	4.742	4.637	4.440	4.091	3.793	3.418	3.021	2.777
200	5.811	5.690	5.572	5.458	5.347	5.241	5.138	5.039	4.944	4.851	4.677	4.361	4.085	3.730	3.431	3.099

$s = 6$

Upper 1% Points

Obere 1% Punkte

m \ n	5	10	15	20	25	30	35	40	45	50	60	80	100	130	160	200
-0.5	2.210	1.606	1.260	1.037	0.881	0.765	0.677	0.607	0.550	0.503	0.429	0.332	0.271	0.212	0.174	0.141
0	2.406	1.764	1.394	1.150	0.980	0.854	0.756	0.678	0.615	0.563	0.481	0.373	0.304	0.238	0.196	0.158
0.5	2.582	1.911	1.518	1.258	1.075	0.938	0.832	0.747	0.679	0.621	0.531	0.412	0.337	0.264	0.218	0.176
1	2.739	2.047	1.636	1.361	1.165	1.018	0.905	0.814	0.739	0.678	0.581	0.451	0.369	0.290	0.239	0.193
1.5	2.880	2.174	1.745	1.458	1.252	1.096	0.975	0.878	0.799	0.732	0.628	0.489	0.400	0.315	0.259	0.210
2	3.013	2.293	1.849	1.550	1.334	1.171	1.043	0.940	0.856	0.786	0.675	0.526	0.431	0.339	0.280	0.227
2.5	3.133	2.405	1.949	1.638	1.414	1.243	1.109	1.001	0.912	0.838	0.720	0.563	0.461	0.364	0.300	0.243
3	3.244	2.510	2.043	1.723	1.490	1.312	1.173	1.060	0.967	0.888	0.765	0.598	0.491	0.387	0.320	0.259
3.5	3.347	2.607	2.133	1.804	1.564	1.380	1.234	1.117	1.020	0.938	0.808	0.633	0.521	0.411	0.339	0.275
4	3.444	2.699	2.219	1.882	1.635	1.445	1.294	1.172	1.071	0.986	0.851	0.668	0.549	0.434	0.359	0.291
4.5	3.533	2.786	2.301	1.957	1.703	1.508	1.352	1.226	1.121	1.033	0.893	0.702	0.578	0.457	0.378	0.307
5	3.618	2.869	2.379	2.030	1.770	1.569	1.409	1.278	1.170	1.079	0.933	0.735	0.606	0.480	0.397	0.323
10	4.217	3.512	3.003	2.625	2.331	2.097	1.904	1.743	1.608	1.492	1.304	1.041	0.867	0.693	0.577	0.472
15	4.571	3.936	3.446	3.062	2.755	2.505	2.296	2.120	1.968	1.840	1.620	1.310	1.100	0.886	0.742	0.610
20	4.809	4.234	3.774	3.400	3.096	2.835	2.618	2.432	2.270	2.129	1.893	1.549	1.311	1.065	0.896	0.740
25	4.984	4.455	4.026	3.668	3.366	3.109	2.888	2.697	2.530	2.382	2.132	1.763	1.503	1.230	1.041	0.863
30	5.111	4.627	4.226	3.885	3.592	3.339	3.119	2.926	2.755	2.603	2.345	1.957	1.678	1.383	1.176	0.980
35	5.206	4.765	4.388	4.064	3.782	3.535	3.317	3.125	2.953	2.799	2.535	2.133	1.840	1.526	1.304	1.093
40	5.286	4.878	4.523	4.215	3.944	3.704	3.490	3.299	3.128	2.974	2.706	2.294	1.990	1.661	1.424	1.197
45	5.350	4.972	4.638	4.344	4.083	3.850	3.642	3.454	3.284	3.130	2.861	2.442	2.130	1.787	1.539	1.298
50	5.404	5.051	4.735	4.455	4.205	3.979	3.776	3.592	3.424	3.271	3.002	2.579	2.259	1.905	1.648	1.395
60	5.489	5.178	4.895	4.639	4.407	4.195	4.003	3.826	3.664	3.515	3.249	2.822	2.494	2.124	1.849	1.577
80	5.602	5.351	5.117	4.899	4.699	4.513	4.340	4.180	4.030	3.891	3.638	3.218	2.885	2.496	2.200	1.899
100	5.673	5.464	5.265	5.077	4.900	4.737	4.580	4.434	4.297	4.167	3.930	3.526	3.197	2.803	2.495	2.176
130	5.743	5.575	5.412	5.256	5.107	4.966	4.832	4.705	4.584	4.468	4.253	3.878	3.562	3.173	2.773	2.528
160	5.788	5.648	5.510	5.377	5.249	5.126	5.009	4.896	4.788	4.684	4.489	4.141	3.843	3.467	3.068	2.820
200	5.829	5.714	5.600	5.488	5.380	5.276	5.175	5.077	4.983	4.892	4.719	4.405	4.130	3.774	3.474	3.141

Tafel 7 (Forts.) Table 7 (cont.)

$s = 7$

Upper 5% Points

Obere 5% Punkte

m \ n	5	10	15	20	25	30	35	40	45	50	60	80	100	130	160	200
-0.5	2.428	1.778	1.403	1.158	0.986	0.859	0.760	0.682	0.619	0.566	0.483	0.374	0.305	0.239	0.197	0.159
0	2.628	1.942	1.541	1.277	1.090	0.951	0.843	0.758	0.688	0.630	0.539	0.418	0.341	0.268	0.220	0.178
0.5	2.810	2.097	1.672	1.391	1.191	1.040	0.924	0.831	0.755	0.692	0.593	0.460	0.376	0.296	0.243	0.197
1	2.978	2.242	1.797	1.500	1.287	1.127	1.002	0.902	0.820	0.752	0.645	0.502	0.411	0.323	0.266	0.215
1.5	3.133	2.378	1.916	1.604	1.380	1.210	1.078	0.972	0.884	0.811	0.697	0.543	0.445	0.350	0.289	0.234
2	3.276	2.507	2.029	1.704	1.469	1.291	1.151	1.039	0.947	0.869	0.747	0.583	0.478	0.377	0.311	0.252
2.5	3.409	2.628	2.137	1.800	1.555	1.369	1.223	1.105	1.007	0.926	0.797	0.623	0.511	0.403	0.333	0.270
3	3.532	2.740	2.240	1.893	1.639	1.445	1.292	1.169	1.067	0.981	0.845	0.662	0.544	0.429	0.354	0.288
3.5	3.648	2.850	2.338	1.982	1.720	1.519	1.360	1.231	1.125	1.035	0.893	0.700	0.576	0.455	0.376	0.305
4	3.756	2.953	2.433	2.070	1.798	1.590	1.426	1.292	1.181	1.088	0.940	0.738	0.608	0.481	0.397	0.323
4.5	3.858	3.051	2.524	2.151	1.874	1.660	1.490	1.352	1.237	1.140	0.986	0.767	0.639	0.506	0.419	0.340
5	3.953	3.144	2.610	2.231	1.947	1.728	1.553	1.410	1.291	1.191	1.031	0.812	0.675	0.531	0.439	0.357
10	4.662	3.880	3.321	2.903	2.578	2.318	2.106	1.930	1.780	1.652	1.444	1.154	0.961	0.768	0.640	0.523
15	5.102	4.380	3.833	3.407	3.065	2.786	2.554	2.357	2.188	2.042	1.801	1.457	1.223	0.986	0.826	0.679
20	5.402	4.741	4.220	3.800	3.456	3.168	2.925	2.716	2.535	2.377	2.102	1.729	1.463	1.188	1.000	0.826
25	5.620	5.014	4.522	4.116	3.775	3.486	3.238	3.023	2.835	2.669	2.388	1.974	1.681	1.376	1.164	0.966
30	5.786	5.228	4.765	4.374	4.041	3.755	3.507	3.289	3.096	2.925	2.634	2.196	1.883	1.551	1.319	1.099
35	5.916	5.401	4.964	4.590	4.267	3.986	3.739	3.521	3.327	3.153	2.854	2.400	2.070	1.716	1.465	1.226
40	6.021	5.542	5.130	4.773	4.461	4.186	3.943	3.726	3.532	3.357	3.054	2.586	2.243	1.870	1.604	1.348
45	6.107	5.661	5.271	4.930	4.628	4.361	4.122	3.908	3.715	3.540	3.234	2.758	2.404	2.016	1.735	1.464
50	6.180	5.762	5.393	5.066	4.775	4.515	4.282	4.071	3.879	3.705	3.399	2.917	2.555	2.153	1.861	1.575
60	6.294	5.924	5.591	5.291	5.020	4.775	4.553	4.349	4.163	3.992	3.689	3.201	2.828	2.406	2.094	1.785
80	6.449	6.147	5.869	5.613	5.377	5.160	4.959	4.773	4.600	4.438	4.147	3.665	3.283	2.839	2.501	2.157
100	6.547	6.293	6.055	5.833	5.625	5.431	5.249	5.079	4.919	4.769	4.494	4.028	3.649	3.197	2.845	2.480
130	6.643	6.438	6.243	6.057	5.881	5.730	5.557	5.408	5.266	5.131	4.880	4.444	4.079	3.632	3.272	2.890
160	6.706	6.533	6.368	6.209	6.057	5.912	5.773	5.640	5.513	5.391	5.163	4.758	4.412	3.977	3.619	3.231
200	6.761	6.620	6.482	6.349	6.220	6.096	5.977	5.862	5.751	5.643	5.440	5.073	4.752	4.340	3.992	3.607

Tafel 7 (Forts.) Table 7 (cont.)

$s = 7$

Upper 1% Points

Obere 1% Punkte

m \ n	5	10	15	20	25	30	35	40	45	50	60	80	100	130	160	200
-0.5	2.631	1.945	1.545	1.280	1.093	0.954	0.846	0.760	0.690	0.632	0.541	0.419	0.343	0.269	0.221	0.179
0	2.832	2.113	1.687	1.404	1.201	1.050	0.933	0.839	0.762	0.698	0.598	0.465	0.380	0.299	0.249	0.199
0.5	3.014	2.270	1.820	1.520	1.305	1.143	1.016	0.915	0.832	0.763	0.654	0.509	0.417	0.328	0.270	0.219
1	3.181	2.417	1.947	1.631	1.404	1.232	1.097	0.989	0.900	0.826	0.709	0.553	0.453	0.357	0.294	0.238
1.5	3.334	2.554	2.068	1.737	1.498	1.317	1.175	1.060	0.966	0.887	0.763	0.596	0.488	0.385	0.317	0.257
2	3.476	2.683	2.182	1.839	1.589	1.400	1.250	1.130	1.030	0.947	0.815	0.637	0.523	0.413	0.341	0.276
2.5	3.607	2.804	2.291	1.937	1.677	1.479	1.323	1.197	1.093	1.005	0.866	0.678	0.558	0.440	0.363	0.295
3	3.729	2.913	2.395	2.030	1.762	1.557	1.394	1.263	1.154	1.062	0.916	0.719	0.591	0.467	0.386	0.313
3.5	3.843	3.025	2.493	2.121	1.844	1.632	1.463	1.326	1.213	1.117	0.965	0.759	0.625	0.494	0.408	0.332
4	3.949	3.126	2.590	2.211	1.924	1.705	1.530	1.389	1.271	1.171	1.013	0.797	0.657	0.520	0.431	0.350
4.5	4.048	3.223	2.680	2.291	2.000	1.775	1.596	1.449	1.327	1.224	1.060	0.824	0.690	0.546	0.452	0.368
5	4.142	3.315	2.766	2.372	2.075	1.844	1.660	1.509	1.383	1.276	1.106	0.873	0.729	0.572	0.474	0.386
10	4.825	4.039	3.468	3.040	2.706	2.438	2.219	2.035	1.879	1.745	1.527	1.222	1.019	0.815	0.680	0.556
15	5.241	4.525	3.972	3.538	3.190	2.904	2.665	2.462	2.288	2.137	1.888	1.530	1.286	1.038	0.870	0.715
20	5.523	4.871	4.350	3.925	3.575	3.282	3.034	2.820	2.635	2.472	2.185	1.804	1.528	1.243	1.047	0.866
25	5.728	5.132	4.643	4.235	3.890	3.597	3.344	3.125	2.933	2.763	2.476	2.050	1.749	1.433	1.214	1.008
30	5.883	5.335	4.876	4.486	4.152	3.862	3.610	3.389	3.193	3.018	2.721	2.273	1.951	1.610	1.370	1.143
35	6.004	5.498	5.067	4.695	4.372	4.089	3.840	3.618	3.421	3.244	2.940	2.476	2.139	1.775	1.517	1.271
40	6.101	5.633	5.227	4.872	4.561	4.285	4.040	3.821	3.624	3.407	3.138	2.662	2.312	1.930	1.657	1.393
45	6.181	5.745	5.362	5.023	4.723	4.456	4.216	4.000	3.805	3.627	3.318	2.834	2.473	2.076	1.789	1.510
50	6.248	5.841	5.477	5.155	4.866	4.606	4.372	4.160	3.967	3.791	3.481	2.992	2.623	2.214	1.915	1.622
60	6.354	5.993	5.666	5.371	5.103	4.859	4.637	4.433	4.246	4.074	3.768	3.275	2.895	2.467	2.148	1.833
80	6.495	6.203	5.931	5.679	5.447	5.231	5.032	4.847	4.674	4.512	4.220	3.735	3.349	2.899	2.555	2.207
100	6.586	6.341	6.108	5.890	5.685	5.494	5.314	5.145	4.986	4.836	4.561	4.093	3.712	3.255	2.899	2.529
130	6.674	6.476	6.286	6.104	5.930	5.782	5.612	5.464	5.323	5.189	4.939	4.504	4.138	3.687	3.324	2.938
160	6.731	6.566	6.405	6.249	6.100	5.957	5.820	5.689	5.563	5.442	5.215	4.812	4.465	4.029	3.669	3.278
200	6.782	6.646	6.513	6.382	6.256	6.134	6.017	5.903	5.793	5.687	5.486	5.121	4.801	4.388	4.039	3.652

$s = 8$

Upper 5% Points

Obere 5% Punkte

m \ n	5	10	15	20	25	30	35	40	45	50	60	80	100	130	160	200
-0.5	2.852	2.127	1.696	1.411	1.208	1.055	0.937	0.843	0.766	0.701	0.601	0.467	0.382	0.300	0.256	0.200
0	3.058	2.301	1.844	1.539	1.320	1.156	1.028	0.925	0.841	0.771	0.662	0.515	0.421	0.331	0.273	0.221
0.5	3.248	2.464	1.984	1.661	1.428	1.253	1.116	1.006	0.915	0.840	0.721	0.562	0.460	0.362	0.299	0.242
1	3.423	2.617	2.117	1.778	1.532	1.346	1.201	1.084	0.987	0.907	0.779	0.608	0.499	0.393	0.324	0.263
1.5	3.586	2.762	2.245	1.891	1.633	1.437	1.284	1.160	1.057	0.972	0.836	0.654	0.537	0.423	0.349	0.283
2	3.738	2.898	2.366	1.999	1.731	1.526	1.364	1.234	1.126	1.035	0.892	0.699	0.574	0.453	0.374	0.304
2.5	3.880	3.028	2.483	2.104	1.825	1.611	1.443	1.306	1.193	1.098	0.947	0.743	0.611	0.482	0.399	0.324
3	4.012	3.151	2.595	2.204	1.916	1.695	1.519	1.377	1.257	1.159	1.001	0.786	0.647	0.512	0.423	0.344
3.5	4.136	3.268	2.701	2.302	2.005	1.776	1.594	1.445	1.323	1.219	1.054	0.829	0.683	0.541	0.447	0.364
4	4.253	3.379	2.804	2.396	2.091	1.855	1.666	1.513	1.385	1.278	1.106	0.871	0.719	0.569	0.471	0.383
4.5	4.363	3.486	2.902	2.486	2.174	1.931	1.737	1.579	1.447	1.335	1.157	0.913	0.754	0.598	0.495	0.403
5	4.467	3.587	2.996	2.574	2.255	2.006	1.807	1.643	1.507	1.392	1.207	0.954	0.788	0.626	0.518	0.422
10	5.251	4.397	3.780	3.314	2.951	2.659	2.420	2.220	2.051	1.905	1.669	1.336	1.115	0.892	0.744	0.609
15	5.748	4.955	4.350	3.875	3.494	3.181	2.919	2.697	2.507	2.341	2.068	1.675	1.410	1.138	0.954	0.785
20	6.094	5.363	4.784	4.316	3.931	3.609	3.335	3.100	2.896	2.717	2.418	1.982	1.679	1.366	1.151	0.952
25	6.347	5.674	5.125	4.671	4.290	3.966	3.688	3.430	3.233	3.046	2.729	2.258	1.926	1.578	1.337	1.110
30	6.541	5.919	5.401	4.964	4.591	4.270	3.991	3.745	3.528	3.335	3.006	2.510	2.154	1.777	1.512	1.261
35	6.694	6.117	5.628	5.209	4.847	4.531	4.254	4.008	3.789	3.592	3.255	2.740	2.365	1.963	1.677	1.405
40	6.818	6.281	5.818	5.417	5.066	4.758	4.484	4.240	4.020	3.823	3.480	2.951	2.561	2.138	1.834	1.543
45	6.920	6.419	5.981	5.496	5.257	4.956	4.687	4.446	4.228	4.029	3.685	3.146	2.744	2.303	1.984	1.674
50	7.006	6.536	6.120	5.752	5.425	5.132	4.869	4.631	4.415	4.218	3.872	3.325	2.915	2.459	2.126	1.801
60	7.143	6.725	6.349	6.010	5.705	5.429	5.177	4.948	4.737	4.544	4.201	3.649	3.225	2.746	2.390	2.039
80	7.328	6.986	6.671	6.382	6.115	5.869	5.642	5.431	5.235	5.052	4.722	4.176	3.743	3.238	2.858	2.463
100	7.447	7.159	6.889	6.636	6.401	6.181	5.975	5.782	5.600	5.430	5.118	4.589	4.159	3.645	3.245	2.830
130	7.564	7.330	7.107	6.897	6.697	6.508	6.328	6.158	5.997	5.844	5.559	5.064	4.650	4.140	3.731	3.297
160	7.639	7.443	7.254	7.073	6.900	6.735	6.577	6.426	6.281	6.143	5.883	5.423	5.029	4.534	4.127	3.686
200	7.707	7.545	7.388	7.236	7.090	6.949	6.812	6.681	6.555	6.433	6.201	5.784	5.418	4.948	4.553	4.114

Tafel 7 (Forts.) Table 7 (cont.)

$s = 8$

Upper 1% Points

Obere 1% Punkte

m \ n	5	10	15	20	25	30	35	40	45	50	60	80	100	130	160	200
-0.5	3.059	2.302	1.846	1.541	1.323	1.158	1.030	0.927	0.843	0.773	0.663	0.516	0.423	0.332	0.283	0.222
0	3.265	2.478	1.996	1.672	1.438	1.262	1.124	1.013	0.922	0.846	0.726	0.566	0.464	0.365	0.301	0.244
0.5	3.455	2.644	2.139	1.797	1.549	1.362	1.215	1.096	0.998	0.917	0.788	0.615	0.505	0.397	0.328	0.266
1	3.630	2.799	2.275	1.916	1.656	1.458	1.302	1.176	1.073	0.986	0.848	0.663	0.545	0.429	0.354	0.287
1.5	3.792	2.944	2.404	2.031	1.758	1.551	1.387	1.254	1.145	1.053	0.907	0.711	0.584	0.461	0.381	0.309
2	3.942	3.081	1.527	2.141	1.858	1.641	1.469	1.330	1.215	1.118	0.965	0.757	0.623	0.492	0.406	0.330
2.5	4.083	3.210	2.645	2.247	1.954	1.728	1.549	1.404	1.284	1.182	1.021	0.802	0.661	0.522	0.432	0.351
3	4.214	3.332	2.757	2.349	2.046	1.813	1.627	1.476	1.349	1.245	1.076	0.847	0.698	0.553	0.457	0.372
3.5	4.337	3.448	2.864	2.447	2.136	1.895	1.703	1.546	1.416	1.306	1.131	0.891	0.735	0.582	0.482	0.392
4	4.452	3.559	2.966	2.542	2.223	1.975	1.777	1.615	1.480	1.366	1.184	0.934	0.772	0.612	0.507	0.413
4.5	4.560	3.664	3.064	2.634	2.307	2.053	1.849	1.682	1.543	1.425	1.236	0.977	0.808	0.641	0.531	0.433
5	4.662	3.764	3.157	2.721	2.389	2.129	1.919	1.748	1.604	1.483	1.287	1.019	0.843	0.670	0.556	0.453
10	5.422	4.562	3.934	3.458	3.085	2.785	2.538	2.331	2.155	2.004	1.757	1.409	1.177	0.943	0.787	0.644
15	5.892	5.107	4.496	4.013	3.625	3.305	3.036	2.809	2.613	2.442	2.160	1.752	1.477	1.193	1.001	0.824
20	6.223	5.500	4.921	4.448	4.057	3.729	3.450	3.210	3.001	2.818	2.511	2.062	1.749	1.425	1.201	0.994
25	6.462	5.799	5.253	4.797	4.412	4.083	3.800	3.554	3.337	3.146	2.822	2.339	1.998	1.639	1.389	1.155
30	6.645	6.033	5.519	5.083	4.708	4.383	4.160	3.851	3.630	3.434	3.098	2.591	2.227	1.839	1.566	1.307
35	6.788	6.222	5.739	5.321	4.958	4.640	4.360	4.111	3.889	3.689	3.346	2.821	2.438	2.026	1.733	1.452
40	6.905	6.378	5.922	5.523	5.173	4.863	4.587	4.340	4.118	3.918	3.570	3.032	2.625	2.202	1.891	1.591
45	7.000	6.509	6.077	5.697	5.359	5.057	4.787	4.544	4.323	4.121	3.774	3.226	2.817	2.367	2.041	1.724
50	7.081	6.620	6.211	5.847	5.522	5.229	4.965	4.725	4.508	4.309	3.959	3.404	2.988	2.523	2.183	1.851
60	7.208	6.800	6.430	6.096	5.794	5.518	5.267	5.037	4.826	4.631	4.285	3.727	3.297	2.810	2.449	2.090
80	7.380	7.047	6.738	6.453	6.190	5.946	5.720	5.510	5.314	5.130	4.800	4.250	3.812	3.302	2.911	2.515
100	7.490	7.210	6.946	6.698	6.466	6.248	6.044	5.852	5.671	5.501	5.189	4.659	4.225	3.707	3.302	2.882
130	7.598	7.372	7.154	6.947	6.751	6.564	6.387	6.219	6.058	5.906	5.622	5.127	4.711	4.199	3.787	3.348
160	7.668	7.478	7.294	7.116	6.946	6.784	6.628	6.478	6.335	6.198	5.939	5.480	5.086	4.589	4.180	3.736
200	7.730	7.574	7.421	7.272	7.128	6.989	6.855	6.726	6.601	6.480	6.250	5.835	5.470	5.000	4.603	4.162

TABLE 8

The T^2-Criterion of H. Hotelling: Tables of D. R. Jensen and R. B. Howe

(a) Content of the tables and definition of the test statistic :

The tables contain the upper percentage points of the test statistic

$$T^2 (p , \nu) = k . D$$

for the treatment of multivariate one- and two sample problems.
The tests statistic is a special case of the T^2-criterion of H. HOTELLING (see Table 6) and the direct multivariate generalization of the univariate t-test.

The definitions are :

(1) For the one sample problem as well as the two sample problem with paired (related) samples :

$$D = (\underline{\bar{x}} - \underline{\mu}_o)' . \underline{S}^{-1} . (\underline{\bar{x}} - \underline{\mu}_o)$$

where $\underline{S}$ = covariance matrix of the sample

and $k = N =$ sample size .

(2) For the two sample problem :

$$D = (\underline{\bar{x}}_1 - \underline{\bar{x}}_2)' . \underline{S}^{-1} . (\underline{\bar{x}}_1 - \underline{\bar{x}}_2)$$

where $\underline{S}$ = pooled covariance matrix of the two samples

$$k = \frac{N_1 . N_2}{N_1 + N_2}$$

with N_1 = size of the first sample

N_2 = size of the second sample

Note that here $\underline{S}$ denotes the covariance matrix while some authors deviate from this notation and use $\underline{S}$ for the SP-matrices.

(b) Extent of the tables and definition of the parameter :

(a) The parameter α :

α = error probability

α = 0.01 ; 0.025 ; 0.05 ; 0.10 .

(b) The parameter p :

p = dimension of the variate

p = 1(1)20(2)30(5)55 .

Condition : $p < \nu$.

(3) The parameter ν :

$$\nu = n_e = \text{degrees of freedom of the error}$$

$$\nu = 2(1)30(5)60(10)120 \ .$$

Condition : $\nu > p$.

For the one sample problem :

$$\nu = N - 1 \ .$$

For the two sample problem :

$$\nu = N_1 + N_2 - 2 \ .$$

(c) Suggestions for applications :

(1) Multivariate one and two sample problems.

(2) Simultaneous confidence regions for the components of multivariate variates.

(d) References :

(1) For the test criterion :

HOTELLING, H.: The generalization of Student's ratio. Annals of Mathematical Statistics 2, 360 - 378 (1931) .

(2) For the source of the tables :

JENSEN, D. R. + HOWE, R. B. : Tables of Hotelling's T^2-Distribution.
Blacksburg / Virginia : Virginia Polytechnic Institute (Technical Report No. 9), March 1968.
Revised Edition, August 1972.

(e) Further remarks :

(1) Given H_o there is the following relationship between the test statistic T^2 and the F-distribution :

(aa) For the one sample problem :

$$F_{\alpha, f_1, f_2} = \frac{N - p}{p(N - 1)} \cdot T^2_{\alpha, p, N - p}$$

where $f_1 = p$

and $f_2 = N - p$.

(bb) For the two sample problem :

$$F_{\alpha, f_1, f_2} = \frac{N_1+N_2-p-1}{(N_1+N_2) \cdot p} \cdot T^2_{\alpha, p, N_1+N_2-p-1}$$

where $f_1 = p$

and $f_2 = N_1 + N_2 - p - 1$.

For percentage points outside the tabulated region, one may switch to the F-distribution.

(2) For $p = 1$ T^2 is identical to $F = t^2$.

Tafel 8 Table 8

$\alpha = 0.010$

ν \ p	1	2	3	4	5
2	98.503				
3	34.116	297.000			
4	21.198	82.177	594.997		
5	16.258	45.000	147.283	992.494	
6	13.745	31.857	75.125	229.679	1489.489
7	12.246	25.491	50.652	111.839	329.433
8	11.259	21.821	39.118	72.908	155.219
9	10.561	19.460	32.598	54.890	98.703
10	10.044	17.826	28.466	44.838	72.882
11	9.646	16.631	25.637	38.533	58.618
12	9.330	15.722	23.588	34.251	49.739
13	9.074	15.008	22.041	31.171	43.745
14	8.862	14.433	20.834	28.857	39.454
15	8.683	13.960	19.867	27.060	36.246
16	8.531	13.566	19.076	25.626	33.762
17	8.400	13.231	18.418	24.458	31.788
18	8.285	12.943	17.861	23.487	30.182
19	8.185	12.694	17.385	22.670	28.852
20	8.096	12.476	16.973	21.972	27.734
21	8.017	12.283	16.613	21.369	26.781
22	7.945	12.111	16.296	20.843	25.959
23	7.881	11.958	16.015	20.381	25.244
24	7.823	11.820	15.763	19.972	24.616
25	7.770	11.695	15.538	19.606	24.060
26	7.721	11.581	15.334	19.279	23.565
27	7.677	11.478	15.149	18.983	23.121
28	7.636	11.383	14.980	18.715	22.721
29	7.598	11.295	14.825	18.471	22.359
30	7.562	11.215	14.683	18.247	22.029
35	7.419	10.890	14.117	17.366	20.743
40	7.314	10.655	13.715	16.750	19.858
45	7.234	10.478	13.414	16.295	19.211
50	7.171	10.340	13.181	15.945	18.718
55	7.119	10.228	12.995	15.667	18.331
60	7.077	10.137	12.843	15.442	18.018
70	7.011	9.996	12.611	15.098	17.543
80	6.963	9.892	12.440	14.849	17.201
90	6.925	9.813	12.310	14.660	16.942
100	6.895	9.750	12.208	14.511	16.740
110	6.871	9.699	12.125	14.391	16.577
120	6.851	9.657	12.057	14.292	16.444
150	6.807	9.565	11.909	14.079	16.156
200	6.763	9.474	11.764	13.871	15.877
400	6.699	9.341	11.551	13.569	15.473
1000	6.660	9.262	11.426	13.392	15.239
∞	6.635	9.210	11.345	13.277	15.086

Tafel 8 (Forts.)

Table 8 (cont.)

α=0.010

ν \ p	6	7	8	9	10
7	2085.984				
8	446.571	2781.978			
9	205.293	581.106	3577.472		
10	128.067	262.076	733.045	4472.464	
11	93.127	161.015	325.576	902.392	5466.956
12	73.969	115.640	197.555	395.797	1089.149
13	62.114	90.907	140.429	237.692	472.742
14	54.150	75.676	109.441	167.499	281.428
15	48.472	65.483	90.433	129.576	196.853
16	44.240	58.241	77.755	106.391	151.316
17	40.975	52.858	68.771	90.969	123.554
18	38.385	48.715	62.109	80.067	105.131
19	36.283	45.435	56.992	71.999	92.134
20	34.546	42.779	52.948	65.813	82.532
21	33.088	40.587	49.679	60.932	75.181
22	31.847	38.750	46.986	56.991	69.389
23	30.779	37.188	44.730	53.748	64.719
24	29.850	35.846	42.816	51.036	60.879
25	29.036	34.680	41.171	48.736	57.671
26	28.316	33.659	39.745	46.762	54.953
27	27.675	32.756	38.496	45.051	52.622
28	27.101	31.954	37.393	43.554	50.604
29	26.584	31.236	36.414	42.234	48.839
30	26.116	30.589	35.538	41.062	47.283
35	24.314	28.135	32.259	36.743	41.651
40	23.094	26.502	30.120	33.984	38.135
45	22.214	25.340	28.617	32.073	35.737
50	21.550	24.470	27.504	30.673	33.998
55	21.030	23.795	26.647	29.603	32.682
60	20.613	23.257	25.967	28.760	31.650
70	19.986	22.451	24.957	27.515	30.139
80	19.536	21.877	24.242	26.642	29.085
90	19.197	21.448	23.710	25.995	28.310
100	18.934	21.115	23.299	25.496	27.714
110	18.722	20.849	22.972	25.101	27.243
120	18.549	20.632	22.705	24.779	26.862
150	18.178	20.167	22.137	24.096	26.054
200	17.819	19.720	21.592	23.446	25.287
400	17.303	19.080	20.818	22.525	24.209
1000	17.006	18.713	20.376	22.003	23.600
∞	16.812	18.475	20.090	21.666	23.209

$\alpha=0.010$

ν \ p	11	12	13	14	15
12	6560.947				
13	1293.319	7754.436			
14	556.413	1514.902	9047.426		
15	328.767	646.811	1753.899	10439.91	
16	228.494	379.710	743.938	2010.310	11931.90
17	174.662	262.423	434.257	847.794	2284.137
18	141.923	199.618	298.642	492.409	958.379
19	120.242	161.501	226.183	337.150	554.167
20	104.973	136.305	182.290	254.358	377.950
21	93.711	118.588	153.320	204.288	284.145
22	85.100	105.538	132.979	171.289	227.499
23	78.323	95.571	118.013	148.147	190.213
24	72.865	87.736	106.596	131.139	164.093
25	68.382	81.432	97.630	118.176	144.916
26	64.639	76.258	90.421	108.005	130.313
27	61.470	71.942	84.509	99.834	118.863
28	58.756	68.291	79.582	93.138	109.671
29	56.406	65.165	75.416	87.560	102.144
30	54.353	62.461	71.851	82.847	95.877
35	47.059	53.053	59.741	67.252	75.749
40	42.617	47.478	52.776	58.578	64.961
45	39.636	43.803	48.272	53.083	58.281
50	37.501	41.203	45.128	49.301	53.752
55	35.898	39.268	42.811	46.543	50.484
60	34.650	37.774	41.034	44.444	48.019
70	32.836	35.617	38.490	41.465	44.549
80	31.581	34.137	36.759	39.453	42.226
90	30.662	33.059	35.504	38.004	40.564
100	29.960	32.238	34.554	36.912	39.316
110	29.406	31.593	33.810	36.059	38.344
120	28.958	31.073	33.210	35.374	37.567
150	28.013	29.980	31.957	33.947	35.952
200	27.122	28.953	30.784	32.619	34.457
400	25.874	27.525	29.163	30.792	32.414
1000	25.174	26.727	28.262	29.782	31.289
∞	24.725	26.217	27.688	29.141	30.578

Tafel 8 (Forts.) Table 8 (cont.)

α=0.010

ν \ p	16	17	18	19	20
17	13523.39				
18	2575.378	15214.37			
19	1075.693	2884.036	17004.86		
20	619.531	1199.738	3210.109	18894.84	
21	421.040	688.501	1330.513	3553.598	20884.33
22	315.543	466.422	761.078	1468.019	3914.503
23	251.921	348.554	514.096	837.263	1612.255
24	210.092	277.556	383.177	564.061	917.054
25	180.818	230.926	304.404	419.413	616.319
26	159.344	198.322	252.716	332.465	457.261
27	143.005	174.425	216.606	275.462	361.739
28	130.204	156.255	190.159	235.669	299.164
29	119.933	142.029	170.063	206.545	255.513
30	111.528	130.622	154.339	184.429	223.584
35	85.434	96.566	109.480	124.618	142.570
40	72.020	79.868	88.644	98.516	109.697
45	63.917	70.051	76.754	84.109	92.215
50	58.511	63.617	69.109	75.035	81.450
55	54.657	59.085	63.795	68.816	74.183
60	51.774	55.726	59.894	64.297	68.958
70	47.753	51.085	54.557	58.179	61.964
80	45.085	48.035	51.084	54.238	57.504
90	43.187	45.880	48.645	51.489	54.417
100	41.769	44.276	46.840	49.465	52.154
110	40.669	43.037	45.451	47.913	50.426
120	39.792	42.051	44.348	46.685	49.063
150	37.976	40.019	42.085	44.173	46.287
200	36.303	38.157	40.021	41.896	43.783
400	34.030	35.641	37.249	38.854	40.459
1000	32.785	34.271	35.749	37.216	38.677
∞	32.000	33.409	34.805	36.191	37.566

Tafel 8 (Forts.)

Table 8 (cont.)

α=0.010

ν \ p	22	24	26	28	30
23	25161.79				
24	4688.561				
25	1920.920	29837.25			
26	1087.459	5532.285			
27	727.712	2256.508	34910.71		
28	537.797	1272.294	6445.674		
29	423.928	848.275	2619.021	40382.16	
30	349.439	624.787	1471.559	7428.729	
35	190.465	264.634	390.968	639.627	1264.993
40	137.128	174.127	226.272	304.124	429.942
45	111.180	134.941	165.471	205.923	261.624
50	96.008	113.433	134.633	160.926	194.282
55	86.107	99.943	116.184	135.500	158.820
60	79.158	90.728	103.970	119.272	137.145
70	70.070	78.992	88.868	99.865	112.188
80	64.405	71.854	79.929	88.717	98.325
90	60.542	67.065	74.034	81.504	89.536
100	57.741	63.632	69.860	76.462	83.478
110	55.619	61.052	66.751	72.742	79.054
120	53.955	59.044	64.347	69.887	75.682
150	50.596	55.023	59.579	64.275	69.123
200	47.599	51.475	55.419	59.438	63.536
400	43.667	46.877	50.085	53.325	56.569
1000	41.581	44.465	47.333	50.187	53.031
∞	40.289	42.980	45.642	48.278	50.892

Tafel 8 (Forts.)

Table 8 (cont.)

α=0.010

ν \ p	35	40	45	50	55
40	1675.315				
45	557.658	2142.966			
50	334.022	701.741	2667.947		
55	244.962	415.009	862.192	3250.258	
60	198.221	301.245	504.591	1039.014	3889.902
70	150.639	205.476	289.266	430.641	709.544
80	126.801	163.979	214.454	286.516	396.666
90	112.571	141.052	177.176	224.418	288.611
100	103.142	126.575	155.011	190.243	234.985
110	96.445	116.626	140.367	168.727	203.193
120	91.448	109.376	129.991	153.973	182.237
150	81.978	96.041	111.529	128.702	147.875
200	74.171	85.434	97.404	110.188	123.887
400	64.756	73.081	81.572	90.249	99.132
1000	60.109	67.161	74.207	81.261	88.332
∞	57.342	63.691	69.957	76.154	82.292

Tafel 8 (Forts.) Table 8 (cont.)

$\alpha=0.025$

ν \ p	1	2	3	4	5
2	38.506				
3	17.443	117.000			
4	12.218	42.784	234.993		
5	10.007	26.623	77.196	392.484	
6	8.813	20.241	44.906	120.808	589.473
7	8.073	16.940	32.607	67.232	173.656
8	7.571	14.952	26.395	47.282	93.645
9	7.209	13.634	22.718	37.363	64.317
10	6.937	12.699	20.310	31.558	49.896
11	6.724	12.004	18.620	27.789	41.527
12	6.554	11.467	17.372	25.163	36.130
13	6.414	11.041	16.416	23.235	32.387
14	6.298	10.694	15.660	21.764	29.653
15	6.200	10.407	15.048	20.606	27.573
16	6.115	10.165	14.543	19.672	25.941
17	6.042	9.959	14.120	18.904	24.628
18	5.978	9.781	13.759	18.261	23.551
19	5.922	9.626	13.449	17.715	22.651
20	5.871	9.490	13.180	17.246	21.888
21	5.827	9.369	12.943	16.839	21.234
22	5.786	9.261	12.734	16.482	20.668
23	5.750	9.164	12.547	16.168	20.172
24	5.717	9.077	12.380	15.888	19.734
25	5.686	8.997	12.230	15.637	19.346
26	5.659	8.925	12.094	15.411	18.998
27	5.633	8.859	11.969	15.207	18.686
28	5.610	8.798	11.856	15.021	18.403
29	5.588	8.743	11.752	14.852	18.146
30	5.568	8.691	11.656	14.697	17.912
35	5.485	8.482	11.273	14.081	16.994
40	5.424	8.330	10.998	13.646	16.356
45	5.377	8.215	10.792	13.324	15.887
50	5.340	8.125	10.632	13.075	15.528
55	5.310	8.052	10.504	12.877	15.244
60	5.286	7.992	10.399	12.715	15.015
70	5.247	7.899	10.237	12.469	14.666
80	5.218	7.831	10.119	12.289	14.413
90	5.196	7.779	10.028	12.152	14.221
100	5.179	7.737	9.956	12.044	14.071
110	5.164	7.703	9.898	11.957	13.950
120	5.152	7.675	9.851	11.886	13.851
150	5.126	7.614	9.747	11.731	13.636
200	5.100	7.554	9.645	11.579	13.427
400	5.062	7.465	9.495	11.358	13.124
1000	5.039	7.412	9.406	11.228	12.948
∞	5.024	7.378	9.348	11.143	12.833

Tafel 8 (Forts.) Table 8 (cont.)

$\alpha=0.025$

ν \ p	6	7	8	9	10
7	825.961				
8	235.755	1101.946			
9	124.164	307.112	1417.429		
10	83.732	158.797	387.730	1772.410	
11	64.017	105.537	197.551	477.612	2166.889
12	52.648	79.737	129.738	240.426	576.758
13	45.354	64.934	97.060	156.337	287.426
14	40.317	55.475	78.389	115.992	185.336
15	36.649	48.966	66.499	93.019	136.533
16	33.868	44.238	58.339	78.430	108.826
17	31.690	40.662	52.427	68.442	91.271
18	29.943	37.868	47.963	61.219	79.277
19	28.511	35.630	44.483	55.776	70.619
20	27.317	33.799	41.698	51.538	64.103
21	26.307	32.275	39.423	48.152	59.037
22	25.443	30.987	37.532	45.389	54.994
23	24.694	29.885	35.935	43.093	51.698
24	24.039	28.932	34.571	41.158	48.963
25	23.463	28.100	33.392	39.505	46.659
26	22.951	27.367	32.363	38.078	44.692
27	22.493	26.718	31.458	36.834	42.996
28	22.082	26.137	30.656	35.740	41.517
29	21.711	25.616	29.940	34.771	40.218
30	21.374	25.146	29.298	33.907	39.068
35	20.067	23.345	26.872	30.693	34.859
40	19.174	22.135	25.272	28.614	32.194
45	18.525	21.266	24.137	27.160	30.357
50	18.032	20.613	23.292	26.088	29.017
55	17.645	20.104	22.638	25.264	27.995
60	17.334	19.696	22.117	24.612	27.191
70	16.863	19.083	21.340	23.645	26.008
80	16.525	18.645	20.788	22.963	25.179
90	16.269	18.316	20.376	22.456	24.566
100	16.070	18.061	20.056	22.065	24.094
110	15.910	17.856	19.801	21.754	23.720
120	15.778	17.688	19.593	21.500	23.416
150	15.496	17.329	19.148	20.960	22.772
200	15.222	16.983	18.721	20.444	22.158
400	14.827	16.485	18.110	19.711	21.291
1000	14.598	16.199	17.761	19.293	20.800
∞	14.449	16.013	17.535	19.023	20.483

Tafel 8 (Forts.) Table 8 (cont.)

$\alpha=0.025$

ν \ p	11	12	13	14	15
12	2600.865				
13	685.169	3074.340			
14	338.551	802.847	3587.313		
15	216.738	393.802	929.791	4139.783	
16	158.686	250.543	453.180	1066.002	4731.751
17	125.810	182.452	286.751	516.684	1211.480
18	105.024	143.974	207.832	325.363	584.317
19	90.847	119.691	163.319	234.825	366.380
20	80.628	103.153	135.271	183.844	263.433
21	72.948	91.248	116.195	151.765	205.550
22	66.983	82.311	102.479	129.974	169.175
23	62.228	75.377	92.193	114.323	144.492
24	58.355	69.855	84.221	102.596	126.780
25	55.144	65.360	77.876	93.515	113.521
26	52.440	61.636	72.716	86.292	103.259
27	50.134	58.503	68.443	80.422	95.104
28	48.146	55.833	64.850	75.564	88.480
29	46.415	53.532	61.790	71.482	83.001
30	44.894	51.529	59.154	68.006	78.399
35	39.429	44.470	50.066	56.316	63.344
40	36.047	40.213	44.737	49.671	55.078
45	33.753	37.373	41.245	45.402	49.879
50	32.097	35.346	38.785	42.433	46.315
55	30.845	33.829	36.959	40.252	43.724
60	29.867	32.650	35.551	38.582	41.756
70	28.437	30.940	33.524	36.197	38.967
80	27.442	29.759	32.135	34.577	37.088
90	26.710	28.895	31.125	33.404	35.737
100	26.150	28.236	30.357	32.516	34.718
110	25.706	27.716	29.753	31.821	33.923
120	25.347	27.296	29.267	31.262	33.285
150	24.588	26.411	28.245	30.093	31.956
200	23.868	25.576	27.287	29.001	30.721
400	22.857	24.410	25.954	27.491	29.023
1000	22.286	23.756	25.210	26.653	28.084
∞	21.920	23.337	24.736	26.119	27.488

Tafel 8 (Forts.)

Table 8 (cont.)

α=0.025

ν \ p	16	17	18	19	20
17	5363.218				
18	1366.226	6034.182			
19	656.076	1530.239	6744.644		
20	409.802	731.963	1703.519	7494.604	
21	293.656	455.629	811.979	1886.067	8284.061
22	228.439	325.494	503.861	896.122	2077.883
23	187.500	252.509	358.947	554.498	984.393
24	159.748	206.742	277.762	394.016	607.541
25	139.851	175.743	226.899	304.197	430.700
26	124.966	153.535	192.477	247.973	331.815
27	113.455	136.934	167.834	209.950	269.963
28	104.312	124.103	149.424	182.748	228.163
29	96.890	113.917	135.203	162.436	198.276
30	90.753	105.652	123.919	146.755	175.972
35	71.303	80.387	90.846	103.002	117.285
40	61.030	67.617	74.944	83.141	92.371
45	54.717	59.966	65.680	71.926	78.780
50	50.457	54.887	59.641	64.755	70.273
55	47.392	51.277	55.400	59.787	64.463
60	45.084	48.582	52.264	56.147	60.251
70	41.842	44.830	47.939	51.179	54.559
80	39.676	42.345	45.101	47.950	50.898
90	38.127	40.579	43.097	45.685	48.348
100	36.965	39.260	41.608	44.010	46.470
110	36.061	38.238	40.457	42.721	45.031
120	35.338	37.423	39.542	41.698	43.893
150	33.837	35.736	37.657	39.599	41.566
200	32.448	34.184	35.929	37.686	39.456
400	30.550	32.074	33.596	35.117	36.637
1000	29.506	30.920	32.325	33.725	35.119
∞	28.845	30.191	31.526	32.852	34.170

Tafel 8 (Forts.) Table 8 (cont.)

α=0.025

ν \ p	22	24	26	28	30
23	9981.471				
24	2489.317				
25	1173.321	11836.87			
26	720.843	2937.822			
27	508.916	1378.761	13850.26		
28	390.599	843.768	3423.399		
29	316.694	593.594	1600.716	16021.65	
30	266.807	454.114	976.317	3946.047	
35	154.767	211.310	304.359	478.892	886.415
40	114.784	144.555	185.741	245.830	340.104
45	94.706	114.455	139.531	172.294	216.659
50	82.734	97.538	115.399	137.340	164.872
55	74.815	86.758	100.691	117.146	136.857
60	69.202	79.312	90.827	104.062	119.429
70	61.787	69.720	78.473	88.187	99.031
80	57.119	63.822	71.071	78.943	87.527
90	53.914	59.834	66.150	72.908	80.162
100	51.579	56.961	62.644	68.662	75.050
110	49.803	54.792	60.022	65.515	71.297
120	48.406	53.099	57.988	63.091	68.426
150	45.575	49.694	53.933	58.302	62.811
200	43.034	46.671	50.374	54.147	57.996
400	39.681	42.727	45.788	48.860	51.947
1000	37.892	40.649	43.394	46.128	48.855
∞	36.781	39.364	41.923	44.461	46.979

Tafel 8 (Forts.)

Table 8 (cont.)

$\alpha=0.025$

p / ν	35	40	45	50	55
40	1174.875				
45	441.806	1503.741			
50	277.177	556.606	1873.014		
55	208.381	344.923	684.504	2282.694	
60	171.270	256.740	419.900	825.502	2732.782
70	132.607	179.873	250.890	368.021	591.556
80	112.837	145.602	189.625	251.665	344.923
90	100.887	126.361	158.440	200.035	255.970
100	92.902	114.084	139.654	171.144	210.846
110	87.196	105.584	127.132	152.753	183.729
120	82.918	99.356	118.200	140.046	165.687
150	74.762	87.820	102.182	118.077	135.785
200	67.986	78.562	89.805	101.801	114.647
400	59.747	67.688	75.793	84.081	92.573
1000	55.651	62.434	69.220	76.022	82.847
∞	53.203	59.342	65.410	71.420	77.380

α=0.050

ν \ p	1	2	3	4	5
2	18.513				
3	10.128	57.000			
4	7.709	25.472	114.986		
5	6.608	17.361	46.383	192.468	
6	5.987	13.887	29.661	72.937	289.446
7	5.591	12.001	22.720	44.718	105.157
8	5.318	10.828	19.028	33.230	62.561
9	5.117	10.033	16.766	27.202	45.453
10	4.965	9.459	15.248	23.545	36.561
11	4.844	9.026	14.163	21.108	31.205
12	4.747	8.689	13.350	19.376	27.656
13	4.667	8.418	12.719	18.086	25.145
14	4.600	8.197	12.216	17.089	23.281
15	4.543	8.012	11.806	16.296	21.845
16	4.494	7.856	11.465	15.651	20.706
17	4.451	7.722	11.177	15.117	19.782
18	4.414	7.606	10.931	14.667	19.017
19	4.381	7.504	10.719	14.283	18.375
20	4.351	7.415	10.533	13.952	17.828
21	4.325	7.335	10.370	13.663	17.356
22	4.301	7.264	10.225	13.409	16.945
23	4.279	7.200	10.095	13.184	16.585
24	4.260	7.142	9.979	12.983	16.265
25	4.242	7.089	9.874	12.803	15.981
26	4.225	7.041	9.779	12.641	15.726
27	4.210	6.997	9.692	12.493	15.496
28	4.196	6.957	9.612	12.359	15.287
29	4.183	6.919	9.539	12.236	15.097
30	4.171	6.885	9.471	12.123	14.924
35	4.121	6.744	9.200	11.674	14.240
40	4.085	6.642	9.005	11.356	13.762
45	4.057	6.564	8.859	11.118	13.409
50	4.034	6.503	8.744	10.934	13.138
55	4.016	6.454	8.652	10.787	12.923
60	4.001	6.413	8.577	10.668	12.748
70	3.978	6.350	8.460	10.484	12.482
80	3.960	6.303	8.375	10.350	12.289
90	3.947	6.267	8.309	10.248	12.142
100	3.936	6.239	8.257	10.167	12.027
110	3.927	6.216	8.215	10.102	11.934
120	3.920	6.196	8.181	10.048	11.858
150	3.904	6.155	8.105	9.931	11.693
200	3.888	6.113	8.031	9.817	11.531
400	3.865	6.052	7.922	9.650	11.297
1000	3.851	6.015	7.857	9.552	11.160
∞	3.841	5.991	7.815	9.488	11.070

Tafel 8 (Forts.) Table 8 (cont.)

α=0.050

ν \ p	6	7	8	9	10
7	405.920				
8	143.050	541.890			
9	83.202	186.622	697.356		
10	59.403	106.649	235.873	872.317	
11	47.123	75.088	132.903	290.806	1066.774
12	39.764	58.893	92.512	161.967	351.421
13	34.911	49.232	71.878	111.676	193.842
14	31.488	42.881	59.612	86.079	132.582
15	28.955	38.415	51.572	70.907	101.499
16	27.008	35.117	45.932	60.986	83.121
17	25.467	32.588	41.775	54.041	71.127
18	24.219	30.590	38.592	48.930	62.746
19	23.189	28.975	36.082	45.023	56.587
20	22.324	27.642	34.054	41.946	51.884
21	21.588	26.525	32.384	39.463	48.184
22	20.954	25.576	30.985	37.419	45.202
23	20.403	24.759	29.798	35.709	42.750
24	19.920	24.049	28.777	34.258	40.699
25	19.492	23.427	27.891	33.013	38.961
26	19.112	22.878	27.114	31.932	37.469
27	18.770	22.388	26.428	30.985	36.176
28	18.463	21.950	25.818	30.149	35.043
29	18.184	21.555	25.272	29.407	34.044
30	17.931	21.198	24.781	28.742	33.156
35	16.944	19.823	22.913	26.252	29.881
40	16.264	18.890	21.668	24.624	27.783
45	15.767	18.217	20.781	23.477	26.326
50	15.388	17.709	20.117	22.627	25.256
55	15.090	17.311	19.600	21.972	24.437
60	14.850	16.992	19.188	21.451	23.790
70	14.485	16.510	18.571	20.676	22.834
80	14.222	16.165	18.130	20.127	22.162
90	14.022	15.905	17.801	19.718	21.663
100	13.867	15.702	17.544	19.401	21.279
110	13.741	15.540	17.340	19.149	20.973
120	13.639	15.407	17.172	18.943	20.725
150	13.417	15.121	16.814	18.504	20.196
200	13.202	14.845	16.469	18.083	19.692
400	12.890	14.447	15.975	17.484	18.976
1000	12.710	14.217	15.692	17.141	18.570
∞	12.592	14.067	15.507	16.919	18.307

Tafel 8 (Forts.) Table 8 (cont.)

$\alpha=0.050$

ν \ p	11	12	13	14	15
12	1280.727				
13	417.719	1514.176			
14	228.529	489.700	1767.120		
15	155.231	266.028	567.364	2039.560	
16	118.138	179.624	306.339	650.712	2331.496
17	96.253	135.998	205.761	349.464	739.744
18	81.996	110.304	155.078	233.643	395.402
19	72.047	93.592	125.276	175.380	263.269
20	64.745	81.945	105.918	141.169	196.903
21	59.177	73.407	92.442	118.974	157.983
22	54.800	66.902	82.573	103.538	132.759
23	51.274	61.793	75.060	92.244	115.234
24	48.378	57.681	69.165	83.653	102.421
25	45.958	54.305	64.423	76.916	92.681
26	43.908	51.487	60.533	71.501	85.048
27	42.149	49.099	57.286	67.061	78.916
28	40.624	47.053	54.538	63.357	73.890
29	39.291	45.280	52.183	60.223	69.700
30	38.115	43.730	50.143	57.539	66.156
35	33.848	38.209	43.030	48.392	54.392
40	31.175	34.833	38.794	43.102	47.807
45	29.346	32.559	35.990	39.665	43.614
50	28.017	30.926	34.000	37.256	40.715
55	27.008	29.696	32.514	35.475	38.593
60	26.216	28.737	31.364	34.106	36.973
70	25.053	27.339	29.699	32.139	34.666
80	24.241	26.370	28.553	30.796	33.103
90	23.642	25.658	27.716	29.820	31.974
100	23.182	25.114	27.079	29.080	31.120
110	22.817	24.683	26.577	28.499	30.453
120	22.521	24.335	26.171	28.030	29.916
150	21.894	23.600	25.317	27.049	28.795
200	21.297	22.904	24.514	26.128	27.749
400	20.457	21.928	23.392	24.851	26.306
1000	19.981	21.379	22.764	24.139	25.505
∞	19.675	21.026	22.362	23.685	24.996

$\alpha=0.050$

ν \ p	16	17	18	19	20
17	2642.928				
18	834.459	2973.855			
19	444.153	934.859	3324.278		
20	294.641	495.717	1040.942	3694.197	
21	219.648	327.758	550.095	1152.710	4083.611
22	175.719	243.615	362.620	607.287	1270.161
23	147.275	194.376	268.804	399.227	667.292
24	127.529	162.522	213.955	295.215	437.581
25	113.103	140.425	178.499	234.456	322.849
26	102.144	124.291	153.920	195.207	255.879
27	93.560	112.042	135.985	168.016	212.647
28	86.668	102.453	122.376	148.186	182.713
29	81.021	94.757	111.727	133.146	160.893
30	76.316	88.455	103.184	121.382	144.352
35	61.152	68.824	77.602	87.737	99.556
40	52.969	58.658	64.961	71.982	79.849
45	47.871	52.476	57.475	62.921	68.879
50	44.398	48.330	52.540	57.058	61.921
55	41.882	45.361	49.047	52.961	57.126
60	39.978	43.131	46.447	49.939	53.624
70	37.287	40.009	42.840	45.787	48.859
80	35.479	37.929	40.457	43.069	45.771
90	34.180	36.444	38.767	41.155	43.610
100	33.203	35.331	37.507	39.733	42.013
110	32.441	34.466	36.530	38.636	40.784
120	31.830	33.775	35.752	37.764	39.811
150	30.559	32.342	34.144	35.968	37.815
200	29.378	31.017	32.665	34.325	35.997
400	27.758	29.209	30.658	32.108	33.558
1000	26.862	28.215	29.561	30.902	32.238
∞	26.296	27.587	28.869	30.144	31.410

Tafel 8 (Forts.) Table 8 (cont.)

α=0.050

ν \ p	22	24	26	28	30
23	4920.928				
24	1522.116				
25	795.744	5836.227			
26	519.524	1796.808			
27	381.784	935.451	6829.509		
28	301.492	608.449	2094.237		
29	249.719	445.609	1086.413	7900.773	
30	213.908	350.794	704.358	2414.402	
35	130.166	175.384	247.780	378.491	666.429
40	98.797	123.657	157.540	206.067	280.417
45	82.644	99.572	120.860	148.362	185.105
50	72.859	85.775	101.254	120.123	143.589
55	66.316	76.873	89.125	103.514	120.641
60	61.642	70.667	80.905	92.623	106.162
70	55.417	62.599	70.504	79.253	88.990
80	51.467	57.594	64.210	71.381	79.185
90	48.739	54.190	59.997	66.204	72.857
100	46.744	51.725	56.983	62.544	68.440
110	45.221	49.860	54.719	59.820	65.184
120	44.022	48.399	52.958	57.715	62.685
150	41.581	45.451	49.435	53.540	57.777
200	39.381	42.822	46.327	49.900	53.545
400	36.462	39.375	42.300	45.239	48.195
1000	34.899	37.547	40.186	42.819	45.446
∞	33.924	36.415	38.885	41.337	43.773

'afel 8 (Forts.) Table 8 (cont.)

α=0.050

ν \ p	35	40	45	50	55
40	884.072				
45	364.829	1132.286			
50	237.278	460.165	1411.072		
55	181.901	295.723	566.426	1720.431	
60	151.364	224.516	360.442	683.615	2060.363
70	118.950	160.681	222.529	322.666	508.710
80	102.098	131.558	170.803	225.527	306.670
90	91.809	114.992	144.016	181.386	231.214
100	84.886	104.330	127.705	156.347	192.247
110	79.914	96.905	116.753	140.267	168.570
120	76.172	91.440	108.900	129.084	152.697
150	69.003	81.259	94.723	109.604	126.158
200	63.009	73.029	83.680	95.040	107.198
400	55.671	63.291	71.074	79.039	87.204
1000	52.002	58.556	65.121	71.707	78.323
∞	49.802	55.758	61.656	67.505	73.311

α=0.100

ν \ p	1	2	3	4	5
2	8.526				
3	5.538	27.000			
4	4.545	14.566	54.971		
5	4.060	10.811	26.954	92.434	
6	3.776	9.071	18.859	42.741	139.389
7	3.589	8.081	15.202	28.751	61.940
8	3.458	7.446	13.155	22.529	40.506
9	3.360	7.005	11.857	19.085	31.077
10	3.285	6.681	10.964	16.917	25.896
11	3.225	6.434	10.314	15.435	22.655
12	3.177	6.239	9.820	14.361	20.448
13	3.136	6.081	9.432	13.548	18.854
14	3.102	5.951	9.119	12.912	17.651
15	3.073	5.842	8.862	12.401	16.713
16	3.048	5.750	8.648	11.981	15.960
17	3.026	5.670	8.465	11.631	15.344
18	3.007	5.600	8.309	11.335	14.830
19	2.990	5.539	8.173	11.081	14.396
20	2.975	5.485	8.053	10.860	14.023
21	2.961	5.437	7.948	10.667	13.701
22	2.949	5.394	7.854	10.497	13.419
23	2.937	5.355	7.770	10.345	13.170
24	2.927	5.320	7.695	10.210	12.949
25	2.918	5.288	7.626	10.088	12.752
26	2.909	5.259	7.564	9.977	12.574
27	2.901	5.232	7.507	9.877	12.414
28	2.894	5.207	7.455	9.785	12.268
29	2.887	5.184	7.407	9.701	12.135
30	2.881	5.163	7.363	9.624	12.013
35	2.855	5.077	7.184	9.316	11.530
40	2.835	5.013	7.054	9.095	11.190
45	2.820	4.965	6.957	8.930	10.937
50	2.809	4.927	6.880	8.802	10.743
55	2.799	4.896	6.818	8.699	10.588
60	2.791	4.871	6.768	8.616	10.462
70	2.779	4.831	6.690	8.487	10.270
80	2.769	4.802	6.632	8.392	10.129
90	2.762	4.780	6.588	8.320	10.023
100	2.756	4.762	6.553	8.263	9.939
110	2.752	4.747	6.524	8.217	9.871
120	2.748	4.735	6.501	8.179	9.815
150	2.739	4.708	6.449	8.096	9.694
200	2.731	4.682	6.399	8.015	9.576
400	2.718	4.643	6.324	7.895	9.403
1000	2.711	4.620	6.280	7.825	9.303
∞	2.706	4.605	6.251	7.779	9.236

Tafel 8 (Forts.) Table 8 (cont.)

α=0.100

ν \ p	6	7	8	9	10
7	195.836				
8	84.556	261.774			
9	54.132	110.590	337.204		
10	40.854	69.632	140.045	422.124	
11	33.600	51.866	87.009	172.920	516.536
12	29.082	42.202	64.114	106.263	209.216
13	26.016	36.204	51.706	77.601	127.396
14	23.808	32.146	44.025	62.113	92.327
15	22.145	29.229	38.840	52.547	73.423
16	20.850	27.036	35.120	46.102	61.772
17	19.814	25.331	32.329	41.486	53.933
18	18.966	23.969	30.161	38.026	48.326
19	18.261	22.857	28.431	35.343	44.129
20	17.665	21.931	27.020	33.203	40.877
21	17.154	21.150	25.847	31.459	38.286
22	16.713	20.482	24.857	30.011	36.175
23	16.327	19.904	24.012	28.790	34.424
24	15.987	19.400	23.281	27.747	32.949
25	15.685	18.955	22.643	26.846	31.690
26	15.415	18.561	22.082	26.061	30.603
27	15.173	18.209	21.584	25.370	29.655
28	14.954	17.893	21.140	24.757	28.821
29	14.755	17.607	20.741	24.211	28.083
30	14.573	17.348	20.380	23.720	27.424
35	13.862	16.343	19.001	21.866	24.972
40	13.369	15.657	18.074	20.642	23.382
45	13.006	15.158	17.408	19.773	22.269
50	12.729	14.779	16.907	19.125	21.446
55	12.510	14.481	16.516	18.623	20.814
60	12.332	14.242	16.202	18.223	20.312
70	12.062	13.880	15.731	17.625	19.568
80	11.867	13.619	15.394	17.200	19.042
90	11.719	13.422	15.141	16.882	18.651
100	11.603	13.268	14.944	16.635	18.348
110	11.509	13.145	14.786	16.438	18.107
120	11.432	13.043	14.657	16.278	17.911
150	11.266	12.826	14.380	15.934	17.493
200	11.105	12.614	14.112	15.603	17.092
400	10.870	12.309	13.727	15.131	16.523
1000	10.734	12.132	13.506	14.859	16.197
∞	10.645	12.017	13.362	14.684	15.987

Tafel 8 (Forts.)

Table 8 (cont.)

α=0.100

ν \ p	11	12	13	14	15
12	620.440				
13	248.935	733.834			
14	150.408	292.075	856.720		
15	108.294	175.299	338.637	989.097	
16	85.639	125.500	202.069	388.620	1130.965
17	71.699	98.760	143.948	230.719	442.027
18	62.334	82.330	112.788	163.637	261.249
19	55.643	71.306	93.666	127.721	184.567
20	50.640	63.437	80.849	105.706	143.561
21	46.766	57.558	71.708	90.963	118.450
22	43.682	53.009	64.885	80.458	101.649
23	41.171	49.391	59.609	72.621	89.686
24	39.090	46.447	55.415	66.565	80.767
25	37.337	44.008	52.005	61.754	73.878
26	35.842	41.955	49.180	57.843	68.408
27	34.551	40.204	46.803	54.606	63.964
28	33.427	38.694	44.778	51.883	60.286
29	32.438	37.378	43.031	49.563	57.194
30	31.563	36.222	41.509	47.563	54.560
35	28.354	32.059	36.137	40.652	45.679
40	26.318	29.475	32.885	36.582	40.605
45	24.911	27.718	30.709	33.905	37.332
50	23.882	26.446	29.151	32.013	35.047
55	23.097	25.483	27.982	30.605	33.363
60	22.478	24.729	27.073	29.517	32.072
70	21.566	23.625	25.750	27.946	30.220
80	20.925	22.855	24.834	26.867	28.958
90	20.451	22.287	24.163	26.080	28.043
100	20.086	21.852	23.649	25.481	27.349
110	19.796	21.507	23.244	25.009	26.805
120	19.560	21.228	22.917	24.629	26.366
150	19.059	20.636	22.225	23.829	25.448
200	18.582	20.075	21.572	23.076	24.587
400	17.906	19.284	20.657	22.027	23.395
1000	17.522	18.837	20.142	21.440	22.731
∞	17.275	18.549	19.812	21.064	22.307

'afel 8 (Forts.) Table 8 (cont.)

α=0.100

ν \ p	16	17	18	19	20
17	1282.325				
18	498.855	1443.175			
19	293.659	559.105	1613.517		
20	206.738	327.949	622.778	1793.351	
21	160.308	230.151	364.119	689.873	1982.675
22	131.900	177.961	254.805	402.169	760.391
23	112.907	146.055	196.521	280.701	442.099
24	99.392	124.737	160.915	215.988	307.838
25	89.322	109.577	137.140	176.481	236.362
26	81.548	98.287	120.241	150.115	192.751
27	75.377	89.575	107.662	131.384	163.663
28	70.367	82.663	97.960	117.447	143.007
29	66.221	77.052	90.265	106.702	127.643
30	62.737	72.412	84.020	98.182	115.803
35	51.311	57.665	64.887	73.166	82.743
40	45.003	49.831	55.156	61.059	67.640
45	41.016	44.990	49.290	53.960	59.050
50	38.272	41.708	45.377	49.306	53.524
55	36.270	39.338	42.584	46.024	49.677
60	34.745	37.548	40.491	43.587	46.848
70	32.578	35.024	37.567	40.212	42.966
80	31.112	33.331	35.622	37.987	40.432
90	30.055	32.118	34.235	36.411	38.648
100	29.256	31.205	33.197	35.236	37.324
110	28.632	30.493	32.391	34.327	36.303
120	28.131	29.924	31.747	33.603	35.492
150	27.084	28.738	30.412	32.106	33.823
200	26.108	27.638	29.179	30.731	32.296
400	24.762	26.129	27.497	28.865	30.236
1000	24.015	25.297	26.573	27.846	29.116
∞	23.542	24.769	25.989	27.204	28.412

Tafel 8 (Forts.) Table 8 (cont.)

α=0.100

ν \ p	22	24	26	28	30
23	2389.797				
24	911.693				
25	527.600	2834.885			
26	365.839	1076.684			
27	279.831	620.621	3317.937		
28	227.410	428.806	1255.365		
29	192.476	326.928	721.164	3838.954	
30	167.689	264.891	496.741	1447.735	
35	107.197	142.519	197.439	292.664	489.993
40	83.352	103.694	130.975	169.280	226.496
45	70.739	84.987	102.719	125.350	155.148
50	62.969	74.052	87.236	103.174	122.805
55	57.713	66.899	77.503	89.880	104.510
60	53.926	61.864	70.832	81.047	92.789
70	48.838	55.253	62.296	70.067	78.688
80	45.581	51.112	57.074	63.524	70.528
90	43.319	48.277	53.554	59.187	65.215
100	41.656	46.215	51.022	56.103	61.485
110	40.383	44.648	49.113	53.798	58.722
120	39.378	43.417	47.624	52.011	56.594
150	37.324	40.924	44.631	48.452	52.394
200	35.464	38.690	41.977	45.329	48.751
400	32.984	35.743	38.517	41.307	44.116
1000	31.648	34.172	36.691	39.206	41.719
∞	30.813	33.196	35.563	37.916	40.256

α=0.100

ν \ p	35	40	45	50	55
40	650.830				
45	295.278	834.351			
50	199.388	373.020	1040.554		
55	156.030	248.991	459.723	1269.442	
60	131.550	193.024	303.959	555.389	1521.013
70	105.035	141.312	194.279	278.326	429.994
80	91.000	117.125	151.606	199.131	268.553
90	82.336	103.169	129.087	162.204	205.947
100	76.462	94.104	115.215	140.944	172.983
110	72.220	87.748	105.826	127.156	152.709
120	69.015	83.048	99.054	117.501	131.787
150	62.839	74.235	86.738	100.542	115.863
200	57.640	67.055	77.061	87.730	99.143
400	51.229	58.488	65.912	73.517	81.316
1000	48.001	54.293	60.606	66.948	73.326
∞	46.059	51.805	57.505	63.167	68.796

PART II

Tables Related to the Multivariate Normal Distribution

Preliminary Remarks:

The distributions and test statistics under consideration in this part of the collection of tables form a substantial addition to the main part I. Here we consider criteria which have received alltogether too little consideration in the literature and in practice up to this time. In part they are test statistics which are to be recommended for a careful preparation of the data.

Other tables relate to aspects of multivariate quality control and related techniques.

For reasons of space, we exclude tables for ordinary, multiple, and partial correlation coefficients, since they are, in any case, tabulated in numerous other places.

Also related to the topic of this part II are the tables 26, 27, and 30, which appear - by technical reasons - in the appendix part IV.

TABLE 9

The Multivariate Normal Distribution with Equal Correlations: Tables of S. S. Gupta

(a) Content of the tables and definition of the tabulated statistic :

The tables give the values of the

$$F_N(H, \rho) = \int_{-\infty}^{\infty} F^N\left(\frac{x.\rho^{1/2} + H}{(1-\rho)^{1/2}}\right) f(x)dx$$

for given values of $\rho \geqq 0$, N , and H. What we have here is the probability that N standardised, normally distributed random variables (variates) with equal correlation coefficient do not exceed the value H .

In the integrand, F and f , respectively, represent the probability function and the density function of a standardised, normally distributed random variable (variate) .

(b) Extent of the tables and definition of the parameters :

(1) The parameter ρ :

ρ = correlation coefficient

ρ = 0.100 ; 0.125 ; 0.200 ; 0.250 ; 0.300

ρ = 1/3 ; 0.375 ; 0.400 ; 0.500 ; 0.600

= 0.625 ; 2/3 ; 0.700 ; 0.750 ; 0.800

= 0.875 ; 0.900 .

(2) The parameter H :

H = bound for normal random variables (variates)

H = -3.50(0.10)3.50 .

(3) The parameter N :

N = number of normal standardised variables (variates)

N = 1(1)12 .

(c) Suggestions for for applications :

The tables allow one to determine the value of the probability $F_N(H,\rho)$ for given values of N, H, and ρ , as discussed in section (a) .

(d) References :

For the source of the tables and for applications :

GUPTA, S. S. : Probability integrals of multivariate normal and multivariate t.
The Annals of Mathematical Statistics 34, 792 - 828 (1963) (Table II).

(e) Further remarks :

Note that $F_N(H , \rho)$ also gives the probability that the maximum of a set of N equally correlated normal standardised random variables do not exceed the value H .
This probability is equal to the one that the minimum of this set is not less than the value $-H$.

Tafel 9 Table 9

ρ = .100

-H \ N	1	2	3	4	5	6	7	8	9	10	11	12
3.50	.00023	.00000	.00000	.00000	.00000	.00000	.00000	.00000	.00000	.00000	.00000	.00000
3.40	.00034	.00000	.00000	.00000	.00000	.00000	.00000	.00000	.00000	.00000	.00000	.00000
3.30	.00048	.00000	.00000	.00000	.00000	.00000	.00000	.00000	.00000	.00000	.00000	.00000
3.20	.00069	.00000	.00000	.00000	.00000	.00000	.00000	.00000	.00000	.00000	.00000	.00000
3.10	.00097	.00000	.00000	.00000	.00000	.00000	.00000	.00000	.00000	.00000	.00000	.00000
3.00	.00135	.00000	.00000	.00000	.00000	.00000	.00000	.00000	.00000	.00000	.00000	.00000
2.90	.00187	.00001	.00000	.00000	.00000	.00000	.00000	.00000	.00000	.00000	.00000	.00000
2.80	.00256	.00002	.00000	.00000	.00000	.00000	.00000	.00000	.00000	.00000	.00000	.00000
2.70	.00347	.00003	.00000	.00000	.00000	.00000	.00000	.00000	.00000	.00000	.00000	.00000
2.60	.00466	.00005	.00000	.00000	.00000	.00000	.00000	.00000	.00000	.00000	.00000	.00000
2.50	.00621	.00008	.00000	.00000	.00000	.00000	.00000	.00000	.00000	.00000	.00000	.00000
2.40	.00820	.00013	.00000	.00000	.00000	.00000	.00000	.00000	.00000	.00000	.00000	.00000
2.30	.01072	.00022	.00001	.00000	.00000	.00000	.00000	.00000	.00000	.00000	.00000	.00000
2.20	.01390	.00035	.00001	.00000	.00000	.00000	.00000	.00000	.00000	.00000	.00000	.00000
2.10	.01786	.00056	.00003	.00000	.00000	.00000	.00000	.00000	.00000	.00000	.00000	.00000
2.00	.02275	.00087	.00005	.00000	.00000	.00000	.00000	.00000	.00000	.00000	.00000	.00000
1.90	.02872	.00134	.00009	.00001	.00000	.00000	.00000	.00000	.00000	.00000	.00000	.00000
1.80	.03593	.00202	.00016	.00002	.00000	.00000	.00000	.00000	.00000	.00000	.00000	.00000
1.70	.04457	.00300	.00028	.00003	.00001	.00000	.00000	.00000	.00000	.00000	.00000	.00000
1.60	.05480	.00440	.00048	.00007	.00001	.00000	.00000	.00000	.00000	.00000	.00000	.00000
1.50	.06681	.00633	.00079	.00012	.00002	.00000	.00000	.00000	.00000	.00000	.00000	.00000
1.40	.08076	.00899	.00129	.00023	.00005	.00001	.00000	.00000	.00000	.00000	.00000	.00000
1.30	.09680	.01256	.00205	.00040	.00009	.00002	.00001	.00000	.00000	.00000	.00000	.00000
1.20	.11507	.01729	.00321	.00071	.00018	.00005	.00002	.00001	.00000	.00000	.00000	.00000
1.10	.13567	.02344	.00491	.00120	.00033	.00010	.00004	.00001	.00001	.00000	.00000	.00000
1.00	.15866	.03132	.00736	.00199	.00061	.00020	.00007	.00003	.00001	.00001	.00000	.00000
0.90	.18406	.04125	.01082	.00322	.00107	.00039	.00015	.00006	.00003	.00001	.00001	.00000
0.80	.21186	.05355	.01559	.00510	.00133	.00072	.00030	.00013	.00006	.00003	.00002	.00001
0.70	.24196	.06854	.02204	.00786	.00306	.00128	.00057	.00027	.00013	.00007	.00004	.00002
0.60	.27425	.08653	.03057	.01186	.00497	.00223	.00106	.00053	.00027	.00015	.00008	.00005
0.50	.30854	.10776	.04162	.01746	.00786	.00375	.00189	.00099	.00054	.00031	.00018	.00011
0.40	.34458	.13242	.05562	.02515	.01210	.00614	.00326	.00181	.00103	.00061	.00037	.00023
0.30	.38209	.16062	.07301	.03543	.01816	.00976	.00547	.00317	.00190	.00117	.00074	.00048
0.20	.42074	.19237	.09418	.04883	.02658	.01508	.00888	.00540	.00337	.00216	.00142	.00095
0.10	.46017	.22755	.11942	.06588	.03794	.02266	.01398	.00887	.00577	.00384	.00261	.00180
0.00	.50000	.26594	.14891	.08709	.05286	.03314	.02137	.01413	.00955	.00659	.00462	.00330

H \ N	1	2	3	4	5	6	7	8	9	10	11	12
0.00	.50000	.26594	.14891	.08709	.05286	.03314	.02137	.01413	.00955	.00659	.00462	.00330
0.10	.53983	.30720	.18271	.11283	.07196	.04720	.03173	.02180	.01528	.01089	.00789	.00580
0.20	.57926	.35089	.22070	.14336	.09574	.06551	.04580	.03263	.02365	.01741	.01299	.00982
0.30	.61791	.39645	.26259	.17875	.12461	.08870	.06431	.04741	.03547	.02689	.02063	.01601
0.40	.65542	.44327	.30791	.21889	.15877	.11724	.08795	.06692	.05157	.04020	.03167	.02519
0.50	.69146	.49068	.35605	.26340	.19820	.15141	.11722	.09186	.07277	.05823	.04701	.03827
0.60	.72575	.53802	.40625	.31173	.24261	.19122	.15244	.12276	.09979	.08180	.06758	.05622
0.70	.75804	.58461	.45769	.36310	.29145	.23642	.19360	.15990	.13310	.11159	.09417	.07995
0.80	.78814	.62984	.50948	.41659	.34393	.28641	.24039	.20320	.17289	.14799	.12737	.11019
0.90	.81594	.67313	.56074	.47120	.39906	.34035	.29216	.25226	.21899	.19105	.16744	.14738
1.00	.84134	.71401	.61064	.52586	.45570	.39715	.34794	.30628	.27080	.24041	.21424	.19159
1.10	.86433	.75211	.65841	.57955	.51267	.45557	.40653	.36416	.32738	.29528	.26716	.24241
1.20	.88493	.78715	.70344	.63131	.56879	.51430	.46657	.42456	.38744	.35451	.32518	.29898
1.30	.90320	.81896	.74522	.68034	.62298	.57205	.52664	.48600	.44951	.41664	.38693	.36002
1.40	.91924	.84747	.78340	.72597	.67430	.62764	.58538	.54698	.51200	.48004	.45078	.42393
1.50	.93319	.87272	.81779	.76774	.72199	.68007	.64155	.60608	.57335	.54307	.51501	.48895
1.60	.94520	.89480	.84831	.80534	.76552	.72855	.69415	.66209	.63215	.60415	.57792	.55331
1.70	.95543	.91387	.87503	.83868	.80458	.77255	.74241	.71402	.68723	.66193	.63800	.61534
1.80	.96407	.93016	.89811	.86777	.83902	.81174	.78583	.76118	.73772	.71535	.69402	.67365
1.90	.97128	.94390	.91777	.89281	.86893	.84607	.82417	.80317	.78302	.76367	.74507	.72718
2.00	.97725	.95537	.93431	.91403	.89449	.87563	.85743	.83986	.82288	.80646	.79058	.77520
2.10	.98214	.96483	.94806	.93179	.91601	.90069	.88580	.87135	.85729	.84362	.83032	.81738
2.20	.98610	.97255	.95933	.94645	.93387	.92160	.90961	.89791	.88647	.87529	.86436	.85367
2.30	.98928	.97877	.96848	.95839	.94850	.93880	.92929	.91995	.91079	.90180	.89298	.88431
2.40	.99180	.98374	.97580	.96800	.96031	.95275	.94530	.93797	.93074	.92363	.91662	.90971
2.50	.99379	.98766	.98161	.97564	.96974	.96391	.95816	.95247	.94686	.94131	.93583	.93042
2.60	.99534	.99072	.98616	.98164	.97716	.97273	.96834	.96399	.95969	.95542	.95120	.94702
2.70	.99653	.99309	.98968	.98630	.98294	.97960	.97630	.97301	.96976	.96652	.96332	.96013
2.80	.99744	.99491	.99238	.98987	.98738	.98490	.98244	.97999	.97756	.97514	.97273	.97034
2.90	.99813	.99628	.99443	.99259	.99076	.98894	.98712	.98532	.98352	.98173	.97995	.97818
3.00	.99865	.99730	.99596	.99463	.99330	.99197	.99065	.98934	.98802	.98672	.98541	.98412
3.10	.99903	.99807	.99710	.99614	.99519	.99423	.99328	.99233	.99139	.99044	.98950	.98856
3.20	.99931	.99863	.99794	.99726	.99658	.99590	.99522	.99454	.99387	.99319	.99252	.99185
3.30	.99952	.99903	.99855	.99807	.99759	.99711	.99663	.99615	.99568	.99520	.99472	.99425
3.40	.99966	.99933	.99899	.99865	.99832	.99798	.99765	.99731	.99698	.99665	.99631	.99598
3.50	.99977	.99953	.99930	.99907	.99884	.99861	.99638	.99814	.99791	.99768	.99745	.99722

Tafel 9 (Forts.) Table 9 (cont.)

ρ = .125

-H \ N	1	2	3	4	5	6	7	8	9	10	11	12
3.50	.00023	.00000	.00000	.00000	.00000	.00000	.00000	.00000	.00000	.00000	.00000	.00000
3.40	.00034	.00000	.00000	.00000	.00000	.00000	.00000	.00000	.00000	.00000	.00000	.00000
3.30	.00048	.00000	.00000	.00000	.00000	.00000	.00000	.00000	.00000	.00000	.00000	.00000
3.20	.00069	.00000	.00000	.00000	.00000	.00000	.00000	.00000	.00000	.00000	.00000	.00000
3.10	.00097	.00000	.00000	.00000	.00000	.00000	.00000	.00000	.00000	.00000	.00000	.00000
3.00	.00135	.00001	.00000	.00000	.00000	.00000	.00000	.00000	.00000	.00000	.00000	.00000
2.90	.00187	.00001	.00000	.00000	.00000	.00000	.00000	.00000	.00000	.00000	.00000	.00000
2.80	.00256	.00002	.00000	.00000	.00000	.00000	.00000	.00000	.00000	.00000	.00000	.00000
2.70	.00347	.00003	.00000	.00000	.00000	.00000	.00000	.00000	.00000	.00000	.00000	.00000
2.60	.00466	.00006	.00000	.00000	.00000	.00000	.00000	.00000	.00000	.00000	.00000	.00000
2.50	.00621	.00009	.00000	.00000	.00000	.00000	.00000	.00000	.00000	.00000	.00000	.00000
2.40	.00820	.00016	.00001	.00000	.00000	.00000	.00000	.00000	.00000	.00000	.00000	.00000
2.30	.01072	.00025	.00001	.00000	.00000	.00000	.00000	.00000	.00000	.00000	.00000	.00000
2.20	.01390	.00040	.00002	.00000	.00000	.00000	.00000	.00000	.00000	.00000	.00000	.00000
2.10	.01786	.00064	.00004	.00000	.00000	.00000	.00000	.00000	.00000	.00000	.00000	.00000
2.00	.02275	.00098	.00007	.00001	.00000	.00000	.00000	.00000	.00000	.00000	.00000	.00000
1.90	.02872	.00149	.00012	.00001	.00000	.00000	.00000	.00000	.00000	.00000	.00000	.00000
1.80	.03593	.00224	.00021	.00003	.00000	.00000	.00000	.00000	.00000	.00000	.00000	.00000
1.70	.04457	.00330	.00036	.00005	.00001	.00000	.00000	.00000	.00000	.00000	.00000	.00000
1.60	.05480	.00480	.00059	.00010	.00002	.00000	.00000	.00000	.00000	.00000	.00000	.00000
1.50	.06681	.00686	.00097	.00018	.00004	.00001	.00000	.00000	.00000	.00000	.00000	.00000
1.40	.08076	.00967	.00155	.00031	.00008	.00002	.00001	.00000	.00000	.00000	.00000	.00000
1.30	.09680	.01343	.00244	.00054	.00014	.00004	.00001	.00001	.00000	.00000	.00000	.00000
1.20	.11507	.01838	.00375	.00093	.00027	.00009	.00003	.00001	.00001	.00000	.00000	.00000
1.10	.13567	.02479	.00566	.00154	.00048	.00017	.00006	.00003	.00001	.00001	.00000	.00000
1.00	.15866	.03295	.00838	.00250	.00085	.00032	.00013	.00006	.00003	.00001	.00001	.00000
0.90	.18406	.04318	.01217	.00396	.00145	.00058	.00025	.00012	.00006	.00003	.00002	.00001
0.80	.21186	.05580	.01734	.00614	.00241	.00103	.00047	.00023	.00012	.00006	.00004	.00002
0.70	.24196	.07112	.02425	.00930	.00392	.00178	.00087	.00045	.00024	.00013	.00008	.00005
0.60	.27425	.08942	.03330	.01377	.00620	.00300	.00154	.00083	.00047	.00027	.00016	.00010
0.50	.30854	.11096	.04491	.01997	.00958	.00490	.00264	.00149	.00088	.00053	.00033	.00021
0.40	.34458	.13589	.05951	.02833	.01444	.00779	.00441	.00260	.00159	.00100	.00065	.00043
0.30	.38209	.16432	.07749	.03936	.02124	.01205	.00714	.00439	.00279	.00182	.00122	.00083
0.20	.42074	.19623	.09923	.05356	.03050	.01817	.01125	.00720	.00474	.00320	.00221	.00155
0.10	.46017	.23152	.12499	.07144	.04281	.02670	.01722	.01144	.00780	.00544	.00387	.00280
0.00	.50000	.26995	.15492	.09344	.05875	.03825	.02566	.01766	.01244	.00894	.00654	.00486

H \ N	1	2	3	4	5	6	7	8	9	10	11	12
0.00	.50000	.26995	.15492	.09344	.05875	.03825	.02566	.01766	.01244	.00894	.00654	.00486
0.10	.53983	.31117	.18905	.11992	.07885	.05346	.03721	.02650	.01925	.01424	.01070	.00815
0.20	.57926	.35475	.22724	.15106	.10360	.07296	.05258	.03865	.02892	.02198	.01694	.01322
0.30	.61791	.40014	.26919	.18692	.13332	.09729	.07242	.05486	.04220	.03291	.02598	.02074
0.40	.65542	.44673	.31443	.22733	.16815	.12685	.09734	.07583	.05987	.04783	.03863	.03150
0.50	.69146	.49388	.36235	.27192	.20804	.16185	.12777	.10217	.08265	.06755	.05573	.04636
0.60	.72575	.54092	.41221	.32011	.25264	.20223	.16390	.13430	.11114	.09279	.07809	.06619
0.70	.75804	.58719	.46321	.37115	.30142	.24769	.20568	.17239	.14570	.12407	.10637	.09177
0.80	.78814	.63209	.51449	.42414	.35357	.29764	.25274	.21629	.18641	.16167	.14103	.12368
0.90	.81594	.67506	.56519	.47811	.40814	.35121	.30441	.26555	.23300	.20554	.18220	.16223
1.00	.84134	.71564	.61450	.53204	.46404	.40737	.35973	.31935	.28487	.25524	.22961	.20733
1.10	.86433	.75346	.66170	.58495	.52013	.46493	.41756	.37663	.34105	.30995	.28263	.25850
1.20	.88493	.78824	.70618	.63593	.57531	.52264	.47659	.43610	.40032	.36856	.34023	.31487
1.30	.90320	.81983	.74746	.68419	.62853	.57929	.53549	.49637	.46127	.42966	.40108	.37517
1.40	.91924	.84816	.78520	.72912	.67891	.63376	.59299	.55603	.52241	.49174	.46366	.43790
1.50	.93319	.87325	.81920	.77025	.72574	.68512	.64792	.61377	.58230	.55325	.52636	.50141
1.60	.94520	.89520	.84940	.80731	.76850	.73261	.69935	.66843	.63963	.61276	.58762	.56407
1.70	.95543	.91417	.87585	.84018	.80688	.77573	.74653	.71911	.69331	.66900	.64605	.62435
1.80	.96407	.93038	.89871	.86890	.84077	.81418	.78902	.76516	.74252	.72099	.70050	.68097
1.90	.97128	.94406	.91821	.89363	.87022	.84790	.82658	.80621	.78671	.76804	.75013	.73295
2.00	.97725	.95548	.93463	.91463	.89542	.87697	.85921	.84212	.82565	.80976	.79443	.77962
2.10	.98214	.96491	.94827	.93221	.91667	.90164	.88708	.87298	.85931	.84605	.83317	.82067
2.20	.98610	.97260	.95948	.94673	.93433	.92226	.91051	.89907	.88791	.87703	.86642	.85606
2.30	.98928	.97880	.96858	.95858	.94881	.93925	.92990	.92075	.91179	.90302	.89442	.88600
2.40	.99180	.98376	.97587	.96813	.96052	.95305	.94572	.93851	.93142	.92446	.91761	.91087
2.50	.99379	.98768	.98165	.97572	.96987	.96411	.95843	.95283	.94731	.94187	.93649	.93119
2.60	.99534	.99073	.98618	.98169	.97725	.97286	.96852	.96422	.95998	.95578	.95163	.94753
2.70	.99653	.99310	.98970	.98633	.98299	.97968	.97641	.97316	.96994	.96675	.96359	.96046
2.80	.99744	.99491	.99239	.98989	.98741	.98495	.98251	.98008	.97767	.97528	.97290	.97054
2.90	.99813	.99628	.99443	.99260	.99078	.98897	.98716	.98537	.98359	.98182	.98005	.97830
3.00	.99865	.99731	.99597	.99464	.99331	.99199	.99068	.98937	.98807	.98677	.98548	.98419
3.10	.99903	.99807	.99711	.99615	.99519	.99424	.99330	.99235	.99141	.99047	.98954	.98860
3.20	.99931	.99863	.99794	.99726	.99658	.99590	.99523	.99455	.99388	.99321	.99254	.99187
3.30	.99952	.99903	.99855	.99807	.99759	.99711	.99664	.99616	.99568	.99521	.99473	.99426
3.40	.99966	.99933	.99899	.99865	.99832	.99799	.99765	.99732	.99699	.99665	.99632	.99599
3.50	.99977	.99953	.99930	.99907	.99884	.99861	.99838	.99815	.99792	.99768	.99745	.99722

Tafel 9 (Forts.) Table 9 (cont.)

ρ = .200

N / − H	1	2	3	4	5	6	7	8	9	10	11	12
3.50	.00023	.00000	.00000	.00000	.00000	.00000	.00000	.00000	.00000	.00000	.00000	.00000
3.40	.00034	.00000	.00000	.00000	.00000	.00000	.00000	.00000	.00000	.00000	.00000	.00000
3.30	.00048	.00000	.00000	.00000	.00000	.00000	.00000	.00000	.00000	.00000	.00000	.00000
3.20	.00069	.00000	.00000	.00000	.00000	.00000	.00000	.00000	.00000	.00000	.00000	.00000
3.10	.00097	.00001	.00000	.00000	.00000	.00000	.00000	.00000	.00000	.00000	.00000	.00000
3.00	.00135	.00001	.00000	.00000	.00000	.00000	.00000	.00000	.00000	.00000	.00000	.00000
2.90	.00187	.00002	.00000	.00000	.00000	.00000	.00000	.00000	.00000	.00000	.00000	.00000
2.80	.00256	.00003	.00000	.00000	.00000	.00000	.00000	.00000	.00000	.00000	.00000	.00000
2.70	.00347	.00006	.00000	.00000	.00000	.00000	.00000	.00000	.00000	.00000	.00000	.00000
2.60	.00466	.00009	.00000	.00000	.00000	.00000	.00000	.00000	.00000	.00000	.00000	.00000
2.50	.00621	.00015	.00001	.00000	.00000	.00000	.00000	.00000	.00000	.00000	.00000	.00000
2.40	.00820	.00024	.00002	.00000	.00000	.00000	.00000	.00000	.00000	.00000	.00000	.00000
2.30	.01072	.00038	.00003	.00000	.00000	.00000	.00000	.00000	.00000	.00000	.00000	.00000
2.20	.01390	.00059	.00005	.00001	.00000	.00000	.00000	.00000	.00000	.00000	.00000	.00000
2.10	.01786	.00091	.00009	.00001	.00000	.00000	.00000	.00000	.00000	.00000	.00000	.00000
2.00	.02275	.00137	.00015	.00002	.00001	.00000	.00000	.00000	.00000	.00000	.00000	.00000
1.90	.02872	.00204	.00025	.00005	.00001	.00000	.00000	.00000	.00000	.00000	.00000	.00000
1.80	.03593	.00298	.00042	.00008	.00002	.00001	.00000	.00000	.00000	.00000	.00000	.00000
1.70	.04457	.00431	.00067	.00015	.00004	.00001	.00000	.00000	.00000	.00000	.00000	.00000
1.60	.05480	.00614	.00107	.00025	.00007	.00003	.00001	.00000	.00000	.00000	.00000	.00000
1.50	.06681	.00861	.00168	.00043	.00014	.00005	.00002	.00001	.00000	.00000	.00000	.00000
1.40	.08076	.01192	.00257	.00072	.00024	.00009	.00004	.00002	.00001	.00001	.00000	.00000
1.30	.09680	.01626	.00387	.00117	.00042	.00017	.00008	.00004	.00002	.00001	.00001	.00000
1.20	.11507	.02189	.00573	.00187	.00072	.00031	.00015	.00008	.00004	.00002	.00001	.00001
1.10	.13567	.02906	.00834	.00293	.00119	.00054	.00027	.00014	.00008	.00005	.00003	.00002
1.00	.15866	.03807	.01192	.00449	.00194	.00093	.00048	.00027	.00016	.00010	.00006	.00004
0.90	.18406	.04921	.01676	.00676	.00308	.00155	.00084	.00048	.00029	.00018	.00012	.00008
0.80	.21186	.06278	.02315	.00996	.00480	.00252	.00142	.00084	.00053	.00034	.00023	.00016
0.70	.24196	.07906	.03146	.01439	.00730	.00401	.00234	.00144	.00093	.00062	.00042	.00030
0.60	.27425	.09830	.04206	.02041	.01087	.00623	.00378	.00240	.00159	.00109	.00076	.00055
0.50	.30854	.12072	.05533	.02840	.01586	.00946	.00595	.00390	.00265	.00186	.00134	.00098
0.40	.34458	.14643	.07165	.03879	.02267	.01405	.00914	.00618	.00432	.00310	.00228	.00171
0.30	.38209	.17552	.09135	.05204	.03174	.02042	.01372	.00954	.00684	.00503	.00378	.00289
0.20	.42074	.20792	.11471	.06856	.04356	.02903	.02011	.01439	.01057	.00794	.00609	.00475
0.10	.46017	.24351	.14192	.08878	.05863	.04041	.02883	.02117	.01593	.01223	.00956	.00759
0.00	.50000	.28205	.17307	.11301	.07741	.05508	.04043	.03044	.02343	.01836	.01463	.01182

N / H	1	2	3	4	5	6	7	8	9	10	11	12
0.00	.50000	.28205	.17307	.11301	.07741	.05508	.04043	.03044	.02343	.01836	.01463	.01182
0.10	.53983	.32317	.20810	.14149	.10033	.07358	.05546	.04277	.03363	.02689	.02181	.01791
0.20	.57926	.36644	.24684	.17430	.12769	.09634	.07448	.05876	.04716	.03842	.03170	.02646
0.30	.61791	.41134	.28893	.21138	.15966	.12375	.09797	.07897	.06463	.05360	.04496	.03810
0.40	.65542	.45728	.33392	.25248	.19624	.15599	.12631	.10390	.08662	.07305	.06224	.05351
0.50	.69146	.50364	.38120	.29720	.23725	.19309	.15970	.13389	.11358	.09734	.08417	.07336
0.60	.72575	.54980	.43008	.34496	.28229	.23486	.19813	.16912	.14584	.12689	.11126	.09824
0.70	.75804	.59513	.47982	.39502	.33077	.28089	.24137	.20952	.18348	.16191	.14385	.12859
0.80	.78814	.63906	.52961	.44658	.38194	.33054	.28895	.25477	.22633	.20240	.18206	.16462
0.90	.81594	.68109	.57869	.49874	.43491	.38301	.34016	.30431	.27398	.24806	.22572	.20632
1.00	.84134	.72076	.62632	.55059	.48871	.43733	.39411	.35733	.32572	.29832	.27439	.25334
1.10	.86433	.75773	.67185	.60127	.54234	.49246	.44975	.41283	.38063	.35234	.32732	.30505
1.20	.88493	.79175	.71472	.64999	.59484	.54732	.50596	.46968	.43760	.40905	.38351	.36052
1.30	.90320	.82266	.75451	.69605	.64532	.60088	.56162	.52669	.49542	.46726	.44177	.41861
1.40	.91924	.85040	.79091	.73891	.69302	.65221	.61565	.58270	.55285	.52568	.50083	.47802
1.50	.93319	.87500	.82373	.77816	.73733	.70051	.66710	.63663	.60872	.58305	.55935	.53740
1.60	.94520	.89654	.85293	.81357	.77782	.74515	.71517	.68753	.66196	.63821	.61609	.59543
1.70	.95543	.91518	.87856	.84504	.81421	.78572	.75929	.73468	.71170	.69017	.66995	.65092
1.80	.96407	.93112	.90074	.87260	.84641	.82197	.79906	.77755	.75728	.73814	.72004	.70287
1.90	.97128	.94460	.91971	.89639	.87448	.85382	.83431	.81582	.79827	.78159	.76569	.75051
2.00	.97725	.95587	.93571	.91664	.89856	.88138	.86502	.84940	.83448	.82019	.80649	.79334
2.10	.98214	.96518	.94904	.93365	.91894	.90486	.89135	.87838	.86590	.85388	.84230	.83111
2.20	.98610	.97279	.96002	.94775	.93594	.92456	.91358	.90297	.89271	.88278	.87315	.86381
2.30	.98928	.97893	.96894	.95928	.94993	.94086	.93206	.92352	.91521	.90714	.89927	.89161
2.40	.99180	.98385	.97612	.96860	.96128	.95416	.94721	.94043	.93381	.92734	.92102	.91484
2.50	.99379	.98773	.98182	.97603	.97038	.96485	.95944	.95413	.94893	.94384	.93884	.93393
2.60	.99534	.99077	.98629	.98189	.97758	.97334	.96918	.96509	.96106	.95710	.95321	.94937
2.70	.99653	.99312	.98977	.98646	.98321	.98000	.97684	.97372	.97065	.96762	.96463	.96168
2.80	.99744	.99492	.99243	.98998	.98755	.98515	.98278	.98044	.97812	.97583	.97357	.97133
2.90	.99813	.99629	.99446	.99265	.99086	.98909	.98733	.98560	.98387	.98217	.98047	.97880
3.00	.99865	.99731	.99598	.99467	.99336	.99207	.99078	.98950	.98824	.98698	.98574	.98450
3.10	.99903	.99807	.99712	.99617	.99523	.99429	.99336	.99243	.99151	.99060	.98969	.98879
3.20	.99931	.99863	.99795	.99727	.99660	.99593	.99526	.99460	.99394	.99329	.99263	.99198
3.30	.99952	.99903	.99856	.99808	.99760	.99713	.99666	.99619	.99572	.99525	.99479	.99433
3.40	.99966	.99933	.99899	.99866	.99833	.99799	.99766	.99733	.99701	.99668	.99635	.99603
3.50	.99977	.99953	.99930	.99907	.99884	.99861	.99838	.99815	.99793	.99770	.99747	.99724

Tafel 9 (Forts.) Table 9 (cont.)

ρ = .250

N / - H	1	2	3	4	5	6	7	8	9	10	11	12
3.50	.00023	.00000	.00000	.00000	.00000	.00000	.00000	.00000	.00000	.00000	.00000	.00000
3.40	.00034	.00000	.00000	.00000	.00000	.00000	.00000	.00000	.00000	.00000	.00000	.00000
3.30	.00048	.00000	.00000	.00000	.00000	.00000	.00000	.00000	.00000	.00000	.00000	.00000
3.20	.00069	.00001	.00000	.00000	.00000	.00000	.00000	.00000	.00000	.00000	.00000	.00000
3.10	.00097	.00001	.00000	.00000	.00000	.00000	.00000	.00000	.00000	.00000	.00000	.00000
3.00	.00135	.00002	.00000	.00000	.00000	.00000	.00000	.00000	.00000	.00000	.00000	.00000
2.90	.00187	.00003	.00000	.00000	.00000	.00000	.00000	.00000	.00000	.00000	.00000	.00000
2.80	.00256	.00005	.00000	.00000	.00000	.00000	.00000	.00000	.00000	.00000	.00000	.00000
2.70	.00347	.00008	.00000	.00000	.00000	.00000	.00000	.00000	.00000	.00000	.00000	.00000
2.60	.00466	.00013	.00001	.00000	.00000	.00000	.00000	.00000	.00000	.00000	.00000	.00000
2.50	.00621	.00020	.00002	.00000	.00000	.00000	.00000	.00000	.00000	.00000	.00000	.00000
2.40	.00820	.00032	.00003	.00000	.00000	.00000	.00000	.00000	.00000	.00000	.00000	.00000
2.30	.01072	.00049	.00005	.00001	.00000	.00000	.00000	.00000	.00000	.00000	.00000	.00000
2.20	.01390	.00075	.00008	.00002	.00000	.00000	.00000	.00000	.00000	.00000	.00000	.00000
2.10	.01786	.00113	.00014	.00003	.00001	.00000	.00000	.00000	.00000	.00000	.00000	.00000
2.00	.02275	.00168	.00024	.00005	.00001	.00000	.00000	.00000	.00000	.00000	.00000	.00000
1.90	.02872	.00247	.00039	.00009	.00003	.00001	.00000	.00000	.00000	.00000	.00000	.00000
1.80	.03593	.00356	.00062	.00015	.00005	.00002	.00001	.00000	.00000	.00000	.00000	.00000
1.70	.04457	.00508	.00098	.00026	.00009	.00003	.00002	.00001	.00000	.00000	.00000	.00000
1.60	.05480	.00715	.00151	.00044	.00015	.00006	.00003	.00002	.00001	.00000	.00000	.00000
1.50	.06681	.00991	.00230	.00071	.00027	.00012	.00006	.00003	.00002	.00001	.00001	.00000
1.40	.08076	.01357	.00345	.00114	.00045	.00021	.00010	.00006	.00003	.00002	.00001	.00001
1.30	.09680	.01832	.00508	.00180	.00075	.00036	.00019	.00010	.00006	.00004	.00003	.00002
1.20	.11507	.02441	.00735	.00278	.00123	.00061	.00033	.00019	.00012	.00007	.00005	.00003
1.10	.13567	.03210	.01047	.00421	.00195	.00101	.00056	.00034	.00021	.00014	.00009	.00007
1.00	.15866	.04168	.01468	.00626	.00305	.00164	.00095	.00058	.00038	.00025	.00017	.00012
0.90	.18406	.05342	.02025	.00914	.00456	.00260	.00156	.00098	.00065	.00045	.00032	.00023
0.80	.21186	.06762	.02749	.01310	.00699	.00405	.00250	.00163	.00110	.00077	.00056	.00041
0.70	.24196	.08453	.03675	.01846	.01028	.00617	.00394	.00263	.00182	.00131	.00096	.00072
0.60	.27425	.10439	.04838	.02556	.01483	.00922	.00606	.00415	.00295	.00216	.00162	.00124
0.50	.30854	.12738	.06274	.03478	.02099	.01350	.00912	.00641	.00465	.00347	.00265	.00206
0.40	.34458	.15360	.08016	.04653	.02916	.01936	.01344	.00968	.00718	.00546	.00424	.00335
0.30	.38209	.18311	.10095	.06121	.03977	.02722	.01941	.01431	.01084	.00839	.00663	.00532
0.20	.42074	.21583	.12532	.07922	.05327	.03755	.02746	.02070	.01599	.01261	.01012	.00825
0.10	.46017	.25162	.15344	.10089	.07010	.05081	.03808	.02932	.02308	.01852	.01510	.01248
0.00	.50000	.29022	.18532	.12648	.09066	.06748	.05176	.04067	.03262	.02660	.02202	.01845

N / H	1	2	3	4	5	6	7	8	9	10	11	12
0.00	.50000	.29022	.18532	.12648	.09066	.06748	.05176	.04067	.03262	.02660	.02202	.01845
0.10	.53983	.33127	.22090	.15615	.11528	.08800	.06900	.05530	.04513	.03739	.03140	.02666
0.20	.57926	.37435	.25994	.18994	.14418	.11274	.09027	.07370	.06116	.05145	.04380	.03768
0.30	.61791	.41893	.30210	.22771	.17745	.14193	.11593	.09634	.08123	.06933	.05981	.05208
0.40	.65542	.46445	.34691	.26918	.21500	.17569	.14623	.12358	.10578	.09154	.07997	.07045
0.50	.69146	.51030	.39377	.31393	.25660	.21393	.18125	.15563	.13514	.11849	.10477	.09331
0.60	.72575	.55588	.44203	.36137	.30181	.25641	.22089	.19252	.16946	.15044	.13453	.12109
0.70	.75804	.60060	.49095	.41080	.35003	.30265	.26484	.23410	.20870	.18745	.16945	.15405
0.80	.78814	.64391	.53980	.46144	.40054	.35202	.31256	.27994	.25259	.22937	.20946	.19223
0.90	.81594	.68530	.58784	.51245	.45248	.40372	.36337	.32946	.30062	.27582	.25429	.23545
1.00	.84134	.72437	.63439	.56299	.50495	.45686	.41638	.38185	.35206	.32616	.30339	.28326
1.10	.86433	.76077	.67884	.61227	.55704	.51046	.47063	.43617	.40607	.37954	.35600	.33497
1.20	.88493	.79427	.72067	.65954	.60787	.56355	.52509	.49137	.46155	.43498	.41115	.38965
1.30	.90320	.82472	.75947	.70419	.65663	.61520	.57874	.54637	.51742	.49135	.46774	.44625
1.40	.91924	.85205	.79498	.74571	.70263	.66457	.63063	.60014	.57258	.54751	.52460	.50357
1.50	.93319	.87630	.82701	.78374	.74533	.71094	.67991	.65173	.62599	.60236	.58058	.56042
1.60	.94520	.89755	.85552	.81806	.78434	.75378	.72590	.70031	.67672	.65489	.63459	.61566
1.70	.95543	.91595	.88057	.84858	.81943	.79270	.76807	.74525	.72403	.70422	.68568	.66826
1.80	.96407	.93170	.90228	.87533	.85050	.82750	.80609	.78609	.76734	.74971	.73308	.71736
1.90	.97128	.94503	.92086	.89847	.87761	.85812	.83982	.82258	.80630	.79089	.77626	.76234
2.00	.97725	.95618	.93656	.91819	.90093	.88465	.86925	.85464	.64074	.82750	.81486	.80276
2.10	.98214	.96540	.94966	.93478	.92069	.90729	.89453	.88234	.87068	.05950	.84877	.83844
2.20	.98610	.97294	.96046	.94856	.93721	.92634	.91592	.90591	.89628	.88699	.87804	.86938
2.30	.98928	.97904	.96925	.95986	.95083	.94214	.93375	.92565	.91782	.91024	.90288	.89575
2.40	.99180	.98392	.97633	.96900	.96191	.95505	.94840	.94194	.93567	.92957	.92363	.91784
2.50	.99379	.98778	.98196	.97630	.97081	.96547	.96026	.95519	.95024	.94541	.94068	.93607
2.60	.99534	.99080	.98638	.98208	.97787	.97376	.96974	.96581	.96196	.95819	.95449	.95086
2.70	.99653	.99314	.98983	.98658	.98340	.98027	.97721	.97420	.97125	.96835	.96549	.96269
2.80	.99744	.99494	.99247	.99005	.98767	.98533	.98302	.98076	.97852	.97632	.97414	.97200
2.90	.99813	.99630	.99449	.99270	.99094	.98920	.98749	.98580	.98413	.98248	.98085	.97924
3.00	.99865	.99732	.99600	.99470	.99341	.99214	.99088	.98963	.98840	.98718	.98597	.98478
3.10	.99903	.99807	.99713	.99619	.99526	.99433	.99342	.99251	.99162	.99072	.98984	.98897
3.20	.99931	.99863	.99795	.99728	.99662	.99596	.99530	.99465	.99400	.99336	.99272	.99209
3.30	.99952	.99904	.99856	.99808	.99761	.99714	.99668	.99622	.99576	.99530	.99484	.99439
3.40	.99966	.99933	.99899	.99866	.99833	.99800	.99768	.99735	.99703	.99670	.99638	.99606
3.50	.99977	.99954	.99930	.99907	.99885	.99862	.99839	.99816	.99794	.99771	.99749	.99727

Tafel 9 (Forts.) Table 9 (cont.)

ρ = .300

−H \ N	1	2	3	4	5	6	7	8	9	10	11	12
3.50	.00023	.00000	.00000	.00000	.00000	.00000	.00000	.00000	.00000	.00000	.00000	.00000
3.40	.00034	.00000	.00000	.00000	.00000	.00000	.00000	.00000	.00000	.00000	.00000	.00000
3.30	.00048	.00000	.00000	.00000	.00000	.00000	.00000	.00000	.00000	.00000	.00000	.00000
3.20	.00069	.00001	.00000	.00000	.00000	.00000	.00000	.00000	.00000	.00000	.00000	.00000
3.10	.00097	.00001	.00000	.00000	.00000	.00000	.00000	.00000	.00000	.00000	.00000	.00000
3.00	.00135	.00002	.00000	.00000	.00000	.00000	.00000	.00000	.00000	.00000	.00000	.00000
2.90	.00187	.00004	.00000	.00000	.00000	.00000	.00000	.00000	.00000	.00000	.00000	.00000
2.80	.00256	.00006	.00000	.00000	.00000	.00000	.00000	.00000	.00000	.00000	.00000	.00000
2.70	.00347	.00010	.00001	.00000	.00000	.00000	.00000	.00000	.00000	.00000	.00000	.00000
2.60	.00466	.00017	.00002	.00000	.00000	.00000	.00000	.00000	.00000	.00000	.00000	.00000
2.50	.00621	.00026	.00003	.00001	.00000	.00000	.00000	.00000	.00000	.00000	.00000	.00000
2.40	.00820	.00041	.00005	.00001	.00000	.00000	.00000	.00000	.00000	.00000	.00000	.00000
2.30	.01072	.00062	.00008	.00002	.00000	.00000	.00000	.00000	.00000	.00000	.00000	.00000
2.20	.01390	.00094	.00014	.00003	.00001	.00000	.00000	.00000	.00000	.00000	.00000	.00000
2.10	.01786	.00139	.00022	.00005	.00002	.00001	.00000	.00000	.00000	.00000	.00000	.00000
2.00	.02275	.00204	.00036	.00009	.00003	.00001	.00001	.00000	.00000	.00000	.00000	.00000
1.90	.02872	.00295	.00057	.00016	.00006	.00002	.00001	.00001	.00000	.00000	.00000	.00000
1.80	.03593	.00422	.00089	.00027	.00010	.00004	.00002	.00001	.00001	.00000	.00000	.00000
1.70	.04457	.00594	.00137	.00044	.00017	.00008	.00004	.00002	.00001	.00001	.00001	.00000
1.60	.05480	.00826	.00207	.00070	.00029	.00014	.00007	.00004	.00002	.00002	.00001	.00001
1.50	.06681	.01133	.00308	.00111	.00048	.00024	.00013	.00008	.00005	.00003	.00002	.00001
1.40	.08076	.01535	.00451	.00172	.00078	.00040	.00022	.00013	.00009	.00006	.00004	.00003
1.30	.09680	.02052	.00650	.00262	.00124	.00066	.00038	.00024	.00015	.00010	.00007	.00005
1.20	.11507	.02709	.00923	.00393	.00195	.00107	.00064	.00040	.00027	.00019	.00013	.00010
1.10	.13567	.03531	.01290	.00580	.00299	.00170	.00104	.00068	.00046	.00032	.00024	.00018
1.00	.15866	.04546	.01777	.00840	.00451	.00265	.00167	.00111	.00077	.00055	.00041	.00031
0.90	.18406	.05781	.02410	.01196	.00668	.00406	.00263	.00179	.00126	.00092	.00069	.00053
0.80	.21186	.07263	.03221	.01676	.00971	.00609	.00405	.00282	.00203	.00151	.00115	.00090
0.70	.24196	.09017	.04243	.02309	.01388	.00896	.00611	.00434	.00320	.00242	.00188	.00148
0.60	.27425	.11063	.05509	.03131	.01948	.01295	.00904	.00657	.00493	.00379	.00298	.00239
0.50	.30854	.13418	.07052	.04177	.02687	.01836	.01314	.00974	.00744	.00582	.00464	.00377
0.40	.34458	.16090	.08902	.05487	.03645	.02558	.01872	.01416	.01100	.00874	.00708	.00582
0.30	.38209	.19082	.11085	.07096	.04861	.03500	.02618	.02019	.01596	.01287	.01056	.00879
0.20	.42074	.22385	.13620	.09039	.06376	.04705	.03595	.02824	.02268	.01856	.01543	.01300
0.10	.46017	.25983	.16516	.11344	.08229	.06217	.04847	.03875	.03162	.02624	.02210	.01883
0.00	.50000	.29849	.19774	.14031	.10453	.08077	.06421	.05221	.04325	.03638	.03101	.02673

H \ N	1	2	3	4	5	6	7	8	9	10	11	12
0.00	.50000	.29849	.19774	.14031	.10453	.08077	.06421	.05221	.04325	.03638	.03101	.02673
0.10	.53983	.33949	.23381	.17109	.13073	.10321	.08358	.06907	.05805	.04948	.04267	.03718
0.20	.57926	.38237	.27314	.20575	.16104	.12975	.10694	.08977	.07650	.06602	.05759	.05070
0.30	.61791	.42664	.31534	.24413	.19547	.16056	.13458	.11466	.09901	.08647	.07625	.06781
0.40	.65542	.47175	.35996	.28591	.23387	.19565	.16662	.14397	.12590	.11122	.09911	.08898
0.50	.69146	.51710	.40641	.33064	.27594	.23487	.20306	.17782	.15738	.14055	.12649	.11461
0.60	.72575	.56213	.45405	.37774	.32124	.27789	.24370	.21613	.19348	.17458	.15861	.14496
0.70	.75804	.60624	.50219	.42653	.36916	.32423	.28817	.25863	.23403	.21325	.19549	.18015
0.80	.78814	.64892	.55011	.47627	.41898	.37325	.33592	.30488	.27869	.25630	.23696	.22008
0.90	.81594	.68969	.59715	.52617	.46991	.42417	.38623	.35424	.32690	.30327	.28263	.26446
1.00	.84134	.72815	.64264	.57545	.52111	.47615	.43829	.40593	.37795	.35349	.33191	.31275
1.10	.86433	.76398	.68602	.62337	.57172	.52828	.49117	.45905	.43095	.40613	.38404	.36423
1.20	.88493	.79695	.72682	.66925	.62093	.57967	.54394	.51264	.48495	.46026	.43807	.41801
1.30	.90320	.82692	.76465	.71252	.66803	.62949	.59568	.56573	.53895	.51484	.49299	.47308
1.40	.91924	.85383	.79926	.75273	.71240	.67698	.64554	.61737	.59195	.56886	.54776	.52838
1.50	.93319	.87771	.83049	.78954	.75354	.72151	.69275	.66674	.64304	.62132	.60132	.58283
1.60	.94520	.89866	.85830	.82277	.79110	.76260	.73674	.71311	.69140	.67135	.65276	.63544
1.70	.95543	.91681	.88275	.85234	.82489	.79991	.77703	.75594	.73640	.71822	.70124	.68532
1.80	.96407	.93236	.90397	.87828	.85483	.83328	.81335	.79483	.77754	.76135	.74612	.73175
1.90	.97128	.94552	.92214	.90073	.88099	.86267	.84558	.82958	.81454	.80035	.78693	.77420
2.00	.97725	.95654	.93751	.91990	.90350	.88816	.87373	.86013	.84726	.83504	.82342	.81233
2.10	.98214	.96566	.95036	.93606	.92262	.90995	.89795	.88657	.87572	.86538	.85548	.84600
2.20	.98610	.97313	.96096	.94949	.93863	.92831	.91848	.90909	.90010	.89148	.88319	.87521
2.30	.98928	.97917	.96961	.96052	.95186	.94357	.93563	.92800	.92067	.91359	.90676	.90015
2.40	.99180	.98401	.97658	.96947	.96264	.95608	.94975	.94365	.93774	.93203	.92648	.92111
2.50	.99379	.98784	.98213	.97663	.97132	.96619	.96122	.95640	.95172	.94717	.94274	.93843
2.60	.99534	.99084	.98650	.98230	.97822	.97426	.97040	.96665	.96299	.95943	.95594	.95254
2.70	.99653	.99317	.98990	.98673	.98363	.98061	.97766	.97478	.97196	.96920	.96650	.96385
2.80	.99744	.99495	.99252	.99015	.98783	.98555	.98333	.98114	.97900	.97690	.97483	.97280
2.90	.99813	.99631	.99452	.99276	.99104	.98935	.98769	.98605	.98445	.98286	.98130	.97977
3.00	.99865	.99732	.99602	.99474	.99347	.99223	.99101	.98980	.98861	.98743	.98627	.98513
3.10	.99903	.99808	.99714	.99621	.99530	.99439	.99350	.99262	.99175	.99088	.99003	.98919
3.20	.99931	.99863	.99796	.99730	.99664	.99599	.99535	.99471	.99408	.99346	.99284	.99223
3.30	.99952	.99904	.99856	.99809	.99763	.99717	.99671	.99625	.99581	.99536	.99492	.99448
3.40	.99966	.99933	.99900	.99867	.99834	.99802	.99769	.99737	.99706	.99674	.99643	.99612
3.50	.99977	.99953	.99930	.99908	.99885	.99862	.99840	.99818	.99796	.99773	.99752	.99730

Tafel 9 (Forts.) Table 9 (cont.)

$\rho = 1/3$

−H \ N	1	2	3	4	5	6	7	8	9	10	11	12
3.50	.00023	.00000	.00000	.00000	.00000	.00000	.00000	.00000	.00000	.00000	.00000	.00000
3.40	.00034	.00000	.00000	.00000	.00000	.00000	.00000	.00000	.00000	.00000	.00000	.00000
3.30	.00048	.00001	.00000	.00000	.00000	.00000	.00000	.00000	.00000	.00000	.00000	.00000
3.20	.00069	.00001	.00000	.00000	.00000	.00000	.00000	.00000	.00000	.00000	.00000	.00000
3.10	.00097	.00002	.00000	.00000	.00000	.00000	.00000	.00000	.00000	.00000	.00000	.00000
3.00	.00135	.00003	.00000	.00000	.00000	.00000	.00000	.00000	.00000	.00000	.00000	.00000
2.90	.00187	.00005	.00000	.00000	.00000	.00000	.00000	.00000	.00000	.00000	.00000	.00000
2.80	.00256	.00008	.00001	.00000	.00000	.00000	.00000	.00000	.00000	.00000	.00000	.00000
2.70	.00347	.00013	.00001	.00000	.00000	.00000	.00000	.00000	.00000	.00000	.00000	.00000
2.60	.00466	.00020	.00002	.00000	.00000	.00000	.00000	.00000	.00000	.00000	.00000	.00000
2.50	.00621	.00031	.00004	.00001	.00000	.00000	.00000	.00000	.00000	.00000	.00000	.00000
2.40	.00820	.00048	.00007	.00002	.00000	.00000	.00000	.00000	.00000	.00000	.00000	.00000
2.30	.01072	.00072	.00011	.00003	.00001	.00000	.00000	.00000	.00000	.00000	.00000	.00000
2.20	.01390	.00108	.00018	.00005	.00002	.00001	.00000	.00000	.00000	.00000	.00000	.00000
2.10	.01786	.00159	.00029	.00008	.00003	.00001	.00001	.00000	.00000	.00000	.00000	.00000
2.00	.02275	.00231	.00046	.00014	.00005	.00002	.00001	.00001	.00000	.00000	.00000	.00000
1.90	.02872	.00331	.00072	.00023	.00009	.00004	.00002	.00001	.00001	.00000	.00000	.00000
1.80	.03593	.00469	.00111	.00037	.00015	.00007	.00004	.00002	.00001	.00001	.00001	.00000
1.70	.04457	.00656	.00168	.00059	.00026	.00013	.00007	.00004	.00003	.00002	.00001	.00001
1.60	.05480	.00906	.00251	.00093	.00042	.00022	.00012	.00007	.00005	.00003	.00002	.00002
1.50	.06681	.01234	.00368	.00145	.00068	.00036	.00021	.00013	.00008	.00006	.00004	.00003
1.40	.08076	.01661	.00532	.00220	.00108	.00059	.00035	.00022	.00015	.00010	.00007	.00006
1.30	.09680	.02207	.00758	.00330	.00168	.00095	.00058	.00038	.00026	.00018	.00013	.00010
1.20	.11507	.02896	.01063	.00486	.00257	.00150	.00094	.00063	.00043	.00031	.00023	.00018
1.10	.13567	.03754	.01469	.00705	.00386	.00232	.00150	.00102	.00072	.00053	.00040	.00031
1.00	.15866	.04807	.02002	.01005	.00571	.00353	.00233	.00162	.00117	.00087	.00066	.00052
0.90	.18406	.06083	.02688	.01411	.00829	.00528	.00357	.00253	.00186	.00141	.00109	.00086
0.80	.21186	.07607	.03558	.01948	.01185	.00776	.00537	.00388	.00290	.00223	.00175	.00140
0.70	.24196	.09402	.04644	.02649	.01664	.01119	.00792	.00583	.00443	.00346	.00275	.00223
0.60	.27425	.11489	.05978	.03547	.02298	.01586	.01147	.00860	.00665	.00526	.00425	.00349
0.50	.30854	.13881	.07591	.04678	.03123	.02209	.01631	.01246	.00978	.00785	.00642	.00533
0.40	.34458	.16586	.09512	.06076	.04176	.03024	.02280	.01772	.01413	.01150	.00952	.00799
0.30	.38209	.19604	.11763	.07778	.05495	.04072	.03130	.02475	.02003	.01651	.01383	.01174
0.20	.42074	.22928	.14360	.09813	.07119	.05393	.04224	.03395	.02787	.02328	.01972	.01692
0.10	.46017	.26538	.17311	.12206	.09082	.07028	.05604	.04576	.03808	.03221	.02760	.02392
0.00	.50000	.30409	.20613	.14974	.11413	.09012	.07311	.06061	.05113	.04375	.03790	.03318

H \ N	1	2	3	4	5	6	7	8	9	10	11	12
0.00	.50000	.30409	.20613	.14974	.11413	.09012	.07311	.06061	.05113	.04375	.03790	.03318
0.10	.53983	.34504	.24251	.18121	.14133	.11377	.09384	.07893	.06744	.05838	.05111	.04516
0.20	.57926	.38780	.28201	.21642	.17250	.14144	.11853	.10109	.08744	.07654	.06768	.06036
0.30	.61791	.43187	.32423	.25516	.20763	.17324	.14739	.12737	.11149	.09863	.08805	.07922
0.40	.65542	.47670	.36871	.29711	.24653	.20913	.18049	.15796	.13983	.12497	.11260	.10217
0.50	.69146	.52173	.41489	.34180	.28887	.24891	.21777	.19288	.17258	.15574	.14157	.12950
0.60	.72575	.56638	.46212	.38866	.33418	.29223	.25897	.23200	.20971	.19100	.17508	.16140
0.70	.75804	.61010	.50974	.43703	.38187	.33858	.30371	.27501	.25100	.23061	.21310	.19789
0.80	.78814	.65236	.55707	.48618	.43124	.38733	.35140	.32143	.29605	.27426	.25536	.23880
0.90	.81594	.69271	.60344	.53536	.48150	.43771	.40135	.37062	.34429	.32145	.30144	.28377
1.00	.84134	.73076	.64824	.58381	.53186	.48892	.45275	.42180	.39499	.37150	.35073	.33223
1.10	.86433	.76621	.69093	.63084	.58150	.54009	.50473	.47412	.44731	.42360	.40245	.38346
1.20	.88493	.79882	.73104	.67581	.62968	.59039	.55641	.52665	.50033	.47683	.45570	.43657
1.30	.90320	.82847	.76822	.71818	.67570	.63902	.60691	.57849	.55310	.53024	.50951	.49062
1.40	.91924	.85509	.80223	.75753	.71900	.68529	.65545	.62876	.60471	.58286	.56291	.54458
1.50	.93319	.87873	.83292	.79354	.75912	.72863	.70134	.67670	.65430	.63380	.61493	.59748
1.60	.94520	.89946	.86026	.82604	.79573	.76858	.74403	.72166	.70115	.68224	.66471	.64841
1.70	.95543	.91743	.88431	.85497	.82866	.80484	.78310	.76312	.74466	.72752	.71153	.69656
1.80	.96407	.93283	.90518	.88036	.85785	.83727	.81831	.80076	.78441	.76913	.75479	.74129
1.90	.97128	.94588	.92307	.90235	.88336	.86583	.84956	.83437	.82013	.80674	.79409	.78213
2.00	.97725	.95681	.93821	.92114	.90533	.89062	.87686	.86392	.85172	.84017	.82921	.81878
2.10	.98214	.96586	.95088	.93698	.92401	.91184	.90037	.88952	.87922	.86942	.86007	.85113
2.20	.98610	.97327	.96134	.95017	.93966	.92973	.92031	.91134	.90279	.89460	.88676	.87922
2.30	.98928	.97927	.96988	.96102	.95261	.94462	.93699	.92969	.92269	.91595	.90947	.90322
2.40	.99180	.98408	.97677	.96982	.96319	.95684	.95074	.94488	.93923	.93378	.92851	.92341
2.50	.99379	.98789	.98226	.97688	.97171	.96673	.96193	.95729	.95280	.94845	.94423	.94012
2.60	.99534	.99088	.98659	.98247	.97849	.97463	.97090	.96728	.96376	.96034	.95701	.95376
2.70	.99653	.99319	.98997	.98684	.98381	.98087	.97801	.97522	.97250	.96984	.96725	.96471
2.80	.99744	.99497	.99256	.99023	.98795	.98573	.98356	.98144	.97937	.97734	.97535	.97339
2.90	.99813	.99632	.99454	.99281	.99112	.98947	.98784	.98625	.98469	.98316	.98165	.98017
3.00	.99865	.99733	.99604	.99477	.99353	.99231	.99111	.98993	.98877	.98763	.98650	.98540
3.10	.99903	.99808	.99715	.99623	.99533	.99444	.99357	.99270	.99185	.99101	.99018	.98937
3.20	.99931	.99863	.99797	.99731	.99666	.99602	.99539	.99477	.99415	.99354	.99294	.99234
3.30	.99951	.99904	.99857	.99810	.99764	.99718	.99673	.99629	.99585	.99541	.99498	.99455
3.40	.99966	.99933	.99900	.99867	.99835	.99803	.99771	.99739	.99708	.99677	.99647	.99616
3.50	.99977	.99953	.99931	.99908	.99885	.99863	.99841	.99819	.99797	.99775	.99754	.99732

$\rho = .375$

H \ N	1	2	3	4	5	6	7	8	9	10	11	12
3.50	.00023	.00000	.00000	.00000	.00000	.00000	.00000	.00000	.00000	.00000	.00000	.00000
3.40	.00034	.00000	.00000	.00000	.00000	.00000	.00000	.00000	.00000	.00000	.00000	.00000
3.30	.00048	.00001	.00000	.00000	.00000	.00000	.00000	.00000	.00000	.00000	.00000	.00000
3.20	.00069	.00001	.00000	.00000	.00000	.00000	.00000	.00000	.00000	.00000	.00000	.00000
3.10	.00097	.00002	.00000	.00000	.00000	.00000	.00000	.00000	.00000	.00000	.00000	.00000
3.00	.00135	.00004	.00000	.00000	.00000	.00000	.00000	.00000	.00000	.00000	.00000	.00000
2.90	.00187	.00006	.00001	.00000	.00000	.00000	.00000	.00000	.00000	.00000	.00000	.00000
2.80	.00256	.00010	.00001	.00000	.00000	.00000	.00000	.00000	.00000	.00000	.00000	.00000
2.70	.00347	.00016	.00002	.00000	.00000	.00000	.00000	.00000	.00000	.00000	.00000	.00000
2.60	.00466	.00025	.00004	.00001	.00000	.00000	.00000	.00000	.00000	.00000	.00000	.00000
2.50	.00621	.00038	.00006	.00001	.00001	.00000	.00000	.00000	.00000	.00000	.00000	.00000
2.40	.00820	.00058	.00010	.00003	.00001	.00000	.00000	.00000	.00000	.00000	.00000	.00000
2.30	.01072	.00086	.00016	.00005	.00002	.00001	.00000	.00000	.00000	.00000	.00000	.00000
2.20	.01390	.00128	.00025	.00008	.00003	.00001	.00001	.00000	.00000	.00000	.00000	.00000
2.10	.01786	.00186	.00040	.00013	.00005	.00002	.00001	.00001	.00000	.00000	.00000	.00000
2.00	.02275	.00268	.00063	.00021	.00009	.00004	.00002	.00001	.00001	.00001	.00000	.00000
1.90	.02872	.00381	.00096	.00034	.00015	.00008	.00004	.00003	.00002	.00001	.00001	.00001
1.80	.03593	.00534	.00145	.00054	.00025	.00013	.00007	.00005	.00003	.00002	.00001	.00001
1.70	.04457	.00740	.00215	.00085	.00040	.00022	.00013	.00008	.00005	.00004	.00003	.00002
1.60	.05480	.01013	.00315	.00130	.00064	.00036	.00022	.00014	.00009	.00007	.00005	.00004
1.50	.06681	.01369	.00455	.00197	.00101	.00058	.00036	.00024	.00016	.00012	.00009	.00007
1.40	.08076	.01828	.00647	.00293	.00155	.00091	.00058	.00039	.00028	.00020	.00015	.00012
1.30	.09680	.02410	.00908	.00431	.00236	.00143	.00093	.00064	.00046	.00034	.00026	.00020
1.20	.11507	.03141	.01256	.00622	.00352	.00219	.00146	.00102	.00074	.00056	.00043	.00034
1.10	.13567	.04044	.01714	.00885	.00518	.00330	.00224	.00160	.00118	.00090	.00071	.00056
1.00	.15866	.05146	.02305	.01239	.00748	.00490	.00340	.00247	.00185	.00144	.00114	.00092
0.90	.18406	.06473	.03059	.01709	.01064	.00714	.00506	.00374	.00285	.00224	.00179	.00146
0.80	.21186	.08049	.04004	.02323	.01489	.01023	.00739	.00556	.00431	.00342	.00278	.00229
0.70	.24196	.09896	.05171	.03111	.02051	.01442	.01063	.00812	.00639	.00514	.00422	.00352
0.60	.27425	.12032	.06591	.04106	.02783	.02000	.01502	.01166	.00930	.00758	.00629	.00530
0.50	.30854	.14470	.08292	.05342	.03717	.02730	.02088	.01647	.01331	.01098	.00921	.00784
0.40	.34458	.17216	.10299	.06852	.04890	.03667	.02854	.02286	.01873	.01563	.01325	.01138
0.30	.38209	.20268	.12633	.08668	.06338	.04850	.03839	.03120	.02589	.02186	.01872	.01622
0.20	.42074	.23616	.15306	.10815	.08096	.06315	.05081	.04188	.03519	.03004	.02598	.02272
0.10	.46017	.27242	.18323	.13315	.10193	.08100	.06620	.05531	.04704	.04058	.03544	.03127
0.00	.50000	.31118	.21677	.16179	.12653	.10235	.08493	.07189	.06185	.05392	.04753	.04229

H \ N	1	2	3	4	5	6	7	8	9	10	11	12
0.00	.50000	.31118	.21677	.16179	.12653	.10235	.08493	.07189	.06185	.05392	.04753	.04229
0.10	.53983	.35208	.25352	.19408	.15491	.12745	.10730	.09199	.08002	.07046	.06267	.05622
0.20	.57926	.39468	.29320	.22992	.18711	.15645	.13356	.11590	.10191	.09059	.08128	.07349
0.30	.61791	.43850	.33544	.26907	.22304	.18940	.16385	.14383	.12777	.11464	.10371	.09450
0.40	.65542	.48300	.37975	.31120	.26249	.22620	.19816	.17589	.15779	.14282	.13025	.11955
0.50	.69146	.52763	.42558	.35582	.30511	.26660	.23637	.21202	.19200	.17526	.16105	.14886
0.60	.72575	.57182	.47230	.40237	.35040	.31021	.27817	.25203	.23027	.21189	.19615	.18252
0.70	.75804	.61503	.51929	.45020	.39778	.35652	.32315	.29555	.27234	.25252	.23540	.22045
0.80	.78814	.65678	.56587	.49862	.44655	.40489	.37070	.34209	.31775	.29677	.27849	.26239
0.90	.81594	.69661	.61143	.54690	.49598	.45458	.42015	.39099	.36592	.34411	.32493	.30792
1.00	.84134	.73415	.65537	.59434	.54530	.50483	.47072	.44150	.41613	.39385	.37411	.35646
1.10	.86433	.76911	.69719	.64028	.59376	.55481	.52158	.49279	.46756	.44521	.42524	.40726
1.20	.88493	.80127	.73645	.68413	.64066	.60376	.57190	.54402	.51934	.49730	.47746	.45948
1.30	.90320	.83050	.77282	.72539	.68537	.65095	.62089	.59433	.57061	.54925	.52988	.51221
1.40	.91924	.85676	.80608	.76367	.72736	.69574	.66783	.64293	.62051	.60017	.58159	.56453
1.50	.93319	.88007	.83610	.79868	.76622	.73761	.71211	.68914	.66829	.64924	.63172	.61554
1.60	.94520	.90053	.86284	.83028	.80166	.77616	.75321	.73237	.71330	.69575	.67951	.66442
1.70	.95543	.91827	.88636	.85840	.83352	.81113	.79079	.77217	.75501	.73912	.72431	.71048
1.80	.96407	.93348	.90679	.88309	.86177	.84239	.82464	.90826	.79306	.77889	.76562	.75315
1.90	.97128	.94637	.92431	.90449	.88646	.86993	.85466	.84047	.82722	.81479	.80309	.79204
2.00	.97725	.95718	.93916	.92278	.90775	.89385	.88091	.86880	.85743	.84670	.83654	.82690
2.10	.98214	.96613	.95159	.93823	.92586	.91433	.90352	.89334	.88372	.97460	.86593	.85765
2.20	.98610	.97347	.96186	.95110	.94106	.93162	.92272	.91429	.90628	.89864	.89135	.88436
2.30	.98928	.97942	.97026	.96170	.95364	.94603	.93880	.93192	.92534	.91904	.91300	.90719
2.40	.99180	.98418	.97704	.97031	.96393	.95787	.95208	.94654	.94122	.93610	.93117	.92642
2.50	.99379	.98796	.98246	.97723	.97224	.96747	.96290	.95850	.95425	.95016	.94620	.94236
2.60	.99534	.99093	.98673	.98271	.97886	.97516	.97159	.96815	.96481	.96158	.95844	.95539
2.70	.99653	.99322	.99006	.98701	.98407	.98124	.97849	.97583	.97324	.97072	.96827	.96588
2.80	.99744	.99499	.99263	.99034	.98813	.98598	.98389	.98186	.97988	.97795	.97606	.97422
2.90	.99813	.99633	.99458	.99289	.99124	.98964	.98807	.98654	.98504	.98358	.98215	.98074
3.00	.99865	.99734	.99606	.99482	.99361	.99242	.99126	.99012	.98901	.98791	.98684	.98578
3.10	.99903	.99809	.99717	.99626	.99538	.99451	.99366	.99283	.99201	.99120	.99041	.98962
3.20	.99931	.99864	.99798	.99733	.99670	.99607	.99545	.99485	.99425	.99366	.99308	.99251
3.30	.99952	.99904	.99857	.99811	.99766	.99721	.99677	.99634	.99591	.99549	.99507	.99466
3.40	.99966	.99933	.99900	.99868	.99836	.99805	.99773	.99743	.99712	.99682	.99652	.99623
3.50	.99977	.99954	.99931	.99908	.99886	.99864	.99843	.99821	.99800	.99778	.99758	.99737

$\rho = .400$

– H \ N	1	2	3	4	5	6	7	8	9	10	11	12
3.50	.00023	.00000	.00000	.00000	.00000	.00000	.00000	.00000	.00000	.00000	.00000	.00000
3.40	.00034	.00001	.00000	.00000	.00000	.00000	.00000	.00000	.00000	.00000	.00000	.00000
3.30	.00048	.00001	.00000	.00000	.00000	.00000	.00000	.00000	.00000	.00000	.00000	.00000
3.20	.00069	.00002	.00000	.00000	.00000	.00000	.00000	.00000	.00000	.00000	.00000	.00000
3.10	.00097	.00003	.00000	.00000	.00000	.00000	.00000	.00000	.00000	.00000	.00000	.00000
3.00	.00135	.00005	.00001	.00000	.00000	.00000	.00000	.00000	.00000	.00000	.00000	.00000
2.90	.00187	.00007	.00001	.00000	.00000	.00000	.00000	.00000	.00000	.00000	.00000	.00000
2.80	.00256	.00012	.00002	.00000	.00000	.00000	.00000	.00000	.00000	.00000	.00000	.00000
2.70	.00347	.00018	.00003	.00001	.00000	.00000	.00000	.00000	.00000	.00000	.00000	.00000
2.60	.00466	.00028	.00004	.00001	.00000	.00000	.00000	.00000	.00000	.00000	.00000	.00000
2.50	.00621	.00043	.00007	.00002	.00001	.00000	.00000	.00000	.00000	.00000	.00000	.00000
2.40	.00820	.00065	.00012	.00004	.00001	.00001	.00000	.00000	.00000	.00000	.00000	.00000
2.30	.01072	.00096	.00020	.00006	.00002	.00001	.00001	.00000	.00000	.00000	.00000	.00000
2.20	.01390	.00141	.00031	.00010	.00004	.00002	.00001	.00001	.00000	.00000	.00000	.00000
2.10	.01786	.00204	.00048	.00017	.00007	.00004	.00002	.00001	.00001	.00001	.00000	.00000
2.00	.02275	.00292	.00074	.00027	.00012	.00006	.00004	.00002	.00001	.00001	.00001	.00001
1.90	.02872	.00413	.00112	.00043	.00020	.00011	.00006	.00004	.00003	.00002	.00001	.00001
1.80	.03593	.00576	.00168	.00067	.00032	.00018	.00011	.00007	.00005	.00003	.00002	.00002
1.70	.04457	.00794	.00247	.00103	.00052	.00029	.00018	.00012	.00008	.00006	.00004	.00003
1.60	.05480	.01081	.00358	.00157	.00081	.00047	.00030	.00020	.00014	.00010	.00008	.00006
1.50	.06681	.01454	.00513	.00234	.00125	.00075	.00048	.00033	.00023	.00017	.00013	.00010
1.40	.08076	.01933	.00724	.00345	.00191	.00117	.00077	.00053	.00039	.00029	.00022	.00018
1.30	.09680	.02538	.01007	.00500	.00285	.00179	.00120	.00085	.00062	.00047	.00037	.00029
1.20	.11507	.03294	.01382	.00714	.00420	.00270	.00185	.00133	.00099	.00076	.00060	.00049
1.10	.13567	.04225	.01872	.01006	.00610	.00401	.00281	.00205	.00155	.00121	.00097	.00079
1.00	.15866	.05356	.02500	.01394	.00870	.00587	.00418	.00311	.00239	.00189	.00152	.00125
0.90	.18406	.06714	.03295	.01906	.01223	.00843	.00612	.00463	.00361	.00288	.00235	.00195
0.80	.21186	.08321	.04285	.02567	.01693	.01193	.00882	.00677	.00535	.00433	.00357	.00299
0.70	.24196	.10199	.05501	.03409	.02308	.01661	.01251	.00975	.00780	.00638	.00532	.00450
0.60	.27425	.12365	.06973	.04463	.03099	.02277	.01744	.01380	.01119	.00926	.00779	.00665
0.50	.30854	.14831	.08726	.05763	.04100	.03074	.02394	.01921	.01577	.01320	.01122	.00966
0.40	.34458	.17600	.10785	.07339	.05347	.04086	.03234	.02631	.02187	.01850	.01588	.01380
0.30	.38209	.20672	.13168	.09222	.06872	.05350	.04302	.03547	.02983	.02550	.02209	.01936
0.20	.42074	.24036	.15885	.11436	.08709	.06902	.05634	.04706	.04004	.03457	.03023	.02671
0.10	.46017	.27671	.18940	.13998	.10884	.08775	.07268	.06148	.05288	.04611	.04068	.03623
0.00	.50000	.31549	.22324	.16917	.13419	.10998	.09238	.07909	.06876	.06053	.05385	.04833

H \ N	1	2	3	4	5	6	7	8	9	10	11	12
0.00	.50000	.31549	.22324	.16917	.13419	.10998	.09238	.07909	.06876	.06053	.05385	.04833
0.10	.53983	.35636	.26020	.20193	.16325	.13592	.11570	.10022	.08803	.07821	.07015	.06344
0.20	.57926	.39887	.29999	.23812	.19603	.16569	.14287	.12514	.11100	.09949	.08996	.08194
0.30	.61791	.44255	.34223	.27750	.23241	.19928	.17396	.15401	.13791	.12466	.11358	.10419
0.40	.65542	.48685	.38643	.31971	.27216	.23657	.20894	.18689	.16887	.15390	.14126	.13044
0.50	.69146	.53123	.43205	.36428	.31491	.27730	.24765	.22368	.20388	.18725	.17309	.16087
0.60	.72575	.57515	.47847	.41063	.36017	.32105	.28977	.26416	.24277	.22463	.20904	.19550
0.70	.75804	.61807	.52508	.45814	.40734	.36730	.33484	.30794	.28523	.26579	.24895	.23419
0.80	.78814	.65950	.57122	.50612	.45575	.41542	.38229	.35450	.33080	.31033	.29244	.27666
0.90	.81594	.69902	.61629	.55387	.50468	.46470	.43141	.40319	.37888	.35770	.33904	.32245
1.00	.84134	.73625	.65972	.60071	.55339	.51436	.48147	.45327	.42876	.40721	.38808	.37095
1.10	.86433	.77091	.70102	.64600	.60115	.56364	.53165	.50394	.47963	.45808	.43881	.42144
1.20	.88493	.80280	.73978	.68919	.64729	.61179	.58117	.55438	.53067	.50949	.49040	.47309
1.30	.90320	.83178	.77567	.72979	.69122	.65813	.62927	.60378	.58104	.56056	.54198	.52502
1.40	.91924	.85781	.80847	.76743	.73243	.70205	.67527	.65141	.62994	.61047	.59269	.57636
1.50	.93319	.88093	.83807	.80185	.77055	.74305	.71859	.69660	.67666	.65845	.64171	.62625
1.60	.94520	.90121	.86445	.83290	.80529	.78078	.75876	.73881	.72058	.70383	.68833	.67394
1.70	.95543	.91881	.88765	.86053	.83652	.81498	.79547	.77764	.76124	.74607	.73195	.71877
1.80	.96407	.93390	.90781	.88480	.86420	.84554	.82850	.81281	.79829	.78476	.77212	.76024
1.90	.97128	.94669	.92511	.90583	.88840	.87247	.85780	.84420	.83153	.81967	.80851	.79798
2.00	.97725	.95742	.93977	.92383	.90927	.89585	.88341	.87180	.86091	.85066	.84098	.83180
2.10	.98214	.96631	.95205	.93903	.92703	.91589	.90548	.89571	.88649	.87777	.86949	.86161
2.20	.98610	.97360	.96221	.95170	.94194	.93282	.92423	.91613	.90844	.90113	.89416	.88750
2.30	.98928	.97951	.97051	.96214	.95431	.94693	.93995	.93332	.92700	.92097	.91519	.90964
2.40	.99180	.98425	.97722	.97063	.96442	.95853	.95293	.94759	.94247	.93756	.93284	.92829
2.50	.99379	.98801	.98258	.97746	.97259	.96796	.96352	.95927	.95518	.95124	.94744	.94376
2.60	.99534	.99096	.98681	.98287	.97911	.97551	.97204	.96871	.96548	.96237	.95935	.95642
2.70	.99653	.99325	.99012	.98712	.98425	.98148	.97881	.97623	.97372	.97129	.96893	.96663
2.80	.99744	.99501	.99267	.99042	.98825	.98615	.98411	.98214	.98022	.97835	.97653	.97475
2.90	.99813	.99634	.99461	.99294	.99132	.98975	.98822	.98673	.98528	.98386	.98247	.98111
3.00	.99865	.99734	.99608	.99485	.99366	.99250	.99136	.99025	.98916	.98810	.98706	.98604
3.10	.99903	.99809	.99718	.99629	.99542	.99456	.99373	.99291	.99211	.99133	.99055	.98979
3.20	.99931	.99864	.99799	.99735	.99672	.99610	.99550	.99491	.99432	.99375	.99318	.99263
3.30	.99952	.99904	.99858	.99812	.99768	.99724	.99680	.99638	.99596	.99554	.99513	.99473
3.40	.99966	.99933	.99901	.99869	.99837	.99806	.99775	.99745	.99715	.99686	.99656	.99628
3.50	.99977	.99954	.99931	.99909	.99887	.99865	.99844	.99822	.99801	.99781	.99760	.99740

Tafel 9 (Forts.) Table 9 (cont.)

ρ = .500

-H \ N	1	2	3	4	5	6	7	8	9	10	11	12
3.50	.00023	.00001	.00000	.00000	.00000	.00000	.00000	.00000	.00000	.00000	.00000	.00000
3.40	.00034	.00001	.00000	.00000	.00000	.00000	.00000	.00000	.00000	.00000	.00000	.00000
3.30	.00048	.00002	.00000	.00000	.00000	.00000	.00000	.00000	.00000	.00000	.00000	.00000
3.20	.00069	.00003	.00001	.00000	.00000	.00000	.00000	.00000	.00000	.00000	.00000	.00000
3.10	.00097	.00005	.00001	.00000	.00000	.00000	.00000	.00000	.00000	.00000	.00000	.00000
3.00	.00135	.00008	.00002	.00000	.00000	.00000	.00000	.00000	.00000	.00000	.00000	.00000
2.90	.00187	.00013	.00003	.00001	.00000	.00000	.00000	.00000	.00000	.00000	.00000	.00000
2.80	.00256	.00020	.00004	.00001	.00001	.00000	.00000	.00000	.00000	.00000	.00000	.00000
2.70	.00347	.00030	.00007	.00002	.00001	.00001	.00000	.00000	.00000	.00000	.00000	.00000
2.60	.00466	.00045	.00011	.00004	.00002	.00001	.00001	.00000	.00000	.00000	.00000	.00000
2.50	.00621	.00067	.00017	.00006	.00003	.00002	.00001	.00001	.00000	.00000	.00000	.00000
2.40	.00820	.00098	.00027	.00011	.00005	.00003	.00002	.00001	.00001	.00001	.00000	.00000
2.30	.01072	.00143	.00041	.00017	.00009	.00005	.00003	.00002	.00001	.00001	.00001	.00001
2.20	.01390	.00204	.00062	.00027	.00014	.00008	.00005	.00004	.00003	.00002	.00001	.00001
2.10	.01786	.00289	.00093	.00041	.00022	.00013	.00009	.00006	.00004	.00003	.00003	.00002
2.00	.02275	.00405	.00137	.00063	.00035	.00021	.00014	.00010	.00007	.00006	.00004	.00004
1.90	.02872	.00561	.00201	.00096	.00054	.00034	.00023	.00016	.00012	.00009	.00007	.00006
1.80	.03593	.00767	.00289	.00143	.00083	.00053	.00037	.00026	.00020	.00016	.00012	.00010
1.70	.04457	.01037	.00411	.00210	.00125	.00082	.00057	.00042	.00032	.00025	.00020	.00017
1.60	.05480	.01386	.00576	.00305	.00185	.00124	.00088	.00066	.00051	.00040	.00033	.00027
1.50	.06681	.01832	.00799	.00436	.00272	.00185	.00133	.00101	.00079	.00063	.00052	.00043
1.40	.08076	.02394	.01093	.00616	.00393	.00272	.00200	.00153	.00121	.00098	.00081	.00068
1.30	.09680	.03094	.01476	.00858	.00560	.00395	.00294	.00228	.00182	.00149	.00124	.00105
1.20	.11507	.03955	.01970	.01179	.00788	.00565	.00427	.00335	.00270	.00223	.00188	.00160
1.10	.13567	.04999	.02596	.01600	.01092	.00797	.00611	.00485	.00395	.00329	.00279	.00241
1.00	.15866	.06251	.03380	.02142	.01494	.01109	.00861	.00692	.00570	.00479	.00410	.00355
0.90	.18406	.07734	.04347	.02832	.02015	.01521	.01197	.00973	.00809	.00686	.00591	.00516
0.80	.21186	.09469	.05526	.03695	.02683	.02058	.01641	.01348	.01133	.00969	.00841	.00739
0.70	.24196	.11472	.06941	.04762	.03525	.02746	.02218	.01842	.01562	.01348	.01179	.01043
0.60	.27425	.13757	.08619	.06060	.04570	.03614	.02957	.02481	.02124	.01847	.01627	.01449
0.50	.30854	.16332	.10580	.07617	.05850	.04694	.03887	.03296	.02847	.02495	.02214	.01984
0.40	.34458	.19198	.12841	.09458	.07393	.06017	.05041	.04318	.03762	.03323	.02969	.02677
0.30	.38209	.22349	.15414	.11606	.09227	.07612	.06451	.05579	.04932	.04363	.03924	.03561
0.20	.42074	.25771	.18303	.14075	.11374	.09508	.08147	.07113	.06302	.05650	.05115	.04668
0.10	.46017	.29442	.21503	.16873	.13850	.11727	.10156	.08948	.07991	.07215	.06574	.06034
0.00	.50000	.33333	.25000	.20000	.16667	.14286	.12500	.11111	.10000	.09091	.08333	.07692

H \ N	1	2	3	4	5	6	7	8	9	10	11	12
0.00	.50000	.33333	.25000	.20000	.16667	.14286	.12500	.11111	.10000	.09091	.08333	.07692
0.10	.53983	.37408	.28772	.23446	.19823	.17192	.15193	.13620	.12350	.11302	.10422	.09673
0.20	.57926	.41623	.32788	.27192	.23308	.20445	.18240	.16487	.15058	.13869	.12863	.12000
0.30	.61791	.45931	.37006	.31206	.27104	.24032	.21637	.19713	.18129	.16801	.15669	.14692
0.40	.65542	.50282	.41379	.35450	.31177	.27930	.25368	.23287	.21559	.20099	.18845	.17757
0.50	.69146	.54624	.45855	.39874	.35486	.32104	.29403	.27187	.25332	.23750	.22384	.21190
0.60	.72575	.58906	.50376	.44425	.39981	.36509	.33704	.31380	.29417	.27732	.26266	.24977
0.70	.75804	.63079	.54885	.49041	.44605	.41091	.38220	.35820	.33775	.32006	.30458	.29089
0.80	.78814	.67098	.59323	.53661	.49293	.45788	.42894	.40451	.38353	.36525	.34915	.33483
0.90	.81594	.70922	.63638	.58224	.53983	.50536	.47660	.45211	.43090	.41231	.39582	.38107
1.00	.84134	.74520	.67778	.62670	.58608	.55257	.52450	.50030	.47920	.46056	.44394	.42898
1.10	.86433	.77866	.71701	.66944	.63107	.59913	.57195	.54839	.52770	.50930	.49279	.47786
1.20	.88493	.80941	.75373	.71000	.67424	.64414	.61828	.59568	.57569	.55780	.54166	.52698
1.30	.90320	.83734	.78766	.74799	.71510	.68713	.66287	.64151	.62248	.60535	.58980	.57559
1.40	.91924	.86243	.81864	.78309	.75326	.72763	.70519	.68529	.66744	.65127	.63652	.62297
1.50	.93319	.88471	.84656	.81513	.78843	.76525	.74480	.72652	.71001	.69498	.68119	.66847
1.60	.94520	.90427	.87143	.84398	.82040	.79973	.78134	.76479	.74975	.73598	.72328	.71151
1.70	.95543	.92124	.89331	.86964	.84908	.83090	.81459	.79982	.78631	.77388	.76236	.75163
1.80	.96407	.93581	.91233	.89218	.87448	.85870	.84444	.83144	.81948	.80841	.79811	.78848
1.90	.97128	.94817	.92867	.91172	.89669	.88317	.87087	.85958	.84914	.83943	.83036	.82183
2.00	.97725	.95855	.94253	.92845	.91585	.90442	.89395	.88429	.87531	.86691	.85902	.85159
2.10	.98214	.96717	.95416	.94260	.93217	.92264	.91385	.90569	.89806	.89090	.88415	.87776
2.20	.98610	.97424	.96380	.95443	.94590	.93805	.93077	.92397	.91758	.91156	.90586	.90045
2.30	.98928	.97998	.97169	.96419	.95730	.95092	.94497	.93938	.93411	.92911	.92437	.91984
2.40	.99180	.98459	.97809	.97215	.96665	.96153	.95673	.95220	.94790	.94382	.93992	.93619
2.50	.99379	.98825	.98321	.97856	.97423	.97017	.96635	.96272	.95927	.95597	.95282	.94979
2.60	.99534	.99113	.98726	.98366	.98030	.97712	.97411	.97125	.96851	.96588	.96337	.96094
2.70	.99653	.99336	.99043	.98768	.98509	.98264	.98030	.97807	.97592	.97386	.97188	.96996
2.80	.99744	.99509	.99288	.99081	.98884	.98697	.98517	.98345	.98180	.98020	.97865	.97716
2.90	.99813	.99640	.99476	.99321	.99173	.99032	.98896	.98765	.98639	.98517	.98398	.98283
3.00	.99865	.99738	.99618	.99504	.99394	.99288	.99187	.99089	.98994	.98901	.98812	.98724
3.10	.99903	.99812	.99724	.99641	.99560	.99483	.99408	.99335	.99264	.99195	.99128	.99063
3.20	.99931	.99866	.99803	.99743	.99684	.99628	.99573	.99520	.99468	.99417	.99367	.99319
3.30	.99952	.99905	.99861	.99817	.99776	.99735	.99695	.99657	.99619	.99582	.99546	.99511
3.40	.99966	.99934	.99902	.99872	.99842	.99813	.99785	.99757	.99730	.99704	.99678	.99652
3.50	.99977	.99954	.99932	.99911	.99890	.99870	.99850	.99830	.99811	.99792	.99774	.99756

Tafel 9 (Forts.) Table 9 (cont.)

$\rho = .600$

-H \ N	1	2	3	4	5	6	7	8	9	10	11	12
3.50	.00023	.00001	.00000	.00000	.00000	.00000	.00000	.00000	.00000	.00000	.00000	.00000
3.40	.00034	.00002	.00001	.00000	.00000	.00000	.00000	.00000	.00000	.00000	.00000	.00000
3.30	.00048	.00004	.00001	.00000	.00000	.00000	.00000	.00000	.00000	.00000	.00000	.00000
3.20	.00069	.00006	.00001	.00001	.00000	.00000	.00000	.00000	.00000	.00000	.00000	.00000
3.10	.00097	.00009	.00002	.00001	.00000	.00000	.00000	.00000	.00000	.00000	.00000	.00000
3.00	.00135	.00014	.00004	.00002	.00001	.00000	.00000	.00000	.00000	.00000	.00000	.00000
2.90	.00187	.00021	.00006	.00003	.00001	.00001	.00001	.00000	.00000	.00000	.00000	.00000
2.80	.00256	.00032	.00010	.00004	.00002	.00001	.00001	.00001	.00001	.00000	.00000	.00000
2.70	.00347	.00047	.00015	.00007	.00004	.00002	.00002	.00001	.00001	.00001	.00001	.00000
2.60	.00466	.00069	.00023	.00011	.00006	.00004	.00003	.00002	.00001	.00001	.00001	.00001
2.50	.00621	.00101	.00035	.00017	.00010	.00006	.00004	.00003	.00002	.00002	.00002	.00001
2.40	.00820	.00145	.00053	.00026	.00016	.00010	.00007	.00005	.00004	.00003	.00003	.00002
2.30	.01072	.00206	.00078	.00040	.00024	.00016	.00012	.00009	.00007	.00005	.00004	.00004
2.20	.01390	.00289	.00115	.00060	.00037	.00025	.00018	.00014	.00011	.00009	.00007	.00006
2.10	.01786	.00401	.00166	.00090	.00056	.00039	.00028	.00022	.00017	.00014	.00012	.00010
2.00	.02275	.00550	.00237	.00132	.00084	.00059	.00044	.00034	.00027	.00022	.00019	.00016
1.90	.02872	.00747	.00335	.00191	.00124	.00088	.00066	.00052	.00042	.00035	.00029	.00025
1.80	.03593	.01003	.00468	.00274	.00182	.00131	.00099	.00078	.00064	.00053	.00045	.00039
1.70	.04457	.01333	.00645	.00387	.00262	.00191	.00147	.00117	.00096	.00081	.00069	.00060
1.60	.05480	.01753	.00880	.00541	.00372	.00275	.00214	.00172	.00143	.00121	.00104	.00091
1.50	.06681	.02279	.01186	.00747	.00523	.00392	.00308	.00251	.00209	.00178	.00154	.00135
1.40	.08076	.02933	.01581	.01018	.00725	.00551	.00438	.00359	.00302	.00259	.00226	.00199
1.30	.09680	.03736	.02082	.01372	.00994	.00765	.00614	.00509	.00431	.00372	.00326	.00290
1.20	.11507	.04708	.02711	.01826	.01345	.01049	.00851	.00711	.00607	.00528	.00465	.00415
1.10	.13567	.05873	.03492	.02403	.01798	.01420	.01164	.00981	.00844	.00738	.00655	.00587
1.00	.15866	.07253	.04447	.03125	.02375	.01898	.01572	.01336	.01158	.01019	.00909	.00819
0.90	.18406	.08866	.05601	.04017	.03099	.02507	.02096	.01796	.01568	.01389	.01246	.01128
0.80	.21186	.10731	.06978	.05104	.03996	.03271	.02761	.02385	.02097	.01869	.01685	.01533
0.70	.24196	.12863	.08600	.06412	.05093	.04216	.03593	.03128	.02769	.02483	.02250	.02057
0.60	.27425	.15270	.10488	.07965	.06416	.05370	.04618	.04052	.03610	.03256	.02966	.02724
0.50	.30854	.17956	.12657	.09786	.07989	.06759	.05864	.05184	.04649	.04217	.03862	.03563
0.40	.34458	.20919	.15118	.11893	.09837	.08410	.07360	.06553	.05915	.05395	.04965	.04602
0.30	.38209	.24150	.17876	.14299	.11978	.10344	.09128	.08187	.07434	.06819	.06305	.05869
0.20	.42074	.27631	.20927	.17012	.14425	.12580	.11192	.10107	.09233	.08514	.07910	.07394
0.10	.46017	.31338	.24262	.20030	.17186	.15130	.13567	.12334	.11334	.10505	.09804	.09204
0.00	.50000	.35242	.27862	.23345	.20259	.17999	.16263	.14882	.13753	.12810	.12009	.11320

H \ N	1	2	3	4	5	6	7	8	9	10	11	12
0.00	.50000	.35242	.27862	.23345	.20259	.17999	.16263	.14882	.13753	.12810	.12009	.11320
0.10	.53983	.39304	.31701	.26941	.23636	.21185	.19282	.17756	.16499	.15442	.14540	.13759
0.20	.57926	.43482	.35742	.30790	.27297	.24674	.22618	.20954	.19573	.18406	.17404	.16532
0.30	.61791	.47732	.39946	.34859	.31214	.28445	.26252	.24463	.22969	.21698	.20600	.19639
0.40	.65542	.52003	.44266	.39105	.35352	.32466	.30160	.28263	.26668	.25302	.24116	.23073
0.50	.69146	.56248	.48650	.43480	.39665	.36699	.34305	.32321	.30641	.29194	.27930	.26815
0.60	.72575	.60419	.53046	.47931	.44104	.41094	.38644	.36596	.34851	.33339	.32012	.30835
0.70	.75804	.64470	.57400	.52403	.48612	.45599	.43123	.41039	.39251	.37693	.36319	.35095
0.80	.78814	.68360	.61660	.56840	.53133	.50155	.47688	.45595	.43788	.42204	.40801	.39545
0.90	.81594	.72054	.65780	.61187	.57608	.54704	.52277	.50204	.48402	.46815	.45402	.44131
1.00	.84134	.75521	.69715	.65392	.61982	.59186	.56830	.54804	.53032	.51463	.50059	.48791
1.10	.86433	.78740	.73429	.69410	.66202	.63545	.61289	.59334	.57616	.56086	.54710	.53462
1.20	.88493	.81694	.76893	.73203	.70222	.67730	.65598	.63738	.62093	.60621	.59291	.58080
1.30	.90320	.84376	.80085	.76739	.74004	.71697	.69707	.67961	.66407	.65010	.63743	.62584
1.40	.91924	.86782	.82993	.79994	.77516	.75408	.73576	.71958	.70510	.69202	.68010	.66916
1.50	.93319	.88918	.85610	.82955	.80739	.78836	.77171	.75691	.74361	.73152	.72046	.71027
1.60	.94520	.90793	.87938	.85616	.83658	.81963	.80469	.79133	.77926	.76825	.75813	.74876
1.70	.95543	.92420	.89984	.87979	.86269	.84778	.83455	.82266	.81185	.80194	.79281	.78432
1.80	.96407	.93817	.91763	.90050	.88577	.87282	.86125	.85079	.84124	.83245	.82431	.81673
1.90	.97128	.95004	.93291	.91846	.90591	.89481	.88482	.87574	.86742	.85972	.85256	.84587
2.00	.97725	.96000	.94588	.93383	.92328	.91387	.90536	.89759	.89042	.88377	.87756	.87173
2.10	.98213	.96828	.95677	.94684	.93808	.93020	.92304	.91647	.91038	.90470	.89938	.89438
2.20	.98610	.97508	.96580	.95772	.95052	.94402	.93807	.93258	.92747	.92269	.91820	.91396
2.30	.98928	.98061	.97321	.96671	.96087	.95556	.95068	.94615	.94192	.93795	.93421	.93066
2.40	.99180	.98505	.97922	.97405	.96937	.96509	.96114	.95745	.95399	.95073	.94765	.94472
2.50	.99379	.98859	.98405	.97998	.97628	.97287	.96970	.96673	.96394	.96130	.95880	.95641
2.60	.99534	.99137	.98787	.98471	.98181	.97913	.97662	.97427	.97204	.96993	.96792	.96600
2.70	.99653	.99354	.99087	.98844	.98620	.98411	.98216	.98031	.97856	.97689	.97530	.97378
2.80	.99744	.99521	.99319	.99135	.98964	.98803	.98652	.98509	.98373	.98243	.98119	.97999
2.90	.99813	.99648	.99498	.99359	.99230	.99108	.98993	.98884	.98779	.98679	.98583	.98490
3.00	.99865	.99744	.99633	.99530	.99433	.99342	.99255	.99173	.99094	.99017	.98944	.98873
3.10	.99903	.99816	.99735	.99659	.99588	.99520	.99455	.99394	.99334	.99277	.99222	.99169
3.20	.99931	.99868	.99810	.99755	.99703	.99653	.99606	.99560	.99516	.99474	.99433	.99393
3.30	.99952	.99907	.99865	.99826	.99788	.99752	.99718	.99685	.99652	.99621	.99591	.99562
3.40	.99966	.99935	.99905	.99877	.99851	.99825	.99800	.99776	.99753	.99730	.99708	.99687
3.50	.99977	.99955	.99934	.99915	.99896	.99877	.99860	.99843	.99826	.99810	.99794	.99779

ρ = .625

N / −H	1	2	3	4	5	6	7	8	9	10	11	12
3.50	.00023	.00002	.00000	.00000	.00000	.00000	.00000	.00000	.00000	.00000	.00000	.00000
3.40	.00034	.00003	.00001	.00000	.00000	.00000	.00000	.00000	.00000	.00000	.00000	.00000
3.30	.00048	.00004	.00001	.00000	.00000	.00000	.00000	.00000	.00000	.00000	.00000	.00000
3.20	.00069	.00007	.00002	.00001	.00000	.00000	.00000	.00000	.00000	.00000	.00000	.00000
3.10	.00097	.00010	.00003	.00001	.00001	.00000	.00000	.00000	.00000	.00000	.00000	.00000
3.00	.00135	.00016	.00005	.00002	.00001	.00001	.00000	.00000	.00000	.00000	.00000	.00000
2.90	.00187	.00024	.00008	.00003	.00002	.00001	.00001	.00001	.00000	.00000	.00000	.00000
2.80	.00256	.00036	.00012	.00006	.00003	.00002	.00001	.00001	.00001	.00001	.00000	.00000
2.70	.00347	.00053	.00018	.00009	.00005	.00003	.00002	.00002	.00001	.00001	.00001	.00001
2.60	.00466	.00077	.00028	.00014	.00008	.00005	.00004	.00003	.00002	.00002	.00001	.00001
2.50	.00621	.00111	.00042	.00021	.00013	.00009	.00006	.00005	.00004	.00003	.00002	.00002
2.40	.00820	.00159	.00062	.00032	.00020	.00013	.00010	.00007	.00006	.00005	.00004	.00003
2.30	.01072	.00224	.00091	.00049	.00031	.00021	.00015	.00012	.00009	.00008	.00006	.00005
2.20	.01390	.00314	.00132	.00073	.00046	.00032	.00024	.00019	.00015	.00012	.00010	.00009
2.10	.01786	.00433	.00190	.00107	.00070	.00049	.00037	.00029	.00023	.00019	.00016	.00014
2.00	.02275	.00592	.00269	.00156	.00103	.00074	.00056	.00044	.00036	.00030	.00026	.00022
1.90	.02872	.00801	.00378	.00224	.00151	.00109	.00084	.00067	.00055	.00046	.00040	.00034
1.80	.03593	.01071	.00524	.00318	.00217	.00160	.00124	.00100	.00083	.00070	.00060	.00053
1.70	.04457	.01417	.00718	.00446	.00310	.00231	.00181	.00147	.00123	.00104	.00090	.00079
1.60	.05480	.01855	.00973	.00618	.00437	.00330	.00261	.00214	.00180	.00154	.00134	.00118
1.50	.06681	.02404	.01303	.00846	.00607	.00465	.00372	.00307	.00260	.00224	.00196	.00174
1.40	.08076	.03083	.01725	.01144	.00835	.00647	.00523	.00436	.00371	.00322	.00284	.00253
1.30	.09680	.03912	.02259	.01530	.01134	.00890	.00726	.00610	.00523	.00457	.00405	.00362
1.20	.11507	.04914	.02926	.02023	.01522	.01208	.00995	.00843	.00728	.00640	.00570	.00512
1.10	.13567	.06111	.03748	.02643	.02018	.01620	.01348	.01150	.01001	.00885	.00792	.00716
1.00	.15866	.07523	.04748	.03414	.02644	.02148	.01803	.01551	.01358	.01208	.01086	.00987
0.90	.18406	.09171	.05951	.04361	.03425	.02812	.02382	.02065	.01821	.01628	.01472	.01343
0.80	.21186	.11070	.07380	.05507	.04385	.03640	.03110	.02716	.02410	.02167	.01969	.01805
0.70	.24196	.13234	.09056	.06879	.05550	.04655	.04013	.03529	.03152	.02850	.02602	.02395
0.60	.27425	.15672	.10997	.08499	.06945	.05886	.05116	.04532	.04072	.03702	.03396	.03140
0.50	.30854	.18387	.13219	.10387	.08595	.07357	.06447	.05751	.05199	.04751	.04379	.04065
0.40	.34458	.21375	.15730	.12561	.10521	.09092	.08032	.07212	.06559	.06024	.05578	.05200
0.30	.38209	.24626	.18534	.15031	.12739	.11112	.09893	.08942	.08178	.07549	.07021	.06572
0.20	.42074	.28122	.21625	.17803	.15260	.13432	.12049	.10961	.10080	.09350	.08734	.08207
0.10	.46017	.31838	.24993	.20875	.18089	.16063	.14514	.13285	.12283	.11448	.10740	.10131
0.00	.50000	.35745	.28618	.24234	.21224	.19007	.17295	.15926	.14802	.13860	.13056	.12362

N / H	1	2	3	4	5	6	7	8	9	10	11	12
0.00	.50000	.35745	.28618	.24234	.21224	.19007	.17295	.15926	.14802	.13860	.13056	.12362
0.10	.53983	.39804	.32471	.27864	.24651	.22257	.20391	.18886	.17643	.16594	.15694	.14913
0.20	.57926	.43974	.36518	.31736	.28351	.25800	.23792	.22160	.20803	.19651	.18658	.17791
0.30	.61791	.48208	.40717	.35815	.32294	.29610	.27479	.25733	.24271	.23023	.21942	.20994
0.40	.65542	.54259	.45021	.50059	.36443	.33656	.31423	.29580	.28027	.26693	.25532	.24510
0.50	.69146	.56680	.49381	.44419	.40753	.37897	.35587	.33668	.32040	.30634	.29404	.28316
0.60	.72575	.60822	.53744	.48843	.45173	.42283	.39927	.37954	.36270	.34808	.33523	.32381
0.70	.75804	.64842	.58058	.53276	.49649	.46763	.44390	.42389	.40670	.39170	.37845	.36663
0.80	.78814	.68699	.62273	.57665	.54124	.51280	.48921	.46918	.45187	.43669	.42321	.41114
0.90	.81594	.72359	.66343	.61956	.58543	.55775	.53460	.51482	.49762	.48246	.46894	.45677
1.00	.84134	.75792	.70225	.66100	.62852	.60192	.57950	.56022	.54336	.52842	.51503	.50294
1.10	.86433	.78978	.73886	.70053	.67000	.64477	.62334	.60479	.58847	.57394	.56087	.54902
1.20	.88493	.81901	.77297	.73779	.70945	.68581	.66560	.64798	.63240	.61846	.60586	.59439
1.30	.90320	.84552	.80438	.77248	.74650	.72463	.70580	.68929	.67461	.66141	.64943	.63848
1.40	.91924	.86931	.83296	.80438	.78086	.76089	.74357	.72829	.71464	.70231	.69107	.68076
1.50	.93319	.89043	.85868	.83337	.81233	.79433	.77861	.76465	.75212	.74074	.73034	.72076
1.60	.94520	.90896	.88154	.85941	.84082	.82479	.81069	.79811	.78675	.77640	.76690	.75811
1.70	.95543	.92504	.90164	.88250	.86628	.85218	.83970	.82851	.81835	.80905	.80048	.79253
1.80	.96407	.93895	.91909	.90275	.88877	.87652	.86561	.85577	.84680	.83856	.83093	.82383
1.90	.97128	.95057	.93409	.92029	.90838	.89788	.88846	.87992	.87211	.86489	.85819	.85194
2.00	.97725	.96042	.94682	.93531	.92529	.91639	.90836	.90105	.89432	.88809	.88228	.87684
2.10	.98214	.96860	.95751	.94802	.93968	.93223	.92548	.91929	.91358	.90826	.90329	.89862
2.20	.98610	.97533	.96637	.95864	.95180	.94563	.94002	.93485	.93006	.92558	.92138	.91742
2.30	.98928	.98080	.97365	.96742	.96186	.95683	.95222	.94796	.94399	.94027	.93676	.93345
2.40	.99180	.98519	.97956	.97459	.97014	.96608	.96234	.95886	.95562	.95256	.94968	.94694
2.50	.99379	.98869	.98429	.98039	.97686	.97362	.97062	.96783	.96520	.96273	.96038	.95815
2.60	.99534	.99145	.98805	.98501	.98224	.97969	.97732	.97510	.97301	.97103	.96914	.96735
2.70	.99653	.99359	.99100	.98866	.98652	.98453	.98268	.98093	.97929	.97772	.97623	.97480
2.80	.99744	.99525	.99329	.99151	.98987	.98834	.98691	.98556	.98427	.98305	.98188	.98076
2.90	.99813	.99651	.99505	.99371	.99247	.99130	.99021	.98917	.98819	.98724	.98634	.98547
3.00	.99865	.99746	.99638	.99538	.99445	.99358	.99276	.99197	.99122	.99050	.98981	.98915
3.10	.99903	.99817	.99738	.99665	.99596	.99531	.99470	.99411	.99355	.99301	.99249	.99199
3.20	.99931	.99869	.99812	.99759	.99709	.99661	.99616	.99572	.99531	.99491	.99452	.99414
3.30	.99952	.99908	.99867	.99828	.99792	.99758	.99725	.99693	.99662	.99633	.99604	.99577
3.40	.99966	.99935	.99906	.99879	.99853	.99828	.99805	.99782	.99760	.99738	.99718	.99697
3.50	.99977	.99955	.99935	.99916	.99897	.99880	.99863	.99847	.99831	.99815	.99801	.99786

Tafel 9 (Forts.) Table 9 (cont.)

$\rho = 2/3$

-H \ N	1	2	3	4	5	6	7	8	9	10	11	12
3.50	.00023	.00002	.00001	.00000	.00000	.00000	.00000	.00000	.00000	.00000	.00000	.00000
3.40	.00034	.00003	.00001	.00000	.00000	.00000	.00000	.00000	.00000	.00000	.00000	.00000
3.30	.00048	.00005	.00002	.00001	.00000	.00000	.00000	.00000	.00000	.00000	.00000	.00000
3.20	.00069	.00008	.00003	.00001	.00001	.00000	.00000	.00000	.00000	.00000	.00000	.00000
3.10	.00097	.00013	.00004	.00002	.00001	.00001	.00001	.00000	.00000	.00000	.00000	.00000
3.00	.00135	.00020	.00007	.00003	.00002	.00001	.00001	.00001	.00001	.00000	.00000	.00000
2.90	.00187	.00029	.00010	.00005	.00003	.00002	.00002	.00001	.00001	.00001	.00001	.00000
2.80	.00256	.00043	.00016	.00008	.00005	.00003	.00002	.00002	.00001	.00001	.00001	.00001
2.70	.00347	.00063	.00024	.00013	.00008	.00005	.00004	.00003	.00002	.00002	.00002	.00001
2.60	.00466	.00091	.00037	.00020	.00013	.00009	.00006	.00005	.00004	.00003	.00003	.00002
2.50	.00621	.00131	.00055	.00030	.00019	.00014	.00010	.00008	.00006	.00005	.00004	.00004
2.40	.00820	.00185	.00080	.00045	.00030	.00021	.00016	.00012	.00010	.00008	.00007	.00006
2.30	.01072	.00259	.00116	.00067	.00044	.00032	.00024	.00019	.00016	.00013	.00011	.00010
2.20	.01390	.00360	.00167	.00099	.00066	.00048	.00037	.00030	.00025	.00021	.00018	.00016
2.10	.01786	.00493	.00236	.00143	.00098	.00072	.00056	.00045	.00038	.00032	.00028	.00024
2.00	.02275	.00669	.00331	.00204	.00142	.00106	.00083	.00068	.00057	.00048	.00042	.00037
1.90	.02872	.00897	.00459	.00289	.00204	.00154	.00122	.00100	.00085	.00073	.00063	.00056
1.80	.03593	.01192	.00630	.00405	.00290	.00222	.00177	.00147	.00124	.00107	.00094	.00084
1.70	.04457	.01567	.00854	.00560	.00407	.00315	.00254	.00212	.00181	.00157	.00139	.00124
1.60	.05480	.02039	.01144	.00765	.00564	.00441	.00360	.00302	.00259	.00227	.00201	.00180
1.50	.06681	.02626	.01517	.01035	.00773	.00611	.00503	.00425	.00368	.00323	.00288	.00259
1.40	.08076	.03347	.01990	.01383	.01047	.00837	.00694	.00592	.00515	.00455	.00407	.00368
1.30	.09680	.04224	.02581	.01827	.01402	.01133	.00948	.00813	.00712	.00632	.00569	.00516
1.20	.11507	.05277	.03313	.02387	.01856	.01515	.01278	.01105	.00973	.00869	.00785	.00716
1.10	.13567	.06528	.04208	.03085	.02430	.02003	.01704	.01483	.01313	.01179	.01070	.00980
1.00	.15866	.07997	.05287	.03943	.03145	.02618	.02245	.01967	.01752	.01581	.01441	.01325
0.90	.18406	.09702	.06575	.04985	.04025	.03383	.02924	.02579	.02310	.02095	.01919	.01771
0.80	.21186	.11659	.08092	.06234	.05094	.04323	.03765	.03342	.03011	.02744	.02523	.02338
0.70	.24196	.13880	.09859	.07715	.06378	.05462	.04793	.04282	.03878	.03551	.03280	.03051
0.60	.27425	.16371	.11892	.09448	.07899	.06825	.06033	.05424	.04939	.04543	.04214	.03935
0.50	.30854	.19134	.14203	.11451	.09679	.08436	.07510	.06793	.06218	.05746	.05351	.05015
0.40	.34458	.22163	.16797	.13737	.11736	.10315	.09248	.08413	.07741	.07185	.06718	.06319
0.30	.38209	.25448	.19677	.16314	.14082	.12479	.11264	.10307	.09530	.08885	.08340	.07871
0.20	.42074	.28970	.22835	.19185	.16727	.14941	.13575	.12491	.11605	.10866	.10238	.09696
0.10	.46017	.32702	.26257	.22343	.19669	.17705	.16189	.14977	.13981	.13145	.12431	.11812
0.00	.50000	.36614	.29921	.25775	.22902	.20769	.19109	.17771	.16665	.15732	.14931	.14234

H \ N	1	2	3	4	5	6	7	8	9	10	11	12
0.00	.50000	.36614	.29921	.25775	.22902	.20769	.19109	.17771	.16665	.15732	.14931	.14234
0.10	.53983	.40668	.33798	.29459	.26411	.24124	.22328	.20871	.19659	.18630	.17744	.16969
0.20	.57926	.44822	.37852	.33366	.30173	.27751	.25833	.24266	.22955	.21837	.20868	.20018
0.30	.61791	.49031	.42041	.37459	.34154	.31622	.29601	.27937	.26537	.25336	.24292	.23372
0.40	.65542	.53248	.46319	.41696	.38317	.35703	.33599	.31856	.30380	.29108	.27996	.27014
0.50	.69146	.57426	.50637	.46027	.42616	.39950	.37788	.35985	.34449	.33119	.31952	.30916
0.60	.72575	.61520	.54944	.50403	.47001	.44317	.42123	.40281	.38704	.37331	.36121	.35043
0.70	.75804	.65487	.59190	.54770	.51418	.48750	.46551	.44694	.43095	.41697	.40459	.39351
0.80	.78814	.69288	.63328	.59076	.55815	.53194	.51020	.49171	.47570	.46164	.44913	.43790
0.90	.81594	.72890	.67313	.63272	.60137	.57596	.55471	.53654	.52072	.50676	.49429	.48305
1.00	.84134	.76266	.71106	.67311	.64335	.61900	.59851	.58087	.56544	.55175	.53948	.52839
1.10	.86433	.79394	.74676	.71154	.68362	.66059	.64106	.62416	.60929	.59605	.58414	.57332
1.20	.88493	.82263	.77996	.74766	.72179	.70026	.68189	.66590	.65176	.63911	.62769	.61728
1.30	.90320	.84863	.81049	.78122	.75753	.73766	.72059	.70565	.69237	.68044	.66963	.65974
1.40	.91924	.87196	.83824	.81202	.79059	.77248	.75681	.74302	.73072	.71961	.70950	.70023
1.50	.93319	.89264	.86317	.83997	.82081	.80450	.79030	.77774	.76647	.75626	.74693	.73835
1.60	.94520	.91079	.88533	.86502	.84810	.83359	.82088	.80957	.79939	.79012	.78163	.77378
1.70	.95543	.92654	.90478	.88722	.87246	.85971	.84847	.83842	.82933	.82102	.81338	.80630
1.80	.96407	.94006	.92168	.90667	.89394	.88287	.87305	.86423	.85621	.84886	.84207	.83577
1.90	.97128	.95154	.93618	.92350	.91266	.90316	.89469	.88704	.88006	.87363	.86768	.86213
2.00	.97725	.96119	.94849	.93790	.92877	.92072	.91350	.90695	.90095	.89541	.89025	.88543
2.10	.98213	.96920	.95883	.95009	.94249	.93574	.92967	.92413	.91903	.91430	.90990	.90576
2.20	.98610	.97579	.96741	.96027	.95402	.94844	.94339	.93876	.93448	.93050	.92678	.92328
2.30	.98928	.98115	.97445	.96869	.96361	.95905	.95489	.95107	.94753	.94422	.94112	.93819
2.40	.99180	.98546	.98016	.97557	.97149	.96780	.96443	.96131	.95841	.95570	.95314	.95072
2.50	.99379	.98889	.98475	.98113	.97789	.97494	.97224	.96973	.96738	.96518	.96310	.96113
2.60	.99534	.99159	.98839	.98557	.98302	.98070	.97855	.97656	.97468	.97292	.97125	.96966
2.70	.99653	.99370	.99125	.98907	.98710	.98529	.98361	.98203	.98056	.97916	.97783	.97657
2.80	.99744	.99532	.99347	.99181	.99029	.98890	.98760	.98638	.98522	.98413	.98309	.98210
2.90	.99813	.99656	.99517	.99392	.99277	.99171	.99071	.98978	.98889	.98805	.98724	.98647
3.00	.99865	.99750	.99647	.99554	.99467	.99387	.99312	.99241	.99173	.99109	.99047	.98988
3.10	.99903	.99819	.99744	.99675	.99612	.99552	.99496	.99442	.99391	.99343	.99296	.99252
3.20	.99931	.99871	.99816	.99766	.99720	.99676	.99634	.99595	.99557	.99521	.99486	.99452
3.30	.99952	.99909	.99870	.99833	.99800	.99768	.99737	.99708	.99681	.99654	.99628	.99603
3.40	.99966	.99936	.99908	.99883	.99858	.99835	.99813	.99792	.99772	.99753	.99734	.99716
3.50	.99977	.99956	.99936	.99918	.99901	.99884	.99869	.99854	.99839	.99825	.99812	.99799

Tafel 9 (Forts.) Table 9 (cont.)

$\rho = .700$

-H \ N	1	2	3	4	5	6	7	8	9	10	11	12
3.50	.00023	.00003	.00001	.00000	.00000	.00000	.00000	.00000	.00000	.00000	.00000	.00000
3.40	.00034	.00004	.00001	.00001	.00000	.00000	.00000	.00000	.00000	.00000	.00000	.00000
3.30	.00048	.00007	.00002	.00001	.00001	.00000	.00000	.00000	.00000	.00000	.00000	.00000
3.20	.00069	.00010	.00004	.00002	.00001	.00001	.00001	.00000	.00000	.00000	.00000	.00000
3.10	.00097	.00015	.00006	.00003	.00002	.00001	.00001	.00001	.00001	.00000	.00000	.00000
3.00	.00135	.00023	.00009	.00005	.00003	.00002	.00001	.00001	.00001	.00001	.00001	.00001
2.90	.00187	.00034	.00014	.00007	.00005	.00003	.00002	.00002	.00001	.00001	.00001	.00001
2.80	.00256	.00050	.00021	.00011	.00007	.00005	.00004	.00003	.00002	.00002	.00002	.00001
2.70	.00347	.00073	.00031	.00017	.00011	.00008	.00006	.00005	.00004	.00003	.00003	.00002
2.60	.00466	.00105	.00046	.00026	.00017	.00013	.00010	.00008	.00006	.00005	.00005	.00004
2.50	.00621	.00149	.00067	.00040	.00027	.00019	.00015	.00012	.00010	.00008	.00007	.00006
2.40	.00820	.00209	.00098	.00059	.00040	.00029	.00023	.00019	.00015	.00013	.00011	.00010
2.30	.01072	.00291	.00141	.00086	.00059	.00044	.00035	.00028	.00024	.00020	.00018	.00016
2.20	.01390	.00401	.00200	.00124	.00087	.00066	.00052	.00043	.00036	.00031	.00027	.00024
2.10	.01786	.00546	.00280	.00178	.00127	.00096	.00077	.00064	.00054	.00047	.00041	.00036
2.00	.02275	.00736	.00389	.00252	.00182	.00140	.00113	.00094	.00080	.00070	.00061	.00055
1.90	.02872	.00983	.00535	.00353	.00258	.00201	.00163	.00137	.00117	.00102	.00091	.00081
1.80	.03593	.01298	.00727	.00488	.00361	.00284	.00233	.00197	.00170	.00149	.00133	.00119
1.70	.04457	.01698	.00977	.00668	.00501	.00398	.00329	.00279	.00243	.00214	.00191	.00173
1.60	.05480	.02198	.01300	.00904	.00687	.00551	.00459	.00392	.00343	.00304	.00273	.00248
1.50	.06681	.02817	.01710	.01210	.00930	.00754	.00633	.00545	.00478	.00426	.00384	.00350
1.40	.08076	.03574	.02226	.01601	.01247	.01020	.00863	.00748	.00660	.00591	.00535	.00489
1.30	.09680	.04490	.02867	.02097	.01652	.01364	.01163	.01014	.00900	.00810	.00737	.00676
1.20	.11507	.05586	.03654	.02716	.02165	.01804	.01549	.01360	.01213	.01097	.01002	.00923
1.10	.13567	.06882	.04610	.03480	.02806	.02359	.02041	.01802	.01616	.01468	.01346	.01245
1.00	.15866	.08398	.05756	.04412	.03598	.03051	.02658	.02362	.02129	.01942	.01788	.01659
0.90	.18406	.10151	.07114	.05534	.04563	.03903	.03424	.03060	.02773	.02541	.02349	.02187
0.80	.21186	.12156	.08705	.06871	.05725	.04939	.04363	.03921	.03571	.03287	.03050	.02850
0.70	.24196	.14423	.10547	.08442	.07109	.06182	.05498	.04969	.04548	.04203	.03915	.03671
0.60	.27425	.16958	.12654	.10268	.08734	.07656	.06853	.06229	.05728	.05316	.04970	.04676
0.50	.30854	.19760	.15037	.12364	.10621	.09383	.08453	.07725	.07136	.06650	.06240	.05889
0.40	.34458	.22824	.17700	.14741	.12784	.11380	.10316	.09477	.08795	.08229	.07749	.07337
0.30	.38209	.26137	.20641	.17404	.15235	.13662	.12460	.11505	.10725	.10073	.09519	.09040
0.20	.42074	.29679	.23852	.20354	.17977	.16237	.14895	.13822	.12941	.12201	.11569	.11021
0.10	.46017	.33425	.27316	.23580	.21009	.19106	.17628	.16438	.15455	.14625	.13913	.13293
0.00	.50000	.37341	.31011	.27069	.24319	.22265	.20656	.19353	.18269	.17351	.16559	.15868

H \ N	1	2	3	4	5	6	7	8	9	10	11	12
0.00	.50000	.37341	.31011	.27069	.24319	.22265	.20656	.19353	.18269	.17351	.16559	.15868
0.10	.53983	.41390	.34906	.30794	.27891	.25700	.23971	.22561	.21382	.20378	.19509	.18747
0.20	.57926	.45531	.38964	.34727	.31697	.29389	.27554	.26048	.24782	.23698	.22756	.21927
0.30	.61791	.49720	.43144	.38828	.35705	.33304	.31380	.29790	.28448	.27293	.26285	.25395
0.40	.65542	.53909	.47400	.43056	.39875	.37407	.35414	.33758	.32352	.31137	.30072	.29128
0.50	.69146	.58053	.51682	.47362	.44161	.41655	.39617	.37913	.36458	.35196	.34085	.33096
0.60	.72575	.62107	.55943	.51695	.48513	.45999	.43941	.42209	.40724	.39428	.38283	.37261
0.70	.75804	.66030	.60133	.56006	.52880	.50389	.48335	.46598	.45099	.43786	.42622	.41579
0.80	.78814	.69785	.64206	.60244	.57210	.54772	.52747	.51025	.49532	.48218	.47049	.45997
0.90	.81594	.73339	.68121	.64361	.61451	.59093	.57122	.55436	.53967	.52669	.51509	.50463
1.00	.84134	.76667	.71842	.68315	.65557	.63304	.61409	.59778	.58350	.57084	.55948	.54920
1.10	.86433	.79749	.75336	.72067	.69484	.67358	.65557	.63999	.62629	.61409	.60310	.59313
1.20	.88493	.82572	.78582	.75586	.73196	.71213	.69523	.68053	.66755	.65593	.64544	.63588
1.30	.90320	.85130	.81563	.78849	.76663	.74836	.73269	.71899	.70684	.69592	.68602	.67698
1.40	.91924	.87422	.84269	.81839	.79863	.78200	.76765	.75504	.74380	.73366	.72444	.71599
1.50	.93319	.89455	.86698	.84547	.82783	.81287	.79988	.78841	.77815	.76885	.76037	.75257
1.60	.94520	.91238	.88854	.86972	.85415	.84085	.82924	.81894	.80968	.80126	.79355	.78644
1.70	.95543	.92785	.90746	.89119	.87760	.86592	.85567	.84652	.83827	.83074	.82382	.81742
1.80	.96407	.94112	.92388	.90997	.89826	.88812	.87917	.87115	.86388	.85723	.85110	.84541
1.90	.97128	.95239	.93798	.92622	.91624	.90754	.89982	.89288	.88656	.88075	.87538	.87039
2.00	.97725	.96186	.94994	.94011	.93170	.92433	.91776	.91181	.90638	.90138	.89673	.89240
2.10	.98214	.96973	.95998	.95186	.94486	.93868	.93315	.92812	.92351	.91925	.91528	.91157
2.20	.98610	.97620	.96831	.96167	.95591	.95080	.94620	.94200	.93813	.93454	.93120	.92806
2.30	.98928	.98146	.97515	.96979	.96510	.96092	.95713	.95367	.95046	.94748	.94469	.94207
2.40	.99180	.98569	.98070	.97642	.97265	.96927	.96619	.96336	.96074	.95829	.95599	.95383
2.50	.99379	.98907	.98515	.98177	.97878	.97608	.97361	.97132	.96920	.96722	.96535	.96358
2.60	.99534	.99172	.98869	.98605	.98370	.98156	.97960	.97779	.97609	.97450	.97300	.97157
2.70	.99653	.99379	.99147	.98943	.98760	.98594	.98440	.98297	.98163	.98037	.97918	.97804
2.80	.99744	.99539	.99363	.99208	.99067	.98938	.98819	.98708	.98603	.98505	.98411	.98322
2.90	.99813	.99661	.99529	.99412	.99305	.99207	.99115	.99030	.98949	.98873	.98800	.98731
3.00	.99865	.99753	.99655	.99567	.99487	.99413	.99344	.99279	.99217	.99159	.99103	.99050
3.10	.99903	.99822	.99750	.99685	.99626	.99570	.99519	.99470	.99423	.99379	.99337	.99297
3.20	.99931	.99873	.99821	.99773	.99729	.99689	.99650	.99614	.99580	.99547	.99515	.99485
3.30	.99952	.99910	.99872	.99838	.99806	.99777	.99749	.99722	.99697	.99672	.99649	.99627
3.40	.99966	.99937	.99910	.99886	.99863	.99842	.99821	.99802	.99783	.99766	.99749	.99732
3.50	.99977	.99956	.99937	.99920	.99904	.99889	.99874	.99860	.99847	.99834	.99822	.99810

Tafel 9 (Forts.) Table 9 (cont.)

ρ = .750

-H \ N	1	2	3	4	5	6	7	8	9	10	11	12
3.50	.00023	.00004	.00001	.00001	.00000	.00000	.00000	.00000	.00000	.00000	.00000	.00000
3.40	.00034	.00006	.00002	.00001	.00001	.00001	.00000	.00000	.00000	.00000	.00000	.00000
3.30	.00048	.00009	.00004	.00002	.00001	.00001	.00001	.00001	.00000	.00000	.00000	.00000
3.20	.00069	.00013	.00006	.00003	.00002	.00001	.00001	.00001	.00001	.00001	.00001	.00000
3.10	.00097	.00020	.00009	.00005	.00003	.00002	.00002	.00001	.00001	.00001	.00001	.00001
3.00	.00135	.00029	.00013	.00008	.00005	.00004	.00003	.00002	.00002	.00002	.00001	.00001
2.90	.00187	.00043	.00020	.00012	.00008	.00006	.00005	.00004	.00003	.00003	.00002	.00002
2.80	.00256	.00062	.00029	.00018	.00012	.00009	.00007	.00006	.00005	.00004	.00004	.00003
2.70	.00347	.00090	.00043	.00027	.00019	.00014	.00011	.00009	.00008	.00007	.00006	.00005
2.60	.00466	.00128	.00063	.00040	.00028	.00021	.00017	.00014	.00012	.00010	.00009	.00008
2.50	.00621	.00179	.00092	.00059	.00042	.00032	.00026	.00022	.00018	.00016	.00014	.00013
2.40	.00820	.00250	.00131	.00085	.00062	.00048	.00039	.00033	.00028	.00024	.00022	.00019
2.30	.01072	.00345	.00186	.00123	.00090	.00070	.00057	.00048	.00042	.00037	.00033	.00029
2.20	.01390	.00470	.00260	.00174	.00129	.00102	.00084	.00071	.00062	.00055	.00049	.00044
2.10	.01786	.00636	.00360	.00245	.00184	.00147	.00122	.00104	.00090	.00080	.00072	.00065
2.00	.02275	.00850	.00494	.00342	.00259	.00209	.00174	.00149	.00131	.00116	.00105	.00095
1.90	.02872	.01125	.00670	.00471	.00361	.00293	.00246	.00213	.00187	.00167	.00151	.00138
1.80	.03593	.01475	.00899	.00641	.00498	.00407	.00345	.00299	.00264	.00237	.00215	.00197
1.70	.04457	.01914	.01193	.00864	.00678	.00559	.00477	.00416	.00370	.00333	.00303	.00279
1.60	.05480	.02460	.01569	.01153	.00914	.00760	.00652	.00572	.00511	.00462	.00422	.00390
1.50	.06681	.03130	.02041	.01521	.01219	.01021	.00882	.00778	.00698	.00634	.00582	.00538
1.40	.08076	.03945	.02627	.01987	.01608	.01358	.01180	.01047	.00943	.00860	.00792	.00735
1.30	.09680	.04924	.03349	.02567	.02099	.01786	.01562	.01393	.01261	.01155	.01067	.00993
1.20	.11507	.06088	.04226	.03283	.02711	.02325	.02046	.01834	.01667	.01533	.01421	.01327
1.10	.13567	.07455	.05280	.04156	.03464	.02993	.02650	.02388	.02180	.02012	.01872	.01753
1.00	.15866	.09046	.06532	.05207	.04382	.03814	.03396	.03076	.02821	.02612	.02439	.02291
0.90	.18406	.10874	.08002	.06459	.05485	.04808	.04307	.03920	.03610	.03356	.03143	.02962
0.80	.21186	.12954	.09709	.07933	.06797	.06000	.05406	.04943	.04571	.04265	.04007	.03787
0.70	.24196	.15293	.11668	.09647	.08338	.07411	.06715	.06169	.05728	.05363	.05055	.04790
0.60	.27425	.17896	.13891	.11618	.10127	.09062	.08255	.07620	.07103	.06674	.06309	.05996
0.50	.30854	.20761	.16386	.13859	.12181	.10970	.10047	.09315	.08718	.08218	.07794	.07427
0.40	.34458	.23878	.19153	.16375	.14509	.13150	.12107	.11274	.10591	.10018	.09528	.09104
0.30	.38209	.27235	.22187	.19170	.17118	.15611	.14445	.13510	.12738	.12088	.11531	.11046
0.20	.42074	.30809	.25478	.22237	.20008	.18356	.17069	.16031	.15170	.14441	.13814	.13267
0.10	.46017	.34575	.29006	.25565	.23172	.21382	.19979	.18840	.17891	.17084	.16388	.15777
0.00	.50000	.38497	.32746	.29135	.26594	.24679	.23167	.21932	.20899	.20017	.19252	.18580

H \ N	1	2	3	4	5	6	7	8	9	10	11	12
0.00	.50000	.38497	.32746	.29135	.26594	.24679	.23167	.21932	.20899	.20017	.19252	.18580
0.10	.53983	.42540	.36666	.32920	.30255	.28229	.26618	.25296	.24184	.23231	.22402	.21671
0.20	.57926	.46661	.40728	.36886	.34123	.32005	.30310	.28912	.27730	.26712	.25824	.25038
0.30	.61791	.50817	.44891	.40995	.38164	.35976	.34214	.32752	.31510	.30437	.29497	.28663
0.40	.65542	.54963	.49109	.45204	.42336	.40103	.38292	.36781	.35493	.34375	.33392	.32517
0.50	.69146	.59053	.53335	.49465	.46594	.44341	.42502	.40960	.39639	.38489	.37473	.36567
0.60	.72575	.63046	.57522	.53731	.50890	.48642	.46790	.45242	.43904	.42734	.41698	.40770
0.70	.75804	.66901	.61623	.57951	.55172	.52958	.51129	.49579	.48239	.47063	.46018	.45080
0.80	.78814	.70583	.65597	.62080	.59394	.57236	.55444	.53918	.52593	.51426	.50385	.49447
0.90	.81594	.74062	.69403	.66074	.63506	.61430	.59694	.58208	.56914	.55769	.54744	.53819
1.00	.84134	.77315	.73009	.69892	.67467	.65491	.63830	.62401	.61151	.60041	.59045	.58143
1.10	.86433	.80322	.76387	.73503	.71237	.69379	.67808	.66451	.65258	.64195	.63238	.62369
1.20	.88493	.83074	.79516	.76877	.74786	.73059	.71590	.70315	.69190	.68185	.67276	.66449
1.30	.90320	.85564	.82383	.79995	.78087	.76500	.75144	.73961	.72912	.71972	.71120	.70342
1.40	.91924	.87794	.84981	.82845	.81123	.79682	.78444	.77359	.76393	.75525	.74735	.74012
1.50	.93319	.89769	.87309	.85419	.83884	.82591	.81473	.80489	.79611	.78818	.78095	.77431
1.60	.94520	.91500	.89372	.87719	.86366	.85218	.84221	.83340	.82550	.81835	.81180	.80578
1.70	.95543	.93001	.91180	.89751	.88571	.87564	.86686	.85905	.85204	.84566	.83981	.83440
1.80	.96407	.94289	.92748	.91525	.90508	.89636	.88870	.88187	.87571	.87009	.86492	.86014
1.90	.97128	.95382	.94092	.93058	.92191	.91443	.90784	.90194	.89659	.89169	.88718	.88299
2.00	.97725	.96300	.95231	.94367	.93637	.93003	.92442	.91937	.91478	.91057	.90668	.90306
2.10	.98214	.97063	.96187	.95472	.94864	.94334	.93861	.93435	.93046	.92688	.92356	.92046
2.20	.98610	.97690	.96980	.96396	.95895	.95456	.95063	.94707	.94381	.94080	.93800	.93538
2.30	.98928	.98200	.97631	.97158	.96751	.96391	.96068	.95774	.95504	.95254	.95021	.94803
2.40	.99180	.98610	.98160	.97781	.97453	.97162	.96899	.96660	.96439	.96233	.96042	.95862
2.50	.99379	.98938	.98584	.98285	.98023	.97790	.97579	.97386	.97207	.97041	.96885	.96738
2.60	.99534	.99195	.98921	.98687	.98481	.98297	.98129	.97975	.97832	.97698	.97573	.97455
2.70	.99653	.99396	.99185	.99004	.98844	.98700	.98568	.98447	.98334	.98228	.98128	.98034
2.80	.99744	.99551	.99391	.99253	.99129	.99018	.98916	.98821	.98733	.98650	.98572	.98497
2.90	.99813	.99670	.99549	.99445	.99351	.99265	.99187	.99114	.99046	.98982	.98921	.98863
3.00	.99865	.99759	.99670	.99591	.99521	.99456	.99397	.99341	.99289	.99240	.99193	.99149
3.10	.99903	.99826	.99760	.99702	.99650	.99601	.99557	.99515	.99476	.99438	.99403	.99369
3.20	.99931	.99876	.99828	.99785	.99746	.99711	.99678	.99647	.99617	.99589	.99563	.99538
3.30	.99952	.99912	.99877	.99847	.99818	.99792	.99768	.99745	.99723	.99703	.99683	.99664
3.40	.99966	.99938	.99914	.99892	.99871	.99852	.99835	.99818	.99802	.99787	.99773	.99759
3.50	.99977	.99957	.99940	.99924	.99910	.99896	.99883	.99871	.99860	.99849	.99839	.99829

Tafel 9 (Forts.) Table 9 (cont.)

ρ = .800

-H \ N	1	2	3	4	5	6	7	8	9	10	11	12
3.50	.00023	.00005	.00002	.00001	.00001	.00001	.00001	.00000	.00000	.00000	.00000	.00000
3.40	.00034	.00008	.00004	.00002	.00001	.00001	.00001	.00001	.00001	.00001	.00000	.00000
3.30	.00048	.00011	.00005	.00003	.00002	.00002	.00001	.00001	.00001	.00001	.00001	.00001
3.20	.00069	.00017	.00008	.00005	.00004	.00003	.00002	.00002	.00002	.00001	.00001	.00001
3.10	.00097	.00025	.00013	.00008	.00006	.00004	.00004	.00003	.00003	.00002	.00002	.00002
3.00	.00135	.00037	.00019	.00012	.00009	.00007	.00006	.00005	.00004	.00004	.00003	.00003
2.90	.00187	.00054	.00028	.00018	.00013	.00011	.00009	.00007	.00006	.00006	.00005	.00004
2.80	.00256	.00078	.00042	.00028	.00020	.00016	.00013	.00011	.00010	.00009	.00008	.00007
2.70	.00347	.00110	.00060	.00041	.00030	.00024	.00020	.00017	.00015	.00013	.00012	.00011
2.60	.00466	.00156	.00087	.00059	.00045	.00036	.00030	.00025	.00022	.00020	.00018	.00016
2.50	.00621	.00217	.00124	.00086	.00065	.00052	.00044	.00038	.00033	.00030	.00027	.00024
2.40	.00820	.00299	.00175	.00122	.00094	.00076	.00064	.00056	.00049	.00044	.00040	.00037
2.30	.01072	.00409	.00244	.00173	.00134	.00110	.00093	.00081	.00072	.00064	.00059	.00054
2.20	.01390	.00553	.00337	.00242	.00190	.00156	.00133	.00116	.00104	.00094	.00085	.00079
2.10	.01786	.00741	.00461	.00336	.00265	.00220	.00189	.00166	.00148	.00134	.00123	.00113
2.00	.02275	.00983	.00624	.00460	.00367	.00307	.00264	.00233	.00209	.00190	.00175	.00162
1.90	.02872	.01290	.00836	.00624	.00502	.00423	.00367	.00325	.00293	.00267	.00246	.00228
1.80	.03593	.01678	.01109	.00838	.00680	.00576	.00503	.00448	.00405	.00370	.00342	.00319
1.70	.04457	.02162	.01455	.01114	.00912	.00778	.00682	.00610	.00554	.00509	.00471	.00440
1.60	.05480	.02758	.01891	.01465	.01209	.01039	.00916	.00823	.00750	.00691	.00642	.00601
1.50	.06681	.03486	.02434	.01907	.01588	.01373	.01217	.01098	.01005	.00929	.00866	.00813
1.40	.08076	.04363	.03101	.02458	.02064	.01796	.01600	.01450	.01332	.01235	.01155	.01087
1.30	.09680	.05411	.03913	.03136	.02655	.02325	.02082	.01896	.01747	.01626	.01524	.01438
1.20	.11507	.06648	.04890	.03962	.03381	.02979	.02682	.02452	.02268	.02117	.01990	.01882
1.10	.13567	.08093	.06052	.04957	.04263	.03779	.03419	.03139	.02914	.02728	.02572	.02439
1.00	.15866	.09764	.07419	.06141	.05322	.04746	.04315	.03978	.03705	.03480	.03290	.03127
0.90	.18406	.11673	.09010	.07535	.06580	.05902	.05391	.04990	.04664	.04394	.04165	.03968
0.80	.21186	.13833	.10841	.09158	.08056	.07267	.06669	.06197	.05813	.05492	.05219	.04984
0.70	.24196	.16250	.12925	.11025	.09768	.08852	.08170	.07621	.07172	.06796	.06475	.06198
0.60	.27425	.18926	.15271	.13151	.11733	.10703	.09912	.09281	.08752	.08326	.07954	.07630
0.50	.30854	.21856	.17883	.15543	.13962	.12804	.11910	.11193	.10601	.10102	.09674	.09301
0.40	.34458	.25031	.20758	.18205	.16462	.15176	.14177	.13371	.12704	.12139	.11652	.11227
0.30	.38209	.28434	.23890	.21134	.19234	.17822	.16718	.15824	.15079	.14447	.13901	.13423
0.20	.42074	.32042	.27262	.24321	.22274	.20740	.19535	.18553	.17733	.17034	.16428	.15895
0.10	.46017	.35828	.30854	.27750	.25569	.23923	.22621	.21556	.20663	.19898	.19234	.18649
0.00	.50000	.39758	.34638	.31399	.29100	.27354	.25965	.24823	.23861	.23034	.22314	.21678

H \ N	1	2	3	4	5	6	7	8	9	10	11	12
0.00	.50000	.39758	.34638	.31399	.29100	.27354	.25965	.24823	.23861	.23034	.22314	.21678
0.10	.53983	.43794	.38580	.35238	.32844	.31011	.29545	.28335	.27311	.26428	.25657	.24973
0.20	.57926	.47894	.42643	.39232	.36766	.34866	.33337	.32068	.30990	.30059	.29242	.28516
0.30	.61791	.52016	.46785	.43342	.40831	.38882	.37305	.35990	.34870	.33898	.33042	.32281
0.40	.65542	.56115	.50960	.47524	.44995	.43019	.41411	.40065	.38913	.37910	.37025	.36235
0.50	.69146	.60149	.55124	.51732	.49214	.47233	.45613	.44250	.43079	.42057	.41151	.40342
0.60	.72575	.64075	.59230	.55920	.53440	.51477	.49862	.48498	.47322	.46292	.45377	.44556
0.70	.75804	.67858	.63236	.60040	.57626	.55702	.54112	.52763	.51594	.50568	.49653	.48831
0.80	.78814	.71462	.67102	.64051	.61727	.59863	.58314	.56994	.55847	.54836	.53933	.53119
0.90	.81594	.74861	.70792	.67911	.65698	.63912	.62421	.61145	.60032	.59047	.58166	.57369
1.00	.84134	.78033	.74275	.71585	.69502	.67811	.66391	.65171	.64103	.63156	.62305	.61534
1.10	.86433	.80960	.77528	.75044	.73105	.71521	.70185	.69032	.68019	.67117	.66306	.65568
1.20	.88493	.83634	.80534	.78264	.76480	.75012	.73769	.72692	.71742	.70894	.70129	.69432
1.30	.90320	.86051	.83279	.81229	.79605	.78261	.77117	.76122	.75242	.74454	.73740	.73089
1.40	.91924	.88212	.85761	.83930	.82467	.81251	.80210	.79302	.78495	.77770	.77113	.76511
1.50	.93319	.90124	.87981	.86362	.85060	.83971	.83035	.82215	.81484	.80826	.80227	.79677
1.60	.94520	.91798	.89943	.88529	.87383	.86419	.85587	.84854	.84199	.83608	.83069	.82573
1.70	.95543	.93249	.91661	.90438	.89440	.88597	.87865	.87218	.86638	.86113	.85633	.85191
1.80	.96407	.94492	.93147	.92102	.91243	.90512	.89876	.89312	.88804	.88343	.87921	.87530
1.90	.97128	.95547	.94420	.93536	.92805	.92179	.91632	.91145	.90706	.90306	.89938	.89598
2.00	.97725	.96432	.95498	.94759	.94143	.93614	.93149	.92733	.92357	.92014	.91698	.91404
2.10	.98214	.97168	.96402	.95790	.95277	.94834	.94443	.94093	.93775	.93483	.93214	.92964
2.20	.98610	.97772	.97151	.96650	.96227	.95861	.95536	.95244	.94978	.94733	.94507	.94296
2.30	.98928	.98264	.97765	.97359	.97015	.96715	.96448	.96207	.95987	.95785	.95597	.95421
2.40	.99180	.98660	.98263	.97939	.97661	.97418	.97201	.97005	.96825	.96659	.96504	.96360
2.50	.99379	.98975	.98663	.98406	.98185	.97991	.97816	.97658	.97512	.97378	.97252	.97135
2.60	.99534	.99223	.98981	.98780	.98605	.98451	.98313	.98186	.98070	.97962	.97861	.97767
2.70	.99653	.99417	.99231	.99074	.98939	.98818	.98709	.98610	.98518	.98432	.98352	.98277
2.80	.99744	.99567	.99425	.99305	.99200	.99107	.99022	.98945	.98873	.98806	.98743	.98684
2.90	.99813	.99681	.99574	.99483	.99403	.99332	.99267	.99207	.99151	.99099	.99051	.99004
3.00	.99865	.99767	.99688	.99619	.99559	.99505	.99456	.99410	.99367	.99328	.99290	.99255
3.10	.99903	.99832	.99773	.99723	.99678	.99637	.99600	.99565	.99533	.99503	.99475	.99448
3.20	.99931	.99880	.99837	.99800	.99767	.99736	.99709	.99683	.99659	.99637	.99615	.99595
3.30	.99952	.99915	.99884	.99857	.99833	.99811	.99790	.99771	.99754	.99737	.99721	.99706
3.40	.99966	.99940	.99918	.99899	.99881	.99865	.99850	.99837	.99824	.99811	.99800	.99789
3.50	.99977	.99958	.99943	.99929	.99916	.99905	.99894	.99884	.99875	.99866	.99858	.99850

Tafel 9 (Forts.) Table 9 (cont.)

$\rho = .875$

N / −H	1	2	3	4	5	6	7	8	9	10	11	12
3.50	.00023	.00008	.00004	.00003	.00002	.00002	.00002	.00001	.00001	.00001	.00001	.00001
3.40	.00034	.00012	.00007	.00005	.00004	.00003	.00003	.00002	.00002	.00002	.00002	.00002
3.30	.00048	.00017	.00010	.00007	.00006	.00005	.00004	.00004	.00003	.00003	.00003	.00002
3.20	.00069	.00026	.00015	.00011	.00009	.00007	.00006	.00006	.00005	.00004	.00004	.00004
3.10	.00097	.00037	.00023	.00017	.00013	.00011	.00010	.00008	.00008	.00007	.00006	.00006
3.00	.00135	.00054	.00034	.00025	.00020	.00017	.00014	.00013	.00012	.00011	.00010	.00009
2.90	.00187	.00077	.00049	.00036	.00029	.00025	.00021	.00019	.00017	.00016	.00015	.00014
2.80	.00256	.00109	.00070	.00053	.00043	.00036	.00032	.00028	.00026	.00024	.00022	.00020
2.70	.00347	.00152	.00100	.00076	.00062	.00053	.00046	.00041	.00038	.00035	.00032	.00030
2.60	.00466	.00211	.00141	.00108	.00088	.00076	.00067	.00060	.00055	.00051	.00047	.00044
2.50	.00621	.00290	.00196	.00151	.00125	.00108	.00096	.00086	.00079	.00073	.00068	.00064
2.40	.00820	.00395	.00271	.00211	.00176	.00152	.00135	.00122	.00112	.00104	.00097	.00092
2.30	.01072	.00533	.00370	.00291	.00244	.00212	.00190	.00172	.00158	.00147	.00138	.00130
2.20	.01390	.00711	.00502	.00398	.00336	.00294	.00263	.00240	.00221	.00206	.00194	.00183
2.10	.01786	.00941	.00673	.00539	.00457	.00402	.00361	.00330	.00306	.00286	.00269	.00255
2.00	.02275	.01232	.00893	.00722	.00616	.00544	.00491	.00451	.00418	.00392	.00370	.00351
1.90	.02872	.01599	.01175	.00957	.00823	.00730	.00662	.00609	.00567	.00532	.00503	.00478
1.80	.03593	.02056	.01531	.01258	.01087	.00969	.00882	.00814	.00760	.00715	.00677	.00645
1.70	.04457	.02619	.01975	.01637	.01423	.01275	.01164	.01078	.01009	.00951	.00903	.00862
1.60	.05480	.03305	.02525	.02109	.01845	.01660	.01522	.01413	.01326	.01253	.01192	.01139
1.50	.06681	.04132	.03197	.02692	.02369	.02141	.01970	.01835	.01726	.01635	.01558	.01492
1.40	.08076	.05120	.04010	.03404	.03012	.02734	.02525	.02359	.02225	.02113	.02017	.01935
1.30	.09680	.06288	.04984	.04263	.03794	.03459	.03205	.03004	.02840	.02703	.02586	.02485
1.20	.11507	.07653	.06138	.05290	.04734	.04335	.04030	.03788	.03590	.03424	.03283	.03160
1.10	.13567	.09232	.07491	.06504	.05852	.05381	.05020	.04732	.04496	.04297	.04127	.03979
1.00	.15866	.11040	.09059	.07924	.07'67	.06617	.06194	.05856	.05577	.05342	.05140	.04964
0.90	.18406	.13089	.10859	.09566	.08698	.08063	.07572	.07179	.06853	.06578	.06341	.06134
0.80	.21186	.15386	.12902	.11446	.10460	.09735	.09173	.08719	.08343	.08024	.07749	.07509
0.70	.24196	.17935	.15198	.13574	.12467	.11649	.11010	.10494	.10064	.09699	.09383	.09107
0.60	.27425	.20734	.17749	.15959	.14729	.13814	.13098	.12516	.12031	.11617	.11259	.10944
0.50	.30854	.23775	.20554	.18602	.17250	.16239	.15444	.14796	.14253	.13790	.13387	.13033
0.40	.34458	.27046	.23607	.21500	.20030	.18925	.18051	.17337	.16737	.16224	.15777	.15382
0.30	.38209	.30527	.26894	.24644	.23063	.21867	.20919	.20140	.19484	.18921	.18430	.17996
0.20	.42074	.34194	.30396	.28018	.26336	.25057	.24038	.23198	.22489	.21878	.21344	.20872
0.10	.46017	.38015	.34086	.31601	.29830	.28476	.27393	.26498	.25739	.25084	.24511	.24001
0.00	.50000	.41957	.37935	.35365	.33521	.32104	.30965	.30020	.29218	.28523	.27913	.27370

N / H	1	2	3	4	5	6	7	8	9	10	11	12
0.00	.50000	.41957	.37935	.35365	.33521	.32104	.30965	.30020	.29218	.28523	.27913	.27370
0.10	.53983	.45981	.41908	.39278	.37378	.35910	.34726	.33740	.32900	.32171	.31529	.30958
0.20	.57926	.50046	.45963	.43301	.41365	.39861	.38643	.37625	.36755	.35999	.35332	.34736
0.30	.61791	.54109	.50061	.47395	.45442	.43918	.42679	.41639	.40749	.39972	.39286	.38671
0.40	.65542	.58130	.54157	.51516	.49568	.48040	.46792	.45742	.44840	.44052	.43353	.42727
0.50	.69146	.62068	.58210	.55620	.53697	.52182	.50939	.49891	.48987	.48195	.47491	.46860
0.60	.72575	.65883	.62176	.59664	.57787	.56300	.55076	.54039	.53144	.52357	.51656	.51027
0.70	.75804	.69542	.66017	.63606	.61793	.60350	.59157	.58144	.57266	.56493	.55803	.55182
0.80	.78814	.73015	.69698	.67409	.65676	.64291	.63141	.62161	.61309	.60558	.59886	.59280
0.90	.81594	.76276	.73189	.71038	.69400	.68084	.66987	.66050	.65233	.64511	.63864	.63278
1.00	.84134	.79309	.76464	.74465	.72933	.71696	.70662	.69775	.69000	.68313	.67696	.67137
1.10	.86433	.82099	.79505	.77667	.76249	.75099	.74134	.73304	.72576	.71930	.71348	.70821
1.20	.88493	.84639	.82299	.80626	.79328	.78271	.77380	.76611	.75936	.75334	.74792	.74299
1.30	.90320	.86928	.84839	.83333	.82157	.81195	.80381	.79677	.79057	.78503	.78003	.77548
1.40	.91924	.88969	.87124	.85782	.84729	.83863	.83128	.82490	.81926	.81422	.80966	.80550
1.50	.93319	.90771	.89158	.87975	.87041	.86270	.85613	.85041	.84535	.84081	.83669	.83293
1.60	.94520	.92345	.90949	.89918	.89099	.88419	.87839	.87332	.86882	.86477	.86110	.85773
1.70	.95543	.93705	.92511	.91621	.90910	.90318	.89810	.89365	.88970	.88613	.88289	.87991
1.80	.96407	.94870	.93858	.93098	.92488	.91977	.91538	.91152	.90808	.90497	.90213	.89953
1.90	.97128	.95856	.95007	.94366	.93847	.93412	.93036	.92704	.92408	.92140	.91895	.91670
2.00	.97725	.96682	.95978	.95442	.95006	.94639	.94320	.94039	.93787	.93558	.93349	.93156
2.10	.98214	.97368	.96790	.96346	.95984	.95677	.95410	.95174	.94962	.94769	.94592	.94428
2.20	.98610	.97930	.97461	.97098	.96800	.96546	.96325	.96129	.95952	.95791	.95643	.95506
2.30	.98928	.98388	.98010	.97716	.97474	.97267	.97085	.96924	.96778	.96645	.96523	.96409
2.40	.99180	.98756	.98455	.98220	.98024	.97857	.97710	.97578	.97460	.97351	.97251	.97158
2.50	.99379	.99048	.98812	.98625	.98469	.98335	.98217	.98112	.98016	.97928	.97847	.97772
2.60	.99534	.99279	.99095	.98948	.98825	.98719	.98626	.98542	.98465	.98395	.98330	.98270
2.70	.99653	.99459	.99317	.99203	.99107	.99024	.98951	.98884	.98824	.98769	.98717	.98669
2.80	.99744	.99598	.99489	.99402	.99328	.99264	.99206	.99155	.99108	.99064	.99024	.98986
2.90	.99813	.99704	.99622	.99555	.99499	.99450	.99406	.99366	.99329	.99296	.99265	.99235
3.00	.99865	.99784	.99722	.99673	.99630	.99593	.99559	.99529	.99501	.99475	.99451	.99429
3.10	.99903	.99844	.99798	.99761	.99729	.99701	.99676	.99653	.99632	.99613	.99595	.99578
3.20	.99931	.99888	.99855	.99828	.99804	.99783	.99764	.99747	.99732	.99717	.99703	.99691
3.30	.99952	.99921	.99897	.99877	.99859	.99844	.99830	.99818	.99806	.99795	.99785	.99776
3.40	.99966	.99944	.99927	.99913	.99900	.99889	.99879	.99870	.99861	.99853	.99846	.99839
3.50	.99977	.99961	.99949	.99939	.99930	.99922	.99915	.99908	.99902	.99896	.99891	.99886

Tafel 9 (Forts.) Table 9 (cont.)

ρ = .900

N / −H	1	2	3	4	5	6	7	8	9	10	11	12
3.50	.00023	.00009	.00006	.00004	.00003	.00003	.00002	.00002	.00002	.00002	.00002	.00002
3.40	.00034	.00014	.00009	.00006	.00005	.00004	.00004	.00003	.00003	.00003	.00003	.00002
3.30	.00048	.00020	.00013	.00010	.00008	.00007	.00006	.00005	.00005	.00004	.00004	.00004
3.20	.00069	.00029	.00019	.00014	.00012	.00010	.00009	.00008	.00007	.00007	.00006	.00006
3.10	.00097	.00043	.00028	.00021	.00018	.00015	.00013	.00012	.00011	.00010	.00009	.00009
3.00	.00135	.00061	.00041	.00031	.00026	.00022	.00020	.00018	.00016	.00015	.00014	.00013
2.90	.00187	.00087	.00059	.00046	.00038	.00033	.00029	.00026	.00024	.00023	.00021	.00020
2.80	.00256	.00122	.00084	.00066	.00055	.00048	.00043	.00039	.00036	.00033	.00031	.00029
2.70	.00347	.00170	.00119	.00094	.00079	.00069	.00062	.00056	.00052	.00048	.00045	.00043
2.60	.00466	.00235	.00166	.00132	.00112	.00098	.00088	.00080	.00074	.00069	.00065	.00062
2.50	.00621	.00322	.00230	.00184	.00157	.00138	.00124	.00114	.00106	.00099	.00093	.00088
2.40	.00820	.00436	.00315	.00255	.00218	.00193	.00174	.00160	.00149	.00140	.00132	.00125
2.30	.01072	.00585	.00428	.00349	.00300	.00266	.00242	.00223	.00208	.00195	.00185	.00176
2.20	.01390	.00778	.00576	.00473	.00409	.00365	.00332	.00307	.00287	.00270	.00256	.00244
2.10	.01786	.01024	.00768	.00635	.00552	.00494	.00452	.00418	.00392	.00370	.00351	.00335
2.00	.02275	.01336	.01013	.00844	.00738	.00663	.00608	.00565	.00530	.00501	.00477	.00456
1.90	.02872	.01727	.01325	.01111	.00976	.00882	.00811	.00755	.00710	.00673	.00642	.00615
1.80	.03593	.02211	.01715	.01450	.01280	.01160	.01071	.01000	.00943	.00895	.00855	.00820
1.70	.04457	.02806	.02201	.01873	.01662	.01513	.01400	.01311	.01239	.01179	.01128	.01084
1.60	.05480	.03527	.02797	.02397	.02138	.01953	.01814	.01703	.01613	.01538	.01473	.01418
1.50	.06681	.04395	.03522	.03039	.02724	.02499	.02327	.02191	.02079	.01986	.01906	.01837
1.40	.08076	.05427	.04395	.03818	.03439	.03166	.02957	.02791	.02655	.02541	.02443	.02358
1.30	.09680	.06641	.05434	.04752	.04300	.03974	.03723	.03523	.03359	.03220	.03101	.02998
1.20	.11507	.08056	.06659	.05860	.05329	.04942	.04644	.04406	.04209	.04043	.03900	.03776
1.10	.13567	.09687	.08086	.07162	.06543	.06090	.05739	.05458	.05225	.05028	.04858	.04710
1.00	.15866	.11549	.09734	.08675	.07961	.07436	.07028	.06700	.06427	.06196	.05996	.05822
0.90	.18406	.13652	.11615	.10415	.09600	.08998	.08529	.08149	.07833	.07565	.07333	.07130
0.80	.21186	.16002	.13740	.12395	.11474	.10791	.10257	.09823	.09462	.09153	.08886	.08652
0.70	.24196	.18602	.16116	.14624	.13595	.12829	.12226	.11736	.11326	.10976	.10672	.10405
0.60	.27425	.21448	.18746	.17107	.15970	.15118	.14447	.13899	.13439	.13046	.12703	.12402
0.50	.30854	.24533	.21624	.19844	.18601	.17665	.16925	.16318	.15808	.15370	.14989	.14652
0.40	.34458	.27841	.24743	.22829	.21484	.20467	.19659	.18995	.18435	.17954	.17533	.17161
0.30	.38209	.31352	.28088	.26051	.24611	.23517	.22645	.21926	.21318	.20794	.20335	.19929
0.20	.42074	.35040	.31636	.29492	.27967	.26803	.25871	.25100	.24447	.23883	.23388	.22949
0.10	.46017	.38875	.35361	.33129	.31530	.30304	.29319	.28502	.27808	.27207	.26680	.26210
0.00	.50000	.42822	.39233	.36931	.35274	.33996	.32967	.32110	.31380	.30747	.30189	.29693

N / H	1	2	3	4	5	6	7	8	9	10	11	12
0.00	.50000	.42822	.39233	.36931	.35274	.33996	.32967	.32110	.31380	.30747	.30189	.29693
0.10	.53983	.46841	.43213	.40866	.39165	.37848	.36783	.35894	.35135	.34475	.33892	.33373
0.20	.57926	.50892	.47263	.44894	.43168	.41825	.40734	.39822	.39040	.38359	.37758	.37220
0.30	.61791	.54934	.51341	.48976	.47241	.45886	.44782	.43855	.43059	.42364	.41749	.41198
0.40	.65542	.58925	.55405	.53068	.51344	.49991	.48884	.47953	.47151	.46450	.45827	.45269
0.50	.69146	.62825	.59412	.57127	.55431	.54095	.52998	.52072	.51273	.50572	.49950	.49390
0.60	.72575	.66597	.63323	.61112	.59462	.58155	.57079	.56168	.55381	.54689	.54072	.53517
0.70	.75804	.70209	.67099	.64982	.63393	.62130	.61086	.60200	.59431	.58754	.58151	.57607
0.80	.78814	.73631	.70708	.68703	.67189	.65980	.64978	.64124	.63383	.62728	.62143	.61615
0.90	.81594	.76840	.74122	.72241	.70814	.69670	.68718	.67905	.67197	.66570	.66009	.65502
1.00	.84134	.79818	.77317	.75572	.74241	.73169	.72274	.71508	.70839	.70246	.69714	.69232
1.10	.86433	.82554	.80276	.78674	.77446	.76452	.75620	.74905	.74280	.73725	.73226	.72773
1.20	.88493	.85042	.82988	.81533	.80411	.79500	.78734	.78075	.77497	.76982	.76519	.76098
1.30	.90320	.87281	.85449	.84141	.83127	.82300	.81603	.81001	.80472	.80000	.79575	.79188
1.40	.91924	.89275	.87658	.86494	.85587	.84845	.84217	.83674	.83195	.82767	.82380	.82028
1.50	.93319	.91033	.89620	.88596	.87793	.87134	.86574	.86089	.85660	.85276	.84929	.84611
1.60	.94520	.92568	.91345	.90453	.89751	.89171	.88678	.88248	.87868	.87527	.87218	.86936
1.70	.95543	.93893	.92847	.92078	.91469	.90965	.90534	.90159	.89826	.89526	.89254	.89005
1.80	.96407	.95025	.94139	.93483	.92962	.92528	.92156	.91831	.91542	.91282	.91045	.90828
1.90	.97128	.95984	.95241	.94687	.94245	.93875	.93558	.93280	.93032	.92808	.92604	.92417
2.00	.97725	.96786	.96170	.95708	.95336	.95025	.94757	.94521	.94310	.94120	.93946	.93786
2.10	.98214	.97451	.96946	.96563	.96255	.95995	.95771	.95573	.95396	.95236	.95090	.94955
2.20	.98610	.97997	.97586	.97273	.97020	.96806	.96620	.96456	.96309	.96176	.96054	.95941
2.30	.98928	.98440	.98110	.97857	.97650	.97476	.97324	.97189	.97068	.96958	.96858	.96764
2.40	.99180	.98796	.98533	.98331	.98165	.98024	.97901	.97791	.97693	.97603	.97521	.97445
2.50	.99379	.99080	.98873	.98712	.98580	.98467	.98368	.98281	.98201	.98129	.98063	.98001
2.60	.99534	.99303	.99142	.99015	.98911	.98822	.98744	.98674	.98611	.98553	.98500	.98451
2.70	.99653	.99477	.99352	.99254	.99173	.99103	.99042	.98987	.98937	.98892	.98850	.98811
2.80	.99744	.99611	.99516	.99441	.99378	.99324	.99276	.99234	.99195	.99159	.99127	.99096
2.90	.99813	.99714	.99642	.99585	.99537	.99495	.99459	.99426	.99396	.99368	.99343	.99319
3.00	.99865	.99791	.99737	.99694	.99658	.99627	.99599	.99574	.99551	.99530	.99510	.99492
3.10	.99903	.99849	.99809	.99777	.99750	.99727	.99706	.99687	.99669	.99654	.99639	.99625
3.20	.99931	.99892	.99863	.99839	.99819	.99802	.99786	.99772	.99759	.99747	.99736	.99726
3.30	.99952	.99923	.99902	.99885	.99870	.99857	.99846	.99836	.99826	.99817	.99809	.99801
3.40	.99966	.99946	.99931	.99919	.99908	.99899	.99890	.99883	.99876	.99869	.99863	.99858
3.50	.99977	.99963	.99952	.99943	.99935	.99929	.99923	.99917	.99912	.99907	.99903	.99899

TABLE 10

The Distribution of the Maximum of N Equally Correlated Normal Standardised Random Variables: Tables of S. S. Gupta, K. Nagel, and S. Panchapakesan

(a) Content of the tables and definition of the tabulated statistic :

Let $X_1, X_2, \ldots, X_N$ be normal standardised random variables with equal correlations $\rho_{ij} = \rho$.

$X_{(1)} \leq X_{(2)} \leq \ldots \leq X_{(N)}$ denote the X_i arranged in increasing order.

Further let

$$F_N(H;\rho) \equiv pr(X_{(N)} \leq H) = \int_{-\infty}^{\infty} \Phi^N\{x\rho^{1/2} + H)/(1-\rho)^{1/2}\}\phi(x)dx .$$

$\Phi(x)$ and $\phi(x)$ respectively, denote the distribution and the density function of a normal standardised randomvariable.

The tables contain the upper percentage points for $X_{(N)}$, that is, they contain those values of H which satisfy the condition

$$F_N(H;\rho) = 1 - \alpha .$$

(b) Extent of the tables and definition of the parameters :

(1) The parameter α :

α = probability of exceeding the bound

α = 0.010 ; 0.025 ; 0.050 ; 0.100 ; 0.250 .

(2) The parameter ρ :

ρ = correlation coefficient

ρ = 0.100 ; 0.125 ; 0.200 ; 0.250 ; 0.300 ;
1/3 ; 0.375 ; 0.400 ; 1/2 ; 0.625 ;
2/3 ; 0.700 ; 0.750 ; 0.800 ; 0.875 ;
0.900 .

(3) The parameter N :

N = number of normal standardised random variables (variates)
= 1(1)10(2)50 .

(c) Suggestions for applications :

(1) To determine the percentage points for $X_{(N)}$ as discussed in section (a).

(2) Certain hypothesis tests (See: GUPTA, S. S. et al.(1973)).

(3) Certain multiple comparisons (See: GUPTA, S. S. et al. (1973)).

(d) References :

For the source of the tables and for applications :

GUPTA, S. S. + NAGEL, K. + PANCHAPAKESAN, S. : On the order statistics from equally correlated normal random variables. Biometrika 60 , 403 - 413 (1973).

(e) Further remarks :

The present table 10 complements table 9 , in which the probabilities , that $X_{(N)}$ does not exceed specified values H , are given .

Tafel 10 | Table 10

N \ α	$\rho = 0·100$: 0·010	0·025	0·050	0·100	0·250	$\rho = 0·125$: 0·010	0·025	0·050	0·100	0·250
1	2·3263	1·9600	1·6449	1·2816	0·6745	2·3263	1·9600	1·6449	1·2816	0·6745
2	2·5739	2·2368	1·9508	1·6258	1·0943	2·5736	2·2361	1·9497	1·6239	1·0905
3	2·7105	2·3878	2·1158	1·8089	1·3120	2·7099	2·3868	2·1141	1·8061	1·3064
4	2·8041	2·4907	2·2276	1·9320	1·4563	2·8034	2·4894	2·2255	1·9286	1·4496
5	2·8750	2·5683	2·3116	2·0240	1·5632	2·8742	2·5668	2·3092	2·0201	1·5555
6	2·9318	2·6303	2·3785	2·0970	1·6474	2·9310	2·6288	2·3759	2·0928	1·6391
7	2·9792	2·6819	2·4340	2·1574	1·7167	2·9783	2·6802	2·4312	2·1529	1·7078
8	3·0197	2·7259	2·4813	2·2087	1·7753	3·0188	2·7241	2·4784	2·2039	1·7659
9	3·0551	2·7642	2·5224	2·2533	1·8260	3·0541	2·7624	2·5194	2·2483	1·8161
10	3·0864	2·7982	2·5587	2·2926	1·8706	3·0855	2·7963	2·5556	2·2874	1·8603
12	3·1401	2·8561	2·6207	2·3594	1·9460	3·1390	2·8541	2·6173	2·3539	1·9352
14	3·1848	2·9044	2·6722	2·4148	2·0083	3·1837	2·9022	2·6686	2·4090	1·9969
16	3·2231	2·9456	2·7161	2·4620	2·0612	3·2219	2·9434	2·7124	2·4560	2·0494
18	3·2566	2·9816	2·7544	2·5031	2·1070	3·2554	2·9793	2·7506	2·4968	2·0948
20	3·2863	3·0135	2·7883	2·5393	2·1474	3·2850	3·0111	2·7844	2·5329	2·1349
22	3·3129	3·0421	2·8187	2·5718	2·1835	3·3117	3·0397	2·8146	2·5652	2·1706
24	3·3371	3·0680	2·8461	2·6012	2·2161	3·3358	3·0655	2·8420	2·5944	2·2029
26	3·3592	3·0917	2·8712	2·6279	2·2457	3·3579	3·0891	2·8670	2·6210	2·2322
28	3·3796	3·1135	2·8943	2·6525	2·2728	3·3782	3·1109	2·8900	2·6454	2·2591
30	3·3984	3·1337	2·9156	2·6752	2·2979	3·3970	3·1310	2·9112	2·6680	2·2840
32	3·4160	3·1524	2·9355	2·6963	2·3212	3·4146	3·1497	2·9310	2·6890	2·3070
34	3·4324	3·1700	2·9540	2·7160	2·3429	3·4310	3·1672	2·9495	2·7086	2·3285
36	3·4478	3·1864	2·9714	2·7345	2·3632	3·4464	3·1836	2·9668	2·7270	2·3486
38	3·4624	3·2019	2·9878	2·7519	2·3823	3·4609	3·1991	2·9831	2·7443	2·3675
40	3·4761	3·2166	3·0032	2·7683	2·4002	3.4746	3·2137	2·9985	2·7606	2·3854
42	3·4891	3·2304	3·0179	2·7838	2·4173	3·4876	3·2275	3·0131	2·7760	2·4022
44	3·5015	3·2436	3·0318	2·7986	2·4334	3·5000	3·2407	3·0269	2·7907	2·4182
46	3·5133	3·2562	3·0450	2·8126	2·4487	3·5117	3·2532	3·0401	2·8046	2·4334
48	3·5245	3·2682	3·0576	2·8260	2·4634	3·5230	3·2651	3·0527	2·8179	2·4479
50	3·5353	3·2796	3·0697	2·8388	2·4773	3·5337	3·2766	3·0647	2·8306	2·4617

N \ α	$\rho = 0·200$: 0·010	0·025	0·050	0·100	0·250	$\rho = 0·250$: 0·010	0·25	0·50	0·100	0·250
1	2·3263	1·9600	1·6449	1·2816	0·6745	2·3263	1·9600	1·6449	1·2816	0·6745
2	2·5722	2·2336	1·9456	1·6175	1·0784	2·5709	2·2314	1·9423	1·6125	1·0696
3	2·7078	2·3829	2·1080	1·7964	1·2882	2·7058	2·3795	2·1029	1·7888	1·2748
4	2·8008	2·4846	2·2180	1·9167	1·4273	2·7983	2·4804	2·2116	1·9072	1·4108
5	2·8712	2·5613	2·3006	2·0065	1·5303	2·8683	2·5564	2·2933	1·9957	1·5115
6	2·9277	2·6227	2·3664	2·0779	1·6114	2·9244	2·6172	2·3584	2·0659	1·5908
7	2·9747	2·6736	2·4210	2·1368	1·6781	2·9712	2·6677	2·4123	2·1240	1·6560
8	3·0149	2·7171	2·4675	2·1869	1·7346	3·0112	2·7109	2·4582	2·1733	1·7111
9	3·0501	2·7551	2·5079	2·2304	1·7834	3·0461	2·7484	2·4982	2·2160	1·7588
10	3·0812	2·7886	2·5437	2·2688	1·8262	3·0770	2·7816	2·5334	2·2538	1·8007
12	3·1344	2·8459	2·6046	2·3340	1·8989	3·1299	2·8383	2·5935	2·3179	1·8716
14	3·1788	2·8935	2·6551	2·3880	1·9588	3·1740	2·8855	2·6434	2·3710	1·9300
16	3·2168	2·9343	2·6983	2·4340	2·0096	3·2117	2·9258	2·6860	2·4162	1·9796
18	3·2501	2·9698	2·7359	2·4741	2·0536	3·2447	2·9610	2·7231	2·4555	2·0226
20	3·2795	3·0013	2·7691	2·5094	2·0924	3·2739	2·9921	2·7559	2·4902	2·0604
22	3·3060	3·0295	2·7989	2·5410	2·1271	3·3002	3·0200	2·7852	2·5212	2·0942
24	3·3299	3·0551	2·8259	2·5696	2·1583	3·3240	3·0453	2·8118	2·5493	2·1247
26	3·3519	3·0784	2·8505	2·5956	2·1867	3·3457	3·0684	2·8361	2·5748	2·1524
28	3·3720	3·0999	2·8731	2·6196	2·2128	3·3657	3·0896	2·8583	2·5983	2·1778
30	3·3907	3·1198	2·8940	2·6417	2·2369	3·3843	3·1093	2·8789	2·6200	2·2012
32	3·4081	3·1383	2·9135	2·6622	2·2592	3·4015	3·1275	2·8981	2·6401	2·2230
34	3·4244	3·1556	2·9316	2·6814	2·2800	3·4177	3·1446	2·9160	2·6589	2·2432
36	3·4397	3·1718	2·9487	2·6993	2·2994	3·4328	3·1606	2·9327	2·6765	2·2622
38	3·4541	3·1870	2·9647	2·7162	2·3177	3·4471	3·1757	2·9485	2·6931	2·2800
40	3·4677	3·2015	2·9798	2·7322	2·3350	3·4606	3·1900	2·9634	2·7087	2·2968
42	3·4806	3·2151	2·9942	2·7472	2·3513	3·4734	3·2035	2·9775	2·7235	2·3127
44	3·4929	3·2281	3·0078	2·7616	2·3667	3·4855	3·2163	2·9909	2·7375	2·3278
46	3·5045	3·2405	3·0207	2·7752	2·3814	3·4971	3·2285	3·0037	2·7509	2·3421
48	3·5157	3·2523	3·0331	2·7882	2·3954	3·5081	3·2401	3·0158	2·7636	2·3557
50	3·5263	3·2636	3·0449	2·8006	2·4088	3·5187	3·2513	3·0274	2·7758	2·3687

N \ α	$\rho = 0{\cdot}300$					$\rho = \frac{1}{3}$				
	0·010	0·025	0·050	0·100	0·250	0·010	0·025	0·050	0·100	0·250
1	2·3263	1·9600	1·6449	1·2816	0·6745	2·3263	1·9600	1·6449	1·2816	0·6745
2	2·5692	2·2287	1·9385	1·6069	1·0601	2·5679	2·2267	1·9356	1·6028	1·0533
3	2·7032	2·3753	2·0969	1·7801	1·2603	2·7011	2·3721	2·0924	1·7738	1·2500
4	2·7950	2·4752	2·2042	1·8965	1·3929	2·7924	2·4711	2·1985	1·8886	1·3802
5	2·8645	2·5504	2·2847	1·9834	1·4911	2·8614	2·5457	2·2782	1·9742	1·4765
6	2·9202	2·6105	2·3488	2·0523	1·5684	2·9167	2·6052	2·3415	2·0422	1·5524
7	2·9666	2·6604	2·4019	2·1092	1·6319	2·9628	2·6547	2·3940	2·0983	1·6147
8	3·0062	2·7030	2·4472	2·1576	1·6856	3·0022	2·6969	2·4387	2·1460	1·6674
9	3·0408	2·7401	2·4865	2·1995	1·7320	3·0365	2·7336	2·4776	2·1873	1·7129
10	3·0715	2·7729	2·5212	2·2365	1·7728	3·0669	2·7661	2·5119	2·2237	1·7529
12	3·1239	2·8289	2·5804	2·2993	1·8418	3·1189	2·8215	2·5703	2·2856	1·8205
14	3·1675	2·8755	2·6294	2·3513	1·8987	3·1622	2·8675	2·6187	2·3367	1·8763
16	3·2049	2·9152	2·6713	2·3956	1·9469	3·1993	2·9069	2·6600	2·3803	1·9236
18	3·2376	2·9499	2·7077	2·4341	1·9887	3·2316	2·9411	2·6959	2·4181	1·9645
20	3·2665	2·9806	2·7399	2·4680	2·0255	3·2603	2·9715	2·7277	2·4515	2·0006
22	3·2925	3·0081	2·7688	2·4984	2·0583	3·2860	2·9987	2·7561	2·4814	2·0327
24	3·3160	3·0330	2·7949	2·5258	2·0879	3·3094	3·0233	2·7819	2·5084	2·0617
26	3·3375	3·0558	2·8187	2·5508	2·1149	3·3307	3·0458	2·8053	2·5330	2·0881
28	3·3573	3·0767	2·8405	2·5737	2·1396	3·3503	3·0664	2·8269	2·5555	2·1123
30	3·3756	3·0960	2·8607	2·5949	2·1624	3·3684	3·0855	2·8468	2·5764	2·1346
32	3·3927	3·1140	2·8795	2·6146	2·1835	3·3853	3·1033	2·8653	2·5957	2·1553
34	3·4086	3·1308	2·8971	2·6330	2·2032	3·4011	3·1199	2·8826	2·6138	2·1746
36	3·4236	3·1466	2·9135	2·6502	2·2216	3·4159	3·1355	2·8988	2·6307	2·1926
38	3·4377	3·1614	2·9290	2·6664	2·2389	3·4299	3·1501	2·9140	2·6466	2·2095
40	3·4510	3·1755	2·9436	2·6817	2·2552	3·4430	3·1640	2·9284	2·6616	2·2255
42	3·4636	3·1887	2·9574	2·6961	2·2706	3·4555	3·1771	2·9420	2·6758	2·2406
44	3·4756	3·2013	2·9706	2·7098	2·2853	3·4674	3·1895	2·9550	2·6892	2·2549
46	3·4870	3·2134	2·9831	2·7229	2·2992	3·4787	3·2014	2·9673	2·7020	2·2685
48	3·4979	3·2248	2·9950	2·7353	2·3124	3·4895	3·2127	2·9790	2·7143	2·2815
50	3·5084	3·2358	3·0064	2·7472	2·3250	3·4998	3·2235	2·9902	2·7259	2·2939

N \ α	$\rho = 0{\cdot}375$					$\rho = 0{\cdot}400$				
	0·010	0·025	0·050	0·100	0·250	0·010	0·025	0·050	0·100	0·250
1	2·3263	1·9600	1·6449	1·2816	0·6745	2·3263	1·9600	1·6449	1·2816	0·6745
2	2·5660	2·2237	1·9316	1·5972	1·0444	2·5647	2·2218	1·9289	1·5936	1·0387
3	2·6981	2·3675	2·0861	1·7651	1·2363	2·6959	2·3644	2·0820	1·7595	1·2277
4	2·7885	2·4653	2·1906	1·8777	1·3633	2·7857	2·4613	2·1854	1·8707	1·3527
5	2·8568	2·5389	2·2690	1·9618	1·4572	2·8536	2·5343	2·2629	1·9537	1·4450
6	2·9116	2·5976	2·3313	2·0284	1·5312	2·9080	2·5925	2·3246	2·0194	1·5178
7	2·9572	2·6464	2·3830	2·0834	1·5919	2·9532	2·6408	2·3756	2·0737	1·5775
8	2·9961	2·6880	2·4269	2·1300	1·6432	2·9919	2·6820	2·4190	2·1197	1·6279
9	3·0301	2·7242	2·4651	2·1705	1·6875	3·0256	2·7178	2·4568	2·1596	1·6715
10	3·0601	2·7562	2·4987	2·2061	1·7265	3·0554	2·7495	2·4900	2·1948	1·7098
12	3·1115	2·8107	2·5561	2·2667	1·7923	3·1063	2·8035	2·5467	2·2544	1·7745
14	3·1543	2·8561	2·6036	2·3168	1·8466	3.1487	2·8483	2·5936	2·3038	1·8278
16	3·1909	2·8948	2·6441	2·3594	1·8926	3·1850	2·8866	2·6336	2·3458	1·8730
18	3·2228	2·9285	2·6794	2·3963	1·9324	3·2166	2·9199	2·6684	2·3822	1·9121
20	3·2511	2·9583	2·7105	2·4290	1·9675	3·2446	2·9494	2·6991	3·4144	1·9466
22	3·2764	2·9851	2·7384	2·4582	1·9988	3·2698	2·9759	2·7267	2·4431	1·9773
24	3·2994	3·0092	2·7636	2·4845	2·0270	3·2925	2·9998	2·7515	2·4691	2·0050
26	3·3204	3·0313	2·7866	2·5085	2·0526	3·3133	3·0216	2·7742	2·4927	2·0302
28	3·3398	3·0516	2·8077	2·5306	2·0761	3·3325	3·0416	2·7950	2·5144	2·0533
30	3·3577	3·0704	2·8272	2·5509	2·0978	3·3502	3·0602	2·8143	2·5344	2·0746
32	3·3743	3·0878	2·8453	2·5698	2·1179	3·3666	3·0774	2·8321	2·5530	2·0943
34	3·3898	3·1041	2·8623	2·5874	2·1366	3·3820	3·0935	2·8488	2·5704	2·1127
36	3·4044	3·1194	2·8781	2·6039	2·1542	3·3965	3·1086	2·8645	2·5866	2·1299
38	3·4182	3·1338	2·8931	2·6195	2·1706	3·4101	3·1228	2·8792	2·6019	2·1461
40	3·4312	3·1474	2·9071	2·6341	2·1862	3·4229	3·1362	2·8931	2·6163	2·1613
42	3·4435	3·1603	2·9205	2·6480	2·2008	3·4351	3·1489	2·9062	2·6299	2·1757
44	3·4552	3·1725	2·9331	2·6611	2·2147	3·4466	3·1610	2·9187	2·6429	2·1894
46	3·4663	3·1841	2·9452	2·6736	2·2280	3·4577	3·1725	2·9306	2·6552	2·2024
48	3·4769	3·1952	2·9566	2·6855	2·2406	3·4682	3·1834	2·9419	2·6669	2·2147
50	3·4871	3·2058	2·9676	2·6969	2·2526	3·4782	3·1939	2·9527	2·6781	2·2265

$N \backslash \alpha$	$\rho = \frac{1}{2}$ 0·010	0·025	0·050	0·100	0·250	$\rho = 0{\cdot}600$ 0·010	0·025	0·050	0·100	0·250
1	2·3263	1·9600	1·6449	1·2816	0·6745	2·3263	1·9600	1·6449	1·2816	0·6745
2	2·5578	2·2121	1·9163	1·5770	1·0139	2·5476	2·1987	1·8997	1·5560	0·9845
3	2·6849	2·3490	2·0621	1·7335	1·1895	2·6684	2·3276	2·0358	1·7008	1·1444
4	2·7716	2·4418	2·1603	1·8383	1·3055	2·7505	2·4146	2·1272	1·7974	1·2499
5	2·8370	2·5115	2·2338	1·9162	1·3911	2·8123	2·4798	2·1954	1·8691	1·3276
6	2·8893	2·5670	2·2922	1·9779	1·4585	2·8615	2·5317	2·2495	1·9258	1·3886
7	2·9327	2·6130	2·3404	2·0288	1·5137	2·9024	2·5745	2·2941	1·9725	1·4387
8	2·9698	2·6522	2·3814	2·0719	1·5604	2·9372	2·6110	2·3320	2·0120	1·4809
9	3·0021	2·6862	2·4170	2·1092	1·6007	2·9674	2·6426	2·3648	2·0462	1·5173
10	3·0306	2·7163	2·4484	2·1421	1·6360	2·9942	2·6705	2·3937	2·0762	1·5493
12	3·0793	2·7675	2·5017	2·1979	1·6958	3·0397	2·7180	2·4428	2·1272	1·6033
14	3·1198	2·8099	2·5458	2·2439	1·7450	3·0775	2·7573	2·4834	2·1692	1·6478
16	3·1544	2·8461	2·5834	2·2831	1·7867	3·1097	2·7907	2·5179	2·2049	1·6854
18	3·1845	2·8776	2·6161	2·3170	1·8227	3·1377	2·8198	2·5479	2·2359	1·7180
20	3·2112	2·9055	2·6449	2·3470	1·8545	3·1625	2·8455	2·5743	2·2632	1·7466
22	3·2351	2·9304	2·6707	2·3737	1·8828	3·1847	2·8685	2·5980	2·2875	1·7721
24	3·2567	2·9529	2·6940	2·3979	1·9083	3·2048	2·8893	2·6193	2·3095	1·7951
26	3·2765	2·9735	2·7152	2·4199	1·9315	3·2231	2·9082	2·6387	2·3295	1·8160
28	3·2946	2·9924	2·7347	2·4400	1·9527	3·2400	2·9256	2·6565	2·3479	1·8352
30	3·3114	3·0098	2·7527	2·4587	1·9723	3·2555	2·9416	2·6730	2·3648	1·8528
32	3·3270	3·0260	2·7694	2·4759	1·9905	3·2700	2·9565	2·6883	2·3805	1·8692
34	3·3416	3·0412	2·7850	2·4921	2·0074	3·2835	2·9704	2·7025	2·3951	1·8844
36	3·3553	3·0554	2·7996	2·5071	2·0233	3·2961	2·9835	2·7159	2·4088	1·8987
38	3·3682	3·0687	2·8134	2·5213	2·0381	3·3080	2·9957	2·7284	2·4217	1·9121
40	3·3803	3·0813	2·8264	2·5347	2·0522	3·3193	3·0073	2·7403	2·4338	1·9247
42	3·3919	3·0933	2·8386	2·5473	2·0654	3·3299	3·0182	2·7515	2·4453	1·9366
44	3·4028	3·1046	2·8503	2·5593	2·0780	3·3400	3·0286	2·7621	2·4562	1·9479
46	3·4132	3·1153	2·8613	2·5707	2·0899	3·3496	3·0385	2·7722	2·4665	1·9587
48	3·4231	3·1256	2·8719	2·5816	2·1012	3·3588	3·0479	2·7818	2·4764	1·9689
50	3·4326	3·1354	2·8820	2·5920	2·1121	3·3675	3·0569	2·7910	2·4858	1·9787

$N \backslash \alpha$	$\rho = 0{\cdot}625$ 0·010	0·025	0·050	0·100	0·250	$\rho = \frac{2}{3}$ 0·010	0·025	0·050	0·100	0·250
1	2·3263	1·9600	1·6449	1·2816	0·6745	2·3263	1·9600	1·6449	1·2816	0·6745
2	2·5444	2·1946	1·8947	1·5500	0·9763	2·5382	2·1869	1·8855	1·5389	0·9617
3	2·6632	2·3210	2·0279	1·6913	1·1318	2·6532	2·3088	2·0135	1·6741	1·1094
4	2·7438	2·4063	2·1173	1·7856	1·2343	2·7312	2·3909	2·0992	1·7641	1·2067
5	2·8044	2·4702	2·1840	1·8555	1·3098	2·7896	2·4522	2·1630	1·8308	1·2783
6	2·8528	2·5209	2·2368	1·9107	1·3692	2·8362	2·5009	2·2136	1·8835	1·3346
7	2·8928	2·5628	2·2804	1·9562	1·4178	2·8747	2·5411	2·2552	1·9268	1·3806
8	2·9269	2·5985	2·3173	1·9947	1·4588	2·9075	2·5752	2·2905	1·9634	1·4195
9	2·9565	2·6294	2·3494	2·0280	1·4941	2·9360	2·6049	2·3211	1·9951	1·4530
10	2·9827	2·6567	2·3776	2·0573	1·5252	2·9612	2·6310	2·3480	2·0230	1·4824
12	3·0273	2·7030	2·4254	2·1069	1·5776	3·0039	2·6753	2·3937	2·0702	1·5320
14	3·0642	2·7414	2·4650	2·1478	1·6208	3·0394	2·7120	2·4314	2·1091	1·5728
16	3·0958	2·7740	2·4986	2·1825	1·6573	3·0696	2·7432	2·4635	2·1421	1·6074
18	3·1232	2·8024	2·5278	2·2126	1·6889	3·0959	2·7703	2·4913	2·1707	1·6373
20	3·1474	2·8275	2·5536	2·2392	1·7166	3·1191	2·7942	2·5158	2·1959	1·6636
22	3·1691	2·8499	2·5766	2·2628	1·7414	3·1398	2·8156	2·5377	2·2183	1·6870
24	3·1887	2·8701	2·5974	2·2842	1·7637	3·1586	2·8349	2·5575	2·2386	1·7081
26	3·2066	2·8886	2·6163	2·3036	1·7840	3·1757	2·8525	2·5755	2·2571	1·7273
28	3·2231	2·9055	2·6337	2·3215	1·8026	3·1914	2·8686	2·5920	2·2740	1·7449
30	3·2383	2·9212	2·6497	2·3379	1·8197	3·2060	2·8836	2·6072	2·2896	1·7611
32	3·2524	2·9357	2·6645	2·3532	1·8356	3·2194	2·8974	2·6214	2·3041	1·7761
34	3·2656	2·9493	2·6784	2·3674	1·8504	3·2320	2·9103	2·6345	2·3176	1·7900
36	3·2779	2·9620	2·6914	2·3807	1·8642	3·2438	2·9224	2·6469	2·3302	1·8031
38	3·2895	2·9739	2·7036	2·3932	1·8772	3·2549	2·9338	2·6585	2·3421	1·8154
40	3·3005	2·9852	2·7151	2·4050	1·8894	3·2654	2·9445	2·6694	2·3532	1·8270
42	3·3109	2·9958	2·7260	2·4162	1·9010	3·2753	2·9547	2·6798	2·3638	1·8379
44	3·3207	3·0059	2·7364	2·4267	1·9120	3·2847	2·9643	2·6896	2·3738	1·8483
46	3·3301	3·0156	2·7462	2·4368	1·9224	3·2936	2·9734	2·6989	2·3834	1·8581
48	3·3390	3·0247	2·7555	2·4464	1·9323	3·3022	2·9822	2·7078	2·3925	1·8675
50	3·3476	3·0335	2·7645	2·4555	1·9417	3·3103	2·9905	2·7163	2·4011	1·8764

Tafel 10 (Forts.) Table 10 (cont.)

N \ α	$\rho = 0{\cdot}700$: 0·010	0·025	0·050	0·100	0·250	$\rho = 0{\cdot}750$: 0·010	0·025	0·050	0·100	0·250
1	2·3263	1·9600	1·6449	1·2816	0·6745	2·3263	1·9600	1·6449	1·2816	0·6745
2	2·5324	2·1799	1·8773	1·5291	0·9490	2·5221	2·1676	1·8631	1·5126	0·9281
3	2·6440	2·2976	2·0005	1·6589	1·0901	2·6274	2·2781	1·9783	1·6332	1·0581
4	2·7194	2·3768	2·0830	1·7452	1·1828	2·6985	2·3523	2·0551	1·7132	1·1435
5	2·7759	2·4359	2·1443	1·8091	1·2511	2·7515	2·4075	2·1122	1·7724	1·2063
6	2·8209	2·4827	2·1928	1·8595	1·3047	2·7937	2·4512	2·1572	1·8191	1·2555
7	2·8580	2·5214	2·2327	1·9009	1·3486	2·8285	2·4872	2·1943	1·8574	1·2958
8	2·8897	2·5542	2·2666	1·9360	1·3856	2·8581	2·5178	1·2257	1·8897	1·3298
9	2·9171	2·5827	2·2959	1·9662	1·4175	2·8837	2·5442	2·2529	1·9177	1·3591
10	2·9413	2·6077	2·3217	1·9929	1·4455	2·9063	2·5676	2·2768	1·9423	1·3848
12	2·9825	2·6503	2·3654	2·0379	1·4927	2·9447	2·6071	2·3173	1·9838	1·4281
14	3·0166	2·6855	2·4015	2·0751	1·5316	2·9765	2·6397	2·3507	2·0181	1·4638
16	3·0456	2·7154	2·4322	2·1066	1·5645	3·0035	2·6675	2·3790	2·0471	1·4939
18	3·0709	2·7414	2·4588	2·1339	1·5929	3·0270	2·6916	2·4036	2·0723	1·5200
20	3·0931	2·7643	2·4822	2·1579	1·6179	3·0477	2·7128	2·4253	2·0944	1·5429
22	3·1131	2·7848	2·5032	2·1794	1·6402	3·0662	2·7317	2·4446	2·1142	1·5633
24	3·1311	2·8032	2·5221	2·1987	1·6603	2·0830	2·7489	2·4620	2·1320	1·5817
26	3·1475	2·8201	2·5393	2·2163	1·67[illegible]5	3·0982	2·7644	2·4779	2·1482	1·5984
28	3·1626	2·8355	2·5550	2·2324	1·6952	3·1122	2·7787	2·4925	2·1630	1·6137
30	3·1765	2·8498	2·5696	2·2473	1·7106	3·1251	2·7919	2·5059	2·17[illegible]	1·6278
32	3·1894	2·8631	2·5831	2·2611	1·7249	3·1371	2·8042	2·5184	2·1894	1·6409
34	3·2015	2·8754	2·5957	2·2740	1·7382	3·1483	2·8156	2·5300	2·2012	1·6531
36	3·2128	2·8870	2·6075	2·2860	1·7506	3·1588	2·8263	2·5408	2·2123	1·6645
38	3·2234	2·8979	2·6186	2·2973	1·7623	3·1686	2·8363	2·5510	2·2227	1·6752
40	3·2335	2·9081	2·6290	2·3080	1·7733	3·1779	2·8458	2·5607	2·2325	1·6852
42	3·2430	2·9178	2·6389	2·3181	1·7837	3·1867	2·8548	2·5698	2·2417	1·6948
44	3·2520	2·9270	2·6483	2·3276	1·7935	3·1951	2·8633	2·5784	2·2505	1·7038
46	3·2605	2·9358	2·6572	2·3367	1·8029	3·2030	2·8714	2·5866	2·2589	1·7123
48	3·2687	2·9441	2·6657	2·3454	1·8118	3·2106	2·8791	2·5944	2·2668	1·7205
50	3·2765	2·9521	2·6738	2·3536	1·8203	3·2178	2·8864	2·6019	2·2744	1·7283

N \ α	$\rho = 0{\cdot}800$: 0·010	0·025	0·050	0·100	0·250	$\rho = 0{\cdot}875$: 0·010	0·025	0·050	0·100	0·250
1	2·3263	1·9600	1·6449	1·2816	0·6745	2·3263	1·9600	1·6449	1·2816	0·6745
2	2·5091	2·1524	1·8460	1·4931	0·9042	2·4818	2·1218	1·8123	1·4556	0·8601
3	2·6068	2·2543	1·9516	1·6031	1·0217	2·5638	2·2065	1·8993	1·5454	0·9544
4	2·6724	2·3224	2·0219	1·6759	1·0987	2·6185	2·2627	1·9569	1·6045	1·0162
5	2·7213	2·3730	2·0740	1·7296	1·1553	2·6591	2·3043	1·9994	1·6481	1·0615
6	2·7600	2·4130	2·1150	1·7719	1·1996	2.6911	2.3371	2·0329	1·6823	1·0970
7	2·7920	2·4460	2·1488	1·8066	1·2359	2·7175	2·3641	2·0603	1·7103	1·1260
8	2·8191	2·4739	2·1773	1·8359	1·2665	2·7398	2·3869	2·0835	1·7340	1·1504
9	2·8426	2·4980	2·2020	1·8613	1·2929	2·7591	2·4066	2·1036	1·7544	1·1714
10	2·8633	2·5192	2·2237	1·8835	1·3160	2·7761	2·4239	2·1212	1·7723	1·1899
12	2·8984	2·5552	2·2605	1·9211	1·3549	2·8049	2·4532	2·1509	1·8026	1·2210
14	2·9274	2·5849	2·2908	1·9520	1·3870	2·8286	2·4773	2·1754	1·8274	1·2465
16	2·9521	2·6102	2·3164	1·9782	1·4141	2·8487	2·4978	2·1961	1·8485	1·2681
18	2·9735	2·6320	2·3387	2·0009	1·4375	2·8662	2·5155	2·2141	1·8667	1·2867
20	2·9924	2·6513	2·3583	2·0209	1·4581	2·8815	2·5311	2·2299	1·8827	1·3031
22	3·0093	2·6685	2·3758	2·0387	1·4764	2·8952	2·5450	2·2440	1·8970	1·3177
24	3·0245	2·6840	2·3916	2·0548	1·4929	2·9076	2·5576	2·2566	1·9099	1·3309
26	3·0383	2·6982	2·4059	2·0694	1·5079	2·9189	2·5690	2·2682	1·9216	1·3428
28	3·0511	2·7111	2·4191	2·0828	1·5217	2·9292	2·5794	2·2788	1·9323	1·3537
30	3·0628	2·7231	2·4312	2·0951	1·5343	2·9387	2·5891	2·2885	1·9422	1·3638
32	3·0737	2·7342	2·4425	2·1066	1·5461	2·9476	2·5980	2·2976	1·9513	1·3731
34	3·0839	2·7445	2·4530	2·1172	1·5570	2·9558	2·6064	2·3060	1·9599	1·3818
36	3·0934	2·7542	2·4628	2·1272	1·5672	2·9635	2·6142	2·3139	1·9678	1·3899
38	3·1023	2·7633	2·4720	2·1366	1·5768	2·9708	2·6215	2·3213	1·9753	1·3976
40	3·1108	2·7719	2·4807	2·1454	1·5859	2·9776	2·6284	2·3283	1·9824	1·4048
42	3·1188	2·7800	2·4889	2·1537	1·5944	2·9840	2·6350	2·3349	1·9891	1·4115
44	3·1263	2·7877	2·4967	2·1616	1·6025	2·9902	2·6412	2·3411	1·9954	1·4180
46	3·1335	2·7950	2·5041	2·1691	1·6102	2·9960	2·6470	2·3471	2·0014	1·4241
48	3·1404	2·8019	2·5112	2·1763	1·6175	3·0015	2·6526	2·3527	2·0071	1·42[illegible]
50	3·1469	2·8086	2·5179	2·1831	1·6245	3·0068	2·6580	2·3581	2·0126	1·4[illegible]

Tafel 10 (Forts.) Table 10 (cont.)

$\rho = 0{\cdot}900$

N \ α	0·010	0·025	0·050	0·100	0·250
1	2·3263	1·9600	1·6449	1·2816	0·6745
2	2·4694	2·1081	1·7976	1·4397	0·8418
3	2·5444	2·1853	1·8767	1·5209	0·9268
4	2·5943	2·2365	1·9289	1·5744	0·9823
5	2·6313	2·2743	1·9674	1·6137	1·0230
6	2·6604	2·3040	1·9977	1·6446	1·0549
7	2·6844	2·3285	2·0225	1·6699	1·0810
8	2·7046	2·3491	2·0434	1·6912	1·1029
9	2·7222	2·3669	2·0615	1·7096	1·1218
10	2·7375	2·3826	2·0774	1·7257	1·1383
12	2·7636	2·4090	2·1042	1·7529	1·1662
14	2·7851	2·4308	2·1263	1·7753	1·1891
16	2·8033	2·4493	2·1450	1·7943	1·2085
18	2·8191	2·4653	2·1612	1·8106	1·2252
20	2·8330	2·4794	2·1754	1·8250	1·2399
22	2·8454	2·4919	2·1881	1·8379	1·2530
24	2·8565	2·5033	2·1995	1·8495	1·2648
26	2·8667	2·5136	2·2099	1·8600	1·2755
28	2·8760	2·5230	2·2194	1·8696	1·2853
30	2·8846	2·5317	2·2282	1·8785	1·2943
32	2·8926	2·5398	2·2364	1·8867	1·3027
34	2·9001	2·5473	2·2440	1·8944	1·3105
36	2·9070	2·5543	2·2510	1·9015	1·3177
38	2·9135	2·5609	2·2577	1·9083	1·3246
40	2·9197	2·5671	2·2640	1·9146	1·3310
42	2·9255	2·5730	2·2699	1·9206	1·3371
44	2·9310	2·5786	2·2755	1·9263	1·3429
46	2·9363	2·5839	2·2809	1·9316	1·3483
48	2·9413	2·5889	2·2860	1·9368	1·3535
50	2·9461	2·5937	2·2908	1·9417	1·3585

TABLE 11

The Sphericity Test of J. W. Mauchly: Tables of B. N. Nagarsenker and K. C. S. Pillai

(a) Content of the tables and definition of the test statistic :

The tables contain the lower percentage points for MAUCHLY's criterion for testing for "sphericity":

A p-dimensional normal distribution is called spherically symmetric if for all p single components the variances are equal and all covariances are zero .

This sphericity , i. e. the null hypothesis

$$H_o : \underline{\Sigma} = \sigma^2 . \underline{I}_p \quad , \text{ where } \quad \sigma^2 > 0 \ ,$$

can be tested on the basis of a sample against the alternative

$$H_1 : \underline{\Sigma} \neq \sigma^2 . \underline{I}_p \quad ,$$

where $\underline{I}_p$ is the identity matrix.

Following J. W. MAUCHLY (1940) , one does this by means of the likelihood criterion

$$W = | \underline{S} | \, / \, (\text{Trace } \underline{S}/p)^p \quad ,$$

where $\underline{S}$ is the SP-matrix (= matrix of sums of the products of the deviations) for a sample of the size N .

(b) Extent of the tables and definition of the parameters :

(1) The parameter α :

α = error probability

α = 0.005 ; 0.01 ; 0.025 ; 0.05
0.1 ; 0.25 .

(2) The parameter p :

p = dimension of the variate

p = 2(1)10 .

(3) The parameter N :

N = size of the p-dimensional sample

N = various values up to 300 .

(c) Suggestions for applications :

As noted in section (a) , the tables give the lower percentage points for the sphericity test criterion W .

(d) References :

(1) For the test criterion :

MAUCHLY, J. W. : Significance test for sphericity of a normal n-variate distribution.
The Annals of Mathematical Statistics 11 , 204 - 209 (1940) .

NAGARSENKER, B. N. + PILLAI, K. C. S. : The distribution of the sphericity test criterion.
Journal of Multivariate Analysis 3, 226 - 235 (1973).

(2) For the source of the tables :

NAGARSENKER, B. N. + PILLAI, K. C. S. : The distribution of the sphericity test criterion.
Mimeograph Series No. 284 , Department of Statistics, Purdue University, 1972 .

(3) For the approximation of section (e) :

DAVIS, A. W. : Percentile approximations for a class of likelihood ratio criteria.
Biometrika 58 , 349 - 356 (1971) .

WILKS, S. S. : The large sample distribution of the likelihood ratio for testing composite hypotheses.
The Annals of Mathematical Statistics 9 , 60 - 62 (1938) .

(e) Further remarks :

According to A. W. DAVIS (1971) , for large N one can dispense with the sphericity hypothesis with an (approximative) error probabiliy α , whenever the χ^2-distribution yields

$$T = -\left(N - \frac{2p^2 + p + 2}{6p} \right) . \ln W > \chi_f^2 (\alpha)$$

with f degrees of freedom , where

$$f = \frac{p(p + 1)}{2} - 1 \quad .$$

Tafel 11

Table 11

p = 2

N \ α	.005	.01	.025	.05	.1	.25
3	$.0^{4}25000$	$.0^{3}10000$	$.0^{3}62500$	$.0^{2}25000$	.010000	.062500
4	$.0^{2}50000$	.010000	.025000	.050000	.10000	.25000
5	.029240	.046416	.08550	.13572	.21544	.39685
6	.070711	.10000	.15811	.22361	.31623	.50000
7	.12011	.15849	.22865	.30171	.39811	.57435
8	.17100	.21544	.29240	.36840	.46416	.62996
9	.22007	.26827	.34855	.42489	.51795	.67295
10	.26591	.31623	.39764	.47287	.56234	.70711
11	.30808	.35938	.44054	.51390	.59948	.73487
12	.34657	.39811	.47818	.54928	.63096	.75786
13	.38162	.43288	.51135	.58003	.65793	.77720
14	.41352	.46416	.54074	.60696	.68129	.79370
15	.44258	.49239	.56693	.63073	.70170	.80793
16	.46912	.51795	.59038	.65184	.71969	.82034
17	.49340	.54117	.61149	.67070	.73564	.83124
18	.51567	.56234	.63058	.68766	.74989	.84090
19	.53616	.58171	.64792	.70297	.76270	.84951
20	.55505	.59945	.66373	.71687	.77426	.85724
22	.58870	.63096	.69150	.74113	.79433	.87055
24	.61775	.65793	.71509	.76160	.81113	.88159
26	.64305	.68129	.73535	.77908	.82540	.89090
28	.66527	.70170	.75295	.79418	.83768	.89885
30	.68492	.71969	.76836	.80736	.84834	.90572
34	.71810	.74989	.79409	.82925	.86596	.91700
38	.74501	.77426	.81470	.84668	.87992	.92587
42	.76727	.79433	.83157	.86089	.89125	.93303
46	.78597	.81113	.84563	.87269	.90063	.93893
50	.80191	.82540	.85753	.88265	.90852	.94387
60	.83302	.85317	.88056	.90186	.92367	.95332
70	.85570	.87333	.89718	.91566	.93452	.96005
80	.87297	.88862	.90975	.92606	.94267	.96508
90	.88655	.90063	.91958	.93418	.94901	.96898
100	.89751	.91030	.92748	.94069	.95410	.97210
120	.91411	.92491	.93939	.95049	.96172	.97678
140	.92609	.93544	.94794	.95751	.96718	.98011
160	.93513	.94337	.95438	.96279	.97127	.98261
180	.94221	.94957	.95940	.96690	.97446	.98454
200	.94789	.95455	.96342	.97019	.97701	.98609
250	.95817	.96354	.97069	.97613	.98160	.98888
300	.96507	.96957	.97555	.98010	.98467	.99074

Tafel 11 (Forts.) Table 11 (cont.)

p = 3

N \ α	.005	.01	.025	.05	.1	.25
4	$.0^{5}39305$	$.0^{4}15228$	$.0^{4}99478$	$.0^{3}40104$	$.0^{2}16700$	.011603
5	$.0^{2}11700$	$.0^{2}23667$	$.0^{2}61070$	.012679	.026853	.076732
6	$.0^{2}88748$	.014398	.027585	.045683	.076928	.16044
7	.025882	.037466	.061687	.090921	.13590	.24004
8	.050467	.068151	.10225	.14026	.19471	.31002
9	.079827	.10285	.14486	.18921	.24970	.37019
10	.11161	.13898	.18696	.23564	.29971	.42176
11	.14418	.17494	.22726	.27876	.34471	.46613
12	.17647	.20981	.26516	.31836	.38503	.50453
13	.20786	.24391	.30048	.35457	.42118	.53800
14	.23799	.27457	.33321	.38762	.45365	.56738
15	.26666	.30417	.36350	.41779	.48290	.59335
16	.29383	.33192	.39149	.44538	.50934	.61645
17	.31948	.35789	.41737	.47065	.53332	.63712
18	.34366	.38219	.44133	.49386	.55516	.65571
19	.36644	.40492	.46355	.51522	.57511	.67251
20	.38789	.42619	.48417	.53493	.59340	.68778
22	.42713	.46482	.52124	.57006	.62573	.71444
24	.46203	.49889	.55354	.60040	.65338	.73694
26	.49319	.52908	.58190	.62684	.67729	.75618
28	.52111	.55598	.60696	.65006	.69816	.77281
30	.54624	.58007	.62926	.67060	.71651	.78732
34	.58958	.62136	.66715	.70529	.74730	.81144
38	.62556	.65540	.69811	.73343	.77210	.83066
42	.65584	.68391	.72386	.75670	.79248	.84634
46	.68166	.70811	.74559	.77626	.80953	.85736
50	.70393	.72891	.76417	.79293	.82400	.87035
60	.74809	.76997	.80064	.82546	.85211	.89155
70	.78086	.80028	.82737	.84918	.87249	.90679
80	.80612	.82356	.84779	.86723	.88794	.91828
90	.82617	.84199	.86390	.88143	.90006	.92725
100	.84247	.85694	.87693	.89289	.90981	.93444
120	.86737	.87972	.89672	.91024	.92454	.94527
140	.88548	.89624	.91103	.92276	.93513	.95303
160	.89925	.90878	.92186	.93221	.94312	.95886
180	.91006	.91861	.93034	.93961	.94936	.96340
200	.91877	.92654	.93716	.94555	.95436	.96704
250	.93462	.94092	.94952	.95629	.96340	.97361
300	.94529	.95059	.95781	.96350	.96945	.97799

Tafel 11 (Forts.) Table 11 (cont.)

p = 4

N \ α	.005	.01	.025	.05	.1	.25
5	$.0^{6}91162$	$.0^{5}36645$	$.0^{4}23265$	$.0^{4}95283$	$.0^{3}40030$	$.0^{2}29305$
6	$.0^{3}33678$	$.0^{3}69040$	$.0^{2}18194$	$.0^{2}38662$	$.0^{2}84730$	.026147
7	$.0^{2}30556$	$.0^{2}50312$	$.0^{2}99040$	.016868	.029512	.066529
8	.010209	.015033	.025485	.038664	.060019	.11410
9	.022162	.030463	.047058	.066398	.095554	.16287
10	.038208	.050095	.072584	.097393	.1396	.20994
11	.057311	.072583	.10033	.12972	.17030	.25404
12	.078477	.096785	.12902	.16211	.20651	.29477
13	.10089	.12183	.15780	.19381	.24102	.33213
14	.12391	.14708	.18610	.22435	.27358	.36631
15	.14708	.17211	.21356	.25352	.30412	.39756
16	.17006	.19663	.23999	.28119	.33269	.42615
17	.19263	.22044	.26528	.30736	.35936	.45236
18	.21462	.24343	.28938	.33205	.38425	.47643
19	.23595	.26553	.31230	.35332	.40749	.49860
20	.25655	.28673	.33406	.37723	.42920	.51905
22	.29546	.32641	.37429	.41734	.46850	.55550
24	.33132	.36261	.41046	.45301	.50304	.58698
26	.36428	.39559	.44305	.48484	.53356	.61440
28	.39455	.42567	.47247	.51337	.56068	.63847
30	.42235	.45313	.49912	.53903	.58492	.65977
34	.47149	.50132	.54542	.58326	.62635	.69571
38	.51337	.54207	.58415	.61995	.66039	.72486
42	.54938	.57689	.61695	.65082	.68883	.74894
46	.58059	.60692	.64506	.67712	.71293	.76918
50	.60788	.63307	.66939	.69978	.73359	.78641
60	.66298	.68558	.71790	.74471	.77429	.82004
70	.70468	.72509	.75409	.77801	.80425	.84454
80	.73729	.75584	.78211	.80366	.82721	.86318
90	.76346	.78045	.80441	.82401	.84536	.87784
100	.78491	.80057	.82259	.84055	.86007	.88966
120	.81798	.83149	.85042	.86580	.88244	.90756
140	.84225	.85413	.87072	.88415	.89865	.92046
160	.86083	.87141	.88617	.89809	.91094	.93021
180	.87550	.88504	.89832	.90904	.92057	.93782
200	.88737	.89606	.90814	.91787	.92832	.94394
250	.90906	.91615	.92599	.93390	.94238	.95501
300	.92375	.92974	.93804	.94470	.95183	.96243

Tafel 11 (Forts.) Table 11 (cont.)

p = 5

N \ α	.005	.01	.025	.05	.1	.25
6	$.0^{6}24579$	$.0^{6}98368$	$.0^{5}72524$	$.0^{4}25776$	$.0^{3}10959$	$.0^{3}83762$
7	$.0^{3}10563$	$.0^{3}21839$	$.0^{3}58374$	$.0^{2}12621$	$.0^{2}28373$	$.0^{2}92522$
8	$.0^{2}10968$	$.0^{2}18281$	$.0^{2}36768$	$.0^{2}64001$	.011530	.027554
9	$.0^{2}40994$	$.0^{2}61227$	.010628	.016501	.026388	.053105
10	$.0^{2}97579$	.013613	.021543	.031104	.046080	.082916
11	.018156	.024161	.035852	.049192	.069047	.11473
12	.0290262	.037303	.052770	.069704	.093963	.14705
13	.041953	.052479	.071536	.091741	.11983	.17893
14	.056485	.069151	.091503	.11460	.14594	.20983
15	.072206	.086848	.11215	.13775	.17180	.23944
16	.088751	.10518	.13309	.16082	.19710	.26760
17	.10582	.12385	.15402	.18354	.22163	.29428
18	.12317	.14261	.17473	.20575	.24527	.31947
19	.14061	.16129	.19507	.22731	.26797	.34324
20	.15799	.17974	.21492	.24817	.28969	.36563
22	.19215	.21560	.25292	.28761	.33025	.40663
24	.22503	.24971	.28847	.32400	.36713	.44311
26	.25634	.28186	.32151	.35746	.40063	.47566
28	.28593	.31200	.35214	.38818	.43110	.50482
30	.31379	.34018	.38049	.41641	.45885	.53106
34	.36449	.39103	.43108	.46628	.50740	.57625
38	.40909	.43536	.47461	.50878	.54831	.61370
42	.44838	.47413	.51231	.54529	.58315	.64520
46	.48312	.50821	.54519	.57692	.61314	.67203
50	.51397	.53834	.57407	.60456	.63919	.69513
60	.57761	.60010	.63275	.66033	.69137	.74089
70	.62690	.64760	.67745	.70250	.73049	.77478
80	.66607	.68517	.71257	.73544	.76088	.80086
90	.69790	.71558	.74086	.76186	.78514	.82155
100	.72425	.74069	.76411	.78351	.80495	.83836
120	.76529	.77966	.80005	.81686	.83535	.86399
140	.79575	.80850	.82652	.84133	.85757	.88262
160	.81924	.83068	.84682	.86004	.87451	.89676
180	.83790	.84827	.86287	.87481	.88786	.90787
200	.85307	.86255	.87588	.88677	.89864	.91682
250	.88095	.8875	.89969	.90860	.91828	.93307
300	.89995	.90656	.91584	.92337	.93155	.94401

p = 6

N \ α	.005	.01	.025	.05	.1	.25
7	$.0^{7}70557$	$.0^{6}29697$	$.0^{5}18030$	$.0^{5}74790$	$.0^{4}31547$	$.0^{3}24844$
8	$.0^{4}34541$	$.0^{4}71870$	$.0^{3}19456$	$.0^{3}42669$	$.0^{3}97879$	$.0^{2}33335$
9	$.0^{3}40126$	$.0^{3}67578$	$.0^{2}13837$	$.0^{2}25527$	$.0^{2}45255$	.011336
10	$.0^{2}16522$	$.0^{2}24979$	$.0^{2}44243$	$.0^{2}70038$	.011482	.024216
11	$.0^{2}42686$	$.0^{2}60326$	$.0^{2}97479$	.014353	.021791	.041033
12	$.0^{2}85127$	.011478	.017390	.024325	.034966	.060679
13	.014444	.018800	.027141	.036529	.050383	.082172
14	.021960	.027821	.038682	.050510	.067439	.10473
15	.030903	.038302	.051661	.065830	.085610	.12776
16	.041061	.049980	.065743	.082100	.104468	.15085
17	.052219	.062606	.080634	.098998	.12368	.17368
18	.064174	.075950	.096078	.11626	.14298	.19605
19	.076743	.089816	.11187	.13367	.16217	.21782
20	.089763	.10403	.12783	.15107	.18112	.23890
22	.11662	.13299	.15975	.18538	.21788	.27885
24	.14388	.16196	.19107	.21850	.25277	.31577
26	.17096	.19041	.22132	.25008	.28556	.34973
28	.19745	.21798	.25027	.27997	.31624	.38093
30	.22313	.24449	.27778	.30812	.34485	.40960
34	.27153	.29397	.32844	.35939	.39633	.46025
38	.31571	.33866	.37355	.40450	.44105	.50339
42	.35576	.37885	.41364	.44424	.48005	.54044
46	.39199	.41497	.44935	.47936	.51425	.57254
50	.42477	.44748	.48125	.51055	.54441	.60056
60	.49406	.51571	.54754	.57485	.60606	.65708
70	.54917	.56955	.59930	.62460	.65331	.69978
80	.59381	.61292	.64067	.66413	.69059	.73311
90	.63060	.64852	.67442	.69622	.72071	.75983
100	.66139	.67822	.70246	.72278	.74553	.78171
120	.70994	.72488	.74628	.76412	.78400	.81540
140	.74642	.75981	.77891	.79479	.81241	.84011
160	.77480	.78691	.80415	.81843	.83423	.85899
180	.79750	.80854	.82423	.83720	.85152	.87389
200	.81605	.82620	.84058	.85246	.86555	.88595
250	.85037	.85879	.87069	.88048	.89124	.90796
300	.87391	.88110	.89124	.89957	.90870	.92285

p = 7

N \ α	.005	.01	.025	.05	.1	.25
8	$.0^{7}21289$	$.0^{7}86044$	$.0^{6}55120$	$.0^{5}22835$	$.0^{5}95942$	$.0^{4}72580$
9	$.0^{4}115809$	$.0^{4}24239$	$.0^{4}66388$	$.0^{3}14730$	$.0^{3}34311$	$.0^{2}12149$
10	$.0^{3}14825$	$.0^{3}25197$	$.0^{3}52402$	$.0^{3}94336$	$.0^{2}17761$	$.0^{2}46290$
11	$.0^{3}66517$	$.0^{2}10165$	$.0^{2}18324$	$.0^{2}29501$	$.0^{2}49404$	.010845
12	$.0^{2}18510$	$.0^{2}26462$	$.0^{2}43552$	$.0^{2}65237$	.010119	.019815
13	$.0^{2}39356$	$.0^{2}53692$	$.0^{2}82864$	.011790	.017307	.031195
14	$.0^{2}70567$	$.0^{2}92955$	.013667	.018704	.026327	.044531
15	.011263	.014435	.020431	.027115	.036919	.059370
16	.016537	.020729	.028448	.036821	.048798	.075298
17	.022812	.028074	.037553	.047610	.061690	.091967
18	.029994	.036345	.047578	.059270	.075347	.10909
19	.037977	.045412	.058355	.071609	.089554	.12644
20	.046648	.055143	.069730	.084457	.10413	.14384
22	.065631	.076124	.093740	.11111	.13380	.17825
24	.086164	.098448	.11870	.13831	.16346	.21158
26	.10761	.12146	.14396	.16541	.19254	.24343
28	.12949	.14467	.16905	.19200	.22067	.27360
30	.15142	.16774	.19367	.21781	.24767	.30205
34	.19449	.21253	.24073	.26653	.29794	.35389
38	.23555	.25472	.28433	.31106	.34320	.39952
42	.27402	.29390	.32429	.35146	.38380	.43972
46	.30974	.33002	.36076	.38801	.42019	.47524
50	.34277	.36321	.39399	.42108	.45287	.50678
60	.41458	.43479	.46486	.49099	.52126	.57177
70	.47344	.49296	.52177	.54656	.57505	.62203
80	.52215	.54082	.56817	.59156	.61826	.66192
90	.56296	.58071	.60661	.62863	.65366	.69431
100	.59755	.61441	.63890	.65965	.68314	.72109
120	.65283	.66805	.69003	.70854	.72937	.76277
140	.69496	.70876	.72861	.74526	.76391	.79366
160	.72806	.74066	.75872	.77381	.79067	.81747
180	.75474	.76630	.78284	.79664	.81201	.83636
200	.77668	.78735	.80260	.81529	.82941	.85172
250	.81754	.82649	.83923	.84979	.86149	.87991
300	.84580	.85349	.86441	.87344	.88343	.89910

Tafel 11 (Forts.) Table 11 (cont.)

p = 8

N \ α	.005	.01	.025	.05	.1	.25
9	$.0^{8}71788$	$.0^{7}27598$	$.0^{6}17155$	$.0^{6}72189$	$.0^{5}31412$	$.0^{4}20202$
10	$.0^{5}39473$	$.0^{5}83064$	$.0^{4}22961$	$.0^{4}51489$	$.0^{3}12180$	$.0^{3}44134$
11	$.0^{4}55082$	$.0^{4}94377$	$.0^{3}19900$	$.0^{3}36314$	$.0^{3}69598$	$.0^{2}18771$
12	$.0^{3}26703$	$.0^{3}41204$	$.0^{3}75447$	$.0^{2}12329$	$.0^{2}21036$	$.0^{2}47822$
13	$.0^{3}79550$	$.0^{2}11491$	$.0^{2}19226$	$.0^{2}29243$	$.0^{2}46224$	$.0^{2}93706$
14	$.0^{2}17954$	$.0^{2}24756$	$.0^{2}38847$	$.0^{2}56126$	$.0^{2}83944$	.015649
15	$.0^{2}33920$	$.0^{2}45162$	$.0^{2}67510$	$.0^{2}93791$	.013445	.023497
16	$.0^{2}56696$	$.0^{2}73433$	.010564	.014227	.019719	.032724
17	$.0^{2}86711$	.010982	.015313	.020106	.027108	.043115
18	.012404	.015421	.020950	.026931	.035479	.054455
19	.016850	.020620	.027401	.034597	.044691	.066544
20	.021969	.026523	.034584	.042993	.054605	.079202
22	.034018	.040171	.050778	.061544	.076025	.10563
24	.048080	.055801	.068835	.081781	.098838	.13274
26	.063675	.072873	.088141	.10304	.12235	.15986
28	.080369	.090921	.10820	.12482	.14605	.18654
30	.097792	.10956	.12861	.14671	.16957	.21248
34	.13368	.14749	.16941	.18984	.21517	.26153
38	.16963	.18497	.20900	.23105	.25801	.30642
42	.20463	.22109	.24659	.26972	.29769	.34717
46	.23812	.25537	.28186	.30567	.33419	.38404
50	.26985	.28764	.31474	.33892	.36766	.41741
60	.34107	.35942	.38699	.41123	.43966	.48794
70	.40149	.41973	.44689	.47052	.49797	.54398
80	.45272	.47053	.49686	.51959	.54581	.58935
90	.49642	.51364	.53895	.56068	.58561	.62671
100	.53399	.55054	.57479	.59551	.61917	.65795
120	.59497	.61020	.63235	.65116	.67250	.70718
140	.64217	.65616	.67642	.69354	.71289	.74412
160	.67967	.79256	.71117	.72684	.74448	.77285
180	.71015	.72207	.73925	.75367	.76986	.79580
200	.73538	.74646	.76238	.77572	.79067	.81456
250	.78277	.79216	.80559	.81680	.82932	.84922
300	.81582	.82394	.83554	.84520	.85595	.87299

p = 9

N \ α	.005	.01	.025	.05	.1	.25
10	$.0^{8}25350$	$.0^{8}92163$	$.0^{7}56095$	$.0^{6}23259$	$.0^{5}10164$	$.0^{5}61014$
11	$.0^{5}13612$	$.0^{5}28789$	$.0^{5}80284$	$.0^{4}18169$	$.0^{4}42570$	$.0^{3}16718$
12	$.0^{4}20532$	$.0^{4}35438$	$.0^{4}75647$	$.0^{3}13971$	$.0^{3}27192$	$.0^{3}75487$
13	$.0^{3}10683$	$.0^{3}16629$	$.0^{3}30880$	$.0^{3}51137$	$.0^{3}88727$	$.0^{2}20826$
14	$.0^{3}33897$	$.0^{3}49428$	$.0^{3}83940$	$.0^{2}12945$	$.0^{2}20813$	$.0^{2}43517$
15	$.0^{3}80903$	$.0^{2}11264$	$.0^{2}17946$	$.0^{2}26291$	$.0^{2}399930$	$.0^{2}76857$
16	$.0^{2}16061$	$.0^{2}21595$	$.0^{2}32774$	$.0^{2}46163$	$.0^{2}67287$	.012112
17	$.0^{2}28055$	$.0^{2}36693$	$.0^{2}53582$	$.0^{2}73143$	.010304	.017596
18	$.0^{2}44630$	$.0^{2}57070$	$.0^{2}80753$	.010744	.014715	.024063
19	$.0^{2}66139$	$.0^{2}8300$	.011439	.014894	.019925	.031413
20	$.0^{2}92748$	.011455	.015437	.019734	.025874	.039535
22	.016116	.019398	.025210	.031285	.039702	.057658
24	.024859	.029330	.037060	.044940	.055599	.077608
26	.035265	.040948	.050587	.060218	.072995	.098694
28	.047052	.053924	.065397	.076672	.091395	.12038
30	.059939	.067945	.081137	.093923	.11040	.14225
34	.088001	.098046	.11426	.12963	.14902	.18545
38	.11775	.12949	.14914	.16553	.18713	.22683
42	.14799	.16109	.18165	.20057	.22377	.26572
46	.17794	.19211	.21411	.23416	.25850	.30189
50	.20710	.22208	.24517	.26601	.29111	.33536
60	.27498	.29117	.31575	.33759	.36349	.40820
70	.33480	.35143	.37641	.39836	.42410	.46791
80	.38693	.40335	.42832	.44991	.47503	.51732
90	.43229	.44866	.47289	.49388	.51815	.55869
100	.47190	.48787	.51140	.53167	.55501	.59374
120	.53733	.55232	.57426	.59302	.61448	.64973
140	.58885	.60282	.62318	.64050	.66020	.69235
160	.63029	.64331	.66220	.67821	.69636	.72583
180	.66429	.67643	.69401	.70886	.72564	.75279
200	.69264	.70400	.72040	.73422	.74981	.77495
250	.74638	.75611	.77011	.78186	.79505	.81621
300	.78422	.79271	.80488	.81507	.82648	.84472

Tafel 11 (Forts.) Table 11 (cont.)

p = 10

N \ α	.005	.010	.025	.05	.1	.25
11	$.0^{8}11246$	$.0^{8}35733$	$.0^{7}19324$	$.0^{7}77218$	$.0^{6}33526$	$.0^{5}21386$
12	$.0^{6}47154$	$.0^{5}10036$	$.0^{5}28256$	$.0^{5}64552$	$.0^{4}15891$	$.0^{4}60952$
13	$.0^{5}76669$	$.0^{4}13324$	$.0^{4}28760$	$.0^{4}53699$	$.0^{3}10610$	$.0^{3}30376$
14	$.0^{4}42583$	$.0^{4}66814$	$.0^{3}12568$	$.0^{3}21066$	$.0^{3}37102$	$.0^{3}89399$
15	$.0^{3}14331$	$.0^{3}21078$	$.0^{3}36286$	$.0^{3}56666$	$.0^{3}92523$	$.0^{2}19892$
16	$.0^{3}36052$	$.0^{3}50647$	$.0^{3}81824$	$.0^{2}12140$	$.0^{2}18755$	$.0^{2}37055$
17	$.0^{3}75024$	$.0^{2}10179$	$.0^{2}15665$	$.0^{2}22346$	$.0^{2}33072$	$.0^{2}61162$
18	$.0^{2}13670$	$.0^{2}18040$	$.0^{2}26712$	$.0^{2}36920$	$.0^{2}52683$	$.0^{2}92661$
19	$.0^{2}22587$	$.0^{2}29142$	$.0^{2}41802$	$.0^{2}56300$	$.0^{2}78244$	.013126
20	$.0^{2}34640$	$.0^{2}43855$	$.0^{2}61253$	$.0^{2}80714$	.010952	.017701
22	$.0^{2}69174$	$.0^{2}84980$	.011376	.014478	.018907	.028781
24	.011834	.014208	.018414	.022817	.028932	.042056
26	.018193	.021449	.027092	.032865	.040709	.057042
28	.025886	.030071	.037194	.044348	.053894	.073283
30	.034761	.039886	.048484	.056983	.068156	.090387
34	.055364	.062312	.073700	.084683	.098783	.12595
38	.078662	.087261	.10111	.11422	.13076	.16184
42	.10353	.11356	.12949	.14436	.16285	.19695
46	.12913	.14036	.15800	.17428	.19429	.23066
50	.15483	.16705	.18607	.20346	.22464	.26265
60	.21714	.23106	.25238	.27151	.29443	.33462
70	.27444	.28925	.31169	.33157	.35512	.39574
80	.32589	.34109	.36390	.38395	.40748	.44759
90	.37168	.38694	.40970	.42955	.45270	.49182
100	.41235	.42747	.44991	.46937	.49195	.52983
120	.48083	.49536	.51674	.53515	.55634	.59151
140	.53578	.54954	.56969	.58695	.60669	.63923
160	.58059	.59356	.61249	.62861	.64700	.67713
180	.61773	.62994	.64770	.66279	.67994	.70791
200	.64895	.66046	.67715	.69129	.70732	.73339
250	.70872	.71871	.73313	.74529	.75903	.78122
300	.75125	.76003	.77268	.78332	.79529	.81455

TABLE 12

The Test Criteria L_{mvc}, L_{vc}, and L_m of S. S. Wilks: Tables of S. S. Wilks and also of J. Roy and V. K. Murthy

(a) Content of the tables and definition of the test statistics :

The tables contain the lower percentage points for WILKS' likelihood ratio test statistics L_{mvc}, L_{vc}, and L_m for p univariate components x_i of a p-dimensional normal distribution

$$\underline{x}' = (x_1 , x_2 , \dots , x_p) \quad .$$

The corresponding hypotheses are :

H_{mvc} : all means are equal and all variances and covariances are equal ,

H_{vc} : all variances and all covariances are equal, independent of the means ,

H_m : all means are equal, and under the assumption that H_{vc} (equality of variances and covariances) obtains.

The test statistics are computed from the values of a p-dimensional sample of size n : $\{ x_{ik} \}$ with $i = 1,2,3, \dots , p$ and $k = 1,2,\dots , n$.

First find :

$$\bar{x}_i = (1/n) \sum_{k=1}^{n} x_{ik} \quad ; \quad \bar{x} = (1/p) . \sum_{i=1}^{p} \bar{x}_i \quad ;$$

$$s_{ij} = \frac{1}{n} . \sum_{k=1}^{n} (x_{ik} - \bar{x}_i)(x_{jk} - \bar{x}_j) = \frac{1}{n} . \sum_{k=1}^{n} x_{ik} . x_{jk} - \bar{x}_i . \bar{x}_j \quad ;$$

$$s^2 = \frac{1}{p} . \sum_{i=1}^{p} s_{ii} \quad ; \quad s^2 . r = \frac{1}{p(p-1)} . \sum_{i \neq j=1}^{p} s_{ij} \quad .$$

Use these to find the test criteria as follows :

$$L_{mvc} = L_{vc} . L_m^{p-1} \qquad \text{for } H_{mvc} \; ;$$

$$L_{vc} = \frac{| s_{ij} |}{(s^2)^p . (1 - r)^{p-1} (1 + (p-1) . r)} \qquad \text{for } H_{vc} \; ;$$

$$L_m = \frac{s^2 . (1 - r)}{s^2 (1 - r) + \frac{1}{p-1} . \sum_{i=1}^{p} (\bar{x}_i - \bar{x})^2} \qquad \text{for } H_m \; .$$

$| s_{ij} | = | \underline{S} |$ is the determinant of the variance-covariance matrix.

Refer to :

Table 12a for percentage points for L_{mvc} and L_{vc}

Table 12b for percentage points for L_m

Table 12c for approximate percentage points for the variates $-n.\ln L_{mvc}$, $-n.\ln L_{vc}$; and $-n(p-1).\ln L_m$

Table 12d for a survey of the accuracy of the approximative percentage points for L_{mvc} , L_{vc} , and L_m given in table 12c .

(b) Extent of the tables and definition of the parameters :

(1) The parameter α :

α = error probsbility

α = 5% and 1% .

(2) The parameter p :

p = dimension of the variates

p = 2(1)6 in table 12a ,

= 2(1)5 in table 12b ,

= 2(1)6 in table 12c ,

= 2(1)5 in table 12d .

(3) The parameter n :

n = sample size

n = 3(1)32,42,62,122, ∞ for p = 2 in table 12a ;

= 4(1)18,23,33,63, ∞ for p = 3 in table 12a ;

= 25(5)60(10)100 for p = 4(1)7 in table 12a ;

= various values of n in tables 12b, 12c , and 12d .

(c) Suggestions for applications :

The tables are used to test the hypotheses H_{mvc} , H_{vc} , and H_m on the basis of samples from a p-dimensional normally distributed population.

(d) References :

(1) For the source of the tables 12a (1st page), 12b , 12c , and 12d :

WILKS, S. S. : Sample criteria for testing equality of means, equality of variances, and equality of covariances in a normal multivariate distribution.
The Annals of Mathematical Statistics 17,257-281(1946).
(Table I, Table II, Table III, Table IV).

(2) For the source of the table 12a (2nd page) :

ROY, J. + Murthy, V. K. : Percentage points of Wilks' L_{mvc} and L_{vc} criteria.
Psychometrika 25,243 - 250 (1960).
(Table 3 ; Table 4) .

(e) Further remarks :

(1) For large sample sizes , the test statistics $-n.\ln L_{mvc}$, $-n.\ln L_{vc}$, and $-n.(p - 1).\ln L_m$ are approximately χ^2 - distributed with respective degrees of freedom

$(1/2).p.(p+3) - 3$; $(1/2).p.(p+1) - 2$; $p - 1$.

Table 12d gives a survey of the accuracy of the χ^2- approximation.

(2) In some sense the test statistic L_{vc} for hypothesis H_{vc} is an extension of the sphericity test of J. W. MAUCHLY in table 11 .

(3) The test statistic L_m is equivalent to a variance analysis test for a $(p \times n)$ experimental design.
The corresponding F-test for hypothesis H_m is

$$F = \frac{(1/2).(n-1).(p-1).(1-L_m)}{(1/2).(p-1).L_m}$$

with $f_1 = p - 1$ and $f_2 = (n-1).(p-1)$ degrees of freedom.

(4) Exact tests for H_{mvc} and H_{vc} are available in dimensions $p = 2, 3$.

(aa) For $p = 2$:

H_{mvc} is to be rejected whenever

$$\frac{(n-2).(1-L_{mvc})}{2 \cdot L_{mvc}} > F_{2\,;\,n-2}(\alpha) .$$

H_{vc} is to be rejected whenever

$$\frac{(n-2).(1-L_{vc})}{L_{vc}} > F_{1\,;\,n-2}(\alpha) .$$

(bb) For $p = 3$:

H_{mvc} is to be rejected whenever

$$\frac{(n-3).(1-\sqrt{L_{mvc}})}{3 \cdot \sqrt{L_{mvc}}} > F_{6\,;\,2n-6}(\alpha) .$$

H_{vc} is to be rejected wheneveer

$$\frac{(n-3).(1-\sqrt{L_{vc}})}{2 \cdot \sqrt{L_{vc}}} > F_{4\,;\,2n-6}(\alpha) .$$

(5) The probability integrals for L_{mvc} and L_{vc} in the case $p = 2$, as well as those for $\sqrt{L_{mvc}}$ and $\sqrt{L_{vc}}$ in the case $p = 3$ and those for L_m for arbitrary p are all incomplete beta functions.

The corresponding parameters of the incomplete beta functions, here denoted by p_1 and q_1, can be found in the following table :

p	Criterion	p_1	q_1
2	L_{mvc}	$\frac{1}{2}(n - 2)$	1
2	L_{vc}	$\frac{1}{2}(n - 2)$	$\frac{1}{2}$
3	$\sqrt{L_{mvc}}$	$n - 3$	3
3	$\sqrt{L_{vc}}$	$n - 3$	2
p	L_m	$\frac{1}{2}(n - 1)(p - 1)$	$\frac{1}{2}(p - 1)$

afel 12a

Table 12a

Untere 5%- und 1%- Punkte von L_{mvc} und L_{vc} /
Lower 5%- and 1%- points of L_{mvc} and L_{vc}

	p = 2			
	L_{mvc}		L_{vc}	
n	5%	1%	5%	1%
3	0.0025	.0001	0.0062	.0002
4	.0500	.0100	.0975	.0199
5	.1357	.0464	.2285	.0808
6	.2236	.1000	.3416	.1588
7	.3017	.1585	.4307	.2352
8	.3684	.2154	.5005	.3039
9	.4249	.2683	.5559	.3637
10	.4729	.3162	.6007	.4154
11	.5139	.3594	.6375	.4601
12	.5493	.3981	.6682	.4989
13	.5800	.4329	.6943	.5328
14	.6070	.4642	.7165	.5626
15	.6307	.4924	.7358	.5889
16	.6518	.5180	.7528	.6124
17	.6707	.5411	.7675	.6334
18	.6877	.5623	.7807	.6522
19	.7030	.5817	.7925	.6693
20	.7169	.5995	.8031	.6848
21	.7294	.6159	.8126	.6989
22	.7411	.6310	.8213	.7119
23	.7518	.6450	.8292	.7237
24	.7616	.6579	.8365	.7347
25	.7707	.6700	.8431	.7448
26	.7791	.6813	.8493	.7542
27	.7869	.6918	.8549	.7629
28	.7942	.7017	.8602	.7710
29	.8010	.7110	.8651	.7786
30	.8074	.7197	.8697	.7857
31	.8133	.7279	.8739	.7924
32	.8190	.7356	.8779	.7987
42	.8609	.7943	.9073	.8454
62	.9050	.8577	.9375	.8945
122	.9513	.9261	.9684	.9460
∞	1.0000	1.0000	1.0000	1.0000

	p = 3			
	L_{mvc}		L_{vc}	
n	5%	1%	5%	1%
4	0.00029	0.00001	0.00064	0.00003
5	.0095	.0018	.0183	.0035
6	.0358	.0112	.0618	.0198
7	.0736	.0300	.1174	.0493
8	.1165	.0559	.1749	.0866
9	.1603	.0860	.2297	.1272
10	.2028	.1181	.2802	.1682
11	.2432	.1508	.3259	.2079
12	.2808	.1829	.3670	.2457
13	.3157	.2141	.4040	.2811
14	.3480	.2439	.4373	.3141
15	.3778	.2722	.4674	.3448
16	.4052	.2990	.4946	.3732
17	.4306	.3243	.5193	.3996
18	.4540	.3482	.5418	.4240
23	.5484	.4482	.6293	.5230
33	.6660	.5811	.7326	.6470
63	.8135	.7591	.8549	.8029
∞	1.0000	1.0000	1.0000	1.0000

Tafel 12a (Forts.) Table 12a (cont.)

5%- und 1%- Punkte des Kriteriums L_{mvc}

.05 and .01 Points of L_{mvc} Criterion

	.05 point				.01 point			
n\p	4	5	6	7	4	5	6	7
25	.4206	.2920	.1923	.1196	.3366	.2251	.1427	.0854
30	.4918	.3658	.2623	.1781	.4098	.2957	.2048	.1356
35	.5482	.4273	.3229	.2339	.4698	.3570	.2623	.1859
40	.5937	.4787	.3759	.2852	.5193	.4098	.3143	.2338
45	.6311	.5222	.4220	.3314	.5607	.4552	.3606	.2783
50	.6623	.5592	.4625	.3730	.5958	.4946	.4019	.3191
55	.6887	.5911	.4979	.4102	.6258	.5290	.4387	.3654
60	.7113	.6187	.5292	.4437	.6518	.5591	.4715	.3903
70	.7480	.6645	.5815	.5011	.6943	.6096	.5274	.4494
80	.7763	.7005	.6239	.5483	.7276	.6498	.5730	.4985
90	.7992	.7296	.6586	.5876	.7545	.6826	.6108	.5403
100	.8177	.7536	.6875	.6208	.7765	.7099	.6426	.5757

5%- und 1%- Punkte des Kriteriums L_{vc}

.05 and .01 Points of L_{vc} Criterion

	.05 point				.01 point			
n\p	4	5	6	7	4	5	6	7
25	.5129	.3768	.2473	.1601	.4209	.2985	.1866	.1163
30	.5773	.4490	.3219	.2273	.4908	.3709	.2563	.1756
35	.6271	.5071	.3853	.2883	.5464	.4313	.3181	.2322
40	.6666	.5546	.4390	.3424	.5913	.4819	.3721	.2841
45	.7002	.5941	.4847	.3899	.6284	.5248	.4191	.3310
50	.7251	.6273	.5239	.4318	.6594	.5613	.4601	.3731
55	.7473	.6555	.5578	.4686	.6857	.5928	.4961	.4108
60	.7663	.6799	.5873	.5013	.7083	.6201	.5278	.4446
70	.7967	.7196	.6362	.5564	.7450	.6653	.5810	.5024
80	.8202	.7506	.6748	.6008	.7736	.7011	.6236	.5499
90	.8394	.7755	.7062	.6374	.7964	.7299	.6586	.5894
100	.8560	.7959	.7321	.6679	.8151	.7538	.6877	.6226

5%- und 1%- Punkte des Kriteriums L_m

5% and 1% points of L_m

n	p = 2		n	p = 3		n	p = 4		n	p = 5	
	5%	1%		5%	1%		5%	1%		5%	1%
2	0.0062	0.0002	2	.0500	0.0100	2	.0973	.0328	2	.1354	.0589
3	.0975	.0199	3	.2236	.1000	3	.2960	.1698	3	.3426	.2221
4	.2285	.0808	4	.3684	.2154	4	.4372	.3002	4	.4793	.3566
5	.3416	.1588	5	.4729	.3162	5	.5340	.4019	5	.5709	.4560
6	.4307	.2352	6	.5493	.6033	6	.6033	.4800	6	.6356	.5302
7	.5005	.3039	7	.6070	.4642	7	.6550	.5409	7	.6837	.5872
8	.5559	.3637	8	.6518	.5180	8	.6950	.5895	8	.7206	.6321
9	.6007	.4154	9	.6877	.5623	9	.7267	.6290	11	.7933	.7232
10	.6375	.4601	10	.7169	.5995	10	.7525	.6617	16	.8559	.8043
11	.6682	.4989	11	.7411	.6310	11	.7739	.6892	31	.9246	.8961
12	.6943	.5328	12	.7616	.6579	21	.8788	.8290	∞	1.0000	1.0000
13	.7165	.5626	13	.7791	.6813	41	.9372	.9101			
14	.7358	.5889	14	.7942	.7017	∞	1.0000	1.0000			
15	.7527	.6124	15	.8074	.7197						
16	.7675	.6334	16	.8190	.7356						
17	.7807	.6522	21	.8609	.7943						
18	.7925	.6693	31	.9050	.8577						
19	.8031	.6848	61	.9513	.9261						
20	.8126	.6989	∞	1.0000	1.0000						
21	.8213	.7119									
22	.8292	.7237									
23	.8365	.7347									
24	.8431	.7448									
25	.8493	.7542									
26	.8549	.7629									
27	.8602	.7710									
28	.8651	.7786									
29	.8697	.7857									
30	.8739	.7924									
31	.8779	.7987									
41	.9073	.8454									
61	.9375	.8945									
121	.9684	.9460									
∞	1.0000	1.0000									

Tafel 12c

Table 12c

Genäherte 5%- und 1%- Punkte / Approximate 5% and 1% points

p	$-n \ln L_{mvc}$			$-n \ln L_{vc}$			$-n\,(p-1) \ln L_m$		
	d.f.	5%	1%	*d.f.*	5%	5%	*d.f.*	5%	1%
2	2	5.99147	9.21034	1	3.84146	6.63490	1	3.84146	6.63490
3	6	12.5916	16.8119	4	9.48773	13.2767	2	5.99147	9.21034
4	11	19.6751	24.7250	8	15.5073	20.0902	3	7.81473	11.3449
5	17	27.5871	33.4087	13	22.3621	27.6883	4	9.48773	13.2767
6	24	36.4151	42.9798	19	30.1435	36.1908	5	11.0705	15.0863

Tafel 12d

Table 12d

Genauigkeit der genäherten Prozentpunkte in Tafel 12c / Accuracy of the approximate percentage points in table 12c

criterion	p	n	5%		1%	
			exact	approx.	exact	approx.
L_{mvc}	2	30	0.8074	0.8190 (5.53)*	0.7197	0.7357 (1.73)*
L_{mvc}	2	62	.9050	.9079 (5.25)	.8577	.8619 (1.36)
L_{mvc}	2	122	.9513	.9521 (5.13)	.9261	.9273 (1.19)
L_{mvc}	3	33	.6660	.6828 (5.79)	.5811	.6008 (1.88)
L_{mvc}	3	63	.8135	.8188 (5.40)	.7591	.7658 (1.49)
L_{vc}	2	30	.8697	.8799 (5.49)	.7857	.8016 (1.76)
L_{vc}	2	62	.9375	.9399 (5.22)	.8945	.8985 (1.37)
L_{vc}	2	122	.9684	.9690 (5.11)	.9460	.9471 (1.20)
L_{vc}	3	33	.7326	.7501 (5.82)	.6470	.6688 (2.01)
L_{vc}	3	63	.8549	.8602 (5.41)	.8029	.8100 (1.55)
L_m	2	31	.8779	.8835 (5.28)	.7987	.8073 (1.43)
L_m	2	61	.9375	.9389 (5.13)	.8945	.8969 (1.20)
L_m	2	121	.9684	.9688 (5.07)	.9460	.9467 (1.13)
L_m	3	31	.9050	.9079 (5.25)	.8577	.8619 (1.36)
L_m	3	61	.9513	.9521 (5.10)	.9261	.9273 (1.14)
L_m	4	41	.9372	.9385 (5.19)	.9101	.9119 (1.26)
L_m	5	31	.9246	.9264 (5.25)	.8961	.8984 (1.32)

***The numbers in the parentheses are approximate percentages (obtained by linear interpolation) to which the approximate percent points correspond.**

**Die Zahlen in Klammern sind genäherte Prozentsätze, die den genäherten Prozentpunkten entsprechen.*

TABLE 13

The Multivariate Outlier Criteria of S. S. Wilks

(a) Content of the tables and definition of the test statistics :

The tables contain the lower percentage points for WILKS' tests statistics r_1 and r_2 fore one , respectively two , outliers in a sample of size n from a p-dimesnional normally distributed population.

Let $(\bar{x}_1, \bar{x}_2, \ldots , \bar{x}_n)$ be the vector of sample means with

$$\bar{x}_i = \frac{1}{n} \cdot \sum_{\xi=1}^{n} x_{i\xi} \qquad \text{and} \qquad a_{ij} = \sum_{\xi=1}^{n} (x_{i\xi}-\bar{x}_i)(x_{j\xi}-\bar{x}_j)$$

$$\text{for} \quad i , j = 1,2, \ldots , p .$$

The sample can be regarded as a cluster of n points in a p-dimensional Euclidian space, R_p . Any p of these points , together with the "center of mass" $(\bar{x}_1, \bar{x}_2, \ldots , \bar{x}_p)$, form a simplex . Square the volume of this simplex, sum over all possible simplexes which can be formed in this way, and one obtains S. S. WILKS' statistic

$$V = (p!) \cdot | a_{ij} |$$

where $| a_{ij} |$ denotes the determinant of the SP-matrix (a_{ij}) .
Following S. S. WILKS , $|a_{ij}|$ is called the internal scatter of the given sample.

Whenever $n > p$, then $| a_{ij} | > 0$ with probability 1 .

In the following, we distinguish two cases :

(1) <u>A single outlier</u> :

Eliminating the ξ-th element of the sample yields a cluster of n-1 points in R_n . The "internal scatter" of these n-1 points will be denoted by $a_{ij\xi}$. Again $|a_{ij\xi}|>0$ with probability 1 whenever $n > p + 1$.
Now we define

$$R_\xi = \frac{|a_{ij\xi}|}{|a_{ij}|} \quad \text{for} \quad \xi = 1,2,3, \dots , n$$

and call the values $R_1, R_2, \dots , R_n$ the scattering ratios for a single outlier from the sample

$$(x_{1\xi}, x_{2\xi}, \dots , x_{p\xi} \ ; \ \xi = 1,2, \dots , n) .$$

Now form a rank ordering $R_{(1)}, R_{(2)}, \dots , R_{(n)}$ of the scattering ratios $R_1, R_2, \dots , R_n$.

As a criterion for the selection of a single extreme observation as well as for testing for a significant outlier , we select, following S. S. WILKS, the smallest scattering ratio $R_{(1)}$ and denote it r_1 .
The most serious candidate for a significant outlier therefore consists of that sample element whose elimination leaves the minimal scattering ratio , $r_1 = \min\{R_\xi\}$.

This element is to be tested as an outlier ; r_1 is the test criterion .

The critical values of r_1 are those which are found at the left end of the distribution. Thus one obtains lower percentage points . Because of the tedious computations for the probabilities $P(r_1 < r)$, one remains satisfied with upper bounds for them.
Therefore, the first four pages of the tables contain those values r_α , for which the upper bound of $P(r_1 < r_\alpha)$ has value α .

(2) Two outliers :

In this case, two elements of the sample are excluded as outliers.

Let $(x_{1\xi}, x_{2\xi}, \ldots, x_{p\xi})$ and $(x_{1\eta}, x_{2\eta}, \ldots, x_{p\eta})$ be given.
The the "internal scatter" of the resulting cluster of n-2 points will be denoted by $|a_{ij\xi\eta}|$; it is positive with probability one whenever $n > p + 2$.

The scattering ratio for two outliers is then defined to be

$$R_{\xi\eta} = \frac{|a_{ij\xi\eta}|}{|a_{ij}|} \quad \text{for} \quad \eta > \xi = 1,2, \ldots , n .$$

As a criterion for the selection of a pair of extreme observations , as well as for testing for a significant pair of outliers one naturally chooses the minimum $r_2 = \min_{\eta>\xi}\{R_{\xi\eta}\}$ of the scattering ratios $R_{\xi\eta}$ for two outliers.

The most serious pair of candidates for a significant pair of outliers therefore consists of those sample elements whose elimination leaves the minimal scattering ratio r_2 .

The critical values of r_2 are those which are found at the left end of the distribution , once again one obtains lower percentage points .

The last four pages therefore contain those values of $\sqrt{r_\alpha}$ for which the upper bounds of $P(r_2 < r_\alpha)$ has value α .

(b) Extent of the tables and definition of the parameters :

(1) The parameter α :

α = error probability

α = 0.100 ; 0.050 ; 0.025 ; 0.010 .

(2) The parameter p :

p = dimension of the variates

p = 1(1)5 .

(3) The parameter n :

n = size of the k-dimensional sample

n = 5(1)30(5)100(100)500 .

(c) Suggestions for applications :

The tables - subject to the discussion of section (a) - allow the testing of the hypothesis that a sample from a multivariate, normally distributed population contains one, respectively two, outliers.

(d) References :

(1) For the test statistic and for the source of the tables :

WILKS, S. S. : Multivariate statistical outliers.
Sankhya, Series A , 25, 407 - 426 (1963).

(2) For further discussion :

WILKS, S. S. : Mathematical Statiatics.
New York : Wiley, 1962.

(e) Further remarks :

S. S. WILKS has extended the above considerations to the general case of t outliers.

For $t = 3$ and $t = 4$ he gives relations from which that value of r, for which the upper bound of

$$P(r_t < r_\alpha) \text{ has value } \alpha,$$

can be determined.

For still larger values of t, he gives several directions for the computation of the upper bound.

Tafel 13 Table 13

Werte von r_α *für* r_1 / *Values of* r_α *for* r_1

$\alpha = 0.100$

	p				
n	1	2	3	4	5
5	0.10000	0.02000	0.00025		
6	.20000	.06525	.01114	0.00012	
7	.26960	.11952	.04172	.00717	0.00007
8	.32610	.17328	.08282	.02959	.00502
9	.37418	.22314	.12675	.06216	.02234
10	.41540	.26827	.16978	.09888	.04901
11	.45106	.30878	.21038	.13629	.08032
12	.48221	.34511	.24801	.17267	.11319
13	.50966	.37776	.28264	.20723	.14593
14	.53405	.40719	.31442	.23967	.17764
15	.55586	.43383	.34358	.26995	.20789
16	.57550	.45804	.37037	.29813	.23651
17	.59328	.48014	.39502	.32433	.26346
18	.60948	.50038	.41777	.34870	.28878
19	.62428	.51899	.43881	.37139	.31254
20	.63789	.53615	.45832	.39255	.33484
21	.65043	.55205	.47645	.41231	.35578
22	.66205	.56680	.49334	.43079	.37545
23	.67282	.58053	.50912	.44812	.39396
24	.68286	.59335	.52389	.46438	.41140
25	.69223	.60535	.53774	.47967	.42784
26	.70101	.61660	.55075	.49408	.44337
27	.70925	.62717	.56301	.50767	.45806
28	.71699	.63713	.57457	.52052	.47197
29	.72429	.64653	.58549	.53268	.48516
30	.73119	.65540	.59583	.54420	.49768
35	.76063	.69342	.64023	.59385	.55182
40	.78375	.72335	.67531	.63326	.59499
45	.80245	.74758	.70379	.66532	.63023
50	.81792	.76763	.72738	.69195	.65955
55	.83094	.78451	.74728	.71443	.68435
60	.84208	.79895	.76430	.73369	.70561
65	.85173	.81145	.77904	.75037	.72404
70	.86017	.82238	.79193	.76497	.74019
75	.86763	.83203	.80331	.77787	.75446
80	.87427	.84061	.81344	.78935	.76717
85	.88023	.84830	.82251	.79963	.77856
90	.88560	.85524	.83069	.80891	.78883
95	.89048	.86152	.83811	.81731	.79814
100	.89492	.86725	.84486	.82497	.80663
200	.94067	.92574	.91371	.90300	.89308
300	.95796	.94751	.93923	.93186	.92503
400	.96722	.95908	.95272	.94711	.94189
500	.97304	.96631	.96113	.95661	.95238

Tafel 13 (Forts.) Table 13 (cont.)

Werte von r_α *für* r_1 / *Values of* r_α *for* r_1

	$\alpha = 0.050$				
	p				
n	1	2	3	4	5
5	0.08083	0.01000			
6	.14529	.04110	0.00556		
7	.20661	.08452	.02620	0.00358	
8	.26161	.13133	.05831	.01856	0.00251
9	.31006	.17711	.09559	.04367	.01400
10	.35261	.22007	.13408	.07438	.03440
11	.39008	.25965	.17171	.10731	.06033
12	.42325	.29584	.20751	.14050	.08896
13	.45277	.32886	.24112	.17285	.11850
14	.47921	.35897	.27245	.20383	.14785
15	.50302	.38650	.30154	.23319	.17642
16	.52457	.41171	.32855	.26086	.20386
17	.54417	.43487	.35361	.28686	.23002
18	.56208	.45620	.37690	.31124	.25486
19	.57852	.47591	.39857	.33412	.27837
20	.59365	.49417	.41876	.35558	.30060
21	.60764	.51113	.43761	.37573	.32160
22	.62061	.52692	.45525	.39467	.34145
23	.63267	.54166	.47178	.41249	.36021
24	.64391	.55545	.48729	.42929	.37796
25	.65443	.56838	.50188	.44513	.39477
26	.66429	.58053	.51563	.46010	.41069
27	.67355	.59197	.52860	.47426	.42580
28	.68226	.60276	.54086	.48767	.44014
29	.69048	.61296	.55247	.50040	.45377
30	.69825	.62260	.56347	.51248	.46674
35	.73146	.66402	.61090	.56478	.52314
40	.75758	.69675	.64857	.60654	.56843
45	.77872	.72330	.67924	.64067	.60557
50	.79621	.74532	.70472	.66909	.63659
55	.81094	.76388	.72624	.69314	.66289
60	.82354	.77975	.74467	.71377	.68548
65	.83444	.79351	.76065	.73167	.70511
70	.84398	.80554	.77464	.74735	.72232
75	.85240	.81616	.78700	.76122	.73754
80	.85989	.82561	.79800	.77356	.75111
85	.86661	.83408	.80786	.78463	.76328
90	.87267	.84172	.81674	.79462	.77426
95	.87816	.84864	.82480	.80367	.78421
100	.88317	.85494	.83214	.81192	.79329
200	.93447	.91924	.90696	.89602	.88591
300	.95372	.94310	.93463	.92711	.92013
400	.96399	.95573	.94924	.94351	.93817
500	.97043	.96361	.95833	.95370	.94938

Tafel 13 (Forts.) Table 13 (cont.)

Werte von r_α *für* r_1 / *Values of* r_α *for* r_1

$\alpha = 0.025$

n	p: 1	2	3	4	5
5	0.05124	0.00500	0.00002		
6	.10353	.02589	.00278	0.00001	
7	.15787	.05976	.01647	.00179	0.00000
8	.20934	.09953	.04111	.01166	.00125
9	.25636	.14057	.07219	.03075	.00879
10	.29873	.18053	.10601	.05606	.02420
11	.33677	.21834	.14030	.08466	.04541
12	.37094	.25361	.17380	.11452	.07008
13	.40170	.28629	.20589	.14441	.09644
14	.42950	.31647	.23627	.17360	.12331
15	.45471	.34433	.26485	.20171	.14998
16	.47768	.37007	.29165	.22854	.17601
17	.49867	.39387	.31674	.25400	.20114
18	.51794	.41593	.34022	.27811	.22525
19	.53569	.43642	.36221	.30089	.24827
20	.55208	.45547	.38281	.32239	.27020
21	.56727	.47324	.40213	.34269	.29106
22	.58139	.48984	.42028	.36187	.31088
23	.59455	.50538	.43735	.37999	.32971
24	.60685	.51996	.45343	.39713	.34760
25	.61836	.53367	.46860	.41336	.36460
26	.62917	.54657	.48292	.42873	.38076
27	.63934	.55874	.49647	.44332	.39614
28	.64891	.57025	.50930	.45717	.41079
29	.65796	.58113	.52147	.47034	.42475
30	.66651	.59144	.53303	.48287	.43806
35	.70317	.63587	.58306	.53737	.49626
40	.73208	.67113	.62301	.58117	.54333
45	.75551	.69982	.65567	.61711	.58213
50	.77492	.72365	.68286	.64715	.61466
55	.79128	.74378	.70589	.67264	.64232
60	.80527	.76102	.72564	.69454	.66613
65	.81738	.77596	.74278	.71357	.68685
70	.82798	.78904	.75780	.73027	.70504
75	.83733	.80060	.77108	.74503	.72116
80	.84566	.81088	.78291	.75820	.73553
85	.85312	.82010	.79352	.77001	.74844
90	.85984	.82841	.80308	.78067	.76009
95	.86594	.83595	.81176	.79034	.77066
100	.87150	.84281	.81967	.79916	.78030
200	.92829	.91280	.90028	.88914	.87885
300	.94949	.93871	.93008	.92240	.91530
400	.96077	.95240	.94579	.93993	.93450
500	.96783	.96093	.95555	.95082	.94642

Werte von r_α *für* r_1 / *Values of* r_α *for* r_1

$\alpha = 0.010$

n	p = 1	2	3	4	5
5	0.02795	0.00200	0.00000		
6	.06592	.01406	.00111	0.00000	
7	.11026	.03780	.00893	.00071	0.00000
8	.15547	.06898	.02593	.00632	.00050
9	.19888	.10358	.04987	.01937	.00476
10	.23942	.13895	.07781	.03866	.01523
11	.27678	.17364	.10755	.06200	.03129
12	.31103	.20689	.13765	.08757	.05126
13	.34238	.23835	.16726	.11407	.07362
14	.37107	.26790	.19590	.14065	.09723
15	.39738	.29556	.22330	.16678	.12128
16	.42156	.32141	.24936	.19215	.14525
17	.44383	.34555	.37404	.21657	.16878
18	.46440	.36810	.29737	.23996	.19167
19	.48344	.38919	.31940	.26228	.21378
20	.50112	.40893	.34019	.28354	.23506
21	.51757	.42743	.35982	.30376	.25547
22	.53292	.44480	.37835	.32298	.27501
23	.54727	.46113	.39588	.34125	.29370
24	.56071	.47651	.41246	.35861	.31155
25	.57334	.49102	.42815	.37513	.32861
26	.58521	.50471	.44304	.39084	.34491
27	.59641	.51767	.45716	.40580	.36048
28	.60698	.52994	.47057	.42006	.37536
29	.61697	.54158	.48333	.43365	.38959
30	.62644	.55263	.49547	.44663	.40320
35	.66716	.60048	.54835	.50344	.46318
40	.69944	.63870	.59091	.54949	.51217
45	.72567	.66994	.62588	.58754	.55284
50	.74745	.69598	.65514	.61947	.58711
55	.76583	.71803	.67997	.64666	.61636
60	.78157	.73694	.70133	.67009	.64162
65	.79521	.75336	.71990	.69050	.66366
70	.80715	.76775	.73620	.70843	.68306
75	.81769	.78048	.75062	.72432	.70026
80	.82708	.79181	.76348	.73851	.71563
85	.83549	.80197	.77503	.75124	.72944
90	.84308	.81115	.78545	.76274	.74192
95	.84995	.81946	.79490	.77319	.75326
100	.85622	.82704	.80352	.78271	.76361
200	.92016	.90435	.89155	.88018	.86361
300	.94392	.93293	.92411	.91625	.90899
400	.95652	.94801	.94125	.93525	.92969
500	.96439	.95739	.95190	.94704	.94254

Tafel 13 (Forts.) Table 13 (cont.)

Werte von $\sqrt{r_\alpha}$ für r_2 / Values of $\sqrt{r_\alpha}$ for r_2

$\alpha = 0.100$

	p				
n	1	2	3	4	5
5	0.10000	0.00501			
6	.18821	.04791	0.00223		
7	.26269	.10904	.02872	0.00119	
8	.32402	.16955	.07368	.01927	0.00072
9	.37493	.22444	.12283	.05397	.01386
10	.41780	.27305	.17039	.09468	.04161
11	.45442	.31592	.21449	.13599	.07599
12	.48609	.35381	.25473	.17565	.11219
13	.51380	.38747	.29126	.21282	.14791
14	.53826	.41753	.32440	.24728	.18212
15	.56006	.44453	.35452	.27909	.21430
16	.57962	.46892	.38196	.30842	.24461
17	.59728	.49106	.40705	.33548	.27282
18	.61332	.51125	.43006	.36047	.29911
19	.62797	.52974	.45124	.38360	.32362
20	.64140	.54676	.47079	.40506	.34649
21	.65378	.56246	.48889	.42501	.36785
22	.66522	.57700	.50571	.44360	.38783
23	.67584	.59051	.52137	.46095	.40655
24	.68572	.60310	.53598	.47720	.42412
25	.69494	.61487	.54966	.49243	.44064
26	.70357	.62589	.56250	.50675	.45619
27	.71167	.63623	.57456	.52023	.47086
28	.71928	.64596	.58592	.53293	.48472
29	.72646	.65513	.59664	.54494	.49784
30	.73323	.66380	.60677	.55631	.51026
35	.76216	.70082	.65015	.60508	.56375
40	.78489	.72990	.68431	.64361	.60614
45	.80328	.75343	.71196	.67486	.64061
50	.81850	.77288	.73485	.70074	.66921
55	.83133	.78926	.75413	.72257	.69335
60	.84231	.80327	.77061	.74124	.71401
65	.85183	.81539	.78488	.75740	.73191
70	.86017	.82600	.79736	.77155	.74758
75	.86754	.83537	.80837	.78403	.76141
80	.87410	.84370	.81817	.79514	.77372
85	.87999	.85118	.82696	.80509	.78475
90	.88531	.85791	.83488	.81407	.79470
95	.89014	.86402	.84205	.82220	.80372
100	.89454	.86960	.84859	.82961	.81193
200	.93994	.92664	.91538	.90518	.89565
300	.95713	.94799	.94025	.93322	.92666
400	.96635	.95936	.95345	.94807	.94305
500	.97215	.96650	.96169	.95733	.95326

Werte von $\sqrt{r_\alpha}$ für r_2 / Values of $\sqrt{r_\alpha}$ for r_2

	$\alpha = 0.050$				
	p				
n	1	2	3	4	5
5	0.07071	0.00250	0.00000		
6	.14938	.03372	.00111		
7	.22090	.08601	.02019	0.00060	
8	.28207	.14167	.05800	.01355	0.00036
9	.33403	.19419	.10237	.04245	.00975
10	.37841	.24188	.14702	.07881	.03273
11	.41670	.28462	.18945	.11716	.06322
12	.45006	.32286	.22884	.15489	.09657
13	.47939	.35712	.26504	.19086	.13029
14	.50539	.38792	.29819	.22462	.16313
15	.52863	.41573	.32853	.25609	.19451
16	.54952	.44096	.35634	.28532	.22417
17	.56842	.46394	.38188	.31244	.25207
18	.58562	.48496	.40540	.33763	.27823
19	.60134	.50426	.42712	.36103	.30274
20	.61578	.52205	.44722	.38282	.32571
21	.62908	.53849	.46588	.40313	.34723
22	.64139	.55374	.48324	.42210	.36743
23	.65282	.56793	.49944	.43986	.38641
24	.66346	.58117	.51459	.45651	.40426
25	.67339	.59354	.52878	.47215	.42108
26	.68269	.60515	.54211	.48687	.43695
27	.69141	.61604	.55465	.50076	.45194
28	.69962	.62630	.56648	.51386	.46612
29	.70735	.63598	.57765	.52625	.47955
30	.71465	.64513	.58821	.53799	.49230
35	.74583	.68424	.63351	.58850	.54730
40	.77032	.71501	.66926	.62850	.59106
45	.79013	.73992	.69824	.66101	.62671
50	.80652	.76052	.72224	.68798	.65635
55	.82032	77787	.74247	.71073	.68138
60	.83213	.79271	.75978	.73021	.70284
65	.84236	.80555	.77476	.74708	.72143
70	.85132	.81678	.78787	.76185	.73772
75	.85922	.82670	.79944	.77489	.75210
80	.86627	.83553	.80974	.78650	.76491
85	.87259	.84343	.81896	.79689	.77639
90	.87829	.85056	.82728	.80627	.78674
95	.88346	.85703	.83482	.81477	.79612
100	.88817	.86292	.84169	.82251	.80467
200	.93664	.92316	.91177	.90145	.89181
300	.95490	.94564	.93780	.93069	.92405
400	.96466	.95759	.95160	.94616	.94107
500	.97080	.96500	.96021	.95580	.95167

Werte von $\sqrt{r_\alpha}$ für r_2 / Values of $\sqrt{r_\alpha}$ for r_2

$\alpha = 0.010$

n	p 1	2	3	4	5
5	0.03162	0.00050			
6	.08736	.01498	0.00022		
7	.14772	.04982	.00896	0.00012	
8	.20444	.09374	.03349	.00601	0.00007
9	.25544	.13926	.06744	.02449	.00433
10	.30069	.18308	.10490	.05181	.01887
11	.34076	.22397	.14265	.08337	.04152
12	.37636	.26160	.17912	.11626	.06861
13	.40812	.29606	.21363	.14889	.09761
14	.43660	.32758	.24595	.18044	.12699
15	.46228	.35641	.27605	.21050	.15589
16	.48553	.38284	.30403	.23893	.18382
17	.50670	.40713	.33002	.26568	.21057
18	.52604	.42952	.35419	.29082	.23601
19	.54380	.45019	.37667	.31441	.26013
20	.56016	.46935	.39764	.33655	.28296
21	.57528	.48714	.41721	.35735	.30454
22	.58930	.50371	.43551	.37689	.32494
23	.60234	.51918	.45266	.39528	.34423
24	.61451	.53365	.46876	.41261	.36248
25	.62588	.54722	.48390	.42895	.37975
26	.63655	.55996	.49816	.44439	.39611
27	.64657	.57196	.51162	.45899	.41164
28	.65599	.58328	.52434	.47282	.42637
29	.66489	.59397	.53637	.48594	.44037
30	.67329	.60409	.54778	.49839	.45370
35	.70925	.64753	.59694	.55228	.51159
40	.73753	.68184	.63596	.59527	.55804
45	.76043	.70968	.66772	.63038	.59611
50	.77937	.73276	.69411	.65962	.62789
55	.79532	.75222	.71638	.68436	.65482
60	.80897	.76887	.73547	.70556	.67796
65	.82077	.78328	.75201	.72396	.69804
70	.83111	.79590	.76649	.74009	.71566
75	.84023	.80704	.77928	.75434	.73124
80	.84835	.81695	.79066	.76703	.74512
85	.85563	.82583	.80086	.77840	.75757
90	.86210	.83384	.81007	.78866	.76880
95	.86814	.84110	.81841	.79797	.77899
100	.87356	.84771	.82601	.80645	.78828
200	.92902	.91518	.90349	.89291	.88304
300	94974	.94023	.93218	.92489	.91808
400	.96076	.95349	.94734	.94176	.93655
500	.96766	.96180	.95679	.95226	.94804

Tafel 13 (Forts.) Table 13 (cont.)

Werte von $\sqrt{r_\alpha}$ *für* r_2 / *Values of* $\sqrt{r_\alpha}$ *for* r_2

$\alpha = 0.025$

	p				
n	1	2	3	4	5
5	0.05000	0.00125			
6	.11856	.02376	0.00056		
7	.18575	.06794	.01422	0.00030	
8	.24556	.11852	.04575	.00954	0.00018
9	.29758	.16820	.08546	.03346	.00687
10	.34274	.21444	.12703	.06572	.02579
11	.38212	.25661	.16755	.10110	.05269
12	.41670	.29479	.20581	.13678	.08327
13	.44728	.32932	.24141	.17138	.11495
14	.47453	.36058	.27432	.20427	.14634
15	.49896	.38897	.30468	.23522	.17670
16	.52099	.41483	.33266	.26419	.20568
17	.54097	.43848	.35849	.29124	.23314
18	.55918	.46017	.38237	.31647	.25905
19	.57585	.48014	.40450	.34003	.28346
20	.59118	.49859	.42504	.36203	.30642
21	.60532	.51567	.44415	.38260	.32802
22	.61842	.53155	.46197	.40187	.34835
23	.63058	.54633	.47863	.41995	.36751
24	.64191	.56014	.49423	.43694	.38557
25	.65250	.57307	.50887	.45292	.40262
26	.66242	.58520	.52263	.46799	.41874
27	.67173	.59660	.53560	.48221	.43399
28	.68049	.60734	.54784	.49566	.44844
29	.68874	.61748	.55941	.50839	.46215
30	.69653	.62707	.57036	.52046	.47517
35	.72985	.66814	.61742	.57252	.53152
40	.75603	.70050	.65465	.61389	.57651
45	.77720	.72671	.68487	.64757	.61326
50	.79471	.74841	.70994	.67555	.64386
55	.80946	.76669	.73108	.69919	.66975
60	.82207	.78233	.74917	.71944	.69195
65	.83300	.79586	.76484	.73699	.71121
70	.84255	.80770	.77855	.75236	.72808
75	.85099	.81815	.79066	.76593	.74299
80	.85851	.82746	.80144	.77801	.75628
85	.86524	.83579	.81109	.78884	.76818
90	.87132	.84330	.81980	.79861	.77892
95	.87683	.85012	.82769	.80746	.78866
100	.88185	.85632	.83487	.81552	.79753
200	.93335	.91971	.90818	.89774	.88801
300	.95267	.94330	.93537	.92818	.92146
400	.96298	.95582	.94976	.94426	.93912
500	.96944	.96350	.95873	.95427	.95010

TABLE 14

Multivariate Tolerance Regions with β-Expectation (Type 2): Tables of D. A. S. Fraser and I. Guttman

(a) Content of the tables and definition of the test statistics :

Here we distinguish between two types of tolerance regions R :

(1) Type 1 : Tolerance regions with β-content : R is constructed that it contains with probability γ at least (100.β)% of the elements of the population. β is that portion of the population, which is to form the tolerance region.

(2) Type 2 : Tolerance regions with β-expectation : R is constructed so that the expectation of that portion of the population , which is to belong to R , is exactly (100.β)% .

For continuated sampling let β_i denote that portion of the population lying in R_i , computed on the basis of the i-th sample , and then

(1') $P(\beta_i \geq \beta) = \gamma$ whenever R is of type 1, and

(2') $E(\beta_i) = \beta$ whenever R is of type 2 .

A (100.β)% tolerance region of type 2 is identical to a (100.β)% predicted region for the next observation .

The tables at hand contain tolerance factors for determining tolerance regions with β-expectation (type 2) for several multivariate normal distribution.

The determination is made on the basis of a given sample od size n .

(aa) Tolerance factors for a univariate normal distribution with unknown mean value μ and unknown variance σ^2:

From a sample $x_1, x_2, \dots, x_n$ with mean $\bar{x}$ and standard deviation s_x one obtains a (minimax and strongest) tolerance region of the form

$$S(x_1,x_2, \dots ,x_n) = [\bar{x} - a_\beta.s_x , \bar{x} + a_\beta.s_x] .$$

Here $a_\beta = (1 + 1/n)^{1/2}.t_{1-\beta/2}$; t_α is that value which is exceeded with probability α in STUDENT's t-distribution with degrees of freedom $n-1$.

The values of a_β are tabulated on the first page of the table. If one wishes, for example, to have the tolerance region include the left end of the distribution , then one sets

$$S(x_1,x_2, \dots ,x_n) =]-\infty , \bar{x} + a'_\gamma.s_x]$$

where $a'_\gamma = a_{2.\gamma-1}$.

(bb) Tolerance factors for a univariate normal distribution with unknown mean value μ and known variance σ^2 :

Here the (minimax and strongest) tolerance region is determined by

$$S(x_1,x_2, \dots ,x_n) = [\bar{x} - b_\beta.\sigma , \bar{x} + b_\beta.\sigma] .$$

The factors b_β are given by

$$b_\beta = (1 + 1/n)^{1/2}. z_{(1-\beta)/2} .$$

z_α is that value which is exceeded with probability α in the standardised normal distribution. The values of b_β are contained in the second page of the tables.

If one wishes, for example, to have a tolerance region include the left end of the distribution, then one sets

$$S(x_1, x_2, \ldots, x_n) =]-\infty, \bar{x} + b'_\gamma \cdot \sigma]$$

where $b'_\gamma = b_{2\cdot\gamma-1}$.

If the mean μ is also known, then one obtains

$$S(x_1, x_2, \ldots, x_n) = [\mu - t'_{(1-\beta)/2} \cdot s_x , \mu + t'_{(1-\beta)/2} \cdot s_x]$$

where $s_x = n^{-1} \cdot \sum_{i=1}^{n} (x_i - \mu)^2$.

t'_α is the right hand α%-point of the t-distribution with n degrees of freedom

When including the left end of the distribution, one obtains

$$S(x_1, x_2, \ldots, x_n) =]-\infty, \mu + t'_{1-\beta} \cdot s_x] \quad .$$

(cc) Tolerance factors c_β for a multivariate normal distribution with unknown mean vector and unknown covariance matrix :

In analogy to the univariate case, one now obtains the (minimax and strongest) tolerance region with β-expectation for a p-variate normal distribution in the form of the ellipsoid shaped region

$$\{ \xi \mid (\xi - \underline{\bar{x}}) \cdot \underline{A}^{-1} \cdot (\xi - \underline{\bar{x}})' \leqq c_\beta \} \quad ,$$

where we define

$\underline{\bar{x}}$ = mean vector for a p-dimensional sample of size n ,

$\underline{A}$ = covariance matrix for a p-dimensional sample of size n ,

$\underline{x}_i$ = i-th (vector) element of this sample .

The values of the factors c_β for dimensions two, three , and four are contained on third through the fifth pages of the tables .

(b) Extent of the tables and definition of the parameters :

(1) The parameter p :

p = dimension of the variates

p = 1(1)4 .

(2) The parameter β :

β = expectation of the tolerance region region as defined in section (a)

β = 0.995 ; 0.990 ; 0.975 ; 0.950 ; 0.900 ; 0.750 .

(3) The parameter n :

n = size of the p-dimensional sample

n = 2(1)31, 41, 61, 121, ∞ when p = 1 ,
= 3(1)32, 42, 62, 122, ∞ when p = 2 ,
= 4(1)33, 43, 63, 123, ∞ when p = 3 ,
= 5(1)34, 44, 64, 124, ∞ when p = 4 .

(c) Suggestions for applications :

The tables, subject to the discussion of section (a) , permit the determination of tolerance factors for setting tolerance regions pf type 2 (with β-expectation).

(d) References :

For the definition of the tolerance factors and the tolerance regions, as well as for the source of the tables :

FRASER, D. A. + GUTTMAN, I. : Tolerance regions.
The Annals of Mathematical Statistics 27, 167 - 179 (1956).

(e) Further remarks :

The tolerance factors c_β in the multinormal case can be computed from

$$c_\beta = (1 + 1/n).(n - 1).(p/n - p).F_{1 - \beta} \quad .$$

By definition, F_α is that point in the F-distribution which is exceeded with probability α . The degrees of freedom in the F-distribution are

$$f_1 = p \quad \text{and} \quad f_2 = n - p \quad .$$

Tafel 14 Table 14

a_β für / for $p = 1$

n	β					
	.995	.99	.975	.95	.90	.75
2	155.9	77.96	31.17	15.56	7.733	2.957
3	16.27	11.46	7.165	4.968	3.372	1.852
4	8.333	6.530	4.669	3.558	2.631	1.591
5	6.132	5.044	3.829	3.041	2.335	1.473
6	5.156	4.355	3.417	2.777	2.176	1.405
7	4.615	3.963	3.174	2.616	2.077	1.361
8	4.274	3.712	3.014	2.508	2.010	1.330
9	4.040	3.537	2.900	2.431	1.960	1.307
10	3.870	3.408	2.816	2.373	1.923	1.290
11	3.741	3.310	2.751	2.327	1.893	1.276
12	3.639	3.233	2.699	2.291	1.869	1.264
13	3.558	3.170	2.657	2.261	1.850	1.255
14	3.491	3.118	2.621	2.236	1.833	1.246
15	3.435	3.074	2.592	2.215	1.819	1.239
16	3.387	3.037	2.567	2.197	1.807	1.234
17	3.346	3.005	2.545	2.181	1.797	1.228
18	3.311	2.978	2.525	2.168	1.787	1.224
19	3.280	2.953	2.509	2.155	1.779	1.220
20	3.252	2.932	2.494	2.145	1.772	1.216
21	3.228	2.912	2.480	2.135	1.765	1.213
22	3.206	2.895	2.468	2.126	1.759	1.210
23	3.186	2.879	2.457	2.119	1.754	1.207
24	3.168	2.865	2.447	2.111	1.749	1.205
25	3.152	2.852	2.438	2.105	1.745	1.202
26	3.137	2.840	2.430	2.099	1.741	1.200
27	3.123	2.830	2.422	2.093	1.737	1.198
28	3.111	2.820	2.415	2.088	1.733	1.197
29	3.099	2.811	2.409	2.083	1.730	1.195
30	3.088	2.802	2.403	2.079	1.727	1.193
31	3.078	2.794	2.397	2.075	1.724	1.192
41	3.007	2.737	2.357	2.046	1.704	1.181
61	2.938	2.682	2.318	2.017	1.684	1.171
121	2.872	2.628	2.279	1.988	1.665	1.161
∞	2.807	2.576	2.241	1.960	1.645	1.150

Tafel 14 (Forts.) Table 14 (cont.)

b_β für / for $p = 1$

n	β					
	.995	.99	.975	.95	.90	.75
2	3.438	3.155	2.745	2.401	2.015	1.409
3	3.241	2.974	2.588	2.263	1.899	1.328
4	3.138	2.880	2.506	2.191	1.839	1.286
5	3.075	2.822	2.455	2.147	1.802	1.260
6	3.032	2.782	2.421	2.117	1.777	1.242
7	3.001	2.754	2.396	2.095	1.758	1.230
8	2.977	2.732	2.377	2.079	1.745	1.220
9	2.959	2.715	2.363	2.066	1.734	1.213
10	2.944	2.702	2.351	2.056	1.725	1.206
11	2.932	2.690	2.341	2.047	1.718	1.201
12	2.922	2.681	2.333	2.040	1.712	1.197
13	2.913	2.673	2.326	2.034	1.707	1.194
14	2.906	2.666	2.320	2.029	1.703	1.191
15	2.899	2.660	2.315	2.024	1.699	1.188
16	2.893	2.655	2.310	2.020	1.696	1.186
17	2.888	2.650	2.306	2.017	1.693	1.184
18	2.884	2.646	2.303	2.014	1.690	1.182
19	2.880	2.643	2.300	2.011	1.688	1.180
20	2.876	2.639	2.297	2.008	1.686	1.179
21	2.873	2.636	2.294	2.006	1.684	1.177
22	2.870	2.634	2.292	2.004	1.682	1.176
23	2.867	2.631	2.290	2.002	1.680	1.175
24	2.865	2.629	2.288	2.000	1.679	1.174
25	2.863	2.627	2.286	1.999	1.677	1.173
26	2.860	2.625	2.284	1.997	1.676	1.172
27	2.859	2.623	2.283	1.996	1.675	1.171
28	2.857	2.621	2.281	1.995	1.674	1.171
29	2.855	2.620	2.280	1.994	1.673	1.170
30	2.853	2.618	2.278	1.992	1.672	1.169
31	2.852	2.617	2.277	1.991	1.671	1.169
41	2.841	2.607	2.269	1.984	1.665	1.164
61	2.830	2.597	2.260	1.976	1.658	1.160
121	2.819	2.586	2.251	1.968	1.652	1.155
∞	2.807	2.576	2.241	1.960	1.645	1.150

Tafel 14 (Forts.) Table 14 (cont.)

c_β für / for $p = 2$

n	β					
	.995	.99	.975	.95	.90	.75
3	106,667	26,664	4,264	1,064	264.0	40.00
4	746.2	371.2	146.2	71.25	33.75	11.25
5	159.4	98.61	51.34	30.57	17.48	7.295
6	76.66	52.50	31.06	20.25	12.61	5.833
7	50.23	36.41	23.13	15.87	10.37	5.082
8	38.18	28.68	19.06	13.50	9.091	4.626
9	31.50	24.25	16.61	12.03	8.273	4.320
10	27.33	21.41	15.00	11.04	7.705	4.101
11	24.50	19.45	13.85	10.32	7.288	3.936
12	22.47	18.02	13.00	9.778	6.970	3.807
13	20.94	16.93	12.35	9.357	6.719	3.705
14	19.75	16.08	11.83	9.019	6.516	3.620
15	18.81	15.40	11.41	8.743	6.348	3.550
16	18.04	14.83	11.06	8.513	6.208	3.491
17	17.39	14.36	10.76	8.318	6.088	3.440
18	16.85	13.97	10.51	8.151	5.985	3.395
19	16.39	13.62	10.30	8.006	5.895	3.356
20	15.99	13.33	10.11	7.879	5.816	3.322
21	15.64	13.07	9.941	7.768	5.747	3.292
22	15.34	12.84	9.795	7.668	5.685	3.265
23	15.07	12.64	9.663	7.580	5.629	3.240
24	14.82	12.46	9.546	7.500	5.579	3.218
25	14.61	12.29	9.440	7.427	5.533	3.198
26	14.41	12.14	9.343	7.362	5.492	3.179
27	14.23	12.01	9.256	7.302	5.454	3.162
28	14.07	11.89	9.176	7.247	5.419	3.147
29	13.92	11.78	9.102	7.197	5.387	3.133
30	13.79	11.67	9.034	7.150	5.357	3.119
31	13.66	11.58	8.971	7.107	5.330	3.107
32	13.54	11.49	8.913	7.067	5.304	3.095
42	12.73	10.87	8.502	6.783	5.122	3.013
62	11.97	10.28	8.110	6.509	4.945	2.931
122	11.26	9.732	7.736	6.246	4.773	2.851
∞	10.60	9.210	7.378	5.991	4.605	2.773

Tafel 14 (Forts.) Table 14 (cont.)

c_β für / for $p = 3$

n	β					
	.995	.99	.975	.95	.90	.75
4	243,169	60,787	9,722	2,427	602.9	92.25
5	1,434	714.0	282.0	138.0	65.96	22.70
6	276.9	171.8	90.06	54.11	31.45	13.74
7	124.8	85.85	51.32	33.90	21.55	10.53
8	78.10	56.98	36.68	25.56	17.10	8.903
9	57.41	43.46	29.33	21.14	14.62	7.931
10	46.17	35.86	24.99	18.44	13.04	7.285
11	39.26	31.05	22.16	16.63	11.96	6.825
12	34.63	27.77	20.17	15.34	11.17	6.481
13	31.33	25.40	18.71	14.38	10.58	6.214
14	28.87	23.62	17.59	13.63	10.11	6.001
15	26.98	22.22	16.70	13.03	9.727	5.827
16	25.47	21.11	15.99	12.54	9.416	5.683
17	24.25	20.20	15.40	12.14	9.156	5.560
18	23.24	19.44	14.90	11.80	8.936	5.456
19	22.39	18.80	14.48	11.51	8.746	5.366
20	21.67	18.25	14.12	11.25	8.581	5.286
21	21.05	17.78	13.81	11.03	8.437	5.216
22	20.51	17.37	13.53	10.84	8.309	5.155
23	20.03	17.00	13.29	10.67	8.196	5.099
24	19.61	16.68	13.07	10.52	8.094	5.049
25	19.24	16.39	12.88	10.38	8.003	5.004
26	18.90	16.14	12.70	10.25	7.920	4.963
27	18.60	15.90	12.54	10.14	7.844	4.926
28	18.33	15.69	12.40	10.04	7.775	4.892
29	18.08	15.50	12.26	9.943	7.712	4.860
30	17.85	15.32	12.14	9.857	7.654	4.831
31	17.64	15.16	12.03	9.777	7.600	4.804
32	17.45	15.01	11.93	9.703	7.550	4.779
33	17.27	14.87	11.83	9.635	7.504	4.756
43	16.04	13.90	11.16	9.150	7.175	4.590
63	14.89	12.99	10.53	8.686	6.857	4.426
123	13.83	12.14	9.922	8.241	6.549	4.266
∞	12.84	11.34	9.348	7.815	6.251	4.108

Tafel 14 (Forts.) Table 14 (cont.)

c_β für / for $p = 4$

n	β					
	.995	.99	.975	.95	.90	.75
5	432,000	107,992	17,272	4,312	1,072	164.8
6	2,325	1,158	457.9	224.5	107.8	37.71
7	422.4	262.5	138.1	83.36	48.85	21.85
8	182.3	125.8	75.64	50.31	32.34	16.26
9	110.6	81.01	52.54	36.92	25.03	13.46
10	79.38	60.38	41.10	29.92	20.99	11.80
11	62.65	48.91	34.43	25.69	18.46	10.70
12	52.46	41.74	30.11	22.87	16.72	9.916
13	45.70	36.89	27.10	20.87	15.47	9.335
14	40.91	33.40	24.89	19.38	14.52	8.886
15	37.37	30.78	23.22	18.23	13.77	8.528
16	34.64	28.75	21.89	17.31	13.18	8.236
17	32.49	27.13	20.83	16.57	12.69	7.994
18	30.75	25.82	19.95	15.96	12.28	7.790
19	29.32	24.72	19.22	15.44	11.93	7.615
20	28.12	23.83	18.60	15.00	11.63	7.464
21	27.10	23.02	18.07	14.62	11.38	7.332
22	26.22	22.34	17.60	14.28	11.15	7.216
23	25.46	21.75	17.20	13.99	10.95	7.113
24	24.79	21.23	16.84	13.73	10.78	7.021
25	24.20	20.77	16.52	13.50	10.62	6.938
26	23.68	20.36	16.24	13.30	10.48	6.863
27	23.21	19.99	15.98	13.11	10.35	6.795
28	22.79	19.66	15.75	12.94	10.23	6.733
29	22.41	19.36	15.54	12.79	10.12	6.676
30	22.06	19.09	15.35	12.64	10.03	6.624
31	21.74	18.84	15.17	12.51	9.935	6.576
32	21.45	18.61	15.01	12.40	9.851	6.532
33	21.19	18.39	14.86	12.28	9.774	6.490
34	20.94	18.20	14.72	12.18	9.703	6.452
44	19.23	16.84	13.75	11.46	9.195	6.177
64	17.66	15.57	12.83	10.77	8.706	5.907
124	16.20	14.38	11.96	10.11	8.234	5.643
∞	14.86	13.28	11.14	9.488	7.780	5.385

TABLE 15

Multivariate Tolerance Regions with β-Content (Type 1): Tables of V. Chew

(a) Content of the tables and definition of the test statistics :

Here we distinguish between two types of tolerance regions R :

(1) Type 1 : Tolerance regions with β-content : R is constructed that it contains with probability γ at least (100.β)% of the elements of the population. β is that portion of the population, which is to form the tolerance region.

(2) Type 2 : Tolerance regions with β-expectation : R is constructed so that the expectation of that portion of the population , which is to belong to R , is exactly (100.β)% .

For continuated sampling let β_i denote that portion of the population lying on R_i , computed on the basis of the i-th sample, and then

(1') $P(\beta_i \geqq \beta) = \gamma$ whenever R is of type 1

(2') $P(\beta\ )\ = \beta$ whenever R is of type 2 .

A (100.β)% tolerance region of type 2 is identical to a (100.β)% predicted region for the next observation.

The tables at hand contain tolerance bounds for determining tolerance regions with β-content (type 1) for bivariate normally distributed populations with unknown mean vector and unknown covariance matrix. The determination is made on the basis of a given sample of size n .

When the mean vector and the covariance matrix are unknown, the tolerance ellipsoid which contains at least (100.β)% of the population with probability approximately (100.γ)% has the form

$$(\underline{x} - \underline{m})'\underline{S}^{-1}(\underline{x} - \underline{m}) = \frac{\chi'^2(1-\beta, p, p/n)}{\{\chi^2(\gamma,\nu p)\}/(\nu p)} = H .$$

This approximation is good whenever $1/(n^2)$ can be disregarded. $\underline{S}$ and $\underline{m}$ are the covariance matrix and the mean vector, respectively, of the given sample of size n .
The remaining terms are defined as follows :

$\nu = n - 1$

$\chi^2(\alpha,\nu)$ = upper (100.α)% points of the χ^2-distribution with degrees of freedom ,

$\chi'^2(\alpha,\nu,\lambda)$ = upper (100.α)% points of the non-central χ'^2-distribution with ν degrees of freedom and non-centrality parameter γ .

Table 15 contains the values of the bound H for different combinations of the initial parameters γ, β, and n .

(b) Extent of the tables and definition of the parameters :

(1) The parameter γ :

γ = confidence level for the prediction

γ = 0.90 ; 0.95 ; 0.99 ;

(2) The parameter β :

β = that portion of the population which is included in the tolerance region ,

β = 0.90 ; 0.95 ; 0.99 .

(3) The parameter p :

p = dimension of the variate

p = 2 .

(4) The parameter n :

n = size of the p-dimensional sample

n = 8(1)20(2)30(5)50, 60, 80, 100, 200, 800, ∞ .

(c) Suggestions for applications :

The table, subject to the discussion of section (a) , gives the constant H and thereby allows the determination of the bivariate tolerance region of type 1 with content β from a given sample of size n .

(d) References :

For the source of the table and for the definition of the tolerance region :

CHEW, V. : Confidence, prediction, and tolerance regions for the multivariate normal distribution. (Table 2). Journal of the American Statistical Association 61 , 605 - 617 (1966) .

(e) Further remarks :

In the case where the mean vector of the population is known and only the covariance matrix remains unknown, one obtains the tolerance ellipsoid

$$(\underline{x} - \underline{\mu})' (\underline{S}')^{-1} (\underline{x} - \underline{\mu}) = \chi^2(1 - \beta, p) / \{\chi^2(\gamma, np)/np\} .$$

Here $\underline{\mu}$ is the known mean vector and $\underline{S}'$ is the covariance matrix of the sample , in the case where the mean vector is known .

H

γ: β: n	.90			.95			.99		
	.90	.95	.99	.90	.95	.99	.90	.95	.99
8	9.31	12.08	18.49	11.03	14.32	21.92	15.54	20.17	30.87
9	8.79	11.41	17.48	10.27	13.34	20.43	14.09	18.30	28.02
10	8.39	10.90	16.70	9.70	12.61	19.32	12.99	16.87	25.86
11	8.08	10.50	16.09	9.25	12.02	18.43	12.17	15.81	24.23
12	7.82	10.16	15.58	8.89	11.55	17.72	11.49	14.94	22.91
13	7.60	9.88	15.17	8.59	11.17	17.14	10.97	14.26	21.88
14	7.42	9.65	14.81	8.34	10.84	16.63	10.52	13.68	20.99
15	7.27	9.45	14.51	8.12	10.56	16.21	10.16	13.20	20.26
16	7.13	9.27	14.22	7.92	10.34	15.86	9.83	12.78	19.62
17	7.01	9.12	14.00	7.78	10.12	15.54	9.55	12.42	19.06
18	6.91	8.98	13.80	7.64	9.93	15.25	9.30	12.09	18.57
19	6.81	8.86	13.60	7.51	9.76	14.99	9.08	11.81	18.14
20	6.72	8.74	13.42	7.38	9.60	14.75	8.88	11.54	17.73
22	6.57	8.54	13.12	7.18	9.34	14.36	8.55	11.12	17.08
24	6.45	8.39	12.89	7.03	9.14	14.04	8.27	10.76	16.53
26	6.35	8.26	12.68	6.89	8.96	13.76	8.06	10.48	16.10
28	6.26	8.14	12.50	6.76	8.80	13.51	7.86	10.23	15.72
30	6.18	8.04	12.35	6.66	8.67	13.32	7.70	10.01	15.38
35	6.02	7.84	12.04	6.44	8.38	12.88	7.35	9.56	14.70
40	5.90	7.67	11.79	6.28	8.17	12.56	7.09	9.23	14.18
45	5.80	7.54	11.60	6.15	8.00	12.29	6.90	8.97	13.79
50	5.71	7.43	11.43	6.04	7.86	12.09	6.73	8.76	13.46
60	5.60	7.28	11.19	5.89	7.66	11.77	6.49	8.44	12.98
80	5.43	7.07	10.87	5.68	7.39	11.36	6.17	8.02	12.33
100	5.33	6.93	10.65	5.54	7.20	11.07	5.96	7.76	11.92
200	5.08	6.61	10.16	5.22	6.79	10.43	5.49	7.14	10.98
800	4.83	6.29	9.66	4.90	6.37	9.78	5.02	6.53	10.02
∞	4.61	5.99	9.21	4.61	5.99	9.21	4.61	5.99	9.21

TABLE 16

Testing a Single Covariance Matrix: Tables of B. P. Korin

(a) Content of the table and definition of the test statistic :

The table contains the upper percentage points of the test statisric

$$L = \nu.\ln|\underline{\Sigma}_o| - \nu.p - \nu.\ln|\underline{S}| + \nu.\text{trace}(\underline{S}.\underline{\Sigma}_o^{-1})$$

$$= \nu . \{ \ln|\underline{\Sigma}_o| - p - \ln|\underline{S}| + \text{trace}(\underline{S} . \underline{\Sigma}_o^{-1}) \} .$$

The statistic L is a likelihood ratio statistic for testing the hypothesis that the condition

$$H_o : \underline{\Sigma} = \underline{\Sigma}_o$$

holds for the covariance matrix $\underline{\Sigma}$ of a multinormal population. $\underline{\Sigma}_o$ is a given matrix. The test is based on the covariance matrix $\underline{S}$ of a p-dimensional sample of size N , where $\nu = N - 1$.

(b) Extent of the table and definition of the parameters :

(1) The parameter α :

α = error probability

α = 5% and 1% .

(2) The parameter p :

p = dimension of the variate

p = 2(1)20 .

(3) The parameter ν :

$\nu = N - 1$ = degrees of freedom of the covariance matrix , where N = size of the p-dimensional sample, for various values of ν between 2 and 75, depending on p .

(c) Suggestions for applications :

The table , as noted in section (a) , serves to test the hypothesis H_0 that the covariance matrix $\underline{\Sigma}$ of a multinormal population is identical to a given matrix $\underline{\Sigma}_0$.

(d) References :

(1) For the test statisric and the computation of the tables:
KORIN, B. P. : On the distribution of a statistic used for testing a covariance matrix.(Table 3). Biometrika 55, 171 - 178 (1968).

(2) For the source of the table in a modified form:
PEARSON, E. S. + HARTLEY, H. O. (eds.) : Biometrika Tables for Statisticians. Vol. II. (Table 53). Cambridge: Cambridge University Press, 1972 (Published for the Biometrika Trustees).

(3) For the approximations of section (e) :

BOX, G. E. P. : A general distribution theory for a class of likelihood criteria. Biometrika 36 , 317 - 346 (1949).

(e) Further remarks :

According to KORIN one can use the following approximations for the distribution of L outside the tabulated parameter regions :

(1) The χ^2-approximation :

$$L \sim \chi^2/(1 - D_1)$$

where $f_1 = (1/2).p.(p + 1)$

$=$ degree of freedom of the χ^2-distribution ,

and $D_1 = \{ 2.p + 1 - \frac{2}{p + 1} \}/(6.\nu)$

(2) The F-approximation :

$$L \sim b \cdot F_{(f_1, f_2)}$$

where $f_1 = (1/2).p.(p + 1)$

$=$ first degree of freedom of the F-distribution ,

$f_2 = (f_1 + 2)/(D_2 - D_1^2)$

$=$ second degree of freedom of the F-distribution ,

D_1 is as in the χ^2-approximation

$$D_2 = (p - 1).(p + 2)/(6.\nu^2) ,$$

and

$$b = \frac{f_1}{1 - D_1 - f_1/f_2} .$$

Tafel 16 — Table 16

Obere Prozentpunkte von L

Upper percentage points of L

p = 2

ν	5%	1%
2	13·50	19·95
3	10·64	15·56
4	9·69	14·13
5	9·22	13·42
6	8·94	13·00
7	8·75	12·73
8	8·62	12·53
9	8·52	12·38
10	8·44	12·26

p = 3

ν	5%	1%
4	18·8	25·6
5	16·82	22·68
6	15·81	21·23
7	15·19	20·36
8	14·77	19·78
9	14·47	19·36
10	14·24	19·04
11	14·06	18·80
12	13·92	18·61
13	13·80	18·45
14	13·70	18·31
15	13·62	18·20

p = 4

ν	5%	1%
7	25·8	30·8
8	24·06	29·33
9	23·00	28·36
10	22·28	27·66
11	21·75	27·13
12	21·35	26·71
13	21·03	26·38
14	20·77	26·10
15	20·56	25·87

p = 5

ν	5%	1%
9	32·5	40·0
10	31·4	38·6
11	30·55	37·51
12	29·92	36·72
13	29·42	36·09
14	29·02	35·57
15	28·68	35·15
16	28·40	34·79
17	28·15	34·49
18	27·94	34·23
19	27·76	34·00
20	27·60	33·79

p = 6

ν	5%	1%
12	40·9	49·0
13	40·0	47·8
14	39·3	47·0
15	38·7	46·2
16	38·22	45·65
17	37·81	45·13
18	37·45	44·70
19	37·14	44·32
20	36·87	43·99
21	36·63	43·69
22	36·41	43·43
24	36·05	42·99
26	35·75	42·63
28	35·49	42·32
30	35·28	42·07

p = 7

ν	5%	1%
18	48·6	56·9
19	48·2	56·3
20	47·7	55·8
21	47·34	55·36
22	47·00	54·96
24	46·43	54·28
26	45·97	53·73
28	45·58	53·27
30	45·25	52·88
32	44·97	52·55
34	44·73	52·27

p = 8

ν	5%	1%
24	58·4	67·1
26	57·7	66·3
28	57·09	65·68
30	56·61	65·12
32	56·20	64·64
34	55·84	64·23
36	55·54	63·87
38	55·26	63·55
40	55·03	63·28

p = 9

ν	5%	1%
28	70·1	79·6
30	69·4	78·8
32	68·8	78·17
34	68·34	77·60
36	(67·91)	(77·08)
38	(67·53)	(76·65)
40	67·21	76·29
45	66·54	75·51
50	66·02	74·92
55	65·61	74·44
60	65·28	74·06

p = 10

ν	5%	1%
34	(82·3)	(92·4)
36	81·7	91·8
38	81·2	91·2
40	80·7	90·7
45	79·83	89·63
50	79·13	88·83
55	78·57	88·20
60	78·13	87·68
65	77·75	87·26
70	77·44	86·89
75	77·18	86·59

Entries in parentheses have been interpolated or extrapolated into Korin's table.

Die Werte in Klammern sind interpoliert oder extrapoliert worden.

TABLE 17

Testing the Equality of k Covariance Matrices: Tables of B. P. Korin

(a) Content of the table and definition of the test statistic :

The table contains the upper percentage points for the test statistic

$$M = N \cdot \ln |\underline{S}| - \sum_{i=1}^{k} \nu_i \cdot \ln |\underline{S}_i|$$

for testing the hypothesis

$$H_o : \underline{\Sigma}_1 = \underline{\Sigma}_2 = \ldots\ldots = \underline{\Sigma}_k ,$$

that the covariance matrices of k multinormal p-dimensional populations are equal. The $\underline{S}_i$ are the $p \times p$ covariance matrices of k given samples.

The sample sizes $n_1, n_2, n_3, \ldots, n_k$ are required to be all equal and are denoted by n_o .

The corresponding degrees of freedom for all samples are therefore all equal to $\nu_o = n_o - 1$.

The sum of the degrees of freedom of the k covariance matrices is thus

$$N = k \cdot \nu_o .$$

Finally one obtains a "pooled" covariance matrix $\underline{S}$ from

$$N \cdot \underline{S} = \nu_o \cdot \underline{S}_1 + \nu_o \cdot \underline{S}_2 + \ldots + \nu_o \cdot \underline{S}_k$$

(b) Extent of the table and definition of the parameters :

(1) The parameter α :

α = error probability

α = 5% .

(2) The parameter p :

p = dimension of the variate

p = 2(1)6 .

(3) The parameter k :

k = the number of samples

k =	2(1)10	when p = 2; 3; 4 ;
=	2(1)7	when p = 5;
=	2(1)5	when p = 6 .

(4) The parameter ν_o :

$\nu_o = n_o - 1$ = degrees of freedom for each covariance matrix , where
n_o = sample size (equal for all k samples)

ν_o =	3(1)10	when p = 2 ;
=	5(1)13	when p = 3 ;
=	6(1)15	when p = 4 ;
=	8(1)16	when p = 5 ;
=	10(1)20	when p = 6 .

(c) Suggestions for applications :

The table contains the upper percentage points for the test statistic M for testing the null hypothesis that the covariance matrices of k p-dimensional normally distributed populations are equal. (See the discussion in section (a)).

(d) References :

(1) For the test criterion and the computation of the table :
KORIN, B. P. : On testing the equality of k covariance matrices. (Table 2).
Biometrika 56, 216 - 218 (1969).

(2) For the source of the table in modified form :
PEARSON, E. S. + HARTLEY, H. O. : Biometrika Tables for Statisticians. Vol. II.(Table 50).
Cambridge: Cambridge University Press, 1972.
(Published for the Biometrika Trusties).

(3) Further literature on the tabulated statistic :
ANDERSON, T. W. : An introduction to multivariate statistcal analysis.
New York : Wiley, 1958.
BOX, G. E. P. : A general distribution theory for a class of likelihood criteria.
Biometrika 36, 317 - 346 (1949)
KORIN, B. P. : On the distribution of a statistic used for testing a covariance matrix.
Biometrika 55, 171 - 178 (1968). (See table 15).
KULLBACK, S. : Information theory and statistics.
New York : Wiley, 1959 .

(e) Further remarks :

Outside the tabulated region of the parameters the following approximations for the distribution of M are available. Here no equality of the k sample sizes n_i nor their corresponding degrees of freedom ν_i is required.

(1) The χ^2-approximation :

We have

$$M \sim \chi^2_{f_1}/(1 - D_1)$$

where $f_1 = (1/2).p.(p + 1).(k - 1)$

is the degree of freedom of the χ^2-distribution ,

and $$D_1 = \frac{2p^2 + 3p - 1}{6(p + 1).(k - 1)}.\{\sum_{t=1}^{k} (1/\nu_t) - (1/N)\} .$$

Whenever $\nu_t = \nu_o$ for $t = 1, 2, \ldots, k$,

that is, when all k sample sizes are equal , then

$$D_1 = \frac{(2p^2 + 3p - 1).(k + 1)}{6.(p + 1).k.\nu_o} .$$

(2) The F-approximation :

Here we have

$$M \sim b . F_{f_1,f_2}$$

where $f_1 = (1/2).p.(p + 1).(k - 1)$

is the first degree of freedom of the F-distribution ,

$$f_2 = (f_1 + 2)/(D_2 - D_1^2)$$

is the second degree of freedom of the F-distribution ,

with

$$D_1 = \text{as in (1) for the } \chi^2\text{-distribution}$$

$$D_2 = \frac{(p-1).(p+2)}{6.(k-1)} .\{\sum_{t=1}^{k}(1/\nu_t^2) - 1/N^2\}$$

or

$$= \frac{(p-1).(p+2).(k^2+k+1)}{6.k^2.\nu_o^2}$$

when $\nu_t = \nu_o$ for $t = 1, 2, \ldots, k$ (i. e. k equal sample sizes)

as well as

$$b = \frac{f_1}{1 - D_1 - f_1/f_2} .$$

5%- Punkte von M

5 per cent points of M

ν_0 \ k	2	3	4	5	6	7	8	9	10
					$p = 2$				
3	12·18	18·70	24·55	30·09	35·45	40·68	45·81	50·87	55·87
4	10·70	16·65	22·00	27·07	31·97	36·76	41·45	46·07	50·64
5	9·97	15·63	20·73	25·56	30·23	34·79	39·26	43·67	48·02
6	9·53	15·02	19·97	24·66	29·19	33·61	37·95	42·22	46·45
7	9·24	14·62	19·46	24·05	28·49	32·82	37·07	41·26	45·40
8	9·04	14·33	19·10	23·62	27·99	32·26	36·45	40·57	44·65
9	8·88	14·11	18·83	23·30	27·62	31·84	35·98	40·06	44·08
10	8·76	13·94	18·61	23·05	27·33	31·51	35·61	36·65	43·64
					$p = 3$				
5	19·2	30·5	41·0	51·0	60·7	70·3	79·7	89·0	98·3
6	17·57	28·24	38·06	47·49	56·68	65·69	74·58	83·37	92·09
7	16·59	26·84	36·29	45·37	54·21	62·89	71·45	79·91	88·29
8	15·93	25·90	35·10	43·93	52·54	60·99	69·33	77·56	85·72
9	15·46	25·22	34·24	42·90	51·34	59·62	67·79	75·86	83·86
10	15·11	24·71	33·59	42·11	50·42	58·58	66·62	74·57	82·45
11	14·83	24·31	33·08	41·50	49·71	57·76	65·71	73·56	81·35
12	14·61	23·99	32·67	41·01	49·13	57·11	64·97	72·75	80·46
13	14·43	23·73	32·33	40·60	48·66	56·57	64·37	72·08	79·72
					$p = 4$				
6	30·07	48·63	65·91	82·6	98·9	115·0	131·0	—	—
7	27·31	44·69	60·90	76·56	91·89	107·0	121·9	137·0	152·0
8	25·61	42·24	57·77	72·78	87·46	101·9	116·2	130·4	144·6
9	24·46	40·56	55·62	70·17	84·42	98·45	112·3	126·1	139·8
10	23·62	39·34	54·05	68·27	82·19	95·91	109·5	122·9	136·3
11	22·98	38·41	52·85	66·81	80·49	93·95	107·3	120·5	133·6
12	22·48	37·67	51·90	65·66	79·14	92·41	105·5	118·5	131·5
13	22·08	37·08	51·13	64·73	78·04	91·16	104·1	117·0	129·7
14	21·75	36·59	50·50	63·96	77·14	90·12	103·0	115·7	128·3
15	21·47	36·17	49·97	63·31	76·38	89·25	102·0	114·6	127·1

ν_0 \ k	2	3	4	5	6	7
			$p = 5$			
8	39·29	65·15	89·46	113·0	—	—
9	36·70	61·40	84·63	107·2	129·3	151·5
10	34·92	58·79	81·25	103·1	124·5	145·7
11	33·62	56·86	78·76	100·0	120·9	141·6
12	32·62	55·37	76·83	97·68	118·2	138·4
13	31·83	54·19	75·30	95·81	116·0	135·9
14	31·19	53·24	74·06	94·29	114·2	133·8
15	30·66	52·44	73·02	93·03	112·7	132·1
16	30·21	51·77	72·14	91·95	111·4	130·6

ν_0 \ k	2	3	4	5
		$p = 6$		
10	49·95	84·43	117·0	—
11	47·43	80·69	112·2	142·9
12	45·56	77·90	108·6	138·4
13	44·11	75·74	105·7	135·0
14	42·96	74·01	103·5	132·2
15	42·03	72·59	101·6	129·9
16	41·25	71·41	100·1	128·0
17	40·59	70·41	98·75	126·4
18	40·02	69·55	97·63	125·0
19	39·53	68·80	96·64	123·8
20	39·11	68·14	95·78	122·7

TABLE 18

Distribution of the Extreme Roots of a Wishart Matrix: Tables of R. Ch. Hanumara and W. A. Thompson

(a) Content of the tables and definition of the tabulatesd statistic :

The tables contain the upper percentage points for the distribution of the largest root of a WISHART matrix and the lower percentage points for the distribution of the smallest root of a WISHART matrix.
Let there be given a sample of size N from a p-variate multinormal population with covariance matrix $\underline{\Sigma}$.

Let the covariance matrix of this sample be

$$\underline{S} = \nu^{-1} . \underline{A} \quad \text{where} \quad \underline{A} = (a_{ij}) \text{ is the SP matrix and}$$

$$\nu = N - 1 .$$

$\underline{A} = \nu . \underline{S}$ belongs to a WISHART distribution $W(\underline{\Sigma}, \nu)$.

The roots of the determinant equation

$$| \underline{A} . \underline{\Sigma}^{-1} - c.\underline{I} | = 0$$

are denoted by

$$0 = c_1 \leq c_2 \leq \ldots \leq c_p .$$

$\underline{I}$ is the identity matrix.
Insofar as the following treats the special case $\underline{\Sigma} = \underline{I}$, the roots c_i are identical to the roots of the WISHART distributed matrix $\underline{A}$.

We now introduce the following notation :

$$c_1 = c_{min} = c(\underline{A}) ,$$

$$c_p = c_{max} = C(\underline{A}) \quad .$$

In this notation, the tables contain the upper percentage points

u_1 for $C(\underline{A})$, determined by

$$P\{ C(\underline{A}) \leqq u_1 \} = 1 - \alpha \quad ,$$

as well as the lower percentage points

l for $c(\underline{A})$, determined by

$$P\{ c(\underline{A}) \geqq l \} = 1 - \alpha \quad .$$

In addition we denote by u the upper bound of

$C(\underline{A})$ determined by the relation

$$P\{ l \leqq c(\underline{A}) \leqq C(\underline{A}) \leqq u \} = 1 - 2.\alpha \quad .$$

In this case it follows from the investigations of HANUMARA and THOMPSON that, with error probabilities $\alpha = 0.05$, one may regard u and u_1 as approximately equal ($u \sim u_1$) so that separate tables for u are not necessary.

(b) Extent of the tables and definition of the parameters :

(1) The parameter α :

α = error probability

α = 0.005 ; 0.010 ; 0.025 ; 0.050 .

(2) The parameter p :

p = dimension of the variantes with multinormal distribution

p = 2(1)10 .

(3) The parameter ν :

ν = degrees of freedom of the WISHART matrix

= N - 1 , where N = sample size ,

ν = p(1)10(5)30(10)100 .

(c) Suggestions for applications :

The tables contain - as discussed in section (a) - the upper percentage points for the largest root and the lower percentage points for the smallest root of a WISHART matrix. They may be used

(1) to determine confidence regions for the roots ,

(2) to construct simultaneous confidence intervals for special experiments.

(d) References :

(1) For the source of the tables :

HANUMARA, R. Ch. + THOMPSON,Jr., W. A. : Percentage points of the extreme roots of a WISHART matrix. Biometrika 55, 505 - 512 (1968).

(2) Further literature :

PEARSON, E. S. + HARTLEY, H. O. (eds.) : Biometrika Tables for Statisticians. Vol. II.(Table 51). Cambridge: Cambridge University Press, 1972. (Published for the Biometrika Trustees).

PILLAI, K. C. S. + CHANG, T. C. : An approximation to the c. d. f. of the largest root of a covariance matrix. Ann. Inst. Statist. Math., Suppl., 6,115 - 124 (1970).

(e) Further remarks :

The modified tables in PEARSON & HARTLEY for the largest root $C(\underline{A})$ include additional parameter values up to $\nu = 200$, although only for

$$\alpha = 5\% \quad \text{and}$$

$$\alpha = 1\% \quad .$$

Tafel 18 Table 18

Obere Prozentpunkte l Obere Prozentpunkte $u_1 \simeq u$

Lower percentage points l *Upper percentage points* $u_1 \simeq u$

ν \ α	0·005	0·010	0·025	0·050	0·005	0·010	0·025	0·050
				$p = 2$				
2	$0{\cdot}0^{4}1518$	$0{\cdot}0^{4}6287$	$0{\cdot}0^{3}3858$	$0{\cdot}0^{2}1500$	13·66	12·16	10·15	8·594
3	$\cdot0^{2}5012$	$\cdot0^{2}1005$	·02532	·05129	16·16	14·57	12·42	10·74
4	·04047	·06477	·1216	·1980	18·40	16·73	14·46	12·68
5	·1264	·1812	·2948	·4314	20·48	18·73	16·36	14·49
6	·2659	·3573	·5340	·7333	22·45	20·64	18·17	16·21
7	·4550	·5858	·8278	1·090	24·33	22·47	19·91	17·88
8	·6880	·8595	1·167	1·489	26·15	24·23	21·59	19·49
9	·9597	1·172	1·544	1·926	27·92	25·95	23·24	21·06
10	1·265	1·518	1·953	2·392	29·65	27·63	24·84	22·60
15	3·184	3·629	4·358	5·059	37·83	35·59	32·48	29·96
20	5·558	6·177	7·166	8·094	45·51	43·08	39·69	36·94
25	8·233	9·009	10·23	11·37	52·86	50·27	46·63	43·67
30	11·13	12·05	13·49	14·80	59·99	57·24	53·39	50·24
40	17·38	18·56	20·38	22·03	73·76	70·75	66·50	63·02
50	24·07	25·48	27·65	29·60	87·08	83·84	79·24	75·46
60	31·07	32·70	35·18	37·39	100·1	96·72	91·72	87·66
70	38·31	40·14	42·91	45·37	112·9	109·2	104·0	99·70
80	45·75	47·76	50·79	53·48	125·4	121·6	116·1	111·6
90	53·34	55·52	58·81	61·71	137·8	133·8	128·1	123·4
100	61·06	63·40	66·93	70·04	150·1	145·9	140·0	135·0
				$p = 3$				
3	$0{\cdot}0^{5}9820$	$0{\cdot}0^{4}3927$	$0{\cdot}0^{3}2454$	$0{\cdot}0^{3}9817$	18·96	17·18	14·90	13·11
4	$0{\cdot}0^{2}3342$	$0{\cdot}0^{2}6701$	·01688	·03420	21·26	19·50	17·12	15·24
5	0·02844	·04550	·08538	·1390	23·45	21·66	19·18	17·22
6	·09224	·1322	·2149	·3142	25·55	23·69	21·13	19·09
7	·1997	·2682	·4004	·5492	27·56	25·64	22·99	20·88
8	·3495	·4497	·6346	·8339	29·49	27·52	24·80	22·62
9	·5383	·6719	·9106	1·160	31·37	29·34	26·55	24·31
10	·7625	·9300	1·223	1·522	33·19	31·12	28·26	25·96
15	2·301	2·638	3·191	3·724	41·79	39·52	36·36	33·80
20	4·338	4·833	5·623	6·364	49·82	47·37	43·95	41·18
25	6·710	7·350	8·356	9·285	57·49	54·89	51·24	48·27
30	9·326	10·10	11·31	12·41	64·90	62·15	58·30	55·15
40	15·08	16·10	17·66	19·07	79·18	76·18	71·96	68·50
50	21·33	22·57	24·45	26·14	92·95	89·73	85·18	81·44
60	27·93	29·37	31·55	33·48	106·4	102·9	98·09	94·09
70	34·81	36·43	38·88	41·04	119·5	115·9	110·8	106·5
80	41·90	43·69	46·39	48·77	132·4	128·6	123·2	118·8
90	49·16	51·12	54·06	56·64	145·2	141·2	135·6	130·9
100	56·57	58·69	61·85	64·63	157·7	153·6	147·8	143·0
				$p = 4$				
4	$0{\cdot}0^{5}7074$	$0{\cdot}0^{4}2830$	$0{\cdot}0^{3}1769$	$0{\cdot}0^{3}7085$	23·78	21·97	19·49	17·52
5	$\cdot0^{2}2506$	$\cdot0^{2}5025$	·01266	·02565	26·11	24·24	21·67	19·63
6	·02197	·03514	·06595	·1073	28·31	26·39	23·74	21·62
7	·07289	·1045	·1698	·2481	30·41	28·43	25·71	23·53
8	·1607	·2158	·3220	·4414	32·43	30·41	27·61	25·37
9	·2854	·3671	·5177	·6798	34·39	32·32	29·45	27·15
10	·4451	·5552	·7519	·9574	36·29	34·18	31·25	28·90
15	1·675	1·935	2·365	2·781	45·25	42·94	39·73	37·13
20	3·435	3·839	4·488	5·096	53·58	51·10	47·64	44·84
25	5·555	6·095	6·946	7·730	61·51	58·88	55·21	52·22
30	7·939	8·607	9·645	10·59	69·16	66·40	62·53	59·37

Tafel 18 (Forts.) Table 18 (cont.)

Untere Prozentpunkte l Obere Prozentpunkte $u_1 \simeq u$

Lower percentage points l Upper percentage points $u_1 \simeq u$

$\nu \backslash \alpha$	0·005	0·010	0·025	0·050	0·005	0·010	0·025	0·050
	Lower				Upper			
				$p = 4$ (*cont.*)				
40	13·28	14·18	15·56	16·79	83·86	80·86	76·64	73·18
50	19·16	20·27	21·95	23·45	98·00	94·79	90·25	86·53
60	25·43	26·73	28·69	30·43	111·8	108·3	103·5	99·55
70	31·99	33·47	35·69	37·65	125·2	121·6	116·5	112·3
80	38·80	40·44	42·90	45·06	138·4	134·7	129·3	124·9
90	45·79	47·59	50·27	52·63	151·4	147·5	142·0	137·4
100	52·94	54·89	57·79	60·33	164·3	160·2	154·4	149·7
				$p = 5$				
5	$0{\cdot}0^{5}5521$	$0{\cdot}0^{4}2209$	$0{\cdot}0^{3}1381$	$0{\cdot}0^{3}5527$	28·85	26·62	23·97	21·85
6	${\cdot}0^{2}2005$	${\cdot}0^{2}4020$	·01013	·02052	31·01	28·86	26·13	23·95
7	·01791	·02865	·05377	·08750	33·11	31·00	28·21	25·96
8	·06035	·08648	·1405	·2054	35·17	33·05	30·19	27·88
9	·1347	·1809	·2698	·3698	37·17	35·04	32·11	29·75
10	·2418	·3110	·4383	·5754	39·14	36·98	33·98	31·57
15	1·210	1·411	1·746	2·073	48·40	46·05	42·79	40·15
20	2·728	3·063	3·602	4·109	56·99	54·49	50·99	48·14
25	4·629	5·092	5·820	6·493	65·15	62·51	58·80	55·78
30	6·811	7·394	8·301	9·128	73·01	70·23	66·34	63·16
40	11·79	12·59	13·82	14·93	88·09	85·08	80·84	77·37
50	17·34	18·35	19·87	21·23	102·6	99·34	94·81	91·08
60	23·32	24·51	26·30	27·88	116·6	113·2	108·4	104·4
70	29·61	30·97	33·01	34·81	130·3	126·8	121·7	117·5
80	36·16	37·68	39·95	41·94	143·8	140·1	134·8	130·4
90	42·91	44·58	47·08	49·25	157·0	153·2	147·6	143·1
100	49·84	51·66	54·36	56·71	170·1	166·1	160·4	155·6
				$p = 6$				
6	$0{\cdot}0^{5}4590$	$0{\cdot}0^{4}1835$	$0{\cdot}0^{3}1148$	$0{\cdot}0^{3}4596$	33·22	31·19	28·39	26·14
7	${\cdot}0^{2}1671$	${\cdot}0^{2}3350$	${\cdot}0^{2}8440$	·01710	35·48	33·40	30·54	28·23
8	·01512	·02420	·04540	·07389	37·65	35·53	32·60	30·24
9	·05153	·07383	·1200	·1753	39·75	37·59	34·60	32·19
10	·1161	·1558	·2325	·3185	41·79	39·59	36·54	34·08
15	·8580	1·012	1·272	1·529	51·33	48·96	45·64	42·96
20	2·162	2·440	2·889	3·313	60·16	57·63	54·09	51·21
25	3·865	4·264	4·893	5·475	68·53	65·87	62·13	59·07
30	5·866	6·379	7·178	7·907	76·58	73·79	69·86	66·66
40	10·51	11·24	12·35	13·25	92·00	88·98	84·72	81·24
50	15·78	16·69	18·09	19·33	106·8	103·6	99·00	95·27
60	21·49	22·58	24·23	25·69	121·1	117·7	112·9	108·9
70	27·53	28·79	30·69	32·35	135·1	131·5	126·4	122·3
80	33·85	35·27	37·39	39·24	148·7	145·0	139·7	135·4
90	40·39	41·95	44·28	46·32	162·2	158·3	152·8	148·3
100	47·12	48·82	51·35	53·56	175·5	171·5	165·8	161·1
				$p = 7$				
7	$0{\cdot}0^{5}3835$	$0{\cdot}0^{4}1534$	$0{\cdot}0^{4}9592$	$0{\cdot}0^{3}3841$	37·82	35·70	32·76	30·40
8	${\cdot}0^{2}1432$	${\cdot}0^{2}2872$	${\cdot}0^{2}7234$	·01466	40·05	37·89	34·90	32·48
9	·01309	·02095	·03930	·06395	42·22	40·02	36·96	34·49
10	·04498	·06444	·1047	·1530	44·31	42·07	38·96	36·45
15	·5909	·7071	·9057	1·105	54·11	51·70	48·34	45·61
20	1·701	1·931	2·306	2·662	63·16	60·60	57·02	54·10

Tafel 18 (Forts.) Table 18 (cont.

Untere Prozentpunkte l Obere Prozentpunkte u_1 ~

Lower percentage points l Upper percentage points u_1 ~

$\nu \backslash \alpha$	0·005	0·010	0·025	0·050	0·005	0·010	0·025	0·050
				$p = 7$ (*cont.*)				
25	3·224	3·569	4·114	4·620	71·72	69·03	65·26	62·17
30	5·058	5·513	6·220	6·867	79·95	77·13	73·18	69·95
40	9·403	10·06	11·07	11·98	95·67	92·64	88·36	84·86
50	14·40	15·24	16·53	17·66	110·7	107·5	102·9	99·18
60	19·86	20·88	22·41	23·77	125·3	121·9	117·1	113·1
70	25·69	26·36	28·63	30·18	139·5	135·9	130·9	126·7
80	31·79	33·12	35·11	36·84	153·4	149·7	144·4	140·0
90	38·13	39·60	41·80	43·71	167·1	163·2	157·7	153·2
100	44·67	46·28	48·67	50·74	180·5	176·5	170·8	166·1
				$p = 8$				
8	0·0^{5}3327	0·0^{4}1331	0·0^{4}8318	0·0^{3}3332	42·35	40·15	37·10	34·63
9	·0^{2}1253	·0^{2}2513	·0^{2}6330	·01283	44·57	42·33	39·22	36·71
10	·01154	·01847	·03465	·05638	46·72	44·45	41·28	38·72
15	·3902	·4753	·6235	·7744	56·76	54·32	50·91	48·15
20	1·323	1·513	1·824	2·122	66·01	63·43	59·81	56·86
25	2·681	2·979	3·452	3·892	74·76	72·04	68·23	65·12
30	4·361	4·764	5·392	5·968	83·15	80·31	76·33	73·07
40	8·426	9·026	9·945	10·77	99·16	96·11	91·81	88·29
50	13·17	13·95	15·14	16·19	114·5	111·2	106·7	102·9
60	18·41	19·36	20·78	22·04	129·3	125·8	121·0	117·0
70	24·02	25·12	26·78	28·23	143·7	140·1	135·0	130·9
80	29·93	31·18	33·05	34·69	157·8	154·0	148·8	144·4
90	36·09	37·48	39·55	41·35	171·6	167·8	162·3	157·8
100	42·45	43·97	46·23	48·20	185·3	181·3	175·6	170·9
				$p = 9$				
9	0·0^{5}2936	0·0^{4}1175	0·0^{4}7343	0·0^{3}2941	46·84	44·57	41·40	38·84
10	·0^{2}1114	·0^{2}2234	·0^{2}5626	·01140	49·05	46·74	43·52	40·91
15	·2426	·3024	·4090	·5202	59·32	56·85	53·39	50·58
20	1·014	1·169	1·425	1·672	68·76	66·14	62·48	59·50
25	2·218	2·475	2·884	3·267	77·67	74·93	71·09	67·94
30	3·753	4·111	4·670	5·183	86·21	83·35	79·34	76·05
40	7·557	8·104	8·944	9·698	102·5	99·43	95·11	91·57
50	12·07	12·79	13·89	14·86	118·0	114·8	110·2	106·4
60	17·09	17·98	19·31	20·49	133·1	129·6	124·8	120·8
70	22·50	23·54	25·10	26·47	147·7	144·1	139·0	134·8
80	28·25	29·41	31·18	32·72	162·0	158·2	152·9	148·6
90	34·21	35·53	37·49	39·20	176·0	172·1	166·6	162·1
100	40·41	41·86	44·01	45·88	189·8	185·8	180·1	175·4
				$p = 10$				
10	0·0^{5}2628	0·0^{4}1051	0·0^{4}6573	0·0^{3}2632	51·32	48·98	45·73	43·12
15	·1382	·1777	·2503	·3284	61·78	59·28	55·78	52·94
20	·7608	·8863	1·095	1·298	71·41	68·77	65·07	62·05
25	1·821	2·042	2·396	2·728	80·48	77·72	73·84	70·67
30	3·221	3·538	4·035	4·493	89·17	86·29	82·24	78·93
40	6·777	7·278	8·047	8·738	105·7	102·6	98·28	94·72
50	11·07	11·74	12·76	13·66	121·5	118·2	113·6	109·8
60	15·89	16·72	17·97	19·07	136·7	133·3	128·4	124·4
70	21·11	22·09	23·56	24·85	151·5	147·9	142·8	138·7
80	26·66	27·78	29·46	30·92	166·0	162·2	157·0	152·6
90	32·48	33·74	35·60	37·23	180·2	176·3	170·8	166·3
100	38·52	39·90	41·95	43·74	194·1	190·1	184·5	179·8

TABLE 19

The Multivariate t-Distribution: Tables of P. R. Krishnaiah and J. V. Armitage

(a) Content of the tables and definition of the test statistic :

The tables contain the upper percentage points of the multivariate t-distribution , which is defined as follows :

Let $\underline{x}' = (x_1, x_2, \ldots, x_p)$ have a p-dimensional normal distribution with mean vector $\underline{\mu}' = (\mu_1, \mu_2, \ldots, \mu_p)$, with common, unknown variance σ^2 , and with known correlation matrix $\underline{\Omega} = (\rho_{ij})$. Further let s^2/σ^2 be a χ^2-distributed random variables with n degrees of freedom which is independent of the variates (= vector components) $x_1, x_2, \ldots, x_p$.

The common distribution of $t_1, t_2, \ldots, t_p$, where $t_i = x_i \cdot \sqrt{n}/s$, is then called a central, respectively non-central, p-variate t-distribution with n degrees of freedom whenever $\underline{\mu}' = \underline{0}'$, respectively $\neq \underline{0}'$.

The tables at hand contain the upper percentage points $a = t_\alpha$ for the special case where

(1) $\underline{\mu}' = \underline{0}'$ (i.e. central p-variate t-distribution)

(2) $\rho_{ij} = \rho$ for $i \neq j = 1, 2, \ldots, p$

hold.

If we denote the density function by $g(t_1, t_2, \ldots, t_p)$, then we have

$$1 - \int_{-\infty}^{a} \cdots \int_{-\infty}^{a} g(t_1, t_2, \ldots, t_p)\,dt_1 dt_2 \ldots dt_p = \alpha .$$

(b) Extent of the tables and definition of the parameters :

(1) The parameter α :

α = error probability

α = 0.05 and 0.01

= 5% and 1% .

(2) The parameter p :

p = dimension of the variates

p = 1(1)10 .

(3) The parameter ρ :

ρ = correlation coefficient

ρ = 0(0.1)0.9 .

(4) The parameter n :

n = degrees of freedom of the multivariate t-distribution

n = 5(1)35 .

(c) Suggestions for applications :

(1) Multiple comparison tests in MANOVA models. (See e. g. KRISHNAIAH, P. R. (1965)).

(2) Selecting a univariate population which is better than a standard population . (See e. g. GUPTA, S. S. + SOBEL, M. (1958)).

(d) References :

(1) For the source of the tables :

KRISHNAIAH, P. R. + ARMITAGE, J. V. : Tables for multivariate t-distribution.
Sankhya, Series B, Vol. 28, 31 - 56 (1966).

(2) For suggestions for use :

GUPTA, S. S. + SOBEL, M. : On selecting a subset which contained all populations better than a standard.
The Annals of Mathematical Statistics 29,235 - 244 (1958) .

KRISHNAIAH, P. R. : Multiple comparison tests in multi-response experiments.
Sankhya, Series A, Vol. 27, 65 - 72 (1965) .

(e) Further remarks :

At the parameter $p = 1$, the tables contain the upper percentage points of the (univariate) STUDENT's t-distribution with n degrees of freedom for the one-sided error probability α .

Tafel 19 Table 19

UPPER 5% POINTS / OBERE 5%-PUNKTE $\rho = 0.0$

n p	1	2	3	4	5	6	7	8	9	10
5	2.01	2.53	2.84	3.06	3.23	3.38	3.50	3.60	3.69	3.77
6	1.94	2.42	2.70	2.89	3.05	3.17	3.28	3.37	3.45	3.53
7	1.89	2.34	2.60	2.78	2.92	3.04	3.14	3.22	3.30	3.36
8	1.86	2.28	2.53	2.70	2.84	2.95	3.04	3.12	3.19	3.25
9	1.83	2.24	2.48	2.64	2.77	2.87	2.96	3.04	3.10	3.16
10	1.81	2.21	2.44	2.60	2.72	2.82	2.90	2.97	3.04	3.10
11	1.79	2.18	2.41	2.56	2.68	2.77	2.86	2.93	2.99	3.04
12	1.78	2.16	2.38	2.53	2.65	2.74	2.82	2.89	2.94	3.00
13	1.77	2.15	2.36	2.51	2.62	2.71	2.79	2.85	2.91	2.96
14	1.76	2.13	2.34	2.48	2.59	2.68	2.76	2.82	2.88	2.93
15	1.75	2.12	2.32	2.47	2.57	2.66	2.74	2.80	2.86	2.91
16	1.75	2.11	2.31	2.45	2.56	2.64	2.72	2.78	2.83	2.88
17	1.74	2.10	2.30	2.44	2.54	2.63	2.70	2.76	2.81	2.86
18	1.73	2.09	2.29	2.42	2.53	2.61	2.68	2.74	2.80	2.84
19	1.73	2.08	2.28	2.41	2.52	2.60	2.67	2.73	2.78	2.83
20	1.72	2.08	2.27	2.40	2.51	2.59	2.66	2.72	2.77	2.81
21	1.72	2.07	2.26	2.39	2.50	2.58	2.65	2.71	2.76	2.80
22	1.72	2.06	2.26	2.39	2.49	2.57	2.64	2.70	2.75	2.79
23	1.71	2.06	2.25	2.38	2.48	2.56	2.63	2.69	2.74	2.78
24	1.71	2.05	2.24	2.37	2.47	2.55	2.62	2.68	2.73	2.77
25	1.71	2.05	2.24	2.37	2.47	2.55	2.61	2.67	2.72	2.76
26	1.71	2.05	2.23	2.36	2.46	2.54	2.60	2.66	2.71	2.75
27	1.70	2.04	2.23	2.36	2.46	2.53	2.60	2.65	2.70	2.75
28	1.70	2.04	2.22	2.35	2.45	2.53	2.59	2.65	2.70	2.74
29	1.70	2.04	2.22	2.35	2.44	2.52	2.59	2.64	2.69	2.73
30	1.70	2.03	2.22	2.34	2.44	2.52	2.58	2.64	2.69	2.73
31	1.70	2.03	2.21	2.34	2.44	2.51	2.58	2.63	2.68	2.72
32	1.69	2.03	2.21	2.34	2.43	2.51	2.57	2.63	2.68	2.72
33	1.69	2.03	2.21	2.33	2.43	2.51	2.57	2.62	2.67	2.71
34	1.69	2.02	2.21	2.33	2.42	2.50	2.56	2.62	2.67	2.71
35	1.69	2.02	2.20	2.33	2.42	2.50	2.56	2.61	2.66	2.71

Tafel 19 (Forts.) Table 19 (cont.)

UPPER 1% POINTS / OBERE 1%-PUNKTE $\rho = 0.0$

n \ p	1	2	3	4	5	6	7	8	9	10
5	3.36	4.00	4.39	4.67	4.90	5.08	5.24	5.38	5.50	5.61
6	3.14	3.68	4.01	4.25	4.44	4.59	4.73	4.84	4.94	5.03
7	3.00	3.48	3.77	3.98	4.15	4.28	4.40	4.50	4.59	4.67
8	2.90	3.34	3.61	3.80	3.95	4.07	4.17	4.26	4.34	4.42
9	2.82	3.24	3.49	3.66	3.80	3.91	4.01	4.09	4.17	4.23
10	2.76	3.16	3.39	3.56	3.69	3.79	3.89	3.96	4.03	4.09
11	2.72	3.10	3.32	3.48	3.60	3.70	3.79	3.86	3.93	3.98
12	2.68	3.05	3.26	3.41	3.53	3.63	3.71	3.78	3.84	3.90
13	2.65	3.00	3.21	3.36	3.47	3.56	3.64	3.71	3.77	3.82
14	2.62	2.97	3.17	3.31	3.42	3.51	3.59	3.65	3.71	3.76
15	2.60	2.94	3.14	3.28	3.38	3.47	3.54	3.60	3.66	3.71
16	2.58	2.92	3.11	3.24	3.35	3.43	3.50	3.56	3.62	3.67
17	2.57	2.89	3.08	3.21	3.32	3.40	3.47	3.53	3.58	3.63
18	2.55	2.87	3.06	3.19	3.29	3.37	3.44	3.50	3.55	3.59
19	2.54	2.86	3.04	3.16	3.26	3.34	3.41	3.47	3.52	3.56
20	2.53	2.84	3.02	3.14	3.24	3.32	3.38	3.44	3.49	3.54
21	2.52	2.83	3.00	3.13	3.22	3.30	3.36	3.42	3.47	3.51
22	2.51	2.81	2.99	3.11	3.20	3.28	3.34	3.40	3.45	3.49
23	2.50	2.80	2.97	3.10	3.19	3.27	3.33	3.38	3.43	3.47
24	2.49	2.79	2.96	3.08	3.17	3.25	3.31	3.36	3.41	3.46
25	2.48	2.78	2.95	3.07	3.16	3.24	3.30	3.35	3.40	3.44
26	2.48	2.77	2.94	3.06	3.15	3.22	3.29	3.34	3.38	3.42
27	2.47	2.77	2.93	3.05	3.14	3.21	3.27	3.32	3.37	3.41
28	2.47	2.76	2.93	3.04	3.13	3.20	3.26	3.31	3.36	3.40
29	2.46	2.75	2.92	3.03	3.12	3.19	3.25	3.30	3.35	3.39
30	2.46	2.75	2.91	3.02	3.11	3.18	3.24	3.29	3.34	3.37
31	2.45	2.74	2.90	3.02	3.10	3.17	3.23	3.28	3.33	3.36
32	2.45	2.73	2.90	3.01	3.10	3.16	3.22	3.27	3.32	3.36
33	2.44	2.73	2.89	3.00	3.09	3.16	3.21	3.27	3.31	3.35
34	2.44	2.72	2.89	2.99	3.08	3.15	3.21	3.26	3.30	3.34
35	2.44	2.72	2.88	2.99	3.08	3.14	3.20	3.25	3.30	3.33

UPPER 5% POINTS / OBERE 5%-PUNKTE $\rho = 0.1$

n \ p	1	2	3	4	5	6	7	8	9	10
5	2.01	2.52	2.82	3.03	3.20	3.33	3.45	3.54	3.63	3.71
6	1.94	2.41	2.68	2.87	3.02	3.14	3.24	3.33	3.41	3.47
7	1.89	2.33	2.58	2.76	2.90	3.01	3.10	3.18	3.26	3.32
8	1.86	2.28	2.52	2.68	2.81	2.92	3.01	3.08	3.15	3.21
9	1.83	2.24	2.47	2.63	2.75	2.85	2.93	3.01	3.07	3.13
10	1.81	2.20	2.43	2.58	2.70	2.80	2.88	2.95	3.01	3.06
11	1.79	2.18	2.39	2.55	2.66	2.75	2.83	2.90	2.96	3.01
12	1.78	2.16	2.37	2.52	2.63	2.72	2.80	2.86	2.92	2.97
13	1.77	2.14	2.35	2.49	2.60	2.69	2.77	2.83	2.89	2.94
14	1.76	2.13	2.33	2.47	2.58	2.67	2.74	2.80	2.86	2.91
15	1.75	2.11	2.31	2.45	2.56	2.65	2.72	2.78	2.83	2.88
16	1.75	2.10	2.30	2.44	2.54	2.63	2.70	2.76	2.81	2.86
17	1.74	2.09	2.29	2.42	2.53	2.61	2.68	2.74	2.79	2.84
18	1.73	2.08	2.28	2.41	2.52	2.60	2.67	2.73	2.78	2.82
19	1.73	2.08	2.27	2.40	2.50	2.58	2.65	2.71	2.76	2.81
20	1.72	2.07	2.26	2.39	2.49	2.57	2.64	2.70	2.75	2.79
21	1.72	2.06	2.25	2.38	2.49	2.56	2.63	2.69	2.74	2.78
22	1.72	2.06	2.25	2.38	2.48	2.55	2.62	2.68	2.73	2.77
23	1.71	2.05	2.24	2.37	2.47	2.55	2.61	2.67	2.72	2.76
24	1.71	2.05	2.24	2.36	2.46	2.54	2.60	2.66	2.71	2.75
25	1.71	2.04	2.23	2.36	2.46	2.53	2.60	2.65	2.70	2.74
26	1.71	2.04	2.23	2.35	2.45	2.53	2.59	2.65	2.70	2.74
27	1.70	2.04	2.22	2.35	2.44	2.52	2.58	2.64	2.69	2.73
28	1.70	2.03	2.22	2.34	2.44	2.52	2.58	2.63	2.68	2.72
29	1.70	2.03	2.21	2.34	2.43	2.51	2.57	2.63	2.68	2.72
30	1.70	2.03	2.21	2.34	2.43	2.51	2.57	2.62	2.67	2.71
31	1.70	2.03	2.21	2.33	2.43	2.50	2.56	2.62	2.67	2.71
32	1.69	2.02	2.20	2.33	2.42	2.50	2.56	2.61	2.66	2.70
33	1.69	2.02	2.20	2.33	2.42	2.49	2.56	2.61	2.66	2.70
34	1.69	2.02	2.20	2.32	2.41	2.49	2.55	2.60	2.65	2.69
35	1.69	2.02	2.20	2.32	2.41	2.49	2.55	2.60	2.65	2.69

Tafel 19 (Forts.) Table 19 (cont.)

UPPER 1% POINTS / OBERE 1%-PUNKTE $\rho = 0.1$

n \ p	1	2	3	4	5	6	7	8	9	10
5	3.36	3.98	4.36	4.64	4.85	5.03	5.18	5.31	5.43	5.53
6	3.14	3.68	4.00	4.23	4.41	4.56	4.68	4.79	4.89	4.98
7	3.00	3.48	3.76	3.96	4.13	4.26	4.37	4.46	4.55	4.62
8	2.90	3.34	3.60	3.78	3.93	4.05	4.15	4.24	4.31	4.38
9	2.82	3.23	3.48	3.65	3.78	3.90	3.99	4.07	4.14	4.20
10	2.76	3.16	3.38	3.55	3.68	3.78	3.87	3.94	4.01	4.07
11	2.72	3.09	3.31	3.47	3.59	3.69	3.77	3.84	3.91	3.96
12	2.68	3.04	3.25	3.40	3.52	3.61	3.69	3.76	3.82	3.88
13	2.65	3.00	3.21	3.35	3.46	3.55	3.63	3.70	3.75	3.80
14	2.62	2.97	3.17	3.31	3.41	3.50	3.57	3.64	3.70	3.75
15	2.60	2.94	3.13	3.27	3.37	3.46	3.53	3.59	3.65	3.70
16	2.58	2.91	3.10	3.23	3.34	3.42	3.49	3.55	3.60	3.65
17	2.57	2.89	3.08	3.21	3.31	3.39	3.46	3.52	3.57	3.61
18	2.55	2.87	3.05	3.18	3.28	3.36	3.43	3.49	3.54	3.58
19	2.54	2.85	3.03	3.16	3.26	3.33	3.40	3.46	3.51	3.55
20	2.53	2.84	3.02	3.14	3.23	3.31	3.38	3.43	3.48	3.53
21	2.52	2.82	3.00	3.12	3.22	3.29	3.36	3.41	3.46	3.51
22	2.51	2.81	2.98	3.11	3.20	3.28	3.34	3.39	3.44	3.48
23	2.50	2.80	2.97	3.09	3.18	3.26	3.32	3.37	3.42	3.46
24	2.49	2.79	2.96	3.08	3.17	3.24	3.31	3.36	3.40	3.45
25	2.48	2.78	2.95	3.07	3.16	3.23	3.29	3.34	3.39	3.43
26	2.48	2.77	2.94	3.06	3.14	3.22	3.28	3.33	3.37	3.42
27	2.47	2.76	2.93	3.05	3.13	3.20	3.27	3.32	3.36	3.40
28	2.47	2.76	2.92	3.04	3.12	3.19	3.25	3.31	3.35	3.39
29	2.46	2.75	2.91	3.03	3.12	3.18	3.24	3.30	3.34	3.38
30	2.46	2.74	2.91	3.02	3.11	3.17	3.23	3.29	3.33	3.37
31	2.45	2.74	2.90	3.01	3.10	3.17	3.23	3.28	3.32	3.36
32	2.45	2.73	2.90	3.00	3.09	3.16	3.22	3.27	3.31	3.35
33	2.44	2.73	2.89	3.00	3.09	3.15	3.21	3.26	3.30	3.34
34	2.44	2.72	2.88	2.99	3.08	3.14	3.20	3.25	3.30	3.33
35	2.44	2.72	2.88	2.99	3.07	3.14	3.19	3.24	3.29	3.33

Tafel 19 (Forts.) Table 19 (cont.)

UPPER 5% POINTS / OBERE 5%-PUNKTE $\rho = 0.2$

n p	1	2	3	4	5	6	7	8	9	10
5	2.01	2.51	2.79	3.00	3.15	3.28	3.39	3.48	3.56	3.63
6	1.94	2.39	2.66	2.84	2.98	3.10	3.19	3.28	3.35	3.41
7	1.89	2.32	2.56	2.73	2.87	2.97	3.06	3.14	3.21	3.27
8	1.86	2.27	2.50	2.66	2.78	2.89	2.97	3.04	3.11	3.16
9	1.83	2.23	2.45	2.60	2.72	2.82	2.90	2.97	3.03	3.08
10	1.81	2.19	2.41	2.56	2.68	2.77	2.85	2.91	2.97	3.02
11	1.79	2.17	2.38	2.53	2.64	2.73	2.80	2.87	2.92	2.97
12	1.78	2.15	2.35	2.50	2.61	2.70	2.77	2.83	2.89	2.94
13	1.77	2.13	2.33	2.47	2.58	2.67	2.74	2.80	2.85	2.90
14	1.76	2.12	2.32	2.45	2.56	2.64	2.71	2.77	2.83	2.87
15	1.75	2.11	2.30	2.44	2.54	2.62	2.69	2.75	2.80	2.85
16	1.75	2.09	2.29	2.42	2.52	2.61	2.67	2.73	2.78	2.83
17	1.74	2.09	2.28	2.41	2.51	2.59	2.66	2.72	2.77	2.81
18	1.73	2.08	2.27	2.40	2.50	2.58	2.64	2.70	2.75	2.80
19	1.73	2.07	2.26	2.39	2.49	2.57	2.63	2.69	2.74	2.78
20	1.72	2.06	2.25	2.38	2.48	2.55	2.62	2.68	2.73	2.77
21	1.72	2.06	2.24	2.37	2.47	2.55	2.61	2.67	2.71	2.76
22	1.72	2.05	2.24	2.36	2.46	2.54	2.60	2.66	2.71	2.75
23	1.71	2.05	2.23	2.36	2.45	2.53	2.59	2.65	2.70	2.74
24	1.71	2.04	2.22	2.35	2.45	2.52	2.58	2.64	2.69	2.73
25	1.71	2.04	2.22	2.34	2.44	2.52	2.58	2.63	2.68	2.72
26	1.71	2.03	2.21	2.34	2.43	2.51	2.57	2.63	2.67	2.71
27	1.70	2.03	2.21	2.33	2.43	2.50	2.57	2.62	2.67	2.71
28	1.70	2.03	2.21	2.33	2.42	2.50	2.56	2.61	2.66	2.70
29	1.70	2.02	2.20	2.33	2.42	2.49	2.55	2.61	2.65	2.70
30	1.70	2.02	2.20	2.32	2.41	2.49	2.55	2.60	2.65	2.69
31	1.70	2.02	2.20	2.32	2.41	2.49	2.55	2.60	2.64	2.69
32	1.69	2.02	2.19	2.32	2.41	2.48	2.54	2.59	2.64	2.68
33	1.69	2.01	2.19	2.31	2.40	2.48	2.54	2.59	2.64	2.68
34	1.69	2.01	2.19	2.31	2.40	2.47	2.53	2.59	2.63	2.67
35	1.69	2.01	2.19	2.31	2.40	2.47	2.53	2.58	2.63	2.67

Tafel 19 (Forts.) Table 19 (cont.)

UPPER 1% POINTS / OBERE 1%-PUNKTE $\rho = 0.2$

n \ p	1	2	3	4	5	6	7	8	9	10
5	3.36	3.97	4.34	4.60	4.80	4.97	5.12	5.24	5.35	5.44
6	3.14	3.66	3.97	4.20	4.37	4.51	4.63	4.74	4.83	4.91
7	3.00	3.47	3.74	3.94	4.10	4.22	4.33	4.42	4.50	4.57
8	2.90	3.33	3.58	3.76	3.90	4.02	4.11	4.20	4.27	4.34
9	2.82	3.23	3.46	3.63	3.76	3.87	3.96	4.04	4.11	4.17
10	2.76	3.15	3.37	3.53	3.66	3.76	3.84	3.91	3.98	4.04
11	2.72	3.09	3.30	3.45	3.57	3.67	3.75	3.82	3.88	3.93
12	2.68	3.04	3.24	3.39	3.50	3.59	3.67	3.74	3.80	3.85
13	2.65	3.00	3.20	3.34	3.45	3.54	3.61	3.67	3.73	3.78
14	2.62	2.96	3.16	3.30	3.40	3.49	3.56	3.62	3.68	3.72
15	2.60	2.93	3.12	3.26	3.36	3.44	3.51	3.57	3.63	3.68
16	2.58	2.91	3.10	3.22	3.33	3.41	3.48	3.53	3.59	3.63
17	2.57	2.89	3.07	3.20	3.30	3.37	3.44	3.50	3.55	3.60
18	2.55	2.87	3.05	3.17	3.27	3.35	3.41	3.47	3.52	3.56
19	2.54	2.85	3.03	3.15	3.25	3.32	3.39	3.44	3.49	3.54
20	2.53	2.83	3.01	3.13	3.22	3.30	3.36	3.42	3.47	3.51
21	2.52	2.82	2.99	3.12	3.21	3.28	3.34	3.40	3.45	3.49
22	2.51	2.81	2.98	3.10	3.19	3.26	3.33	3.38	3.43	3.47
23	2.50	2.80	2.97	3.09	3.17	3.25	3.31	3.36	3.41	3.45
24	2.49	2.79	2.95	3.07	3.16	3.23	3.30	3.35	3.39	3.43
25	2.48	2.78	2.94	3.06	3.15	3.22	3.28	3.33	3.38	3.42
26	2.48	2.77	2.93	3.05	3.14	3.21	3.27	3.32	3.36	3.40
27	2.47	2.76	2.93	3.04	3.13	3.20	3.26	3.31	3.35	3.39
28	2.47	2.75	2.92	3.03	3.12	3.19	3.24	3.30	3.34	3.38
29	2.46	2.75	2.91	3.02	3.11	3.18	3.23	3.29	3.33	3.37
30	2.46	2.74	2.90	3.01	3.10	3.17	3.22	3.28	3.32	3.36
31	2.45	2.74	2.90	3.01	3.09	3.16	3.22	3.27	3.31	3.35
32	2.45	2.73	2.89	3.00	3.09	3.15	3.21	3.26	3.30	3.34
33	2.44	2.73	2.88	2.99	3.08	3.14	3.20	3.25	3.29	3.33
34	2.44	2.72	2.88	2.99	3.07	3.14	3.19	3.24	3.29	3.32
35	2.44	2.72	2.87	2.98	3.06	3.13	3.19	3.23	3.28	3.32

Tafel 19 (Forts.) Table 19 (cont.)

UPPER 5% POINTS / OBERE 5%-PUNKTE $\rho = 0.3$

n \ p	1	2	3	4	5	6	7	8	9	10
5	2.01	2.49	2.76	2.95	3.10	3.22	3.32	3.41	3.49	3.55
6	1.94	2.38	2.63	2.80	2.94	3.05	3.14	3.22	3.28	3.34
7	1.89	2.31	2.54	2.70	2.83	2.93	3.01	3.09	3.15	3.21
8	1.86	2.25	2.48	2.63	2.75	2.85	2.93	2.99	3.05	3.11
9	1.83	2.21	2.43	2.58	2.69	2.78	2.86	2.92	2.98	3.03
10	1.81	2.18	2.39	2.54	2.65	2.73	2.81	2.87	2.93	2.98
11	1.79	2.16	2.36	2.50	2.61	2.70	2.77	2.83	2.88	2.93
12	1.78	2.14	2.34	2.47	2.58	2.66	2.73	2.79	2.85	2.89
13	1.77	2.12	2.32	2.45	2.55	2.64	2.71	2.76	2.81	2.86
14	1.76	2.11	2.30	2.43	2.53	2.61	2.68	2.74	2.79	2.83
15	1.75	2.10	2.29	2.42	2.52	2.59	2.66	2.72	2.77	2.81
16	1.75	2.09	2.27	2.40	2.50	2.58	2.64	2.70	2.75	2.79
17	1.74	2.08	2.26	2.39	2.49	2.56	2.63	2.68	2.73	2.77
18	1.73	2.07	2.25	2.38	2.47	2.55	2.61	2.67	2.72	2.76
19	1.73	2.06	2.24	2.37	2.46	2.54	2.60	2.66	2.71	2.75
20	1.72	2.05	2.23	2.36	2.45	2.53	2.59	2.65	2.69	2.73
21	1.72	2.05	2.23	2.35	2.45	2.52	2.58	2.64	2.68	2.72
22	1.72	2.04	2.22	2.34	2.44	2.51	2.57	2.63	2.67	2.71
23	1.71	2.04	2.22	2.34	2.43	2.51	2.57	2.62	2.66	2.71
24	1.71	2.03	2.21	2.33	2.42	2.50	2.56	2.61	2.66	2.70
25	1.71	2.03	2.21	2.33	2.42	2.49	2.55	2.60	2.65	2.69
26	1.71	2.03	2.20	2.32	2.41	2.49	2.55	2.60	2.64	2.68
27	1.70	2.02	2.20	2.32	2.41	2.48	2.54	2.59	2.64	2.68
28	1.70	2.02	2.19	2.31	2.40	2.48	2.54	2.59	2.63	2.67
29	1.70	2.02	2.19	2.31	2.40	2.47	2.53	2.58	2.63	2.67
30	1.70	2.01	2.19	2.31	2.39	2.47	2.53	2.58	2.62	2.66
31	1.70	2.01	2.18	2.30	2.39	2.46	2.52	2.57	2.62	2.66
32	1.69	2.01	2.18	2.30	2.39	2.46	2.52	2.57	2.61	2.65
33	1.69	2.01	2.18	2.30	2.38	2.46	2.51	2.56	2.61	2.65
34	1.69	2.00	2.18	2.29	2.38	2.45	2.51	2.56	2.60	2.64
35	1.69	2.00	2.17	2.29	2.38	2.45	2.51	2.56	2.60	2.64

Tafel 19 (Forts.) Table 19 (cont.)

UPPER 1% POINTS / OBERE 1%-PUNKTE $\rho = 0.3$

n \ p	1	2	3	4	5	6	7	8	9	10
5	3.36	3.95	4.30	4.55	4.75	4.91	5.04	5.15	5.26	5.35
6	3.14	3.65	3.95	4.16	4.33	4.46	4.58	4.67	4.76	4.84
7	3.00	3.45	3.72	3.91	4.06	4.18	4.28	4.37	4.44	4.51
8	2.90	3.32	3.56	3.74	3.87	3.98	4.07	4.15	4.22	4.28
9	2.82	3.22	3.45	3.61	3.74	3.84	3.92	4.00	4.06	4.12
10	2.76	3.14	3.36	3.51	3.63	3.73	3.81	3.88	3.94	3.99
11	2.72	3.08	3.29	3.44	3.55	3.64	3.72	3.78	3.84	3.90
12	2.68	3.03	3.23	3.37	3.48	3.57	3.64	3.71	3.76	3.81
13	2.65	2.99	3.19	3.32	3.43	3.51	3.58	3.65	3.70	3.75
14	2.62	2.96	3.15	3.28	3.38	3.46	3.53	3.59	3.65	3.69
15	2.60	2.93	3.11	3.24	3.34	3.42	3.49	3.55	3.60	3.65
16	2.58	2.90	3.09	3.21	3.31	3.39	3.45	3.51	3.56	3.60
17	2.57	2.88	3.06	3.18	3.28	3.36	3.42	3.48	3.53	3.57
18	2.55	2.86	3.04	3.16	3.25	3.33	3.39	3.45	3.50	3.54
19	2.54	2.84	3.02	3.14	3.23	3.31	3.37	3.42	3.47	3.51
20	2.53	2.83	3.00	3.12	3.21	3.29	3.35	3.40	3.45	3.49
21	2.52	2.82	2.98	3.10	3.19	3.27	3.33	3.38	3.42	3.47
22	2.51	2.80	2.97	3.09	3.18	3.25	3.31	3.36	3.41	3.45
23	2.50	2.79	2.96	3.07	3.16	3.23	3.29	3.34	3.39	3.43
24	2.49	2.78	2.95	3.06	3.15	3.22	3.28	3.33	3.37	3.41
25	2.48	2.77	2.94	3.05	3.14	3.21	3.27	3.32	3.36	3.40
26	2.48	2.76	2.93	3.04	3.13	3.19	3.25	3.30	3.34	3.38
27	2.47	2.76	2.92	3.03	3.12	3.18	3.24	3.29	3.33	3.37
28	2.47	2.75	2.91	3.02	3.11	3.17	3.23	3.28	3.32	3.36
29	2.46	2.74	2.90	3.01	3.10	3.16	3.22	3.27	3.31	3.35
30	2.46	2.74	2.90	3.00	3.09	3.15	3.21	3.26	3.30	3.34
31	2.45	2.73	2.89	3.00	3.08	3.15	3.20	3.25	3.29	3.33
32	2.45	2.73	2.88	2.99	3.07	3.14	3.19	3.24	3.29	3.32
33	2.44	2.72	2.88	2.98	3.07	3.13	3.19	3.23	3.28	3.31
34	2.44	2.72	2.87	2.98	3.06	3.13	3.18	3.23	3.27	3.31
35	2.44	2.71	2.87	2.97	3.05	3.12	3.17	3.22	3.26	3.30

Tafel 19 (Forts.) Table 19 (cont.

UPPER 5% POINTS / OBERE 5%-PUNKTE $\rho = 0.4$

n p	1	2	3	4	5	6	7	8	9	10
5	2.01	2.47	2.72	2.91	3.04	3.16	3.25	3.33	3.40	3.46
6	1.94	2.36	2.60	2.76	2.89	2.99	3.07	3.15	3.21	3.27
7	1.89	2.29	2.51	2.66	2.78	2.88	2.96	3.02	3.08	3.13
8	1.86	2.24	2.45	2.60	2.71	2.80	2.87	2.94	2.99	3.04
9	1.83	2.20	2.40	2.54	2.65	2.74	2.81	2.87	2.92	2.97
10	1.81	2.17	2.37	2.50	2.61	2.69	2.76	2.82	2.87	2.92
11	1.79	2.14	2.34	2.47	2.57	2.65	2.72	2.78	2.83	2.87
12	1.78	2.12	2.32	2.45	2.54	2.62	2.69	2.75	2.80	2.84
13	1.77	2.11	2.30	2.42	2.52	2.60	2.66	2.72	2.77	2.81
14	1.76	2.09	2.28	2.40	2.50	2.58	2.64	2.70	2.74	2.78
15	1.75	2.08	2.26	2.39	2.48	2.56	2.62	2.68	2.72	2.76
16	1.75	2.07	2.25	2.37	2.47	2.54	2.60	2.66	2.71	2.74
17	1.74	2.06	2.24	2.36	2.46	2.53	2.59	2.64	2.69	2.73
18	1.73	2.06	2.23	2.35	2.44	2.52	2.58	2.63	2.68	2.72
19	1.73	2.05	2.22	2.34	2.43	2.51	2.57	2.62	2.66	2.70
20	1.72	2.04	2.22	2.33	2.42	2.50	2.56	2.61	2.65	2.69
21	1.72	2.04	2.21	2.33	2.42	2.49	2.55	2.60	2.64	2.68
22	1.72	2.03	2.20	2.32	2.41	2.48	2.54	2.59	2.63	2.67
23	1.71	2.03	2.20	2.31	2.40	2.47	2.53	2.58	2.62	2.66
24	1.71	2.02	2.19	2.31	2.40	2.47	2.52	2.57	2.62	2.66
25	1.71	2.02	2.19	2.30	2.39	2.46	2.52	2.57	2.61	2.65
26	1.71	2.01	2.18	2.30	2.38	2.45	2.51	2.56	2.60	2.64
27	1.70	2.01	2.18	2.30	2.38	2.45	2.51	2.56	2.60	2.64
28	1.70	2.01	2.18	2.29	2.38	2.44	2.50	2.55	2.59	2.63
29	1.70	2.00	2.17	2.29	2.37	2.44	2.50	2.55	2.59	2.63
30	1.70	2.00	2.17	2.28	2.37	2.44	2.49	2.54	2.58	2.62
31	1.70	2.00	2.17	2.28	2.36	2.43	2.49	2.54	2.58	2.62
32	1.69	2.00	2.16	2.28	2.36	2.43	2.49	2.53	2.57	2.61
33	1.69	1.99	2.16	2.27	2.36	2.43	2.48	2.53	2.57	2.61
34	1.69	1.99	2.16	2.27	2.35	2.42	2.48	2.53	2.57	2.60
35	1.69	1.99	2.16	2.27	2.35	2.42	2.48	2.52	2.56	2.60

Tafel 19 (Forts.) Table 19 (cont.)

UPPER 1% POINTS / OBERE 1%-PUNKTE $\rho = 0.4$

n *p*	1	2	3	4	5	6	7	8	0	10
5	3.36	3.03	4.26	4.50	4.68	4.83	4.95	5.06	5.15	5.23
6	3.14	3.63	3.92	4.12	4.27	4.40	4.51	4.60	4.68	4.75
7	3.00	3.44	3.69	3.87	4.01	4.13	4.22	4.30	4.37	4.44
8	2.90	3.30	3.54	3.71	3.83	3.94	4.02	4.10	4.16	4.22
9	2.82	3.20	3.43	3.58	3.70	3.80	3.88	3.95	4.01	4.06
10	2.76	3.13	3.34	3.49	3.60	3.69	3.77	3.83	3.89	3.94
11	2.72	3.07	3.27	3.41	3.52	3.61	3.68	3.74	3.80	3.85
12	2.68	3.02	3.21	3.35	3.46	3.54	3.61	3.67	3.72	3.77
13	2.65	2.98	3.17	3.30	3.40	3.48	3.55	3.61	3.66	3.71
14	2.62	2.95	3.13	3.26	3.36	3.44	3.50	3.56	3.61	3.65
15	2.60	2.92	3.10	3.22	3.32	3.40	3.46	3.52	3.56	3.61
16	2.58	2.89	3.07	3.19	3.29	3.36	3.43	3.48	3.53	3.57
17	2.57	2.87	3.05	3.17	3.26	3.33	3.39	3.45	3.50	3.54
18	2.55	2.85	3.02	3.14	3.23	3.31	3.37	3.42	3.47	3.51
19	2.54	2.84	3.00	3.12	3.21	3.28	3.34	3.39	3.44	3.48
20	2.53	2.82	2.99	3.10	3.19	3.26	3.32	3.37	3.42	3.46
21	2.52	2.81	2.97	3 9	3.17	3.24	3.30	3.35	3.40	3.44
22	2.51	2.80	2.96	3.07	3.16	3.23	3.29	3.33	3.38	3.42
23	2.50	2.78	2.95	3.06	3.14	3.21	3.27	3.32	3.36	3.40
24	2.49	2.77	2.94	3.05	3.13	3.20	3.26	3.30	3.35	3.38
25	2.48	2.77	2.93	3.03	3.12	3.18	3.24	3.29	3.33	3.37
26	2.48	2.76	2.92	3.02	3.11	3.17	3.23	3.28	3.32	3.36
27	2.47	2.75	2.91	3.01	3.10	3.16	3.22	3.27	3.31	3.34
28	2.47	2.74	2.90	3.01	3.09	3.15	3.21	3.26	3.30	3.33
29	2.46	2.74	2.89	3.00	3.08	3.14	3.20	3.25	3.20	3.32
30	2.46	2.73	2.89	2.99	3.07	3.14	3.19	3.24	3.28	3.31
31	2.45	2.72	2.88	2.98	3.06	3.13	3.18	3.23	3.27	3.31
32	2.45	2.72	2.87	2.98	3.06	3.12	3.17	3.22	3.26	3.30
33	2.44	2.72	2.87	2.97	3.05	3.11	3.17	3.21	3.25	3.29
34	2.44	2.71	2.86	2.96	3.04	3.11	3.16	3.20	3.25	3.28
35	2.44	2.71	2.86	2.96	3.04	3.10	3.15	3.20	3.24	3.28

Tafel 19 (Forts.) Table 19 (cont.)

UPPER 5% POINTS / OBERE 5%-PUNKTE $\rho = 0.5$

n *p*	1	2	3	4	5	6	7	8	9	10
5	2.01	2.44	2.68	2.85	2.98	3.08	3.16	3.24	3.30	3.36
6	1.94	2.34	2.56	2.71	2.83	2.92	3.00	3.06	3.12	3.17
7	1.89	2.27	2.48	2.62	2.73	2.81	2.89	2.95	3.00	3.05
8	1.86	2.22	2.42	2.55	2.66	2.74	2.81	2.87	2.92	2.96
9	1.83	2.18	2.37	2.50	2.60	2.68	2.75	2.81	2.86	2.90
10	1.81	2.15	2.34	2.47	2.56	2.64	2.70	2.76	2.81	2.85
11	1.79	2.13	2.31	2.43	2.53	2.60	2.67	2.72	2.77	2.81
12	1.78	2.11	2.29	2.41	2.50	2.58	2.64	2.69	2.73	2.77
13	1.77	2.09	2.27	2.39	2.48	2.55	2.61	2.66	2.71	2.75
14	1.76	2.08	2.25	2.37	2.46	2.53	2.59	2.64	2.69	2.72
15	1.75	2.07	2.24	2.36	2.44	2.51	2.57	2.62	2.67	2.71
16	1.75	2.06	2.23	2.34	2.43	2.50	2.56	2.61	2.65	2.69
17	1.74	2.05	2.22	2.33	2.42	2.49	2.54	2.59	2.63	2.67
18	1.73	2.04	2.21	2.32	2.41	2.48	2.53	2.58	2.62	2.66
19	1.73	2.03	2.20	2.31	2.40	2.46	2.52	2.57	2.61	2.65
20	1.72	2.03	2.19	2.30	2.39	2.46	2.51	2.56	2.60	2.64
21	1.72	2.02	2.18	2.30	2.38	2.45	2.50	2.55	2.59	2.63
22	1.72	2.02	2.18	2.29	2.37	2.44	2.50	2.54	2.58	2.62
23	1.71	2.01	2.17	2.28	2.37	2.43	2.49	2.53	2.57	2.61
24	1.71	2.01	2.17	2.28	2.36	2.43	2.48	2.53	2.57	2.60
25	1.71	2.00	2.16	2.27	2.36	2.42	2.48	2.52	2.56	2.60
26	1.71	2.00	2.16	2.27	2.35	2.42	2.47	2.52	2.56	2.59
27	1.70	2.00	2.16	2.26	2.35	2.41	2.46	2.51	2.55	2.58
28	1.70	1.99	2.15	2.26	2.34	2.41	2.46	2.51	2.54	2.58
29	1.70	1.99	2.15	2.26	2.34	2.40	2.46	2.50	2.54	2.57
30	1.70	1.99	2.15	2.25	2.33	2.40	2.45	2.50	2.54	2.57
31	1.70	1.99	2.14	2.25	2.33	2.39	2.45	2.49	2.53	2.57
32	1.69	1.98	2.14	2.25	2.33	2.39	2.44	2.49	2.53	2.56
33	1.69	1.98	2.14	2.24	2.32	2.39	2.44	2.49	2.52	2.56
34	1.69	1.98	2.14	2.24	2.32	2.38	2.44	2.48	2.52	2.56
35	1.69	1.98	2.13	2.24	2.32	2.38	2.43	2.48	2.52	2.55

Tafel 19 (Forts.) Table 19 (cont.)

UPPER 1% POINTS / OBERE 1%-PUNKTE $\rho = 0.5$

n \ p	1	2	3	4	5	6	7	8	9	10
5	3.36	3.90	4.21	4.43	4.60	4.73	4.85	4.94	5.03	5.11
6	3.14	3.61	3.88	4.06	4.21	4.32	4.42	4.51	4.58	4.64
7	3.00	3.42	3.66	3.83	3.96	4.06	4.15	4.22	4.29	4.35
8	2.90	3.29	3.51	3.66	3.78	3.88	3.96	4.03	4.09	4.14
9	2.82	3.19	3.40	3.54	3.66	3.75	3.82	3.89	3.94	3.99
10	2.76	3.11	3.31	3.45	3.56	3.64	3.72	3.78	3.83	3.88
11	2.72	3.06	3.25	3.38	3.48	3.56	3.63	3.69	3.74	3.79
12	2.68	3.01	3.19	3.32	3.42	3.50	3.56	3.62	3.67	3.71
13	2.65	2.97	3.15	3.27	3.37	3.44	3.51	3.56	3.61	3.65
14	2.62	2.93	3.11	3.23	3.32	3.40	3.46	3.51	3.56	3.60
15	2.60	2.91	3.08	3.20	3.29	3.36	3.42	3.47	3.52	3.56
16	2.58	2.88	3.05	3.17	3.26	3.33	3.39	3.44	3.48	3.52
17	2.57	2.86	3.03	3.14	3.23	3.30	3.36	3.41	3.45	3.49
18	2.55	2.84	3.01	3.12	3.20	3.27	3.33	3.38	3.42	3.46
19	2.54	2.83	2.99	3.10	3.18	3.25	3.31	3.36	3.40	3.44
20	2.53	2.81	2.97	3.08	3.16	3.23	3.29	3.34	3.38	3.41
21	2.52	2.80	2.96	3.07	3.15	3.21	3.27	3.32	3.36	3.39
22	2.51	2.79	2.94	3.05	3.13	3.20	3.25	3.30	3.34	3.38
23	2.50	2.77	2.93	3.04	3.12	3.18	3.24	3.28	3.32	3.36
24	2.49	2.77	2.92	3.02	3.11	3.17	3.22	3.27	3.31	3.34
25	2.48	2.76	2.91	3.01	3.10	3.16	3.21	3.26	3.30	3.33
26	2.48	2.75	2.90	3.00	3.08	3.15	3.20	3.24	3.29	3.32
27	2.47	2.74	2.89	2.99	3.07	3.14	3.19	3.23	3.27	3.31
28	2.47	2.73	2.88	2.99	3.06	3.13	3.18	3.22	3.26	3.30
29	2.46	2.73	2.88	2.98	3.06	3.12	3.17	3.21	3.25	3.29
30	2.46	2.72	2.87	2.97	3.05	3.11	3.16	3.20	3.24	3.28
31	2.45	2.72	2.86	2.96	3.04	3.10	3.15	3.20	3.23	3.27
32	2.45	2.71	2.86	2.96	3.03	3.10	3.15	3.19	3.23	3.26
33	2.44	2.71	2.85	2.95	3.03	3.09	3.14	3.18	3.22	3.25
34	2.44	2.70	2.85	2.95	3.02	3.08	3.13	3.17	3.21	3.25
35	2.44	2.70	2.84	2.94	3.01	3.08	3.13	3.17	3.21	3.24

Tafel 19 (Forts.) Table 19 (cont.)

UPPER 5% POINTS / OBERE 5%-PUNKTE $\rho = 0.6$

n p	1	2	3	4	5	6	7	8	9	10
5	2.01	2.41	2.63	2.78	2.89	2.99	3.06	3.13	3.18	3.23
6	1.94	2.31	2.51	2.65	2.75	2.84	2.91	2.97	3.02	3.07
7	1.89	2.24	2.43	2.56	2.66	2.74	2.81	2.86	2.91	2.95
8	1.86	2.19	2.37	2.50	2.59	2.67	2.73	2.78	2.83	2.87
9	1.83	2.16	2.33	2.45	2.54	2.62	2.68	2.73	2.77	2.81
10	1.81	2.13	2.30	2.42	2.51	2.58	2.63	2.68	2.73	2.76
11	1.79	2.10	2.27	2.39	2.47	2.54	2.60	2.65	2.69	2.73
12	1.78	2.09	2.25	2.36	2.45	2.52	2.57	2.62	2.66	2.70
13	1.77	2.07	2.23	2.34	2.43	2.49	2.55	2.59	2.64	2.67
14	1.76	2.06	2.22	2.33	2.41	2.47	2.53	2.57	2.61	2.65
15	1.75	2.05	2.20	2.31	2.39	2.46	2.51	2.56	2.60	2.63
16	1.75	2.04	2.19	2.30	2.38	2.44	2.50	2.54	2.58	2.62
17	1.74	2.03	2.18	2.29	2.37	2.43	2.48	2.53	2.57	2.60
18	1.73	2.02	2.17	2.28	2.36	2.42	2.47	2.52	2.55	2.59
19	1.73	2.01	2.17	2.27	2.35	2.41	2.46	2.51	2.54	2.58
20	1.72	2.01	2.16	2.26	2.34	2.40	2.45	2.50	2.53	2.57
21	1.72	2.00	2.15	2.26	2.33	2.39	2.45	2.49	2.53	2.56
22	1.72	2.00	2.15	2.25	2.33	2.39	2.44	2.48	2.52	2.55
23	1.71	1.99	2.14	2.24	2.32	2.38	2.43	2.47	2.51	2.54
24	1.71	1.99	2.14	2.24	2.32	2.38	2.43	2.47	2.51	2.54
25	1.71	1.98	2.13	2.23	2.31	2.37	2.42	2.46	2.50	2.53
26	1.71	1.98	2.13	2.23	2.31	2.37	2.41	2.46	2.49	2.53
27	1.70	1.98	2.13	2.23	2.30	2.36	2.41	2.45	2.49	2.52
28	1.70	1.97	2.12	2.22	2.30	2.36	2.41	2.45	2.48	2.52
29	1.70	1.97	2.12	2.22	2.29	2.35	2.40	2.44	2.48	2.51
30	1.70	1.97	2.12	2.22	2.29	2.35	2.40	2.44	2.48	2.51
31	1.70	1.97	2.11	2.21	2.29	2.35	2.39	2.44	2.47	2.50
32	1.69	1.96	2.11	2.21	2.28	2.34	2.39	2.43	2.47	2.50
33	1.69	1.96	2.11	2.21	2.28	2.34	2.39	2.43	2.47	2.50
34	1.69	1.96	2.11	2.20	2.28	2.34	2.38	2.43	2.46	2.49
35	1.69	1.96	2.11	2.20	2.28	2.33	2.38	2.42	2.46	2.49

Tafel 19 (Forts.) Table 19 (cont.)

UPPER 1% POINTS / OBERE 1%-PUNKTE $\rho = 0.6$

n \ p	1	2	3	4	5	6	7	8	9	10
5	3.36	3.86	4.15	4.35	4.50	4.62	4.72	4.81	4.89	4.95
6	3.14	3.58	3.82	4.00	4.13	4.23	4.32	4.40	4.46	4.52
7	3.00	3.39	3.61	3.77	3.89	3.98	4.06	4.13	4.19	4.24
8	2.90	3.26	3.47	3.61	3.72	3.81	3.88	3.94	4.00	4.05
9	2.82	3.17	3.36	3.50	3.60	3.68	3.75	3.81	3.86	3.91
10	2.76	3.09	3.28	3.41	3.51	3.58	3.65	3.70	3.75	3.80
11	2.72	3.04	3.21	3.34	3.43	3.51	3.57	3.62	3.67	3.71
12	2.68	2.99	3.16	3.28	3.37	3.44	3.51	3.56	3.60	3.64
13	2.65	2.95	3.12	3.23	3.32	3.39	3.45	3.50	3.55	3.58
14	2.62	2.92	3.08	3.20	3.28	3.35	3.41	3.46	3.50	3.54
15	2.60	2.89	3.05	3.16	3.25	3.31	3.37	3.42	3.46	3.50
16	2.58	2.87	3.02	3.13	3.22	3.28	3.34	3.38	3.42	3.46
17	2.57	2.85	3.00	3.11	3.19	3.25	3.31	3.35	3.39	3.43
18	2.55	2.83	2.98	3.09	3.17	3.23	3.28	3.33	3.37	3.40
19	2.54	2.81	2.96	3.07	3.15	3.21	3.26	3.31	3.34	3.38
20	2.53	2.80	2.95	3.05	3.13	3.19	3.24	3.29	3.32	3.36
21	2.52	2.78	2.93	3.03	3.11	3.17	3.22	3.27	3.31	3.34
22	2.51	2.77	2.92	3.02	3.10	3.16	3.21	3.25	3.29	3.32
23	2.50	2.76	2.91	3.01	3.08	3.14	3.19	3.24	3.27	3.31
24	2.49	2.75	2.90	2.99	3.07	3.13	3.18	3.22	3.26	3.29
25	2.48	2.74	2.89	2.98	3.06	3.12	3.17	3.21	3.25	3.28
26	2.48	2.74	2.88	2.97	3.05	3.11	3.16	3.20	3.24	3.27
27	2.47	2.73	2.87	2.97	3.04	3.10	3.15	3.19	3.22	3.26
28	2.47	2.72	2.86	2.96	3.03	3.09	3.14	3.18	3.21	3.25
29	2.46	2.72	2.85	2.95	3.02	3.08	3.13	3.17	3.20	3.24
30	2.46	2.71	2.85	2.94	3.01	3.07	3.12	3.16	3.20	3.23
31	2.45	2.70	2.84	2.94	3.01	3.07	3.11	3.15	3.19	3.22
32	2.45	2.70	2.84	2.93	3.00	3.06	3.11	3.15	3.18	3.21
33	2.44	2.69	2.83	2.92	2.99	3.05	3.10	3.14	3.17	3.20
34	2.44	2.69	2.82	2.92	2.99	3.05	3.10	3.13	3.17	3.20
35	2.44	2.69	2.82	2.91	2.98	3.04	3.09	3.13	3.16	3.19

Tafel 19 (Forts.) Table 19 (cont.)

UPPER 5% POINTS / OBERE 5%-PUNKTE $\rho = 0.7$

n \ p	1	2	3	4	5	6	7	8	9	10
5	2.01	2.37	2.56	2.70	2.80	2.88	2.94	3.00	3.05	3.09
6	1.94	2.27	2.45	2.57	2.67	2.74	2.80	2.85	2.90	2.94
7	1.89	2.21	2.38	2.49	2.58	2.65	2.71	2.75	2.80	2.83
8	1.86	2.16	2.32	2.43	2.52	2.58	2.64	2.68	2.72	2.76
9	1.83	2.12	2.28	2.39	2.47	2.53	2.59	2.63	2.67	2.71
10	1.81	2.10	2.25	2.36	2.43	2.50	2.55	2.59	2.63	2.66
11	1.79	2.08	2.23	2.33	2.40	2.47	2.52	2.56	2.59	2.63
12	1.78	2.06	2.21	2.31	2.38	2.44	2.49	2.53	2.57	2.60
13	1.77	2.04	2.19	2.29	2.36	2.42	2.47	2.51	2.54	2.58
14	1.76	2.03	2.17	2.27	2.34	2.40	2.45	2.49	2.53	2.56
15	1.75	2.02	2.16	2.26	2.33	2.39	2.43	2.47	2.51	2.54
16	1.75	2.01	2.15	2.25	2.32	2.37	2.42	2.46	2.50	2.52
17	1.74	2.00	2.14	2.24	2.31	2.36	2.41	2.45	2.48	2.51
18	1.73	1.99	2.13	2.23	2.30	2.35	2.40	2.44	2.47	2.50
19	1.73	1.99	2.13	2.22	2.29	2.34	2.39	2.43	2.46	2.49
20	1.72	1.98	2.12	2.21	2.28	2.34	2.38	2.42	2.45	2.48
21	1.72	1.98	2.11	2.20	2.27	2.33	2.37	2.41	2.44	2.47
22	1.72	1.97	2.11	2.20	2.27	2.32	2.37	2.40	2.44	2.47
23	1.71	1.97	2.10	2.19	2.26	2.32	2.36	2.40	2.43	2.46
24	1.71	1.96	2.10	2.19	2.26	2.31	2.35	2.39	2.42	2.45
25	1.71	1.96	2.09	2.18	2.25	2.31	2.35	2.39	2.42	2.45
26	1.71	1.96	2.09	2.18	2.25	2.30	2.34	2.38	2.41	2.44
27	1.70	1.95	2.09	2.18	2.24	2.30	2.34	2.38	2.41	2.44
28	1.70	1.95	2.08	2.17	2.24	2.29	2.34	2.37	2.40	2.43
29	1.70	1.95	2.08	2.17	2.24	2.29	2.33	2.37	2.40	2.43
30	1.70	1.94	2.08	2.17	2.23	2.29	2.33	2.37	2.40	2.43
31	1.70	1.94	2.08	2.16	2.23	2.28	2.33	2.36	2.39	2.42
32	1.69	1.94	2.07	2.16	2.23	2.28	2.32	2.36	2.39	2.42
33	1.69	1.94	2.07	2.16	2.22	2.28	2.32	2.36	2.39	2.42
34	1.69	1.94	2.07	2.16	2.22	2.27	2.32	2.35	2.38	2.41
35	1.69	1.93	2.07	2.15	2.22	2.27	2.32	2.35	2.38	2.41

UPPER 1% POINTS / OBERE 1%-PUNKTE $\rho = 0.7$

n \ p	1	2	3	4	5	6	7	8	9	10
5	3.36	3.82	4.07	4.25	4.38	4.48	4.57	4.65	4.71	4.77
6	3.14	3.54	3.76	3.91	4.03	4.12	4.19	4.26	4.32	4.37
7	3.00	3.36	3.56	3.70	3.80	3.88	3.95	4.01	4.06	4.11
8	2.90	3.23	3.42	3.54	3.64	3.72	3.78	3.84	3.89	3.93
9	2.82	3.14	3.31	3.43	3.53	3.60	3.66	3.71	3.75	3.79
10	2.76	3.07	3.23	3.35	3.44	3.51	3.56	3.61	3.66	3.69
11	2.72	3.01	3.17	3.28	3.37	3.43	3.49	3.54	3.58	3.61
12	2.68	2.96	3.12	3.23	3.31	3.37	3.43	3.47	3.51	3.55
13	2.65	2.93	3.08	3.18	3.26	3.33	3.38	3.42	3.46	3.50
14	2.62	2.90	3.04	3.15	3.22	3.29	3.34	3.38	3.42	3.45
15	2.60	2.87	3.01	3.11	3.19	3.25	3.30	3.34	3.38	3.41
16	2.58	2.84	2.99	3.09	3.16	3.22	3.27	3.31	3.35	3.38
17	2.57	2.82	2.97	3.06	3.14	3.19	3.24	3.28	3.32	3.35
18	2.55	2.81	2.95	3.04	3.11	3.17	3.22	3.26	3.29	3.33
19	2.54	2.79	2.93	3.02	3.10	3.15	3.20	3.24	3.27	3.30
20	2.53	2.78	2.91	3.01	3.08	3.13	3.18	3.22	3.25	3.28
21	2.52	2.76	2.90	2.99	3.06	3.12	3.16	3.20	3.23	3.27
22	2.51	2.75	2.89	2.98	3.05	3.10	3.15	3.18	3.22	3.25
23	2.50	2.74	2.88	2.97	3.03	3.09	3.13	3.17	3.20	3.23
24	2.49	2.73	2.87	2.95	3.02	3.08	3.12	3.16	3.19	3.22
25	2.48	2.72	2.86	2.94	3.01	3.07	3.11	3.15	3.18	3.21
26	2.48	2.72	2.85	2.94	3.00	3.06	3.10	3.14	3.17	3.20
27	2.47	2.71	2.84	2.93	2.99	3.05	3.09	3.13	3.16	3.19
28	2.47	2.70	2.83	2.92	2.98	3.04	3.08	3.12	3.15	3.18
29	2.46	2.70	2.82	2.91	2.98	3.03	3.07	3.11	3.14	3.17
30	2.46	2.69	2.82	2.91	2.97	3.02	3.07	3.10	3.13	3.16
31	2.45	2.69	2.81	2.90	2.96	3.01	3.06	3.10	3.13	3.15
32	2.45	2.68	2.81	2.89	2.96	3.01	3.05	3.09	3.12	3.15
33	2.44	2.68	2.80	2.89	2.95	3.00	3.04	3.08	3.11	3.14
34	2.44	2.67	2.79	2.88	2.94	3.00	3.04	3.08	3.11	3.13
35	2.44	2.67	2.79	2.88	2.94	2.99	3.03	3.07	3.10	3.13

Tafel 19 (Forts.) Table 19 (cont.)

UPPER 5% POINTS / OBERE 5%-PUNKTE $\rho = 0.8$

n \ p	1	2	3	4	5	6	7	8	9	10
5	2.01	2.32	2.48	2.59	2.67	2.74	2.79	2.84	2.88	2.91
6	1.94	2.22	2.37	2.48	2.55	2.61	2.66	2.71	2.74	2.77
7	1.89	2.16	2.31	2.40	2.47	2.53	2.58	2.62	2.65	2.68
8	1.86	2.12	2.25	2.35	2.42	2.47	2.52	2.55	2.59	2.62
9	1.83	2.08	2.22	2.31	2.37	2.43	2.47	2.51	2.54	2.57
10	1.81	2.06	2.19	2.27	2.34	2.39	2.43	2.47	2.50	2.53
11	1.79	2.04	2.16	2.25	2.31	2.36	2.40	2.44	2.47	2.50
12	1.78	2.02	2.14	2.23	2.29	2.34	2.38	2.42	2.45	2.47
13	1.77	2.00	2.13	2.21	2.27	2.32	2.36	2.40	2.43	2.45
14	1.76	1.99	2.11	2.20	2.26	2.31	2.35	2.38	2.41	2.43
15	1.75	1.98	2.10	2.18	2.24	2.29	2.33	2.36	2.39	2.42
16	1.75	1.97	2.09	2.17	2.23	2.28	2.32	2.35	2.38	2.41
17	1.74	1.96	2.08	2.16	2.22	2.27	2.31	2.34	2.37	2.39
18	1.73	1.96	2.08	2.15	2.21	2.26	2.30	2.33	2.36	2.38
19	1.73	1.95	2.07	2.15	2.21	2.25	2.29	2.32	2.35	2.37
20	1.72	1.95	2.06	2.14	2.20	2.25	2.28	2.32	2.34	2.37
21	1.72	1.94	2.06	2.13	2.19	2.24	2.28	2.31	2.34	2.36
22	1.72	1.94	2.05	2.13	2.19	2.23	2.27	2.30	2.33	2.35
23	1.71	1.93	2.05	2.12	2.18	2.23	2.26	2.30	2.32	2.35
24	1.71	1.93	2.04	2.12	2.18	2.22	2.26	2.29	2.32	2.34
25	1.71	1.92	2.04	2.12	2.17	2.22	2.25	2.29	2.31	2.34
26	1.71	1.92	2.04	2.11	2.17	2.21	2.25	2.28	2.31	2.33
27	1.70	1.92	2.03	2.11	2.16	2.21	2.25	2.28	2.31	2.33
28	1.70	1.92	2.03	2.11	2.16	2.21	2.24	2.27	2.30	2.32
29	1.70	1.91	2.03	2.10	2.16	2.20	2.24	2.27	2.30	2.32
30	1.70	1.91	2.02	2.10	2.16	2.20	2.24	2.27	2.29	2.32
31	1.70	1.91	2.02	2.10	2.15	2.20	2.23	2.26	2.29	2.31
32	1.69	1.91	2.02	2.10	2.15	2.19	2.23	2.26	2.29	2.31
33	1.69	1.91	2.02	2.09	2.15	2.19	2.23	2.26	2.29	2.31
34	1.69	1.90	2.01	2.09	2.15	2.19	2.23	2.26	2.28	2.31
35	1.69	1.90	2.01	2.09	2.14	2.19	2.22	2.25	2.28	2.30

Tafel 19 (Forts.) Table 19 (cont.)

UPPER 1% POINTS / OBERE 1%-PUNKTE $\rho = 0.8$

n \ p	1	2	3	4	5	6	7	8	9	10
5	3.36	3.75	3.97	4.11	4.22	4.31	4.38	4.44	4.50	4.54
6	3.14	3.48	3.67	3.80	3.89	3.97	4.03	4.09	4.13	4.17
7	3.00	3.31	3.48	3.59	3.68	3.75	3.81	3.86	3.90	3.94
8	2.90	3.19	3.35	3.45	3.53	3.60	3.65	3.70	3.74	3.77
9	2.82	3.10	3.25	3.35	3.42	3.49	3.54	3.58	3.62	3.65
10	2.76	3.03	3.17	3.27	3.34	3.40	3.45	3.49	3.52	3.56
11	2.72	2.97	3.11	3.21	3.28	3.33	3.38	3.42	3.45	3.48
12	2.68	2.93	3.06	3.15	3.22	3.28	3.32	3.36	3.39	3.42
13	2.65	2.89	3.02	3.11	3.18	3.23	3.28	3.31	3.34	3.37
14	2.62	2.86	2.99	3.08	3.14	3.19	3.24	3.27	3.30	3.33
15	2.60	2.83	2.96	3.05	3.11	3.16	3.20	3.24	3.27	3.30
16	2.58	2.81	2.94	3.02	3.08	3.13	3.17	3.21	3.24	3.27
17	2.57	2.79	2.92	3.00	3.06	3.11	3.15	3.18	3.21	3.24
18	2.55	2.77	2.90	2.98	3.04	3.09	3.13	2.16	3.19	3.22
19	2.54	2.76	2.88	2.96	3.02	3.07	3.11	3.14	3.17	3.20
20	2.53	2.75	2.86	2.94	3.00	3.05	3.09	3.12	3.15	3.18
21	2.52	2.73	2.85	2.93	2.99	3.04	3.08	3.11	3.14	3.16
22	2.51	2.72	2.84	2.92	2.98	3.02	3.06	3.10	3.12	3.15
23	2.50	2.71	2.83	2.91	2.96	3.01	3.05	3.08	3.11	3.13
24	2.49	2.70	2.82	2.90	2.95	3.00	3.04	3.07	3.10	3.12
25	2.48	2.70	2.81	2.89	2.94	2.99	3.03	3.06	3.09	3.11
26	2.48	2.69	2.80	2.88	2.93	2.98	3.02	3.05	3.08	3.10
27	2.47	2.68	2.79	2.87	2.93	2.97	3.01	3.04	3.07	3.09
28	2.47	2.67	2.78	2.86	2.92	2.96	3.00	3.03	3.06	3.08
29	2.46	2.67	2.78	2.85	2.91	2.95	2.99	3.02	3.05	3.07
30	2.46	2.66	2.77	2.85	2.90	2.95	2.98	3.02	3.04	3.07
31	2.45	2.66	2.77	2.84	2.90	2.94	2.98	3.01	3.04	3.06
32	2.45	2.65	2.76	2.84	2.89	2.94	2.97	3.00	3.03	3.05
33	2.44	2.65	2.76	2.83	2.89	2.93	2.97	3.00	3.02	3.05
34	2.44	2.64	2.75	2.83	2.88	2.92	2.96	2.99	3.02	3.04
35	2.44	2.64	2.75	2.82	2.88	2.92	2.95	2.98	3.01	3.04

Tafel 19 (Forts.) Table 19 (cont.)

UPPER 5% POINTS / OBERE 5%-PUNKTE $\rho = 0.9$

n \ p	1	2	3	4	5	6	7	8	9	10
5	2.01	2.24	2.36	2.44	2.50	2.54	2.58	2.61	2.64	2.67
6	1.94	2.15	2.26	2.34	2.39	2.43	2.47	2.50	2.53	2.55
7	1.89	2.09	2.20	2.27	2.32	2.36	2.40	2.42	2.45	2.47
8	1.86	2.05	2.15	2.22	2.27	2.31	2.34	2.37	2.39	2.41
9	1.83	2.02	2.12	2.18	2.23	2.27	2.30	2.33	2.35	2.37
10	1.81	2.00	2.09	2.16	2.20	2.24	2.27	2.30	2.32	2.34
11	1.79	1.98	2.07	2.13	2.18	2.22	2.25	2.27	2.29	2.31
12	1.78	1.96	2.05	2.12	2.16	2.20	2.23	2.25	2.27	2.29
13	1.77	1.95	2.04	2.10	2.14	2.18	2.21	2.23	2.25	2.27
14	1.76	1.93	2.03	2.09	2.13	2.17	2.19	2.22	2.24	2.26
15	1.75	1.93	2.01	2.08	2.12	2.15	2.18	2.21	2.23	2.25
16	1.75	1.92	2.01	2.07	2.11	2.14	2.17	2.19	2.22	2.23
17	1.74	1.91	2.00	2.06	2.10	2.13	2.16	2.19	2.21	2.22
18	1.73	1.90	1.99	2.05	2.09	2.13	2.15	2.18	2.20	2.22
19	1.73	1.90	1.98	2.04	2.09	2.12	2.15	2.17	2.19	2.21
20	1.72	1.89	1.98	2.04	2.08	2.11	2.14	1.16	2.18	2.20
21	1.72	1.89	1.97	2.03	2.07	2.11	2.13	2.16	2.18	2.19
22	1.72	1.88	1.97	2.03	2.07	2.10	2.13	2.15	2.17	2.19
23	1.71	1.88	1.96	2.02	2.06	2.10	2.12	2.15	2.17	2.18
24	1.71	1.88	1.96	2.02	2.06	2.09	2.12	2.14	2.16	2.18
25	1.71	1.87	1.96	2.01	2.06	2.09	2.12	2.14	2.16	2.17
26	1.71	1.87	1.95	2.01	2.05	2.09	2.11	2.13	2.15	2.17
27	1.70	1.87	1.95	2.01	2.05	2.08	2.11	2.13	2.15	2.17
28	1.70	1.86	1.95	2.00	2.05	2.08	2.11	2.13	2.15	2.16
29	1.70	1.86	1.95	2.00	2.04	2.08	2.10	2.12	2.14	2.16
30	1.70	1.86	1.94	2.00	2.04	2.07	2.10	2.12	2.14	2.16
31	1.70	1.86	1.94	2.00	2.04	2.07	2.10	2.12	2.14	2.15
32	1.69	1.85	1.94	1.99	2.04	2.07	2.09	2.12	2.14	2.15
33	1.69	1.85	1.94	1.99	2.03	2.07	2.09	2.11	2.13	2.15
34	1.69	1.85	1.93	1.99	2.03	2.06	2.09	2.11	2.13	2.15
35	1.69	1.85	1.93	1.99	2.03	2.06	2.09	2.11	2.13	2.15

Tafel 19 (Forts.) Table 19 (cont.)

UPPER 1% POINTS / OBERE 1%-PUNKTE $\rho = 0.9$

n \ p	1	2	3	4	5	6	7	8	9	10
5	3.36	3.66	3.82	3.92	4.00	4.06	4.11	4.16	4.19	4.23
6	3.14	3.40	3.54	3.63	3.70	3.75	3.80	3.83	3.87	3.89
7	3.00	3.23	3.36	3.44	3.51	3.56	3.60	3.63	3.66	3.69
8	2.90	3.12	3.24	3.32	3.37	3.42	3.46	3.49	3.52	3.55
9	2.82	3.03	3.14	3.22	3.28	3.32	3.35	3.39	3.41	3.43
10	2.76	2.97	3.07	3.15	3.20	3.24	3.28	3.31	3.33	3.35
11	2.72	2.91	3.02	3.09	3.14	3.18	3.21	3.24	3.27	3.29
12	2.68	2.87	2.97	3.04	3.09	3.13	3.16	3.19	3.21	3.24
13	2.65	2.84	2.94	3.00	3.05	3.09	3.12	3.15	3.17	3.19
14	2.62	2.81	2.90	2.97	3.02	3.05	3.09	3.11	3.14	3.15
15	2.60	2.78	2.88	2.94	2.99	3.02	3.06	3.08	3.10	3.12
16	2.58	2.76	2.85	2.92	2.96	3.00	3.03	3.06	3.08	3.10
17	2.57	2.74	2.83	2.90	2.94	2.98	3.01	3.03	3.05	3.07
18	2.55	2.72	2.82	2.88	2.92	2.96	2.99	3.01	3.03	3.05
19	2.54	2.71	2.80	2.86	2.91	2.94	2.97	2.99	3.02	3.03
20	2.53	2.70	2.79	2.85	2.89	2.93	2.95	2.98	3.00	3.02
21	2.52	2.69	2.77	2.83	2.88	2.91	2.94	2.96	2.98	3.00
22	2.51	2.67	2.76	2.82	2.87	2.90	2.93	2.95	2.97	2.99
23	2.50	2.67	2.75	2.81	2.85	2.89	2.92	2.94	2.96	2.98
24	2.49	2.66	2.74	2.80	2.84	2.88	2.91	2.93	2.95	2.97
25	2.48	2.65	2.73	2.79	2.83	2.87	2.90	2.92	2.94	2.96
26	2.48	2.64	2.73	2.78	2.83	2.86	2.89	2.91	2.93	2.95
27	2.47	2.63	2.72	2.78	2.82	2.85	2.88	2.90	2.92	2.94
28	2.47	2.63	2.71	2.77	2.81	2.84	2.87	2.90	2.92	2.93
29	2.46	2.62	2.71	2.76	2.80	2.84	2.87	2.89	2.91	2.93
30	2.46	2.62	2.70	2.76	2.80	2.83	2.86	2.88	2.90	2.92
31	2.45	2.61	2.70	2.75	2.79	2.83	2.85	2.88	2.90	2.91
32	2.45	2.61	2.69	2.75	2.79	2.82	2.85	2.87	2.89	2.91
33	2.44	2.60	2.69	2.74	2.78	2.81	2.84	2.86	2.89	2.90
34	2.44	2.60	2.68	2.74	2.78	2.81	2.84	2.86	2.88	2.90
35	2.44	2.59	2.68	2.73	2.77	2.80	2.83	2.85	2.88	2.89

PART III

Further Tables for Multivariate Problems

P r e l i m i n a r y   R e m a r k s :

The distributions and test statistics under consideration in this part of the collection of tables are concerned with multivariate problems, which have entered just now into text books of multivariate analysis.

Of great interest for the practitioner are the criteria for sequential multivariate comparisons of means.

Moreover, the selection of tables for this part of the collection must have an individual character for reasons of space.

Also related to the topics of this part III are the tables 28, and 29, which appear - by technical reasons - in the appendix (=part IV).

TABLE 20

The Gamma Distribution: Tables of M. B. Wilk, R. Gnanadesikan, and M. J. Huyette

(a) Content of the tables and definition of the test statistic :

The tables contain the quantiles of the gamma distribution with density function

$$f(y; a; \lambda; \eta) \quad ,$$

where the parameters are defined as follows :

a = initial parameter with $-\infty < a < \infty$;

λ = scale parameter with $\lambda > 0$;

η = profile parameter with $\eta > 0$.

Here we consider the standardised case with $a = 0$ and $\lambda = 1$, so we have the special density function

$$f(y; 0; 1; \eta) = \frac{1}{\Gamma(\eta)} \cdot y^{\eta-1} . e^{-\eta} \quad \text{for} \quad 0 \leq y < \infty \quad ,$$

$$= 0 \quad \text{for} \quad y < 0 \ .$$

The corresponding distribution function is

$$P(y) = F(y; 0; 1; \eta) = \frac{1}{\Gamma(\eta)} \cdot \int_0^y f(t; 0; 1; \eta) dt \quad .$$

The tables cover a broad region of η-values.

(b) Extent of the tables and definition of the parameters :

(1) The parameter P(y) :

P(y) = percentage = probability of the event ($t \leq y$)

for the percentages
0.1 ; 0.5 ; 0.7 ; 1.0(0.5)3(1)5 ; 7.5 ;
10(5)30(10)70(5)90(2.5)97.5 ; 98 ; 99 ;
99.5 ; 99.9 .

(2) The parameter η :

η = profile parameter of the gamma distribution

η = 0.1(0.1)0.6(0.2)5.0(0.5)10.0(1.0)22.0 .

(c) Suggestions for applications :

(1) Graphic representation of the gamma distribution and estimation of its parameters.

(2) Graphic internal comparison procedures for multivariate experiments. (See e. g. : ROY, S. N. + GNANADESIKAN, R. + SRIVASTAVA, J. N. (1971)).

(d) References :

(1) For the source of the tables :

WILK, M. B. + GNANADESIKAN, R. + HUYETTE, M. J. : Probability plots for the gamma distribution. Technometrics 4, 1 - 20 (1962).

(2) For applications :

ROY, S. N. + GNANADESIKAN, R. + SRIVASTAVA, J. N. : Analysis and design of certain quantitative multiresponse experiments.
Oxford + New York a. o. : Pergamon Press, 1971 .

(e) Further remarks :

For the special case $\lambda = 1/2$, one obtains the chi-square distribution from the tabulated gamma distribution. The quantiles given in the tables, therefore are to be multiplied by two.
The parameter $2.\eta$ corresponds to the degrees of freedom n of the chi-square distribution.

Tafel 20 Table 20

QUANTILE DER GAMMA-VERTEILUNG

QUANTILES ØF THE GAMMA DISTRIBUTIØN

(NØTE. 6.07E-31 IS EQUIVALENT TØ 6.07 X 10^{-31})

PER CENT	ETA= 0.1	ETA= 0.2	ETA= 0.3	ETA= 0.4	ETA= 0.5
0.1	6.0730398E-31	6.5254914E-16	6.9727015E-11	2.3449403E-08	7.8539858E-07
0.5	5.9307050E-24	2.0392159E-12	1.4903943E-08	1.3108620E-06	1.9635211E-05
0.7	3.4199421E-22	1.5485293E-11	5.7580040E-08	3.6123166E-06	4.4179955E-05
1.0	6.0730419E-21	6.5254908E-11	1.5022232E-07	7.4153871E-06	7.8543941E-05
1.5	3.5020208E-19	4.9552939E-10	5.8037043E-07	2.0434596E-05	1.7673543E-04
2.0	6.2187947E-18	2.0881566E-09	1.5141467E-06	4.1948807E-05	3.1422512E-04
2.5	5.7917027E-17	6.3725492E-09	3.1856788E-06	7.3283179E-05	4.9103468E-04
3.0	3.5860676E-16	1.5856933E-08	5.8497902E-06	1.1560341E-04	7.0719174E-04
4.0	6.3680429E-15	6.6821009E-08	1.5261804E-05	2.3733141E-04	1.2576912E-03
5.0	5.9307026E-14	2.0392152E-07	3.2110344E-05	4.1465349E-04	1.9660703E-03
7.5	3.4199416E-12	1.5485298E-06	1.2406423E-04	1.1432451E-03	4.4309273E-03
10.0	6.0730380E-11	6.5255202E-06	3.2372462E-04	2.3488758E-03	7.8953891E-03
15.0	3.5020194E-09	4.9554952E-05	1.2515738E-03	6.4918979E-03	1.7882892E-02
20.0	6.2187893E-08	2.0885191E-04	3.2703397E-03	1.3392227E-02	3.2092386E-02
25.0	5.7917008E-07	6.3759328E-04	6.8998047E-03	2.3564933E-02	5.0765538E-02
30.0	3.5860783E-06	1.5877920E-03	1.2726660E-02	3.7541870E-02	7.4235954E-02
40.0	6.3684073E-05	6.7195739E-03	3.3739803E-02	7.9361846E-02	1.3749799E-01
50.0	5.9339001E-04	2.0746364E-02	7.3131159E-02	1.4507806E-01	2.2746834E-01
60.0	3.6844445E-03	5.3010665E-02	1.4125254E-01	2.4475218E-01	3.5416332E-01
70.0	1.7427737E-02	1.2103773E-01	2.5656503E-01	3.9725703E-01	5.3709736E-01
75.0	3.5306306E-02	1.7885940E-01	3.4289970E-01	5.0480585E-01	6.6165223E-01
80.0	6.9389746E-02	2.6354398E-01	4.6007422E-01	6.4557067E-01	8.2118779E-01
85.0	1.3466307E-01	3.9239831E-01	6.2662798E-01	8.3910112E-01	1.0361264E 00
90.0	2.6615398E-01	6.0490358E-01	8.8481154E-01	1.1298418E 00	1.3527737E 00
92.5	3.8439176E-01	7.7335898E-01	1.0811301E 00	1.3461192E 00	1.5850286E 00
95.0	5.8043370E-01	1.0305303E 00	1.3723524E 00	1.6619615E 00	1.9207334E 00
97.5	9.7790323E-01	1.5111122E 00	1.9002707E 00	2.2247126E 00	2.5119509E 00
98.0	1.1190291E 00	1.6744477E 00	2.0765711E 00	2.4107092E 00	2.7059585E 00
99.0	1.5884692E 00	2.2023303E 00	2.6394268E 00	3.0000888E 00	3.3174701E 00
99.5	2.0945469E 00	2.7547526E 00	3.2206054E 00	3.6035306E 00	3.9397731E 00
99.9	3.3636681E 00	4.1023043E 00	4.6191710E 00	5.0425654E 00	5.4140809E 00

PER CENT	ETA= 0.6	ETA= 0.8	ETA= 1.0	ETA= 1.2	ETA= 1.4
0.1	8.2890664E-06	1.6272338E-04	1.0005003E-03	3.4337143E-03	8.4321789E-03
0.5	1.2119545E-04	1.2173541E-03	5.0125419E-03	1.3187497E-02	2.6824095E-02
0.7	2.3823386E-04	2.0217386E-03	7.5282661E-03	1.8533525E-02	3.5971068E-02
1.0	3.8483497E-04	2.8980758E-03	1.0050336E-02	2.3608727E-02	4.4330284E-02
1.5	7.5659043E-04	4.8159973E-03	1.5113637E-02	3.3243716E-02	5.9597079E-02
2.0	1.2224146E-03	6.9081841E-03	2.0202707E-02	4.2425602E-02	7.3617880E-02
2.5	1.7737191E-03	9.1419638E-03	2.5317809E-02	5.1301112E-02	8.6809725E-02
3.0	2.4044995E-03	1.1496962E-02	3.0459208E-02	5.9952244E-02	9.9398144E-02
4.0	3.8873902E-03	1.6518275E-02	4.0821993E-02	7.6773139E-02	1.2327700E-01
5.0	5.6448357E-03	2.1897531E-02	5.1293299E-02	9.3145212E-02	1.4592649E-01
7.5	1.1133291E-02	3.6647758E-02	7.7961543E-02	1.3293248E-01	1.9922959E-01
10.0	1.8060443E-02	5.2981821E-02	1.0536052E-01	1.7189840E-01	2.4973644E-01
15.0	3.5894110E-02	8.9739645E-02	1.6251895E-01	2.4937285E-01	3.4689076E-01
20.0	5.8803372E-02	1.3152850E-01	2.2314357E-01	3.2797601E-01	4.4246886E-01
25.0	8.6771754E-02	1.7827535E-01	2.8768212E-01	4.0903381E-01	5.3886871E-01
30.0	1.1998889E-01	2.3019176E-01	3.5667497E-01	4.9357830E-01	6.3769279E-01
40.0	2.0382268E-01	3.5143535E-01	5.1082570E-01	6.7712659E-01	8.4793257E-01
50.0	3.1570220E-01	5.0135124E-01	6.9314729E-01	8.8793657E-01	1.0843713E 00
60.0	4.6590956E-01	6.9127123E-01	9.1629100E-01	1.1400349E 00	1.3623938E 00
70.0	6.7474850E-01	9.4321526E-01	1.2039732E 00	1.4587471E 00	1.7088393E 00
75.0	8.1365435E-01	1.1058806E 00	1.3862949E 00	1.6581306E 00	1.9234758E 00
80.0	9.8899279E-01	1.3073711E 00	1.6094384E 00	1.9000906E 00	2.1822733E 00
85.0	1.2219616E 00	1.5702344E 00	1.8971206E 00	2.2093879E 00	2.5109470E 00
90.0	1.5605061E 00	1.9452591E 00	2.3025857E 00	2.6414636E 00	2.9669431E 00
92.5	1.8063390E 00	2.2138442E 00	2.5902694E 00	2.9459091E 00	3.2864993E 00
95.0	2.1590177E 00	2.5951444E 00	2.9957345E 00	3.3726729E 00	3.7325141E 00
97.5	2.7747879E 00	3.2526998E 00	3.6888885E 00	4.0973864E 00	4.4858795E 00
98.0	2.9757511E 00	3.4656007E 00	3.9120296E 00	4.3296577E 00	4.7264682E 00
99.0	3.6066221E 00	4.1298904E 00	4.6051941E 00	5.0486563E 00	5.4690573E 00
99.5	4.2455626E 00	4.7977945E 00	5.2983655E 00	5.7645890E 00	6.2058056E 00
99.9	5.7512864E 00	6.3586366E 00	6.9080574E 00	7.4187547E 00	7.9005137E 00

QUANTILE DER GAMMA-VERTEILUNG

QUANTILES ØF THE GAMMA DISTRIBUTIØN

(NØTE. 6.C7E-31 IS EQUIVALENT TØ 6.07 X 10^{-31})

PER CENT	ETA= 1.6	ETA= 1.8	ETA= 2.0	ETA= 2.2	ETA= 2.4
0.1	1.6780863E-02	2.9006379E-02	4.5402019E-02	6.6083795E-02	9.1045250E-02
0.5	4.6409061E-02	7.2019480E-02	1.0349455E-01	1.4055196E-01	1.8285795E-01
0.7	6.0108387E-02	9.0817968E-02	1.2777440E-01	1.7057086E-01	2.1878191E-01
1.0	7.2283818E-02	1.0717552E-01	1.4855474E-01	1.9592441E-01	2.4879515E-01
1.5	9.3902475E-02	1.3560738E-01	1.8407833E-01	2.3869266E-01	2.9887579E-01
2.0	1.1322857E-01	1.6050999E-01	2.1469912E-01	2.7508877E-01	3.4104964E-01
2.5	1.3105810E-01	1.8314475E-01	2.4220929E-01	3.0748374E-01	3.7830102E-01
3.0	1.4781163E-01	2.0416656E-01	2.6752685E-01	3.3707901E-01	4.1212894E-01
4.0	1.7902662E-01	2.4280678E-01	3.1357260E-01	3.9044885E-01	4.7270738E-01
5.0	2.0808401E-01	2.7826796E-01	3.5536154E-01	4.3845224E-01	5.2679405E-01
7.5	2.7489323E-01	3.5837546E-01	4.4846799E-01	5.4422134E-01	6.4488062E-01
10.0	3.3669067E-01	4.3112658E-01	5.3181166E-01	6.3780034E-01	7.4835367E-01
15.0	4.5272915E-01	5.6524168E-01	6.8323869E-01	8.0583392E-01	9.3234974E-01
20.0	5.6432575E-01	6.9199598E-01	8.2438848E-01	9.6070601E-01	1.1003482E 00
25.0	6.7504999E-Q1	8.1617533E-01	9.6127882E-01	1.1096652E 00	1.2608163E 00
30.0	7.8710858E-01	9.4060314E-01	1.0973494E 00	1.2567599E 00	1.4184017E 00
40.0	1.0219104E 00	1.1982486E 00	1.3764216E 00	1.5560676E 00	1.7369269E 00
50.0	1.2817967E 00	1.4798565E 00	1.6783474E 00	1.8771417E 00	2.0761571E 00
60.0	1.5834684E 00	1.8033957E 00	2.0223137E 00	2.2403417E 00	2.4575810E 00
70.0	1.9551916E 00	2.1984807E 00	2.4392173E 00	2.6777864E 00	2.9144873E 00
75.0	2.1837358E 00	2.4398846E 00	2.6926356E 00	2.9425202E 00	3.1899446E 00
80.0	2.4579001E 00	2.7282669E 00	2.9943091E 00	3.2567191E 00	3.5160193E 00
85.0	2.8042824E 00	3.0910513E 00	3.3724436E 00	3.6493274E 00	3.9223586E 00
90.0	3.2821878E 00	3.5892844E 00	3.8897218E 00	4.1845905E 00	4.4746965E 00
92.5	3.6156165E 00	3.9356059E 00	4.2481450E 00	4.5544488E 00	4.8554263E 00
95.0	4.0793485E 00	4.4158303E 00	4.7438694E 00	5.0648451E 00	5.3797783E 00
97.5	4.8591827E 00	5.2203591E 00	5.5716554E 00	5.9146804E 00	6.2506155E 00
98.0	5.1074690E 00	5.4758451E 00	5.8339376E 00	6.1834139E 00	6.5255030E 00
99.0	5.8719870E 00	6.2608809E 00	6.6383883E 00	7.0063273E 00	7.3660060E 00
99.5	6.6282540E 00	7.0353775E 00	7.4301966E 00	7.8146312E 00	8.1900520E 00
99.9	8.3615569E 00	8.8C45225E 00	9.2336768E 00	9.6511943E 00	1.0057676E 01

PER CENT	ETA= 2.6	ETA= 2.8	ETA= 3.0	ETA= 3.2	ETA= 3.4
0.1	1.2020084E-01	1.5341695E-01	1.9053338E-01	2.3137730E-01	2.7577227E-01
0.5	2.3006679E-01	2.8184123E-01	3.3786340E-01	3.9783849E-01	4.6149604E-01
0.7	2.7199522E-01	3.2982536E-01	3.9191958E-01	4.5795804E-01	5.2765206E-01
1.0	3.0670981E-01	3.6925137E-01	4.3604521E-01	5.0675595E-01	5.8108456E-01
1.5	3.6411288E-01	4.3394811E-01	5.0798078E-01	5.8585852E-01	6.6727104E-01
2.0	4.1203167E-01	4.8755657E-01	5.6720963E-01	6.5063010E-01	7.3750284E-01
2.5	4.5408805E-01	5.3435255E-01	6.1867216E-01	7.0668188E-01	7.9806521E-01
3.0	4.9208797E-01	5.7645527E-01	6.6480418E-01	7.5676816E-01	8.5203115E-01
4.0	5.5974186E-01	6.5104351E-01	7.4618340E-01	8.4479663E-01	9.4657009E-01
5.0	6.1977374E-01	7.1688152E-01	8.1769154E-01	9.2184297E-01	1.0290283E 00
7.5	7.4983816E-01	8.5859696E-01	9.7074668E-01	1.0859437E 00	1.2038973E 00
10.0	8.6288491E-01	9.8091840E-01	1.1020654E 00	1.2260021E 00	1.3524560E 00
15.0	1.0622576E 00	1.1951340E 00	1.3306367E 00	1.4684834E 00	1.6084377E 00
20.0	1.2428518E 00	1.3878487E 00	1.5350443E 00	1.6841965E 00	1.8351031E 00
25.0	1.4143367E 00	1.5699138E 00	1.7272996E 00	1.8862906E 00	2.0467182E 00
30.0	1.5819476E 00	1.7471400E 00	1.9137761E 00	2.0816902E 00	2.2507454E 00
40.0	1.9188084E 00	2.1015632E 00	2.2850772E 00	2.4692581E 00	2.6540295E 00
50.0	2.2753401E 00	2.4746498E 00	2.6740610E 00	2.8735522E 00	3.0731070E 00
60.0	2.6741217E 00	2.8900326E 00	3.1053791E 00	3.3202146E 00	3.5345823E 00
70.0	3.1495635E 00	3.3832035E 00	3.6155689E 00	3.8467910E 00	4.0769786E 00
75.0	3.4352389E 00	3.6786528E 00	3.9204034E 00	4.1606653E 00	4.3995793E 00
80.0	3.7726309E 00	4.0268778E 00	4.2790313E 00	4.5293148E 00	4.7779093E 00
85.0	4.1920675E 00	4.4588519E 00	4.7230534E 00	4.9849489E 00	5.2447592E 00
90.0	4.7607076E 00	5.0431069E 00	5.3223249E 00	5.5986977E 00	5.8724992E 00
92.5	5.1518242E 00	5.4441808E 00	5.7329780E 00	6.0185937E 00	6.3013325E 00
95.0	5.6895199E 00	5.9946764E 00	6.2958025E 00	6.5933195E 00	6.8875671E 00
97.5	6.5804864E 00	6.9049583E 00	7.2247059E 00	7.5402121E 00	7.8518687E 00
98.0	6.8612753E 00	7.1914166E 00	7.5166242E 00	7.8374205E 00	8.1542076E 00
99.0	7.7186885E 00	8.0650784E 00	8.4059950E 00	8.7419776E 00	9.0734380E 00
99.5	8.5579089E 00	8.9188460E 00	9.2738498E 00	9.6235664E 00	9.9682106E 00
99.9	1.0456185E C1	1.0846008E 01	1.1229337E 01	1.1606503E 01	1.1977438E 01

Tafel 20 (Forts.) Table 20 (cont.)

QUANTILE DER GAMMA-VERTEILUNG

QUANTILES ØF THE GAMMA DISTRIBUTIØN

(NØTE. 6.07E-31 IS EQUIVALENT TØ 6.07 X 10^{-31})

PER CENT	ETA= 3.6	ETA= 3.8	ETA= 4.0	ETA= 4.2	ETA= 4.4
0.1	3.2354432E-01	3.7452469E-01	4.2855244E-01	4.8547497E-01	5.4514865E-01
0.5	5.2858958E-01	5.9889438E-01	6.7220659E-01	7.4834049E-01	8.2712702E-01
0.7	6.0074238E-01	6.7699502E-01	7.5619978E-01	8.3816664E-01	9.2272364E-01
1.0	6.5876450E-01	7.3955699E-01	8.2324872E-01	9.0964804E-01	9.9858204E-01
1.5	7.5194496E-01	8.3963767E-01	9.3013401E-01	1.0232418E 00	1.1187889E 00
2.0	8.2755259E-01	9.2053593E-01	1.0162386E 00	1.1144695E 00	1.2150586E 00
2.5	8.9254687E-01	9.8988446E-01	1.0898654E 00	1.1923011E 00	1.2970235E 00
3.0	9.5031885E-01	1.0513911E 00	1.1550374E 00	1.2610717E 00	1.3693284E 00
4.0	1.0512345E 00	1.1585543E 00	1.2683247E 00	1.3803650E 00	1.4945144E 00
5.0	1.1389841E 00	1.2514805E 00	1.3663186E 00	1.4833227E 00	1.6023380E 00
7.5	1.3243603E 00	1.4471184E 00	1.5719868E 00	1.6988037E 00	1.8274260E 00
10.0	1.4811965E 00	1.6120247E 00	1.7447698E 00	1.8792817E 00	2.0154290E 00
15.0	1.7503018E 00	1.8939042E 00	2.0390997E 00	2.1857612E 00	2.3337768E 00
20.0	1.9875967E 00	2.1415323E 00	2.2967870E 00	2.4532543E 00	2.6108400E 00
25.0	2.2084427E 00	2.3713439E 00	2.5353205E 00	2.7002840E 00	2.8661566E 00
30.0	2.4208299E 00	2.5918456E 00	2.7637115E 00	2.9363569E 00	3.1097192E 00
40.0	2.8393317E 00	3.0251106E 00	3.2113231E 00	3.3979324E 00	3.5849041E 00
50.0	3.2727174E 00	3.4723707E 00	3.6720615E 00	3.8717843E 00	4.0715334E 00
60.0	3.7485254E 00	3.9620748E 00	4.1752641E 00	4.3881181E 00	4.6006591E 00
70.0	4.3062314E 00	4.5346230E 00	4.7622305E 00	4.9891121E 00	5.2153195E 00
75.0	4.6372802E 00	4.8738642E 00	5.1094297E 00	5.3440557E 00	5.5778067E 00
80.0	5.0249804E 00	5.2706529E 00	5.5150478E 00	5.7582676E 00	6.0003931E 00
85.0	5.5026938E 00	5.7589013E 00	6.0135390E 00	6.2667284E 00	6.5185704E 00
90.0	6.1439911E 00	6.4133576E 00	6.6807875E 00	6.9464370E 00	7.2104183E 00
92.5	6.5814911E 00	6.8592682E 00	7.1348786E 00	7.4084946E 00	7.6802465E 00
95.0	7.1788929E 00	7.4675052E 00	7.7536630E 00	8.0375611E 00	8.3193413E 00
97.5	8.1601093E 00	8.4651448E 00	8.7672890E 00	9.0667806E 00	9.3637615E 00
98.0	8.4674107E 00	8.7772766E 00	9.0841264E 00	9.3882200E 00	9.6896919E 00
99.0	9.4009742E 00	9.7247509E 00	1.0045182E 01	1.0362505E 01	1.0676853E 01
99.5	1.0308698E 01	1.0644946E 01	1.0977605E 01	1.1306876E 01	1.1632821E 01
99.9	1.2344066E 01	1.2705508E 01	1.3062754E 01	1.3416387E 01	1.3765564E 01

PER CENT	ETA= 4.6	ETA= 4.8	ETA= 5.0	ETA= 5.5	ETA= 6.0
0.1	6.0743883E-01	6.7221912E-01	7.3937174E-01	9.1692643E-01	1.1071047E 00
0.5	9.0841244E-01	9.9205569E-01	1.0779283E 00	1.3016111E 00	1.5369119E 00
0.7	1.0097155E 00	1.0990001E 00	1.1904491E 00	1.4277901E 00	1.6763566E 00
1.0	1.0898952E 00	1.1834459E 00	1.2791062E 00	1.5267421E 00	1.7852847E 00
1.5	1.2166211E 00	1.3165983E 00	1.4185948E 00	1.6816918E 00	1.9551829E 00
2.0	1.3178539E 00	1.4227173E 00	1.5295258E 00	1.8043437E 00	2.0891438E 00
2.5	1.4038829E 00	1.5127442E 00	1.6234866E 00	1.9078744E 00	2.2018944E 00
3.0	1.4796606E 00	1.5919354E 00	1.7060345E 00	1.9985808E 00	2.3004521E 00
4.0	1.6106316E 00	1.7285883E 00	1.8482710E 00	2.1543736E 00	2.4692747E 00
5.0	1.7232276E 00	1.8458680E 00	1.9701498E 00	2.2874067E 00	2.6130151E 00
7.5	1.9577285E 00	2.0895982E 00	2.2229352E 00	2.5621264E 00	2.9087574E 00
10.0	2.1530969E 00	2.2921812E 00	2.4325913E 00	2.7888928E 00	3.1518984E 00
15.0	2.4830503E 00	2.6334934E 00	2.7850300E 00	3.1682177E 00	3.5569183E 00
20.0	2.7694638E 00	2.9290513E 00	3.0895401E 00	3.4943373E 00	3.9036642E 00
25.0	3.0328729E 00	3.2003714E 00	3.3686008E 00	3.7920720E 00	4.2192098E 00
30.0	3.2837474E 00	3.4583909E 00	3.6336099E 00	4.0739347E 00	4.5171390E 00
40.0	3.7722123E 00	3.9598298E 00	4.1477370E 00	4.6186435E 00	5.0909867E 00
50.0	4.2713065E 00	4.4710989E 00	4.6709096E 00	5.1705011E 00	5.6701630E 00
60.0	4.8129125E 00	5.0248932E 00	5.2366202E 00	5.7649193E 00	6.2919204E 00
70.0	5.4409055E 00	5.6659053E 00	5.8903626E 00	6.4493365E 00	7.0055531E 00
75.0	5.8107550E 00	6.0429444E 00	6.2744328E 00	6.8503502E 00	7.4227052E 00
80.0	6.2415134E 00	6.4816867E 00	6.7209831E 00	7.3157148E 00	7.9059964E 00
85.0	6.7691775E 00	7.0186161E 00	7.2669716E 00	7.8835530E 00	8.4946588E 00
90.0	7.4728816E 00	7.7339032E 00	7.9935977E 00	8.6375158E 00	9.2746798E 00
92.5	7.9503019E 00	8.2187408E 00	8.4856923E 00	9.1471189E 00	9.8009979E 00
95.0	8.5991930E 00	8.8772030E 00	9.1535279E 00	9.8375876E 00	1.0513056E 01
97.5	9.6584998E 00	9.9510543E 00	1.0241620E 01	1.0960063E 01	1.1668369E 01
98.0	9.9888175E 00	1.0285654E 01	1.0580429E 01	1.1309022E 01	1.2027013E 01
99.0	1.0988642E 01	1.1297808E 01	1.1604693E 01	1.2362570E 01	1.3108568E 01
99.5	1.1956081E 01	1.2276368E 01	1.2594225E 01	1.3378652E 01	1.4150016E 01
99.9	1.4112402E 01	1.4454964E 01	1.4794938E 01	1.5632967E 01	1.6455929E 01

Tafel 20 (Forts.) Table 20 (cont.)

QUANTILE DER GAMMA-VERTEILUNG

QUANTILES ØF THE GAMMA DISTRIBUTIØN

(NØTE. 6.07E-31 IS EQUIVALENT TØ 6.07 X 10^{-31})

PER CENT	ETA= 6.5	ETA= 7.0	ETA= 7.5	ETA= 8.0	ETA= 8.5
0.1	1.3086092E 00	1.5203364E 00	1.7413424E 00	1.9708140E 00	2.2080464E 00
0.5	1.7825174E 00	2.0373377E 00	2.3004580E 00	2.5711030E 00	2.8486089E 00
0.7	1.9348533E 00	2.2022113E 00	2.4775370E 00	2.7600756E 00	3.0491823E 00
1.0	2.0534580E 00	2.3302127E 00	2.6146747E 00	2.9061064E 00	3.2038803E 00
1.5	2.2378305E 00	2.5286208E 00	2.8267128E 00	3.1313983E 00	3.4420770E 00
2.0	2.3827228E 00	2.6840991E 00	2.9924585E 00	3.3071189E 00	3.6275021E 00
2.5	2.5043755E 00	2.8143633E 00	3.1310691E 00	3.4538326E 00	3.7820935E 00
3.0	2.6105067E 00	2.9278158E 00	3.2516130E 00	3.5812568E 00	3.9162049E 00
4.0	2.7918850E 00	3.1213215E 00	3.4568573E 00	3.7978841E 00	4.1438895E 00
5.0	2.9459324E 00	3.2853161E 00	3.6304726E 00	3.9808231E 00	4.3358805E 00
7.5	3.2618875E 00	3.6207577E 00	3.9847463E 00	4.3533348E 00	4.7260877E 00
10.0	3.5207531E 00	3.8947672E 00	4.2733786E 00	4.6561186E 00	5.0425941E 00
15.0	3.9504190E 00	4.3481487E 00	4.7496417E 00	5.1545107E 00	5.5624310E 00
20.0	4.3169312E 00	4.7336645E 00	5.1534806E 00	5.5760590E 00	6.0011336E 00
25.0	4.6495332E 00	5.0826579E 00	5.5182703E 00	5.9561104E 00	6.3959645E 00
30.0	4.9628425E 00	5.4107397E 00	5.8605856E 00	6.3121755E 00	6.7653399E 00
40.0	5.5645716E 00	6.0392421E 00	6.5148761E 00	6.9913707E 00	7.4686382E 00
50.0	6.1698797E 00	6.6696393E 00	7.1694322E 00	7.6692510E 00	8.1690941E 00
60.0	6.8177888E 00	7.3426500E 00	7.8666144E 00	8.3897712E 00	8.9121968E 00
70.0	7.5593646E 00	8.1110528E 00	8.6608512E 00	9.2089498E 00	9.7555171E 00
75.0	7.9919587E 00	8.5584704E 00	9.1225476E 00	9.6844350E 00	1.0244344E 01
80.0	8.4924042E 00	9.0753887E 00	9.6553365E 00	1.0232545E 01	1.0807289E 01
85.0	9.1009954E 00	9.7031254E 00	1.0301515E 01	1.0896538E 01	1.1488524E 01
90.0	9.9059781E 00	1.0532086E 01	1.1153581E 01	1.1770923E 01	1.2384529E 01
92.5	1.0448295E 01	1.1089785E 01	1.1726100E 01	1.2357771E 01	1.2985264E 01
95.0	1.1181042E 01	1.1842418E 01	1.2497918E 01	1.3148141E 01	1.3793597E 01
97.5	1.2367854E 01	1.3059525E 01	1.3744270E 01	1.4422725E 01	1.5095555E 01
98.0	1.2735805E 01	1.3436437E 01	1.4129837E 01	1.4816652E 01	1.5497598E 01
99.0	1.3844243E 01	1.4570737E 01	1.5289087E 01	1.6000054E 01	1.6704526E 01
99.5	1.4909999E 01	1.5659867E 01	1.6401025E 01	1.7133823E 01	1.7859475E 01
99.9	1.7265614E 01	1.8062641E 01	1.8849757E 01	1.9627376E 01	2.0397113E 01

PER CENT	ETA= 9.0	ETA= 9.5	ETA=10.0	ETA=11.0	ETA=12.0
0.1	2.4524246E 00	2.7034082E 00	2.9605206E 00	3.4914847E 00	4.0424412E 00
0.5	3.1324024E 00	3.4219864E 00	3.7169226E 00	4.3213587E 00	4.9431173E 00
0.7	3.3443029E 00	3.6449553E 00	3.9507180E 00	4.5761350E 00	5.2180602E 00
1.0	3.5074561E 00	3.8163654E 00	4.1301997E 00	4.7712465E 00	5.4281815E 00
1.5	3.7582328E 00	4.0794198E 00	4.4052488E 00	5.0695062E 00	5.7487117E 00
2.0	3.9531113E 00	4.2835185E 00	4.6183496E 00	5.3000156E 00	5.9959114E 00
2.5	4.1153736E 00	4.4532587E 00	4.7953894E 00	5.4911615E 00	6.2005758E 00
3.0	4.2559932E 00	4.6002209E 00	4.9485406E 00	5.6562685E 00	6.3771362E 00
4.0	4.4944343E 00	4.8491418E 00	5.2076828E 00	5.9351506E 00	6.6749172E 00
5.0	4.6952280E 00	5.0585075E 00	5.4254066E 00	6.1690078E 00	6.9242137E 00
7.5	5.1026338E 00	5.4826530E 00	5.8658682E 00	6.6409487E 00	7.4262587E 00
10.0	5.4324691E 00	5.8254560E 00	6.2213050E 00	7.0207483E 00	7.8293430E 00
15.0	5.9731265E 00	6.3863619E 00	6.8019311E 00	7.6393782E 00	8.4842802E 00
20.0	6.4284776E 00	6.8578969E 00	7.2892214E 00	8.1570214E 00	9.0309031E 00
25.0	6.8376465E 00	7.2810003E 00	7.7258881E 00	8.6198114E 00	9.5186276E 00
30.0	7.2199330E 00	7.6758316E 00	8.1329297E 00	9.0503625E 00	9.9716172E 00
40.0	7.9466078E 00	8.4252195E 00	8.9044168E 00	9.8643986E 00	1.0826247E 01
50.0	8.6689538E 00	9.1688287E 00	9.6687176E 00	1.0668527E 01	1.1668365E 01
60.0	9.4339553E 00	9.9551048E 00	1.0475689E 01	1.1515335E 01	1.2553178E 01
70.0	1.0300684E 01	1.0844570E 01	1.1387277E 01	1.2469514E 01	1.3547990E 01
75.0	1.0802450E 01	1.1358912E 01	1.1913852E 01	1.3019639E 01	1.4120585E 01
80.0	1.1379781E 01	1.1950216E 01	1.2518760E 01	1.3650738E 01	1.4776670E 01
85.0	1.2077742E 01	1.2664436E 01	1.3248803E 01	1.4411246E 01	1.5566245E 01
90.0	1.2994726E 01	1.3601808E 01	1.4206015E 01	1.5406661E 01	1.6598145E 01
92.5	1.3608924E 01	1.4229087E 01	1.4846038E 01	1.6071234E 01	1.7286167E 01
95.0	1.4434687E 01	1.5071797E 01	1.5705250E 01	1.6962263E 01	1.8207551E 01
97.5	1.5763251E 01	1.6426253E 01	1.7084891E 01	1.8390440E 01	1.9682139E 01
98.0	1.6173158E 01	1.6843826E 01	1.7509872E 01	1.8829848E 01	2.0135297E 01
99.0	1.7402804E 01	1.8095583E 01	1.8783283E 01	2.0144920E 01	2.1490142E 01
99.5	1.8578442E 01	1.9291487E 01	1.9998780E 01	2.1398395E 01	2.2779675E 01
99.9	2.1158045E 01	2.1912130E 01	2.2658903E 01	2.4136056E 01	2.5590844E 01

QUANTILE DER GAMMA-VERTEILUNG

QUANTILES ØF THE GAMMA DISTRIBUTIØN

(NØTE. 6.07E-31 IS EQUIVALENT TØ 6.07×10^{-31})

PER CENT	ETA=13.0	ETA=14.0	ETA=15.0	ETA=16.0	ETA=17.0
0.1	4.6110638E 00	5.1954403E 00	5.7939761E 00	6.4053278E 00	7.0283506E 00
0.5	5.5801195E 00	6.2306684E 00	6.8933615E 00	7.5670171E 00	8.2506377E 00
0.7	5.8744963E 00	6.5438208E 00	7.2246892E 00	7.9159749E 00	8.6167239E 00
1.0	6.0990738E 00	6.7823563E 00	7.4767294E 00	8.1811088E 00	8.8945751E 00
1.5	6.4410392E 00	7.1450050E 00	7.8593836E 00	8.5831507E 00	9.3154399E 00
2.0	6.7042936E 00	7.4237416E 00	8.1530889E 00	8.8913561E 00	9.6377208E 00
2.5	6.9219536E 00	7.6539314E 00	8.3953875E 00	9.1453834E 00	9.9031279E 00
3.0	7.1095212E 00	7.8521072E 00	8.6038142E 00	9.3637351E 00	1.0131110E 01
4.0	7.4254577E 00	8.1855372E 00	8.9541411E 00	9.7304205E 00	1.0513662E 01
5.0	7.6895794E 00	8.4639391E 00	9.2463311E 00	1.0035958E 01	1.0832143E 01
7.5	8.2205200E 00	9.0226990E 00	9.8319496E 00	1.0647562E 01	1.1468939E 01
10.0	8.6459442E 00	9.4696220E 00	1.0299619E 01	1.1135299E 01	1.1976129E 01
15.0	9.3356943E 00	1.0192867E 01	1.1055175E 01	1.1922099E 01	1.2793207E 01
20.0	9.9100994E 00	1.0793988E 01	1.1682060E 01	1.2573894E 01	1.3469138E 01
25.0	1.0421718E 01	1.1328581E 01	1.2238806E 01	1.3152057E 01	1.4068045E 01
30.0	1.0896202E 01	1.1823732E 01	1.2753882E 01	1.3686390E 01	1.4621032E 01
40.0	1.1789720E 01	1.2754630E 01	1.3720816E 01	1.4688148E 01	1.5656520E 01
50.0	1.2668234E 01	1.3668121E 01	1.4668023E 01	1.5667934E 01	1.6667863E 01
60.0	1.3589448E 01	1.4624319E 01	1.5657936E 01	1.6690443E 01	1.7721924E 01
70.0	1.4623175E 01	1.5695444E 01	1.6765128E 01	1.7832464E 01	1.8897699E 01
75.0	1.5217292E 01	1.6310262E 01	1.7399882E 01	1.8486499E 01	1.9570403E 01
80.0	1.5897320E 01	1.7013296E 01	1.8125106E 01	1.9233168E 01	2.0337850E 01
85.0	1.6714752E 01	1.7857509E 01	1.8995142E 01	2.0128173E 01	2.1257023E 01
90.0	1.7781609E 01	1.8957984E 01	2.0128054E 01	2.1292423E 01	2.2451631E 01
92.5	1.8492109E 01	1.9690108E 01	2.0881007E 01	2.2065500E 01	2.3244234E 01
95.0	1.9442629E 01	2.0668632E 01	2.1886569E 01	2.3097180E 01	2.4301234E 01
97.5	2.0961681E 01	2.2230526E 01	2.3489723E 01	2.4740289E 01	2.5983167E 01
98.0	2.1428066E 01	2.2709562E 01	2.3981039E 01	2.5243490E 01	2.6497842E 01
99.0	2.2821100E 01	2.4139324E 01	2.5446398E 01	2.6743183E 01	2.8030844E 01
99.5	2.4145321E 01	2.5497245E 01	2.6836654E 01	2.8164678E 01	2.9482640E 01
99.9	2.7028923E 01	2.8449401E 01	2.9854549E 01	3.1246408E 01	3.2626712E 01

PER CENT	ETA=18.0	ETA=19.0	ETA=20.0	ETA=21.0	ETA=22.0
0.1	7.6620569E 00	8.3055943E 00	8.9582144E 00	9.6192597E 00	1.0288149E 01
0.5	8.9433638E 00	9.6444577E 00	1.0353269E 01	1.1069233E 01	1.1791848E 01
0.7	9.3261169E 00	1.0043450E 01	1.0768106E 01	1.1499544E 01	1.2237289E 01
1.0	9.6163391E 00	1.0345723E 01	1.1082133E 01	1.1825048E 01	1.2574014E 01
1.5	1.0055510E 01	1.0802721E 01	1.1556513E 01	1.2316403E 01	1.3081958E 01
2.0	1.0391474E 01	1.1152007E 01	1.1918788E 01	1.2691355E 01	1.3469300E 01
2.5	1.0667943E 01	1.1439243E 01	1.2216521E 01	1.2999333E 01	1.3787286E 01
3.0	1.0905281E 01	1.1685690E 01	1.2471847E 01	1.3263322E 01	1.4059740E 01
4.0	1.1303252E 01	1.2098665E 01	1.2899443E 01	1.3705187E 01	1.4515540E 01
5.0	1.1634307E 01	1.2441954E 01	1.3254654E 01	1.4072029E 01	1.4893742E 01
7.5	1.2295574E 01	1.3127027E 01	1.3962919E 01	1.4802916E 01	1.5646724E 01
10.0	1.2821652E 01	1.3671478E 01	1.4525265E 01	1.5382713E 01	1.6243567E 01
15.0	1.3668128E 01	1.4546540E 01	1.5428164E 01	1.6312768E 01	1.7200125E 01
20.0	1.4367484E 01	1.5268673E 01	1.6172481E 01	1.7078708E 01	1.7987179E 01
25.0	1.4986524E 01	1.5907287E 01	1.6830154E 01	1.7754958E 01	1.8681570E 01
30.0	1.5557613E 01	1.6495978E 01	1.7435974E 01	1.8377486E 01	1.9320403E 01
40.0	1.6625837E 01	1.7596011E 01	1.8566986E 01	1.9538694E 01	2.0511088E 01
50.0	1.7667794E 01	1.8667731E 01	1.9667683E 01	2.0667638E 01	2.1667592E 01
60.0	1.8752478E 01	1.9782188E 01	2.0811108E 01	2.1839308E 01	2.2866829E 01
70.0	1.9961006E 01	2.1022545E 01	2.2082456E 01	2.3140855E 01	2.4197863E 01
75.0	2.0651832E 01	2.1730978E 01	2.2808022E 01	2.3883145E 01	2.4956471E 01
80.0	2.1439427E 01	2.2538165E 01	2.3634289E 01	2.4728014E 01	2.5819494E 01
85.0	2.2382071E 01	2.3503613E 01	2.4621950E 01	2.5737329E 01	2.6849966E 01
90.0	2.3606131E 01	2.4756340E 01	2.5902573E 01	2.7045162E 01	2.8184337E 01
92.5	2.4417700E 01	2.5586345E 01	2.6750551E 01	2.7910633E 01	2.9066888E 01
95.0	2.5499299E 01	2.6691864E 01	2.7879356E 01	2.9062158E 01	3.0240542E 01
97.5	2.7218828E 01	2.8447969E 01	2.9671091E 01	3.0888593E 01	3.2100914E 01
98.0	2.7744701E 01	2.8984689E 01	3.0218336E 01	3.1446061E 01	3.2668638E 01
99.0	2.9310028E 01	3.0581402E 01	3.1845937E 01	3.3103699E 01	3.4355160E 01
99.5	3.0791157E 01	3.2091351E 01	3.3383925E 01	3.4668762E 01	3.5946929E 01
99.9	3.3995731E 01	3.5355368E 01	3.6703910E 01	3.8046123E 01	3.9379733E 01

TABLE 21

The Bargmann Test for Simple Structure of a Factor Pattern: Tables of R. Bargmann

(a) Content of the tables and definition of the test statistic :

The goal of rotating an (n×r) factor pattern (matrix) of r factors for n variates is to obtain the simplest structure possible , so as to allow for a meaningful interpretation.

The primarily descriptive definition of a simple structure for a factor pattern , originally given by THURSTONE (see section (e)), was made precise by BARGMANN (1955) , who also provided a significance test for factors.

Rotation to a simple structure is a matter of grouping to one factor all those variates whose partial correlation matrix, obtained by eliminating the influence of the remaining variates, has a rank of one. To identify such groups of variates, one determines the (k-1)-dimensional hyperplanes in which these variates lie. In the geometric model, the simple structure for a k-dimensional factor study corresponds to all statistically sufficiently overdetermined (k-1)-dimensional subspaces. The decision that such a hyperplane is statistically sufficiently overdetermined is based on the number of "zero contributions", following BARGMANN. Therefore, for a given number of variates n and a given number of factors r ,the tables give the number of zero contributions , $N(n,r,\alpha)$, for the factor being tested, which could appear purely randomly under H_0 with given probability α . If one regards the variates as vectors in the space of factors, then the interpretation of the condition of the null hypothesis H_0 , is that the vectors are uniformly distributed in the space about the hyperplane , and thus no simple structure is present.

The contribution a_{ij} of the i-th factor to the j-th variate is , by convention and experience , a zero contribution if its value lies between -0.09 and +0.09 .
When the contribution is at these bounds, the decision is based on the communality h_i^2 :
The number of vectors satisfying the condition $|a_{ij}/h_i| > 0.10$ is subtracted from the previously determined total.
One can also briefly define :

$$N_B(n;r;\alpha) = \{ \text{number of } a_{ij} \text{ with } |a_{ij}/h_i| \leq 0.10 \text{ ; given } H_o \}$$

If the given number of zero contributions, $N_{exp}(n;r)$, is greater than the corresponding tabulated value $N_B(n;r;\alpha)$, then the simple structure of the factor is considered statistically certain, and the factor may be interpreted.
Boldface numbers in the tables indicate that the given number of zero contributions reaches the corresponding significance bound exactly.
Italic numbers indicate that the given number of zero contributions does not quite reach the corresponding significance bound. In particular , they are used when the value is closest to the significance bound, but somewhat below it.

(b) Extent of the tables and definition of the parameters :

(1) The parameter α :

α = error probability = probability of the occurrence of more than N_B zero contributions under H_o

α = 0.50 ; 0.25 ; 0.10 ; 0.05 ; 0.01 ; 0.001 .

The values 0.50 and 0.25 are included so that one can see which factors may not be interpreted under any circumstances.

(2) The parameter k :

k = number of factors

k = 2(1)12 .

(3) The parameter n :

n = number of variates

n = between 5 and 70 , depending on the number k of factors .

(c) Suggestions for applications :

The tables provide the critical bounds for the minimum number of zero contributions required for a significant simple structure, as described in section (a) .

(d) References :

(1) For the test statistic and the source of the tables :
BARGMANN, R. : Signifikanzuntersuchungen der Einfachen Struktur in der Faktoren-Analyse. Mitteilungsblatt für mathematische Statiatik Vol. 7, 1 - 24 (1955) .

(2) For simple structure :
HARMAN, H. H. : Modern Factor Analysis.(Second Edition). Chicago : The University of Chicago Press, 1967 .
THURSTONE, L. L. : Multiple Factor Analysis. Chicago : The University of Chicago Press, 1947 .

(e) Further remarks :

(1) At the head of each table - that is, for each factor number k - is given the probability P_k that a variate vector falls into the critical region.

(2) BARGMANN gives precise computational methods for factor analysis with parameters outside the tabulated range.

(3) One tends to characterise simple structure by the following conditions (going back to THURSTONE, 1947) :

(1') Each row of the factor pattern (matrix) should have at least one zero .

(2') For r common factors, each column of the matrix should have at least r zeros .

(3') For each pair of column vectors in the factor pattern there should be several variates which have only zero contributions in one column, but preferably large contributions in the other column.

(4') When more than four factors are present, then for each pair of columns there should preferably be many variates with zero contributions in both columns.

(5') For each pair of columns of the factor pattern there should be only a few variates having large contributions in both columns.

2 Faktoren (Dimensionen)

$P_2 = 0{,}0638$

2 factors (dimensions)

$\left(\frac{a}{h} = \pm\ 0{,}10\right)$

n	(.50)	(.25)	.10	.05	.01	.001
5	2	2	3	3	3	4
6	2	2	3	3	4	5
7	2	3	3	3	4	5
8	2	3	3	4	4	5
9	2	3	3	4	4	5
10	2	3	3	4	5	6
11	2	3	3	4	5	6
12	2	3	4	4	5	6
13	2	3	4	4	5	6
14	3	3	4	4	5	6
15	3	3	4	4	5	6
16	3	3	4	5	5	7
17	3	3	4	5	6	7
18	3	4	4	5	6	7
19	3	4	4	5	6	7
20	3	4	4	5	6	7
21	3	4	5	5	6	7
22	3	4	5	5	6	8
23	3	4	5	5	6	8
24	3	4	5	5	6	8
25	3	4	5	6	7	8
26	3	4	5	6	7	8
27	3	4	5	6	7	8
28	3	4	5	6	7	8
29	4	4	5	6	7	9
30	4	4	5	6	7	9

3 Faktoren (Dimensionen)

$P_3 = 0,1000$

3 factors (dimensions)

$(\frac{a}{h} = \pm 0,10)$

n	(.50)	(.25)	.10	.05	.01	.001
6	3	4	4	4	5	6
7	3	4	4	5	5	6
8	3	4	4	5	6	6
9	3	4	5	5	6	7
10	4	4	5	5	6	7
11	4	4	5	5	6	7
12	4	4	5	6	6	7
13	4	5	5	6	7	8
14	4	5	5	6	7	8
15	4	5	6	6	7	8
16	4	5	6	6	7	8
17	4	5	6	6	7	9
18	4	5	6	7	7	9
19	4	5	6	7	8	9
20	4	5	6	7	8	9
21	5	5	6	7	8	9
22	5	6	7	7	8	10
23	5	6	7	7	8	10
24	5	6	7	7	9	10
25	5	6	7	8	9	10
26	5	6	7	8	9	10
27	5	6	7	8	9	11
28	5	6	7	8	9	11
29	5	6	8	8	9	11
30	5	7	8	8	10	11
31	6	7	8	9	10	11
32	6	7	8	9	10	12
33	6	7	8	9	10	12

Tafel 21 (Forts.) Table 21 (cont.)

4 Faktoren (Dimensionen)

$P_4 = 0{,}1271$

4 factors (dimensions)

$(\frac{a}{h} = \pm 0{,}10)$

n	(.50)	(.25)	.10	.05	.01	.001
10	5	5	6	**6**	7	8
11	5	**5**	6	7	**7**	**8**
12	5	6	**6**	7	8	9
13	5	6	6	7	8	9
14	5	6	7	7	8	**9**
15	5	6	7	**7**	**8**	10
16	**5**	6	7	8	9	10
17	*5*	**6**	7	8	9	10
18	6	6	**7**	8	9	**10**
19	6	7	7	8	**9**	11
20	6	7	8	**8**	10	11
21	6	7	8	9	10	**11**
22	6	7	8	9	10	11
23	6	**7**	**8**	9	**10**	12
24	**6**	7	*8*	9	10	12
25	6	8	9	**9**	11	**12**
26	7	8	9	10	11	*12*
27	7	8	9	10	11	13
28	7	8	9	10	**11**	13
29	7	8	**9**	10	*11*	**13**
30	7	**8**	10	**10**	12	13
31	**7**	8	10	11	12	**14**
32	7	9	10	11	12	14
33	*7*	9	10	11	**12**	14
34	8	9	10	11	13	**14**
35	8	9	**10**	11	13	*14*
36	8	9	*10*	**11**	13	15
37	8	**9**	11	*11*	13	15
38	8	*9*	11	12	**13**	**15**
39	**8**	10	11	12	13	15
40	8	10	11	12	**14**	16
41	*8*	10	**11**	12	14	16
42	9	10	11	**12**	14	16
43	9	**10**	*11*	*12*	**14**	**16**
44	9	10	12	13	14	16
45	9	*10*	12	13	15	17

Tafel 21 (Forts.) Table 21 (cont.)

5 Faktoren (Dimensionen)
$P_5 = 0{,}1495$

5 factors (dimensions)

$(\frac{a}{h} = \pm 0{,}10)$

n	(.50)	(.25)	.10	.05	.01	.001
10	6	6	7	7	8	9
11	6	7	7	8	8	9
12	6	7	7	8	9	10
13	6	7	8	8	9	10
14	6	7	8	8	9	10
15	6	7	8	9	9	11
16	7	7	8	9	10	11
17	7	8	8	9	10	11
18	7	8	9	9	10	12
19	7	8	9	10	11	12
20	7	8	9	10	11	12
21	7	8	9	10	11	13
22	7	8	9	10	11	13
23	8	9	10	10	12	13
24	8	9	10	11	12	13
25	8	9	10	11	12	14
26	8	9	10	11	12	14
27	8	9	10	11	13	14
28	8	9	11	11	13	14
29	8	10	11	12	13	15
30	9	10	11	12	13	15
31	9	10	11	12	13	15
32	9	10	11	12	14	15
33	9	10	12	12	14	16
34	9	11	12	13	14	16
35	9	11	12	13	14	16
36	9	11	12	13	15	16
37	10	11	12	13	15	17
38	10	11	12	13	15	17
39	10	11	13	14	15	17
40	10	12	13	14	15	17
41	10	12	13	14	16	18
42	10	12	13	14	16	18
43	11	12	13	14	16	18
44	11	12	14	15	16	18
45	11	12	14	15	17	19
46	11	12	14	15	17	19
47	11	13	14	15	17	19
48	11	13	14	15	17	19
49	11	13	15	16	17	20
50	12	13	15	16	18	20
51	12	13	15	16	18	20
52	12	13	15	16	18	20
53	12	14	15	16	18	21
54	12	14	15	17	18	21
55	12	14	16	17	19	21

Tafel 21 (Forts.) Table 21 (cont.)

6 Faktoren (Dimensionen) $P_6 = 0,1689$

6 factors (dimensions) $(\frac{a}{h} = \pm 0,10)$

n	(.50)	(.25)	.10	.05	.01	.001
12	7	8	8	9	10	10
13	7	8	9	9	10	11
14	7	8	9	9	10	11
15	7	8	9	10	11	12
16	8	8	9	10	11	12
17	8	9	10	10	11	12
18	8	9	10	10	11	13
19	8	9	10	11	12	13
20	8	9	10	11	12	13
21	8	10	11	11	12	14
22	9	10	11	11	13	14
23	9	10	11	12	13	14
24	9	10	11	12	13	15
25	9	10	11	12	13	15
26	9	11	12	12	14	15
27	9	11	12	13	14	16
28	10	11	12	13	14	16
29	10	11	12	13	14	16
30	10	11	12	13	15	16
31	10	11	13	14	15	17
32	10	12	13	14	15	17
33	10	12	13	14	15	17
34	11	12	13	14	16	18
35	11	12	14	14	16	18
36	11	12	14	15	16	18
37	11	12	14	15	16	18
38	11	13	14	15	17	19
39	11	13	14	15	17	19
40	12	13	15	16	17	19
41	12	13	15	16	17	20
42	12	14	15	16	18	20
43	12	14	15	16	18	20
44	12	14	15	16	18	20
45	12	14	16	17	18	21
46	13	14	16	17	19	21
47	13	14	16	17	19	21
48	13	15	16	17	19	21
49	13	15	16	18	19	22
50	13	15	17	18	20	22
51	13	15	17	18	20	22
52	14	15	17	18	20	22
53	14	16	17	18	20	23
54	14	16	17	19	20	23
55	14	16	18	19	21	23
56	14	16	18	19	21	23
57	14	16	18	19	21	24
58	15	16	18	19	21	24
59	15	17	18	20	22	24
60	15	17	19	20	22	24

Tafel 21 (Forts.) Table 21 (cont.)

7 Faktoren (Dimensionen) $P_7 = 0{,}1863$

7 factors (dimensions)

$\left(\frac{a}{h} = \pm 0{,}10\right)$

n	(.50)	(.25)	.10	.05	.01	.001
15	*8*	**9**	10	11	12	13
16	9	*9*	**10**	11	12	**13**
17	9	10	11	11	**12**	*13*
18	9	10	11	**11**	13	14
19	9	10	11	12	13	**14**
20	**9**	**10**	**11**	12	**13**	*14*
21	10	11	12	**12**	14	15
22	10	11	12	13	14	**15**
23	10	11	12	13	14	16
24	10	**11**	**12**	13	14	16
25	**10**	*11*	13	**13**	15	**16**
26	11	12	13	14	15	17
27	11	12	13	14	**15**	17
28	11	12	**13**	14	*15*	**17**
29	11	**12**	14	**14**	16	18
30	11	13	14	15	16	18
31	**11**	13	14	15	**16**	**18**
32	12	13	**14**	15	17	*18*
33	12	13	15	**15**	17	19
34	12	**13**	15	16	**17**	19
35	12	14	15	16	*17*	**19**
36	**12**	14	**15**	16	18	20
37	13	14	*15*	**16**	18	20
38	13	14	16	17	**18**	**20**
39	13	**14**	16	17	*18*	21
40	13	15	16	17	19	21
41	13	15	**16**	**17**	19	**21**
42	**13**	15	17	18	**19**	*21*
43	14	**15**	17	18	20	22
44	14	*15*	17	18	20	22
45	14	16	**17**	**18**	20	**22**
46	14	16	*17*	19	**20**	23
47	**14**	16	18	19	21	23
48	15	**16**	18	19	21	**23**
49	15	17	18	**19**	21	24
50	15	17	**18**	20	**21**	24
51	15	17	19	20	22	24
52	**15**	17	19	20	22	**24**
53	*15*	**17**	19	20	**22**	25
54	16	18	**19**	**20**	*22*	25
55	16	18	20	21	23	**25**
56	16	18	20	21	23	*25*
57	16	18	20	21	**23**	26
58	**16**	**18**	**20**	**21**	*23*	26
59	**17**	19	*20*	22	24	**26**
60	17	19	21	22	24	27
61	17	19	21	22	**24**	27
62	17	19	21	**22**	*24*	27
63	**17**	**19**	**21**	23	25	**27**
64	18	20	*21*	23	25	28
65	18	20	22	23	25	28

Tafel 21 (Forts.) Table 21 (cont.)

8 Faktoren (Dimensionen) $P_8 = 0{,}2020$

8 factors (dimensions) $(\frac{a}{h} = \pm 0{,}10)$

n	(.50)	(.25)	.10	.05	.01	.001
15	**9**	**10**	11	12	13	**14**
16	10	*10*	**11**	12	13	14
17	10	11	12	12	**13**	**14**
18	10	11	12	**12**	*13*	15
19	10	**11**	12	13	14	**15**
20	**10**	*11*	**12**	13	**14**	*15*
21	11	12	13	**13**	*14*	16
22	11	12	13	14	15	**16**
23	11	12	**13**	14	**15**	17
24	11	**12**	14	**14**	*15*	17
25	**11**	13	14	*14*	16	**17**
26	12	13	14	15	16	18
27	12	13	**14**	15	**16**	18
28	12	**13**	15	**15**	17	**18**
29	12	14	15	16	17	19
30	**12**	14	15	16	**17**	19
31	13	14	**15**	16	18	**19**
32	13	**14**	16	**16**	18	20
33	13	*14*	16	17	**18**	20
34	13	15	16	17	19	**20**
35	**13**	15	**16**	17	19	21
36	14	15	17	**17**	19	21
37	14	**15**	17	18	**19**	**21**
38	14	16	17	18	20	22
39	14	16	**17**	**18**	20	22
40	**14**	16	18	19	**20**	**22**
41	15	**16**	18	19	21	23
42	15	17	18	19	21	23
43	15	17	**18**	**19**	21	**23**
44	15	17	19	20	**21**	24
45	**15**	17	19	20	22	24
46	16	**17**	19	20	22	24
47	16	18	**19**	**20**	**22**	**24**
48	16	18	20	21	*22*	25
49	16	18	20	21	23	25
50	**16**	**18**	20	21	23	**25**
51	17	19	**20**	**21**	**23**	26
52	17	19	*20*	22	24	26
53	17	19	21	22	24	**26**
54	17	19	21	22	24	27
55	**17**	**19**	21	**22**	**24**	27
56	18	20	**21**	23	25	27
57	18	20	22	23	25	**27**
58	18	20	22	23	25	28
59	18	**20**	22	**23**	**25**	28
60	**18**	21	**22**	24	26	28
61	19	21	23	24	26	**28**
62	19	21	23	24	**26**	29
63	19	21	23	**24**	27	29
64	**19**	**21**	**23**	25	27	**29**
65	*19*	22	24	25	27	30

Tafel 21 (Forts.) Table 21 (cont.)

9 Faktoren (Dimensionen) $P_9 = 0{,}2166$

9 factors (dimensions)

$(\frac{a}{h} = \pm\, 0{,}10)$

n	(.50)	(.25)	.10	.05	.01	.001
15	**10**	11	12	**12**	**13**	**14**
16	11	**11**	12	13	14	15
17	11	12	**12**	13	14	**15**
18	11	12	13	**13**	**14**	16
19	11	12	13	**14**	15	16
20	**11**	**12**	**13**	14	15	**16**
21	12	13	14	**14**	**15**	17
22	12	13	14	15	16	**17**
23	12	13	**14**	15	16	18
24	**12**	**13**	*14*	**15**	**16**	18
25	13	14	15	*15*	17	**18**
26	13	**14**	15	16	17	19
27	13	14	**15**	16	**17**	19
28	13	**14**	16	**16**	18	**19**
29	**13**	15	16	17	18	20
30	**14**	15	16	17	**18**	**20**
31	**14**	15	**16**	**17**	19	*20*
32	**14**	**15**	17	*17*	19	21
33	**14**	16	17	18	**19**	**21**
34	**14**	16	**17**	18	20	22
35	15	16	18	**18**	20	22
36	15	**16**	18	19	**20**	**22**
37	15	17	18	19	21	23
38	**15**	17	**18**	**19**	21	23
39	16	17	19	20	**21**	**23**
40	16	**17**	19	20	*21*	24
41	16	18	19	20	22	24
42	16	18	**19**	**20**	22	**24**
43	**16**	**18**	20	21	**22**	25
44	17	*18*	20	21	23	25
45	17	19	20	21	23	**25**
46	17	19	**20**	**21**	**23**	26
47	**17**	**19**	21	22	24	26
48	*17*	*19*	21	22	24	**26**
49	18	20	21	**22**	**24**	27
50	18	20	**21**	23	*24*	27
51	18	20	22	23	25	**27**
52	**18**	**20**	22	23	25	*27*
53	19	*20*	**22**	**23**	**25**	28
54	19	21	*22*	24	26	28
55	19	21	23	**24**	26	**28**
56	19	21	23	24	**26**	29
57	**19**	**21**	**23**	**24**	27	29
58	20	22	*23*	25	27	**29**
59	20	22	24	25	**27**	30
60	20	22	24	25	*27*	30
61	**20**	**22**	**24**	**25**	28	**30**
62	*20*	23	25	26	28	31
63	21	23	25	26	**28**	31
64	21	23	25	**26**	29	**31**
65	**21**	**23**	**25**	27	29	32

Tafel 21 (Forts.) Table 21 (cont.)

10 Faktoren (Dimensionen) $P_{10} = 0{,}2301$

10 factors (dimensions) $(\frac{a}{h} = \pm 0{,}10)$

n	(.50)	(.25)	.10	.05	.01	.001
18	12	13	14	**14**	**15**	**16**
19	12	13	**14**	15	16	17
20	12	**13**	**14**	15	16	**17**
21	**12**	14	15	**15**	**16**	18
22	13	14	15	16	17	**18**
23	13	14	**15**	16	17	*18*
24	13	**14**	16	**16**	**17**	19
25	**13**	15	16	*16*	18	**19**
26	14	15	16	17	**18**	20
27	14	**15**	**16**	17	*18*	**20**
28	14	16	17	**17**	19	*20*
29	**14**	16	17	18	**19**	21
30	15	16	**17**	18	*19*	**21**
31	15	**16**	18	**18**	20	22
32	15	17	18	19	**20**	22
33	**15**	17	18	19	*20*	**22**
34	16	17	**18**	**19**	21	23
35	16	**17**	19	20	**21**	23
36	16	18	19	20	*21*	**23**
37	**16**	18	**19**	20	22	24
38	17	18	20	**20**	**22**	**24**
39	17	**18**	20	21	*22*	25
40	17	19	20	21	23	25
41	17	19	**20**	**21**	23	**25**
42	**17**	19	21	22	**23**	26
43	18	**19**	21	22	24	26
44	18	20	21	**22**	24	**26**
45	18	20	**21**	23	**24**	27
46	**18**	20	22	23	25	27
47	19	**20**	22	23	25	**27**
48	19	21	**22**	**23**	**25**	28
49	19	21	23	24	26	28
50	**19**	21	23	24	26	**28**
51	*19*	**21**	23	**24**	**26**	29
52	20	22	**23**	25	27	29
53	20	22	24	25	27	**29**
54	20	22	24	25	27	30
55	**20**	**22**	24	**25**	**27**	30
56	21	23	**24**	26	28	**30**
57	21	23	25	26	28	31
58	21	23	25	**26**	**28**	31
59	**21**	**23**	**25**	*26*	29	**31**
60	22	24	26	27	29	32
61	22	24	26	27	**29**	32
62	22	24	26	**27**	30	32
63	**22**	**24**	**26**	28	30	**32**
64	23	25	27	28	30	33
65	23	25	27	28	**30**	33
66	23	25	27	**28**	31	**33**
67	23	**25**	**27**	29	31	34
68	**23**	26	28	29	**31**	34
69	24	26	28	**29**	32	**34**
70	24	26	28	*29*	32	35

afel 21 (Forts.) Table 21 (cont.)

11 Faktoren (Dimensionen) $P_{11} = 0{,}2428$

11 factors (dimensions)

n	(.50)	(.25)	.10	.05	.01	.001
18	13	14	*14*	15	16	17
19	13	14	15	**15**	**16**	**17**
20	**13**	**14**	**15**	16	17	18
21	14	15	*15*	16	**17**	**18**
22	14	15	16	**16**	18	19
23	14	15	16	17	18	**19**
24	**14**	**15**	**16**	17	**18**	20
25	15	16	17	**17**	19	**20**
26	15	16	17	18	19	21
27	15	**16**	**17**	18	**19**	21
28	15	*16*	18	**18**	20	**21**
29	**15**	17	18	19	**20**	22
30	16	17	**18**	19	*20*	**22**
31	16	**17**	19	**19**	21	23
32	16	18	19	20	**21**	23
33	**16**	18	19	20	*21*	**23**
34	17	18	**19**	**20**	22	24
35	17	**18**	20	21	**22**	**24**
36	17	19	20	21	23	*24*
37	**17**	19	**20**	**21**	23	25
38	18	**19**	21	22	**23**	**25**
39	18	*19*	21	22	24	26
40	18	20	**21**	**22**	24	26
41	**18**	20	22	23	**24**	**26**
42	19	**20**	22	23	25	27
43	19	21	22	23	25	27
44	19	21	**22**	**23**	**25**	**27**
45	**19**	21	23	24	26	28
46	20	**21**	23	24	26	**28**
47	20	22	**23**	**24**	**26**	29
48	20	22	24	25	27	29
49	**20**	22	24	25	27	**29**
50	21	**22**	24	**25**	**27**	30
51	21	23	**24**	26	28	30
52	21	23	25	26	28	**30**
53	**21**	23	25	26	**28**	31
54	22	**23**	**25**	**26**	29	31
55	22	24	26	27	29	**31**
56	22	24	26	27	**29**	32
57	**22**	**24**	26	**27**	30	32
58	23	25	**26**	28	30	**32**
59	23	25	27	28	30	33
60	23	25	27	**28**	**30**	33
61	23	**25**	**27**	29	31	**33**
62	**23**	26	28	29	31	34
63	24	26	28	29	**31**	34
64	24	26	28	**29**	32	**34**
65	24	**26**	**28**	30	32	35
66	**24**	27	29	30	**32**	35
67	25	27	29	**30**	33	**35**
68	25	27	**29**	31	33	36
69	25	**27**	30	31	**33**	36
70	**25**	28	30	31	*33*	**36**

12 Faktoren (Dimensionen) $P_{12} = 0{,}2548$

12 factors (dimensions)

n	(.50)	(.25)	.10	.05	.01	.001
20	**14**	**15**	16	17	*17*	19
21	**14**	*15*	**16**	17	18	**19**
22	15	16	17	**17**	**18**	20
23	15	16	17	18	19	**20**
24	15	**16**	**17**	18	**19**	*20*
25	**15**	17	18	**18**	*19*	21
26	16	17	18	19	20	**21**
27	16	17	**18**	19	**20**	22
28	16	**17**	19	**19**	21	**22**
29	**16**	18	19	20	21	23
30	17	18	**19**	20	**21**	**23**
31	17	**18**	20	**20**	22	*23*
32	17	19	20	21	**22**	24
33	**17**	19	**20**	21	*22*	**24**
34	18	19	21	**21**	23	25
35	18	**19**	21	22	**23**	**25**
36	**18**	20	21	22	24	*25*
37	*18*	20	**21**	22	24	26
38	19	**20**	22	23	**24**	**26**
39	19	21	22	23	25	27
40	**19**	21	**22**	**23**	25	27
41	*19*	21	23	24	**25**	**27**
42	20	**21**	23	24	26	28
43	20	22	**23**	**24**	26	**28**
44	**20**	22	24	25	**26**	29
45	21	**22**	24	25	27	29
46	21	23	**24**	**25**	27	**29**
47	21	23	*24*	26	**27**	30
48	**21**	23	25	26	28	30
49	22	**23**	25	26	28	**30**
50	22	24	**25**	**26**	**28**	31
51	22	24	26	27	29	**31**
52	**22**	**24**	26	27	29	*31*
53	23	25	**26**	**27**	**29**	32
54	23	25	27	28	30	**32**
55	23	25	27	28	30	33
56	**23**	**25**	**27**	**28**	**30**	33
57	24	26	*27*	29	31	**33**
58	24	26	28	29	31	34
59	24	26	28	**29**	**31**	34
60	**24**	**26**	**28**	30	32	**34**
61	25	27	29	30	32	35
62	25	27	29	**30**	**32**	35
63	25	**27**	**29**	*30*	33	**35**
64	**25**	28	30	31	33	36
65	26	28	30	31	**33**	**36**
66	26	28	30	**31**	34	*36*
67	26	**28**	**30**	32	34	37
68	**26**	29	31	32	**34**	**37**
69	27	29	31	**32**	35	38
70	27	29	**31**	33	35	38

TABLE 22

Upper Percentage Points of the Bonferroni Chi-Square Statistic: Tables of G. B. Beus and D. R. Jensen

(a) Content of the tables and definition of the test statistic :

Let $X_1, X_2, \ldots, X_\tau$ be random variables (variates) of a common distribution of a type for which the marginal distributions of the X_i are central chi-square distributions with ν_i degrees of freedom ($i = 1, 2, \ldots, \tau$). The common distribution of the X_i needs not be known; also the ν_i need not all be equal.

The tables contain the upper percentage points $U(\nu; \alpha; \tau)$ of the BONFERRONI-chi-square-statistic $\chi^2(\alpha/\tau;\nu)$, defined by

$$P\{U(\nu_i; \alpha;\tau) < X_i\} = \alpha/\tau .$$

BONFERRONI chi-square statistics are suited for satisfying a BONFERRONI inequality (See section (e)).

(b) Extent of the tables and definition of the parameters :

(1) The parameter α :

α = error probability

= probability of an error of the 1-st type

α = 0.100 ; 0.050 ; 0.025 ; 0.010 ; 0.005 .

(2) The parameter τ :

τ = number of variates X_i

τ = 1(1)20(2)30 .

(3) The parameter ν :

ν_i = degrees of freedom of the variate X_i , distributed after chi-square

ν_i = 1(1)30(5)60(10)120 .

(c) Suggestions for applications :

(1) To handle multivariate nominal data :

(1.1) Simultaneous tests for fitting an empirical distribution to a hypothetical one.

(1.2) Simultaneous tests for independence in a contingency table.

(2) To determine simultaneous confidence intervals for the variates of a multivariate normal distribution on the basis of samples with missing values.
Here: The determination of upper confidence limits.
For lower confidence limits : See table 23 .

(d) References :

For the source of the tables and for applications :

BEUS, G. B. + JENSEN, D. R. : Percentage points of the BONFERRONI chi-square statistics. (Table 1).
Technical Report No. 3).
Blacksburg, Virginia : Department of Statistics, Virginia Polytechnic Institute and State University, September 1967.

(e) Further remarks :

(1) For $\tau = 1$, the tables contain the upper percentage points of the usual chi-square distribution. The same holds for $\tau > 1$, if one chooses $\alpha' = \alpha/\tau$ and reads the column for τ .

(2) The BONFERRONI inequality :
Events are denoted by $E_1, E_2, \ldots, E_\tau$. The corresponding complements are denoted by $E_1', E_2', \ldots, E_\tau'$.

The BONFERRONI inequality used here reads :

$$\Pr\{\bigcap_{i=1}^{\tau} E_i\} \geqq 1 - \sum_{i=1}^{\tau} \Pr\{E_i'\} \quad .$$

This form is used to construct simultaneous confidence intervals.

One can also write the above BONFERRONI inequality in the form

$$\Pr\{\bigcup_{i=1}^{\tau} E_i'\} \leq \sum_{i=1}^{\tau} \Pr\{E_i'\} \quad .$$

In this form it serves for simultaneous hypothesis testing.

Tafel 22 Table 22

UPPER PERCENTAGE POINTS / OBERE PROZENTPUNKTE

$\chi^2(\alpha/\tau,\nu)$ $\alpha=0.100$

ν \ τ	1	2	3	4	5
1	2.706	3.841	4.529	5.024	5.412
2	4.605	5.991	6.802	7.378	7.824
3	6.251	7.815	8.715	9.348	9.837
4	7.779	9.488	10.461	11.143	11.668
5	9.236	11.070	12.108	12.833	13.388
6	10.645	12.592	13.687	14.449	15.033
7	12.017	14.067	15.216	16.013	16.622
8	13.362	15.507	16.705	17.535	18.168
9	14.684	16.919	18.163	19.023	19.679
10	15.987	18.307	19.594	20.483	21.161
11	17.275	19.675	21.004	21.920	22.618
12	18.549	21.026	22.394	23.337	24.054
13	19.812	22.362	23.768	24.736	25.472
14	21.064	23.685	25.127	26.119	26.873
15	22.307	24.996	26.473	27.488	28.259
16	23.542	26.296	27.808	28.845	29.633
17	24.769	27.587	29.131	30.191	30.995
18	25.989	28.869	30.446	31.526	32.346
19	27.204	30.144	31.751	32.852	33.687
20	28.412	31.410	33.048	34.170	35.020
21	29.615	32.671	34.338	35.479	36.343
22	30.813	33.924	35.620	36.781	37.659
23	32.007	35.172	36.897	38.076	38.968
24	33.196	36.415	38.167	39.364	40.270
25	34.382	37.652	39.431	40.646	41.566
26	35.563	38.885	40.690	41.923	42.856
27	36.741	40.113	41.944	43.195	44.140
28	37.916	41.337	43.194	44.461	45.419
29	39.087	42.557	44.439	45.722	46.693
30	40.256	43.773	45.679	46.979	47.962
35	46.059	49.802	51.825	53.203	54.244
40	51.805	55.758	57.891	59.342	60.436
45	57.505	61.656	63.891	65.410	66.555
50	63.167	67.505	69.837	71.420	72.613
55	68.796	73.311	75.736	77.380	78.619
60	74.397	79.082	81.594	83.298	84.580
70	85.527	90.531	93.209	95.023	96.388
80	96.578	101.879	104.712	106.629	108.069
90	107.565	113.145	116.123	118.136	119.648
100	118.498	124.342	127.457	129.561	131.142
110	129.385	135.480	138.725	140.917	142.562
120	140.233	146.567	149.937	152.211	153.918

UPPER PERCENTAGE POINTS / OBERE PROZENTPUNKTE

$\chi^2(\alpha/\tau,\nu)$ $\alpha=0.100$

ν \ τ	6	7	8	9	10
1	5.731	6.002	6.239	6.447	6.635
2	8.189	8.497	8.764	9.000	9.210
3	10.236	10.571	10.861	11.117	11.345
4	12.094	12.452	12.762	13.034	13.277
5	13.839	14.217	14.544	14.831	15.086
6	15.506	15.903	16.245	16.545	16.812
7	17.115	17.529	17.885	18.197	18.475
8	18.680	19.109	19.478	19.802	20.090
9	20.209	20.653	21.034	21.368	21.666
10	21.707	22.165	22.558	22.903	23.209
11	23.181	23.651	24.056	24.410	24.725
12	24.632	25.115	25.530	25.894	26.217
13	26.064	26.560	26.985	27.357	27.688
14	27.480	27.987	28.422	28.803	29.141
15	28.880	29.398	29.843	30.232	30.578
16	30.267	30.796	31.250	31.647	32.000
17	31.642	32.181	32.644	33.049	33.409
18	33.005	33.555	34.027	34.439	34.805
19	34.358	34.919	35.399	35.818	36.191
20	35.702	36.272	36.761	37.187	37.566
21	37.038	37.617	38.113	38.547	38.932
22	38.365	38.954	39.458	39.898	40.289
23	39.685	40.282	40.794	41.242	41.638
24	40.998	41.604	42.123	42.577	42.980
25	42.304	42.919	43.446	43.906	44.314
26	43.604	44.228	44.762	45.228	45.642
27	44.898	45.530	46.071	46.544	46.963
28	46.187	46.827	47.375	47.854	48.278
29	47.471	48.119	48.674	49.158	49.588
30	48.749	49.406	49.967	50.457	50.892
35	55.077	55.771	56.365	56.883	57.342
40	61.312	62.041	62.665	63.209	63.691
45	67.472	68.234	68.885	69.453	69.957
50	73.567	74.361	75.039	75.630	76.154
55	79.609	80.433	81.136	81.749	82.292
60	85.605	86.456	87.184	87.818	88.379
70	97.477	98.383	99.155	99.829	100.425
80	109.220	110.175	110.990	111.700	112.329
90	120.856	121.858	122.713	123.457	124.116
100	132.403	133.449	134.342	135.119	135.807
110	143.874	144.962	145.891	146.699	147.414
120	155.279	156.408	157.371	158.209	158.950

UPPER PERCENTAGE POINTS / OBERE PROZENTPUNKTE

$\chi^2(\alpha/\tau,\nu)$ α=0.100

ν \ τ	11	12	13	14	15
1	6.805	6.960	7.104	7.237	7.361
2	9.401	9.575	9.735	9.883	10.021
3	11.551	11.739	11.912	12.071	12.220
4	13.496	13.695	13.879	14.048	14.206
5	15.317	15.527	15.719	15.898	16.063
6	17.053	17.272	17.473	17.659	17.832
7	18.726	18.954	19.163	19.356	19.535
8	20.350	20.586	20.802	21.002	21.187
9	21.934	22.177	22.401	22.607	22.798
10	23.485	23.736	23.966	24.178	24.374
11	25.008	25.266	25.502	25.720	25.922
12	26.508	26.772	27.014	27.237	27.444
13	27.986	28.256	28.504	28.732	28.944
14	29.446	29.722	29.975	30.209	30.425
15	30.889	31.171	31.429	31.668	31.889
16	32.317	32.605	32.869	33.111	33.337
17	33.732	34.025	34.294	34.541	34.771
18	35.134	35.433	35.707	35.959	36.192
19	36.526	36.830	37.108	37.364	37.602
20	37.907	38.216	38.498	38.759	39.000
21	39.278	39.592	39.879	40.144	40.389
22	40.641	40.959	41.251	41.520	41.769
23	41.995	42.318	42.614	42.887	43.139
24	43.341	43.669	43.970	44.246	44.502
25	44.681	45.013	45.317	45.598	45.857
26	46.013	46.350	46.658	46.942	47.205
27	47.339	47.681	47.993	48.280	48.547
28	48.659	49.005	49.321	49.612	49.882
29	49.974	50.323	50.643	50.938	51.211
30	51.283	51.636	51.960	52.258	52.534
35	57.754	58.128	58.469	58.784	59.075
40	64.123	64.515	64.874	65.203	65.509
45	70.408	70.818	71.192	71.536	71.855
50	76.624	77.049	77.438	77.796	78.127
55	82.779	83.220	83.623	83.994	84.337
60	88.883	89.339	89.755	90.139	90.493
70	100.960	101.444	101.886	102.292	102.668
80	112.892	113.402	113.868	114.296	114.693
90	124.707	125.241	125.729	126.178	126.593
100	136.423	136.981	137.490	137.958	138.391
110	148.055	148.635	149.164	149.650	150.100
120	159.614	160.215	160.763	161.268	161.734

UPPER PERCENTAGE POINTS / OBERE PROZENTPUNKTE

$\chi^2(\alpha/\tau,\nu)$ α=0.100

ν \ τ	16	17	18	19	20
1	7.477	7.586	7.689	7.787	7.879
2	10.150	10.272	10.386	10.494	10.597
3	12.359	12.489	12.612	12.728	12.838
4	14.353	14.491	14.621	14.744	14.860
5	16.217	16.362	16.499	16.628	16.750
6	17.993	18.144	18.286	18.420	18.548
7	19.702	19.859	20.007	20.146	20.278
8	21.360	21.522	21.675	21.819	21.955
9	22.976	23.143	23.301	23.449	23.589
10	24.558	24.730	24.891	25.044	25.188
11	26.110	26.286	26.452	26.609	26.757
12	27.637	27.817	27.987	28.148	28.300
13	29.141	29.326	29.500	29.664	29.819
14	30.627	30.816	30.993	31.161	31.319
15	32.095	32.287	32.469	32.640	32.801
16	33.547	33.743	33.928	34.102	34.267
17	34.985	35.185	35.373	35.551	35.718
18	36.410	36.614	36.805	36.986	37.156
19	37.823	38.030	38.225	38.409	38.582
20	39.225	39.436	39.634	39.820	39.997
21	40.618	40.831	41.032	41.222	41.401
22	42.001	42.218	42.422	42.614	42.796
23	43.375	43.595	43.802	43.997	44.181
24	44.741	44.964	45.174	45.372	45.559
25	46.099	46.325	46.538	46.738	46.928
26	47.450	47.680	47.895	48.098	48.290
27	48.795	49.027	49.245	49.451	49.645
28	50.133	50.368	50.589	50.797	50.993
29	51.465	51.703	51.926	52.137	52.336
30	52.791	53.032	53.258	53.471	53.672
35	59.346	59.600	59.838	60.063	60.275
40	65.793	66.059	66.308	66.544	66.766
45	72.151	72.429	72.689	72.934	73.166
50	78.435	78.724	78.994	79.249	79.490
55	84.657	84.955	85.235	85.499	85.749
60	90.823	91.132	91.421	91.694	91.952
70	103.018	103.346	103.653	103.942	104.215
80	115.061	115.406	115.729	116.033	116.321
90	126.980	127.340	127.679	127.998	128.299
100	138.794	139.170	139.523	139.855	140.169
110	150.519	150.910	151.277	151.622	151.948
120	162.167	162.573	162.953	163.310	163.648

UPPER PERCENTAGE POINTS / OBERE PROZENTPUNKTE

$\chi^2(\alpha/\tau,\nu)$ α=0.100

ν \ τ	22	24	26	28	30
1	8.052	8.210	8.355	8.490	8.615
2	10.787	10.961	11.121	11.270	11.408
3	13.043	13.229	13.400	13.559	13.706
4	15.076	15.273	15.454	15.622	15.777
5	16.976	17.182	17.372	17.547	17.710
6	18.783	18.998	19.195	19.377	19.547
7	20.522	20.745	20.949	21.138	21.313
8	22.208	22.438	22.648	22.843	23.024
9	23.850	24.086	24.304	24.504	24.690
10	25.456	25.699	25.922	26.128	26.320
11	27.031	27.281	27.510	27.721	27.917
12	28.581	28.836	29.070	29.287	29.487
13	30.107	30.368	30.608	30.829	31.034
14	31.613	31.880	32.124	32.350	32.559
15	33.101	33.373	33.622	33.853	34.066
16	34.572	34.850	35.104	35.338	35.556
17	36.029	36.312	36.570	36.809	37.030
18	37.473	37.760	38.023	38.266	38.491
19	38.904	39.196	39.463	39.710	39.939
20	40.323	40.620	40.892	41.142	41.375
21	41.733	42.034	42.310	42.564	42.800
22	43.132	43.438	43.718	43.976	44.215
23	44.523	44.833	45.116	45.378	45.621
24	45.904	46.219	46.506	46.772	47.018
25	47.278	47.597	47.888	48.157	48.406
26	48.645	48.967	49.263	49.535	49.787
27	50.004	50.331	50.630	50.905	51.161
28	51.357	51.688	51.990	52.269	52.527
29	52.704	53.038	53.344	53.626	53.887
30	54.044	54.382	54.692	54.977	55.241
35	60.667	61.023	61.349	61.649	61.928
40	67.177	67.550	67.891	68.206	68.497
45	73.595	73.984	74.339	74.667	74.971
50	79.935	80.339	80.709	81.049	81.364
55	86.210	86.628	87.011	87.363	87.689
60	92.428	92.860	93.255	93.619	93.956
70	104.720	105.177	105.596	105.981	106.338
80	116.852	117.334	117.775	118.180	118.555
90	128.855	129.360	129.821	130.245	130.638
100	140.750	141.275	141.756	142.198	142.607
110	152.551	153.097	153.596	154.055	154.481
120	164.272	164.838	165.354	165.830	166.270

Tafel 22 (Forts.) Table 22 (cont.

UPPER PERCENTAGE POINTS / OBERE PROZENTPUNKTE

$\chi^2(\alpha/\tau,\nu)$ α=0.050

ν \ τ	1	2	3	4	5
1	3.841	5.024	5.731	6.239	6.635
2	5.991	7.378	8.189	8.764	9.210
3	7.815	9.348	10.236	10.861	11.345
4	9.488	11.143	12.094	12.762	13.277
5	11.070	12.833	13.839	14.544	15.086
6	12.592	14.449	15.506	16.245	16.812
7	14.067	16.013	17.115	17.885	18.475
8	15.507	17.535	18.680	19.478	20.090
9	16.919	19.023	20.209	21.034	21.666
10	18.307	20.483	21.707	22.558	23.209
11	19.675	21.920	23.181	24.056	24.725
12	21.026	23.337	24.632	25.530	26.217
13	22.362	24.736	26.064	26.985	27.688
14	23.685	26.119	27.480	28.422	29.141
15	24.996	27.488	28.880	29.843	30.578
16	26.296	28.845	30.267	31.250	32.000
17	27.587	30.191	31.642	32.644	33.409
18	28.869	31.526	33.005	34.027	34.805
19	30.144	32.852	34.358	35.399	36.191
20	31.410	34.170	35.702	36.761	37.566
21	32.671	35.479	37.038	38.113	38.932
22	33.924	36.781	38.365	39.458	40.289
23	35.172	38.076	39.685	40.794	41.638
24	36.415	39.364	40.998	42.123	42.980
25	37.652	40.646	42.304	43.446	44.314
26	38.885	41.923	43.604	44.762	45.642
27	40.113	43.195	44.898	46.071	46.963
28	41.337	44.451	46.187	47.375	48.278
29	42.557	45.722	47.471	48.674	49.588
30	43.773	46.979	48.749	49.967	50.892
35	49.802	53.203	55.077	56.365	57.342
40	55.758	59.342	61.312	62.665	63.691
45	61.656	65.410	67.472	68.885	69.957
50	67.505	71.420	73.567	75.039	76.154
55	73.311	77.380	79.609	81.136	82.292
60	79.082	83.298	85.605	87.184	88.379
70	90.531	95.023	97.477	99.155	100.425
80	101.879	106.629	109.220	110.990	112.329
90	113.145	118.136	120.856	122.713	124.116
100	124.342	129.561	132.403	134.342	135.807
110	135.480	140.917	143.874	145.891	147.414
120	146.567	152.211	155.279	157.371	158.950

UPPER PERCENTAGE POINTS / OBERE PROZENTPUNKTE

$\chi^2(\alpha/\tau,\nu)$ α=0.050

ν \ τ	6	7	8	9	10
1	6.960	7.237	7.477	7.689	7.879
2	9.575	9.883	10.150	10.386	10.597
3	11.739	12.071	12.359	12.612	12.838
4	13.695	14.048	14.353	14.621	14.860
5	15.527	15.898	16.217	16.499	16.750
6	17.272	17.659	17.993	18.286	18.548
7	18.954	19.356	19.702	20.007	20.278
8	20.586	21.002	21.360	21.675	21.955
9	22.177	22.607	22.976	23.301	23.589
10	23.736	24.178	24.558	24.891	25.188
11	25.266	25.720	26.110	26.452	26.757
12	26.772	27.237	27.637	27.987	28.300
13	28.256	28.732	29.141	29.500	29.819
14	29.722	30.209	30.627	30.993	31.319
15	31.171	31.668	32.095	32.469	32.801
16	32.605	33.111	33.547	33.928	34.267
17	34.025	34.541	34.985	35.373	35.718
18	35.433	35.959	36.410	36.805	37.156
19	36.830	37.364	37.823	38.225	38.582
20	38.216	38.759	39.225	39.634	39.997
21	39.592	40.144	40.618	41.032	41.401
22	40.959	41.520	42.001	42.422	42.796
23	42.318	42.887	43.375	43.802	44.181
24	43.669	44.246	44.741	45.174	45.559
25	45.013	45.598	46.099	46.538	46.928
26	46.350	46.942	47.450	47.895	48.290
27	47.681	48.280	48.795	49.245	49.645
28	49.005	49.612	50.133	50.589	50.993
29	50.323	50.938	51.465	51.926	52.336
30	51.636	52.258	52.791	53.258	53.672
35	58.128	58.784	59.346	59.838	60.275
40	64.515	65.203	65.793	66.308	66.766
45	70.818	71.536	72.151	72.689	73.166
50	77.049	77.796	78.435	78.994	79.490
55	83.220	83.994	84.657	85.235	85.749
60	89.339	90.139	90.823	91.421	91.952
70	101.444	102.292	103.018	103.653	104.215
80	113.402	114.296	115.061	115.729	116.321
90	125.241	126.178	126.980	127.679	128.299
100	136.981	137.958	138.794	139.523	140.169
110	148.635	149.650	150.519	151.277	151.948
120	160.215	161.268	162.167	162.953	163.648

Tafel 22 (Forts.) Table 22 (cont.)

UPPER PERCENTAGE POINTS / OBERE PROZENTPUNKTE

$\chi^2(\alpha/\tau,\nu)$ α=0.050

ν \ τ	11	12	13	14	15
1	8.052	8.210	8.355	8.490	8.615
2	10.787	10.961	11.121	11.270	11.408
3	13.043	13.229	13.400	13.559	13.706
4	15.076	15.273	15.454	15.622	15.777
5	16.976	17.182	17.372	17.547	17.710
6	18.783	18.998	19.195	19.377	19.547
7	20.522	20.745	20.949	21.138	21.313
8	22.208	22.438	22.648	22.843	23.024
9	23.850	24.086	24.304	24.504	24.690
10	25.456	25.699	25.922	26.128	26.320
11	27.031	27.281	27.510	27.721	27.917
12	28.581	28.836	29.070	29.287	29.487
13	30.107	30.368	30.608	30.829	31.034
14	31.613	31.880	32.124	32.350	32.559
15	33.101	33.373	33.622	33.853	34.066
16	34.572	34.850	35.104	35.338	35.556
17	36.029	36.312	36.570	36.809	37.030
18	37.473	37.760	38.023	38.266	38.491
19	38.904	39.196	39.463	39.710	39.939
20	40.323	40.620	40.892	41.142	41.375
21	41.733	42.034	42.310	42.564	42.800
22	43.132	43.438	43.718	43.976	44.215
23	44.523	44.833	45.116	45.378	45.621
24	45.904	46.219	46.506	46.772	47.018
25	47.278	47.597	47.888	48.157	48.406
26	48.645	48.967	49.263	49.535	49.787
27	50.004	50.331	50.630	50.905	51.161
28	51.357	51.688	51.990	52.269	52.527
29	52.704	53.038	53.344	53.626	53.887
30	54.044	54.382	54.692	54.977	55.241
35	60.667	61.023	61.349	61.649	61.928
40	67.177	67.550	67.891	68.206	68.497
45	73.595	73.984	74.339	74.667	74.971
50	79.935	80.339	80.709	81.049	81.364
55	86.210	86.628	87.011	87.363	87.689
60	92.428	92.860	93.255	93.619	93.956
70	104.720	105.177	105.596	105.981	106.338
80	116.852	117.334	117.775	118.180	118.555
90	128.855	129.360	129.821	130.245	130.638
100	140.750	141.275	141.756	142.198	142.607
110	152.551	153.097	153.596	154.055	154.481
120	164.272	164.838	165.354	165.830	166.270

UPPER PERCENTAGE POINTS / OBERE PROZENTPUNKTE

$\chi^2(\alpha/\tau,\nu)$ α=0.050

ν \ τ	16	17	18	19	20
1	8.733	8.844	8.948	9.047	9.141
2	11.537	11.658	11.772	11.880	11.983
3	13.844	13.974	14.096	14.211	14.320
4	15.922	16.059	16.187	16.309	16.424
5	17.862	18.004	18.138	18.265	18.386
6	19.705	19.853	19.993	20.124	20.249
7	21.477	21.630	21.775	21.911	22.040
8	23.193	23.351	23.500	23.641	23.774
9	24.864	25.027	25.181	25.325	25.462
10	26.498	26.665	26.823	26.971	27.112
11	28.100	28.272	28.433	28.585	28.729
12	29.675	29.850	30.015	30.171	30.318
13	31.225	31.405	31.573	31.732	31.883
14	32.755	32.938	33.110	33.272	33.426
15	34.265	34.452	34.627	34.793	34.950
16	35.759	35.949	36.128	36.296	36.456
17	37.237	37.430	37.612	37.784	37.946
18	38.701	38.898	39.083	39.257	39.422
19	40.152	40.352	40.540	40.717	40.885
20	41.592	41.795	41.985	42.165	42.336
21	43.020	43.226	43.420	43.602	43.775
22	44.438	44.647	44.844	45.029	45.204
23	45.847	46.059	46.258	46.446	46.623
24	47.247	47.462	47.663	47.854	48.034
25	48.638	48.856	49.060	49.253	49.435
26	50.022	50.242	50.449	50.644	50.829
27	51.399	51.621	51.831	52.028	52.215
28	52.768	52.994	53.205	53.405	53.594
29	54.131	54.359	54.573	54.775	54.967
30	55.488	55.718	55.935	56.139	56.332
35	62.187	62.430	62.658	62.873	63.076
40	68.769	69.023	69.261	69.486	69.699
45	75.254	75.518	75.767	76.001	76.223
50	81.658	81.933	82.191	82.434	82.664
55	87.993	88.278	88.545	88.797	89.035
60	94.269	94.563	94.839	95.098	95.344
70	106.670	106.981	107.273	107.548	107.808
80	118.905	119.232	119.539	119.828	120.102
90	131.004	131.346	131.667	131.970	132.256
100	142.989	143.345	143.680	143.995	144.293
110	154.876	155.246	155.594	155.921	156.230
120	166.680	167.063	167.423	167.762	168.082

Tafel 22 (Forts.) Table 22 (cont.)

UPPER PERCENTAGE POINTS / OBERE PROZENTPUNKTE

$\chi^2(\alpha/\tau,\nu)$ $\alpha=0.050$

ν \ τ	22	24	26	28	30
1	9.315	9.475	9.622	9.758	9.885
2	12.174	12.348	12.508	12.656	12.794
3	14.523	14.709	14.879	15.037	15.183
4	16.638	16.832	17.011	17.177	17.331
5	18.609	18.812	18.999	19.172	19.332
6	20.481	20.692	20.886	21.065	21.232
7	22.280	22.498	22.699	22.884	23.056
8	24.022	24.247	24.453	24.644	24.821
9	25.717	25.948	26.160	26.357	26.539
10	27.373	27.610	27.828	28.030	28.216
11	28.997	29.240	29.463	29.669	29.860
12	30.592	30.841	31.069	31.279	31.475
13	32.162	32.416	32.649	32.864	33.064
14	33.711	33.970	34.208	34.427	34.631
15	35.240	35.504	35.746	35.970	36.177
16	36.751	37.020	37.267	37.494	37.706
17	38.247	38.520	38.771	39.003	39.217
18	39.728	40.006	40.261	40.496	40.714
19	41.196	41.478	41.737	41.976	42.198
20	42.651	42.938	43.201	43.443	43.668
21	44.095	44.386	44.653	44.899	45.127
22	45.529	45.824	46.094	46.344	46.575
23	46.952	47.252	47.526	47.778	48.013
24	48.367	48.670	48.948	49.204	49.441
25	49.773	50.080	50.361	50.620	50.861
26	51.171	51.482	51.766	52.029	52.272
27	52.561	52.876	53.164	53.429	53.676
28	53.944	54.262	54.554	54.822	55.071
29	55.321	55.642	55.937	56.208	56.460
30	56.690	57.015	57.313	57.588	57.843
35	63.453	63.795	64.108	64.397	64.664
40	70.093	70.450	70.778	71.080	71.360
45	76.633	77.005	77.346	77.661	77.952
50	83.090	83.476	83.830	84.156	84.458
55	89.475	89.875	90.240	90.578	90.890
60	95.798	96.211	96.588	96.936	97.259
70	108.289	108.726	109.125	109.493	109.834
80	120.607	121.066	121.486	121.873	122.231
90	132.784	133.264	133.703	134.108	134.483
100	144.844	145.343	145.801	146.222	146.612
110	156.802	157.321	157.795	158.232	158.637
120	168.674	169.210	169.701	170.154	170.573

UPPER PERCENTAGE POINTS / OBERE PROZENTPUNKTE

$\chi^2(\alpha/\tau,\nu)$ $\alpha=0.025$

ν \ τ	1	2	3	4	5
1	5.024	6.239	6.960	7.477	7.879
2	7.378	8.764	9.575	10.150	10.597
3	9.348	10.861	11.739	12.359	12.838
4	11.143	12.762	13.695	14.353	14.860
5	12.833	14.544	15.527	16.217	16.750
6	14.449	16.245	17.272	17.993	18.548
7	16.013	17.885	18.954	19.702	20.278
8	17.535	19.478	20.586	21.360	21.955
9	19.023	21.034	22.177	22.976	23.589
10	20.483	22.558	23.736	24.558	25.188
11	21.920	24.056	25.266	26.110	26.757
12	23.337	25.530	26.772	27.637	28.300
13	24.736	26.985	28.256	29.141	29.819
14	26.119	28.422	29.722	30.627	31.319
15	27.488	29.843	31.171	32.095	32.801
16	28.845	31.250	32.605	33.547	34.267
17	30.191	32.644	34.025	34.985	35.718
18	31.526	34.027	35.433	36.410	37.156
19	32.852	35.399	36.830	37.823	38.582
20	34.170	36.761	38.216	39.225	39.997
21	35.479	38.113	39.592	40.618	41.401
22	36.781	39.458	40.959	42.001	42.796
23	38.076	40.794	42.318	43.375	44.181
24	39.364	42.123	43.669	44.741	45.559
25	40.646	43.446	45.013	46.099	46.928
26	41.923	44.762	46.350	47.450	48.290
27	43.195	46.071	47.681	48.795	49.645
28	44.461	47.375	49.005	50.133	50.993
29	45.722	48.674	50.323	51.465	52.336
30	46.979	49.967	51.636	52.791	53.672
35	53.203	56.365	58.128	59.346	60.275
40	59.342	62.665	64.515	65.793	66.766
45	65.410	68.885	70.818	72.151	73.166
50	71.420	75.039	77.049	78.435	79.490
55	77.380	81.136	83.220	84.657	85.749
60	83.298	87.184	89.339	90.823	91.952
70	95.023	99.155	101.444	103.018	104.215
80	106.629	110.990	113.402	115.061	116.321
90	118.136	122.713	125.241	126.980	128.299
100	129.561	134.342	136.981	138.794	140.169
110	140.917	145.891	148.635	150.519	151.948
120	152.211	157.371	160.215	162.167	163.648

UPPER PERCENTAGE POINTS / OBERE PROZENTPUNKTE

$\chi^2(\alpha/\tau,\nu)$ $\alpha=0.025$

ν \ τ	6	7	8	9	10
1	8.210	8.490	8.733	8.948	9.141
2	10.961	11.270	11.537	11.772	11.983
3	13.229	13.559	13.844	14.096	14.320
4	15.273	15.622	15.922	16.187	16.424
5	17.182	17.547	17.862	18.138	18.386
6	18.998	19.377	19.705	19.993	20.249
7	20.745	21.138	21.477	21.775	22.040
8	22.438	22.843	23.193	23.500	23.774
9	24.086	24.504	24.864	25.181	25.462
10	25.699	26.128	26.498	26.823	27.112
11	27.281	27.721	28.100	28.433	28.729
12	28.836	29.287	29.675	30.015	30.318
13	30.368	30.829	31.225	31.573	31.883
14	31.880	32.350	32.755	33.110	33.426
15	33.373	33.853	34.265	34.627	34.950
16	34.850	35.338	35.759	36.128	36.456
17	36.312	36.809	37.237	37.612	37.946
18	37.760	38.266	38.701	39.083	39.422
19	39.196	39.710	40.152	40.540	40.885
20	40.620	41.142	41.592	41.985	42.336
21	42.034	42.564	43.020	43.420	43.775
22	43.438	43.976	44.438	44.844	45.204
23	44.833	45.378	45.847	46.258	46.623
24	46.219	46.772	47.247	47.663	48.034
25	47.597	48.157	48.638	49.060	49.435
26	48.967	49.535	50.022	50.449	50.829
27	50.331	50.905	51.399	51.831	52.215
28	51.688	52.269	52.768	53.205	53.594
29	53.038	53.626	54.131	54.573	54.967
30	54.382	54.977	55.488	55.935	56.332
35	61.023	61.649	62.187	62.658	63.076
40	67.550	68.206	68.769	69.261	69.699
45	73.984	74.667	75.254	75.767	76.223
50	80.339	81.049	81.658	82.191	82.664
55	86.628	87.363	87.993	88.545	89.035
60	92.860	93.619	94.269	94.839	95.344
70	105.177	105.981	106.670	107.273	107.808
80	117.334	118.180	118.905	119.539	120.102
90	129.360	130.245	131.004	131.667	132.256
100	141.275	142.198	142.989	143.680	144.293
110	153.097	154.055	154.876	155.594	156.230
120	164.838	165.830	166.680	167.423	168.082

UPPER PERCENTAGE POINTS / OBERE PROZENTPUNKTE

$\chi^2(\alpha/\tau,\nu)$ α=0.025

ν \ τ	11	12	13	14	15
1	9.315	9.475	9.622	9.758	9.885
2	12.174	12.348	12.508	12.656	12.794
3	14.523	14.709	14.879	15.037	15.183
4	16.638	16.832	17.011	17.177	17.331
5	18.609	18.812	18.999	19.172	19.332
6	20.481	20.692	20.886	21.065	21.232
7	22.280	22.498	22.699	22.884	23.056
8	24.022	24.247	24.453	24.644	24.821
9	25.717	25.948	26.160	26.357	26.539
10	27.373	27.610	27.828	28.030	28.216
11	28.997	29.240	29.463	29.669	29.860
12	30.592	30.841	31.069	31.279	31.475
13	32.162	32.416	32.649	32.864	33.064
14	33.711	33.970	34.208	34.427	34.631
15	35.240	35.504	35.746	35.970	36.177
16	36.751	37.020	37.267	37.494	37.706
17	38.247	38.520	38.771	39.003	39.217
18	39.728	40.006	40.261	40.496	40.714
19	41.196	41.478	41.737	41.976	42.198
20	42.651	42.938	43.201	43.443	43.668
21	44.095	44.386	44.653	44.899	45.127
22	45.529	45.824	46.094	46.344	46.575
23	46.952	47.252	47.526	47.778	48.013
24	48.367	48.670	48.948	49.204	49.441
25	49.773	50.080	50.361	50.620	50.861
26	51.171	51.482	51.766	52.029	52.272
27	52.561	52.876	53.164	53.429	53.676
28	53.944	54.262	54.554	54.822	55.071
29	55.321	55.642	55.937	56.208	56.460
30	56.690	57.015	57.313	57.588	57.843
35	63.453	63.795	64.108	64.397	64.664
40	70.093	70.450	70.778	71.080	71.360
45	76.633	77.005	77.346	77.661	77.952
50	83.090	83.476	83.830	84.156	84.458
55	89.475	89.875	90.240	90.578	90.890
60	95.798	96.211	96.588	96.936	97.259
70	108.289	108.726	109.125	109.493	109.834
80	120.607	121.066	121.486	121.873	122.231
90	132.784	133.264	133.703	134.108	134.483
100	144.844	145.343	145.801	146.222	146.612
110	156.802	157.321	157.795	158.232	158.637
120	168.674	169.210	169.701	170.154	170.573

Tafel 22 (Forts.) Table 22 (cont.)

UPPER PERCENTAGE POINTS / OBERE PROZENTPUNKTE

$\chi^2(\alpha/\tau,\nu)$ α=0.025

ν \ τ	16	17	18	19	20
1	10.003	10.115	10.220	10.320	10.415
2	12.923	13.044	13.159	13.267	13.369
3	15.320	15.449	15.570	15.685	15.794
4	17.475	17.610	17.737	17.858	17.972
5	19.482	19.623	19.755	19.881	20.000
6	21.387	21.533	21.671	21.801	21.924
7	23.217	23.368	23.510	23.644	23.771
8	24.987	25.142	25.288	25.426	25.557
9	26.709	26.868	27.019	27.160	27.295
10	28.391	28.554	28.708	28.854	28.991
11	30.039	30.206	30.364	30.513	30.654
12	31.657	31.829	31.990	32.142	32.286
13	33.250	33.425	33.590	33.745	33.892
14	34.821	34.999	35.166	35.325	35.475
15	36.371	36.552	36.723	36.884	37.037
16	37.903	38.087	38.261	38.425	38.580
17	39.418	39.606	39.782	39.949	40.107
18	40.918	41.109	41.288	41.458	41.618
19	42.404	42.598	42.781	42.953	43.115
20	43.878	44.075	44.260	44.434	44.600
21	45.340	45.540	45.727	45.904	46.072
22	46.791	46.993	47.184	47.363	47.533
23	48.232	48.437	48.630	48.811	48.984
24	49.663	49.871	50.066	50.250	50.424
25	51.085	51.296	51.493	51.680	51.856
26	52.499	52.712	52.912	53.101	53.279
27	53.905	54.121	54.323	54.514	54.695
28	55.304	55.522	55.726	55.919	56.102
29	56.695	56.915	57.122	57.318	57.502
30	58.080	58.303	58.512	58.709	58.896
35	64.914	65.148	65.368	65.575	65.771
40	71.621	71.865	72.095	72.311	72.516
45	78.224	78.478	78.717	78.942	79.156
50	84.740	85.003	85.251	85.485	85.706
55	91.181	91.454	91.710	91.952	92.181
60	97.559	97.840	98.105	98.354	98.589
70	110.152	110.449	110.729	110.992	111.241
80	122.565	122.878	123.171	123.448	123.710
90	134.832	135.158	135.465	135.754	136.028
100	146.975	147.315	147.635	147.936	148.221
110	159.014	159.367	159.699	160.011	160.306
120	170.963	171.328	171.671	171.994	172.299

UPPER PERCENTAGE POINTS / OBERE PROZENTPUNKTE

$\chi^2(\alpha/\tau,\nu)$ $\alpha=0.025$

ν \ τ	22	24	26	28	30
1	10.591	10.752	10.900	11.038	11.165
2	13.560	13.734	13.894	14.042	14.180
3	15.996	16.180	16.349	16.506	16.652
4	18.183	18.376	18.554	18.718	18.871
5	20.220	20.421	20.605	20.776	20.935
6	22.152	22.360	22.551	22.728	22.892
7	24.006	24.221	24.418	24.601	24.770
8	25.800	26.021	26.224	26.411	26.586
9	27.544	27.771	27.979	28.172	28.351
10	29.247	29.479	29.693	29.890	30.073
11	30.915	31.153	31.371	31.573	31.760
12	32.553	32.796	33.019	33.225	33.416
13	34.164	34.412	34.640	34.850	35.045
14	35.752	36.005	36.237	36.451	36.650
15	37.319	37.577	37.813	38.031	38.233
16	38.868	39.130	39.370	39.592	39.798
17	40.399	40.666	40.910	41.135	41.344
18	41.915	42.186	42.434	42.663	42.875
19	43.417	43.692	43.943	44.176	44.392
20	44.906	45.184	45.440	45.676	45.894
21	46.382	46.665	46.924	47.163	47.385
22	47.848	48.134	48.396	48.639	48.864
23	49.302	49.593	49.858	50.104	50.332
24	50.747	51.041	51.310	51.559	51.790
25	52.183	52.481	52.753	53.005	53.238
26	53.610	53.911	54.187	54.441	54.678
27	55.029	55.334	55.613	55.870	56.109
28	56.441	56.748	57.030	57.291	57.532
29	57.845	58.156	58.441	58.704	58.948
30	59.242	59.556	59.844	60.110	60.357
35	66.134	66.464	66.767	67.046	67.305
40	72.896	73.241	73.557	73.848	74.118
45	79.550	79.909	80.238	80.541	80.822
50	86.115	86.487	86.828	87.142	87.433
55	92.604	92.988	93.340	93.664	93.965
60	99.026	99.422	99.785	100.119	100.430
70	111.702	112.121	112.505	112.858	113.186
80	124.194	124.634	125.036	125.407	125.752
90	136.534	136.994	137.414	137.802	138.162
100	148.747	149.226	149.664	150.067	150.441
110	160.853	161.349	161.803	162.221	162.609
120	172.865	173.378	173.847	174.280	174.681

UPPER PERCENTAGE POINTS / OBERE PROZENTPUNKTE

$\chi^2(\alpha/\tau,\nu)$ $\alpha=0.010$

ν \ τ	1	2	3	4	5
1	6.635	7.879	8.615	9.141	9.550
2	9.210	10.597	11.408	11.983	12.429
3	11.345	12.838	13.706	14.320	14.796
4	13.277	14.860	15.777	16.424	16.924
5	15.086	16.750	17.710	18.386	18.907
6	16.812	18.548	19.547	20.249	20.791
7	18.475	20.278	21.313	22.040	22.601
8	20.090	21.955	23.024	23.774	24.352
9	21.666	23.589	24.690	25.462	26.056
10	23.209	25.188	26.320	27.112	27.722
11	24.725	26.757	27.917	28.729	29.354
12	26.217	28.300	29.487	30.318	30.957
13	27.688	29.819	31.034	31.883	32.535
14	29.141	31.319	32.559	33.426	34.091
15	30.578	32.801	34.066	34.950	35.628
16	32.000	34.267	35.556	36.456	37.146
17	33.409	35.718	37.030	37.946	38.648
18	34.805	37.156	38.491	39.422	40.136
19	36.191	38.582	39.939	40.885	41.610
20	37.566	39.997	41.375	42.336	43.072
21	38.932	41.401	42.800	43.775	44.522
22	40.289	42.796	44.215	45.204	45.962
23	41.638	44.181	45.621	46.623	47.391
24	42.980	45.559	47.018	48.034	48.812
25	44.314	46.928	48.406	49.435	50.223
26	45.642	48.290	49.787	50.829	51.627
27	46.963	49.645	51.161	52.215	53.023
28	48.278	50.993	52.527	53.594	54.411
29	49.588	52.336	53.887	54.967	55.792
30	50.892	53.672	55.241	56.332	57.167
35	57.342	60.275	61.928	63.076	63.955
40	63.691	66.766	68.497	69.699	70.618
45	69.957	73.166	74.971	76.223	77.179
50	76.154	79.490	81.364	82.664	83.657
55	82.292	85.749	87.689	89.035	90.061
60	88.379	91.952	93.956	95.344	96.404
70	100.425	104.215	106.338	107.808	108.929
80	112.329	116.321	118.555	120.102	121.280
90	124.116	128.299	130.638	132.256	133.489
100	135.807	140.169	142.607	144.293	145.577
110	147.414	151.948	154.481	156.230	157.563
120	158.950	163.648	166.270	168.082	169.461

UPPER PERCENTAGE POINTS / OBERE PROZENTPUNKTE

$\chi^2(\alpha/\tau,\nu)$ α=0.010

ν \ τ	6	7	8	9	10
1	9.885	10.169	10.415	10.633	10.828
2	12.794	13.102	13.369	13.605	13.816
3	15.183	15.510	15.794	16.043	16.266
4	17.331	17.675	17.972	18.233	18.467
5	19.332	19.690	20.000	20.272	20.515
6	21.232	21.603	21.924	22.206	22.458
7	23.056	23.440	23.771	24.062	24.322
8	24.821	25.216	25.557	25.857	26.124
9	26.539	26.945	27.295	27.603	27.877
10	28.216	28.633	28.991	29.307	29.588
11	29.860	30.286	30.654	30.976	31.264
12	31.475	31.910	32.286	32.615	32.909
13	33.064	33.509	33.892	34.228	34.528
14	34.631	35.084	35.475	35.818	36.123
15	36.177	36.639	37.037	37.386	37.697
16	37.706	38.175	38.580	38.936	39.252
17	39.217	39.695	40.107	40.468	40.790
18	40.714	41.200	41.618	41.985	42.312
19	42.198	42.691	43.115	43.488	43.820
20	43.668	44.169	44.600	44.978	45.315
21	45.127	45.635	46.072	46.455	46.797
22	46.575	47.090	47.533	47.922	48.268
23	48.013	48.535	48.984	49.377	49.728
24	49.441	49.970	50.424	50.823	51.179
25	50.861	51.396	51.856	52.260	52.620
26	52.272	52.814	53.279	53.688	54.052
27	53.676	54.223	54.695	55.108	55.476
28	55.071	55.625	56.102	56.520	56.892
29	56.460	57.020	57.502	57.925	58.301
30	57.843	58.409	58.896	59.323	59.703
35	64.664	65.259	65.771	66.220	66.619
40	71.360	71.982	72.516	72.985	73.402
45	77.952	78.599	79.156	79.643	80.077
50	84.458	85.129	85.706	86.211	86.661
55	90.890	91.584	92.181	92.703	93.168
60	97.259	97.974	98.589	99.128	99.607
70	109.834	110.591	111.241	111.811	112.317
80	122.231	123.027	123.710	124.308	124.839
90	134.483	135.314	136.028	136.653	137.208
100	146.612	147.478	148.221	148.871	149.449
110	158.637	159.536	160.306	160.981	161.581
120	170.573	171.502	172.299	172.997	173.617

UPPER PERCENTAGE POINTS / OBERE PROZENTPUNKTE

$\chi^2(\alpha/\tau,\nu)$ α=0.010

ν \ τ	11	12	13	14	15
1	11.004	11.165	11.314	11.452	11.580
2	14.006	14.180	14.340	14.488	14.626
3	16.468	16.652	16.821	16.978	17.123
4	18.678	18.871	19.048	19.211	19.364
5	20.735	20.935	21.119	21.289	21.447
6	22.685	22.892	23.083	23.259	23.422
7	24.556	24.770	24.967	25.148	25.317
8	26.366	26.586	26.788	26.974	27.148
9	28.125	28.351	28.558	28.749	28.927
10	29.842	30.073	30.285	30.481	30.664
11	31.524	31.760	31.977	32.177	32.364
12	33.175	33.416	33.638	33.842	34.032
13	34.799	35.045	35.271	35.480	35.673
14	36.399	36.650	36.880	37.093	37.290
15	37.978	38.233	38.468	38.684	38.885
16	39.538	39.798	40.036	40.256	40.460
17	41.080	41.344	41.587	41.810	42.018
18	42.607	42.875	43.121	43.348	43.559
19	44.119	44.392	44.641	44.872	45.086
20	45.618	45.894	46.148	46.382	46.599
21	47.105	47.385	47.642	47.879	48.099
22	48.580	48.864	49.124	49.364	49.587
23	50.044	50.332	50.595	50.839	51.064
24	51.499	51.790	52.056	52.303	52.532
25	52.944	53.238	53.508	53.757	53.989
26	54.380	54.678	54.951	55.203	55.437
27	55.807	56.109	56.385	56.640	56.877
28	57.227	57.532	57.812	58.070	58.309
29	58.640	58.948	59.230	59.491	59.733
30	60.045	60.357	60.642	60.905	61.150
35	66.978	67.305	67.604	67.881	68.137
40	73.777	74.118	74.431	74.720	74.987
45	80.467	80.822	81.147	81.447	81.725
50	87.065	87.433	87.770	88.080	88.369
55	93.586	93.965	94.313	94.634	94.932
60	100.038	100.430	100.789	101.119	101.426
70	112.772	113.186	113.565	113.914	114.238
80	125.317	125.752	126.149	126.516	126.856
90	137.708	138.162	138.577	138.960	139.315
100	149.969	150.441	150.873	151.272	151.641
110	162.120	162.609	163.057	163.471	163.854
120	174.175	174.681	175.145	175.572	175.968

UPPER PERCENTAGE POINTS / OBERE PROZENTPUNKTE

$\chi^2(\alpha/\tau,\nu)$ α=0.010

ν \ τ	16	17	18	19	20
1	11.700	11.813	11.919	12.020	12.116
2	14.756	14.877	14.991	15.099	15.202
3	17.259	17.387	17.508	17.622	17.730
4	19.506	19.640	19.765	19.884	19.997
5	21.595	21.734	21.865	21.988	22.105
6	23.575	23.719	23.854	23.982	24.103
7	25.474	25.622	25.761	25.893	26.018
8	27.310	27.462	27.605	27.740	27.868
9	29.093	29.249	29.396	29.534	29.666
10	30.834	30.993	31.143	31.285	31.420
11	32.538	32.701	32.854	32.999	33.137
12	34.210	34.376	34.533	34.681	34.821
13	35.854	36.024	36.184	36.335	36.478
14	37.475	37.647	37.810	37.964	38.109
15	39.073	39.249	39.414	39.571	39.719
16	40.651	40.830	40.999	41.158	41.308
17	42.212	42.394	42.565	42.726	42.879
18	43.756	43.941	44.115	44.278	44.434
19	45.286	45.473	45.649	45.816	45.973
20	46.801	46.991	47.170	47.339	47.498
21	48.304	48.497	48.678	48.849	49.011
22	49.795	49.990	50.174	50.347	50.511
23	51.275	51.473	51.659	51.834	52.000
24	52.745	52.945	53.133	53.311	53.479
25	54.205	54.407	54.598	54.777	54.947
26	55.656	55.860	56.053	56.235	56.407
27	57.098	57.305	57.500	57.684	57.858
28	58.532	58.741	58.938	59.124	59.300
29	59.959	60.170	60.369	60.557	60.735
30	61.378	61.592	61.793	61.982	62.162
35	68.376	68.601	68.811	69.010	69.199
40	75.237	75.471	75.691	75.898	76.095
45	81.984	82.227	82.456	82.672	82.876
50	88.637	88.889	89.126	89.349	89.561
55	95.210	95.470	95.714	95.945	96.163
60	101.712	101.980	102.232	102.470	102.695
70	114.541	114.823	115.089	115.340	115.578
80	127.173	127.470	127.749	128.012	128.261
90	139.646	139.956	140.248	140.522	140.782
100	151.986	152.308	152.611	152.897	153.167
110	164.211	164.545	164.859	165.155	165.435
120	176.337	176.682	177.007	177.313	177.603

UPPER PERCENTAGE POINTS / OBERE PROZENTPUNKTE

$\chi^2(\alpha/\tau,\nu)$ α=0.010

ν \ τ	22	24	26	28	30
1	12.293	12.456	12.605	12.744	12.873
2	15.392	15.566	15.727	15.875	16.013
3	17.931	18.114	18.282	18.438	18.583
4	20.207	20.398	20.574	20.736	20.888
5	22.323	22.521	22.703	22.872	23.029
6	24.328	24.532	24.721	24.895	25.056
7	26.249	26.460	26.654	26.833	27.000
8	28.106	28.322	28.521	28.705	28.876
9	29.910	30.132	30.335	30.524	30.699
10	31.669	31.896	32.105	32.298	32.477
11	33.391	33.624	33.837	34.034	34.217
12	35.081	35.318	35.536	35.736	35.923
13	36.743	36.984	37.206	37.411	37.601
14	38.379	38.625	38.851	39.059	39.253
15	39.993	40.243	40.473	40.685	40.882
16	41.587	41.841	42.075	42.290	42.490
17	43.163	43.421	43.658	43.877	44.080
18	44.722	44.984	45.224	45.446	45.653
19	46.265	46.531	46.775	47.000	47.210
20	47.795	48.064	48.311	48.540	48.752
21	49.311	49.584	49.835	50.066	50.281
22	50.815	51.092	51.346	51.580	51.798
23	52.308	52.588	52.845	53.083	53.303
24	53.790	54.074	54.334	54.574	54.797
25	55.263	55.550	55.813	56.056	56.282
26	56.726	57.016	57.282	57.528	57.756
27	58.180	58.473	58.743	58.991	59.222
28	59.626	59.923	60.194	60.446	60.679
29	61.064	61.364	61.638	61.892	62.128
30	62.495	62.797	63.075	63.331	63.569
35	69.547	69.865	70.156	70.424	70.673
40	76.458	76.789	77.093	77.373	77.633
45	83.254	83.597	83.913	84.203	84.473
50	89.952	90.308	90.634	90.935	91.214
55	96.567	96.935	97.272	97.583	97.871
60	103.111	103.490	103.837	104.157	104.454
70	116.017	116.417	116.783	117.120	117.434
80	128.722	129.141	129.525	129.879	130.208
90	141.264	141.701	142.102	142.471	142.814
100	153.667	154.122	154.539	154.923	155.279
110	165.954	166.425	166.857	167.255	167.624
120	178.139	178.626	179.072	179.484	179.865

UPPER PERCENTAGE POINTS / OBERE PROZENTPUNKTE

$\chi^2(\alpha/\tau,\nu)$ α=0.005

ν \ τ	1	2	3	4	5
1	7.879	9.141	9.885	10.415	10.828
2	10.597	11.983	12.794	13.369	13.816
3	12.838	14.320	15.183	15.794	16.266
4	14.860	16.424	17.331	17.972	18.467
5	16.750	18.386	19.332	20.000	20.515
6	18.548	20.249	21.232	21.924	22.458
7	20.278	22.040	23.056	23.771	24.322
8	21.955	23.774	24.821	25.557	26.124
9	23.589	25.462	26.539	27.295	27.877
10	25.188	27.112	28.216	28.991	29.588
11	26.757	28.729	29.860	30.654	31.264
12	28.300	30.318	31.475	32.286	32.909
13	29.819	31.883	33.064	33.892	34.528
14	31.319	33.426	34.631	35.475	36.123
15	32.801	34.950	36.177	37.037	37.697
16	34.267	36.456	37.706	38.580	39.252
17	35.718	37.946	39.217	40.107	40.790
18	37.156	39.422	40.714	41.618	42.312
19	38.582	40.885	42.198	43.115	43.820
20	39.997	42.336	43.668	44.600	45.315
21	41.401	43.775	45.127	46.072	46.797
22	42.796	45.204	46.575	47.533	48.268
23	44.181	46.623	48.013	48.984	49.728
24	45.559	48.034	49.441	50.424	51.179
25	46.928	49.435	50.861	51.856	52.620
26	48.290	50.829	52.272	53.279	54.052
27	49.645	52.215	53.676	54.695	55.476
28	50.993	53.594	55.071	56.102	56.892
29	52.336	54.967	56.460	57.502	58.301
30	53.672	56.332	57.843	58.896	59.703
35	60.275	63.076	64.664	65.771	66.619
40	66.766	69.699	71.360	72.516	73.402
45	73.166	76.223	77.952	79.156	80.077
50	79.490	82.664	84.458	85.706	86.661
55	85.749	89.035	90.890	92.181	93.168
60	91.952	95.344	97.259	98.589	99.607
70	104.215	107.808	109.834	111.241	112.317
80	116.321	120.102	122.231	123.710	124.839
90	128.299	132.256	134.483	136.028	137.208
100	140.169	144.293	146.612	148.221	149.449
110	151.948	156.230	158.637	160.306	161.581
120	163.648	168.082	170.573	172.299	173.617

UPPER PERCENTAGE POINTS / OBERE PROZENTPUNKTE

$\chi^2(\alpha/\tau, \nu)$ $\alpha=0.005$

ν \ τ	6	7	8	9	10
1	11.165	11.452	11.700	11.919	12.116
2	14.180	14.488	14.756	14.991	15.202
3	16.652	16.978	17.259	17.508	17.730
4	18.871	19.211	19.506	19.765	19.997
5	20.935	21.289	21.595	21.865	22.105
6	22.892	23.259	23.575	23.854	24.103
7	24.770	25.148	25.474	25.761	26.018
8	26.586	26.974	27.310	27.605	27.868
9	28.351	28.749	29.093	29.396	29.666
10	30.073	30.481	30.834	31.143	31.420
11	31.760	32.177	32.538	32.854	33.137
12	33.416	33.842	34.210	34.533	34.821
13	35.045	35.480	35.854	36.184	36.478
14	36.650	37.093	37.475	37.810	38.109
15	38.233	38.684	39.073	39.414	39.719
16	39.798	40.256	40.651	40.999	41.308
17	41.344	41.810	42.212	42.565	42.879
18	42.875	43.348	43.756	44.115	44.434
19	44.392	44.872	45.286	45.649	45.973
20	45.894	46.382	46.801	47.170	47.498
21	47.385	47.879	48.304	48.678	49.011
22	48.864	49.364	49.795	50.174	50.511
23	50.332	50.839	51.275	51.659	52.000
24	51.790	52.303	52.745	53.133	53.479
25	53.238	53.757	54.205	54.598	54.947
26	54.678	55.203	55.656	56.053	56.407
27	56.109	56.640	57.098	57.500	57.858
28	57.532	58.070	58.532	58.938	59.300
29	58.948	59.491	59.959	60.369	60.735
30	60.357	60.905	61.378	61.793	62.162
35	67.305	67.881	68.376	68.811	69.199
40	74.118	74.720	75.237	75.691	76.095
45	80.822	81.447	81.984	82.456	82.876
50	87.433	88.080	88.637	89.126	89.561
55	93.965	94.634	95.210	95.714	96.163
60	100.430	101.119	101.712	102.232	102.695
70	113.186	113.914	114.541	115.089	115.578
80	125.752	126.516	127.173	127.749	128.261
90	138.162	138.960	139.646	140.248	140.782
100	150.441	151.272	151.986	152.611	153.167
110	162.609	163.471	164.211	164.859	165.435
120	174.681	175.572	176.337	177.007	177.603

UPPER PERCENTAGE POINTS / OBERE PROZENTPUNKTE

$\chi^2(\alpha/\tau,\nu)$ α=0.005

ν \ τ	11	12	13	14	15
1	12.293	12.456	12.605	12.744	12.873
2	15.392	15.566	15.727	15.875	16.013
3	17.931	18.114	18.282	18.438	18.583
4	20.207	20.398	20.574	20.736	20.888
5	22.323	22.521	22.703	22.872	23.029
6	24.328	24.532	24.721	24.895	25.056
7	26.249	26.460	26.654	26.833	27.000
8	28.106	28.322	28.521	28.705	28.876
9	29.910	30.132	30.335	30.524	30.699
10	31.669	31.896	32.105	32.298	32.477
11	33.391	33.624	33.837	34.034	34.217
12	35.081	35.318	35.536	35.736	35.923
13	36.743	36.984	37.206	37.411	37.601
14	38.379	38.625	38.851	39.059	39.253
15	39.993	40.243	40.473	40.685	40.882
16	41.587	41.841	42.075	42.290	42.490
17	43.163	43.421	43.658	43.877	44.080
18	44.722	44.984	45.224	45.446	45.653
19	46.265	46.531	46.775	47.000	47.210
20	47.795	48.064	48.311	48.540	48.752
21	49.311	49.584	49.835	50.066	50.281
22	50.815	51.092	51.346	51.580	51.798
23	52.308	52.588	52.845	53.083	53.303
24	53.790	54.074	54.334	54.574	54.797
25	55.263	55.550	55.813	56.056	56.282
26	56.726	57.016	57.282	57.528	57.756
27	58.180	58.473	58.743	58.991	59.222
28	59.626	59.923	60.194	60.446	60.679
29	61.064	61.364	61.638	61.892	62.128
30	62.495	62.797	63.075	63.331	63.569
35	69.547	69.865	70.156	70.424	70.673
40	76.458	76.789	77.093	77.373	77.633
45	83.254	83.597	83.913	84.203	84.473
50	89.952	90.308	90.634	90.935	91.214
55	96.567	96.935	97.272	97.583	97.871
60	103.111	103.490	103.837	104.157	104.454
70	116.017	116.417	116.783	117.120	117.434
80	128.722	129.141	129.525	129.879	130.208
90	141.264	141.701	142.102	142.471	142.814
100	153.667	154.122	154.539	154.923	155.279
110	165.954	166.425	166.857	167.255	167.624
120	178.139	178.626	179.072	179.484	179.865

Tafel 22 (Forts.) Table 22 (cont.)

UPPER PERCENTAGE POINTS / OBERE PROZENTPUNKTE

$\chi^2(\alpha/\tau,\nu)$ α=0.005

ν \ τ	16	17	18	19	20
1	12.994	13.107	13.215	13.316	13.412
2	16.142	16.263	16.377	16.486	16.588
3	18.719	18.847	18.967	19.080	19.188
4	21.029	21.162	21.287	21.405	21.517
5	23.175	23.313	23.442	23.565	23.681
6	25.208	25.350	25.483	25.610	25.730
7	27.155	27.301	27.439	27.569	27.692
8	29.036	29.186	29.327	29.461	29.587
9	30.863	31.016	31.161	31.298	31.427
10	32.644	32.801	32.949	33.089	33.221
11	34.388	34.548	34.699	34.842	34.977
12	36.098	36.261	36.415	36.561	36.698
13	37.778	37.945	38.102	38.250	38.390
14	39.434	39.603	39.763	39.914	40.056
15	41.066	41.238	41.401	41.554	41.699
16	42.677	42.852	43.017	43.173	43.321
17	44.270	44.448	44.615	44.773	44.923
18	45.845	46.026	46.196	46.356	46.508
19	47.405	47.588	47.760	47.923	48.077
20	48.950	49.136	49.310	49.475	49.632
21	50.482	50.670	50.847	51.014	51.173
22	52.001	52.192	52.371	52.540	52.701
23	53.509	53.702	53.883	54.055	54.217
24	55.006	55.201	55.384	55.558	55.722
25	56.492	56.690	56.875	57.051	57.217
26	57.969	58.169	58.357	58.534	58.702
27	59.437	59.639	59.829	60.008	60.178
28	60.896	61.100	61.292	61.473	61.645
29	62.348	62.554	62.748	62.931	63.104
30	63.791	63.999	64.195	64.380	64.555
35	70.906	71.124	71.329	71.523	71.706
40	77.875	78.102	78.316	78.518	78.709
45	84.725	84.961	85.183	85.392	85.591
50	91.475	91.719	91.949	92.166	92.371
55	98.140	98.392	98.629	98.853	99.064
60	104.731	104.991	105.235	105.465	105.683
70	117.726	118.000	118.257	118.500	118.730
80	130.514	130.801	131.071	131.326	131.567
90	143.134	143.433	143.715	143.980	144.231
100	155.611	155.923	156.215	156.491	156.752
110	167.969	168.291	168.594	168.880	169.151
120	180.221	180.554	180.867	181.163	181.442

UPPER PERCENTAGE POINTS / OBERE PROZENTPUNKTE

$\chi^2(\alpha/\tau,\nu)$ $\alpha=0.005$

ν \ τ	22	24	26	28	30
1	13.591	13.754	13.905	14.044	14.174
2	16.779	16.953	17.113	17.261	17.399
3	19.388	19.570	19.738	19.894	20.038
4	21.726	21.916	22.090	22.252	22.402
5	23.897	24.093	24.274	24.441	24.597
6	25.952	26.155	26.341	26.513	26.674
7	27.921	28.129	28.321	28.498	28.663
8	29.822	30.036	30.232	30.414	30.583
9	31.667	31.886	32.088	32.273	32.446
10	33.467	33.691	33.896	34.086	34.263
11	35.227	35.456	35.666	35.860	36.040
12	36.954	37.187	37.401	37.598	37.782
13	38.651	38.888	39.106	39.307	39.494
14	40.321	40.563	40.784	40.989	41.179
15	41.969	42.214	42.439	42.647	42.841
16	43.594	43.844	44.073	44.284	44.480
17	45.201	45.454	45.686	45.901	46.101
18	46.790	47.047	47.283	47.500	47.703
19	48.363	48.624	48.863	49.083	49.288
20	49.922	50.185	50.428	50.651	50.859
21	51.466	51.733	51.979	52.205	52.416
22	52.998	53.269	53.517	53.746	53.959
23	54.518	54.792	55.043	55.275	55.491
24	56.026	56.304	56.558	56.793	57.011
25	57.525	57.805	58.062	58.300	58.520
26	59.013	59.297	59.557	59.797	60.020
27	60.492	60.779	61.042	61.284	61.510
28	61.963	62.252	62.518	62.763	62.991
29	63.425	63.718	63.986	64.233	64.464
30	64.879	65.175	65.446	65.696	65.928
35	72.046	72.355	72.638	72.900	73.143
40	79.063	79.385	79.680	79.953	80.206
45	85.958	86.292	86.599	86.882	87.144
50	92.751	93.097	93.414	93.707	93.978
55	99.457	99.814	100.141	100.443	100.723
60	106.087	106.454	106.791	107.102	107.390
70	119.156	119.543	119.898	120.225	120.529
80	132.013	132.419	132.790	133.134	133.452
90	144.697	145.120	145.508	145.866	146.198
100	157.236	157.676	158.079	158.451	158.796
110	169.652	170.107	170.524	170.910	171.267
120	181.960	182.430	182.862	183.259	183.628

TABLE 23

Lower Percentage Points of the Bonferroni Chi-Square Statistic: Tables of G. B. Beus and D. R. Jensen

(a) Content of the tables and definition of the test statistic :

Let $X_1, X_2, \ldots, X_\tau$ be random variables (variates) of a common distribution of a type for which the marginal distributions of the X_i are central chi-square distributions with ν_i degrees of freedom ($i = 1, 2, \ldots, \tau$). The common distribution of the X_i needs not be known; also the ν_i need not all be equal.

The tables contain the lower percentage points $L(\nu; \alpha; \tau)$ of the BONFERRONI-chi-square-statistic $\chi^2(\alpha/\tau;\nu)$, defined by

$$P\{U(\nu_i; \alpha;\tau) < X_i\} = \alpha/\tau \quad .$$

BONFERRONI chi-square statistics are suited for satisfying a BONFERRONI inequality. (See section (e)).

(b) Extent of the tables and definition of the parameters :

(1) The parameter α :

α = error probability

= probability of an error of the 1-st kind

α = 0.050 ; 0.025 ; 0.010 ; 0.005 .

(2) The parameter τ :

τ = number of variates X_i

τ = 1(1)20(2)30 .

(3) The parameter ν :

ν_i = degrees of freedom of the variate X_i , distributed after chi-square ,

ν_i = 1(1)30(5)60(10)120 .

(c) Suggestions for applications :

The tables serve to determine simultaneous confidence intervals for variates of a multivariate normal distribution on the basis of samples with missing values.

Here : The determination of lower confidence limits.
For upper confidence limits : See table 22 .

(d) References :

For the source of the tables and for applications :

BEUS, G. B. + JENSEN, D. R. : Percentage points of the BONFERRONI chi-square statistics. (Table 2). (Technical Report No. 3). Blacksburg / Virginia: Department of Statistics, Virginia Polytechnic Institute and State University,September 1967.

(e) Further remarks :

(1) For $\tau = 1$, the tables contain the lower percentage points of the usual chi-square distribution. The same holds for $\tau > 1$, if one chooses $\alpha' = \alpha/\tau$ and reads the column for τ .

(2) The BONFERRONI inequality :
Events are denoted by $E_1, E_2, \ldots, E_\tau$. The corresponding complements are denoted by $E'_1, E'_2, \ldots, E'_\tau$.

The BONFERRONI inequality used here reads :

$$\Pr\left\{ \bigcap_{i=1}^{\tau} E_i \right\} \geqq 1 - \sum_{i=1}^{\tau} \Pr\{ E'_i \} \quad .$$

This form is used to construct simultaneous confidence intervals.

One can also write the above BONFERRONI inequality in the form

$$\Pr\left\{ \bigcup_{i=1}^{\tau} E'_i \right\} \leqq \sum_{i=1}^{\tau} \Pr\{ E'_i \}$$

In this form it serves for simultaneous hypothesis testing.

Tafel 23

Table 23

LOWER PERCENTAGE POINTS / UNTERE PROZENTPUNKTE

$\chi^2\ (\alpha/\tau,\nu)$ $\alpha=0.050$

ν \ τ	1	2	3	4	5
1	$.0^2 3932$	$.0^3 9821$	$.0^3 4364$	$.0^3 2455$	$.0^3 1571$
2	.1026	.05064	.03361	.02516	.02010
3	.3518	.2158	.1630	.1338	.1148
4	.7107	.4844	.3894	.3342	.2971
5	1.145	.8312	.6933	.6109	.5543
6	1.635	1.237	1.058	.9482	.8721
7	2.167	1.690	1.470	1.334	1.239
8	2.733	2.180	1.921	1.760	1.646
9	3.325	2.700	2.405	2.220	2.088
10	3.940	3.247	2.916	2.707	2.558
11	4.575	3.816	3.451	3.219	3.053
12	5.226	4.404	4.006	3.752	3.571
13	5.892	5.009	4.579	4.305	4.107
14	6.571	5.629	5.168	4.873	4.660
15	7.261	6.262	5.772	5.457	5.229
16	7.962	6.908	6.388	6.054	5.812
17	8.672	7.564	7.017	6.664	6.408
18	9.390	8.231	7.656	7.284	7.015
19	10.117	8.907	8.305	7.916	7.633
20	10.851	9.591	8.963	8.556	8.260
21	11.591	10.283	9.629	9.206	8.897
22	12.338	10.982	10.304	9.864	9.542
23	13.091	11.689	10.986	10.529	10.196
24	13.848	12.401	11.674	11.202	10.856
25	14.611	13.120	12.369	11.881	11.524
26	15.379	13.844	13.071	12.567	12.198
27	16.151	14.573	13.777	13.258	12.879
28	16.928	15.308	14.490	13.956	13.565
29	17.708	16.047	15.207	14.658	14.256
30	18.493	16.791	15.929	15.366	14.953
35	22.465	20.569	19.605	18.973	18.509
40	26.509	24.433	23.373	22.677	22.164
45	30.612	28.366	27.216	26.459	25.901
50	34.764	32.357	31.122	30.307	29.707
55	38.958	36.398	35.081	34.212	33.570
60	43.188	40.482	39.087	38.166	37.485
70	51.739	48.758	47.216	46.196	45.442
80	60.391	57.153	55.475	54.363	53.540
90	69.126	65.647	63.841	62.642	61.754
100	77.929	74.222	72.294	71.014	70.065
110	86.792	82.867	80.824	79.466	78.458
120	95.705	91.573	89.419	87.987	86.923

Tafel 23 (Forts.) Table 23 (cont.)

LOWER PERCENTAGE POINTS / UNTERE PROZENTPUNKTE

$\chi^2\ (\alpha/\tau,\nu)$ $\alpha=0.050$

ν \ τ	6	7	8	9	10
1	$.0^31091$	$.0^48014$	$.0^46136$	$.0^44848$	$.0^43927$
2	.01674	.01434	.01254	.01114	.01003
3	.1014	.09133	.08342	.07702	.07172
4	.2700	.2491	.2324	.2186	.2070
5	.5123	.4794	.4528	.4306	.4117
6	.8150	.7699	.7330	.7022	.6757
7	1.167	1.110	1.063	1.023	.9893
8	1.560	1.491	1.434	1.386	1.344
9	1.987	1.907	1.840	1.784	1.735
10	2.444	2.352	2.276	2.212	2.156
11	2.926	2.824	2.738	2.666	2.603
12	3.430	3.317	3.223	3.143	3.074
13	3.954	3.831	3.729	3.641	3.565
14	4.496	4.363	4.252	4.157	4.075
15	5.053	4.911	4.791	4.690	4.601
16	5.625	5.473	5.346	5.237	5.142
17	6.209	6.048	5.913	5.798	5.697
18	6.805	6.635	6.493	6.371	6.265
19	7.413	7.234	7.084	6.956	6.844
20	8.030	7.843	7.686	7.551	7.434
21	8.657	8.461	8.297	8.157	8.034
22	9.292	9.089	8.918	8.771	8.643
23	9.936	9.724	9.546	9.394	9.260
24	10.587	10.367	10.183	10.025	9.886
25	11.245	11.018	10.827	10.663	10.520
26	11.910	11.676	11.478	11.309	11.160
27	12.582	12.339	12.136	11.961	11.808
28	13.259	13.010	12.800	12.619	12.461
29	13.942	13.685	13.469	13.284	13.121
30	14.631	14.367	14.145	13.954	13.787
35	18.145	17.848	17.597	17.381	17.192
40	21.762	21.433	21.156	20.916	20.707
45	25.463	25.104	24.801	24.540	24.311
50	29.234	28.847	28.520	28.238	27.991
55	33.066	32.652	32.302	32.000	31.735
60	36.949	36.509	36.137	35.816	35.534
70	44.847	44.359	43.946	43.589	43.275
80	52.891	52.357	51.906	51.515	51.172
90	61.053	60.477	59.989	59.567	59.196
100	69.315	68.699	68.177	67.725	67.328
110	77.662	77.007	76.453	75.972	75.550
120	86.083	85.391	84.805	84.298	83.852

afel 23 (Forts.) Table 23 (cont.)

LOWER PERCENTAGE POINTS / UNTERE PROZENTPUNKTE

$\chi^2\ (\alpha/\tau, \nu)$ $\alpha = 0.050$

ν \ τ	11	12	13	14	15
1	$.0^4 3245$	$.0^4 2727$	$.0^4 2324$	$.0^4 2004$	$.0^4 1745$
2	$.0^2 9112$	$.0^2 8351$	$.0^2 7707$	$.0^2 7156$	$.0^2 6678$
3	.06725	.06341	.06007	.05714	.05455
4	.1970	.1884	.1808	.1740	.1679
5	.3954	.3811	.3685	.3571	.3469
6	.6528	.6325	.6145	.5983	.5837
7	.9596	.9333	.9099	.8888	.8696
8	1.308	1.276	1.247	1.221	1.197
9	1.692	1.654	1.620	1.589	1.561
10	2.107	2.063	2.024	1.988	1.956
11	2.548	2.499	2.455	2.414	2.378
12	3.013	2.958	2.909	2.864	2.824
13	3.498	3.438	3.384	3.335	3.291
14	4.002	3.937	3.878	3.825	3.777
15	4.523	4.453	4.390	4.332	4.280
16	5.059	4.984	4.916	4.855	4.799
17	5.608	5.529	5.457	5.392	5.332
18	6.171	6.087	6.011	5.941	5.878
19	6.745	6.656	6.576	6.503	6.436
20	7.330	7.237	7.153	7.076	7.006
21	7.925	7.827	7.739	7.659	7.585
22	8.529	8.427	8.335	8.252	8.175
23	9.142	9.036	8.941	8.853	8.773
24	9.764	9.654	9.554	9.463	9.380
25	10.392	10.278	10.175	10.081	9.995
26	11.029	10.911	10.804	10.707	10.617
27	11.672	11.550	11.440	11.339	11.246
28	12.321	12.196	12.082	11.978	11.882
29	12.977	12.847	12.730	12.623	12.525
30	13.638	13.505	13.385	13.274	13.173
35	17.024	16.873	16.736	16.611	16.496
40	20.520	20.353	20.201	20.062	19.934
45	24.107	23.924	23.758	23.606	23.466
50	27.771	27.572	27.393	27.228	27.077
55	31.499	31.287	31.094	30.918	30.756
60	35.284	35.058	34.853	34.665	34.492
70	42.996	42.744	42.516	42.307	42.115
80	50.866	50.591	50.341	50.112	49.901
90	58.866	58.568	58.297	58.049	57.821
100	66.973	66.654	66.364	66.099	65.854
110	75.173	74.834	74.526	74.243	73.982
120	83.454	83.095	82.769	82.470	82.194

LOWER PERCENTAGE POINTS / UNTERE PROZENTPUNKTE

$\chi^2\ (\alpha/\tau,\nu)$ $\alpha = 0.050$

ν \ τ	16	17	18	19	20
1	$.0^4 1534$	$.0^4 1359$	$.0^4 1212$	$.0^4 1088$	$.0^5 9818$
2	$.0^2 6260$	$.0^2 5891$	$.0^2 5563$	$.0^2 5270$	$.0^2 5006$
3	.05223	.05014	.04824	.04652	.04494
4	.1624	.1575	.1529	.1487	.1449
5	.3376	.3291	.3213	.3142	.3075
6	.5703	.5580	.5467	.5363	.5266
7	.8521	.8360	.8211	.8073	.7945
8	1.176	1.156	1.137	1.120	1.104
9	1.535	1.512	1.490	1.469	1.450
10	1.926	1.899	1.873	1.850	1.827
11	2.344	2.313	2.284	2.257	2.232
12	2.786	2.751	2.719	2.689	2.661
13	3.249	3.211	3.176	3.143	3.112
14	3.732	3.690	3.652	3.616	3.582
15	4.232	4.187	4.145	4.106	4.070
16	4.747	4.699	4.654	4.613	4.573
17	5.277	5.226	5.178	5.133	5.092
18	5.820	5.765	5.715	5.668	5.623
19	6.375	6.317	6.264	6.214	6.167
20	6.941	6.881	6.824	6.772	6.723
21	7.517	7.454	7.396	7.341	7.289
22	8.104	8.038	7.976	7.919	7.865
23	8.699	8.630	8.566	8.506	8.450
24	9.303	9.232	9.165	9.103	9.044
25	9.915	9.841	9.772	9.707	9.646
26	10.534	10.458	10.386	10.319	10.256
27	11.161	11.082	11.008	10.938	10.873
28	11.794	11.712	11.636	11.565	11.497
29	12.434	12.350	12.271	12.197	12.128
30	13.080	12.993	12.912	12.836	12.765
35	16.390	16.291	16.199	16.113	16.032
40	19.816	19.706	19.604	19.508	19.417
45	23.337	23.216	23.104	22.999	22.900
50	26.937	26.807	26.685	26.571	26.464
55	30.605	30.466	30.335	30.212	30.097
60	34.332	34.184	34.045	33.914	33.791
70	41.936	41.770	41.615	41.469	41.332
80	49.705	49.523	49.353	49.193	49.043
90	57.609	57.412	57.228	57.055	56.892
100	65.627	65.416	65.218	65.032	64.857
110	73.741	73.516	73.306	73.108	72.922
120	81.939	81.701	81.478	81.269	81.072

LOWER PERCENTAGE POINTS / UNTERE PROZENTPUNKTE

χ^2 $(\alpha/\tau,\nu)$ $\alpha=0.050$

ν \ τ	22	24	26	28	30
1	$.0^5 8114$	$.0^5 6818$	$.0^5 5809$	$.0^5 5009$	$.0^5 4363$
2	$.0^2 4551$	$.0^2 4171$	$.0^2 3850$	$.0^2 3575$	$.0^2 3336$
3	.04215	.03976	.03768	.03585	.03422
4	.1380	.1320	.1267	.1220	.1178
5	.2955	.2849	.2756	.2672	.2597
6	.5090	.4935	.4797	.4673	.4560
7	.7712	.7505	.7321	.7154	.7003
8	1.075	1.049	1.026	1.005	.9863
9	1.415	1.384	1.357	1.332	1.309
10	1.787	1.751	1.719	1.689	1.663
11	2.186	2.145	2.109	2.075	2.044
12	2.610	2.564	2.523	2.485	2.451
13	3.055	3.005	2.959	2.918	2.880
14	3.520	3.465	3.416	3.370	3.329
15	4.003	3.943	3.890	3.841	3.796
16	4.502	4.438	4.380	4.327	4.279
17	5.015	4.947	4.885	4.829	4.777
18	5.542	5.470	5.404	5.344	5.289
19	6.082	6.005	5.936	5.872	5.814
20	6.633	6.552	6.479	6.412	6.351
21	7.194	7.110	7.033	6.963	6.898
22	7.766	7.677	7.597	7.524	7.456
23	8.347	8.255	8.171	8.094	8.024
24	8.937	8.841	8.753	8.674	8.600
25	9.535	9.435	9.344	9.261	9.185
26	10.141	10.037	9.943	9.857	9.778
27	10.754	10.647	10.549	10.461	10.379
28	11.374	11.263	11.163	11.071	10.987
29	12.001	11.886	11.783	11.688	11.601
30	12.634	12.516	12.409	12.312	12.222
35	15.883	15.749	15.627	15.516	15.413
40	19.251	19.102	18.966	18.842	18.727
45	22.717	22.553	22.404	22.268	22.143
50	26.266	26.088	25.927	25.779	25.643
55	29.885	29.694	29.521	29.362	29.215
60	33.565	33.362	33.177	33.007	32.851
70	41.080	40.853	40.646	40.457	40.282
80	48.766	48.517	48.290	48.081	47.890
90	56.592	56.322	56.076	55.850	55.642
100	64.536	64.245	63.981	63.739	63.515
110	72.579	72.270	71.989	71.731	71.492
120	80.710	80.382	80.085	79.811	79.559

Tafel 23 (Forts.) Table 23 (cont.)

LOWER PERCENTAGE POINTS / UNTERE PROZENTPUNKTE

$\chi^2\ (\alpha/\tau,\nu)$ $\alpha=0.025$

ν \ τ	1	2	3	4	5
1	$.0^3 9821$	$.0^3 2455$	$.0^3 1091$	$.0^4 6136$	$.0^4 3927$
2	.05064	.02516	.01674	.01254	.01003
3	.2158	.1338	.1014	.08342	.07172
4	.4844	.3342	.2700	.2324	.2070
5	.8312	.6109	.5123	.4528	.4117
6	1.237	.9482	.8150	.7330	.6757
7	1.690	1.334	1.167	1.063	.9893
8	2.180	1.760	1.560	1.434	1.344
9	2.700	2.220	1.987	1.840	1.735
10	3.247	2.707	2.444	2.276	2.156
11	3.816	3.219	2.926	2.738	2.603
12	4.404	3.752	3.430	3.223	3.074
13	5.009	4.305	3.954	3.729	3.565
14	5.629	4.873	4.496	4.252	4.075
15	6.262	5.457	5.053	4.791	4.601
16	6.908	6.054	5.625	5.346	5.142
17	7.564	6.664	6.209	5.913	5.697
18	8.231	7.284	6.805	6.493	6.265
19	8.907	7.916	7.413	7.084	6.844
20	9.591	8.556	8.030	7.686	7.434
21	10.283	9.206	8.657	8.297	8.034
22	10.982	9.864	9.292	8.918	8.643
23	11.689	10.529	9.936	9.546	9.260
24	12.401	11.202	10.587	10.183	9.886
25	13.120	11.881	11.245	10.827	10.520
26	13.844	12.567	11.910	11.478	11.160
27	14.573	13.258	12.582	12.136	11.808
28	15.308	13.956	13.259	12.800	12.461
29	16.047	14.658	13.942	13.469	13.121
30	16.791	15.366	14.631	14.145	13.787
35	20.569	18.973	18.145	17.597	17.192
40	24.433	22.677	21.762	21.156	20.707
45	28.366	26.459	25.463	24.801	24.311
50	32.357	30.307	29.234	28.520	27.991
55	36.398	34.212	33.066	32.302	31.735
60	40.482	38.166	36.949	36.137	35.534
70	48.758	46.196	44.847	43.946	43.275
80	57.153	54.363	52.891	51.906	51.172
90	65.647	62.642	61.053	59.989	59.196
100	74.222	71.014	69.315	68.177	67.328
110	82.867	79.466	77.662	76.453	75.550
120	91.573	87.987	86.083	84.805	83.852

LOWER PERCENTAGE POINTS / UNTERE PROZENTPUNKTE

$\chi^2\ (\alpha/\tau,\nu)$ $\alpha=0.025$

ν \ τ	6	7	8	9	10
1	$.0^{4}2727$	$.0^{4}2004$	$.0^{4}1534$	$.0^{4}1212$	$.0^{5}9818$
2	$.0^{2}8351$	$.0^{2}7156$	$.0^{2}6260$	$.0^{2}5563$	$.0^{2}5006$
3	.06341	.05714	.05223	.04824	.04494
4	.1884	.1740	.1624	.1529	.1449
5	.3811	.3571	.3376	.3213	.3075
6	.6325	.5983	.5703	.5467	.5266
7	.9333	.8888	.8521	.8211	.7945
8	1.276	1.221	1.176	1.137	1.104
9	1.654	1.589	1.535	1.490	1.450
10	2.063	1.988	1.926	1.873	1.827
11	2.499	2.414	2.344	2.284	2.232
12	2.958	2.864	2.786	2.719	2.661
13	3.438	3.335	3.249	3.176	3.112
14	3.937	3.825	3.732	3.652	3.582
15	4.453	4.332	4.232	4.145	4.070
16	4.984	4.855	4.747	4.654	4.573
17	5.529	5.392	5.277	5.178	5.092
18	6.087	5.941	5.820	5.715	5.623
19	6.656	6.503	6.375	6.264	6.167
20	7.237	7.076	6.941	6.824	6.723
21	7.827	7.659	7.517	7.396	7.289
22	8.427	8.252	8.104	7.976	7.865
23	9.036	8.853	8.699	8.566	8.450
24	9.654	9.463	9.303	9.165	9.044
25	10.278	10.081	9.915	9.772	9.646
26	10.911	10.707	10.534	10.386	10.256
27	11.550	11.339	11.161	11.008	10.873
28	12.196	11.978	11.794	11.636	11.497
29	12.847	12.623	12.434	12.271	12.128
30	13.505	13.274	13.080	12.912	12.765
35	16.873	16.611	16.390	16.199	16.032
40	20.353	20.062	19.816	19.604	19.417
45	23.924	23.606	23.337	23.104	22.900
50	27.572	27.228	26.937	26.685	26.464
55	31.287	30.918	30.605	30.335	30.097
60	35.058	34.665	34.332	34.045	33.791
70	42.744	42.307	41.936	41.615	41.332
80	50.591	50.112	49.705	49.353	49.043
90	58.568	58.049	57.609	57.228	56.892
100	66.654	66.099	65.627	65.218	64.857
110	74.834	74.243	73.741	73.306	72.922
120	83.095	82.470	81.939	81.478	81.072

Tafel 23 (Forts.) Table 23 (cont.)

LOWER PERCENTAGE POINTS / UNTERE PROZENTPUNKTE

$\chi^2 \ (\alpha/\tau,\nu)$ $\alpha=0.025$

ν \ τ	11	12	13	14	15
1	$.0^5 8114$	$.0^5 6818$	$.0^5 5809$	$.0^5 5009$	$.0^5 4363$
2	$.0^2 4551$	$.0^2 4171$	$.0^2 3850$	$.0^2 3575$	$.0^2 3336$
3	.04215	.03976	.03768	.03585	.03422
4	.1380	.1320	.1267	.1220	.1178
5	.2955	.2849	.2756	.2672	.2597
6	.5090	.4935	.4797	.4673	.4560
7	.7712	.7505	.7321	.7154	.7003
8	1.075	1.049	1.026	1.005	.9863
9	1.415	1.384	1.357	1.332	1.309
10	1.787	1.751	1.719	1.689	1.663
11	2.186	2.145	2.109	2.075	2.044
12	2.610	2.564	2.523	2.485	2.451
13	3.055	3.005	2.959	2.918	2.880
14	3.520	3.465	3.416	3.370	3.329
15	4.003	3.943	3.890	3.841	3.796
16	4.502	4.438	4.380	4.327	4.279
17	5.015	4.947	4.885	4.829	4.777
18	5.542	5.470	5.404	5.344	5.289
19	6.082	6.005	5.936	5.872	5.814
20	6.633	6.552	6.479	6.412	6.351
21	7.194	7.110	7.033	6.963	6.898
22	7.766	7.677	7.597	7.524	7.456
23	8.347	8.255	8.171	8.094	8.024
24	8.937	8.841	8.753	8.674	8.600
25	9.535	9.435	9.344	9.261	9.185
26	10.141	10.037	9.943	9.857	9.778
27	10.754	10.647	10.549	10.461	10.379
28	11.374	11.263	11.163	11.071	10.987
29	12.001	11.886	11.783	11.688	11.601
30	12.634	12.516	12.409	12.312	12.222
35	15.883	15.749	15.627	15.516	15.413
40	19.251	19.102	18.966	18.842	18.727
45	22.717	22.553	22.404	22.268	22.143
50	26.266	26.088	25.927	25.779	25.643
55	29.885	29.694	29.521	29.362	29.215
60	33.565	33.362	33.177	33.007	32.851
70	41.080	40.853	40.646	40.457	40.282
80	48.766	48.517	48.290	48.081	47.890
90	56.592	56.322	56.076	55.850	55.642
100	64.536	64.245	63.981	63.739	63.515
110	72.579	72.270	71.989	71.731	71.492
120	80.710	80.382	80.085	79.811	79.559

Tafel 23 (Forts.) Table 23 (cont.)

LOWER PERCENTAGE POINTS / UNTERE PROZENTPUNKTE

$\chi^2\ (\alpha/\tau,\nu)$ $\alpha=0.025$

ν \ τ	16	17	18	19	20
1	$.0^{5}3835$	$.0^{5}3397$	$.0^{5}3030$	$.0^{5}2720$	$.0^{5}2454$
2	$.0^{2}3127$	$.0^{2}2943$	$.0^{2}2780$	$.0^{2}2633$	$.0^{2}2502$
3	.03277	.03147	.03028	.02920	.02822
4	.1139	.1105	.1073	.1044	.1017
5	.2528	.2465	.2408	.2354	.2305
6	.4457	.4363	.4276	.4196	.4121
7	.6865	.6737	.6620	.6511	.6409
8	.9689	.9528	.9379	.9240	.9111
9	1.288	1.268	1.250	1.233	1.218
10	1.638	1.615	1.594	1.574	1.556
11	2.016	1.990	1.966	1.944	1.923
12	2.420	2.390	2.363	2.338	2.314
13	2.845	2.813	2.783	2.755	2.728
14	3.291	3.255	3.222	3.191	3.163
15	3.754	3.716	3.680	3.647	3.615
16	4.234	4.193	4.154	4.118	4.085
17	4.729	4.685	4.644	4.605	4.569
18	5.239	5.191	5.148	5.107	5.068
19	5.760	5.711	5.664	5.621	5.580
20	6.294	6.242	6.193	6.147	6.104
21	6.839	6.784	6.732	6.684	6.639
22	7.394	7.336	7.282	7.232	7.184
23	7.959	7.898	7.842	7.789	7.740
24	8.533	8.470	8.411	8.356	8.304
25	9.115	9.050	8.989	8.931	8.878
26	9.705	9.638	9.574	9.515	9.459
27	10.303	10.233	10.168	10.106	10.048
28	10.909	10.836	10.768	10.705	10.645
29	11.521	11.446	11.376	11.310	11.248
30	12.139	12.062	11.990	11.922	11.859
35	15.318	15.230	15.148	15.070	14.998
40	18.621	18.523	18.431	18.344	18.263
45	22.026	21.918	21.817	21.722	21.632
50	25.516	25.399	25.289	25.186	25.089
55	29.080	28.953	28.835	28.724	28.620
60	32.706	32.571	32.445	32.327	32.215
70	40.120	39.969	39.828	39.695	39.570
80	47.712	47.546	47.391	47.245	47.107
90	55.449	55.269	55.101	54.942	54.793
100	63.308	63.115	62.934	62.764	62.604
110	71.272	71.066	70.873	70.692	70.521
120	79.325	79.107	78.903	78.711	78.530

Tafel 23 (Forts.) Table 23 (cont.)

LOWER PERCENTAGE POINTS / UNTERE PROZENTPUNKTE

χ^2 $(\alpha/\tau,\nu)$ α=0.025

ν \ τ	22	24	26	28	30
1	$.0^5 2028$	$.0^5 1704$	$.0^5 1452$	$.0^5 1252$	$.0^5 1091$
2	$.0^2 2274$	$.0^2 2084$	$.0^2 1924$	$.0^2 1787$	$.0^2 1667$
3	.02647	.02497	.02367	.02252	.02150
4	.09690	.09271	.08901	.08573	.08278
5	.2216	.2138	.2068	.2006	.1950
6	.3985	.3866	.3759	.3663	.3576
7	.6224	.6060	.5914	.5781	.5661
8	.8876	.8667	.8480	.8310	.8156
9	1.189	1.164	1.141	1.120	1.101
10	1.522	1.493	1.466	1.441	1.419
11	1.884	1.850	1.819	1.791	1.765
12	2.271	2.232	2.197	2.165	2.136
13	2.680	2.637	2.598	2.563	2.530
14	3.110	3.063	3.020	2.981	2.945
15	3.558	3.506	3.460	3.418	3.379
16	4.023	3.967	3.917	3.871	3.829
17	4.503	4.444	4.390	4.341	4.296
18	4.997	4.934	4.877	4.825	4.777
19	5.505	5.438	5.377	5.322	5.271
20	6.025	5.954	5.890	5.831	5.777
21	6.556	6.481	6.414	6.352	6.295
22	7.097	7.019	6.948	6.884	6.824
23	7.649	7.567	7.493	7.425	7.363
24	8.210	8.124	8.047	7.977	7.912
25	8.779	8.690	8.610	8.537	8.469
26	9.357	9.265	9.181	9.105	9.035
27	9.942	9.847	9.761	9.681	9.609
28	10.535	10.437	10.347	10.265	10.190
29	11.135	11.033	10.941	10.856	10.778
30	11.742	11.637	11.541	11.454	11.374
35	14.864	14.744	14.634	14.534	14.442
40	18.114	17.979	17.857	17.745	17.641
45	21.468	21.320	21.185	21.061	20.948
50	24.910	24.749	24.602	24.468	24.344
55	28.427	28.254	28.096	27.951	27.818
60	32.010	31.825	31.656	31.501	31.359
70	39.340	39.133	38.944	38.771	38.611
80	46.855	46.626	46.418	46.228	46.051
90	54.519	54.271	54.045	53.838	53.646
100	62.309	62.042	61.799	61.576	61.371
110	70.206	69.922	69.663	69.425	69.206
120	78.197	77.895	77.621	77.369	77.136

Tafel 23 (Forts.) Table 23 (cont.)

LOWER PERCENTAGE POINTS / UNTERE PROZENTPUNKTE

$\chi^2\ (\alpha/\tau,\nu)$ $\alpha=0.010$

ν \ τ	1	2	3	4	5
1	$.0^{3}1571$	$.0^{4}3927$	$.0^{4}1745$	$.0^{5}9818$	$.0^{5}6283$
2	.02010	.01003	$.0^{2}6678$	$.0^{2}5006$	$.0^{2}4004$
3	.1148	.07172	.05455	.04494	.03868
4	.2971	.2070	.1679	.1449	.1292
5	.5543	.4117	.3469	.3075	.2801
6	.8721	.6757	.5837	.5266	.4864
7	1.239	.9893	.8696	.7945	.7411
8	1.646	1.344	1.197	1.104	1.038
9	2.088	1.735	1.561	1.450	1.370
10	2.558	2.156	1.956	1.827	1.734
11	3.053	2.603	2.378	2.232	2.126
12	3.571	3.074	2.824	2.661	2.543
13	4.107	3.565	3.291	3.112	2.982
14	4.660	4.075	3.777	3.582	3.440
15	5.229	4.601	4.280	4.070	3.916
16	5.812	5.142	4.799	4.573	4.408
17	6.408	5.697	5.332	5.092	4.915
18	7.015	6.265	5.878	5.623	5.436
19	7.633	6.844	6.436	6.167	5.969
20	8.260	7.434	7.006	6.723	6.514
21	8.897	8.034	7.585	7.289	7.070
22	9.542	8.643	8.175	7.865	7.636
23	10.196	9.260	8.773	8.450	8.212
24	10.856	9.886	9.380	9.044	8.796
25	11.524	10.520	9.995	9.646	9.389
26	12.198	11.160	10.617	10.256	9.989
27	12.879	11.808	11.246	10.873	10.597
28	13.565	12.461	11.882	11.497	11.212
29	14.256	13.121	12.525	12.128	11.833
30	14.953	13.787	13.173	12.765	12.461
35	18.509	17.192	16.496	16.032	15.686
40	22.164	20.707	19.934	19.417	19.032
45	25.901	24.311	23.466	22.900	22.477
50	29.707	27.991	27.077	26.464	26.006
55	33.570	31.735	30.756	30.097	29.605
60	37.485	35.534	34.492	33.791	33.267
70	45.442	43.275	42.115	41.332	40.747
80	53.540	51.172	49.901	49.043	48.400
90	61.754	59.196	57.821	56.892	56.196
100	70.065	67.328	65.854	64.857	64.110
110	78.458	75.550	73.982	72.922	72.126
120	86.923	83.852	82.194	81.072	80.230

Tafel 23 (Forts.) Table 23 (cont.)

LOWER PERCENTAGE POINTS / UNTERE PROZENTPUNKTE

$\chi^2\ (\alpha/\tau,\nu)$ $\alpha = 0.010$

ν \ τ	6	7	8	9	10
1	$.0^5 4363$	$.0^5 3206$	$.0^5 2454$	$.0^5 1939$	$.0^5 1571$
2	$.0^2 3336$	$.0^2 2859$	$.0^2 2502$	$.0^2 2223$	$.0^2 2001$
3	.03422	.03086	.02822	.02607	.02430
4	.1178	.1089	.1017	.09580	.09080
5	.2597	.2436	.2305	.2196	.2102
6	.4560	.4319	.4121	.3954	.3811
7	.7003	.6678	.6409	.6181	.5985
8	.9863	.9452	.9111	.8822	.8571
9	1.309	1.259	1.218	1.182	1.152
10	1.663	1.604	1.556	1.515	1.479
11	2.044	1.978	1.923	1.875	1.834
12	2.451	2.377	2.314	2.261	2.214
13	2.880	2.797	2.728	2.669	2.617
14	3.329	3.238	3.163	3.098	3.041
15	3.796	3.698	3.615	3.545	3.483
16	4.279	4.173	4.085	4.008	3.942
17	4.777	4.664	4.569	4.488	4.416
18	5.289	5.169	5.068	4.981	4.905
19	5.814	5.687	5.580	5.488	5.407
20	6.351	6.217	6.104	6.006	5.921
21	6.898	6.758	6.639	6.536	6.447
22	7.456	7.309	7.184	7.077	6.983
23	8.024	7.870	7.740	7.628	7.529
24	8.600	8.440	8.304	8.188	8.085
25	9.185	9.019	8.878	8.756	8.649
26	9.778	9.605	9.459	9.333	9.222
27	10.379	10.200	10.048	9.918	9.803
28	10.987	10.802	10.645	10.510	10.391
29	11.601	11.410	11.248	11.109	10.986
30	12.222	12.025	11.859	11.715	11.588
35	15.413	15.188	14.998	14.833	14.688
40	18.727	18.476	18.263	18.079	17.916
45	22.143	21.867	21.632	21.429	21.251
50	25.643	25.343	25.089	24.868	24.674
55	29.215	28.893	28.620	28.382	28.173
60	32.851	32.507	32.215	31.962	31.738
70	40.282	39.897	39.570	39.287	39.036
80	47.890	47.467	47.107	46.795	46.520
90	55.642	55.183	54.793	54.454	54.155
100	63.515	63.023	62.604	62.239	61.918
110	71.492	70.968	70.521	70.132	69.789
120	79.559	79.003	78.530	78.119	77.755

Tafel 23 (Forts.) Table 23 (cont.)

LOWER PERCENTAGE POINTS / UNTERE PROZENTPUNKTE

χ^2 $(\alpha/\tau,\nu)$ $\alpha=0.010$

ν \ τ	11	12	13	14	15
1	$.0^{5}1298$	$.0^{5}1091$	$.0^{6}9295$	$.0^{6}8014$	$.0^{6}6981$
2	$.0^{2}1819$	$.0^{2}1667$	$.0^{2}1539$	$.0^{2}1429$	$.0^{2}1334$
3	.02280	.02150	.02038	.01940	.01852
4	.08652	.08278	.07949	.06756	.07393
5	.2021	.1950	.1887	.1830	.1779
6	.3686	.3576	.3477	.3389	.3308
7	.5813	.5661	.5525	.5402	.5290
8	.8351	.8156	.7980	.7822	.7677
9	1.125	1.101	1.080	1.060	1.042
10	1.447	1.419	1.393	1.370	1.349
11	1.797	1.765	1.736	1.709	1.684
12	2.173	2.136	2.103	2.073	2.045
13	2.571	2.530	2.493	2.460	2.429
14	2.990	2.945	2.905	2.867	2.833
15	3.428	3.379	3.334	3.294	3.257
16	3.882	3.829	3.782	3.738	3.698
17	4.353	4.296	4.244	4.197	4.154
18	4.837	4.777	4.722	4.671	4.625
19	5.335	5.271	5.212	5.159	5.110
20	5.845	5.777	5.716	5.659	5.607
21	6.367	6.295	6.230	6.171	6.116
22	6.899	6.824	6.756	6.694	6.636
23	7.442	7.363	7.292	7.227	7.167
24	7.994	7.912	7.837	7.769	7.706
25	8.554	8.469	8.391	8.321	8.255
26	9.123	9.035	8.954	8.881	8.813
27	9.701	9.609	9.525	9.449	9.378
28	10.285	10.190	10.103	10.024	9.952
29	10.877	10.778	10.689	10.607	10.532
30	11.475	11.374	11.282	11.197	11.120
35	14.559	14.442	14.336	14.240	14.150
40	17.772	17.641	17.523	17.414	17.314
45	21.091	20.948	20.817	20.697	20.587
50	24.501	24.344	24.202	24.072	23.951
55	27.986	27.818	27.665	27.524	27.395
60	31.539	31.359	31.195	31.045	30.906
70	38.813	38.611	38.427	38.259	38.103
80	46.274	46.051	45.849	45.663	45.492
90	53.888	53.646	53.426	53.224	53.038
100	61.630	61.371	61.134	60.917	60.716
110	69.483	69.206	68.953	68.721	68.507
120	77.430	77.136	76.869	76.623	76.396

Tafel 23 (Forts.) Table 23 (cont.)

LOWER PERCENTAGE POINTS / UNTERE PROZENTPUNKTE

$\chi^2\ (\alpha/\tau,\nu)$ $\alpha = 0.010$

ν \ τ	16	17	18	19	20
1	$.0^{6}6136$	$.0^{6}5435$	$.0^{6}4848$	$.0^{6}4351$	$.0^{6}3927$
2	$.0^{2}1250$	$.0^{2}1177$	$.0^{2}1111$	$.0^{2}1053$	$.0^{2}1000$
3	.01774	.01703	.01639	.01581	.01528
4	.07156	.06940	.06742	.06560	.06392
5	.1733	.1690	.1651	.1615	.1581
6	.3235	.3168	.3105	.3048	.2994
7	.5187	.5093	.5006	.4924	.4849
8	.7544	.7422	.7308	.7202	.7104
9	1.026	1.011	.9970	.9839	.9717
10	1.330	1.312	1.295	1.280	1.265
11	1.662	1.641	1.622	1.604	1.587
12	2.020	1.996	1.974	1.954	1.934
13	2.400	2.374	2.350	2.327	2.305
14	2.802	2.773	2.746	2.721	2.697
15	3.223	3.191	3.161	3.134	3.108
16	3.661	3.626	3.594	3.564	3.536
17	4.114	4.077	4.043	4.011	3.980
18	4.583	4.543	4.506	4.472	4.439
19	5.065	5.023	4.984	4.947	4.912
20	5.560	5.515	5.474	5.435	5.398
21	6.066	6.019	5.975	5.934	5.896
22	6.583	6.534	6.488	6.445	6.404
23	7.111	7.059	7.011	6.966	6.924
24	7.648	7.595	7.544	7.497	7.453
25	8.195	8.139	8.086	8.037	7.991
26	8.750	8.692	8.637	8.586	8.538
27	9.313	9.253	9.196	9.143	9.093
28	9.884	9.822	9.763	9.708	9.656
29	10.463	10.398	10.337	10.280	10.227
30	11.048	10.981	10.918	10.860	10.804
35	14.068	13.991	13.919	13.851	13.787
40	17.221	17.135	17.054	16.978	16.906
45	20.484	20.389	20.299	20.216	20.137
50	23.840	23.736	23.638	23.547	23.461
55	27.274	27.162	27.057	26.959	26.866
60	30.778	30.658	30.545	30.440	30.340
70	37.958	37.824	37.698	37.579	37.467
80	45.333	45.184	45.045	44.914	44.791
90	52.865	52.703	52.552	52.410	52.276
100	60.530	60.356	60.193	60.040	59.896
110	68.308	68.123	67.949	67.785	67.631
120	76.185	75.988	75.804	75.630	75.467

Tafel 23 (Forts.) Table 23 (cont.)

LOWER PERCENTAGE POINTS / UNTERE PROZENTPUNKTE

$\chi^2 \ (\alpha/\tau,\nu)$ α=0.010

ν \ τ	22	24	26	28	30
1	$.0^6 3246$	$.0^6 2727$	$.0^6 2324$	$.0^6 2004$	$.0^6 1745$
2	$.0^3 9093$	$.0^3 8335$	$.0^3 7694$	$.0^3 7144$	$.0^3 6668$
3	.01434	.01353	.01282	.01220	.01165
4	.06092	.05830	.05599	.05393	.05209
5	.1521	.1468	.1421	.1378	.1340
6	.2897	.2811	.2735	.2666	.2603
7	.4711	.4590	.4480	.4382	.4292
8	.6924	.6765	.6621	.6491	.6373
9	.9494	.9296	.9118	.8956	.8808
10	1.238	1.215	1.193	1.174	1.156
11	1.556	1.528	1.503	1.481	1.460
12	1.899	1.868	1.839	1.813	1.790
13	2.266	2.230	2.199	2.169	2.143
14	2.653	2.614	2.579	2.547	2.517
15	3.060	3.017	2.978	2.943	2.911
16	3.484	3.438	3.396	3.357	3.322
17	3.925	3.875	3.829	3.788	3.750
18	4.380	4.326	4.278	4.234	4.193
19	4.849	4.792	4.741	4.693	4.650
20	5.331	5.271	5.216	5.166	5.120
21	5.825	5.761	5.704	5.651	5.602
22	6.330	6.263	6.202	6.147	6.096
23	6.846	6.776	6.712	6.654	6.600
24	7.371	7.298	7.231	7.170	7.114
25	7.906	7.830	7.760	7.696	7.638
26	8.450	8.370	8.298	8.232	8.171
27	9.001	8.919	8.844	8.775	8.712
28	9.561	9.476	9.398	9.327	9.261
29	10.128	10.040	9.959	9.886	9.818
30	10.703	10.611	10.528	10.452	10.382
35	13.670	13.565	13.469	13.381	13.300
40	16.775	16.656	16.548	16.449	16.358
45	19.991	19.860	19.741	19.631	19.530
50	23.302	23.159	23.029	22.909	22.799
55	26.695	26.540	26.399	26.270	26.151
60	30.157	29.992	29.841	29.703	29.575
70	37.262	37.076	36.906	36.750	36.607
80	44.564	44.359	44.171	44.000	43.841
90	52.029	51.805	51.601	51.414	51.241
100	59.629	59.389	59.169	58.967	58.781
110	67.347	67.089	66.855	66.639	66.440
120	75.165	74.892	74.643	74.414	74.203

Tafel 23 (Forts.) Table 23 (cont.)

LOWER PERCENTAGE POINTS / UNTERE PROZENTPUNKTE

$\chi^2\ (\alpha/\tau,\nu)$ $\alpha=0.005$

ν \ τ	1	2	3	4	5
1	$.0^4 3927$	$.0^5 9818$	$.0^5 4363$	$.0^5 2454$	$.0^5 1571$
2	.01003	$.0^2 5006$	$.0^2 3336$	$.0^2 2502$	$.0^2 2001$
3	.07172	.04494	.03422	.02822	.02430
4	.2070	.1449	.1178	.1017	.09080
5	.4117	.3075	.2597	.2305	.2102
6	.6757	.5266	.4560	.4121	.3811
7	.9893	.7945	.7003	.6409	.5985
8	1.344	1.104	.9863	.9111	.8571
9	1.735	1.450	1.309	1.218	1.152
10	2.156	1.827	1.663	1.556	1.479
11	2.603	2.232	2.044	1.923	1.834
12	3.074	2.661	2.451	2.314	2.214
13	3.565	3.112	2.880	2.728	2.617
14	4.075	3.582	3.329	3.163	3.041
15	4.601	4.070	3.796	3.615	3.483
16	5.142	4.573	4.279	4.085	3.942
17	5.697	5.092	4.777	4.569	4.416
18	6.265	5.623	5.289	5.068	4.905
19	6.844	6.167	5.814	5.580	5.407
20	7.434	6.723	6.351	6.104	5.921
21	8.034	7.289	6.898	6.639	6.447
22	8.643	7.865	7.456	7.184	6.983
23	9.260	8.450	8.024	7.740	7.529
24	9.886	9.044	8.600	8.304	8.085
25	10.520	9.646	9.185	8.878	8.649
26	11.160	10.256	9.778	9.459	9.222
27	11.808	10.873	10.379	10.048	9.803
28	12.461	11.497	10.987	10.645	10.391
29	13.121	12.128	11.601	11.248	10.986
30	13.787	12.765	12.222	11.859	11.588
35	17.192	16.032	15.413	14.998	14.688
40	20.707	19.417	18.727	18.263	17.916
45	24.311	22.900	22.143	21.632	21.251
50	27.991	26.464	25.643	25.089	24.674
55	31.735	30.097	29.215	28.620	28.173
60	35.534	33.791	32.851	32.215	31.738
70	43.275	41.332	40.282	39.570	39.036
80	51.172	49.043	47.890	47.107	46.520
90	59.196	56.892	55.642	54.793	54.155
100	67.328	64.857	63.515	62.604	61.918
110	75.550	72.922	71.492	70.521	69.789
120	83.852	81.072	79.559	78.530	77.755

Tafel 23 (Forts.) Table 23 (cont.)

LOWER PERCENTAGE POINTS / UNTERE PROZENTPUNKTE

$\chi^2\ (\alpha/\tau,\nu)$ $\alpha = 0.005$

ν \ τ	6	7	8	9	10
1	$.0^5 1091$	$.0^6 8014$	$.0^6 6136$	$.0^6 4848$	$.0^6 3927$
2	$.0^2 1667$	$.0^2 1429$	$.0^2 1250$	$.0^2 1111$	$.0^2 1000$
3	.02150	.01940	.01774	.01639	.01528
4	.08278	.07656	.07156	.06742	.06392
5	.1950	.1830	.1733	.1651	.1581
6	.3576	.3389	.3235	.3105	.2994
7	.5661	.5402	.5187	.5006	.4849
8	.8156	.7822	.7544	.7308	.7104
9	1.101	1.060	1.026	.9970	.9717
10	1.419	1.370	1.330	1.295	1.265
11	1.765	1.709	1.662	1.622	1.587
12	2.136	2.073	2.020	1.974	1.934
13	2.530	2.460	2.400	2.350	2.305
14	2.945	2.867	2.802	2.746	2.697
15	3.379	3.294	3.223	3.161	3.108
16	3.829	3.738	3.661	3.594	3.536
17	4.296	4.197	4.114	4.043	3.980
18	4.777	4.671	4.583	4.506	4.439
19	5.271	5.159	5.065	4.984	4.912
20	5.777	5.659	5.560	5.474	5.398
21	6.295	6.171	6.066	5.975	5.896
22	6.824	6.694	6.583	6.488	6.404
23	7.363	7.227	7.111	7.011	6.924
24	7.912	7.769	7.648	7.544	7.453
25	8.469	8.321	8.195	8.086	7.991
26	9.035	8.881	8.750	8.637	8.538
27	9.609	9.449	9.313	9.196	9.093
28	10.190	10.024	9.884	9.763	9.656
29	10.778	10.607	10.463	10.337	10.227
30	11.374	11.197	11.048	10.918	10.804
35	14.442	14.240	14.068	13.919	13.787
40	17.641	17.414	17.221	17.054	16.906
45	20.948	20.697	20.484	20.299	20.137
50	24.344	24.072	23.840	23.638	23.461
55	27.818	27.524	27.274	27.057	26.866
60	31.359	31.045	30.778	30.545	30.340
70	38.611	38.259	37.958	37.698	37.467
80	46.051	45.663	45.333	45.045	44.791
90	53.646	53.224	52.865	52.552	52.276
100	61.371	60.917	60.530	60.193	59.896
110	69.206	68.721	68.308	67.949	67.631
120	77.136	76.623	76.185	75.804	75.467

Tafel 23 (Forts.) Table 23 (cont.)

LOWER PERCENTAGE POINTS / UNTERE PROZENTPUNKTE

χ^2 ($\alpha/\tau,\nu$) α=0.005

ν \ τ	11	12	13	14	15
1	$.0^63246$	$.0^62727$	$.0^62324$	$.0^62004$	$.0^61745$
2	$.0^39093$	$.0^38335$	$.0^37694$	$.0^37144$	$.0^36668$
3	.01434	.01353	.01282	.01220	.01165
4	.06092	.05830	.05599	.05393	.05209
5	.1521	.1468	.1421	.1378	.1340
6	.2897	.2811	.2735	.2666	.2603
7	.4711	.4590	.4480	.4382	.4292
8	.6924	.6765	.6621	.6491	.6373
9	.9494	.9296	.9118	.8956	.8808
10	1.238	1.215	1.193	1.174	1.156
11	1.556	1.528	1.503	1.481	1.460
12	1.899	1.868	1.839	1.813	1.790
13	2.266	2.230	2.199	2.169	2.143
14	2.653	2.614	2.579	2.547	2.517
15	3.060	3.017	2.978	2.943	2.911
16	3.484	3.438	3.396	3.357	3.322
17	3.925	3.875	3.829	3.788	3.750
18	4.380	4.326	4.278	4.234	4.193
19	4.849	4.792	4.741	4.693	4.650
20	5.331	5.271	5.216	5.166	5.120
21	5.825	5.761	5.704	5.651	5.602
22	6.330	6.263	6.202	6.147	6.096
23	6.846	6.776	6.712	6.654	6.600
24	7.371	7.298	7.231	7.170	7.114
25	7.906	7.830	7.760	7.696	7.638
26	8.450	8.370	8.298	8.232	8.171
27	9.001	8.919	8.844	8.775	8.712
28	9.561	9.476	9.398	9.327	9.261
29	10.128	10.040	9.959	9.886	9.818
30	10.703	10.611	10.528	10.452	10.382
35	13.670	13.565	13.469	13.381	13.300
40	16.775	16.656	16.548	16.449	16.358
45	19.991	19.860	19.741	19.631	19.530
50	23.302	23.159	23.029	22.909	22.799
55	26.695	26.540	26.399	26.270	26.151
60	30.157	29.992	29.841	29.703	29.575
70	37.262	37.076	36.906	36.750	36.607
80	44.564	44.359	44.171	44.000	43.841
90	52.029	51.805	51.601	51.414	51.241
100	59.629	59.389	59.169	58.967	58.781
110	67.347	67.089	66.855	66.639	66.440
120	75.165	74.892	74.643	74.414	74.203

Tafel 23 (Forts.) Table 23 (cont.)

LOWER PERCENTAGE POINTS / UNTERE PROZENTPUNKTE

$\chi^2\ (\alpha/\tau,\nu)$ $\alpha=0.005$

ν \ τ	16	17	18	19	20
1	$.0^{6}1534$	$.0^{6}1359$	$.0^{6}1212$	$.0^{6}1088$	$.0^{7}9818$
2	$.0^{3}6251$	$.0^{3}5883$	$.0^{3}5556$	$.0^{3}5264$	$.0^{3}5001$
3	.01116	.01072	.01032	$.0^{2}9949$	$.0^{2}9614$
4	.05042	.04890	.04752	.04624	.04506
5	.1305	.1273	.1244	.1217	.1192
6	.2546	.2493	.2445	.2400	.2358
7	.4210	.4134	.4064	.3999	.3938
8	.6264	.6164	.6071	.5984	.5903
9	.8672	.8546	.8429	.8320	.8219
10	1.140	1.125	1.111	1.097	1.085
11	1.441	1.423	1.407	1.391	1.377
12	1.768	1.748	1.729	1.711	1.695
13	2.118	2.095	2.074	2.054	2.036
14	2.490	2.465	2.441	2.419	2.398
15	2.881	2.853	2.827	2.803	2.780
16	3.290	3.260	3.231	3.205	3.180
17	3.715	3.682	3.652	3.624	3.597
18	4.155	4.120	4.088	4.057	4.029
19	4.610	4.573	4.538	4.505	4.475
20	5.077	5.038	5.001	4.966	4.934
21	5.557	5.515	5.476	5.440	5.405
22	6.048	6.004	5.963	5.925	5.888
23	6.550	6.504	6.461	6.421	6.382
24	7.062	7.014	6.969	6.926	6.887
25	7.584	7.533	7.486	7.442	7.400
26	8.114	8.062	8.013	7.967	7.923
27	8.653	8.599	8.548	8.500	8.455
28	9.200	9.144	9.091	9.041	8.994
29	9.755	9.696	9.641	9.590	9.542
30	10.317	10.256	10.200	10.146	10.096
35	13.225	13.155	13.089	13.028	12.970
40	16.273	16.194	16.120	16.051	15.985
45	19.436	19.349	19.267	19.190	19.118
50	22.697	22.601	22.512	22.428	22.349
55	26.041	25.937	25.841	25.750	25.664
60	29.457	29.346	29.243	29.145	29.054
70	36.473	36.349	36.232	36.122	36.019
80	43.693	43.556	43.427	43.306	43.191
90	51.081	50.931	50.790	50.658	50.533
100	58.608	58.446	58.295	58.152	58.018
110	66.255	66.082	65.921	65.768	65.624
120	74.006	73.823	73.651	73.489	73.337

Tafel 23 (Forts.) Table 23 (cont.)

LOWER PERCENTAGE POINTS / UNTERE PROZENTPUNKTE

χ^2 $(\alpha/\tau,\nu)$ α=0.005

ν \ τ	22	24	26	28	30
1	$.0^7 8114$	$.0^7 6818$	$.0^7 5810$	$.0^7 5009$	$.0^7 4363$
2	$.0^3 4546$	$.0^3 4167$	$.0^3 3847$	$.0^3 3572$	$.0^3 3334$
3	$.0^2 9021$	$.0^2 8512$	$.0^2 8069$	$.0^2 7680$	$.0^2 7334$
4	.04295	.04111	.03948	.03804	.03674
5	.1147	.1107	.1071	.1040	.1011
6	.2282	.2215	.2155	.2101	.2052
7	.3827	.3729	.3642	.3562	.3490
8	.5756	.5625	.5507	.5400	.5302
9	.8033	.7867	.7718	.7583	.7459
10	1.063	1.043	1.025	1.008	.9930
11	1.351	1.327	1.306	1.286	1.269
12	1.664	1.637	1.613	1.590	1.570
13	2.002	1.971	1.943	1.918	1.895
14	2.360	2.326	2.295	2.267	2.242
15	2.739	2.701	2.667	2.636	2.608
16	3.135	3.094	3.057	3.023	2.992
17	3.548	3.504	3.464	3.427	3.393
18	3.976	3.929	3.886	3.846	3.810
19	4.418	4.368	4.322	4.280	4.241
20	4.874	4.820	4.772	4.727	4.686
21	5.342	5.285	5.234	5.186	5.143
22	5.822	5.762	5.707	5.658	5.612
23	6.312	6.249	6.192	6.140	6.092
24	6.813	6.747	6.687	6.633	6.582
25	7.324	7.255	7.192	7.135	7.082
26	7.843	7.772	7.706	7.647	7.592
27	8.372	8.297	8.229	8.167	8.110
28	8.908	8.831	8.760	8.696	8.636
29	9.453	9.372	9.299	9.232	9.171
30	10.004	9.921	9.846	9.777	9.713
35	12.863	12.767	12.679	12.599	12.525
40	15.865	15.756	15.657	15.566	15.482
45	18.984	18.864	18.754	18.653	18.560
50	22.203	22.071	21.951	21.841	21.739
55	25.507	25.364	25.234	25.115	25.005
60	28.884	28.732	28.592	28.464	28.346
70	35.828	35.656	35.498	35.354	35.220
80	42.980	42.790	42.616	42.456	42.308
90	50.303	50.095	49.906	49.731	49.570
100	57.770	57.546	57.341	57.153	56.979
110	65.359	65.119	64.900	64.699	64.513
120	73.055	72.800	72.567	72.353	72.155

TABLE 24

The Sequential Chi-square Criterion for Multivariate Comparisons of Means: Tables of R. J. Freund and J. E. Jackson

(a) Content of the tables and definition of the test statistic :

JACKSON and BRADLEY (1959 ; 1961) extended the well known univariate sequential probability ratio test for testing hypotheses related to means (sequential t-test) to the multivariate case of testing mean vectors.

Consider a multivariate normally distributed population with mean vector $\underline{\mu}' = (\mu_1, \mu_2, \ldots, \mu_p)$ and covariance matrix $\underline{\Sigma}$.

In the following considerations we assume $\underline{\Sigma}$ to be known. The case of an unknown covariance matrix $\underline{\Sigma}$ is treated in table 25.

For sequential testing of the null hypothesis

$$H_o \; : \; (\underline{\mu} - \underline{\mu}_o)' \cdot \underline{\Sigma}^{-1} \cdot (\underline{\mu} - \underline{\mu}_o) = 0$$

(equivalent to $H_o : \underline{\mu} = \underline{\mu}_o$)

against the alternative hypothesis

$$H_1 \; : \; (\underline{\mu} - \underline{\mu}_o)' \cdot \underline{\Sigma}^{-1} \cdot (\underline{\mu} - \underline{\mu}_o) = \lambda^2$$

(λ^2 = non-centrality parameter)

with a sample of size n , one obtains the probability ratio from

$$p_{on} = L(\underline{x}_n ; \Theta_o) = \text{likelihood function for the case that } H_o \text{ is correct},$$

and

$$p_{1n} = L(\underline{x}_n ; \Theta_1) = \text{likelihood function for the case that } H_1 \text{ is correct}.$$

One obtains

$$p_{1n}/p_{on} = e^{-(1/2).n.\lambda^2} \, {}_oF_1(p/2 \; ; \; n.\lambda^2.\chi_n^2/4)$$

where ${}_oF_1$ is the <u>confluent hypergeometric function</u>

$${}_oF_1(a;x) = \sum_{i=o}^{\infty} \frac{x^i}{i!a(a+1) \; ... \; (a+i+1)} \quad .$$

The decision process is as usual :

(1) In case $p_{1n}/p_{on} \leqq \frac{\beta}{1-\alpha}$, accept H_o .

(2) In case $p_{1n}/p_{on} = \frac{1-\beta}{\alpha}$, reject H_o in favor of H_1 .

(3) In case $\frac{\beta}{1-\alpha} < p_{1n}/p_{on} < \frac{1-\beta}{\alpha}$, continue sampling .

In the usual notation ,

α = probability of rejecting H_o , although H_o is correct ,

β = probability of accepting H_o , although H_1 is correct .

The computation of the probability ratio requires the value χ_n^2 , defined by

$$\chi_n^2 = n.({}_n\underline{\bar{x}} - \underline{\mu}_o)'.\,\underline{\Sigma}^{-1}.\,({}_n\underline{\bar{x}} - \underline{\mu}_o) \quad ,$$

where ${}_n\underline{\bar{x}}$ is the mean vector of the sample of size n and $\underline{\Sigma}$ is the covariance matrix , assumed known , of the p-dimensional population.

The tedious computation of the probability ratio by means of the confluent hypergeometric function for each sample of size n can be avoided by taking the inequality from the decision process for accepting H_o ,

$$e^{-(1/2).n.\lambda^2} {}_oF_1(p/2 \; ; \; n.\lambda^2.\chi_n^2/4) \leqq \frac{\alpha}{1-\alpha}$$

substituting various values of α, β, λ^2 , and n , and solving for χ_n^2 , which then becomes a function of the sample size alone.

Similarly for the inequality for the rejection of H_o. In the first case, the solution gives a value for χ_n^2 , which is an upper bound for the acceptance region for H_o , and which is denoted by $\underline{\chi}_n^2$. In the second case, the solution of the inequality for the rejection of H_o gives a value for χ_n^2 , which is a lower bound for the rejection region for H_o , and which is denoted by $\overline{\chi}_n^2$.

For every χ_n^2 satisfying the inequality $\underline{\chi}_n^2 < \chi_n^2 < \overline{\chi}_n^2$, the testing - and hence the sampling - is to be continued.

In summary , the modified decision process is :

(1') In case $\chi_n^2 \leqq \underline{\chi}_n^2$, accept H_o ,

(2') In case $\chi_n^2 \geqq \overline{\chi}_n^2$, reject H_o ,

(3') In case $\underline{\chi}_n^2 < \chi_n^2 < \overline{\chi}_n^2$, continue testing .

The tables consequently contain the upper bounds $\underline{\chi}_n^2$ of the acceptance region for H_o as well as the lower bounds $\overline{\chi}_n^2$ of the rejection region for H_o , all for various values of the parameters α, β, λ^2, and n .

(b) Extent of the tables and definition of the parameters :

(1) The parameter α :

α = error probability
= probability of an error of the first kind

α = 0.05 .

(2) The parameter β :

β = probability of an error of the second kind

β = 0.05 .

(3) The parameter p :

p = dimension of the variates

p = 2(1)9 .

(4) The parameter λ^2 :

λ^2 = non-centrality parameter

λ^2 = 0.25 ; 0.50 ; 1.00 .

(5) The parameter n :

n = sample size

n = 1(1)150 when p = 2; 3;
= 1(1)100 when p = 4(1)6 ;
= 1(1)60 when p = 7(1)9 .

(c) Suggestions for applications :

As explained in section (a) , the tables serve for a simplified multivariate sequential testing for means with a known covariance matrix .

(d) References :

(1) For the source of the tables and for applications :

FREUND, R. J. + JACKSON, J. E. : Tables to facilitate multivariate sequential testing for means. (Technical Report No. 12). (Table 7). Blacksburg , Virginia : Department of Statistics and Statistical Laboratory, Virginia Agricultural Experiment Station, Virginia Polytechnic Institute, September 1960.

(2) For further suggestions :

JACKSON, J. E. + BRADLEY, E. A. : Multivariate sequential procedures for testing means. (Technical Report No. 10). Blacksburg , Virginia : Virginia Polytechnic Institute, 1959 .

JACKSON, J. E. + BRADLEY, R. A. : Sequential χ^2- and T^2-Tests.The Annals of Mathematical Statistics 32, 1063 - 1077 (1961) .

(e) Further remarks :

(1) On the basis of the following considerations, the tabulated parameter region for the non-centrality parameter λ^2 can be expanded considerably :

For given values of the ratios $(1 - \beta)/\alpha$ and $\alpha/(1 - \beta)$ as well as for a given number p of variates one has the relation

$$\chi^2(n ; \lambda^2) = \chi^2(k.n ; \lambda^2/k)$$

for positive values of k .

Thus if one, for example, needs the χ^2-values for n=10 observations in order to test against the alternative $\lambda^2 = 2$, one can obtain these values (which are not contained in the tables) by finding the χ^2-values for $\lambda = 1$ and $n = 20$, or for $\lambda = 1/2$ and $n = 40$ in the tables.

(2) One can prove that the sequential test process discribed here, leads to a decision with probability one for finite n .

Tafel 24 Table 24

$\alpha = \beta = .05, \ \lambda^2 = .25$

n	p = 2 $\underline{\chi}^2$	p = 2 $\overline{\chi}^2$	p = 3 $\underline{\chi}^2$	p = 3 $\overline{\chi}^2$		n
1		8970		11924		1
2		4755		6300		2
3		3355		4430		3
4		2658		3498		4
5		2243		2942		5
6		1968		2574		6
7		1774		2313		7
8		1630		2119		8
9		1519		1970		9
10		1432		1852		10
11		1362		1757		11
12		1305		1679		12
13		1258		1613		13
14		1218		1559		14
15		1185		1512		15
16		1156		1472		16
17		1132		1438		17
18		1111		1408		18
19		1093		1382		19
20		1078		1360		20
21		1064		1340		21
22		1053		1323		22
23		1043		1308		23
24	3	1034	5	1294		24
25	12	1027	17	1283		25
26	20	1020	29	1272		26
27	28	1015	41	1263		27
28	36	1010	52	1255		28
29	44	1007	64	1248		29
30	51	1004	74	1242		30
31	59	1001	85	1237		31
32	67	999	96	1233		32
33	74	998	106	1229		33
34	82	997	116	1226		34
35	90	997	126	1224		35
36	97	997	136	1222		36
37	105	997	146	1220		37
38	112	998	155	1219		38
39	119	999	165	1218		39
40	127	1000	174	1218		40
41	134	1001	183	1218		41
42	141	1003	193	1219		42
43	149	1005	202	1219		43
44	156	1007	211	1220		44
45	163	1009	220	1221		45
46	171	1012	229	1223		46
47	178	1015	238	1224		47
48	185	1018	246	1226		48
49	192	1021	255	1228		49
50	200	1024	264	1231		50

Tafel 24 (Forts.) Table 24 (cont.)

$\alpha = \beta = .05, \quad \lambda^2 = .25$

n	p = 2		p = 3			n
	$\underline{\chi}^2$	$\bar{\chi}^2$	$\underline{\chi}^2$	$\bar{\chi}^2$		
51	207	1027	272	1233		51
52	214	1030	281	1236		52
53	221	1034	289	1239		53
54	228	1038	298	1241		54
55	235	1041	306	1245		55
56	242	1045	315	1248		56
57	250	1049	323	1251		57
58	257	1053	331	1254		58
59	264	1057	339	1258		59
60	271	1061	348	1262		60
61	278	1066	356	1265		61
62	285	1070	364	1269		62
63	292	1074	372	1273		63
64	299	1079	380	1277		64
65	306	1083	388	1281		65
66	313	1088	396	1285		66
67	320	1093	404	1290		67
68	327	1097	412	1294		68
69	334	1102	420	1298		69
70	341	1107	428	1303		70
71	348	1112	436	1307		71
72	355	1117	443	1312		72
73	362	1122	451	1317		73
74	369	1127	459	1321		74
75	376	1132	467	1326		75
76	383	1137	475	1331		76
77	389	1142	482	1336		77
78	396	1147	490	1341		78
79	403	1152	498	1345		79
80	410	1157	505	1350		80
81	417	1162	513	1355		81
82	424	1168	520	1360		82
83	431	1173	528	1366		83
84	438	1178	536	1371		84
85	444	1184	543	1376		85
86	451	1189	551	1381		86
87	458	1194	558	1386		87
88	465	1200	566	1392		88
89	472	1205	573	1397		89
90	479	1211	581	1402		90
91	485	1216	588	1407		91
92	492	1222	596	1413		92
93	499	1227	603	1418		93
94	506	1233	610	1424		94
95	513	1238	618	1429		95
96	519	1244	625	1435		96
97	526	1249	633	1440		97
98	533	1255	640	1446		98
99	540	1261	647	1451		99
100	546	1266	655	1457		100

$\alpha = \beta = .05, \quad \lambda^2 = .25$

n	p = 2 $\underline{\chi}^2$	p = 2 $\overline{\chi}^2$	p = 3 $\underline{\chi}^2$	p = 3 $\overline{\chi}^2$		n
101	553	1272	662	1462		101
102	560	1278	669	1468		102
103	567	1283	677	1473		103
104	573	1289	684	1479		104
105	580	1295	691	1485		105
106	587	1300	698	1490		106
107	594	1306	706	1496		107
108	600	1312	713	1502		108
109	607	1318	720	1508		109
110	614	1323	727	1513		110
111	620	1329	735	1519		111
112	627	1335	742	1525		112
113	634	1341	749	1531		113
114	641	1347	756	1536		114
115	647	1353	763	1542		115
116	654	1358	771	1548		116
117	661	1364	778	1554		117
118	667	1370	785	1560		118
119	674	1376	792	1566		119
120	681	1382	799	1571		120
121	687	1388	806	1577		121
122	694	1394	813	1583		122
123	701	1399	821	1589		123
124	707	1405	828	1595		124
125	714	1411	835	1601		125
126	721	1417	842	1607		126
127	727	1423	849	1613		127
128	734	1429	856	1619		128
129	741	1435	863	1625		129
130	747	1441	870	1631		130
131	754	1447	877	1637		131
132	761	1453	884	1643		132
133	767	1459	891	1649		133
134	774	1465	898	1655		134
135	780	1471	905	1661		135
136	787	1477	912	1667		136
137	794	1483	919	1673		137
138	800	1489	926	1679		138
139	807	1495	933	1685		139
140	814	1501	940	1691		140
141	820	1507	947	1697		141
142	827	1513	954	1703		142
143	833	1519	961	1709		143
144	840	1525	968	1715		144
145	847	1531	975	1721		145
146	853	1537	982	1727		146
147	860	1543	989	1733		147
148	866	1549	996	1739		148
149	873	1555	1003	1745		149
150	880	1561	1010	1751		150

Tafel 24 (Forts.) Table 24 (cont.

$\alpha = \beta = .05, \quad \lambda^2 = .5$

n	p = 4 $\underline{\chi}^2$	p = 4 $\overline{\chi}^2$	p = 5 $\underline{\chi}^2$	p = 5 $\overline{\chi}^2$	p = 6 $\underline{\chi}^2$	p = 6 $\overline{\chi}^2$	n
1		7763		9180		10569	1
2		4293		5062		5814	2
3		3146		3699		4240	3
4		2580		3026		3461	4
5		2247		2628		2999	5
6		2029		2367		2697	6
7		1878		2186		2486	7
8		1769		2054		2331	8
9		1687		1954		2215	9
10		1624		1877		2124	10
11		1575		1817		2053	11
12		1537	9	1770	11	1996	12
13		1507	49	1732	58	1951	13
14	69	1483	85	1702	101	1914	14
15	97	1465	119	1677	141	1884	15
16	124	1450	151	1658	178	1859	16
17	149	1439	181	1642	213	1839	17
18	173	1430	210	1630	246	1823	18
19	197	1424	237	1621	277	1811	19
20	220	1420	264	1614	307	1801	20
21	242	1418	289	1609	336	1793	21
22	263	1417	314	1606	364	1788	22
23	284	1418	338	1604	390	1784	23
24	304	1420	361	1604	416	1781	24
25	325	1422	383	1604	441	1781	25
26	344	1426	405	1606	465	1781	26
27	364	1430	427	1609	489	1782	27
28	383	1435	448	1613	512	1784	28
29	401	1440	469	1617	535	1787	29
30	420	1446	489	1622	557	1791	30
31	438	1453	509	1627	579	1796	31
32	456	1460	529	1633	600	1801	32
33	474	1467	548	1640	621	1807	33
34	491	1475	568	1647	642	1813	34
35	509	1483	587	1654	662	1819	35
36	526	1492	605	1662	682	1827	36
37	543	1500	624	1670	702	1834	37
38	560	1509	642	1679	722	1842	38
39	577	1519	660	1687	741	1850	39
40	594	1528	678	1696	760	1858	40
41	610	1538	696	1706	779	1867	41
42	627	1547	714	1715	798	1876	42
43	643	1557	731	1725	816	1885	43
44	659	1568	748	1734	835	1895	44
45	675	1578	766	1744	853	1904	45
46	691	1588	783	1754	871	1914	46
47	707	1599	800	1765	889	1924	47
48	723	1610	816	1775	907	1934	48
49	739	1620	833	1786	924	1945	49
50	755	1631	850	1796	942	1955	50

$\alpha = \beta = .05, \quad \lambda^2 = .5$

n	p = 4 $\underline{\chi}^2$	p = 4 $\overline{\chi}^2$	p = 5 $\underline{\chi}^2$	p = 5 $\overline{\chi}^2$	p = 6 $\underline{\chi}^2$	p = 6 $\overline{\chi}^2$	n
51	770	1642	866	1807	959	1966	51
52	786	1653	883	1818	977	1976	52
53	801	1665	899	1829	994	1987	53
54	817	1676	916	1840	1011	1998	54
55	832	1687	932	1851	1028	2009	55
56	847	1699	948	1863	1045	2020	56
57	863	1710	964	1874	1061	2032	57
58	878	1722	980	1886	1078	2043	58
59	893	1733	996	1897	1095	2054	59
60	908	1745	1012	1909	1111	2066	60
61	923	1757	1027	1920	1128	2077	61
62	938	1768	1043	1932	1144	2089	62
63	953	1780	1059	1944	1160	2101	63
64	968	1792	1074	1956	1177	2112	64
65	983	1804	1090	1967	1193	2124	65
66	998	1816	1105	1979	1209	2136	66
67	1013	1828	1121	1991	1225	2148	67
68	1027	1840	1136	2003	1241	2160	68
69	1042	1852	1152	2015	1257	2172	69
70	1057	1864	1167	2028	1273	2184	70
71	1071	1876	1182	2040	1289	2196	71
72	1086	1888	1197	2052	1304	2208	72
73	1101	1901	1213	2064	1320	2221	73
74	1115	1913	1228	2076	1336	2233	74
75	1130	1925	1243	2089	1351	2245	75
76	1144	1937	1258	2101	1367	2258	76
77	1159	1950	1273	2113	1383	2270	77
78	1173	1962	1288	2125	1398	2282	78
79	1188	1974	1303	2138	1413	2295	79
80	1202	1987	1318	2150	1429	2307	80
81	1216	1999	1333	2163	1444	2320	81
82	1231	2011	1347	2175	1460	2332	82
83	1245	2024	1362	2188	1475	2345	83
84	1259	2036	1377	2200	1490	2357	84
85	1273	2049	1392	2213	1505	2370	85
86	1288	2061	1406	2225	1520	2382	86
87	1302	2074	1421	2238	1536	2395	87
88	1316	2086	1436	2250	1551	2407	88
89	1330	2099	1451	2263	1566	2420	89
90	1344	2111	1465	2275	1581	2433	90
91	1359	2124	1480	2288	1596	2445	91
92	1373	2136	1494	2301	1611	2458	92
93	1387	2149	1509	2313	1626	2471	93
94	1401	2161	1523	2326	1641	2483	94
95	1415	2174	1538	2339	1656	2496	95
96	1429	2186	1552	2351	1671	2509	96
97	1443	2199	1567	2364	1685	2522	97
98	1457	2212	1581	2377	1700	2534	98
99	1471	2224	1596	2389	1715	2547	99
100	1485	2237	1610	2402	1730	2560	100

Tafel 24 (Forts.) Table 24 (cont.)

$\alpha = \beta = .05, \quad \lambda^2 = 1$

n	p = 7		p = 8		p = 9		n
	$\underline{\chi}^2$	$\overline{\chi}^2$	$\underline{\chi}^2$	$\overline{\chi}^2$	$\underline{\chi}^2$	$\overline{\chi}^2$	
1		6555		7288		8014	1
2		3889		4311		4730	2
3		3021		3341		3658	3
4		2604		2873		3139	4
5		2366		2605		2840	5
6	13	2218	14	2436	16	2652	6
7	117	2121	133	2325	149	2527	7
8	205	2056	232	2250	259	2441	8
9	282	2012	318	2198	353	2381	9
10	350	1984	393	2163	436	2340	10
11	413	1965	461	2140	510	2312	11
12	470	1955	524	2125	578	2293	12
13	524	1951	583	2118	640	2282	13
14	575	1952	637	2116	699	2277	14
15	624	1956	689	2118	754	2277	15
16	670	1964	739	2124	807	2281	16
17	715	1974	786	2132	857	2287	17
18	758	1986	832	2143	905	2297	18
19	800	2000	876	2156	952	2308	19
20	840	2016	919	2170	997	2321	20
21	880	2033	961	2186	1041	2336	21
22	919	2051	1002	2203	1084	2352	22
23	957	2069	1042	2221	1125	2370	23
24	995	2089	1081	2240	1166	2388	24
25	1032	2109	1119	2259	1206	2407	25
26	1068	2130	1157	2280	1245	2427	26
27	1104	2151	1194	2301	1284	2447	27
28	1139	2173	1231	2322	1322	2469	28
29	1174	2195	1267	2344	1359	2490	29
30	1208	2218	1303	2367	1396	2512	30
31	1242	2241	1338	2390	1433	2535	31
32	1276	2264	1373	2413	1469	2558	32
33	1310	2288	1408	2436	1504	2581	33
34	1343	2312	1442	2460	1540	2605	34
35	1376	2336	1476	2484	1575	2628	35
36	1408	2360	1510	2508	1609	2653	36
37	1441	2385	1543	2532	1644	2677	37
38	1473	2409	1576	2557	1678	2701	38
39	1505	2434	1609	2581	1711	2726	39
40	1537	2459	1642	2606	1745	2751	40
41	1568	2484	1674	2631	1778	2776	41
42	1600	2509	1707	2656	1811	2801	42
43	1631	2534	1739	2682	1844	2826	43
44	1662	2559	1770	2707	1877	2852	44
45	1693	2585	1802	2733	1909	2877	45
46	1724	2610	1834	2758	1942	2903	46
47	1754	2636	1865	2784	1974	2928	47
48	1785	2661	1896	2809	2006	2954	48
49	1815	2687	1928	2835	2037	2980	49
50	1846	2713	1959	2861	2069	3006	50

Tafel 24 (Forts.) Table 24 (cont.)

$\alpha = \beta = .05, \quad \lambda^2 = 1$

n	p = 7		p = 8		p = 9		n
	$\underline{\chi}^2$	$\overline{\chi}^2$	$\underline{\chi}^2$	$\overline{\chi}^2$	$\underline{\chi}^2$	$\overline{\chi}^2$	
51	1876	2738	1989	2887	2101	3032	51
52	1906	2764	2020	2913	2132	3058	52
53	1936	2790	2051	2939	2163	3084	53
54	1966	2816	2081	2965	2194	3110	54
55	1996	2842	2112	2991	2225	3136	55
56	2025	2868	2142	3017	2256	3163	56
57	2055	2894	2172	3043	2287	3189	57
58	2084	2920	2202	3069	2318	3215	58
59	2114	2946	2232	3095	2348	3241	59
60	2143	2972	2262	3122	2379	3268	60

TABLE 25

The Sequential T^2-Criterion for Multivariate Testing for Means: Tables of R. J. Freund and J. E. Jackson

(a) Content of the tables and definition of the parameters :

JACKSON and BRADLEY (1959 ; 1961) extended the well known univariate sequential probability ratio test for testing hypotheses related to means (sequential t-test) to the multivariate case of testing mean vectors.

Consider a multivariate normally distributed population with mean vector $\underline{\mu}' = (\mu_1, \mu_2, \dots, \mu_p)$ and covariance matrix $\underline{\Sigma}$.

In the following considerations we assume $\underline{\Sigma}$ to be unknown. The case of a known covariance matrix $\underline{\Sigma}$ is treated in table 24 .

For sequential testing of the null hypothesis

$$H_o \quad : \quad (\underline{\mu} - \underline{\mu}_o)' \cdot \underline{\Sigma}^{-1} \cdot (\underline{\mu} - \underline{\mu}_o) = 0$$

(equivalent to $H_o : \underline{\mu} = \underline{\mu}_o$)

against the alternative hypothesis

$$H_1 \quad : \quad (\underline{\mu} - \underline{\mu}_o)' \cdot \underline{\Sigma}^{-1} \cdot (\underline{\mu} - \underline{\mu}_o) = \lambda^2$$

(λ^2 = non-centrality parameter)

with a sample of size n , one obtains the probability ratio from

$$p_{on} = L(\underline{x}_n ; \Theta_o) = \text{likelihood function for the case that } H_o \text{ is correct ,}$$

and

$$p_{1n} = L(\underline{x}_n ; \Theta_1) = \text{likelihood function for the case that } H_1 \text{ is correct .}$$

One obtains

$$p_{1n}/p_{on} = e^{-(1/2).n.\lambda^2} \, {}_1F_1(n/2 \; ; \; p/2 \; ; \; \frac{n.\lambda^2.T_n^2}{2(n - 1 + T_n^2)}) \; ,$$

where ${}_1F_1$ is the <u>confluent hypergeometric function</u>

$${}_1F_1(a,b,x) = \sum_{i=0}^{\infty} \frac{a(a+1) \ldots (a+n-1).x^i}{b(b+1) \ldots (b+n-1).i!} \; .$$

The decision process is as usual

(1) In case $p_{1n}/p_{on} \leq \dfrac{\beta}{1-\alpha}$, accept H_o .

(2) In case $p_{1n}/p_{on} \geq \dfrac{1-\beta}{\alpha}$, reject H_o in favour of H_1.

(3) In case $\dfrac{\beta}{1-\alpha} < p_{1n}/p_{on} < \dfrac{1-\beta}{\alpha}$, continue sampling.

In the usual notation ,

α = probability of rejecting H_o , although H_o is correct ,

β = probability of accepting H_o , although H_1 is correct .

The computation of the probability ratio requires the value T_n^2 , defined by

$$T_n^2 = n.({}_n\underline{\bar{x}} - \underline{\mu}_o)' . \underline{S}^{-1} . ({}_n\underline{\bar{x}} - \underline{\mu}_o) \qquad ,$$

where $_n\underline{\bar{x}}$ is the mean vector of the sample of size n and $\underline{S}$ is the covariance matrix of the p-dimensional sample, used as an approximation for $\underline{\Sigma}$.

The tedious computation of the probability ratio by means of the confluent hypergeometric function for each sample of size n can be avoided by taking the <u>inequality</u> from the decision process for accepting H_o ,

$$e^{-(1/2).n.\lambda^2} {}_1F_1(n/2 \ ; \ p/2 \ ; \ \frac{n.\lambda^2.T_n^2}{2(n-1+T_n^2)}) \leqq \frac{\beta}{1-\alpha} \quad ,$$

substituting various values of α, β, λ^2, and n, and solving for T_n^2, which then becomes a function of the sample size alone (similarly to what was done for the sequential χ^2-test described in <u>Table 24</u>).

Similarly for the inequality for the rejection of H_o. In the <u>first case</u> , the solution gives a value for T_n^2 , which is an <u>upper bound</u> for the acceptance region for H_o , and which is denoted by $\underline{T}_n^2$.

In the <u>second case</u>, the solution of the inequality for the rejection of H_o gives a value for T_n^2 , which is a <u>lower bound</u> for the rejection region for H_o, and which is denoted by $\bar{T}_n^2$.

For every T_n^2 satisfying the inequality $\underline{T}^2 < T_n^2 < \bar{T}_n^2$, the testing - and hence the sampling - is to be continued.

In summary , the <u>modified decision process</u> is :

(1') In case $T_n^2 \leqq \underline{T}_n^2$, accept H_o ,

(2') In case $T_n^2 \geqq \bar{T}_n^2$, reject H_o ,

(3') In case $\underline{T}_n^2 < T_n^2 < \bar{T}_n^2$, continue testing .

The tables consequently contain the <u>upper bounds</u> $\underline{T}_n^2$ of the acceptance region for H_o as well as the <u>lower bounds</u> of the rejection region for H_o , all for various values of the parameters α, β, λ^2, and n .

(b) Extent of the tables and definition of the parameters :

(1) The parameter α :

α = eror probability
= probability of an error of the first kind

α = 0.05 .

(2) The parameter β :

β = probability of an error of the second kind

β = 0.05 .

(3) The parameter p :

p = dimension of the variates

p = 2(1)9 .

(4) The parameter λ^2 :

λ^2 = non-centrality parameter

λ^2 = 0.5(0.5)3(1)6;10 .

(5) The parameter n :

n = sample size

n = values between 1 and 75 , depending on the parameters p and λ^2 .

(c) Suggestions for applications :

As explained in section (a) , the tables serve for a simplified multivariate sequential testing for means with an unknown covariance matrix .

(d) References :

(1) For the source of the tables and for applications :

FREUND, R. J. + JACKSON, J. E. : Tables to facilitate multivariate sequential testing for means. (Table 8). (Technical Report No. 12). Blacksburg / Virginia : Department of Statistics and Statistical Laboratory, Virginia Agricultural Experiment Station , Virginia Polytechnic Institute, 1960 .

(2) For further suggestions :

JACKSON, J. E. + BRADLEY, R. A. : Multivariate sequential procedures for testing means. (Technical Report No. 10). Blacksburg / Virginia : Virginia Polytechnic Institute , 1959 .

JACKSON, J. E. + BRADLEY, R. A. : Sequential χ^2- and T^2-Tests. The Annals of Mathematical Statistics 32,1063 - 1077 (1961) .

(e) Further remarks :

(1) In contrast to the sequential χ^2-test (Table 24) there is no simple relation between the parameters λ^2 and n in the present case , so one needs substantially more extensive tables for the sequential T^2-Test.

(2) One can prove that the sequential test process described here, leads to a decision with probability one for finite n .

Tafel 25 Table 25

$\alpha = \beta = .05$

$p = 2$

n	$\lambda^2=0.5$ $\underline{T}^2$	$\lambda^2=0.5$ $\overline{T}^2$	$\lambda^2=1.0$ $\underline{T}^2$	$\lambda^2=1.0$ $\overline{T}^2$	$\lambda^2=1.5$ $\underline{T}^2$	$\lambda^2=1.5$ $\overline{T}^2$	$\lambda^2=2.0$ $\underline{T}^2$	$\lambda^2=2.0$ $\overline{T}^2$	$\lambda^2=2.5$ $\underline{T}^2$	$\lambda^2=2.5$ $\overline{T}^2$	n
3							2		35		3
4					2		52		99		4
5					42		105		166		5
6			3		83	51295	159	7251	234	4947	6
7			31	14633	124	4182	210	3194	302	2901	7
8			59	3899	166	2642	269	2388	370	2345	8
9		12266	88	2580	208	2130	324	2069	438	2112	9
10		4420	116	2079	249	1888	379	1913	505	2002	10
11		2942	144	1736	291	1757	433	1832	572	1951	11
12	3	2322	172	1675	332	1681	487	1791	639	1934	12
13	18	1985	201	1581	373	1638	541	1775	705	1937	13
14	33	1775	229	1520	414	1615	594	1774	771	1953	14
15	48	1635	257	1480	455	1606	648	1784	836	1980	15
16	63	1533	284	1454	496	1606	701	1802	902	2012	16
17	78	1459	312	1438	536	1613	754	1826	967	2050	17
18	92	1404	340	1429	576	1625	807	1854	1032	2092	18
19	107	1361	367	1426	617	1641	859	1885	1097	2137	19
20	121	1328	395	1428	657	1660	912	1919	1161	2184	20
21	136	1302	422	1432	697	1682	964	1955	1226	2234	21
22	150	1282	450	1440	736	1706	1016	1993	1290	2284	22
23	165	1266	477	1449	776	1731	1068	2032	1354	2336	23
24	179	1254	504	1461	816	1758	1120	2072	1418	2390	24
25	193	1245	531	1474	855	1786	1172	2114	1482	2444	25
26	207	1239	558	1488	895	1815	1224	2156	1546	2499	26
27	222	1234	585	1504	934	1845	1275	2199	1610	2554	27
28	236	1232	612	1520	974	1876	1327	2243	1674	2610	28
29	250	1231	639	1537	1013	1907	1379	2287	1737	2667	29
30	264	1231	665	1555	1052	1939	1430	2332	1801	2724	30
31	278	1232	692	1574	1091	1971					31
32	292	1234	719	1593	1130	2004					32
33	306	1237	745	1612	1169	2037					33
34	320	1241	772	1632	1208	2071					34
35	334	1246	798	1653	1247	2104					35
36	348	1251	825	1673	1286	2138					36

Tafel 25 (Forts.) Table 25 (cont.)

p = 2

n	$\lambda^2 = .5$		$\lambda^2 = 1.0$		$\lambda^2 = 1.5$		n
	$\underline{T}^2$	$\overline{T}^2$	$\underline{T}^2$	$\overline{T}^2$	$\underline{T}^2$	$\overline{T}^2$	
37	361	1257	851	1694	1325	2173	37
38	375	1263	878	1716	1363	2207	38
39	389	1269	904	1737	1402	2242	39
40	403	1276	930	1759	1441	2277	40
41	416	1284	956	1781			41
42	430	1291	983	1803			42
43	444	1299	1009	1826			43
44	457	1308	1035	1848			44
45	471	1316	1061	1871			45
46	485	1325	1087	1894			46
47	498	1334	1114	1917			47
48	512	1343	1140	1940			48
49	525	1352	1166	1963			49
50	539	1362	1192	1986			50
51	552	1371					51
52	566	1381					52
53	579	1391					53
54	593	1401					54
55	606	1411					55
56	620	1421					56
57	633	1432					57
58	647	1442					58
59	660	1453					59
60	673	1463					60
61	687	1474					61
62	700	1485					62
63	713	1496					63
64	727	1507					64
65	740	1518					65
66	753	1529					66
67	767	1540					67
68	780	1551					68
69	793	1562					69
70	806	1574					70

Tafel 25 (Forts.)

Table 25 (cont.)

p = 2

n	$\lambda^2 = .5$ $\underline{T}^2$	$\overline{T}^2$		n
71	8.20	15.85		71
72	8.33	15.96		72
73	8.46	16.08		73
74	8.59	16.19		74
75	8.73	16.31		75

$p = 3$

n	$\lambda^2 = .5$ $\underline{T}^2$	$\overline{T}^2$	$\lambda^2 = 1.0$ $\underline{T}^2$	$\overline{T}^2$	$\lambda^2 = 1.5$ $\underline{T}^2$	$\overline{T}^2$	$\lambda^2 = 2.0$ $\underline{T}^2$	$\overline{T}^2$	$\lambda^2 = 2.5$ $\overline{T}^2$	$\overline{T}^2$	n
4					4		78		149		4
5					63		155		242		5
6			4		122		229		331		6
7			46		179		300	23972	416	10621	7
8			87		234	8945	369	5694	498	4855	8
9			127	8453	287	4467	435	3787	577	3603	9
10			165	4439	339	3306	500	3090	654	3087	10
11		11580	202	3275	389	2790	563	2746	729	2824	11
12	5	5544	238	2731	438	2509	624	2554	803	2678	12
13	27	3886	274	2423	486	2340	685	2440	876	2597	13
14	49	3116	308	2230	533	2233	745	2372	948	2554	14
15	71	2674	342	2101	580	2164	803	2333	1019	2536	15
16	91	2390	375	2011	625	2120	862	2314	1090	2535	16
17	112	2193	408	1948	671	2093	919	2308	1159	2547	17
18	131	2051	440	1903	715	2078	976	2313	1228	2567	18
19	151	1943	472	1871	759	2073	1033	2326	1297	2595	19
20	170	1861	503	1849	803	2074	1089	2344	1365	2627	20
21	188	1796	534	1835	847	2081	1145	2367	1433	2664	21
22	207	1744	565	1826	890	2093	1200	2394	1501	2705	22
23	225	1702	596	1822	933	2107	1255	2424	1568	2748	23
24	242	1669	626	1822	975	2125	1310	2456	1635	2794	24
25	260	1641	656	1825	1018	2145	1365	2491	1702	2841	25
26	277	1619	686	1831	1060	2167	1419	2527	1768	2890	26
27	295	1601	715	1838	1102	2191	1473	2564	1835	2941	27
28	312	1587	744	1848	1144	2217	1527	2603	1901	2992	28
29	328	1575	774	1859	1185	2243	1581	2643	1967	3045	29
30	345	1566	803	1872	1226	2271	1635	2684	2032	3099	30
31	361	1559	832	1886	1268	2299					31
32	378	1554	860	1901	1309	2328					32
33	394	1550	889	1916	1350	2358					33
34	410	1548	917	1933	1391	2389					34
35	426	1547	946	1950	1431	2420					35
36	442	1548	974	1968	1472	2452					36
37	458	1549	1002	1986	1512	2484					37

Tafel 25 (Forts.) Table 25 (cont.)

p = 3

n	$\lambda^2 = .5$ $\underline{T}^2$	$\overline{T}^2$	$\lambda^2 = 1.0$ $\underline{T}^2$	$\overline{T}^2$	$\lambda^2 = 1.5$ $\underline{T}^2$	$\overline{T}^2$		n
38	473	1551	1030	2005	1553	2517		38
39	489	1554	1058	2025	1593	2550		39
40	505	1557	1086	2045	1633	2583		40
41	520	1561	1114	2065				41
42	535	1566	1142	2085				42
43	551	1571	1169	2106				43
44	566	1577	1197	2127				44
45	581	1583	1225	2148				45
46	596	1589	1252	2170				46
47	611	1596	1279	2192				47
48	626	1604	1307	2214				48
49	641	1611	1334	2236				49
50	656	1619	1361	2258				50
51	670	1627						51
52	685	1635						52
53	700	1644						53
54	714	1652						54
55	729	1661						55
56	744	1670						56
57	758	1680						57
58	773	1689						58
59	787	1699						59
60	801	1708						60
61	816	1718						61
62	830	1728						62
63	844	1738						63
64	859	1748						64
65	873	1759						65
66	887	1769						66
67	901	1780						67
68	916	1790						68
69	930	1801						69
70	944	1812						70
71	958	1822						71

Tafel 25 (Forts.)

Table 25 (cont.)

$p = 3$

	$\lambda^2 = .5$			
n	$\underline{T}^2$	$\overline{T}^2$		n
72	972	1833		72
73	986	1844		73
74	1000	1855		74
75	1014	1866		75

p = 4

n	$\lambda^2 = 1.0$ $\underline{T}^2$	$\lambda^2 = 1.0$ $\overline{T}^2$	$\lambda^2 = 2.0$ $\underline{T}^2$	$\lambda^2 = 2.0$ $\overline{T}^2$	$\lambda^2 = 3.0$ $\underline{T}^2$	$\lambda^2 = 3.0$ $\overline{T}^2$	$\lambda^2 = 5.0$ $\underline{T}^2$	$\lambda^2 = 5.0$ $\overline{T}^2$	n
6	7		403		769		1459		6
7	78		513		899		1609		7
8	146		611		1015		1752	59100	8
9	211		700		1124	21585	1891	13101	9
10	270		783	13919	1226	9331	2028	8642	10
11	327	46042	862	7590	1324	6660	2163	7031	11
12	379	10630	936	5673	1419	5528	2297	6245	12
13	429	6597	1007	4769	1511	4927	2429	5809	13
14	477	5056	1077	4257	1602	4570	2561	5553	14
15	522	4252	1144	3937	1690	4345	2692	5404	15
16	565	3764	1209	3726	1778	4201	2822	5322	16
17	607	3441	1274	3582	1865	4109	2952	5285	17
18	648	3215	1337	3482	1950	4052	3082	5280	18
19	688	3050	1399	3414	2035	4021	3211	5298	19
20	726	2926	1460	3367	2119	4009	3339	5333	20
21	764	2832	1521	3338	2203	4011	3467	5382	21
22	800	2760	1580	3321	2286	4024	3595	5442	22
23	836	2704	1640	3315	2368	4046	3723	5510	23
24	872	2660	1699	3316	2450	4076	3850	5586	24
25	907	2627	1757	3324	2532	4111	3977	5666	25
26	941	2601	1815	3337	2613	4151	4104	5752	26
27	975	2582	1872	3355	2694	4194	4230	5841	27
28	1008	2569	1930	3376	2775	4242	4356	5934	28
29	1042	2559	1986	3400	2855	4292	4483	6030	29
30	1074	2554	2043	3427	2935	4344	4609	6128	30
31	1107	2552	2099	3457					31
32	1139	2552	2155	3488					32
33	1171	2555	2211	3521					33
34	1202	2560	2267	3556					34
35	1234	2567	2322	3592					35
36	1265	2575	2377	3629					36
37	1296	2585	2432	3667					37

Tafel 25 (Forts.)

Table 25 (cont.)

p = 4

n	$\lambda^2 = 1.0$ $\underline{T}^2$	$\overline{T}^2$	$\lambda^2 = 2.0$ $\underline{T}^2$	$\overline{T}^2$	n
38	1326	2596	2487	3707	38
39	1357	2608	2542	3747	39
40	1387	2621	2597	3787	40
41	1418	2636	2651	3829	41
42	1448	2650	2706	3871	42
43	1478	2666	2760	3914	43
44	1507	2682	2814	3957	44
45	1537	2699	2868	4001	45
46	1566	2717	2922	4045	46
47	1596	2735	2975	4089	47
48	1625	2753	3029	4134	48
49	1654	2772	3083	4179	49
50	1684	2791	3136	4224	50
51	1713	2810			51
52	1741	2830			52
53	1770	2851			53
54	1799	2871			54
55	1828	2892			55
56	1856	2913			56
57	1885	2934			57
58	1913	2956			58
59	1942	2977			59
60	1970	2999			60

p = 5

n	$\lambda^2 = 1.0$ $\underline{T}^2$	$\lambda^2 = 1.0$ $\overline{T}^2$	$\lambda^2 = 2.0$ $\underline{T}^2$	$\lambda^2 = 2.0$ $\overline{T}^2$	$\lambda^2 = 3.0$ $\underline{T}^2$	$\lambda^2 = 3.0$ $\overline{T}^2$	$\lambda^2 = 5.0$ $\underline{T}^2$	$\lambda^2 = 5.0$ $\overline{T}^2$	n
7	95		652		1168		2141		7
8	179		766		1289		2247		8
9	257		868		1398		2359	104906	9
10	329		959		1500	32447	2476	17626	10
11	396		1043	20148	1597	12416	2596	11155	11
12	458	274091	1122	10039	1691	8542	2717	8857	12
13	516	15557	1198	7252	1783	6935	2840	7725	13
14	570	8813	1270	5970	1873	6080	2964	7082	14
15	621	6505	1339	5248	1961	5565	3089	6689	15
16	670	5349	1407	4795	2048	5234	3214	6442	16
17	716	4662	1473	4492	2134	5013	3339	6286	17
18	761	4211	1538	4281	2219	4863	3465	6192	18
19	804	3896	1601	4130	2303	4761	3591	6142	19
20	845	3667	1664	4021	2387	4693	3717	6123	20
21	886	3494	1725	3943	2471	4652	3842	6128	21
22	925	3361	1786	3887	2554	4629	3968	6151	22
23	964	3257	1847	3848	2636	4621	4094	6189	23
24	1002	3175	1907	3823	2718	4625	4220	6238	24
25	1039	3110	1966	3808	2800	4639	4346	6297	25
26	1075	3058	2025	3802	2882	4661	4472	6363	26
27	1111	3017	2083	3803	2963	4689	4597	6436	27
28	1146	2984	2141	3810	3044	4722	4723	6515	28
29	1180	2959	2199	3822	3124	4760	4848	6597	29
30	1215	2939	2256	3838	3205	4802	4974	6684	30
31	1249	2924	2313	3858					31
32	1282	2914	2370	3880					32
33	1315	2907	2426	3905					33
34	1348	2903	2483	3933					34
35	1380	2902	2539	3963					35
36	1413	2903	2595	3994					36
37	1445	2906	2650	4027					37
38	1476	2912	2706	4061					38

afel 25 (Forts.)

Table 25 (cont.)

p = 5

n	$\lambda^2 = 1.0$ $\underline{T}^2$	$\lambda^2 = 1.0$ $\overline{T}^2$	$\lambda^2 = 2.0$ $\underline{T}^2$	$\lambda^2 = 2.0$ $\overline{T}^2$	n
39	1508	2918	2761	4097	39
40	1539	2927	2816	4134	40
41	1570	2936	2871	4171	41
42	1601	2947	2926	4210	42
43	1632	2959	2981	4249	43
44	1663	2971	3036	4290	44
45	1693	2985	3090	4331	45
46	1724	2999	3145	4372	46
47	1754	3014	3199	4414	47
48	1784	3030	3253	4457	48
49	1814	3047	3307	4500	49
50	1844	3063	3362	4543	50
51	1873	3081			51
52	1903	3099			52
53	1933	3117			53
54	1962	3136			54
55	1991	3155			55
56	2020	3174			56
57	2050	3194			57
58	2079	3214			58
59	2108	3234			59
60	2137	3255			60

$p = 6$

n	$\lambda^2 = 1.0$ $\underline{T}^2$	$\overline{T}^2$	$\lambda^2 = 2.0$ $\underline{T}^2$	$\overline{T}^2$	$\lambda^2 = 3.0$ $\underline{T}^2$	$\overline{T}^2$	$\lambda^2 = 5.0$ $\underline{T}^2$	$\overline{T}^2$	n
5			212		452		908		5
6	6		309		578		1076		6
7	62		398		693		1233	32070	7
8	116		481	192013	800	13641	1383	9282	8
9	167		559	9191	901	6714	1527	6445	9
10	216	18311	633	5512	998	5009	1668	5407	10
11	262	7039	704	4288	1092	4273	1806	4911	11
12	307	4766	772	3696	1184	3884	1942	4650	12
13	349	3803	839	3360	1274	3661	2077	4511	13
14	390	3279	903	3153	1362	3527	2210	4445	14
15	430	2955	967	3020	1449	3449	2342	4426	15
16	468	2739	1029	2934	1535	3406	2474	4437	16
17	505	2587	1090	2878	1620	3388	2604	4470	17
18	542	2476	1151	2844	1704	3389	2734	4520	18
19	578	2397	1211	2825	1787	3402	2863	4581	19
20	613	2337	1270	2818	1870	3425	2992	4652	20
21	647	2292	1328	2821	1952	3456	3120	4730	21
22	681	2258	1386	2830	2034	3494	3248	4815	22
23	714	2233	1444	2846	2116	3536	3375	4903	23
24	747	2216	1501	2865	2197	3583	3502	4996	24
25	780	2204	1558	2889	2277	3632	3629	5093	25
26	812	2197	1614	2916	2357	3685	3755	5192	26
27	844	2194	1670	2945	2437	3740	3881	5293	27
28	875	2194	1726	2977	2517	3797	4007	5396	28
29	907	2196	1782	3010	2597	3856	4133	5501	29
30	938	2201	1837	3046	2676	3917	4258	5608	30
31	969	2208	1892	3082					31
32	999	2217	1947	3120					32
33	1029	2227	2002	3159					33
34	1060	2239	2056	3199					34
35	1090	2252	2111	3240					35
36	1119	2266	2165	3281					36

Tafel 25 (Forts.) Table 25 (cont.)

p = 6

n	$\lambda^2 = 1.0$		$\lambda^2 = 2.0$			n
	$\underline{T}^2$	$\overline{T}^2$	$\underline{T}^2$	$\overline{T}^2$		
37	1149	2280	2219	3323		37
38	1178	2296	2273	3366		38
39	1208	2312	2327	3410		39
40	1237	2329	2380	3453		40
41	1266	2347	2434	3498		41
42	1295	2365	2487	3542		42
43	1324	2383	2541	3587		43
44	1353	2402	2594	3633		44
45	1381	2422	2647	3679		45
46	1410	2442	2700	3725		46
47	1439	2462	2753	3771		47
48	1467	2482	2806	3817		48
49	1495	2503	2859	3864		49
50	1524	2524	2912	3911		50
51	1552	2545				51
52	1580	2567				52
53	1608	2588				53
54	1636	2610				54
55	1664	2632				55
56	1692	2655				56
57	1719	2677				57
58	1747	2700				58
59	1775	2722				59
60	1802	2745				60

p = 7

n	$\lambda^2 = 2.0$ $\underline{T}^2$	$\overline{T}^2$	$\lambda^2 = 4.0$ $\underline{T}^2$	$\overline{T}^2$	$\lambda^2 = 6.0$ $\underline{T}^2$	$\overline{T}^2$	$\lambda^2 = 10.0$ $\underline{T}^2$	$\overline{T}^2$	n
8	958		2311		3582		5993		8
9	1070		2383		3578		5801		9
10	1168		2463		3626	74739	5767	42709	10
11	1257		2549	26991	3702	21536	5818	22043	11
12	1338	28165	2639	14233	3798	14195	5921	16486	12
13	1415	12878	2732	10530	3906	11416	6057	14006	13
14	1488	9028	2828	8811	4023	9979	6215	12668	14
15	1559	7299	2925	7845	4147	9141	6388	11877	15
16	1627	6333	3023	7247	4275	8615	6574	11391	16
17	1693	5727	3122	6854	4407	8276	6768	11092	17
18	1758	5319	3222	6588	4541	8054	6968	10915	18
19	1822	5031	3323	6406	4678	7912	7174	10823	19
20	1885	4822	3423	6282	4816	7827	7384	10793	20
21	1946	4668	3525	6200	4956	7784			21
22	2008	4553	3626	6150	5097	7771			22
23	2068	4467	3728	6123	5238	7783			23
24	2128	4403	3829	6115	5381	7815			24
25	2187	4357	3931	6122	5524	7861			25
26	2246	4324	4033	6141	5668	7920			26
27	2305	4302	4135	6170	5812	7990			27
28	2363	4289	4237	6207	5956	8068			28
29	2421	4284	4339	6252	6101	8153			29
30	2478	4285	4441	6302	6246	8244			30
31	2536	4292	4543	6356					31
32	2593	4302	4645	6416					32
33	2649	4317	4747	6479					33
34	2706	4336	4849	6545					34
35	2762	4357	4951	6614					35
36	2818	4381	5053	6685					36
37	2874	4407	5155	6759					37
38	2930	4435	5256	6834					38
39	2986	4465	5358	6912					39

Tafel 25 (Forts.) Table 25 (cont.)

p = 7

n	$\lambda^2 = 2.0$		$\lambda^2 = 4.0$		n
	$\underline{T}^2$	$\overline{T}^2$	$\underline{T}^2$	$\overline{T}^2$	
40	3041	4496	5460	6991	40
41	3097	4529			41
42	3152	4563			42
43	3207	4598			43
44	3262	4635			44
45	3317	4672			45
46	3372	4710			46
47	3426	4749			47
48	3481	4789			48
49	3535	4829			49
50	3590	4870			50

Tafel 25 (Forts.) Table 25 (cont.)

p = 8

n	$\lambda^2 = 2.0$		$\lambda^2 = 4.0$		$\lambda^2 = 6.0$		$\lambda^2 = 10.0$		n
	$\underline{T}^2$	$\overline{T}^2$	$\underline{T}^2$	$\overline{T}^2$	$\underline{T}^2$	$\overline{T}^2$	$\underline{T}^2$	$\overline{T}^2$	
9	1322		3067		4694		7816		9
10	1423		3076		4578		7371		10
11	1512		3110		4541	104705	7176	54438	11
12	1593		3162	34736	4559	26913	7117	27184	12
13	1669	38396	3226	17565	4609	17358	7139	19959	13
14	1740	16120	3298	12747	4682	13715	7213	16721	14
15	1809	11004	3377	10522	4771	11845	7322	14951	15
16	1875	8760	3460	9270	4872	10740	7457	13880	16
17	1940	7515	3546	8486	4981	10036	7610	13199	17
18	2003	6735	3635	7964	5097	9567	7777	12755	18
19	2065	6208	3727	7603	5217	9248	7955	12468	19
20	2126	5834	3820	7348	5342	9031	8141	12288	20
21	2187	5560	3914	7167	5471	8885			21
22	2247	5355	4010	7038	5601	8792			22
23	2307	5199	4106	6949	5734	8738			23
24	2366	5079	4203	6890	5869	8715			24
25	2424	4987	4301	6855	6006	8716			25
26	2483	4917	4399	6839	6143	8737			26
27	2540	4864	4498	6837	6282	8774			27
28	2598	4824	4597	6848	6422	8823			28
29	2655	4796	4697	6870	6562	8883			29
30	2713	4777	4796	6900	6703	8953			30
31	2769	4766	4896	6937					31
32	2826	4762	4996	6981					32
33	2883	4763	5096	7030					33
34	2939	4769	5197	7083					34
35	2995	4780	5297	7141					35
36	3051	4794	5398	7202					36
37	3107	4811	5498	7267					37
38	3163	4831	5599	7334					38
39	3218	4854	5700	7403					39
40	3274	4878	5801	7475					40

$p = 8$

n	$\lambda^2 = 2.0$ $\underline{T}^2$	$\overline{T}^2$	n
41	3329	4905	41
42	3384	4934	42
43	3439	4964	43
44	3494	4995	44
45	3549	5028	45
46	3604	5062	46
47	3659	5097	47
48	3713	5134	48
49	3768	5171	49
50	3822	5209	50

Tafel 25 (Forts.) Table 25 (cont.)

p = 9

n	$\lambda^2 = 2.0$ $\underline{T}^2$	$\lambda^2 = 2.0$ $\overline{T}^2$	$\lambda^2 = 4.0$ $\underline{T}^2$	$\lambda^2 = 4.0$ $\overline{T}^2$	$\lambda^2 = 6.0$ $\underline{T}^2$	$\lambda^2 = 6.0$ $\overline{T}^2$	$\lambda^2 = 10.0$ $\underline{T}^2$	$\lambda^2 = 10.0$ $\overline{T}^2$	
10	1743		3930		5977		9880		10
11	1826		3856		5701		9129		11
12	1900		3832		5562	139921	8739	67669	12
13	1969		3840	43614	5503	32907	8545	32865	13
14	2034	51210	3870	21256	5500	20800	8468	23761	14
15	2097	19779	3916	15172	5526	16220	8467	19672	15
16	2158	13184	3972	12381	5580	13864	8517	17416	16
17	2218	10351	4037	10808	5652	12463	8604	16033	17
18	2277	8794	4108	9817	5737	11558	8718	15132	18
19	2336	7819	4184	9152	5833	10946	8853	14527	19
20	2394	7160	4264	8685	5937	10518	9003	14116	20
21	2452	6690	4348	8349	6047	10216			21
22	2509	6344	4434	8103	6163	10003			22
23	2566	6083	4522	7923	6282	9854			23
24	2623	5861	4612	7791	6405	9754			24
25	2680	5724	4703	7696	6531	9692			25
26	2736	5601	4795	7630	6660	9660			26
27	2793	5504	4889	7588	6790	9652			27
28	2849	5428	4983	7563	6922	9663			28
29	2905	5369	5079	7554	7056	9690			29
30	2961	5324	5175	7558	7191	9731			30
31	3017	5290	5271	7572					31
32	3072	5266	5368	7596					32
33	3128	5250	5466	7627					33
34	3183	5240	5563	7664					34
35	3239	5237	5661	7707					35
36	3294	5238	5760	7756					36
37	3349	5244	5858	7808					37
38	3405	5254	5957	7865					38
39	3460	5268	6056	7925					39
40	3515	5284	6156	7988					40
41	3570	5303							41

Tafel 25 (Forts.) | Table 25 (cont.)

p = 9

n	$\lambda^2 = 2.0$ $\underline{T}^2$	$\overline{T}^2$	n
42	3624	5325	42
43	3679	5349	43
44	3734	5374	44
45	3788	5402	45
46	3843	5431	46
47	3898	5461	47
48	3952	5493	48
49	4006	5526	49
50	4061	5560	50

PART IV (APPENDIX)
Supplement

Preliminary Remarks:

The distributions and test statistics under consideration in this part of the collection of tables have been published in recent years from 1975 to 1982 , namely , after establishing the content of the German edition of this work in 1975 .

By technical reasons they have not been included in part II or part III, respectively , but have been compiled in this appendix.

Especially , the tables 26, 27 , and 30 are related to part II , and the tables 28, and 29 are related to part III .

TABLE 26

The Mardia-Test for Multivariate Normality, Skewness, and Kurtosis: Tables by K. V. Mardia

(a) Content of the tables and definition of the test statistics :

The test for multinormality proposed by K. V. MARDIA (1970 ; 1971) is based on the simultaneous testing for multivariate skewness and multivariate kurtosis .

Let $\underline{x}' = (x_1, x_2, \ldots, x_p)$ and $\underline{y}' = (y_1, y_2, \ldots, y_p)$ be two independent random vectors belonging to the same distribution. The mean vector of the corresponding population will be denoted by

$$E(\underline{x}) = E(\underline{y}) = \underline{\mu} = (\mu_1, \mu_2, \ldots, \mu_p)$$

and the covariance matrix by $\underline{\Sigma}$.

As a measure of the multivariate skewness of the p-dimensional distribution of the population one may follow MARDIA (1974) and choose the formula

$$\beta_{1,p} = E\{(\underline{x} - \underline{\mu})' \cdot \underline{\Sigma}^{-1} \cdot (\underline{y} - \underline{\mu})\}^3 \quad .$$

The corresponding formula for the skewness of a p-dimensional sample of size n is

$$b_{1,p} = \frac{1}{n^2} \cdot \sum_{i=1}^{n} \sum_{j=1}^{n} \{ (\underline{x}_i - \underline{\bar{x}})' \cdot \underline{S}^{-1} \cdot (\underline{x}_j - \underline{\bar{x}}) \}^3 \quad .$$

Here $\underline{\bar{x}}' = (\bar{x}_1, \bar{x}_2, \ldots, \bar{x}_p)$ represents the mean vector and $\underline{S}$ the covariance matrix of a p-dimensional sample of size n consisting of elements $\underline{x}_i = (x_1, x_2, \ldots, x_p)$.

As a measure of multivariate kurtosis of a p-dimensional distribution of the population one chooses , again following MARDIA (1974) , the formula

$$\beta_{2,p} = E \{ (\underline{x} - \underline{\mu})' \cdot \underline{\Sigma}^{-1} \cdot (\underline{x} - \underline{\mu}) \}^2 \quad .$$

The analog for the multivariate kurtosis of a p-dimensional sample of size n is

$$b_{2,p} = \frac{1}{n} \cdot \sum_{i=1}^{n} \{ (\underline{x}_i - \underline{\bar{x}}) \cdot \underline{S}^{-1} \cdot (\underline{x}_i - \underline{\bar{x}}) \}^2 \quad .$$

The test for multivariate normality on the basis of a sample then consists of simultaneous testing for multivariate skewness by

$$H_{o1} : \beta_{1,p} = 0 \quad ,$$

and for multivariate kurtosis by

$$H_{o2} : \beta_{2,p} = p \cdot (p + 2)$$

by means of the corresponding test statistics $b_{1,p}$ and $b_{2,p}$. The tables therefore contain the upper percentage points for $b_{1,p}$ and the upper and lower percentage points for $b_{2,p}$.

(b) Extent of the tables and definition of the parameters :

(1) The parameter α :

α = error probability

(1') for $b_{1,p}$:

α = 0.001 ; 0.01 ; 0.025 ; 0.05 ;
0.075 ; 0.10 .

(2') for $b_{2,p}$ and $p = 2$:

lower and upper percentage points for

α = 0.005 ; 0.0125 ; 0.025 ; 0.05 ,

or for simultaneous testing

$2.\alpha$ = 0.01 ; 0.025 ; 0.05 ; 0.10 .

(3') for $b_{2,p}$ and $p = 3 , 4$:

lower and upper percentage points for

α = 0.01 ; 0.025 ; 0.05 ; 0.10 ,
or for simultaneous testing

$2.\alpha$ = 0.02 ; 0.05 ; 0.10 ; 0.20 .

(2) The parameter p :

p = dimension of the variates

p = 2(1)4 .

(3) Thr parameter n :

n = size of the p-dimensional sample

n = 10(2)20(5)30(10)100(50)200(100)400(200)1000;
1500; ... ;5000 .

(c) Suggestions for applications :

As explained in section (a) , the tables contain the upper percentage points for the criterion $b_{1,p}$ for testing for multivariate skewness as well as the upper and lower percentage points for the criterion $b_{2,p}$ for testing for multivariate kurtosis.
The simultaneous testing of both criteria serves as a test for multivariate normality .

(d) References :

(1) For the source of the tables :

MARDIA, K. V. : Personal communication , August 1975 . Department of Statistics, School of Mathematics, The University of Leeds,Leeds, England .

(2) For the test criteria and for further suggestions :

MARDIA, K. V. : Measures of multivariate skewness and kurtosis with applications. Biometrika 57, 519 - 530 (1970) .

MARDIA, K. V. : Applications of some measures of multivariate skewness and kurtosis in testing normality and robustness studies. Sankhya, Series B, Vol. 36, 115 - 128 (1974) .

MARDIA, K. V. : Assessment of multinormality and robustness of HOTELLING's T^2 test. Applied Statistics 24, 163 - 171 (1975) .

MARDIA, K. V. + ZEMROCH, P. J. : Algorithm AS 84 : Measures of multivariate skewness and kurtosis. Applied Statistics 24, 262 - 265 (1975) .

PEARSON, E. S. + HARTLEY, H. O. (eds.) : Biometrika Tables for Statisticians , Volume 1 (Third Edition). Cambridge : Cambridge University Press, 1966 . (Published for the Biometrika Trustees).

(e) Further remarks :

(1) For p = 1 . the critical values for $b_{1,p}$ and $b_{2,p}$ can already be found in PEARSON + HARTLEY (1966), (pp. 207-208).

(2) A FORTRAN subroutine for computing the test criteria $b_{1,p}$ and $b_{2,p}$ can be found in MARDIA + ZEMROCH (1975) .

(3) For dimensions of $p > 2$, MARDIA (1974) provides the following approximations :

(1') $A = n.b_{1,p}/6 \sim \chi^2_f$-distributed ,

where $f = p.(p+1)(p+2)/6$.

(2') $B = \{ b_{2,p} - p.(p+2)\}/\{8.p(p+2)/n\}^{1/2}$

is asymptotically N(0;1)-distributed .

(3') $A' = n . K . b_{1,p}/6$, where

$K = (p + 1).(n + 1).(n + 3)/\{n((n+1)(p+1)-6)\}$.

instead of A as in (1') .

(4') $$B' = \frac{\{(n+1).b_{2,p} - p(p+2).(n-1)\}.\{(n+3).(n+5)\}^{1/2}}{\{8.p(p+2).(n-3).(n-p-1).(n-p+1)\}^{1/2}}$$

is asympltotically N(0;1)-distributed .

(4) Extensive Monte Carlo studies led MARDIA to the following recommendations for $n \geqq 50$:

(1') For the upper 5% points of $b_{1,p}$ use the approximations in (3)/(3') .

(2') For the lower 2.5% points of $b_{2,p}$ regard $b_{2,p}$ as normally distributed with mean $p(p+2)(n+p+1)/n$ and variance $8.p(p+2)/(n-1)$, whenever $50 \leqq n \leqq 400$. For $n > 400$, use (3)/(2') .

(3') For the upper 2.5% points of $b_{2,p}$ also use (3)/(2').

Tafel 26 Table 26

$b_{1,p}$ $p = 2$

Obere Prozentpunkte /

Upper percentage points

n ↓ α →	0.001	0.01	0.025	0.05	0.075	0.10
10	6.994	5.194	4.294	3.694	3.263	2.994
12	6.744	4.938	3.931	3.319	2.944	2.681
14	6.419	4.581	3.619	3.031	2.669	2.419
16	6.062	4.231	3.337	2.775	2.444	2.219
18	5.737	3.962	3.100	2.556	2.256	2.050
20	5.425	3.669	2.881	2.356	2.081	1.894
25	4.719	3.106	2.438	1.969	1.744	1.581
30	4.238	2.681	2.094	1.687	1.513	1.363
40	3.369	2.087	1.606	1.319	1.181	1.050
50	2.706	1.744	1.306	1.069	0.969	0.862
60	2.200	1.444	1.094	0.906	0.819	0.731
70	1.863	1.244	0.937	0.794	0.725	0.631
80	1.587	1.056	0.812	0.694	0.637	0.544
90	1.400	0.919	0.725	0.638	0.569	0.487
100	1.231	0.831	0.656	0.581	0.506	0.438
150	0.794	0.531	0.444	0.400	0.344	0.281
200	0.569	0.394	0.331	0.300	0.269	0.219
300	0.369	0.256	0.225	0.209	0.169	0.144
400	0.275	0.197	0.166	0.141	0.129	0.116
600	0.183	0.131	0.110	0.094	0.085	0.077
800	0.137	0.099	0.083	0.071	0.064	0.058
1000	0.110	0.079	0.066	0.057	0.051	0.046
1500	0.074	0.053	0.044	0.038	0.034	0.031
2500	0.044	0.032	0.027	0.023	0.021	0.019
3000	0.037	0.027	0.022	0.019	0.017	0.016
4000	0.028	0.020	0.017	0.014	0.013	0.012
5000	0.022	0.016	0.013	0.011	0.010	0.009

afel 26 (Forts.) Table 26 (cont.)

$b_{2,p}$ $p = 2$

n ↓	Obere Prozentpunkte / *Upper percentage points* α → 0.005	0.0125	0.025	0.05	Untere Prozentpunkte / *Lower percentage points* 0.05	0.025	0.0125	0.005
10	10.378	9.781	9.203	8.606	5.057	4.887	4.722	4.580
12	10.881	10.150	9.593	8.947	5.232	5.053	4.899	4.732
14	11.159	10.375	9.769	9.162	5.358	5.179	5.015	4.842
16	11.387	10.562	9.941	9.331	5.482	5.318	5.149	4.977
18	11.478	10.628	10.005	9.403	5.555	5.382	5.219	5.045
20	11.609	10.691	10.114	9.469	5.717	5.533	5.262	5.175
25	11.628	10.584	10.159	9.503	5.871	5.689	5.525	5.351
30	11.594	10.556	10.156	9.516	6.038	5.855	5.692	5.518
40	11.453	10.563	10.109	9.497	6.229	6.139	5.871	5.703
50	11.181	10.372	9.987	9.453	6.403	6.239	6.083	5.909
60	10.994	10.250	9.889	9.401	6.505	6.335	6.189	6.015
70	10.753	10.106	9.781	9.356	6.602	6.437	6.290	6.139
80	10.537	9.981	9.694	9.309	6.683	6.539	6.372	6.223
90	10.325	9.885	9.688	9.256	6.749	6.622	6.475	6.332
100	10.188	9.806	9.556	9.210	6.793	6.665	6.521	6.389
150	10.253	9.475	9.300	9.027	6.972	6.858	6.749	6.615
200	9.506	9.269	9.141	8.919	7.083	6.979	6.889	6.761
300	9.219	9.031	8.916	8.766	7.245	7.142	7.052	6.949
400	9.061	8.917	8.787	8.664	7.342	7.252	7.171	7.079
600	8.874	8.749	8.647	8.547	7.464	7.369	7.295	7.232
800	8.747	8.641	8.562	8.472	7.536	7.451	7.372	7.304
1000	8.656	8.569	8.497	8.419	7.585	7.504	7.433	7.367
1500	8.532	8.463	8.405	8.339	7.661	7.595	7.537	7.460
2000	8.461	8.401	8.351	8.293	7.707	7.649	7.599	7.535
2500	8.412	8.359	8.314	8.262	7.738	7.686	7.641	7.588
3000	8.376	8.327	8.286	8.240	7.760	7.714	7.673	7.624
4000	8.326	8.284	8.248	8.207	7.793	7.752	7.716	7.674
5000	8.291	8.254	8.222	8.186	7.814	7.778	7.746	7.709

Tafel 26 (Forts.) Table 26 (cont.)

$b_{1,p}$ $p = 3$

Obere Prozentpunkte / *Upper percentage points*

n ↓ \ α →	0.001	0.01	0.025	0.05	0.075	0.1
10	11.5	8.8	7.7	6.9	6.5	6.0
12	10.5	8.1	7.1	6.4	5.9	5.5
14	9.7	7.4	6.5	5.9	5.4	5.0
16	8.9	6.8	6.1	5.4	4.9	4.6
18	8.3	6.4	5.6	5.1	4.6	4.2
20	7.7	6.0	5.3	4.7	4.2	3.9
25	6.5	5.2	4.5	3.9	3.5	3.3
30	5.6	4.4	3.9	3.3	3.0	2.8
40	4.2	3.5	3.0	2.7	2.4	2.2
50	3.4	2.8	2.4	2.2	1.9	1.7
60	2.9	2.4	2.0	1.8	1.6	1.5
70	2.5	2.0	1.7	1.5	1.4	1.3
80	2.2	1.7	1.5	1.3	1.2	1.13
90	1.9	1.5	1.3	1.16	1.08	1.01
100	1.7	1.3	1.18	1.05	0.97	0.92
150	1.15	0.90	0.80	0.71	0.66	0.62
200	0.87	0.68	0.60	0.54	0.50	0.47
300	0.58	0.46	0.40	0.36	0.33	0.32
400	0.44	0.34	0.30	0.272	0.252	0.237
600	0.294	0.230	0.203	0.182	0.168	0.159
800	0.221	0.173	0.153	0.137	0.127	0.119
1000	0.177	0.139	0.122	0.109	0.101	0.095
1500	0.118	0.093	0.082	0.073	0.068	0.064
2000	0.089	0.069	0.061	0.055	0.051	0.048
3000	0.059	0.046	0.041	0.037	0.034	0.032
4000	0.044	0.035	0.031	0.027	0.025	0.024
5000	0.035	0.028	0.025	0.022	0.020	0.019

Tafel 26 (Forts.) Table 26 (cont.)

$b_{2,p}$ p = 3

	Obere Prozentpunkte / *Upper percentage points*				Untere Prozentpunkte / *Lower percentage points*			
n ↓ α →	0.01	0.025	0.05	0.10	0.10	0.05	0.025	0.01
10	15.6	15.0	14.4	14.0	10.7	10.4	10.2	10.0
12	16.4	15.9	15.2	14.7	11.0	10.7	10.4	10.2
14	17.1	16.5	15.8	15.1	11.3	10.9	10.6	10.4
16	17.5	16.8	16.1	15.4	11.5	11.1	10.8	10.5
18	17.8	17.1	16.4	15.5	11.6	11.3	11.0	10.7
20	18.0	17.2	16.5	15.7	11.8	11.4	11.1	10.8
25	18.2	17.4	16.7	15.9	12.1	11.8	11.4	11.1
30	18.3	17.5	16.7	16.0	12.3	12.0	11.6	11.3
40	18.2	17.4	16.7	16.1	12.7	12.4	12.0	11.7
50	18.0	17.3	16.7	16.1	12.9	12.6	12.3	11.9
60	17.9	17.2	16.6	16.1	13.1	12.8	12.5	12.1
70	17.7	17.1	16.6	16.1	13.2	13.0	12.6	12.3
80	17.6	17.0	16.5	16.1	13.3	13.1	12.8	12.4
90	17.5	16.9	16.5	16.0	13.5	13.2	12.9	12.5
100	17.4	16.8	16.4	16.0	13.5	13.3	13.0	12.6
150	17.0	16.5	16.2	15.9	13.8	13.6	13.3	13.0
200	16.8	16.3	16.1	15.8	14.0	13.8	13.5	13.2
300	16.5	16.1	15.9	15.7	14.2	14.0	13.8	13.6
400	16.3	16.0	15.8	15.6	14.3	14.1	13.9	13.7
600	15.97	15.81	15.67	15.51	14.4	14.3	14.1	13.9
800	15.85	15.71	15.59	15.45	14.5	14.3	14.2	14.1
1000	15.77	15.64	15.53	15.41	14.53	14.41	14.30	14.17
1500	15.63	15.53	15.44	15.34	14.62	14.52	14.43	14.33
2000	15.55	15.46	15.39	15.30	14.67	14.58	14.51	14.42
3000	15.45	15.38	15.32	15.25	14.73	14.66	14.60	14.53
4000	15.39	15.33	15.28	15.21	14.77	14.71	14.65	14.59
5000	15.35	15.30	15.25	15.19	14.80	14.74	14.69	14.63

Tafel 26 (Forts.) Table 26 (cont.)

$b_{1,p}$ $p = 4$

Obere Prozentpunkte / *Upper percentage points*

n↓ \ α→	0.001	0.01	0.025	0.05	0.075	0.1
10	17.9	15.3	13.3	12.2	11.6	11.1
12	16.2	13.9	12.2	11.2	10.6	10.1
14	14.8	12.7	11.2	10.2	9.7	9.2
16	13.6	11.6	10.3	9.4	8.8	8.4
18	12.6	10.7	9.5	8.7	8.0	7.7
20	11.6	9.9	8.8	8.0	7.4	7.0
25	9.7	8.1	7.1	6.6	6.2	5.9
30	8.1	6.8	6.0	5.6	5.3	5.0
40	6.2	5.2	4.6	4.3	4.1	3.9
50	5.0	4.2	3.8	3.5	3.3	3.1
60	4.2	3.5	3.2	2.9	2.8	2.7
70	3.7	3.0	2.8	2.5	2.4	2.3
80	3.2	2.7	2.4	2.2	2.1	2.0
90	2.9	2.4	2.2	2.0	1.89	1.81
100	2.6	2.2	1.97	1.81	1.71	1.64
150	1.76	1.46	1.33	1.22	1.16	1.11
200	1.33	1.10	1.00	0.92	0.87	0.84
300	0.89	0.74	0.67	0.62	0.59	0.56
400	0.67	0.56	0.51	0.47	0.44	0.42
600	0.45	0.37	0.34	0.31	0.295	0.282
800	0.34	0.280	0.255	0.234	0.222	0.212
1000	0.271	0.224	0.204	0.188	0.177	0.170
1500	0.181	0.150	0.136	0.125	0.118	0.113
2000	0.136	0.112	0.102	0.094	0.089	0.085
3000	0.091	0.075	0.068	0.063	0.059	0.057
4000	0.068	0.056	0.051	0.047	0.045	0.043
5000	0.054	0.045	0.041	0.038	0.036	0.034

Tafel 26 (Forts.) Table 26 (cont.)

$b_{2,p}$ $p = 4$

	Obere Prozentpunkte / *Upper percentage points*				Untere Prozentpunkte / *Lower percentage points*			
n ↓ α →	0.01	0.025	0.05	0.10	0.10	0.05	0.025	0.01
10	24.0	23.0	22.4	21.5	17.8	17.6	17.3	17.0
12	25.4	24.2	23.3	22.3	18.3	18.0	17.7	17.4
14	26.1	25.0	24.0	23.0	18.6	18.3	18.0	17.7
16	26.6	25.4	24.4	23.4	18.9	18.6	18.2	18.0
18	26.9	25.8	24.7	23.8	19.2	18.8	18.4	18.2
20	27.1	26.1	25.0	24.0	19.4	19.0	18.6	18.4
25	27.3	26.4	25.4	24.5	19.8	19.5	19.1	18.8
30	27.4	26.6	25.5	24.7	20.2	19.8	19.4	19.1
40	27.4	26.7	25.7	25.0	21.0	20.3	19.9	19.6
50	27.3	26.6	25.7	25.1	21.0	20.6	20.3	20.0
60	27.2	26.6	25.7	25.14	21.3	20.9	20.5	20.2
70	27.0	26.5	25.7	25.15	21.5	21.0	20.7	20.4
80	26.9	26.4	25.6	25.15	21.7	21.2	21.0	20.6
90	26.8	26.3	25.6	25.14	21.8	21.4	21.1	20.8
100	26.7	26.2	25.6	25.12	21.9	21.5	21.2	20.9
150	26.3	25.9	25.42	25.03	22.33	22.0	21.7	21.4
200	26.0	25.6	25.29	24.95	22.57	22.2	22.0	21.7
300	25.7	25.3	25.11	24.83	22.85	22.57	22.33	22.1
400	25.46	25.20	24.99	24.75	23.02	22.77	22.56	22.3
600	25.21	25.01	24.83	24.63	23.21	23.01	22.83	22.63
800	25.06	24.89	24.74	24.56	23.32	23.15	22.99	22.82
1000	24.96	24.80	24.67	24.51	23.40	23.24	23.10	22.94
1500	24.79	24.66	24.55	24.42	23.51	23.38	23.27	23.14
2000	24.69	24.58	24.48	24.37	23.58	23.47	23.37	23.26
3000	24.57	24.48	24.40	24.31	23.66	23.57	23.49	23.40
4000	24.50	24.42	24.35	24.27	23.71	23.63	23.56	23.48
5000	24.45	24.37	24.31	24.24	23.74	23.67	23.61	23.54

TABLE 27

Sample Size Requirements for the T^2-Test of Manova in One-way Classifications: Tables of J. Läuter

(a) Content of the table and definition of the tabulated quantity :

This table contains the quantity K , the sample size , as needed for a multivariate experiment which is to be analyzed by a one-way multivariate analysis of variance (MANOVA). It is assumed that there are J classes with (an equal number of) K p-dimensional observations which are distributed according to $N(\underline{\mu}_j, \underline{\Sigma})$ with $j = 1, 2, \ldots, J$ (stochastically independent).

The hypothesis

$$H_o : \quad \underline{\mu}_1 = \underline{\mu}_2 = \ldots\ldots = \underline{\mu}_J$$

shall be tested (against certain alternatives) by Hotelling's Criterion

$$T^2 = \text{trace}\ (\underline{H} \cdot \underline{G}^{-1}) \ .$$

Here $\underline{H}$ and $\underline{G}$ are the known independent Wishart matrices with the degrees of freedom $f_1 = J - 1$ and $f_2 = J.(K - 1)$ respectively.
The matrix $\underline{H}$ has a noncentrality $\underline{\gamma}$ when H_o is false.
For the determination of the sample size K we assume the risk of the first kind α , the risk of the second kind β , and a certain minimum deviation from H_o , which can be given in three different forms H_1 , H_2 , and H_3 .

Under these assumptions the table contains the minimal sample size K for a T^2- test of H_o against hte alternatives H_1 , H_2 , and H_3.

K depends on p , J , α , β , and H_1 , H_2 , H_3 respectively.

The deviations from the hypothesis H_0 are given now in the following different forms :

(1) H_1 : There exists a variate, so that

$$\frac{1}{\sigma^2} \sum_{j=1}^{J} (\mu_{ij} - \mu_{i.}) \geqq q^2 \quad ,$$

where μ_{ij} is the mean of the special i^{th} variate in the jth class , and

$$\mu_{i.} = \frac{1}{J} \sum_{j=1}^{J} \mu_{ij}$$ is the total mean and

σ^2 is the corresponding variance.

(2) H_2 : There exists a variate and two classes , so that

$$\frac{1}{\sigma_i} \mid \mu_{ij_1} \quad \mu_{ij_2} \mid \geqq d \quad ,$$ where

the index i characterizes the special variate and the indices j_1 and j_2 characterize the both classes.

(3) H_3 : There exists a variate, so that for all pairs j_1 and j_2 of classes

$$\frac{1}{\sigma_i} \mid \mu_{ij_1} - \mu_{ij_2} \mid \geqq c \quad .$$

Now it follows for the noncentrality parameter γ_1 of the matrix $\underline{H}$:

$$\gamma_1 \geqq K.q^2 \quad \text{in the case of } H_1 \quad ,$$

$$\gamma_1 \geqq (K/2).d^2 \quad \text{in the case of } H_2 \quad ,$$

$$\gamma_1 \geqq \frac{(J-1).J.(J+1).K}{12} \; . \; c^2 \quad \text{in the case of } H_3 \; .$$

The other noncentrality parameters γ_2 , γ_3 , γ_3 , ... can be zero under the hypotheses H_1 , H_2 , or H_3 .

As J. V. Srivastava (1964) has shown it is sufficient for the determination of the sample size to know the distribution of T^2 in its dependence of the first noncentrality parameter γ_1 alone and to set the other noncentrality parameters zero (the so called linear case). (See: J. Läuter (1978)).

(b) Extent of the table and definition of the parameters :

(1) The parameter α :

α = error probability (level of significance ; risk of the first kind)

α = 0.05 and 0.01
= 5% and 1% .

(2) The parameter β :

β = error of the second kind. ($1 - \beta$ = power).

β = 0.3 ; 0.2 ; 0.1 and 0.05 .

(3) The parameter p :

p = dimension of the variate

p = 1(1)10(5)20,30,50,100.

(4) The parameter J :

J = number of classes in the one-way classification

J = 2; 3; 4; 5; 6; 8; 10 .

(5) The parameter K :

K = number of p-dimensional observation vectors for each of the J classes. K is to be assumed equal for all classes.

(6) The parameter d :

d = minimum deviation from the hypothesis H_o (See under (a)).

d = 3; 2; 1.5; 1.0; 0.75; 0.5 .

(7) The parameter q^2 :

q^2 = a parameter to characterize (together with d and c) the minimum deviation from the hypothesis H_o as defined already under (a).

q^2 = the specific values are not the same for all combinations of the other parameters. They are printed in the tables.

(8) The parameter c :

c = a parameter to characterize (together with q^2 and c) the minimum deviation from the hypothesis H_o as defined already under (a).

c = the specific values are not the same for all combinations of the other parameters. They are printed in the tables.

(c) Suggestions for applications :

The tables can be used to determine the sample size of a multivariate experiment which is designed according to a one-way classification and which is to be analyzed and tested by Hotelling's T^2- criterion.

(d) References :

(1) For the source of the tables :

Läuter, J.: Sample Size Requirements for the T^2 Test of MANOVA (Tables for One-way Classification). Biometrical Journal 20, 389 - 406 (1978).

(2) For further information concerning theory, computations, and use :

Läuter, J.: (1978). (As cited before!!).

Srivastava, J. N. : On the Monotonicity Property of the Three Main Tests for Multivariate Analysis of Variance. Journal of The Royal Statistical Society, Series B, Vol. 26, 77 - 81 (1964).

(e) Further remarks :

The tabulated sample size K is rounded to the next higher multiple

of 1 in the interval [1 , 40]

of 2 in the interval [40 , 100]

of 5 in the interval [100 , 200]

of 10 in the interval [200 , 400]

of 20 in the interval [400 , 1000]

of 50 in the interval [1000 , 1500] .

Tafel 27 Table 27

J = 2 classes

	p	$\alpha=0.05$ $\beta=.3$	$\beta=.2$	$\beta=.1$	$\beta=.05$	$\alpha=0.01$ $\beta=.3$	$\beta=.2$	$\beta=.1$	$\beta=.05$
$q^2=4.5$	1	3	4	4	5	5	5	6	6
$d=3$	2	4	5	5	6	6	6	7	8
$c=3$	3	5	5	6	7	7	7	8	9
	4	6	6	7	8	7	8	9	10
	5	6	7	8	8	8	9	10	10
	6	7	7	8	9	9	9	10	11
	8	8	9	10	10	10	11	12	13
	10	9	10	11	12	11	12	13	14
	15	12	13	14	15	14	15	16	17
	20	15	15	16	17	17	18	19	20
	30	20	21	22	23	22	23	25	26
	50	30	31	32	33	33	34	35	37
	100	56	56	58	60	58	60	62	62
$q^2=2$	1	5	6	7	8	7	8	10	11
$d=2$	2	6	7	9	10	9	10	12	13
$c=2$	3	7	8	10	11	10	11	13	15
	4	8	9	11	12	11	12	14	16
	5	9	10	12	14	12	13	16	17
	6	10	11	13	14	13	14	17	19
	8	11	12	14	16	15	16	19	21
	10	13	14	16	18	16	18	20	22
	15	16	17	20	22	20	22	24	27
	20	19	20	23	25	23	25	28	30
	30	24	26	29	31	29	31	34	37
	50	35	37	40	44	42	44	48	50
	100	62	64	68	70	68	70	74	78
$q^2=1.125$	1	7	9	11	13	11	13	15	18
$d=1.5$	2	9	11	13	16	13	15	18	21
$c=1.5$	3	10	12	15	18	15	17	21	24
	4	12	14	17	20	17	19	22	26
	5	13	15	18	21	18	20	24	27
	6	14	16	19	22	19	22	26	29
	8	16	18	22	25	21	24	28	32
	10	18	20	24	27	24	26	31	34
	15	21	24	28	32	28	31	36	40
	20	25	28	32	36	32	36	42	46
	30	32	35	40	44	40	44	50	54
	50	44	48	54	58	54	58	64	70
	100	72	76	84	90	84	88	96	105

Tafel 27 (Forts.) | Table 27 (cont.)

J = 2 classes

	p	$\alpha=0.05$ $\beta=.3$	$\beta=.2$	$\beta=.1$	$\beta=.05$	$\alpha-0.01$ $\beta=.3$	$\beta=.2$	$\beta=.1$	$\beta=.05$
$q^2=0.5$	1	14	17	23	27	21	26	32	38
$d=1$	2	17	21	27	33	26	31	38	44
$c=1$	3	20	24	31	37	29	34	42	50
	4	22	27	34	40	32	37	46	54
	5	24	29	36	44	35	40	48	56
	6	26	31	39	46	37	44	52	60
	8	29	35	44	50	42	48	56	64
	10	32	38	46	54	44	52	60	70
	15	38	46	54	64	52	60	70	80
	20	44	52	62	70	60	68	78	88
	30	54	62	74	84	72	80	92	105
	50	70	80	94	105	92	105	115	130
	100	105	120	135	150	135	145	165	180
$q^2=0.2813$	1	23	29	39	48	36	44	56	66
$d=0.75$	2	29	36	48	58	44	52	66	76
$c=0.75$	3	34	42	54	64	50	58	72	84
	4	37	46	58	70	54	64	78	90
	5	42	50	62	74	58	68	84	96
	6	44	52	66	78	62	72	88	105
	8	48	58	72	86	68	80	96	110
	10	54	64	78	92	74	86	105	120
	15	64	74	92	110	86	100	120	135
	20	72	84	105	120	98	110	130	150
	30	86	100	120	140	115	130	155	175
	50	110	130	150	175	145	165	190	220
	100	160	180	210	240	210	230	260	290
$q^2=0.125$	1	52	64	86	105	80	96	125	145
$d=0.5$	2	64	80	105	130	96	115	145	170
$c=0.5$	3	74	90	120	140	110	130	160	185
	4	80	98	130	155	120	140	170	200
	5	88	110	135	165	125	150	180	210
	6	94	115	145	175	135	160	190	230
	8	105	125	160	190	150	175	210	240
	10	115	135	170	200	160	185	230	260
	15	135	160	195	230	185	220	260	300
	20	150	180	220	260	210	240	290	330
	30	180	210	260	300	250	280	330	370
	50	230	260	320	370	300	340	400	460
	100	310	360	440	500	420	460	540	600

Tafel 27 (Forts.) Table 27 (cont.)

J = 3 classes

	p	$\alpha=0.05$ $\beta=.3$	$\beta=.2$	$\beta=.1$	$\beta=.05$	$\alpha=0.01$ $\beta=.3$	$\beta=.2$	$\beta=.1$	$\beta=.05$
$q^2=4.5$	1	3	4	5	5	5	5	6	7
$d=3$	2	4	5	6	6	6	6	7	8
$c=1.5$	3	5	5	6	7	7	7	8	9
	4	6	6	7	8	7	8	9	10
	5	6	7	8	8	8	9	10	11
	6	7	7	8	9	9	9	10	11
	8	8	8	9	10	10	10	12	13
	10	8	9	10	11	11	12	13	14
	15	11	11	13	14	13	14	15	17
	20	13	13	15	16	15	16	18	19
	30	16	17	19	20	19	20	22	23
	50	23	25	26	28	27	28	30	32
	100	42	42	44	46	44	46	48	50
$q^2=2$	1	5	6	8	9	8	9	11	12
$d=2$	2	7	8	10	11	10	11	13	15
$c=1$	3	8	9	11	13	11	12	15	16
	4	9	10	12	14	12	14	16	18
	5	10	11	13	15	13	15	17	19
	6	10	12	14	16	14	16	18	20
	8	12	13	16	18	16	18	20	23
	10	13	15	17	19	17	19	22	24
	15	16	18	21	23	21	23	26	29
	20	18	21	24	26	24	26	29	32
	30	23	25	29	32	29	32	36	39
	50	32	34	38	42	39	42	46	50
	100	50	54	60	64	60	64	70	74
$q^2=1.125$	1	8	10	13	15	12	14	18	20
$d=1.5$	2	11	13	16	18	15	17	21	24
$c=0.75$	3	12	14	18	21	17	20	24	27
	4	14	16	19	23	19	22	26	29
	5	15	17	21	24	20	23	28	31
	6	16	18	22	26	22	25	29	33
	8	18	21	25	29	24	28	32	36
	10	20	23	27	31	27	30	35	39
	15	24	27	32	37	32	35	42	46
	20	27	31	37	42	36	40	46	52
	30	33	38	44	50	44	48	56	62
	50	44	50	58	64	56	62	70	78
	100	68	74	84	92	84	90	100	110

Tafel 27 (Forts.) Table 27 (cont.)

J = 3 classes

	p	$\alpha=0.05$ $\beta=.3$	$\beta=.2$	$\beta=.1$	$\beta=.05$	$\alpha=0.01$ $\beta=.3$	$\beta=.2$	$\beta=.1$	$\beta=.05$
$q^2=0.5$	1	17	21	27	32	25	30	37	44
$d=1$	2	21	26	33	39	31	36	44	52
$c=0.5$	3	25	29	37	44	35	42	50	58
	4	27	33	42	48	38	44	54	62
	5	30	35	44	52	42	48	58	66
	6	32	38	48	56	44	52	62	70
	8	36	42	52	60	50	56	68	78
	10	39	46	56	66	54	62	74	84
	15	46	54	66	76	64	72	84	96
	20	54	62	74	86	72	80	94	110
	30	64	74	88	105	86	96	115	125
	50	82	94	115	130	110	120	140	155
	100	120	135	155	175	155	170	195	220
$q^2=0.2813$	1	29	36	46	56	44	52	64	76
$d=0.75$	2	36	44	58	68	54	62	76	90
$c=0.375$	3	42	52	64	76	60	70	86	100
	4	46	56	70	84	66	78	94	110
	5	50	60	76	90	72	82	100	115
	6	54	66	82	96	76	88	105	125
	8	60	72	90	105	84	98	120	135
	10	66	78	98	115	92	105	125	145
	15	78	92	115	135	110	125	145	165
	20	90	105	130	150	125	140	165	185
	30	110	125	150	175	145	165	190	220
	50	135	160	190	220	180	210	240	270
	100	195	220	260	300	250	280	330	360
$q^2=0.125$	1	64	80	105	125	96	115	145	170
$d=0.5$	2	80	98	125	150	115	140	170	200
$c=0.25$	3	92	115	145	170	135	155	190	220
	4	105	125	155	185	145	170	210	240
	5	110	135	170	200	155	185	220	260
	6	120	145	180	210	165	195	240	270
	8	135	160	200	240	185	220	260	300
	10	145	175	220	250	200	230	280	320
	15	170	210	250	290	240	270	320	370
	20	195	230	280	330	270	300	360	420
	30	230	270	330	380	310	360	420	480
	50	290	340	420	480	390	440	520	580
	100	400	480	560	640	540	600	700	780

Tafel 27 (Forts.) Table 27 (cont.)

J = 4 classes

	p	$\alpha=0.05$ $\beta=.3$	$\beta=.2$	$\beta=.1$	$\beta=.05$	$\alpha=0.01$ $\beta=.3$	$\beta=.2$	$\beta=.1$	$\beta=.05$
$q^2=4.5$	1	4	4	5	5	5	5	6	7
$d=3$	2	4	5	6	7	6	7	8	8
$c=0.9487$	3	5	6	7	7	7	7	8	9
	4	6	6	7	8	7	8	9	10
	5	6	7	8	9	8	9	10	11
	6	7	7	8	9	9	9	11	12
	8	7	8	9	10	10	11	12	13
	10	8	9	10	11	11	12	13	14
	15	10	11	12	14	13	14	15	17
	20	12	13	14	16	15	16	18	19
	30	15	16	18	19	18	20	21	23
	50	21	22	24	26	24	26	28	30
	100	34	36	38	42	38	40	44	46
$q^2=2$	1	6	7	9	10	8	10	12	13
$d=2$	2	7	9	11	12	10	12	14	16
$c=0.6325$	3	9	10	12	14	12	13	16	18
	4	9	11	13	15	13	15	17	19
	5	10	12	14	16	14	16	18	21
	6	11	13	15	17	15	17	20	22
	8	13	14	17	19	17	19	22	24
	10	14	16	19	21	18	21	24	26
	15	17	19	22	25	22	24	28	31
	20	19	22	25	28	25	28	31	35
	30	24	26	30	34	30	33	38	42
	50	31	35	39	44	40	44	48	54
	100	48	52	58	64	58	64	70	76
$q^2=1.125$	1	9	11	14	17	13	16	19	22
$d=1.5$	2	12	14	17	20	17	19	23	26
$c=0.4743$	3	14	16	20	23	19	22	26	29
	4	15	18	22	25	21	24	28	32
	5	16	19	23	27	23	26	30	34
	6	18	21	25	29	24	27	32	37
	8	20	23	28	32	27	30	36	40
	10	22	25	30	35	29	33	39	44
	15	26	30	36	42	35	39	46	52
	20	30	34	40	46	40	44	52	58
	30	36	42	48	56	48	54	62	68
	50	48	54	62	70	62	68	78	86
	100	70	78	88	98	88	96	110	120

Tafel 27 (Forts.)

Table 27 (cont.)

J = 4 classes

	p	$\alpha=0.05$ $\beta=.3$	$\beta=.2$	$\beta=.1$	$\beta=.05$	$\alpha=0.01$ $\beta=.3$	$\beta=.2$	$\beta=.1$	$\beta=.05$
$q^2=0.5$	1	19	23	30	36	28	33	40	48
$d=1$	2	24	29	37	44	34	40	50	56
$c=0.3162$	3	28	33	42	50	39	46	56	64
	4	31	37	46	54	44	50	60	70
	5	34	40	50	58	48	54	64	74
	6	36	44	54	62	50	58	70	78
	8	42	48	60	68	56	64	76	86
	10	46	52	64	74	62	70	82	94
	15	54	62	76	88	72	82	96	110
	20	60	70	86	98	82	92	110	125
	30	74	84	105	115	98	110	130	145
	50	94	110	130	145	125	140	160	180
	100	135	150	180	200	175	195	220	250
$q^2=0.2813$	1	33	40	52	64	48	58	70	84
$d=0.75$	2	42	50	64	76	60	70	86	98
$c=0.2372$	3	48	58	72	86	68	80	96	110
	4	54	64	80	94	76	88	105	120
	5	58	70	86	105	82	94	115	130
	6	62	74	92	110	86	100	120	140
	8	70	84	105	120	96	115	135	155
	10	78	92	115	130	105	120	145	165
	15	92	110	130	155	125	145	170	190
	20	105	125	150	170	140	160	190	210
	30	125	145	175	200	170	190	220	250
	50	160	185	220	250	210	240	280	310
	100	220	260	300	340	290	330	380	420
$q^2=0.125$	1	72	90	115	140	110	130	160	185
$d=0.5$	2	92	115	145	170	130	155	190	220
$c=0.1581$	3	105	130	165	195	150	175	220	250
	4	120	145	180	210	165	195	240	270
	5	130	155	195	230	180	210	250	290
	6	140	165	210	240	190	220	270	310
	8	155	185	230	270	220	250	300	340
	10	170	200	250	290	240	270	320	370
	15	200	240	290	340	280	320	370	420
	20	230	270	330	380	310	350	420	480
	30	270	320	390	460	370	420	500	560
	50	340	400	480	560	460	520	620	680
	100	480	560	660	760	640	720	820	920

Tafel 27 (Forts.) Table 27 (cont.)

J = 5 classes

	p	$\alpha=0.05$ $\beta=.3$	$\beta=.2$	$\beta=.1$	$\beta=.05$	$\alpha=0.01$ $\beta=.3$	$\beta=.2$	$\beta=.1$	$\beta=.05$
$q^2=4.5$	1	4	4	5	6	5	6	7	7
$d=3$	2	5	5	6	7	6	7	8	9
$c=0.6708$	3	5	6	7	8	7	8	9	10
	4	6	6	8	8	8	8	10	11
	5	6	7	8	9	8	9	10	11
	6	7	8	9	10	9	10	11	12
	8	8	8	10	11	10	11	12	13
	10	8	9	11	12	11	12	13	15
	15	10	11	13	14	13	14	16	17
	20	12	13	15	16	15	16	18	19
	30	14	16	18	20	18	19	22	24
	50	20	21	24	26	24	25	28	30
	100	31	33	36	39	36	38	42	44
$q^2=2$	1	6	7	9	11	9	10	12	14
$d=2$	2	8	9	11	13	11	12	15	17
$c=0.4472$	3	9	11	13	15	13	14	17	19
	4	10	12	14	16	14	16	18	21
	5	11	13	15	18	15	17	20	22
	6	12	14	16	19	16	18	21	24
	8	13	15	18	21	18	20	23	26
	10	15	17	20	23	20	22	25	28
	15	18	20	24	27	23	26	30	33
	20	20	23	27	30	27	29	34	37
	30	25	28	32	36	32	35	40	44
	50	32	36	42	46	42	46	52	56
	100	48	54	60	66	60	66	72	80
$q^2=1.125$	1	10	12	15	18	14	17	20	23
$d=1.5$	2	13	15	19	22	18	20	25	28
$c=0.3354$	3	15	17	21	25	20	23	28	32
	4	16	19	23	27	22	26	30	35
	5	18	21	25	29	24	28	33	37
	6	19	22	27	31	26	30	35	40
	8	22	25	30	35	29	33	39	44
	10	24	27	33	38	32	36	42	48
	15	28	33	39	46	38	44	50	56
	20	32	37	44	50	44	48	56	62
	30	39	46	54	60	52	58	66	74
	50	50	58	68	76	66	74	84	92
	100	74	82	96	105	94	105	120	130

Tafel 27 (Forts.) Table 27 (cont.)

J = 5 classes

	p	$\alpha=0.05$ $\beta=.3$	$\beta=.2$	$\beta=.1$	$\beta=.05$	$\alpha=0.01$ $\beta=.3$	$\beta=.2$	$\beta=.1$	$\beta=.05$
$q^2=0.5$	1	21	25	32	39	30	35	44	50
$d=1$	2	26	32	40	48	37	44	54	62
$c=0.2236$	3	31	37	46	54	44	50	60	68
	4	34	42	50	60	48	56	66	76
	5	37	44	54	64	52	60	70	80
	6	40	48	58	68	56	64	76	86
	8	46	54	66	76	62	70	84	94
	10	50	58	72	82	68	78	90	105
	15	60	70	84	96	80	90	110	120
	20	68	78	94	110	90	105	120	135
	30	82	94	115	130	110	125	145	160
	50	105	120	145	160	140	155	180	200
	100	145	165	195	220	190	220	250	270
$q^2=0.2813$	1	36	44	56	68	52	62	76	88
$d=0.75$	2	46	56	70	84	66	76	92	110
$c=0.1677$	3	54	64	80	94	74	86	105	120
	4	60	72	88	105	82	96	115	135
	5	64	78	96	115	90	105	125	145
	6	70	82	105	120	96	110	135	150
	8	78	92	115	135	110	125	145	170
	10	86	105	125	145	120	135	160	180
	15	105	120	145	170	140	160	185	210
	20	120	135	165	190	160	180	210	240
	30	140	165	195	230	190	210	250	280
	50	180	210	250	280	240	270	310	350
	100	250	290	340	390	330	370	440	480
$q^2=0.125$	1	80	98	125	150	115	140	170	200
$d=0.5$	2	100	125	155	185	145	170	210	240
$c=0.1118$	3	120	145	180	210	165	195	240	270
	4	130	160	195	230	185	210	260	300
	5	145	170	220	250	200	230	280	320
	6	155	185	230	270	220	250	300	340
	8	175	210	260	300	240	280	330	370
	10	190	230	280	320	260	300	360	400
	15	230	270	330	380	310	350	420	480
	20	260	300	370	420	350	400	480	520
	30	310	360	440	500	420	480	560	620
	50	390	460	540	620	520	600	680	760
	100	560	640	760	860	740	820	940	1050

Tafel 27 (Forts.) Table 27 (cont.)

J = 6 classes

	p	$\alpha=0.05$ $\beta=.3$	$\beta=.2$	$\beta=.1$	$\beta=.05$	$\alpha=0.01$ $\beta=.3$	$\beta=.2$	$\beta=.1$	$\beta=.05$
$q^2=4.5$	1	4	4	5	6	5	6	7	8
$d=3$	2	5	5	6	7	6	7	8	9
$c=0.5071$	3	5	6	7	8	7	8	9	10
	4	6	7	8	9	8	9	10	11
	5	6	7	8	9	9	9	11	12
	6	7	8	9	10	9	10	11	13
	8	8	9	10	11	10	11	13	14
	10	9	10	11	12	11	12	14	15
	15	10	11	13	14	13	15	16	18
	20	12	13	15	16	15	16	18	20
	30	14	16	18	20	18	20	22	24
	50	19	21	24	26	24	26	28	31
	100	29	32	35	38	35	38	42	46
$q^2=2$	1	7	8	10	11	9	11	13	15
$d=2$	2	8	10	12	14	12	13	16	18
$c=0.3381$	3	10	11	14	16	13	15	18	20
	4	11	12	15	17	15	16	19	22
	5	12	14	16	19	16	18	21	23
	6	13	14	17	20	17	19	22	25
	8	14	16	19	22	19	21	25	28
	10	16	18	21	24	21	23	27	30
	15	19	21	25	28	25	28	32	35
	20	21	24	28	32	28	31	36	40
	30	26	29	34	38	34	37	44	48
	50	33	38	44	50	44	48	54	60
	100	50	54	62	70	62	68	76	84
$q^2=1.125$	1	11	13	16	19	15	18	21	25
$d=1.5$	2	14	16	20	23	19	22	26	30
$c=0.2535$	3	16	18	23	26	22	25	29	34
	4	18	21	25	29	24	27	32	37
	5	19	22	27	32	26	30	35	40
	6	21	24	29	34	28	32	37	4
	8	23	27	33	37	31	35	42	48
	10	25	30	36	42	34	39	46	52
	15	30	35	42	48	42	46	54	60
	20	35	40	48	54	46	52	60	68
	30	42	48	58	64	56	62	72	80
	50	54	62	72	82	72	78	90	100
	100	76	88	105	115	98	110	125	140

Tafel 27 (Forts.) Table 27 (cont.)

J = 6 classes

	p	$\alpha=0.05$ $\beta=.3$	$\beta=.2$	$\beta=.1$	$\beta=.05$	$\alpha=0.01$ $\beta=.3$	$\beta=.2$	$\beta=.1$	$\beta=.05$
$q^2=0.5$	1	22	27	34	42	32	38	46	54
$d=1$	2	28	34	44	50	40	46	56	66
$c=0.1690$	3	33	39	50	58	46	54	64	74
	4	37	44	54	64	52	60	70	80
	5	40	48	60	68	56	64	76	86
	6	44	52	64	74	60	68	82	92
	8	50	58	70	82	68	76	90	105
	10	54	64	78	90	74	84	98	115
	15	64	76	90	105	88	98	115	130
	20	74	86	105	120	98	115	130	145
	30	88	105	125	140	120	135	155	175
	50	115	130	155	175	150	170	195	220
	100	160	180	220	240	210	240	270	300
$q^2=0.2813$	1	39	48	60	72	56	66	80	94
$d=0.75$	2	50	60	76	88	70	82	98	115
$c=0.1268$	3	58	70	86	105	80	94	115	130
	4	64	76	96	115	90	105	125	145
	5	70	84	105	120	98	115	135	155
	6	76	90	110	130	105	120	145	165
	8	86	100	125	145	120	135	160	180
	10	94	110	135	160	130	145	175	195
	15	115	135	160	185	155	175	210	230
	20	130	150	180	210	175	195	230	260
	30	155	180	220	250	210	230	270	310
	50	195	230	270	310	260	290	340	380
	100	280	320	370	420	360	420	480	520
$q^2=0.125$	1	86	105	135	160	125	145	180	210
$d=0.5$	2	110	135	170	200	155	180	220	260
$c=0.0845$	3	130	155	190	230	180	210	250	290
	4	145	170	220	250	200	230	280	320
	5	155	185	230	270	220	250	300	340
	6	170	200	250	290	230	270	320	360
	8	190	230	280	320	260	300	350	400
	10	210	250	300	350	290	330	390	440
	15	250	290	360	420	340	380	460	520
	20	280	330	400	460	380	440	520	580
	30	340	400	480	560	460	520	600	680
	50	440	500	600	680	580	660	760	840
	100	600	700	820	940	800	900	1050	1150

Tafel 27 (Forts.) Table 27 (cont.)

J = 8 classes

	p	$\alpha=0.05$ $\beta=.3$	$\beta=.2$	$\beta=.1$	$\beta=.05$	$\alpha=0.01$ $\beta=.3$	$\beta=.2$	$\beta=.1$	$\beta=.05$
$q^2=4.5$	1	4	5	6	6	5	6	7	8
$d=3$	2	5	6	7	8	7	7	9	10
$c=0.3273$	3	6	6	8	9	8	8	10	11
	4	6	7	8	9	8	9	11	12
	5	7	8	9	10	9	10	11	13
	6	7	8	10	11	10	11	12	14
	8	8	9	11	12	11	12	14	15
	10	9	10	12	13	12	13	15	16
	15	11	12	14	15	14	15	17	19
	20	12	14	16	18	16	18	20	22
	30	15	17	19	21	19	21	24	26
	50	19	21	25	27	24	27	30	33
	100	29	31	36	39	35	38	42	46
$q^2=2$	1	7	9	11	12	10	12	14	16
$d=2$	2	9	11	13	15	13	14	17	19
$c=0.2182$	3	11	12	15	17	14	16	19	22
	4	12	14	17	19	16	18	21	24
	5	13	15	18	21	17	20	23	26
	6	14	16	19	22	19	21	24	27
	8	16	18	21	25	21	24	27	31
	10	17	20	23	27	23	26	30	33
	15	21	23	28	32	27	30	35	39
	20	23	27	32	36	31	35	40	44
	30	28	32	38	44	37	42	48	52
	50	36	42	48	54	48	54	60	66
	100	52	58	68	76	66	74	84	92
$q^2=1.125$	1	12	14	18	21	17	19	23	27
$d=1.5$	2	15	18	22	26	21	24	29	33
$c=0.1637$	3	18	21	25	29	24	27	33	37
	4	20	23	28	32	27	30	36	42
	5	21	25	30	35	29	33	39	44
	6	23	27	33	38	31	35	42	48
	8	26	30	37	42	35	40	46	52
	10	29	33	40	46	39	44	52	58
	15	34	40	48	54	46	52	60	68
	20	39	46	54	62	52	58	68	76
	30	48	54	64	74	64	70	80	90
	50	60	70	82	92	80	88	105	115
	100	84	96	115	130	110	125	140	155

Tafel 27 (Forts.) Table 27 (cont.)

J = 8 classes

	p	$\alpha=0.05$ $\beta=.3$	$\beta=.2$	$\beta=.1$	$\beta=.05$	$\alpha=0.01$ $\beta=.3$	$\beta=.2$	$\beta=.1$	$\beta=.05$
$q^2=0.5$	1	25	30	38	46	35	42	50	58
$d=1$	2	32	38	48	56	46	52	62	72
$c=0.1091$	3	37	44	56	64	52	60	72	82
	4	42	50	62	72	58	66	78	90
	5	46	54	66	78	64	72	86	96
	6	50	58	72	82	68	78	92	105
	8	56	66	80	92	76	86	105	115
	10	62	72	88	100	84	94	115	125
	15	74	86	105	120	100	115	135	150
	20	84	98	120	135	115	130	150	165
	30	105	120	140	160	135	155	175	200
	50	130	150	175	200	175	195	230	250
	100	180	210	250	280	240	270	310	340
$q^2=0.2813$	1	44	52	66	80	62	72	88	105
$d=0.75$	2	56	68	84	98	78	90	110	125
$c=0.0818$	3	66	78	96	115	90	105	125	145
	4	74	88	110	125	100	115	140	160
	5	80	96	120	135	110	130	150	170
	6	86	105	125	145	120	135	160	185
	8	98	115	140	165	135	155	180	210
	10	110	130	155	180	145	165	195	220
	15	130	150	185	210	175	200	230	260
	20	145	170	210	240	200	230	260	300
	30	175	210	250	280	240	270	310	350
	50	230	260	310	360	310	340	400	440
	100	320	370	440	500	420	480	540	600
$q^2=0.125$	1	96	120	150	180	140	160	195	230
$d=0.5$	2	125	150	185	220	175	200	250	280
$c=0.0546$	3	145	175	220	260	200	240	280	320
	4	165	195	240	280	230	260	310	360
	5	180	210	260	310	250	280	340	380
	6	195	230	280	330	270	300	360	420
	8	220	260	320	370	300	340	400	460
	10	240	280	350	400	330	370	440	500
	15	290	340	420	480	390	440	520	580
	20	330	380	460	540	440	500	580	660
	30	390	460	560	640	540	600	700	780
	50	520	580	700	800	680	760	880	980
	100	700	820	960	1100	940	1050	1200	1350

Tafel 27 (Forts.) Table 27 (cont.)

J = 10 classes

	p	$\alpha=0.05$ $\beta=.3$	$\beta=.2$	$\beta=.1$	$\beta=.05$	$\alpha=0.01$ $\beta=.3$	$\beta=.2$	$\beta=.1$	$\beta=.05$
$q^2=4.5$	1	4	5	6	7	6	6	7	8
$d=3$	2	5	6	7	8	7	8	9	10
$c=0.2335$	3	6	7	8	9	8	9	10	11
	4	7	8	9	10	9	10	11	13
	5	7	8	10	11	10	11	12	14
	6	8	9	10	12	10	11	13	14
	8	9	10	11	13	11	13	14	16
	10	10	11	13	14	13	14	16	17
	15	11	13	15	17	15	16	19	21
	20	13	15	17	19	17	19	21	23
	30	16	17	20	23	20	22	25	28
	50	20	22	26	29	26	28	32	35
	100	29	32	37	42	36	39	44	48
$q^2=2$	1	8	9	11	13	11	12	15	17
$d=2$	2	10	12	14	16	13	15	18	20
$c=0.1557$	3	11	13	16	19	16	18	21	23
	4	13	15	18	21	17	20	23	26
	5	14	16	19	22	19	21	25	28
	6	15	17	21	24	20	23	27	30
	8	17	20	23	27	23	26	30	33
	10	19	21	26	29	25	28	32	36
	15	22	26	30	34	30	33	38	44
	20	25	29	34	39	34	38	44	48
	30	31	35	42	48	42	46	52	58
	50	39	46	52	60	52	58	66	72
	100	56	64	74	82	72	80	90	100
$q^2=1.125$	1	13	15	19	22	18	21	25	29
$d=1.5$	2	16	19	24	28	23	26	31	35
$c=0.1168$	3	19	22	27	32	26	30	35	40
	4	21	25	31	35	29	33	39	44
	5	23	27	33	38	32	36	42	48
	6	25	30	36	42	34	39	46	52
	8	29	33	40	46	38	44	52	58
	10	31	37	44	50	42	48	56	62
	15	38	44	52	60	50	58	66	74
	20	44	50	60	68	58	64	74	84
	30	52	60	72	80	70	78	90	100
	50	66	76	90	105	88	98	115	125
	100	92	110	125	140	125	135	155	170

Tafel 27 (Forts.) Table 27 (cont.)

J = 10 classes

	p	$\alpha=0.05$ $\beta=.3$	$\beta=.2$	$\beta=.1$	$\beta=.05$	$\alpha=0.01$ $\beta=.3$	$\beta=.2$	$\beta=.1$	$\beta=.05$
$q^2=0.5$	1	27	33	42	50	38	44	54	62
$d=1$	2	35	42	52	62	50	56	68	78
$c=0.0778$	3	42	50	60	70	56	66	78	88
	4	46	54	68	78	64	72	86	98
	5	52	60	74	84	70	80	94	105
	6	56	64	78	90	74	86	100	115
	8	62	72	88	105	84	96	115	130
	10	68	80	96	115	92	105	125	140
	15	82	96	115	135	110	125	145	165
	20	94	110	130	150	125	140	165	185
	30	115	130	155	180	150	170	200	220
	50	145	165	200	230	195	220	250	280
	100	200	230	280	310	270	300	340	380
$q^2=0.2813$	1	48	58	72	86	68	78	94	110
$d=0.75$	2	62	74	92	110	86	98	120	135
$c=0.0584$	3	72	86	105	125	100	115	140	155
	4	82	96	120	140	110	130	150	175
	5	90	105	130	150	125	140	165	190
	6	96	115	140	160	130	150	175	200
	8	110	130	155	180	150	170	200	230
	10	120	140	170	195	165	185	220	250
	15	145	170	210	240	195	220	260	290
	20	165	190	230	270	220	250	290	330
	30	200	230	280	320	270	300	350	390
	50	260	300	350	400	340	380	440	500
	100	350	420	480	560	480	520	600	680
$q^2=0.125$	1	105	130	160	190	150	175	210	250
$d=0.5$	2	135	165	210	240	190	220	270	310
$c=0.0389$	3	160	190	240	280	220	260	310	350
	4	180	220	270	310	250	290	340	390
	5	200	240	290	340	270	310	370	420
	6	220	260	310	360	290	340	400	460
	8	250	290	350	400	330	380	440	500
	10	270	320	380	440	360	420	480	560
	15	320	380	460	520	440	500	580	640
	20	370	440	520	600	500	560	660	740
	30	440	520	620	700	600	660	780	860
	50	580	660	780	880	760	860	980	1100
	100	780	900	1100	1250	1050	1200	1350	1500

TABLE 28

Critical Values for Simultaneous and Sequential Bonferroni z-Tests: Tables of G. A. Lienert, O. Ludwig, and K. Rockenfeller

(a) Content of the table and definition of the test statistic :

The use of sequential Bonferroni tests for testing R simultaneous hypotheses

$$H_{o1} \; ; \; H_{o2} \; ; \; \ldots\ldots \; ; \; H_{oR}$$

against specific alternatives

$$H_{a1} \; ; \; H_{a2} \; ; \; \ldots\ldots \; ; \; H_{aR}$$

requires critical limits of a test statistic at levels

$$\alpha/R \; ; \; \alpha/(R-1) \; ; \; \alpha/(R-2) \; ; \; \ldots\ldots \; ; \; \alpha/1 \; .$$

This table contains such necessary upper limits of the standardized normal distribution z (Gauß - distribution).

(b) Extent of the table and definition of the parameters :

(1) The parameter α :

α = error probability
α = 0.10 ; 0.05 ; 0.025 ; 0.01 ; 0.005
for one-sided testing

2α = 0.20 ; 0.10 ; 0.05 ; 0.02 ; 0.01

for two-sided testing.

(2) The parameter R :

R = number of hypotheses to be tested

R = 1(1)100(2)300(5)800(10)1600(20)3200(50)10000 .

(c) Suggestions for applications :

(1) Multidimensional contingency tables :

(1.1) Configural frequency analysis.
(1.2) Residual analysis.

(2) Multiple comparisons :

(2.1) of k independent samples.
(2.2) of k dependent samples resp. repeated measurements.

(d) References :

(1) For the source of the tables :

Lienert, G. A. + Ludwig, O. + Rockenfeller, K.: Tables of the critical values for simultaneous and sequential Bonferroni-z-tests. Biometrical Journal 24, 239 - 255 (1982) .

(2) For further informations concerning theory, computations , and applications :

Lienert, G. A. et al. (1982). (As cited before !!)

Holm, S. A.: A simple sequentially rejective multiple test

procedure.
Scandinavian Journal of Statistics 6, 65 - 70 (1979).

Krauth, J. + Lienert, G. A. : Die Konfigurationsfrequenzanalyse und ihre Anwendungen in Psychologie und Medizin.
Freiburg/Br.: Alber, 1973.

Krauth, J. + Lienert, G. A. : Die Konfigurationsfrequenzanalyse: XII. Symptommusterfolgen (Durchgangssyndrome).
Z. Klin. Psychol. u. Psychother. 28, 302 - 315 (1980).

(e) Further remarks :

(1) For the Bonferroni χ^2- statistic see table 22 and table 23 of this volume.

(2) For the Bonferroni t - statistic see table 29 of this volume.

Tafel 28

Table 28

R	Z(0.10/R)	Z(0.05/R)	Z(0.025/R)	Z(0.01/R)	Z(0.005/R)
1-SIDED 2-SIDED	10% 20%	5% 10%	2.5% 5%	1% 2%	0.5% 1%
1	1.282	1.645	1.960	2.326	2.576
2	1.645	1.960	2.241	2.576	2.807
3	1.834	2.128	2.394	2.713	2.935
4	1.960	2.241	2.498	2.807	3.023
5	2.054	2.326	2.576	2.878	3.090
6	2.128	2.394	2.638	2.935	3.144
7	2.189	2.450	2.690	2.983	3.189
8	2.241	2.498	2.734	3.023	3.227
9	2.287	2.539	2.773	3.059	3.261
10	2.326	2.576	2.807	3.090	3.291
11	2.362	2.609	2.838	3.118	3.317
12	2.394	2.638	2.865	3.144	3.341
13	2.423	2.665	2.891	3.167	3.364
14	2.450	2.690	2.914	3.189	3.384
15	2.475	2.713	2.935	3.209	3.403
16	2.498	2.734	2.955	3.227	3.421
17	2.519	2.754	2.974	3.245	3.437
18	2.539	2.773	2.991	3.261	3.452
19	2.558	2.790	3.008	3.276	3.467
20	2.576	2.807	3.023	3.291	3.481
21	2.593	2.823	3.038	3.304	3.494
22	2.609	2.838	3.052	3.317	3.506
23	2.624	2.852	3.065	3.330	3.518
24	2.638	2.865	3.078	3.341	3.529
25	2.652	2.878	3.090	3.353	3.540
26	2.665	2.891	3.102	3.364	3.550
27	2.678	2.902	3.113	3.374	3.560
28	2.690	2.914	3.124	3.384	3.570
29	2.702	2.925	3.134	3.394	3.579
30	2.713	2.935	3.144	3.403	3.588
31	2.724	2.945	3.154	3.412	3.596
32	2.734	2.955	3.163	3.421	3.605
33	2.744	2.965	3.172	3.429	3.613
34	2.754	2.974	3.180	3.437	3.620
35	2.764	2.983	3.189	3.445	3.628
36	2.773	2.991	3.197	3.452	3.635
37	2.782	3.000	3.205	3.460	3.642
38	2.790	3.008	3.213	3.467	3.649
39	2.799	3.016	3.220	3.474	3.656
40	2.807	3.023	3.227	3.481	3.662
41	2.815	3.031	3.234	3.487	3.669
42	2.823	3.038	3.241	3.494	3.675
43	2.830	3.045	3.248	3.500	3.681
44	2.838	3.052	3.254	3.506	3.687
45	2.845	3.059	3.261	3512	3.692
46	2.852	3.065	3.267	3.518	3.698
47	2.859	3.072	4.273	3.524	3.703
48	2.865	3.078	3.279	3.529	3.709
49	2.872	3.084	3.285	3.535	3.714
50	2.878	3.090	3.291	3.540	3.719
51	2.884	3.096	3.296	3.545	3.724

Tafel 28 (Forts.) Table 28 (cont.)

R	$Z(0.10/R)$	$Z(0.05/R)$	$Z(0.025/R)$	$Z(0.01/R)$	$Z(0.005/R)$
1-SIDED	10%	5%	2.5%	1%	0.5%
2-SIDED	20%	10%	5%	2%	1%
52	2.891	3.102	3.302	3.550	3.729
53	2.896	3.107	3.307	3.555	3.734
54	2.902	3.113	3.312	3.560	3.738
55	2.908	3.118	3.317	3.565	3.743
56	2.914	3.124	3.322	3.570	3.748
57	2.919	3.129	3.327	3.575	3.752
58	2.925	3.134	3.332	3.579	3.756
59	2.930	3.139	3.337	3.584	3.761
60	2.935	3.144	3.341	3.588	3.765
61	2.940	3.149	3.346	3.592	3.769
62	2.945	3.154	3.351	3.596	3.773
63	2.950	3.158	3.355	3.601	3.777
64	2.955	3.163	3.359	3.605	3.781
65	2.960	3.167	3.364	3.609	3.785
66	2.965	3.172	3.368	3.613	3.789
67	2.969	3.176	3.372	3.617	3.792
68	2.974	3.180	3.376	3.620	3.796
69	2.978	3.185	3.380	3.624	3.800
70	2.983	3.189	3.384	3.628	3.803
71	2.987	3.193	3.388	3.632	3.807
72	2.991	3.197	3.392	3.635	3.810
73	2.996	3.201	3.396	3.639	3.814
74	3.000	3.205	3.399	3.642	3.817
75	3.004	3.209	3.403	3.646	3.820
76	3.008	3.213	3.407	3.649	3.823
77	3.012	3.216	3.410	3.652	3.827
78	3.016	3.220	3.414	3.656	3.830
79	3.020	3.224	3.417	3.659	3.833
80	3.023	3.227	3.421	3.662	3.836
81	3.027	3.231	3.424	3.665	3.839
82	3.031	3.234	3.427	3.669	3.842
83	3.034	3.238	3.431	3.672	3.845
84	3.038	3.241	3.434	3.675	3.848
85	3.042	3.245	3.437	3.678	3.851
86	3.045	3.248	3.440	3.681	3.854
87	3.049	3.251	3.443	3.684	3.857
88	3.052	3.254	3.446	3.687	3.859
89	3.055	3.258	3.449	3.689	3.862
90	3.059	3.261	3.452	3.692	3.865
91	3.062	3.264	3.455	3.695	3.868
92	3.065	3.267	3.458	3.698	3.870
93	3.069	3.270	3.461	3.701	3.873
94	3.072	3.273	3.464	3.703	3.876
95	3.075	3.276	3.467	3.706	3.878
96	3.078	3.279	3.470	3.709	3.881
97	3.081	3.282	3.473	3.711	3.883
98	3.084	3.285	3.475	3.714	3.886
99	3.087	3.288	3.478	3.716	3.888
100	3.090	3.291	3.481	3.719	3.891
102	3.096	3.296	3.486	3.724	3.895
104	3.102	3.302	3.491	3.729	3.900
106	3.107	3.307	3.496	3.734	3.905

Tafel 28 (Forts.) Table 28 (cont.)

R	$Z(0.10/R)$	$Z(0.05/R)$	$Z(0.025/R)$	$Z(0.01/R)$	$Z(0.005/R)$
1-SIDED	10%	5%	2.5%	1%	0.5%
2-SIDED	20%	10%	5%	2%	1%
108	3.113	3.312	3.501	3.738	3.909
110	3.118	3.317	3.506	3.743	3.914
112	3.124	3.322	3.511	3.748	3.918
114	3.129	3.327	3.516	3.752	3.922
116	3.134	3.332	3.520	3.756	3.926
118	3.139	3.337	3.525	3.761	3.931
120	3.144	3.341	3.529	3.765	3.935
122	3.149	3.346	3.534	3.769	3.939
124	3.154	3.351	3.538	3.773	3.942
126	3.158	3.355	3.542	3.777	3.946
128	3.163	3.359	3.546	3.781	3.950
130	3.167	3.364	3.550	3.785	3.954
132	3.172	3.368	3.554	3.789	3.957
134	3.176	3.372	3.558	3.792	3.961
136	3.180	3.376	3.562	3.796	3.965
138	3.185	3.380	3.566	3.800	3.968
140	3.189	3.384	3.570	3.803	3.971
142	3.193	3.388	3.574	3.807	3.975
144	3.197	3.392	3.577	3.810	3.978
146	3.201	3.396	3.581	3.814	3.981
148	3.205	3.399	3.584	3.817	3.985
150	3.209	3.403	3.588	3.820	3.988
152	3.213	3.407	3.591	3.823	3.991
154	3.216	3.410	3.595	3.827	3.994
156	3.220	3.414	3.598	3.830	3.997
158	3.224	3.417	3.601	3.833	4.000
160	3.227	3.421	3.605	3.836	4.003
162	3.231	3.424	3.608	3.839	4.006
164	3.234	3.427	3.611	3.842	4.009
166	3.238	3.431	3.614	3.845	4.012
168	3.241	3.434	3.617	3.848	4.015
170	3.245	3.437	3.620	3.851	4.017
172	3.248	3.440	3.623	3.854	4.020
174	3.251	3.443	3.626	3.857	4.023
176	3.254	3.446	3.629	3.859	4.026
178	3.258	3.449	3.632	3.862	4.028
180	3.261	3.452	3.635	3.865	4.031
182	3.264	3.455	3.638	3.868	4.034
184	3.267	3.458	3.641	3.870	4.036
186	3.270	3.461	3.644	3.873	4.039
188	3.273	3.464	3.646	3.876	4.041
190	3.276	3.467	3.649	3.878	4.044
192	3.279	3.470	3.652	3.881	4.046
194	3.282	3.473	3.654	3.883	4.049
196	3.285	3.475	3.657	3.886	4.051
198	3.288	3.478	3.660	3.888	4.053
200	3.291	3.481	3.662	3.891	4.056
202	3.293	3.483	3.665	3.893	4.058
204	3.296	3.486	3.667	3.895	4.060
206	3.299	3.489	3.670	3.898	4.063
208	3.302	3.491	3.672	3.900	4.065
210	3.304	3.494	3.675	3.902	4.067

Tafel 28 (Forts.) Table 28 (cont.)

R	$Z(0.10/R)$	$Z(0.05/R)$	$Z(0.025/R)$	$Z(0.01/R)$	$Z(0.005/R)$
1-SIDED 2-SIDED	10% 20%	5% 10%	2.5% 5%	1% 2%	0.5% 1%
212	3.307	3.496	3.677	3.905	4.069
214	3.310	3.499	3.680	3.907	4.071
216	3.312	3.501	3.682	3.909	4.074
218	3.315	3.504	3.684	3.911	4.076
220	3.317	3.506	3.687	3.914	4.078
222	3.320	3.509	3.689	3.916	4.080
224	3.322	3.511	3.691	3.918	4.082
226	3.325	3.513	3.693	3.920	4.084
228	3.327	3.516	3.696	3.922	4.086
230	3.330	3.518	3.698	3.924	4.088
232	3.332	3.520	3.700	3.926	4.090
234	3.334	3.523	3.702	3.929	4.092
236	3.337	3.525	3.704	3.931	4.094
238	3.339	3.527	3.707	3.933	4.096
240	3.341	3.529	3.709	3.935	4.098
242	3.344	3.531	3.711	3.937	4.100
244	3.346	3.534	3.713	3.939	4.102
246	3.348	3.536	3.715	3.941	4.104
248	3.351	3.538	3.717	3.942	4.106
250	3.353	3.540	3.719	3.944	4.107
252	3.355	3.542	3.721	3.946	4.109
254	3.357	3.544	3.723	3.948	4.111
256	3.359	3.546	3.725	3.950	4.113
258	3.362	3.548	3.727	3.952	4.115
260	3.364	3.550	3.729	3.954	4.117
262	3.366	3.552	3.731	3.956	4.118
264	3.368	3.554	3.733	3.957	4.120
266	3.370	3.556	3.735	3.959	4.122
268	3.372	3.558	3.737	3.961	4.124
270	3.374	3.560	3.738	3.963	4.125
272	3.376	3.562	3.740	3.965	4.127
274	3.378	3.564	3.742	3.966	4.129
276	3.380	3.566	3.744	3.968	4.130
278	3.382	3.568	3.746	3.970	4.132
280	3.384	3.570	3.748	3.971	4.134
282	3.386	3.572	3.749	3.973	4.135
284	3.388	3.574	3.751	3.975	4.137
286	3.390	3.575	3.753	3.977	4.138
288	3.392	3.577	3.755	3.978	4.140
290	3.394	3.579	3.756	3.980	4.142
292	3.396	3.581	3.758	3.981	4.143
294	3.397	3.583	3.760	3.983	4.145
296	3.399	3.584	3.761	3.985	4.146
298	3.401	3.586	3.763	3.986	4.148
300	3.403	3.588	3.765	3.988	4.149
305	3.407	3.592	3.769	3.992	4.153
310	3.412	3.596	3.773	3.996	4.157
315	3.416	3.601	3.777	3.999	4.161
320	3.421	3.605	3.781	4.003	4.164
325	3.425	3.609	3.785	4.007	4.168
330	3.429	3.613	3.789	4.010	4.171
335	3.433	3.617	3.792	4.014	4.175

Tafel 28 (Forts.) Table 28 (cont.)

R	$Z(0.10/R)$	$Z(0.05/R)$	$Z(0.025/R)$	$Z(0.01/R)$	$Z(0.005/R)$
1-SIDED 2-SIDED	10% 20%	5% 10%	2.5% 5%	1% 2%	0.5% 1%
340	3.437	3.620	3.796	4.017	4.178
345	3.441	3.624	3.800	4.021	4.181
350	3.445	3.628	3.803	4.024	4.185
355	3.449	3.632	3.807	4.028	4.188
360	3.452	3.635	3.810	4.031	4.191
365	3.456	3.639	3.814	4.034	4.194
370	3.460	3.642	3.817	4.037	4.197
375	3.463	3.646	3.820	4.041	4.200
380	3.467	3.649	3.823	4.044	4.203
385	3.471	3.652	3.827	4.047	4.206
390	3.474	3.656	3.830	4.050	4.209
395	3.477	3.659	3.833	4.053	4.212
400	3.481	3.662	3.836	4.056	4.215
405	3.484	3.665	3.839	4.059	4.218
410	3.487	3.669	3.842	4.061	4.220
415	3.491	3.672	3.845	4.064	4.223
420	3.494	3.675	3.848	4.067	4.226
425	3.497	3.678	3.851	4.070	4.228
430	3.500	3.681	3.854	4.072	4.231
435	3.503	3.684	3.857	4.075	4.234
440	3.506	3.687	3.859	4.078	4.236
445	3.509	3.689	3.862	4.080	4.239
450	3.512	3.692	3.865	4.083	4.241
455	3.515	3.695	3.868	4.086	4.244
460	3.518	3.698	3.870	4.088	4.246
465	3.521	3.701	3.873	4.091	4.249
470	3.524	3.703	3.876	4.093	4.251
475	3.527	3.706	3.878	4.096	4.253
480	3.529	3.709	3.881	4.098	4.256
485	3.532	3.711	3.883	4.100	4.258
490	3.535	3.714	3.886	4.103	4.260
495	3.537	3.716	3.888	4.105	4.263
500	3.540	3.719	3.891	4.107	4.265
505	3.543	3.722	3.893	4.110	4.267
510	3.545	3.724	3.895	4.112	4.269
515	3.548	3.726	3.898	4.114	4.271
520	3.550	3.729	3.900	4.117	4.274
525	3.553	3.731	3.902	4.119	4.276
530	3.555	3.734	3.905	4.121	4.278
535	3.558	3.736	3.907	4.123	4.280
540	3.560	3.738	3.909	4.125	4.282
545	3.563	3.741	3.911	4.127	4.284
550	3.565	3.743	3.914	4.129	4.286
555	3.568	3.745	3.916	4.132	4.288
560	3.570	3.748	3.918	4.134	4.290
565	3.572	3.750	3.920	4.136	4.292
570	3.575	3.752	3.922	4.138	4.294
575	3.577	3.754	3.924	4.140	4.296
580	3.579	3.756	3.926	4.142	4.298
585	3.581	3.758	3.929	4.144	4.300
590	3.584	3.761	3.931	4.146	4.302
595	3.586	3.763	3.933	4.147	4.304

Tafel 28 (Forts.) Table 28 (cont.)

R	$Z(0.10/R)$	$Z(0.05/R)$	$Z(0.025)/R$	$Z(0.01)/R$	$Z(0.005/R)$
1-SIDED	10%	5%	2.5%	1%	0.5%
2-SIDED	20%	10%	5%	2%	1%
600	3.588	3.765	3.935	4.149	4.305
605	3.590	3.767	3.937	4.151	4.307
610	3.592	3.769	3.939	4.153	4.309
615	3.594	3.771	3.941	4.155	4.311
620	3.596	3.773	3.942	4.157	4.313
625	3.599	3.775	3.944	4.159	4.314
630	3.601	3.777	3.946	4.161	4.316
635	3.603	3.779	3.948	4.162	4.318
640	3.605	3.781	3.950	4.164	4.320
645	3.607	3.783	3.952	4.166	4.321
650	3.609	3.785	3.954	4.168	4.323
655	3.611	3.787	3.956	4.169	4.325
660	3.613	3.789	3.957	4.171	4.326
665	3.615	3.790	3.959	4.173	4.328
670	3.617	3.792	3.961	4.175	4.330
675	3.619	3.794	3.963	4.176	4.331
680	3.620	3.796	3.965	4.178	4.333
685	3.622	3.798	3.966	4.180	4.335
690	3.624	3.800	3.968	4.181	4.336
695	3.626	3.801	3.970	4.183	4.338
700	3.628	3.803	3.971	4.185	4.339
705	3.630	3.805	3.973	4.186	4.341
710	3.632	3.807	3.975	4.188	4.343
715	3.633	3.808	3.977	4.189	4.344
720	3.635	3.810	3.978	4.191	4.346
725	3.637	3.812	3.980	4.193	4.347
730	3.639	3.814	3.981	4.194	4.349
735	3.641	3.815	3.983	4.196	4.350
740	3.642	3.817	3.985	4.197	4.352
745	3.644	3.819	3.986	4.199	4.353
750	3.646	3.820	3.988	4.200	4.355
755	3.647	3.822	3.989	4.202	4.356
760	3.649	3.823	3.991	4.203	4.357
765	3.651	3.825	3.993	4.205	4.359
770	3.652	3.827	3.994	4.206	4.360
775	3.654	3.828	3.996	4.208	4.362
780	3.656	3.830	3.997	4.209	4.363
785	3.657	3.831	3.999	4.211	4.365
790	3.659	3.833	4.000	4.212	4.366
795	3.661	3.835	4.002	4.213	4.367
800	3.662	3.836	4.003	4.215	4.369
810	3.665	3.839	4.006	4.218	4.371
820	3.669	3.842	4.009	4.220	4.374
830	3.672	3.845	4.012	4.223	4.377
840	3.675	3.848	4.015	4.226	4.379
850	3.678	3.851	4.017	4.228	4.382
860	3.681	3.854	4.020	4.231	4.384
870	3.684	3.857	4.023	4.234	4.387
880	3.687	3.859	4.026	4.236	4.389
890	3.689	3.862	4.028	4.239	4.392
900	3.692	3.865	4.031	4.241	4.394
910	3.695	3.868	4.034	4.244	4.397

Tafel 28 (Forts.) Table 28 (cont.)

R	$Z(0.10/R)$	$Z(0.05/R)$	$Z(0.025/R)$	$Z(0.01)/R$	$Z(0.005/R)$
1-SIDED	10%	5%	2.5%	1%	0.5%
2-SIDED	20%	10%	5%	2%	1%
920	3.698	3.870	4.036	4.246	4.399
930	3.701	3.873	4.039	4.249	4.401
940	3.703	3.876	4.041	4.251	4.404
950	3.706	3.878	4.044	4.253	4.406
960	3.709	3.881	4.046	4.256	4.408
970	3.711	3.883	4.049	4.258	4.411
980	3.714	3.886	4.051	4.260	4.413
990	3.716	3.888	4.053	4.263	4.415
1000	3.719	3.891	4.056	4.265	4.417
1010	3.722	3.893	4.058	4.267	4.419
1020	3.724	3.895	4.060	4.269	4.421
1030	3.726	3.898	4.063	4.271	4.424
1040	3.729	3.900	4.065	4.274	4.426
1050	3.731	3.902	4.067	4.276	4.428
1060	3.734	3.905	4.069	4.278	4.430
1070	3.736	3.907	4.071	4.280	4.432
1080	3.738	3.909	4.074	4.282	4.434
1090	3.741	3.911	4.076	4.284	4.436
1100	3.743	3.914	4.078	4.286	4.438
1110	3.745	3.916	4.080	4.288	4.440
1120	3.748	3.918	4.082	4.290	4.442
1130	3.750	3.920	4.084	4.292	4.444
1140	3.752	3.922	4.086	4.294	4.445
1150	3.754	3.924	4.088	4.296	4.447
1160	3.756	3.926	4.090	4.298	4.449
1170	3.758	3.929	4.092	4.300	4.451
1180	3.761	3.931	4.094	4.302	4.453
1190	3.763	3.933	4.096	4.304	4.455
1200	3.765	3.935	4.098	4.305	4.456
1210	3.767	3.937	4.100	4.307	4.458
1220	3.769	3.939	4.102	4.309	4.460
1230	3.771	3.941	4.104	4.311	4.462
1240	3.773	3.942	4.106	4.313	4.463
1250	3.775	3.944	4.107	4.314	4.465
1260	3.777	3.946	4.109	4.316	4.467
1270	3.779	3.948	4.111	4.318	4.469
1280	3.781	3.950	4.113	4.320	4.470
1290	3.783	3.952	4.115	4.321	4.472
1300	3.785	3.954	4.117	4.323	4.474
1310	3.787	3.956	4.118	4.325	4.475
1320	3.789	3.957	4.120	4.326	4.477
1330	3.790	3.959	4.122	4.328	4.478
1340	3.792	3.961	4.124	4.330	4.480
1350	3.794	3.963	4.125	4.331	4.482
1360	3.796	3.965	4.127	4.333	4.483
1370	3.798	3.966	4.129	4.335	4.485
1380	3.800	3.968	4.130	4.336	4.486
1390	3.801	3.970	4.132	4.338	4.488
1400	3.803	3.971	4.134	4.339	4.489
1410	3.805	3.973	4.135	4.341	4.491
1420	3.807	3.975	4.137	4.343	4.492
1430	3.808	3.977	4.138	4.344	4.494

Tafel 28 (Forts.)

Table 28 (cont.)

R	$Z(0.10/R)$	$Z(0.05/R)$	$Z(0.025/R)$	$Z(0.01/R)$	$Z(0.005/R)$
1-SIDED	10%	5%	2.5%	1%	0.5%
2-SIDED	20%	10%	5%	2%	1%
1440	3.810	3.978	4.140	4.346	4.495
1450	3.812	3.980	4.142	4.347	4.497
1460	3.814	3.981	4.143	4.349	4.498
1470	3.815	3.983	4.145	4.350	4.500
1480	3.817	3.985	4.146	4.352	4.501
1490	3.819	3.986	4.148	4.353	4.503
1500	3.820	3.988	4.149	4.355	4.504
1510	3.822	3.989	4.151	4.356	4.505
1520	3.823	3.991	4.152	4.357	4.507
1530	3.825	3.993	4.154	4.359	4.508
1540	3.827	3.994	4.155	4.360	4.510
1550	3.828	3.996	4.157	4.362	4.511
1560	3.830	3.997	4.158	4.363	4.512
1570	3.831	3.999	4.160	4.365	4.514
1580	3.833	4.000	4.161	4.366	4.515
1590	3.835	4.002	4.163	4.367	4.516
1600	3.836	4.003	4.164	4.369	4.518
1620	3.839	4.006	4.167	4.371	4.520
1640	3.842	4.009	4.170	4.374	4.523
1660	3.845	4.012	4.173	4.377	4.526
1680	3.848	4.015	4.175	4.379	4.528
1700	3.851	4.017	4.178	4.382	4.531
1720	3.854	4.020	4.181	4.384	4.533
1740	3.857	4.023	4.183	4.387	4.535
1760	3.859	4.026	4.186	4.389	4.538
1780	3.862	4.028	4.188	4.392	4.540
1800	3.865	4.031	4.191	4.394	4.543
1820	3.868	4.034	4.193	4.397	4.545
1840	3.870	4.036	4.196	4.399	4.547
1860	3.873	4.039	4.198	4.401	4.550
1880	3.876	4.041	4.201	4.404	4.552
1900	3.878	4.044	4.203	4.406	4.554
1920	3.881	4.046	4.206	4.408	4.556
1940	3.883	4.049	4.208	4.411	4.558
1960	3.886	4.051	4.210	4.413	4.561
1980	3.888	4.053	4.213	4.415	4.563
2000	3.891	4.056	4.215	4.417	4.565
2020	3.893	4.058	4.217	4.419	4.567
2040	3.895	4.060	4.219	4.421	4.569
2060	3.898	4.063	4.221	4.424	4.571
2080	3.900	4.065	4.224	4.426	4.573
2100	3.902	4.067	4.226	4.428	4.575
2120	3.905	4.069	4.228	4.430	4.577
2140	3.907	4.071	4.230	4.432	4.579
2160	3.909	4.074	4.232	4.434	4.581
2180	3.911	4.076	4.234	4.436	4.583
2200	3.914	4.078	4.236	4.438	4.585
2220	3.916	4.080	4.238	4.440	4.587
2240	3.918	4.082	4.240	4.442	4.589
2260	3.920	4.084	4.242	4.444	4.590
2280	3.922	4.086	4.244	4.445	4.592
2300	3.924	4.088	4.246	4.447	4.594

Tafel 28 (Forts.) Table 28 (cont.)

R	$Z(0.10/R)$	$Z(0.05/R)$	$Z(0.025/R)$	$Z(0.01/R)$	$Z(0.005/R)$
1-SIDED	10%	5%	2.5%	1%	0.5%
2-SIDED	20%	10%	5%	2%	1%
2320	3.926	4.090	4.248	4.449	4.596
2340	3.929	4.092	4.250	4.451	4.598
2360	3.931	4.094	4.252	4.453	4.599
2380	3.933	4.096	4.254	4.455	4.601
2400	3.935	4.098	4.256	4.456	4.603
2420	3.937	4.100	4.258	4.458	4.605
2440	3.939	4.102	4.259	4.460	4.606
2460	3.941	4.104	4.261	4.462	4.608
2480	3.942	4.106	4.263	4.463	4.610
2500	3.944	4.107	4.265	4.465	4.611
2520	3.946	4.109	4.267	4.467	4.613
2540	3.948	4.111	4.268	4.469	4.615
2560	3.950	4.113	4.270	4.470	4.616
2580	3.952	4.115	4.272	4.472	4.618
2600	3.954	4.117	4.274	4.474	4.620
2620	3.956	4.118	4.275	4.475	4.621
2640	3.957	4.120	4.277	4.477	4.623
2660	3.959	4.122	4.279	4.478	4.624
2680	3.961	4.124	4.280	4.480	4.626
2700	3.963	4.125	4.282	4.482	4.627
2720	3.965	4.127	4.284	4.483	4.629
2740	3.966	4.129	4.285	4.485	4.630
2760	3.968	4.130	4.287	4.486	4.632
2780	3.970	4.132	4.289	4.488	4.633
2800	3.971	4.134	4.290	4.489	4.635
2820	3.973	4.135	4.292	4.491	4.636
2840	3.975	4.137	4.293	4.492	4.638
2860	3.977	4.138	4.295	4.494	4.639
2880	3.978	4.140	4.296	4.495	4.641
2900	3.980	4.142	4.298	4.497	4.642
2920	3.981	4.143	4.299	4.498	4.644
2940	3.983	4.145	4.301	4.500	4.645
2960	3.985	4.146	4.302	4.501	4.646
2980	3.986	4.148	4.304	4.503	4.648
3000	3.988	4.149	4.305	4.504	4.649
3020	3.989	4.151	4.307	4.505	4.651
3040	3.991	4.152	4.308	4.507	4.652
3060	3.993	4.154	4.310	4.508	4.653
3080	3.994	4.155	4.311	4.510	4.655
3100	3.996	4.157	4.313	4.511	4.656
3120	3.997	4.158	4.314	4.512	4.657
3140	3.999	4.160	4.316	4.514	4.659
3160	4.000	4.161	4.317	4.515	4.660
3180	4.002	4.163	4.318	4.516	4.661
3200	4.003	4.164	4.320	4.518	4.662
3250	4.007	4.168	4.323	4.521	4.666
3300	4.010	4.171	4.326	4.524	4.669
3350	4.014	4.175	4.330	4.527	4.672
3400	4.017	4.178	4.333	4.531	4.675
3450	4.021	4.181	4.336	4.534	4.678
3500	4.024	4.185	4.339	4.537	4.681
3550	4.028	4.188	4.343	4.540	4.684

Tafel 28 (Forts.) Table 28 (cont.)

R	$Z(0.10/R)$	$Z(0.05/R)$	$Z(0.025/R)$	$Z(0.01/R)$	$Z(0.005/R)$
1-SIDED 2-SIDED	10% 20%	5% 10%	2.5% 5%	1% 2%	0.5% 2%
3600	4.031	4.191	4.346	4.543	4.687
3650	4.034	4.194	4.349	4.546	4.689
3700	4.037	4.197	4.352	4.548	4.692
3750	4.041	4.200	4.355	4.551	4.695
3800	4.044	4.203	4.357	4.554	4.698
3850	4.047	4.206	4.360	4.557	4.700
3900	4.050	4.209	4.363	4.559	4.703
3950	4.053	4.212	4.366	4.562	4.706
4000	4.056	4.215	4.369	4.565	4.708
4050	4.059	4.218	4.371	4.567	4.711
4100	4.061	4.220	4.374	4.570	4.713
4150	4.064	4.223	4.377	4.573	4.716
4200	4.067	4.226	4.379	4.575	4.718
4250	4.070	4.228	4.382	4.577	4.720
4300	4.072	4.231	4.384	4.580	4.723
4350	4.075	4.234	4.387	4.582	4.723
4400	4.078	4.236	4.389	4.585	4.728
4450	4.080	4.239	4.392	4.587	4.730
4500	4.083	4.241	4.394	4.589	4.732
4550	4.086	4.244	4.397	4.592	4.734
4600	4.088	4.246	4.399	4.594	4.737
4650	4.091	4.249	4.401	4.596	4.739
4700	4.093	4.251	4.404	4.599	4.741
4750	4.096	4.253	4.406	4.601	4.743
4800	4.098	4.256	4.408	4.603	4.745
4850	4.100	4.258	4.411	4.605	4.747
4900	4.103	4.260	4.413	4.607	4.749
4950	4.105	4.263	4.415	4.609	4.751
5000	4.007	4.265	4.417	4.611	4.753
5050	4.110	4.267	4.419	4.613	4.755
5100	4.112	4.269	4.421	4.615	4.757
5150	4.114	4.271	4.424	4.618	4.759
5200	4.117	4.274	4.426	4.620	4.761
5250	4.119	4.276	4.428	4.622	4.763
5300	4.121	4.278	4.430	4.623	4.765
5350	4.123	4.280	4.432	4.625	4.767
5400	4.125	4.282	4.434	4.627	4.769
5450	4.127	4.284	4.436	4.629	4.771
5500	4.129	4.286	4.438	4.631	4.773
5550	4.132	4.288	4.440	4.633	4.774
5600	4.134	4.290	4.442	4.635	4.776
5650	4.136	4.292	4.444	4.637	4.778
5700	4.138	4.294	4.445	4.639	4.780
5750	4.140	4.296	4.447	4.640	4.782
5800	4.142	4.298	4.449	4.642	4.783
5850	4.144	4.300	4.451	4.644	4.785
5900	4.146	4.302	4.453	4.646	4.787
5950	4.147	4.304	4.455	4.647	4.788
6000	4.149	4.305	4.456	4.649	4.790
6050	4.151	4.307	4.458	4.651	4.792
6100	4.153	4.309	4.460	4.653	4.793
6150	4.155	4.311	4.462	4.654	4.795

Tafel 28 (Forts.) Table 28 (cont.)

R	$Z(0.10/R)$	$Z(0.05/R)$	$Z(0.025/R)$	$Z(0.01/R)$	$Z(0.005/R)$
1-SIDED 2-SIDED	10% 20%	5% 10%	2.5% 5%	1% 2%	0.5% 1%
6200	4.157	4.313	4.463	4.656	4.797
6250	4.159	4.314	4.465	4.658	4.798
6300	4.161	4.316	4.467	4.659	4.800
6350	4.162	4.318	4.469	4.661	4.802
6400	4.164	4.320	4.470	4.662	4.803
6450	4.166	4.321	4.472	4.664	4.805
6500	4.168	4.323	4.474	4.666	4.806
6550	4.169	4.325	4.475	4.667	4.808
6600	4.171	4.326	4.477	4.669	4.809
6650	4.173	4.328	4.478	4.670	4.811
6700	4.175	4.330	4.480	4.672	4.812
6750	4.176	4.331	4.482	4.673	4.814
6800	4.178	4.333	4.483	4.675	4.815
6850	4.180	4.335	4.485	4.676	4.817
6900	4.181	4.336	4.486	4.678	4.818
6950	4.183	4.338	4.488	4.679	4.820
7000	4.185	4.339	4.489	4.681	4.821
7050	4.186	4.341	4.491	4.682	4.822
7100	4.188	4.343	4.492	4.684	4.824
7150	4.189	4.344	4.494	4.685	4.825
7200	4.191	4.346	4.495	4.687	4.827
7250	4.193	4.347	4.497	4.688	4.828
7300	4.194	4.349	4.498	4.689	4.829
7350	4.196	4.350	4.500	4.691	4.831
7400	4.197	4.352	4.501	4.692	4.832
7450	4.199	4.353	4.503	4.694	4.833
7500	4.200	4.355	4.504	4.695	4.835
7550	4.202	4.356	4.505	4.696	4.836
7600	4.203	4.357	4.507	4.698	4.837
7650	4.205	4.359	4.508	4.699	4.839
7700	4.206	4.360	4.510	4.700	4.840
7750	4.208	4.362	4.511	4.702	4.841
7800	4.209	4.363	4.512	4.703	4.843
7850	4.211	4.365	4.514	4.704	4.844
7900	4.212	4.366	4.515	4.706	4.845
7950	4.213	4.367	4.516	4.707	4.846
8000	4.215	4.369	4.518	4.708	4.848
8050	4.216	4.370	4.519	4.709	4.849
8100	4.218	4.371	4.520	4.711	4.850
8150	4.219	4.373	4.522	4.712	4.851
8200	4.220	4.374	4.523	4.713	4.852
8250	4.222	4.375	4.524	4.714	4.854
8300	4.223	4.377	4.526	4.716	4.855
8350	4.224	4.378	4.527	4.717	4.856
8400	4.226	4.379	4.528	4.718	4.857
8450	4.227	4.381	4.529	4.719	4.858
8500	4.228	4.382	4.531	4.720	4.860
8550	4.230	4.383	4.532	4.722	4.861
8600	4.231	4.384	4.533	4.723	4.862
8650	4.232	4.386	4.534	4.724	4.863
8700	4.234	4.387	4.535	4.725	4.864
8750	4.235	4.388	4.537	4.726	4.865

Tafel 28 (Forts.) Table 28 (cont.)

R	$Z(0.10/R)$	$Z(0.05/R)$	$Z(0.025/R)$	$Z(0.01/R)$	$Z(0.005/R)$
1-SIDED 2-SIDED	10% 20%	5% 10%	2.5% 5%	1% 2%	0.5% 1%
8800	4.236	4.389	4.538	4.728	4.866
8850	4.238	4.391	4.539	4.729	4.868
8900	4.239	4.392	4.540	4.730	4.869
8950	4.240	4.393	4.541	4.731	4.870
9000	4.241	4.394	4.543	4.732	4.871
9050	4.243	4.396	4.544	4.733	4.872
9100	4.244	4.397	4.545	4.734	4.873
9150	4.245	4.398	4.546	4.735	4.874
9200	4.246	4.399	4.547	4.737	4.875
9250	4.247	4.400	4.548	4.738	4.876
9300	4.249	4.401	4.550	4.739	4.877
9350	4.250	4.403	4.551	4.740	4.878
9400	4.251	4.404	4.552	4.741	4.879
9450	4.252	4.405	4.553	4.742	4.880
9500	4.253	4.406	4.554	4.743	4.882
9550	4.255	4.407	4.555	4.744	4.883
9600	4.256	4.408	4.556	4.745	4.884
9650	4.257	4.409	4.557	4.746	4.885
9700	4.258	4.411	4.558	4.747	4.886
9750	4.259	4.412	4.559	4.748	4.887
9800	4.260	4.413	4.561	4.749	4.888
9850	4.262	4.414	4.562	4.750	4.889
9900	4.263	4.415	4.563	4.751	4.890
9950	4.264	4.416	4.564	4.752	4.891
10000	4.265	4.417	4.565	4.753	4.892

TABLE 29

Upper Percentage Points of the Bonferroni t-Statistic: Tables of B. J. R. Bailey

(a) Content of the table and definition of the statistic :

This table contains upper $(1/2).\alpha/k$ percentage points $t_\nu^{\alpha/2k}$ of the t - distribution :

$$P (T_\nu > t_\nu^{\alpha/2k}) = (1/2).\alpha/k \quad ,$$

where α is the given error probability , k the number of means to be compared, and ν the degree of freedom.

These percentage points are used for multiple comparisons of means and for simultaneous testing.

For a given value of α the so called Bonferroni inequality insures that the resulting intervals have a probability of at least $1 - \alpha$ of simultaneously containing the respective population means.

(b) Extent of the table and definition of the parameters :

(1) The parameter α :

α = error probability (risk of the first kind).

$\alpha = 0.05$ and 0.01

(2) The parameter k :

k = number of means to be compared.

$k = 1(1)20$ and a proportion $(1/2).\alpha/k$,

$k = 7(1)20$ and a proportion $(1/2).\alpha/\binom{k}{2}$.

(3) The parameter ν :

ν = degree of freedom of the t-distribution

$\nu = 2(1)30(5)60(10)120;\ 250;\ 500;\ 1000;\ 1000;\ \infty$.

(c) Suggestions for applications :

(1) Multiple comparisons of means.

(2) Simultaneous confidence intervals for means of dependent normal distributions with unknown variances.

(3) Tests to detect outliers in experiments.

(d) References :

(1) For the source of the tables :

Bailey, B. J. R. : Tables of the Bonferroni t Statistic. Journal of the American Statistical Association 72, 469 - 478 (1977).

(2) For further informations concerning theory, computations, and applications :

Bailey, B. J. R. (1977):(As cited before !!).

Prescott, Ph. : An Approximative Test for Outliers in Linear Models.
Technometrics 17, 129 - 132 (1975).

Stefansky, W. : Rejecting Outliers in Factorial Designs.
Technometrics 14, 469 - 479 (1972).

(e) Further remarks :

Some of the tests to detect outliers require extreme deviations of the F-distribution , in which the numerator has one degree of freedom. They correspond to the squares of deviates of the t-distribution.

Tafel 29 Table 29

	$\alpha = 0.05$									
k:	1	2	$3 = \binom{3}{2}$	4	5	$6 = \binom{4}{2}$	7	8	9	$10 = \binom{5}{2}$
ν $100\alpha/k$:	5.0000	2.5000	1.6667	1.2500	1.0000	0.8333	0.7143	0.6250	0.5556	0.5000
2	4.3027	6.2053	7.6488	8.8602	9.9248	10.8859	11.7687	12.5897	13.3604	14.0890
3	3.1824	4.1765	4.8567	5.3919	5.8409	6.2315	6.5797	6.8952	7.1849	7.4533
4	2.7764	3.4954	3.9608	4.3147	4.6041	4.8510	5.0675	5.2611	5.4366	5.5976
5	2.5706	3.1634	3.5341	3.8100	4.0321	4.2193	4.3818	4.5257	4.6553	4.7733
6	2.4469	2.9687	3.2875	3.5212	3.7074	3.8630	3.9971	4.1152	4.2209	4.3168
7	2.3646	2.8412	3.1276	3.3353	3.4995	3.6358	3.7527	3.8552	3.9467	4.0293
8	2.3060	2.7515	3.0158	3.2060	3.3554	3.4789	3.5844	3.6766	3.7586	3.8325
9	2.2622	2.6850	2.9333	3.1109	3.2498	3.3642	3.4616	3.5465	3.6219	3.6897
10	2.2281	2.6338	2.8701	3.0382	3.1693	3.2768	3.3682	3.4477	3.5182	3.5814
11	2.2010	2.5931	2.8200	2.9809	3.1058	3.2081	3.2949	3.3702	3.4368	3.4966
12	2.1788	2.5600	2.7795	2.9345	3.0545	3.1527	3.2357	3.3078	3.3714	3.4284
13	2.1604	2.5326	2.7459	2.8961	3.0123	3.1070	3.1871	3.2565	3.3177	3.3725
14	2.1448	2.5096	2.7178	2.8640	2.9768	3.0688	3.1464	3.2135	3.2727	3.3257
15	2.1314	2.4899	2.6937	2.8366	2.9467	3.0363	3.1118	3.1771	3.2346	3.2860
16	2.1199	2.4729	2.6730	2.8131	2.9208	3.0083	3.0821	3.1458	3.2019	3.2520
17	2.1098	2.4581	2.6550	2.7925	2.8982	2.9840	3.0563	3.1186	3.1735	3.2224
18	2.1009	2.4450	2.6391	2.7745	2.8784	2.9627	3.0336	3.0948	3.1486	3.1966
19	2.0930	2.4334	2.6251	2.7586	2.8609	2.9439	3.0136	3.0738	3.1266	3.1737
20	2.0860	2.4231	2.6126	2.7444	2.8453	2.9271	2.9958	3.0550	3.1070	3.1534
21	2.0796	2.4138	2.6013	2.7316	2.8314	2.9121	2.9799	3.0382	3.0895	3.1352
22	2.0739	2.4055	2.5912	2.7201	2.8188	2.8985	2.9655	3.0231	3.0737	3.1188
23	2.0687	2.3979	2.5820	2.7097	2.8073	2.8863	2.9525	3.0095	3.0595	3.1040
24	2.0639	2.3909	2.5736	2.7002	2.7969	2.8751	2.9406	2.9970	3.0465	3.0905
25	2.0595	2.3846	2.5660	2.6916	2.7874	2.8649	2.9298	2.9856	3.0346	3.0782
26	2.0555	2.3788	2.5589	2.6836	2.7787	2.8555	2.9199	2.9752	3.0237	3.0669
27	2.0518	2.3734	2.5525	2.6763	2.7707	2.8469	2.9107	2.9656	3.0137	3.0565
28	2.0484	2.3685	2.5465	2.6695	2.7633	2.8389	2.9023	2.9567	3.0045	3.0469
29	2.0452	2.3638	2.5409	2.6632	2.7564	2.8316	2.8945	2.9485	2.9959	3.0380
30	2.0423	2.3596	2.5357	2.6574	2.7500	2.8247	2.8872	2.9409	2.9880	3.0298
35	2.0301	2.3420	2.5145	2.6334	2.7238	2.7966	2.8575	2.9097	2.9554	2.9960
40	2.0211	2.3289	2.4989	2.6157	2.7045	2.7759	2.8355	2.8867	2.9314	2.9712
45	2.0141	2.3189	2.4868	2.6021	2.6896	2.7599	2.8187	2.8690	2.9130	2.9521
50	2.0086	2.3109	2.4772	2.5913	2.6778	2.7473	2.8053	2.8550	2.8984	2.9370
55	2.0040	2.3044	2.4694	2.5825	2.6682	2.7370	2.7944	2.8436	2.8866	2.9247
60	2.0003	2.2990	2.4630	2.5752	2.6603	2.7286	2.7855	2.8342	2.8768	2.9146
70	1.9944	2.2906	2.4529	2.5639	2.6479	2.7153	2.7715	2.8195	2.8615	2.8987
80	1.9901	2.2844	2.4454	2.5554	2.6387	2.7054	2.7610	2.8086	2.8502	2.8870
90	1.9867	2.2795	2.4395	2.5489	2.6316	2.6978	2.7530	2.8002	2.8414	2.8779
100	1.9840	2.2757	2.4349	2.5437	2.6259	2.6918	2.7466	2.7935	2.8344	2.8707
110	1.9818	2.2725	2.4311	2.5394	2.6213	2.6868	2.7414	2.7880	2.8287	2.8648
120	1.9799	2.2699	2.4280	2.5359	2.6174	2.6827	2.7370	2.7835	2.8240	2.8599
250	1.9695	2.2550	2.4102	2.5159	2.5956	2.6594	2.7124	2.7577	2.7972	2.8322
500	1.9647	2.2482	2.4021	2.5068	2.5857	2.6488	2.7012	2.7460	2.7850	2.8195
1000	1.9623	2.2448	2.3980	2.5022	2.5808	2.6435	2.6957	2.7402	2.7790	2.8133
∞	1.9600	2.2414	2.3940	2.4977	2.5758	2.6383	2.6901	2.7344	2.7729	2.8070

Tafel 29 (Forts.) Table 29 (cont.)

	$\alpha = 0.05$								
k:	11	12	13	14	$15 = \binom{6}{2}$	16	17	18	19
ν $100\,\alpha/k$:	0.4545	0.4167	0.3846	0.3571	0.3333	0.3125	0.2941	0.2778	0.2632
2	14.7818	15.4435	16.0780	16.6883	17.2772	17.8466	18.3984	18.9341	19.4551
3	7.7041	7.9398	8.1625	8.3738	8.5752	8.7676	8.9521	9.1294	9.3001
4	5.7465	5.8853	6.0154	6.1380	6.2541	6.3643	6.4693	6.5697	6.6659
5	4.8819	4.9825	5.0764	5.1644	5.2474	5.3259	5.4005	5.4715	5.5393
6	4.4047	4.4858	4.5612	4.6317	4.6979	4.7604	4.8196	4.8759	4.9295
7	4.1048	4.1743	4.2388	4.2989	4.3553	4.4084	4.4586	4.5062	4.5514
8	3.8999	3.9618	4.0191	4.0724	4.1224	4.1693	4.2137	4.2556	4.2955
9	3.7513	3.8079	3.8602	3.9088	3.9542	3.9969	4.0371	4.0752	4.1114
10	3.6388	3.6915	3.7401	3.7852	3.8273	3.8669	3.9041	3.9394	3.9728
11	3.5508	3.6004	3.6462	3.6887	3.7283	3.7654	3.8004	3.8335	3.8648
12	3.4801	3.5274	3.5709	3.6112	3.6489	3.6842	3.7173	3.7487	3.7783
13	3.4221	3.4674	3.5091	3.5478	3.5838	3.6176	3.6493	3.6793	3.7076
14	3.3736	3.4173	3.4576	3.4949	3.5296	3.5621	3.5926	3.6214	3.6487
15	3.3325	3.3749	3.4139	3.4501	3.4837	3.5151	3.5447	3.5725	3.5989
16	3.2973	3.3386	3.3765	3.4116	3.4443	3.4749	3.5036	3.5306	3.5562
17	3.2667	3.3070	3.3440	3.3783	3.4102	3.4400	3.4680	3.4944	3.5193
18	3.2399	3.2794	3.3156	3.3492	3.3804	3.4095	3.4369	3.4626	3.4870
19	3.2163	3.2550	3.2906	3.3235	3.3540	3.3826	3.4094	3.4347	3.4585
20	3.1952	3.2333	3.2683	3.3006	3.3306	3.3587	3.3850	3.4098	3.4332
21	3.1764	3.2139	3.2483	3.2802	3.3097	3.3373	3.3632	3.3876	3.4106
22	3.1595	3.1965	3.2304	3.2618	3.2909	3.3181	3.3436	3.3676	3.3903
23	3.1441	3.1807	3.2142	3.2451	3.2739	3.3007	3.3259	3.3495	3.3719
24	3.1302	3.1663	3.1994	3.2300	3.2584	3.2849	3.3097	3.3331	3.3552
25	3.1175	3.1532	3.1859	3.2162	3.2443	3.2705	3.2950	3.3181	3.3400
26	3.1058	3.1412	3.1736	3.2035	3.2313	3.2572	3.2815	3.3044	3.3260
27	3.0951	3.1301	3.1622	3.1919	3.2194	3.2451	3.2691	3.2918	3.3132
28	3.0852	3.1199	3.1517	3.1811	3.2084	3.2339	3.2577	3.2801	3.3013
29	3.0760	3.1105	3.1420	3.1712	3.1982	3.2235	3.2471	3.2694	3.2904
30	3.0675	3.1017	3.1330	3.1620	3.1888	3.2138	3.2373	3.2594	3.2802
35	3.0326	3.0658	3.0962	3.1242	3.1502	3.1744	3.1971	3.2185	3.2386
40	3.0069	3.0393	3.0690	3.0964	3.1218	3.1455	3.1676	3.1884	3.2081
45	2.9872	3.0191	3.0482	3.0751	3.1000	3.1232	3.1450	3.1654	3.1846
50	2.9716	3.0030	3.0318	3.0582	3.0828	3.1057	3.1271	3.1472	3.1661
55	2.9589	2.9900	3.0184	3.0446	3.0688	3.0914	3.1125	3.1324	3.1511
60	2.9485	2.9792	3.0074	3.0333	3.0573	3.0796	3.1005	3.1202	3.1387
70	2.9321	2.9624	2.9901	3.0156	3.0393	3.0613	3.0818	3.1012	3.1194
80	2.9200	2.9500	2.9773	3.0026	3.0259	3.0476	3.0679	3.0870	3.1050
90	2.9106	2.9403	2.9675	2.9924	3.0156	3.0371	3.0572	3.0761	3.0939
100	2.9032	2.9327	2.9596	2.9844	3.0073	3.0287	3.0487	3.0674	3.0851
110	2.8971	2.9264	2.9532	2.9778	3.0007	3.0219	3.0417	3.0604	3.0779
120	2.8921	2.9212	2.9479	2.9724	2.9951	3.0162	3.0360	3.0545	3.0720
250	2.8635	2.8919	2.9178	2.9416	2.9637	2.9842	3.0034	3.0213	3.0383
500	2.8505	2.8785	2.9041	2.9276	2.9494	2.9696	2.9885	3.0063	3.0230
1000	2.8440	2.8719	2.8973	2.9207	2.9423	2.9624	2.9812	2.9988	3.0154
∞	2.8376	2.8653	2.8905	2.9137	2.9352	2.9552	2.9738	2.9913	3.0078

Tafel 29 (Forts.)

Table 29 (cont.)

	$\alpha = 0.05$							
k:	20	$21 = \binom{7}{2}$	$28 = \binom{8}{2}$	$36 = \binom{9}{2}$	$45 = \binom{10}{2}$	$55 = \binom{11}{2}$	$66 = \binom{12}{2}$	$78 = \binom{13}{2}$
ν $100\alpha/k$:	0.2500	0.2381	0.1786	0.1389	0.1111	0.0909	0.0758	0.0641
2	19.9625	20.4573	23.6326	26.8049	29.9750	33.1436	36.3112	39.4778
3	9.4649	9.6242	10.6166	11.5632	12.4715	13.3471	14.1943	15.0165
4	6.7583	6.8471	7.3924	7.8998	8.3763	8.8271	9.2558	9.6655
5	5.6042	5.6665	6.0447	6.3914	6.7126	7.0128	7.2952	7.5625
6	4.9807	5.0297	5.3255	5.5937	5.8399	6.0680	6.2810	6.4813
7	4.5946	4.6359	4.8839	5.1068	5.3101	5.4973	5.6712	5.8339
8	4.3335	4.3699	4.5869	4.7810	4.9570	5.1183	5.2675	5.4065
9	4.1458	4.1786	4.3744	4.5485	4.7058	4.8494	4.9818	5.1048
10	4.0045	4.0348	4.2150	4.3747	4.5184	4.6492	4.7695	4.8810
11	3.8945	3.9229	4.0913	4.2400	4.3735	4.4947	4.6059	4.7087
12	3.8065	3.8334	3.9925	4.1327	4.2582	4.3719	4.4761	4.5722
13	3.7345	3.7602	3.9118	4.0452	4.1643	4.2721	4.3706	4.4614
14	3.6746	3.6992	3.8448	3.9725	4.0865	4.1894	4.2833	4.3698
15	3.6239	3.6477	3.7882	3.9113	4.0209	4.1198	4.2099	4.2928
16	3.5805	3.6036	3.7398	3.8589	3.9649	4.0604	4.1473	4.2272
17	3.5429	3.5654	3.6980	3.8137	3.9165	4.0091	4.0933	4.1706
18	3.5101	3.5321	3.6614	3.7742	3.8744	3.9644	4.0463	4.1214
19	3.4812	3.5027	3.6292	3.7395	3.8373	3.9251	4.0050	4.0781
20	3.4554	3.4765	3.6006	3.7087	3.8044	3.8903	3.9683	4.0398
21	3.4325	3.4532	3.5751	3.6812	3.7750	3.8593	3.9357	4.0056
22	3.4118	3.4322	3.5522	3.6564	3.7487	3.8314	3.9064	3.9750
23	3.3931	3.4132	3.5314	3.6341	3.7249	3.8062	3.8800	3.9474
24	3.3761	3.3960	3.5126	3.6139	3.7033	3.7834	3.8560	3.9223
25	3.3606	3.3803	3.4955	3.5954	3.6836	3.7626	3.8342	3.8995
26	3.3464	3.3659	3.4797	3.5785	3.6656	3.7436	3.8142	3.8787
27	3.3334	3.3526	3.4653	3.5629	3.6491	3.7261	3.7959	3.8595
28	3.3214	3.3404	3.4520	3.5486	3.6338	3.7101	3.7790	3.8419
29	3.3102	3.3291	3.4397	3.5354	3.6198	3.6952	3.7634	3.8256
30	3.2999	3.3186	3.4282	3.5231	3.6067	3.6814	3.7489	3.8105
35	3.2577	3.2758	3.3816	3.4730	3.5534	3.6252	3.6900	3.7490
40	3.2266	3.2443	3.3473	3.4362	3.5143	3.5840	3.6468	3.7040
45	3.2028	3.2201	3.3211	3.4081	3.4845	3.5525	3.6138	3.6696
50	3.1840	3.2010	3.3003	3.3858	3.4609	3.5277	3.5878	3.6425
55	3.1688	3.1856	3.2836	3.3679	3.4418	3.5076	3.5668	3.6206
60	3.1562	3.1728	3.2697	3.3530	3.4260	3.4910	3.5494	3.6025
70	3.1366	3.1529	3.2481	3.3299	3.4015	3.4652	3.5224	3.5744
80	3.1220	3.1381	3.2321	3.3127	3.3833	3.4460	3.5024	3.5536
90	3.1108	3.1267	3.2197	3.2995	3.3693	3.4313	3.4870	3.5375
100	3.1018	3.1176	3.2099	3.2890	3.3582	3.4196	3.4747	3.5248
110	3.0945	3.1102	3.2018	3.2804	3.3491	3.4100	3.4648	3.5144
120	3.0885	3.1041	3.1952	3.2733	3.3416	3.4021	3.4565	3.5058
250	3.0543	3.0694	3.1577	3.2332	3.2991	3.3575	3.4099	3.4573
500	3.0387	3.0537	3.1406	3.2150	3.2798	3.3373	3.3887	3.4354
1000	3.0310	3.0459	3.1322	3.2059	3.2703	3.3272	3.3783	3.4245
∞	3.0233	3.0381	3.1237	3.1970	3.2608	3.3172	3.3678	3.4136

Tafel 29 (Forts.) Table 29 (cont.)

	$\alpha = 0.05$						
ν k:	$91 = \binom{14}{2}$	$105 = \binom{15}{2}$	$120 = \binom{16}{2}$	$136 = \binom{17}{2}$	$153 = \binom{18}{2}$	$171 = \binom{19}{2}$	$190 = \binom{20}{2}$
$100\alpha/k$:	0.0549	0.0476	0.0417	0.0368	0.0327	0.0292	0.0263
2	42.6439	45.8094	48.9745	52.1392	55.3037	58.4679	61.6320
3	15.8165	16.5964	17.3582	18.1035	18.8336	19.5497	20.2528
4	10.0585	10.4367	10.8016	11.1545	11.4966	11.8288	12.1519
5	7.8166	8.0591	8.2913	8.5143	8.7290	8.9362	9.1365
6	6.6705	6.8500	7.0210	7.1844	7.3410	7.4914	7.6363
7	5.9868	6.1313	6.2684	6.3990	6.5236	6.6430	6.7577
8	5.5368	5.6594	5.7755	5.8857	5.9906	6.0909	6.1869
9	5.2197	5.3276	5.4295	5.5260	5.6177	5.7051	5.7888
10	4.9849	5.0823	5.1740	5.2608	5.3431	5.4215	5.4963
11	4.8044	4.8939	4.9781	5.0576	5.1330	5.2046	5.2729
12	4.6615	4.7450	4.8233	4.8972	4.9672	5.0336	5.0969
13	4.5457	4.6243	4.6981	4.7675	4.8332	4.8956	4.9549
14	4.4500	4.5247	4.5947	4.6606	4.7228	4.7818	4.8379
15	4.3695	4.4410	4.5079	4.5708	4.6302	4.6865	4.7400
16	4.3011	4.3698	4.4341	4.4946	4.5516	4.6056	4.6568
17	4.2421	4.3085	4.3706	4.4289	4.4839	4.5360	4.5853
18	4.1907	4.2551	4.3154	4.3719	4.4251	4.4755	4.5232
19	4.1456	4.2083	4.2669	4.3218	4.3736	4.4225	4.4688
20	4.1057	4.1669	4.2240	4.2776	4.3280	4.3756	4.4208
21	4.0701	4.1300	4.1858	4.2381	4.2874	4.3339	4.3780
22	4.0382	4.0969	4.1516	4.2028	4.2510	4.2966	4.3397
23	4.0095	4.0671	4.1207	4.1710	4.2183	4.2629	4.3052
24	3.9834	4.0400	4.0928	4.1422	4.1886	4.2325	4.2739
25	3.9597	4.0154	4.0674	4.1160	4.1616	4.2047	4.2455
26	3.9380	3.9929	4.0441	4.0920	4.1370	4.1794	4.2196
27	3.9181	3.9723	4.0228	4.0700	4.1144	4.1562	4.1958
28	3.8997	3.9533	4.0032	4.0498	4.0936	4.1349	4.1739
29	3.8828	3.9357	3.9850	4.0311	4.0744	4.1151	4.1537
30	3.8671	3.9195	3.9682	4.0138	4.0566	4.0969	4.1350
35	3.8032	3.8533	3.8999	3.9434	3.9842	4.0226	4.0590
40	3.7564	3.8049	3.8499	3.8919	3.9314	3.9684	4.0035
45	3.7208	3.7680	3.8118	3.8527	3.8911	3.9271	3.9612
50	3.6926	3.7389	3.7818	3.8218	3.8594	3.8946	3.9279
55	3.6699	3.7154	3.7576	3.7969	3.8337	3.8684	3.9010
60	3.6511	3.6960	3.7376	3.7763	3.8126	3.8467	3.8789
70	3.6220	3.6658	3.7065	3.7444	3.7798	3.8131	3.8445
80	3.6004	3.6435	3.6835	3.7207	3.7555	3.7883	3.8191
90	3.5837	3.6263	3.6658	3.7025	3.7369	3.7691	3.7995
100	3.5705	3.6127	3.6517	3.6880	3.7220	3.7539	3.7840
110	3.5598	3.6016	3.6403	3.6763	3.7100	3.7416	3.7714
120	3.5509	3.5924	3.6308	3.6665	3.7000	3.7313	3.7609
250	3.5007	3.5405	3.5774	3.6117	3.6437	3.6737	3.7020
500	3.4779	3.5170	3.5532	3.5868	3.6182	3.6477	3.6754
1000	3.4666	3.5054	3.5412	3.5745	3.6056	3.6348	3.6622
∞	3.4554	3.4938	3.5293	3.5623	3.5931	3.6219	3.6491

Tafel 29 (Forts.) Table 29 (cont.)

ν		$\alpha = 0.01$									
	k:	1	2	$3 = \binom{3}{2}$	4	5	$6 = \binom{4}{2}$	7	8	9	$10 = \binom{5}{2}$
	$100\alpha/k$:	1.0000	0.5000	0.3333	0.2500	0.2000	0.1667	0.1429	0.1250	0.1111	0.1000
2		9.9248	14.0890	17.2772	19.9625	22.3271	24.4643	26.4292	28.2577	29.9750	31.5991
3		5.8409	7.4533	8.5752	9.4649	10.2145	10.8688	11.4532	11.9838	12.4715	12.9240
4		4.6041	5.5976	6.2541	6.7583	7.1732	7.5287	7.8414	8.1216	8.3763	8.6103
5		4.0321	4.7733	5.2474	5.6042	5.8934	6.1384	6.3518	6.5414	6.7126	6.8688
6		3.7074	4.3168	4.6979	4.9807	5.2076	5.3982	5.5632	5.7090	5.8399	5.9588
7		3.4995	4.0293	4.3553	4.5946	4.7853	4.9445	5.0815	5.2022	5.3101	5.4079
8		3.3554	3.8325	4.1224	4.3335	4.5008	4.6398	4.7590	4.8636	4.9570	5.0413
9		3.2498	3.6897	3.9542	4.1458	4.2968	4.4219	4.5288	4.6224	4.7058	4.7809
10		3.1693	3.5814	3.8273	4.0045	4.1437	4.2586	4.3567	4.4423	4.5184	4.5869
11		3.1058	3.4966	3.7283	3.8945	4.0247	4.1319	4.2232	4.3028	4.3735	4.4370
12		3.0545	3.4284	3.6489	3.8065	3.9296	4.0308	4.1169	4.1918	4.2582	4.3178
13		3.0123	3.3725	3.5838	3.7345	3.8520	3.9484	4.0302	4.1013	4.1643	4.2208
14		2.9768	3.3257	3.5296	3.6746	3.7874	3.8798	3.9582	4.0263	4.0865	4.1405
15		2.9467	3.2860	3.4837	3.6239	3.7328	3.8220	3.8975	3.9630	4.0209	4.0728
16		2.9208	3.2520	3.4443	3.5805	3.6862	3.7725	3.8456	3.9089	3.9649	4.0150
17		2.8982	3.2224	3.4102	3.5429	3.6458	3.7297	3.8007	3.8623	3.9165	3.9651
18		2.8784	3.1966	3.3804	3.5101	3.6105	3.6924	3.7616	3.8215	3.8744	3.9216
19		2.8609	3.1737	3.3540	3.4812	3.5794	3.6595	3.7271	3.7857	3.8373	3.8834
20		2.8453	3.1534	3.3306	3.4554	3.5518	3.6303	3.6966	3.7539	3.8044	3.8495
21		2.8314	3.1352	3.3097	3.4325	3.5272	3.6043	3.6693	3.7255	3.7750	3.8193
22		2.8188	3.1188	3.2909	3.4118	3.5050	3.5808	3.6448	3.7000	3.7487	3.7921
23		2.8073	3.1040	3.2739	3.3931	3.4850	3.5597	3.6226	3.6770	3.7249	3.7676
24		2.7969	3.0905	3.2584	3.3761	3.4668	3.5405	3.6025	3.6561	3.7033	3.7454
25		2.7874	3.0782	3.2443	3.3606	3.4502	3.5230	3.5842	3.6371	3.6836	3.7251
26		2.7787	3.0669	3.2313	3.3464	3.4350	3.5069	3.5674	3.6197	3.6656	3.7066
27		2.7707	3.0565	3.2194	3.3334	3.4210	3.4922	3.5520	3.6037	3.6491	3.6896
28		2.7633	3.0469	3.2084	3.3214	3.4082	3.4786	3.5378	3.5889	3.6338	3.6739
29		2.7564	3.0380	3.1982	3.3102	3.3962	3.4660	3.5247	3.5753	3.6198	3.6594
30		2.7500	3.0298	3.1888	3.2999	3.3852	3.4544	3.5125	3.5626	3.6067	3.6460
35		2.7238	2.9960	3.1502	3.2577	3.3400	3.4068	3.4628	3.5110	3.5534	3.5911
40		2.7045	2.9712	3.1218	3.2266	3.3069	3.3718	3.4263	3.4732	3.5143	3.5510
45		2.6896	2.9521	3.1000	3.2028	3.2815	3.3451	3.3984	3.4442	3.4845	3.5203
50		2.6778	2.9370	3.0828	3.1840	3.2614	3.3239	3.3763	3.4214	3.4609	3.4960
55		2.6682	2.9247	3.0688	3.1688	3.2451	3.3068	3.3585	3.4029	3.4418	3.4764
60		2.6603	2.9146	3.0573	3.1562	3.2317	3.2927	3.3437	3.3876	3.4260	3.4602
70		2.6479	2.8987	3.0393	3.1366	3.2108	3.2707	3.3208	3.3638	3.4015	3.4350
80		2.6387	2.8870	3.0259	3.1220	3.1953	3.2543	3.3037	3.3462	3.3833	3.4163
90		2.6316	2.8779	3.0156	3.1108	3.1833	3.2417	3.2906	3.3326	3.3693	3.4019
100		2.6259	2.8707	3.0073	3.1018	3.1737	3.2317	3.2802	3.3218	3.3582	3.3905
110		2.6213	2.8648	3.0007	3.0945	3.1660	3.2235	3.2717	3.3130	3.3491	3.3812
120		2.6174	2.8599	2.9951	3.0885	3.1595	3.2168	3.2646	3.3057	3.3416	3.3735
250		2.5956	2.8322	2.9637	3.0543	3.1232	3.1785	3.2248	3.2644	3.2991	3.3299
500		2.5857	2.8195	2.9494	3.0387	3.1066	3.1612	3.2067	3.2457	3.2798	3.3101
1000		2.5808	2.8133	2.9423	3.0310	3.0984	3.1526	3.1977	3.2365	3.2703	3.3003
∞		2.5758	2.8070	2.9352	3.0233	3.0902	3.1440	3.1888	3.2272	3.2608	3.2905

Tafel 29 (Forts.) Table 29 (cont.)

	$\alpha = 0.01$								
k:	11	12	13	14	$15 = \binom{6}{2}$	16	17	18	19
ν $100\alpha/k$:	0.0909	0.0833	0.0769	0.0714	0.0667	0.0625	0.0588	0.0556	0.0526
2	33.1436	34.6194	36.0347	37.3965	38.7105	39.9812	41.2129	42.4087	43.5718
3	13.3471	13.7450	14.1214	14.4787	14.8194	15.1451	15.4575	15.7577	16.0471
4	8.8271	9.0294	9.2192	9.3983	9.5679	9.7291	9.8828	10.0298	10.1708
5	7.0128	7.1464	7.2712	7.3884	7.4990	7.6037	7.7032	7.7981	7.8888
6	6.0680	6.1690	6.2630	6.3510	6.4338	6.5121	6.5862	6.6568	6.7240
7	5.4973	5.5799	5.6565	5.7282	5.7954	5.8588	5.9188	5.9757	6.0300
8	5.1183	5.1892	5.2549	5.3162	5.3737	5.4278	5.4789	5.5274	5.5735
9	4.8494	4.9124	4.9706	5.0249	5.0757	5.1235	5.1686	5.2114	5.2520
10	4.6492	4.7065	4.7594	4.8087	4.8547	4.8980	4.9388	4.9774	5.0141
11	4.4947	4.5477	4.5966	4.6420	4.6845	4.7244	4.7620	4.7975	4.8312
12	4.3719	4.4215	4.4673	4.5099	4.5496	4.5868	4.6219	4.6551	4.6865
13	4.2721	4.3191	4.3624	4.4026	4.4401	4.4752	4.5083	4.5396	4.5693
14	4.1894	4.2342	4.2755	4.3138	4.3495	4.3829	4.4144	4.4442	4.4724
15	4.1198	4.1628	4.2024	4.2391	4.2733	4.3054	4.3355	4.3640	4.3910
16	4.0604	4.1018	4.1400	4.1754	4.2084	4.2393	4.2683	4.2958	4.3217
17	4.0091	4.0493	4.0863	4.1205	4.1525	4.1823	4.2104	4.2369	4.2620
18	3.9644	4.0035	4.0394	4.0727	4.1037	4.1327	4.1600	4.1857	4.2101
19	3.9251	3.9632	3.9983	4.0307	4.0609	4.0892	4.1157	4.1408	4.1645
20	3.8903	3.9276	3.9618	3.9935	4.0230	4.0506	4.0765	4.1010	4.1241
21	3.8593	3.8958	3.9293	3.9603	3.9892	4.0162	4.0416	4.0655	4.0881
22	3.8314	3.8672	3.9001	3.9306	3.9589	3.9854	4.0103	4.0337	4.0559
23	3.8062	3.8414	3.8738	3.9037	3.9316	3.9576	3.9820	4.0050	4.0268
24	3.7834	3.8181	3.8499	3.8794	3.9068	3.9324	3.9564	3.9790	4.0004
25	3.7626	3.7968	3.8282	3.8572	3.8842	3.9094	3.9331	3.9554	3.9765
26	3.7436	3.7773	3.8083	3.8369	3.8635	3.8884	3.9118	3.9337	3.9545
27	3.7261	3.7595	3.7900	3.8183	3.8446	3.8692	3.8922	3.9139	3.9344
28	3.7101	3.7430	3.7732	3.8012	3.8271	3.8514	3.8742	3.8956	3.9159
29	3.6952	3.7278	3.7577	3.7853	3.8110	3.8350	3.8575	3.8787	3.8987
30	3.6814	3.7136	3.7433	3.7706	3.7961	3.8198	3.8421	3.8631	3.8829
35	3.6252	3.6561	3.6845	3.7108	3.7352	3.7579	3.7792	3.7993	3.8183
40	3.5840	3.6140	3.6415	3.6670	3.6906	3.7126	3.7333	3.7527	3.7710
45	3.5525	3.5818	3.6087	3.6335	3.6565	3.6780	3.6982	3.7171	3.7350
50	3.5277	3.5564	3.5828	3.6071	3.6297	3.6508	3.6705	3.6890	3.7066
55	3.5076	3.5359	3.5618	3.5858	3.6080	3.6287	3.6481	3.6664	3.6836
60	3.4910	3.5189	3.5445	3.5682	3.5901	3.6105	3.6297	3.6477	3.6646
70	3.4652	3.4926	3.5176	3.5408	3.5622	3.5822	3.6010	3.6186	3.6352
80	3.4460	3.4730	3.4977	3.5205	3.5416	3.5613	3.5797	3.5970	3.6134
90	3.4313	3.4579	3.4823	3.5048	3.5257	3.5451	3.5633	3.5804	3.5966
100	3.4196	3.4460	3.4701	3.4924	3.5131	3.5323	3.5503	3.5673	3.5832
110	3.4100	3.4362	3.4602	3.4823	3.5028	3.5219	3.5398	3.5565	3.5724
120	3.4021	3.4281	3.4520	3.4739	3.4943	3.5132	3.5310	3.5477	3.5634
250	3.3575	3.3826	3.4055	3.4267	3.4462	3.4645	3.4815	3.4976	3.5127
500	3.3373	3.3619	3.3845	3.4052	3.4245	3.4424	3.4591	3.4749	3.4897
1000	3.3272	3.3517	3.3740	3.3946	3.4137	3.4314	3.4480	3.4636	3.4783
∞	3.3172	3.3415	3.3636	3.3840	3.4029	3.4205	3.4370	3.4524	3.4670

Tafel 29 (Forts.) Table 29 (cont.)

ν		$\alpha = 0.01$							
	k:	20	$21 = \binom{7}{2}$	$28 = \binom{8}{2}$	$36 = \binom{9}{2}$	$45 = \binom{10}{2}$	$55 = \binom{11}{2}$	$66 = \binom{12}{2}$	$78 = \binom{13}{2}$
	$100\alpha/k$:	0.0500	0.0476	0.0357	0.0278	0.0222	0.0182	0.0152	0.0128
2		44.7046	45.8094	52.9009	59.9875	67.0709	74.1519	81.2312	88.3091
3		16.3263	16.5964	18.2806	19.8889	21.4337	22.9239	24.3667	25.7675
4		10.3063	10.4367	11.2378	11.9851	12.6881	13.3540	13.9882	14.5946
5		7.9757	8.0591	8.5667	9.0332	9.4665	9.8722	10.2546	10.6168
6		6.7883	6.8500	7.2226	7.5617	7.8737	8.1636	8.4348	8.6901
7		6.0818	6.1313	6.4295	6.6987	6.9448	7.1721	7.3837	7.5819
8		5.6174	5.6594	5.9114	6.1375	6.3432	6.5323	6.7076	6.8712
9		5.2907	5.3276	5.5484	5.7458	5.9245	6.0883	6.2395	6.3803
10		5.0490	5.0823	5.2810	5.4578	5.6175	5.7634	5.8978	6.0225
11		4.8633	4.8939	5.0761	5.2378	5.3833	5.5160	5.6379	5.7509
12		4.7165	4.7450	4.9144	5.0644	5.1991	5.3216	5.4340	5.5380
13		4.5975	4.6243	4.7837	4.9244	5.0506	5.1651	5.2700	5.3670
14		4.4992	4.5247	4.6759	4.8091	4.9284	5.0364	5.1354	5.2266
15		4.4166	4.4410	4.5854	4.7125	4.8261	4.9289	5.0229	5.1094
16		4.3463	4.3698	4.5086	4.6305	4.7393	4.8377	4.9275	5.0102
17		4.2858	4.3085	4.4425	4.5600	4.6648	4.7594	4.8457	4.9251
18		4.2332	4.2551	4.3850	4.4987	4.6001	4.6915	4.7748	4.8514
19		4.1869	4.2083	4.3345	4.4450	4.5434	4.6320	4.7127	4.7868
20		4.1460	4.1669	4.2900	4.3976	4.4933	4.5795	4.6579	4.7299
21		4.1096	4.1300	4.2503	4.3554	4.4487	4.5328	4.6092	4.6794
22		4.0769	4.0969	4.2147	4.3175	4.4089	4.4910	4.5657	4.6342
23		4.0474	4.0671	4.1826	4.2835	4.3730	4.4534	4.5265	4.5935
24		4.0207	4.0400	4.1536	4.2527	4.3405	4.4194	4.4911	4.5567
25		3.9964	4.0154	4.1272	4.2246	4.3109	4.3885	4.4589	4.5233
26		3.9742	3.9929	4.1031	4.1990	4.2840	4.3602	4.4295	4.4928
27		3.9538	3.9723	4.0809	4.1755	4.2592	4.3344	4.4025	4.4649
28		3.9351	3.9533	4.0606	4.1539	4.2365	4.3106	4.3778	4.4392
29		3.9177	3.9357	4.0418	4.1339	4.2155	4.2886	4.3549	4.4155
30		3.9016	3.9195	4.0243	4.1154	4.1960	4.2683	4.3337	4.3936
35		3.8362	3.8533	3.9534	4.0403	4.1170	4.1857	4.2479	4.3047
40		3.7884	3.8049	3.9017	3.9855	4.0594	4.1256	4.1854	4.2399
45		3.7519	3.7680	3.8622	3.9437	4.0156	4.0798	4.1378	4.1907
50		3.7231	3.7389	3.8311	3.9108	3.9811	4.0438	4.1004	4.1520
55		3.6999	3.7154	3.8060	3.8843	3.9532	4.0147	4.0702	4.1208
60		3.6807	3.6960	3.7853	3.8624	3.9303	3.9908	4.0454	4.0951
70		3.6509	3.6658	3.7531	3.8284	3.8946	3.9537	4.0069	4.0553
80		3.6288	3.6435	3.7293	3.8033	3.8683	3.9262	3.9784	4.0259
90		3.6118	3.6263	3.7110	3.7839	3.8480	3.9051	3.9565	4.0032
100		3.5983	3.6127	3.6964	3.7686	3.8319	3.8883	3.9391	3.9853
110		3.5874	3.6016	3.6846	3.7561	3.8189	3.8747	3.9250	3.9707
120		3.5783	3.5924	3.6748	3.7458	3.8080	3.8634	3.9133	3.9586
250		3.5270	3.5405	3.6196	3.6875	3.7471	3.8000	3.8475	3.8907
500		3.5037	3.5170	3.5946	3.6612	3.7195	3.7713	3.8179	3.8601
1000		3.4922	3.5054	3.5822	3.6481	3.7059	3.7571	3.8032	3.8449
∞		3.4808	3.4938	3.5699	3.6352	3.6923	3.7430	3.7886	3.8299

Tafel 29 (Forts.) Table 29 (cont.)

	$\alpha = 0.01$						
ν k:	$91 = \binom{14}{2}$	$105 = \binom{15}{2}$	$120 = \binom{16}{2}$	$136 = \binom{17}{2}$	$153 = \binom{18}{2}$	$171 = \binom{19}{2}$	$190 = \binom{20}{2}$
$100\alpha/k$:	0.0110	0.0095	0.0083	0.0074	0.0065	0.0058	0.0053
2	95.3861	102.4622	109.5377	116.6126	123.6871	130.7612	137.8350
3	27.1309	28.4606	29.7598	31.0310	32.2766	33.4985	34.8984
4	15.1768	15.7375	16.2788	16.8026	17.3105	17.8040	18.2841
5	10.9616	11.2910	11.6067	11.9102	12.2025	12.4848	12.7578
6	8.9317	9.1612	9.3800	9.5893	9.7901	9.9831	10.1692
7	7.7685	7.9452	8.1130	8.2729	8.4258	8.5724	8.7132
8	7.0248	7.1696	7.3069	7.4373	7.5617	7.6806	7.7947
9	6.5121	6.6361	6.7533	6.8645	6.9703	7.0713	7.1679
10	6.1391	6.2485	6.3517	6.4495	6.5423	6.6308	6.7154
11	5.8562	5.9550	6.0480	6.1359	6.2193	6.2987	6.3745
12	5.6348	5.7254	5.8107	5.8911	5.9674	6.0399	6.1091
13	5.4571	5.5413	5.6204	5.6951	5.7658	5.8329	5.8969
14	5.3113	5.3904	5.4647	5.5347	5.6009	5.6637	5.7235
15	5.1897	5.2647	5.3350	5.4011	5.4637	5.5230	5.5794
16	5.0868	5.1583	5.2252	5.2882	5.3478	5.4042	5.4578
17	4.9986	5.0671	5.1313	5.1916	5.2486	5.3025	5.3538
18	4.9222	4.9882	5.0500	5.1080	5.1628	5.2146	5.2639
19	4.8554	4.9192	4.9789	5.0350	5.0879	5.1379	5.1854
20	4.7965	4.8584	4.9163	4.9707	5.0219	5.0704	5.1163
21	4.7442	4.8044	4.8607	4.9136	4.9633	5.0104	5.0551
22	4.6974	4.7562	4.8111	4.8625	4.9111	4.9569	5.0004
23	4.6553	4.7128	4.7664	4.8167	4.8641	4.9089	4.9513
24	4.6173	4.6736	4.7261	4.7753	4.8217	4.8654	4.9070
25	4.5828	4.6380	4.6894	4.7377	4.7831	4.8261	4.8667
26	4.5513	4.6055	4.6560	4.7034	4.7480	4.7901	4.8300
27	4.5224	4.5757	4.6255	4.6721	4.7159	4.7573	4.7965
28	4.4959	4.5484	4.5974	4.6432	4.6864	4.7271	4.7657
29	4.4714	4.5232	4.5714	4.6166	4.6591	4.6993	4.7372
30	4.4487	4.4998	4.5475	4.5921	4.6340	4.6735	4.7110
35	4.3569	4.4053	4.4503	4.4924	4.5320	4.5694	4.6047
40	4.2901	4.3365	4.3797	4.4201	4.4580	4.4937	4.5275
45	4.2393	4.2843	4.3261	4.3651	4.4018	4.4363	4.4689
50	4.1994	4.2432	4.2840	4.3220	4.3577	4.3913	4.4230
55	4.1672	4.2101	4.2500	4.2872	4.3222	4.3550	4.3860
60	4.1408	4.1829	4.2221	4.2586	4.2929	4.3252	4.3556
70	4.0997	4.1407	4.1788	4.2143	4.2476	4.2790	4.3085
80	4.0694	4.1096	4.1469	4.1816	4.2142	4.2449	4.2738
90	4.0461	4.0856	4.1223	4.1565	4.1886	4.2187	4.2471
100	4.0276	4.0666	4.1028	4.1366	4.1682	4.1979	4.2260
110	4.0126	4.0512	4.0870	4.1204	4.1517	4.1811	4.2088
120	4.0001	4.0384	4.0739	4.1070	4.1380	4.1671	4.1946
250	3.9303	3.9667	4.0004	4.0318	4.0612	4.0889	4.1149
500	3.8987	3.9343	3.9673	3.9980	4.0266	4.0536	4.0790
1000	3.8831	3.9183	3.9509	3.9812	4.0095	4.0362	4.0612
∞	3.8676	3.9024	3.9346	3.9646	3.9926	4.0189	4.0436

TABLE 30

Upper Percentage Points of Statistics for Testing Covariance Matrices: Tables of J. C. Lee, T. C. Chang, and P. R. Krishnaiah

(a) Content of the tables and definition of the statistics :

(0) Preliminaries :

Let $\underline{X}' = (\underline{X}_1' , \underline{X}_2' , \ldots , \underline{X}_q')$ be distributed as a multivariate normal with mean vector $\underline{\mu}'$ and covariance matrix $\underline{\Sigma}$. Also let

$$E (\underline{X}_i) = \underline{\mu}_i \qquad \text{and}$$

$$E \{(\underline{X}_i - \underline{\mu}_i)(\underline{X}_j - \underline{\mu}_j)'\} = \underline{\Sigma}_{ij}$$

where $\underline{X}_i$ is of order $p_i x1$ with

$$s = \sum_{i=1}^{q} p_i \quad .$$

Now we consider the distributions of the likelihood ratio statistics for testing the following five hypothese :

$$H_1 : \underline{\Sigma}_{ij} = \underline{0} \qquad \text{for } i = j = 1,2, \ldots , q .$$

(= multiple independence of several sets of variates)

$H_2 \quad : \quad \underline{\Sigma} = \sigma^2 \cdot \underline{\Sigma}_o$, where σ^2 is unknown ,

(sphericity)

$H_3 \quad : \quad \underline{\Sigma} = \underline{\Sigma}_o$,

(equality to a special matrix)

$H_4 \quad : \quad \underline{\Sigma}_{11} = \underline{\Sigma}_{22} = \ldots = \underline{\Sigma}_{qq}$ under the assumption that H_1 is true and

$p_1 = p_2 = \ldots = p_q$

(homogeneity of covariance matrices)

$$H_5 \quad : \quad \begin{cases} \underline{\Sigma}_{11} = \ldots = \underline{\Sigma}_{q_1, q_1} \\ \\ \underline{\Sigma}_{q_1+1, q_1+1} = \ldots = \underline{\Sigma}_{q_2', q_2'} \\ \\ \underline{\Sigma}_{q_{k-1}'+1, q_{k-1}'+1} = \ldots = \underline{\Sigma}_{q,q} \\ \text{where } q_o' = 0 \ , \ q_j' = \sum_{i=1}^{j} q_i \ , \\ \text{and } q_k' = q \ . \end{cases}$$

(multiple homogeneity of covariance matrices)

(1) Testing H_1 : Multiple independence of several sets of variates :

Let $\underline{X}_j' = (\underline{X}_1' , \ldots , \underline{X}_q')$ with $j = 1,2, \ldots , N$

be the j-th independent observation on $\underline{X}'$ and let

$$\underline{A} = \begin{bmatrix} A_{11} & \cdots\cdots & A_{1q} \\ A_{21} & \cdots\cdots & A_{2q} \\ \cdot\cdot & \cdots\cdots & \cdot\cdot \\ A_{q1} & \cdots\cdots & A_{qq} \end{bmatrix}$$

with

$$A_{lm} = \sum_{j=1}^{N} (\underline{X}_{lj} - \overline{\underline{X}}_{l.}) \cdot (\underline{X}_{mj} - \overline{\underline{X}}_{m.})'$$

and $$\overline{\underline{X}}_{l.} = \frac{1}{N} \sum_{j=1}^{N} \underline{X}_{lj} \quad .$$

Then the likelihood ratio statistic for testing H_1 is

$$V_1 = \frac{| \underline{A} |}{\prod_{j=1}^{q} | A_{jj} |} \quad .$$

Furthermore we define

$$V_1' = -2.\log V_1$$

and accept or reject H_1 accordingly as

$$V_1' \lessgtr C_1 \quad ,$$

where C_1 is given by

$$P (V_1' \leqq C_1 \mid H_1) = 1 - \alpha \quad .$$

The values of C_1 for certain values of α , p , q , and $M = n - s - 3$ are given in table 30a .

(2) Testing H_2 : Sphericity :

The likelihood ratio statistic for testing H_2 is given by

$$V_2 = \frac{| A \cdot \underline{\Sigma}_o^{-1} |}{\{ \mathrm{tr}\, A \cdot \underline{\Sigma}_o^{-1} / s \}^s}$$

with A as defined under (1) .

Critical values for this test are defined by

$$P (V_2 \geqq C_2 \mid H_2) = 1 - \alpha \quad .$$

This is the sphericity test of J. W. Mauchly which is handled in table 11 of this volume .

(3) Testing H_3 : The covariance matrix is equal to a specified matrix :

The hypothesis H_3 can be tested by the likelihood ratio

statistic

$$V_3 = (e/n)^{s.n/2} \mid \underline{A} \cdot \underline{\Sigma}_o \mid^{n/2} \operatorname{etr} (\tfrac{1}{2} \cdot A \cdot \underline{\Sigma}_o^{-1}) \quad .$$

Critical values for this test are defined by

$$P (V_3 \geqq C_3 \mid H_3) = 1 - \alpha \quad .$$

For testing H_3 see <u>table 16</u> of this volume.

(4) <u>Testing H_4 : The homogeneity of covariance matrices</u> :

Let $\underline{X}' = (\underline{X}_1' , \ldots , \underline{X}_q')$ be as defined above

and let $\underline{X}_{ij}$ $(j = 1,2,3, \ldots , N_i)$ be the j-th observation

on $\underline{X}_i$.

Furthermore let be $p_i = p$ for $i = 1,2,3, \ldots , q$

and

$$A_{ii} = \sum_{j=1}^{N_i} (\underline{X}_{ij} - \overline{\underline{X}}_i) (\underline{X}_{ij} - \overline{\underline{X}}_i)' \quad .$$

The likelihood ratio statistic for testing H_4 (as derived by S. S. WILKS , 1932) can be modified by setting

$$n_i = N_i - 1 \qquad \text{and} \qquad n = \sum_{i=1}^{q} n_i \quad .$$

Then H_4 can be tested by the likelihood ratio statistic

$$V_4 = \frac{\prod_{g=1}^{q} | A_{gg} |^{n_g/2}}{| \sum_{g=1}^{q} A_{gg} |^{n/2}} \cdot \frac{n^{p.n/2}}{\prod_{g=1}^{q} n_g^{p.n_g/2}} .$$

Now we define

$$V_4' = - 2 . \log V_4$$

and accept or reject H_4 accordingly as

$$V_4' \lesseqgtr C_4 \quad , \text{ where } C_4 \text{ is given by}$$

$$P (V_4' \leq C_4 \mid H_4) = 1 - \alpha .$$

The values of C_4 for certain α , p , q , and n_o are given in table 30b .

(5) Testing H_5 : Multiple homogeneity of covariance matrices :

The likelihood ratio statistic for testing H_5 (as defined above) is given by

$$V_5 = \frac{\prod_{i=1}^{q} | A_{ii} / n_i |^{n_i/2}}{\prod_{j=1}^{k} | \sum_{i=q'_{j-1}+1}^{q'_j} A_{ii}/n'_j |^{n'_j/2}} ,$$

where n_i and A_{ii} are defined as before and

$$n'_j = \sum_{i=q_{j-1}+1}^{q'_j} n_i \quad .$$

Now we define

$$V'_5 = -2 \cdot \log V_5$$

and accept or reject H_5 accordingly as

$$V'_5 \lessgtr C_5 \quad , \quad \text{where} \quad C_5 \text{ is given}$$

by

$$P(V'_5 \leqq C_5 \mid H_5) = 1 - \alpha \quad .$$

The values of C_5 for certain α , p , q , k , and $M = n_o - p$ are given in table 30c .

(b) Extent of the tables and definitions of the parameters :

(ba) Table 30a for testing H_1 :

(1) The parameter α :

α = error probability

α = 0.10 ; 0.05 ; 0.01 .

(2) The parameter p :

p = dimension of the p_i-th variate

p = 1 ; 2 ; 3 .

(3) The parameter q :

q = number of sets of variates

q = 3 ; 4 ; 5 .

(4) The parameter M :

$M = n - s - 3$ with $n = N - 1$, where

N = number of observations

$s = \sum_{i=1}^{q} p_i$, for

M = 1(1)20(2)30 .

(bb) Table 30b for testing H_{4-}:

(1) The parameter α :

α = error probability

α = 0.10 ; 0.05 ; 0.01 .

(2) The parameter p :

p = dimension of the variate

(all $p_i = p$ for $i = 1,2, \ldots , q$)

= 2(1)6 .

(3) The parameter q :

q = number of variates

q = 2(1)10 .

(4) <u>The parameter n_o</u>:

$n_o = n_i = N_i - 1$

with N_i = number of observations

$n_o = (p+1)(1)20(5)30$.

(bc) <u>Table 30c for testing H_5</u>:

(1) <u>The parameter α</u>:

α = error probability

α = 0.05 ; 0.01 .

(2) <u>The parameter p</u>:

p = dimension of the variate

p = 1 ; 2 ; 3 ; 4 ;

(3) <u>The parameter q</u>:

q = number of sets of variates

q = 4 ; 6 for k = 2

q = 6 ; 9 for k = 3

(4) <u>The parameter k</u>:

k = as defined with H_5 under (a)/(1)

k = 2 ; 3 .

(5) <u>The parameter M</u>:

$M = n_o - p$ with $n_o = n_i = N_i - 1$.

M = 1(1)20(2)30 .

(c) Suggestions for applications :

The tables can be used for testing the hypotheses H_1, H_4, and H_5 respectively, on the structures of covariance matrices as defined under (a)/(0).

(d) References :

(1) For the source of the tables :

Lee, J. C. + Chang, T. C. + Krishnaiah, P. R.: Approximations to the distributions of the likelihood ratio statistics for testing certain structures on the covariance matrices of real multivariate normal populations.
ARL - report 75 - 0167 (1975).
(Table III ; table VII ; Table IX).

(2) See also :

Lee, J. C. + Chang, T. C. + Krishnaiah, P. R.: Approximations to the distributions of the likelihood ratio statistics for testing certain structures on the covariance matrices of real multivariate normal populations.
= pp. 105 - 118 in
Krishnaiah, P.R.(ed.): Multivariate Analysis-IV. Amsterdam : North-Holland, 1977.

(e) Further remarks :

For theory and computations of the tables see

Lee, J. C. + Chang, T. C. + Krishnaiah, P. R. : ARL 75-/167(1975) as cited above under (d)/(1).

PERCENTAGE POINTS OF $-2 \log V_1$

$q=3$

	p=1			p=2			p=3		
M \ α	10%	5%	1%	10%	5%	1%	10%	5%	1%
1	2.419	3.023	4.390	5.606	6.371	7.991	9.161	10.037	11.838
2	2.027	2.534	3.678	4.051	5.509	6.899	8.091	8.857	10.427
3	1.744	2.180	3.165	4.279	4.857	6.076	7.255	7.937	9.334
4	1.531	1.913	2.778	3.829	4.345	5.432	6.582	7.198	8.457
5	1.363	1.704	2.475	3.466	3.933	4.914	6.026	6.587	7.735
6	1.229	1.537	2.231	3.167	3.593	4.487	5.559	6.075	7.130
7	1.119	1.400	2.031	2.916	3.307	4.129	5.160	5.639	6.615
8	1.027	1.284	1.864	2.701	3.064	3.825	4.815	5.262	6.171
9	.949	1.187	1.723	2.517	2.854	3.563	4.515	4.933	5.784
10	.882	1.103	1.602	2.356	2.672	3.335	4.250	4.643	5.443
11	.824	1.030	1.496	2.215	2.511	3.134	4.016	4.386	5.142
12	.773	.967	1.403	2.089	2.369	2.956	3.805	4.157	4.871
13	.728	.910	1.322	1.977	2.242	2.798	3.616	3.950	4.629
14	.688	.860	1.249	1.877	2.128	2.655	3.445	3.763	4.409
15	.652	.815	1.183	1.786	2.025	2.527	3.289	3.593	4.210
16	.620	.775	1.125	1.704	1.932	2.410	3.148	3.438	4.028
17	.591	.739	1.072	1.629	1.847	2.304	3.018	3.296	3.061
18	.564	.705	1.024	1.561	1.769	2.207	2.898	3.165	3.707
19	.540	.675	.979	1.497	1.697	2.117	2.788	3.044	3.566
20	.517	.647	.939	1.439	1.631	2.035	2.685	2.932	3.434
22	.478	.597	.867	1.335	1.513	1.888	2.501	2.732	3.199
24	.444	.555	.805	1.245	1.412	1.761	2.341	2.557	2.995
26	.414	.518	.752	1.167	1.323	1.650	2.200	2.403	2.814
28	.388	.486	.705	1.098	1.244	1.552	2.076	2.267	2.655
30	.366	.458	.664	1.036	1.175	1.465	1.965	2.145	2.512

* The entries in this table are the values of c_1 where $M = n - s - 3$ and

$$p[-2 \log V_1 \leq c_1] = (1 - \alpha).$$

Tafel 30a (Forts.)

Table 30a (cont.)

q = 4

M \ α	p=1 10%	p=1 5%	p=1 1%	p=2 10%	p=2 5%	p=2 1%	p=3 10%	p=3 5%	p=3 1%
1	3.694	4.376	5.861	8.557	9.417	11.195	13.949	14.932	16.905
2	3.143	3.721	4.980	7.534	8.285	9.831	12.539	13.408	15.155
3	2.736	3.238	4.330	6.739	7.406	8.777	11.407	12.190	13.762
4	2.422	2.867	3.832	6.100	6.701	7.936	10.473	11.189	12.618
5	2.174	2.572	3.438	5.574	6.122	7.245	9.689	10.347	11.660
6	1.971	2.333	3.118	5.134	5.637	6.669	9.017	9.627	10.846
7	1.804	2.134	2.852	4.769	5.225	6.179	8.436	9.004	10.140
8	1.663	1.967	2.628	4.436	4.870	5.757	7.927	8.460	9.524
9	1.542	1.825	2.437	4.154	4.560	5.391	7.477	7.980	8.981
10	1.438	1.701	2.272	3.907	4.289	5.068	7.078	7.552	8.498
11	1.347	1.593	2.128	3.688	4.048	4.783	6.719	7.169	8.066
12	1.266	1.496	2.001	3.493	3.833	4.528	6.395	6.823	7.676
13	1.195	1.414	1.888	3.317	3.639	4.299	6.103	6.510	7.323
14	1.132	1.338	1.788	3.158	3.465	4.093	5.835	6.225	7.001
15	1.074	1.271	1.697	3.013	3.306	3.905	5.591	5.964	6.708
16	1.023	1.210	1.616	2.882	3.162	3.734	5.367	5.725	6.437
17	.976	1.154	1.541	2.761	3.030	3.578	5.160	5.503	6.188
18	.933	1.104	1.474	2.650	2.908	3.434	4.968	5.299	5.958
19	.894	1.057	1.412	2.548	2.796	3.302	4.790	5.110	5.745
20	.858	1.015	1.355	2.453	2.692	3.179	4.626	4.933	5.546
22	.794	.939	1.254	2.284	2.506	2.959	4.327	4.615	5.188
24	.739	.874	1.167	2.137	2.344	2.768	4.066	4.336	4.874
26	.691	.817	1.091	2.007	2.202	2.600	3.833	4.088	4.596
28	.649	.767	1.025	1.893	2.076	2.451	3.627	3.868	4.348
30	.612	.723	.966	1.790	1.964	2.318	3.442	3.671	4.125

q = 5

M	α 10% (p=1)	5% (p=1)	1% (p=1)	10% (p=2)	5% (p=2)	1% (p=2)	10% (p=3)	5% (p=3)	1% (p=3)
1	5.039	5.764	7.368	11.665	12.600	14.500	18.945	20.006	22.106
2	4.339	4.978	6.333	10.405	11.228	12.899	17.242	18.192	20.067
3	3.813	4.371	5.557	9.405	10.146	11.639	15.852	16.716	18.421
4	3.402	3.900	4.954	8.589	9.261	10.616	14.687	15.477	17.043
5	3.071	3.520	4.470	7.909	8.525	9.767	13.691	14.425	15.876
6	2.801	3.208	4.073	7.331	7.900	9.047	12.828	13.515	14.864
7	2.573	2.949	3.742	6.835	7.364	8.430	12.073	12.717	13.981
8	2.381	2.727	3.461	6.402	6.897	7.893	11.408	12.011	13.204
9	2.215	2.537	3.219	6.023	6.487	7.423	10.811	11.386	12.513
10	2.071	2.372	3.009	5.686	6.125	7.006	10.279	10.821	11.890
11	1.944	2.227	2.826	5.386	5.801	6.635	9.798	10.313	11.332
12	1.832	2.099	2.663	5.116	5.510	6.301	9.360	9.85[illegible]	10.822
13	1.733	1.985	2.518	4.873	5.247	6.000	8.960	9.432	10.359
14	1.644	1.882	2.388	4.652	5.009	5.727	8.594	9.047	9.935
15	1.563	1.790	2.270	4.449	4.791	5.477	8.259	8.691	9.545
16	1.490	1.706	2.165	4.265	4.592	5.249	7.948	8.364	9.185
17	1.424	1.631	2.068	4.095	4.409	5.039	7.661	8.061	8.850
18	1.363	1.561	1.980	3.938	4.240	4.846	7.393	7.779	8.541
19	1.307	1.497	1.898	3.793	4.084	4.667	7.144	7.518	8.252
20	1.256	1.438	1.823	3.658	3.938	4.501	6.911	7.272	7.983
22	1.164	1.[illegible]33	1.690	3.415	3.677	4.202	6.490	6.828	7.495
24	1.085	1.2[illegible]3	1.576	3.203	3.448	3.940	6.116	6.435	7.063
26	1.016	1.163	1.475	3.016	3.247	3.709	5.784	6.086	6.679
28	.955	1.094	1.387	2.849	3.067	3.504	5.486	5.772	6.335
30	.901	1.032	1.309	2.700	2.906	3.321	5.219	5.491	6.025

Tafel 30b

Table 30b

PERCENTAGE POINTS OF THE DISTRIBUTION OF $-2 \log V_4$

$\alpha = 0.10 \qquad p = 2$

n_o	q=2	q=3	q=4	q=5	q=6	q=7	q=8	q=9	q=10
3	9.75	15.30	21.29	26.52	31.50	36.58	41.49	46.32	51.12
4	8.56	14.07	19.08	23.87	28.52	33.07	37.55	41.97	46.36
5	7.97	13.22	17.99	22.55	26.97	31.31	35.58	39.80	43.97
6	7.63	12.70	17.33	21.75	26.05	30.25	34.39	38.49	42.53
7	7.39	12.36	16.89	21.22	25.42	29.54	33.60	37.61	41.57
8	7.23	12.11	16.57	20.84	24.98	29.04	33.03	36.98	40.88
9	7.10	11.93	16.34	20.55	24.65	28.66	32.61	36.51	40.37
10	[illegible]	11.78	16.16	20.33	24.39	28.36	32.28	36.14	39.97
11	6.93	11.67	16.01	20.15	24.18	28.12	32.01	35.85	39.65
12	6.87	11.58	15.88	20.01	24.01	27.93	31.79	35.61	39.38
13	6.82	11.50	15.79	19.89	23.87	27.77	31.62	35.41	39.16
14	6.77	11.43	15.70	19.78	23.75	27.63	31.46	35.24	38.98
15	6.74	11.37	15.63	19.69	23.65	27.52	31.33	35.09	38.82
16	6.71	11.33	15.57	19.62	23.56	27.42	31.21	34.97	38.68
17	6.68	11.28	15.51	19.55	23.48	27.32	31.11	34.86	38.56
18	6.65	11.25	15.46	19.49	23.41	27.25	31.03	34.76	38.45
19	6.63	11.21	15.42	19.44	23.35	27.18	30.95	34.68	38.36
20	6.61	11.18	15.38	19.40	23.30	27.12	30.88	34.60	38.28
25	6.53	11.07	15.24	19.22	23.09	26.88	30.61	34.30	37.95
30	6.48	11.00	15.14	19.10	22.96	26.73	30.45	34.12	37.74

$\alpha = 0.10 \qquad p = 3$

n_o	q=2	q=3	q=4	q=5	q=6	q=7	q=8	q=9	q=10
4	18.87	30.76	41.78	52.41	62.90	73.03	83.14	93.16	103.12
5	16.19	26.97	36.79	46.36	55.73	64.95	74.05	83.09	92.05
6	14.83	24.87	34.22	43.24	52.06	60.76	69.35	77.86	86.30
7	14.01	23.66	32.64	41.33	49.51	58.18	66.44	74.64	82.76
8	13.46	22.83	31.58	40.03	48.29	56.44	64.48	72.45	80.37
9	13.06	22.24	30.81	39.09	47.19	55.17	63.05	70.88	78.64
10	12.77	21.79	30.23	38.38	46.36	54.21	61.95	69.68	77.31
11	12.53	21.44	29.77	37.82	45.70	53.46	61.13	68.74	76.26
12	12.35	21.16	29.40	37.37	45.17	52.86	60.45	67.98	75.45
13	12.19	20.93	29.10	37.00	44.74	52.36	59.89	67.35	74.77
14	12.07	20.73	28.85	36.69	44.38	51.94	59.42	66.84	74.19
15	11.96	20.57	28.63	36.43	44.06	51.59	59.03	66.39	73.70
16	11.87	20.43	28.45	36.21	43.80	51.28	58.67	66.01	73.29
17	11.78	20.30	28.29	36.01	43.57	51.02	58.38	65.67	72.92
18	11.71	20.20	28.15	35.83	43.37	50.79	58.11	65.38	72.59
19	11.65	20.10	28.02	35.68	43.19	50.58	57.88	65.12	72.32
20	11.60	20.02	27.91	35.55	43.02	50.39	57.68	64.90	72.07
25	11.39	19.70	27.50	35.05	42.43	49.71	56.91	64.04	71.12
30	11.26	19.50	27.23	34.72	42.05	49.27	56.42	63.49	70.52

Tafel 30b (Forts.) Table 30b (cont.)

$\alpha = 0.10 \quad p = 4$

n_o	q=2	q=3	q=4	q=5	q=6	q=7	q=8	q=9	q=10
5	30.70	50.45	68.96	86.91	104.53	121.93	139.15	156.25	173.24
6	26.16	43.85	60.44	76.56	92.38	107.98	123.46	138.81	154.09
7	23.79	40.34	55.91	71.02	85.37	100.51	115.03	129.45	143.75
8	22.33	38.15	53.06	67.55	81.76	95.81	109.73	123.52	137.25
9	21.33	36.65	51.11	65.15	78.34	92.55	106.06	119.46	132.76
10	20.60	35.56	49.67	63.39	76.17	90.17	103.37	116.45	129.48
11	20.06	34.72	48.58	62.04	75.28	88.36	101.30	114.15	126.93
12	19.62	34.06	47.71	60.98	74.02	86.90	99.68	112.34	124.95
13	19.27	33.53	47.00	60.12	73.00	85.73	98.36	110.88	123.31
14	18.93	33.08	46.42	59.40	72.17	84.76	97.26	109.65	121.98
15	18.74	32.71	45.94	58.80	71.45	83.95	96.32	108.61	120.85
16	18.53	32.40	45.52	58.30	70.34	83.25	95.55	107.73	119.87
17	18.36	32.12	45.16	57.65	70.33	82.65	94.86	106.98	119.03
18	18.21	31.89	44.85	57.47	69.57	82.13	94.26	106.32	118.30
19	18.07	31.68	44.58	57.13	69.47	81.66	93.75	105.73	117.67
20	17.96	31.49	44.33	56.83	69.12	81.25	93.28	105.22	117.10
25	17.51	30.81	43.43	55.72	67.30	79.75	91.57	103.33	115.00
30	17.24	30.37	42.86	55.02	66.37	78.79	90.49	102.11	113.67

$\alpha = 0.10 \quad p = 5$

n_o	q=2	q=3	q=4	q=5	q=6	q=7	q=8	q=9	q=10
6	45.25	74.90	102.86	130.09	156.89	183.39	209.65	235.79	261.72
7	38.56	65.12	90.21	114.66	138.70	162.50	186.15	209.58	232.90
8	34.95	59.77	83.27	106.16	128.70	150.99	173.12	195.06	216.94
9	32.69	56.37	78.81	100.70	122.23	143.54	164.69	185.70	206.61
10	31.12	54.00	75.70	96.88	117.72	138.36	158.82	179.16	199.37
11	29.97	52.24	73.40	94.03	114.37	134.47	154.42	174.28	194.00
12	29.08	50.88	71.62	91.85	111.78	131.47	151.03	170.52	189.94
13	28.39	49.81	70.20	90.09	109.70	129.10	148.33	167.49	186.53
14	27.82	48.93	69.04	88.66	108.01	127.14	146.14	165.04	183.78
15	27.35	48.21	68.07	87.48	106.60	125.51	144.31	162.94	181.54
16	26.95	47.59	67.26	86.47	105.42	124.14	142.75	161.24	179.62
17	26.62	47.07	66.57	85.62	104.40	122.98	141.42	159.75	177.97
18	26.32	46.61	65.96	84.87	103.50	121.95	140.27	158.45	176.56
19	26.07	46.21	65.44	84.22	102.74	121.06	139.23	157.32	175.30
20	25.85	45.85	64.97	83.65	102.05	120.28	138.36	156.32	174.20
25	25.03	44.59	63.28	81.56	99.57	117.41	135.10	152.67	170.19
30	24.52	43.78	62.22	80.23	98.01	115.60	133.04	150.39	167.69

$\alpha = 0.10 \quad p = 6$

n_o	q=2	q=3	q=4	q=5	q=6	q=7	q=8	q=9	q=10
7	62.54	104.12	143.53	181.97	219.37	257.40	294.64	331.76	368.58
8	53.42	90.76	126.20	160.82	194.94	228.72	262.28	295.59	328.81
9	48.40	83.29	116.44	148.84	180.79	212.41	243.93	275.04	306.13
10	45.18	78.42	110.07	141.01	171.49	201.73	231.72	261.56	291.26
11	42.92	74.99	105.56	135.44	164.91	194.13	223.14	251.93	280.69
12	41.23	72.42	102.16	131.26	159.38	188.42	216.65	244.70	272.69
13	39.94	70.43	99.54	128.00	156.11	183.93	211.57	239.10	266.48
14	38.91	68.83	97.42	125.39	153.01	180.35	207.54	234.56	261.46
	[illegible]	67.51	95.67	123.24	150.46	177.39	204.15	230.78	257.33
16	37.35	66.42	94.23	121.46	148.31	174.95	201.33	227.67	253.85
17	36.76	65.49	93.00	119.93	146.51	172.84	199.02	225.05	250.97
18	36.26	64.69	91.93	118.60	144.95	171.03	196.97	222.74	248.44
19	35.82	64.00	91.01	117.47	143.59	169.49	195.17	220.73	246.20
20	35.43	63.39	90.21	116.47	142.40	166.11	193.61	219.00	244.29
25	34.06	61.22	87.30	112.86	138.11	163.13	187.98	212.70	237.33
30	33.21	59.97	85.49	110.63	135.45	160.06	184.51	208.78	233.04

$\alpha = 0.05 \quad p = 2$

n_o	q=2	q=3	q=4	q=5	q=6	q=7	q=8	q=9	q=10
3	12.18	18.70	24.55	30.09	35.45	40.68	45.81	50.87	55.86
4	10.70	16.55	22.00	27.07	31.97	36.75	41.45	46.07	50.64
5	9.97	15.63	20.73	25.57	30.23	34.79	39.26	43.67	46.02
6	9.53	15.02	19.97	24.66	29.19	33.61	37.95	42.22	46.45
7	9.24	14.62	19.46	24.05	28.49	32.83	37.08	41.26	45.40
8	9.04	14.33	19.10	23.62	27.99	32.26	36.44	40.57	44.64
9	8.88	14.11	18.83	23.30	27.62	31.84	35.98	40.05	44.08
10	8.75	13.94	18.61	23.05	27.33	31.51	35.61	39.65	43.64
11	8.67	13.81	18.44	22.85	27.10	31.25	35.32	39.33	43.29
12	8.59	13.70	18.30	22.68	26.90	31.03	35.03	39.07	43.00
13	8.52	13.60	18.19	22.54	26.75	30.85	34.87	38.84	42.76
14	8.47	13.53	18.10	22.42	26.61	30.70	34.71	38.66	42.56
15	8.42	13.46	18.01	22.33	26.50	30.57	34.57	38.50	42.38
16	8.38	13.40	17.94	22.24	26.40	30.45	34.43	38.36	42.23
17	8.35	13.35	17.87	22.17	26.31	30.35	34.32	38.24	42.10
18	8.32	13.30	17.82	22.10	26.23	30.27	34.23	38.13	41.99
19	8.28	13.26	17.77	22.04	26.16	30.19	34.14	38.04	41.88
20	8.26	13.23	17.72	21.98	26.10	30.12	34.07	37.95	41.79
25	8.17	13.10	17.55	21.79	25.87	29.86	33.78	37.63	41.44
30	8.11	13.01	17.44	21.65	25.72	29.69	33.59	37.42	41.21

Tafel 30b (Forts.) Table 30b (cont.)

$\alpha = 0.05 \quad p = 3$

n_o	q=2	q=3	q=4	q=5	q=6	q=7	q=8	q=9	q=10
4	22.41	35.00	46.58	57.68	68.30	79.11	89.60	99.94	110.21
5	19.19	30.52	40.95	50.95	60.59	70.26	79.69	89.03	98.27
6	17.57	28.24	38.06	47.49	56.57	65.69	74.58	83.39	92.09
7	16.59	26.84	36.29	45.37	54.20	62.89	71.44	79.90	88.30
8	15.93	25.90	35.10	43.93	52.[illegible]	60.99	69.32	77.57	8[illegible].73
9	15.46	25.22	34.24	42.90	51.33	59.62	67.78	75.86	83.87
10	15.11	24.71	33.59	42.11	50.42	58.57	66.62	74.58	82.46
11	14.83	24.31	33.08	41.50	49.71	57.76	65.71	73.57	81.[illegible]
12	[illegible]	23.99	32.67	41.00	49.13	57.11	64.97	72.75	80.45
13	14.43	23.73	32.33	40.60	48.65	56.56	64.36	72.09	79.72
14	14.28	23.50	32.05	40.26	48.26	56.11	63.86	71.53	79.11
15	14.15	23.32	31.81	39.97	47.92	55.73	63.43	71.05	78.60
16	14.04	23.16	31.60	39.72	47.63	55.40	63.06	70.64	78.14
17	13.94	23.02	31.43	39.50	47.38	55.11	62.73	70.27	77.76
18	13.86	22.89	31.26	39.31	47.16	54.86	62.45	69.97	77.41
19	13.79	22.76	31.13	39.15	46.96	54.64	62.21	69.69	77.11
20	13.72	22.69	31.01	39.00	46.79	54.44	61.98	69.45	76.84
25	13.48	22.33	30.55	38.44	46.15	53.70	61.16	68.54	75.84
30	13.32	22.10	30.25	38.09	45.73	53.22	60.62	67.94	75.18

$\alpha = 0.05 \quad p = 4$

n_o	q=2	q=3	q=4	q=5	q=6	q=7	q=8	q=9	q=10
5	35.39	56.10	75.36	93.97	112.17	130.11	147.81	165.39	182.80
6	30.06	48.62	65.90	82.60	98.93	115.03	130.94	146.69	162.34
7	27.31	44.69	60.89	76.56	91.88	106.98	121.90	136.71	151.39
8	25.61	42.24	57.77	72.77	87.46	101.94	116.23	130.43	144.50
9	24.45	40.57	55.62	70.17	84.42	98.46	112.32	126.08	139.74
10	23.62	39.34	5[illegible].04	68.26	82.19	95.90	109.46	122.91	136.24
11	22.98	38.41	52.84	66.81	80.48	93.95	107.27	120.46	133.57
12	[illegible]	37.67	[illegible]	65.66	79.14	92.41	105.54	118.55	131.45
13	22.08	37.08	51.13	64.73	78.04	91.15	104.12	116.98	129.74
14	21.75	36.59	50.50	63.95	77.13	90.12	102.97	115.69	128.32
15	21.47	36.17	49.97	63.30	76.37	89.26	101.99	114.59	127.14
16	21.24	35.82	49.51	62.76	75.73	88.51	101.14	113.67	126.10
17	21.03	35.52	49.12	62.28	75.16	87.87	100.42	112.87	125.22
18	20.86	35.26	48.78	61.86	74.68	87.31	99.80	112.17	124.46
19	20.70	35.02	48.47	61.50	74.25	86.82	99.25	111.56	123.79
20	20.56	34.82	48.21	61.17	73.87	86.38	98.75	111.02	123.18
25	20.06	34.06	47.23	59.98	72.47	84.78	96.95	109.01	120.99
30	19.74	33.59	46.61	59.21	71.58	83.74	95.79	107.71	119.57

Tafel 30b (Forts.) Table 30b (cont.)

$\alpha = 0.05 \quad p = 5$

n_o	q=2	q=3	q=4	q=5	q=6	q=7	q=8	q=9	q=10
6	51.11	81.99	110.92	138.98	166.54	193.71	220.66	247.37	273.88
7	43.40	71.06	97.03	122.22	146.95	171.34	195.49	219.47	243.30
8	39.29	65.15	89.45	113.03	136.18	159.04	181.65	204.14	226.48
9	36.71	61.39	84.62	107.17	129.30	151.17	172.80	194.27	215.64
10	34.93	58.78	81.25	103.06	124.48	145.64	166.56	187.37	208.02
11	33.62	56.95	78.75	100.02	120.92	141.54	161.93	182.24	202.37
12	32.62	55.37	76.83	97.68	118.15	138.38	158.38	176.23	198.03
13	31.83	54.19	75.30	95.82	115.36	135.86	155.54	175.10	194.51
14	31.19	53.23	74.05	94.29	114.16	133.80	143.21	172.49	191.66
[illegible]	[illegible]	52.64	73.01	93.02	112.56	132.07	151.29	170.36	189.30
16	30.22	51.76	72.14	91.94	111.41	130.61	149.56	168.53	167.32
17	29.83	51.19	71.39	91.03	110.34	129.38	148.25	166.99	185.61
18	29.51	50.69	70.74	90.23	109.39	128.29	147.03	165.65	184.10
19	29.22	50.26	70.17	89.54	108.57	127.36	145.97	164.45	182.81
20	28.97	49.88	69.67	88.93	107.35	126.52	145.02	163.38	181.65
25	28.05	48.48	67.86	86.70	105.21	123.51	141.62	159.60	177.49
30	27.48	47.61	66.71	85.29	103.56	121.60	139.47	157.22	174.67

$\alpha = 0.05 \quad p = 6$

n_o	q=2	q=3	q=4	q=5	q=6	q=7	q=8	q=9	q=10
7	69.62	112.69	153.28	192.75	231.60	269.98	308.03	345.81	383.45
8	59.24	97.92	134.42	169.94	204.36	239.37	273.52	307.61	341.44
9	53.56	89.72	123.37	157.11	189.33	222.14	254.18	286.00	317.69
10	49.95	84.42	117.03	148.77	179.37	210.85	241.48	271.84	302.13
11	47.43	80.69	112.18	142.85	173.10	202.85	232.46	261.80	291.03
12	45.56	77.90	108.56	138.40	167.79	196.84	225.65	254.25	282.76
13	44.11	75.74	105.72	134.94	163.72	192.15	220.37	248.39	276.24
14	42.96	74.30	103.46	132.19	160.44	188.39	216.09	243.67	271.04
15	42.03	72.59	101.61	129.90	157.76	185.31	212.63	239.72	266.75
16	41.24	71.41	100.07	128.01	155.52	182.73	209.70	236.48	263.15
17	40.58	70.40	98.76	126.38	153.60	180.51	207.23	233.73	260.07
18	40.02	69.54	97.63	125.01	151.36	178.63	205.09	231.36	257.49
19	39.53	68.80	96.64	123.80	150.55	176.99	203.21	229.26	255.19
20	39.11	68.13	95.77	122.73	149.29	175.53	201.60	227.45	253.19
25	37.58	65.79	92.69	118.93	144.78	170.35	195.69	220.92	246.00
30	36.63	64.33	90.77	116.57	141.37	167.13	192.04	216.82	241.49

Tafel 30b (Forts.) Table 30b (cont.)

$\alpha = 0.01 \quad p = 2$

n_o	q=2	q=3	q=4	q=5	q=6	q=7	q=8	q=9	q=10
3	17.67	24.99	31.49	37.60	43.45	49.14	54.70	60.16	65.52
4	15.53	22.25	28.20	33.80	39.15	44.36	49.44	54.43	59.35
5	14.46	20.38	26.56	31.90	37.01	41.97	46.82	51.57	56.26
6	13.84	20.07	25.58	30.76	35.72	40.54	45.24	49.86	54.40
7	13.42	19.52	24.93	30.00	34.36	39.59	44.20	48.71	53.16
8	13.12	19.13	24.46	29.46	34.25	38.90	43.44	47.89	52.28
9	12.90	18.94	24.11	29.05	33.80	38.39	42.89	47.26	51.61
10	12.72	18.61	23.84	28.74	33.44	37.99	42.44	46.80	51.10
11	12.58	18.43	23.62	28.49	33.15	37.68	42.09	46.42	50.69
12	12.47	18.29	23.44	28.28	32.92	37.41	41.83	46.11	50.35
13	[illegible]	18.15	23.30	28.11	32.72	37.20	41.57	45.85	50.07
14	12.30	18.06	23.17	27.97	32.56	37.01	41.36	45.63	49.83
15	12.23	17.97	23.07	27.84	32.42	36.85	41.19	45.44	49.62
16	12.17	17.89	22.97	27.73	32.29	36.72	41.04	45.28	49.45
17	12.12	17.83	22.89	27.64	32.18	36.60	40.91	45.13	49.29
18	12.07	17.77	22.82	27.56	32.09	36.49	40.79	45.01	49.16
19	12.03	17.71	22.75	27.48	32.01	36.40	40.69	44.89	49.03
20	11.99	17.67	22.69	27.42	31.93	36.32	40.60	44.79	48.93
25	11.86	17.49	22.48	27.16	31.65	36.00	40.25	44.42	48.52
30	11.77	17.37	22.34	27.00	31.47	35.80	40.02	44.17	48.25

$\alpha = 0.01 \quad p = 3$

n_o	q=2	q=3	q=4	q=5	q=6	q=7	q=8	q=9	q=10
4	30.17	44.00	56.58	68.55	80.14	91.49	102.60	113.57	124.41
5	25.75	38.23	49.56	60.36	70.30	81.01	91.03	100.91	110.68
6	23.53	35.30	46.00	56.17	66.02	75.65	85.10	94.42	103.61
7	22.20	33.53	43.83	53.62	63.11	72.38	81.47	90.45	99.30
8	21.31	32.34	42.37	51.91	61.14	70.17	79.03	87.76	96.38
9	20.67	31.48	41.32	50.67	59.73	68.58	77.25	85.82	94.27
10	20.19	30.86	40.52	49.73	58.65	67.37	75.91	84.34	92.69
11	[illegible]	30.33	39.93	49.01	57.81	66.42	74.86	83.20	91.43
12	19.52	29.93	39.40	48.42	57.14	65.66	74.03	82.28	90.42
13	19.28	29.60	39.00	47.93	56.58	65.03	73.33	81.51	89.59
14	19.07	29.32	38.65	47.53	56.13	64.51	72.75	80.87	88.90
15	18.90	29.06	38.36	47.18	55.73	64.07	72.27	80.34	88.31
16	18.75	28.88	38.11	46.89	55.39	63.69	71.83	79.86	87.79
17	18.62	28.71	37.89	46.63	55.10	63.36	71.47	79.47	87.36
18	18.51	28.55	37.70	46.41	54.84	63.07	71.14	79.12	86.97
19	18.41	28.42	37.54	46.21	54.61	62.80	70.87	78.80	86.64
20	18.32	28.30	37.39	46.04	54.41	62.58	70.60	78.52	86.33
25	17.99	27.85	36.83	45.38	53.65	61.74	69.65	77.48	85.20
30	17.79	27.56	36.47	44.96	53.16	61.18	69.05	76.81	84.48

Tafel 30b (Forts.) Table 30b (cont.)

$\alpha = 0.01 \quad p = 4$

n_o	q=2	q=3	q=4	q=5	q=6	q=7	q=8	q=9	q=10
5	45.48	67.39	88.52	108.33	127.50	146.51	165.13	183.54	201.82
6	38.41	58.52	77.05	94.81	112.11	129.07	145.79	162.32	178.70
7	34.80	53.66	71.05	87.72	103.35	119.86	135.53	151.06	166.42
8	32.58	50.06	67.33	83.30	98.86	114.12	129.14	143.99	158.73
9	31.08	48.61	64.78	80.27	95.36	110.17	124.74	139.17	153.43
10	30.00	47.12	62.92	78.07	92.52	107.28	121.54	135.60	149.57
11	29.18	45.99	61.50	76.38	90.97	105.07	119.07	132.89	146.62
12	28.54	45.10	60.39	75.06	89.33	103.34	117.13	130.77	144.28
13	28.03	44.38	59.48	73.98	88.09	101.92	115.56	129.03	142.39
14	27.60	43.79	58.74	73.09	87.05	100.75	114.26	127.60	140.81
15	27.24	43.29	58.12	72.34	86.19	99.78	113.16	126.40	139.49
[illegible]	[illegible]	42.88	57.59	71.71	85.45	98.94	112.24	125.35	138.36
17	26.63	42.50	57.13	71.17	84.82	98.23	111.42	124.46	137.41
18	26.46	42.18	56.72	70.68	84.26	97.59	110.71	123.70	136.56
19	26.26	41.90	56.38	70.26	83.79	97.04	110.11	123.02	135.81
20	26.08	41.65	56.06	69.89	83.36	96.55	109.56	122.42	135.13
25	25.44	40.74	54.92	68.52	81.77	94.75	107.55	120.19	132.73
30	25.03	40.17	54.19	67.64	80.75	93.60	106.26	118.78	131.15

$\alpha = 0.01 \quad p = 5$

n_o	q=2	q=3	q=4	q=5	q=6	q=7	q=8	q=9	q=10
6	63.59	96.63	127.33	156.94	185.34	214.26	242.31	270.07	297.64
7	53.60	83.24	110.79	137.35	163.28	188.83	214.03	238.93	263.64
8	48.35	76.11	101.91	126.76	151.07	174.97	198.58	221.92	245.07
9	45.09	71.61	96.28	120.07	143.30	166.13	188.68	211.02	233.21
10	42.86	68.51	92.37	115.39	137.88	159.97	181.82	203.43	224.84
11	41.23	66.22	89.50	111.93	133.85	155.44	176.71	197.81	218.68
12	39.99	64.47	87.29	109.29	130.76	151.90	172.78	193.43	213.96
13	39.00	63.08	85.53	107.18	128.32	149.13	169.68	190.01	210.18
[illegible]	[illegible]	[illegible]	84.10	105.46	125.31	[illegible]	167.08	187.15	207.35
15	37.55	61.01	82.90	104.02	124.66	144.93	164.97	184.83	204.48
16	37.00	60.23	81.30	102.82	123.26	143.34	163.17	182.85	202.34
17	36.52	59.55	81.04	101.78	122.04	141.98	161.57	181.15	200.45
18	36.12	58.97	80.30	100.88	120.39	140.77	160.31	179.65	198.86
19	35.77	58.46	79.65	100.10	120.08	139.76	159.14	178.35	197.42
20	35.46	58.01	79.09	99.41	119.29	138.82	158.11	177.22	196.18
25	34.33	56.38	77.00	96.91	116.36	135.49	154.38	173.09	191.64
30	33.62	55.37	75.70	95.34	114.52	133.40	152.02	170.48	188.82

Tafel 30b (Forts.) Table 30b (cont.)

$\alpha = 0.01 \qquad p = 6$

n_o	q=2	q=3	q=4	q=5	q=6	q=7	q=8	q=9	q=10
7	84.51	130.23	173.00	214.41	254.35	294.86	334.36	373.45	412.35
8	71.34	112.45	150.89	188.07	224.50	260.43	295.38	331.06	365.98
9	64.27	102.74	138.71	173.53	207.62	241.23	274.41	307.34	340.08
10	59.81	96.51	130.87	164.10	196.59	228.79	260.53	291.97	323.20
11	56.71	92.15	125.34	157.47	188.32	219.93	250.64	280.98	311.19
12	54.43	88.92	121.24	152.51	183.14	213.36	243.19	272.79	302.24
13	52.68	86.41	118.04	148.67	178.64	208.24	237.44	266.46	295.22
14	51.29	84.41	115.48	145.56	175.06	204.06	232.85	261.33	289.58
15	50.16	82.77	113.38	143.04	172.11	200.75	229.06	257.12	284.98
16	49.22	81.41	111.65	140.92	169.51	197.92	225.90	253.59	281.09
17	48.41	80.26	110.16	139.13	167.53	195.49	223.20	250.62	277.86
18	47.74	79.27	108.91	137.60	165.73	193.46	220.87	248.03	275.04
19	47.15	78.40	107.80	136.26	164.17	191.64	218.95	245.80	272.61
20	46.63	77.55	106.83	135.09	162.80	190.07	217.07	243.87	270.47
25	44.80	74.37	103.36	130.88	157.87	184.44	210.76	236.82	262.71
30	43.67	73.30	101.21	128.26	154.80	180.93	206.79	232.45	257.88

PERCENTAGE POINTS OF V_5^*

	p = 1 α = 0.05				p = 1 α = 0.01			
	k = 2		k = 3		k = 2		k = 3	
M	q = 4	q = 6	q = 6	q = 9	q = 4	q = 6	q = 6	q = 9
1	7.23	9.42	11.26	14.96	10.90	13.44	15.55	19.73
2	6.89	8.95	10.76	14.29	10.48	12.92	14.95	18.95
3	6.69	8.72	10.48	13.91	10.22	12.59	14.60	18.49
4	6.56	8.56	10.29	13.66	10.05	12.38	14.36	18.19
5	6.47	8.44	10.17	13.49	9.92	12.22	14.20	17.98
6	6.41	8.35	10.07	13.37	9.83	12.11	14.08	17.83
7	6.36	8.29	10.00	13.28	9.76	12.02	13.98	17.70
8	6.32	8.24	9.95	13.20	9.70	11.95	13.91	17.61
9	6.29	8.20	9.90	13.14	9.65	11.89	13.85	17.54
10	6.26	8.16	9.87	13.10	9.61	11.84	13.80	17.47
11	6.24	8.14	9.84	13.05	9.58	11.80	13.76	17.42
12	6.22	8.11	9.81	13.02	9.55	11.77	13.72	17.37
13	6.20	8.09	9.79	12.99	9.53	11.74	13.69	17.33
14	6.19	8.07	9.77	12.96	9.51	11.71	13.66	17.30
15	6.18	8.06	9.75	12.94	9.49	11.69	13.64	17.27
16	6.17	8.04	9.73	12.92	9.47	11.67	13.62	17.25
17	6.16	8.03	9.72	12.90	9.46	11.65	13.60	17.22
18	6.15	8.02	9.71	12.89	9.45	11.64	13.58	17.20
19	6.14	8.01	9.70	12.87	9.44	11.62	13.57	17.18
20	6.13	8.00	9.69	12.86	9.43	11.61	13.55	17.16
22	6.12	7.98	9.67	12.84	9.41	11.59	13.53	17.13
24	6.11	7.97	9.66	12.82	9.39	11.57	13.51	17.11
26	6.10	7.96	9.64	12.80	9.38	11.55	13.49	17.09
28	6.09	7.95	9.63	12.79	9.37	11.54	13.48	17.07
30	6.09	7.94	9.62	12.78	9.36	11.53	13.47	17.05

Tafel 30c (Forts.) Table 30c (cont.)

	p = 2 α = 0.05				p = 2 α = 0.01			
	k = 2		k = 3		k = 2		k = 3	
M	q = 4	q = 6	q = 6	q = 9	q = 4	q = 6	q = 6	q = 9
1	19.64	26.38	31.22	42.86	26.19	33.76	38.95	51.69
2	17.24	23.17	27.80	38.16	23.01	29.66	34.69	46.03
3	16.06	21.58	26.11	35.84	21.44	27.63	32.56	43.21
4	15.36	20.64	25.09	34.44	20.50	26.43	31.29	41.53
5	14.89	20.01	24.41	33.51	19.88	25.63	30.44	40.40
6	14.56	19.57	23.92	32.85	19.44	25.05	29.84	39.60
7	14.31	19.23	23.56	32.35	19.11	24.63	29.38	39.00
8	14.12	18.97	23.28	31.96	18.85	24.30	29.03	38.53
9	13.97	18.77	23.05	31.66	18.65	24.03	28.75	38.16
10	13.84	18.60	22.87	31.40	18.48	23.81	28.52	37.85
11	13.73	18.46	22.72	31.19	18.34	23.63	28.32	37.60
12	13.65	18.34	22.59	31.01	18.22	23.48	28.16	37.39
13	13.57	18.24	22.48	30.86	18.12	23.35	28.02	37.20
14	13.50	18.15	22.38	30.73	18.03	23.24	27.90	37.04
15	13.45	18.07	22.29	30.61	17.95	23.14	27.80	36.90
16	13.40	18.00	22.22	30.51	17.89	23.05	27.71	36.78
17	13.35	17.94	22.15	30.42	17.83	22.97	27.62	36.67
18	13.31	17.89	22.09	30.33	17.77	22.91	27.55	36.57
19	13.28	17.84	22.04	30.26	17.73	22.84	27.48	36.48
20	13.24	17.79	21.99	30.20	17.68	22.79	27.42	36.40
22	13.19	17.72	21.91	30.08	17.61	22.69	27.32	36.26
24	13.14	17.65	21.84	29.98	17.54	22.61	27.23	36.15
26	13.10	17.60	21.78	29.90	17.49	22.54	27.15	36.05
28	13.06	17.55	21.72	29.83	17.44	22.48	27.09	35.96
30	13.03	17.51	21.68	29.77	17.40	22.43	27.03	35.89

Tafel 30c (Forts.)

Table 30c (cont.)

	p = 3	α 0.05			p = [illegible]	α = 0.01		
	k = 2		k = 3		k = [illegible]		k = 3	
M	q = 4	q = 6	q = 6	q = 9	q = 4	q = 6	q = 6	q = 9
1	37.36	51.25	60.49	84.63	46.88	62.09	71.79	97.60
2	32.02	43.95	52.80	73.90	40.07	53.13	62.50	85.09
3	29.32	40.25	48.87	68.42	36.65	48.61	57.78	78.71
4	27.70	38.02	46.47	65.06	34.58	45.88	54.90	74.83
5	26.60	36.52	44.84	62.79	33.20	44.07	52.97	72.21
6	25.82	35.44	43.67	61.16	32.22	42.76	51.58	70.32
7	25.23	34.63	42.79	59.93	31.47	41.77	50.53	68.89
8	24.77	34.00	42.10	58.96	30.90	41.01	49.70	67.77
9	24.40	33.49	41.55	58.18	30.43	40.40	49.05	66.88
10	24.09	33.08	41.09	57.55	30.05	39.89	48.50	66.14
11	23.84	32.73	40.71	57.01	29.74	39.47	48.05	65.52
12	23.62	32.44	40.39	56.56	29.47	39.11	47.67	65.00
13	23.44	32.19	40.10	56.16	29.24	38.81	47.34	64.55
14	23.28	31.96	39.86	55.82	29.04	38.54	47.05	64.17
15	23.14	31.76	39.65	55.53	28.86	38.31	46.80	63.83
16	23.02	31.60	39.46	55.27	28.71	38.11	46.58	63.51
17	22.91	31.45	39.30	55.03	28.57	37.93	46.39	63.25
18	22.81	31.32	39.15	54.83	28.44	37.76	46.21	63.01
19	22.72	31.20	39.01	54.63	28.33	37.61	46.05	62.80
20	22.64	31.09	38.89	54.47	28.23	37.48	45.90	62.59
22	22.50	30.90	38.68	54.17	28.06	37.25	45.65	62.25
24	22.39	30.74	38.50	53.91	27.92	37.06	45.44	61.97
26	22.29	30.60	38.35	53.71	27.79	36.89	45.26	61.73
28	22.20	30.48	38.22	53.52	27.68	36.75	45.10	61.52
30	22.12	30.38	38.10	53.35	27.53	36.62	44.97	61.32

Tafel 30c (Forts.)

Table 30c (cont.)

	$p = 4$ $\alpha = 0.05$				$p = 4$ $\alpha = 0.01$			
	$k = 2$		$k = 3$		$k = 2$		$k = 3$	
M	$q = 4$	$q = 6$	$q = 6$	$q = 9$	$q = 4$	$q = 6$	$q = 6$	$q = 9$
1	50.49	84.13	99.17	140.36	73.06	98.57	114.18	157.80
2	51.49	71.57	86.14	122.00	61.91	83.68	98.79	136.76
3	46.80	65.17	79.22	112.27	56.16	75.97	90.72	125.72
4	43.30	61.15	74.92	106.21	52.63	71.23	85.73	118.83
5	41.93	58.41	71.95	102.04	50.24	68.00	82.29	114.11
6	40.50	56.43	69.79	98.97	48.51	65.68	79.81	110.66
7	39.41	54.92	68.15	96.63	47.19	63.91	77.91	108.04
8	38.55	53.73	66.85	94.81	46.16	62.52	76.41	105.98
9	37.38	52.78	65.80	93.31	45.33	61.39	75.20	104.30
10	37.31	51.99	64.94	92.08	44.65	60.47	74.20	102.93
11	36.83	51.33	64.21	91.06	44.08	59.70	73.36	101.77
12	35.43	50.76	63.58	90.16	43.59	59.03	72.65	100.79
13	36.08	50.28	63.04	89.42	43.17	58.47	72.03	99.94
14	35.78	49.86	62.58	88.75	42.80	57.98	71.49	99.20
15	35.51	49.49	62.17	88.16	42.48	57.55	71.02	98.53
16	35.23	49.16	61.81	87.66	42.21	57.17	70.60	97.95
17	35.07	48.87	61.48	87.20	41.95	56.82	70.23	97.44
18	34.38	48.61	61.19	86.78	41.73	56.52	69.90	96.99
19	34.71	48.37	60.92	86.40	41.52	56.25	69.60	96.56
20	34.55	48.15	60.68	86.06	41.33	55.99	69.32	96.17
22	34.29	47.78	60.26	85.47	41.01	55.56	68.85	95.53
24	34.00	47.47	59.92	84.97	40.74	55.19	68.45	94.96
26	33.87	47.20	59.61	84.55	40.51	54.88	68.09	94.48
28	33.70	46.37	59.35	84.19	40.31	54.61	67.80	94.07
30	33.55	46.76	59.13	83.87	40.14	54.37	67.54	93.70

FINAL REMARKS:

Univariate Test Distributions as a Special Case of Their Multivariate Analogs

The usual test distributions of univariate statistics are obtained in this collection of tables as special cases of their multivariate analogs. The respective details for the tables can be found under the "Further Remarks" in "Section (e)" for each table.

(a) The t-distribution :

This distribution is contained in Table 19 for $p = 1$. It is also contained in Table 8 , for , when $p = 1$, the variate T^2 is identical to $F = t^2$.

(b) The F-distribution :

This distribution is contained in Table 3 for $s = p = 1$.

(c) The beta-distribution :

This distribution is contained in Table 1 for $p = 1$.

(d) The chi-square distribution :

(1) The upper percentage points of this distribution are contained in Table 22 for $\tau = 1$.

(2) The lower percentage points of this distribution are contained in Table 23 for $\tau = 1$.